Progress in Mathematics
Volume 133

The Floer Memorial Volume

Helmut Hofer
Clifford H. Taubes
Alan Weinstein
Eduard Zehnder
Editors

Birkhäuser Verlag
Basel · Boston · Berlin

Editors:

Helmut Hofer
Dept. of Mathematics
ETH-Zentrum
8092 Zürich
Switzerland

Clifford H. Taubes
Dept. of Mathematics
Harvard University
Cambridge, MA 02138
USA

Alan Weinstein
Dept. of Mathematics
University of California
Berkeley, CA 94720
USA

Eduard Zehnder
Dept. of Mathematics
ETH-Zentrum
8092 Zürich
Switzerland

A CIP catalogue record for this book is available from the Library of Congress, Washington D.C., USA

Deutsche Bibliothek Cataloging-in-Publication Data

The **Floer memorial volume** / Helmut Hofer ... ed. – Basel ; Boston ; Berlin : Birkhäuser, 1995
(Progress in mathematics ; Vol. 133)

NE: Hofer, Helmut [Hrsg.]; Floer, Andreas: Festschrift; GT

Softcover reprint of the hardcover 1st edition 1995

Printed on acid-free paper produced of chlorine-free pulp ∞

ISBN-13: 978-3-0348-9948-2 e-ISBN-13: 978-3-0348-9217-9
DOI: 10.1007/978-3-0348-9217-9

9 8 7 6 5 4 3 2 1

Contents

Preface

The death of Andreas Floer in Spring 1991 has left us shaken and sad. He was an exceptionally gifted mathematician, who during his short life has significantly influenced the development of mathematics. His untimely death was a personal tragedy and a great loss for mathematics.

Mathematics was one of the focal points in Andreas' life, giving him pleasure and satisfaction. We decided to collect essays for a memorial volume in his honour related to his fields of interest: gauge theory, dynamical systems, symplectic geometry and topology. Mathematicians with a special relationship to Andreas Floer, to his work and his ideas were invited to contribute. We did not set any conditions as to style and contents and welcomed new research results as well as surveys, speculations and personal reminiscences. The response was overwhelming, bearing testimony to the esteem and appreciation of this colleague.

We would like to thank all the contributors. We also owe thanks to Peter Braam for preparing two unfinished manuscripts by Andreas Floer for inclusion in this volume and adding a short commentary. Finally, our thanks go to Birkhäuser Verlag for support and professional help.

H. Hofer
C. Taubes
A. Weinstein
E. Zehnder

Andreas Floer 1956–1991

Photo courtesy Prof. George Bergmann, Berkeley

Andreas Floer was born on August 23, 1956 in Duisburg, Germany. He studied at the Ruhr-University Bochum, where he specialized in algebraic topology and dynamical systems and received his Diploma in mathematics in 1982. He then spent a year and a half at Berkeley, working with C. Taubes and A. Weinstein. After his return to Bochum, he received his doctoral degree under E. Zehnder in 1984 with a thesis on V.I. Arnold's fixed point problem for global Hamiltonian mappings. There followed research positions at SUNY at Stony Brook (1985–1986) and at the Courant Institute of NYU (1986–1988) and a period as assistant professor in Berkeley. In the fall of 1990 he was appointed to the chair of Analysis and Geometry at the Ruhr-University in Bochum. He died on May 15, 1991 in tragic circumstances.

Andreas Floer worked in the fields of dynamical systems, symplectic geometry, Yang-Mills theory and low dimensional topology. Motivated by the global existence problem of periodic solutions for Hamiltonian systems and starting from ideas of Conley, Gromov and Witten, he developed his Floer homology. Floer's homology theory is based on the combinatorial study of solutions of certain elliptic partial differential equations on manifolds. They occur as connecting orbits in variational principles like the action principle of classical mechanics and that for the Chern-Simons functional in 3-dimensional topology. Applied to the highly degenerate action functional on the loop space, his construction recovers the homology of the underlying compact symplectic manifold and leads to a solution of the V.I. Arnold conjecture in many cases. For other functionals, it produces previously unknown invariants of the underlying structures. The best known example is the Floer homology of the homology 3-sphere, which refines the Casson invariant and is related to the Donaldson invariants for 4-manifolds. In joint work with H. Hofer, the combination of the Floer homology and the symplectic capacity theory led to a symplectic homology theory and to new symplectic invariants. In his last work, Andreas Floer defined Yang-Mills type invariants for knots and gave an axiomatic characterization of his 3-manifold- and knot invariants.

Andreas Floer's work was tragically interrupted, but his visions and deep contributions have provided new, powerful methods which can be applied to problems inaccessible only a few years ago. Andreas Floer was a brilliant mathematician, an exceptionally gifted and a very sensitive man, and a friend. We miss him.

Publications of Andreas Floer

[1] Integrality of the monopole number in $SU(2)$ Yang-Mills-Higgs theories on $\mathbb{R}^3$. Comm. Math. Phys., 93:367–378, 1984.

[2] Fixed point results for symplectic maps related to the Arnold conjecture (with E. Zehnder). In: Dynamical Systems and Bifurcation, Proc. Groningen 1984, Lecture Notes in Math. 1125, Springer 1985, 47–63.

[3] Proof of the Arnold conjecture for surfaces and generalizations to certain Kähler manifolds. Duke math. J. 53 (1986), 1–32.

[4] A refinement of the Conley index and an application to the stability of hyperbolic invariant sets. Ergod. Th. and Dynam. Sys. (1987), 7, 93–103.

[5] The equivariant Conley index and bifurcations of periodic solutions of Hamiltonian systems (with E. Zehnder). Charles Conley Memorial Vol., Ergod. Th. Dynamical Systems 8* (1988), 87–98.

[6] Nonspreading wave packets for the cubic Schrödinger equation with a bounded potential (with A. Weinstein). J. Funct. Anal. 69 (1986), 397–408.

[7] Monopoles on asymptotically Euclidean 3-manifolds. Bull. Amer. Math. Soc. 16 (1987), 125–127.

[8] A topological persistance theorem for normally hyperbolic manifolds via the Conley index. Transact. Amer. Math. Soc. 321 (1990) No. 2, 647–657.

[9] Viterbo's index and the Morse index for the symplectic action, in: Periodic Solutions of Hamiltonian Systems and related topics, edited by A. Ambrosetti, I. Ekeland, P. Rabinowitz and E. Zehnder, NATO ASI Series C, Vol. 209, (1987) 147–152.

[10] Holomorphic curves and a Morse theory for exact symplectomorphisms. In: Aspect dynamiques et topologiques des group infinis de transformation de la mécanique classique, Traveaux en Cours 25, 49–61.

[11] Morse theory for fixed points of symplectic diffeomorphisms. Bull. Amer. Math. Soc. 16 (1987), 279–281.

[12] Morse theory for Lagrangian intersections. J. Diff. Geom. 28, (1988) 513–547.

[13] A relative Morse index for the symplectic action. Comm. Pure Appl. Math., Vol. XLI (1988), 393–407.

[14] The unregularized gradient flow of the symplectic action. Comm. Pure Appl. Math., Vol. XLI (1988), 775–813.

[15] Witten's complex and infinite dimensional Morse theory. Journal Differential Geom. 30 (1989), 207–221.

[16] Cuplength estimates for Lagrangian intersections. Comm. Pure Appl. Math., Vol. XLII (1989), 335–356.

[17] Selfdual conformal structures on lCP^2. Journal Differential Geometry 33 (1991), 551–573.

[18] An instanton-invariant for 3-manifolds. Comm. Math. Phys. 118 (1988), 215–240.

[19] Symplectic fixed points and holomorphic spheres. Comm. Math. Phs. 120 (1989), 575–611.

[20] A proof of the Weinstein conjecture in $P \times \mathbf{C}^l$ (with H. Hofer and C. Viterbo). Math. Zeit. 203 (1990), 469–482.

[21] Instanton homology, surgery and knots, in Geometry of low-dimensional Manifolds, London Mathematical Society Lecture Note Series, Edited by K. Donaldson and C.B. Thomas, Vol. 150, (1989), 97–114.

[22] Elliptic methods in variational problems. Proceedings of the ICM Kyoto 1990, Vol. 1, 87–97.

[23] Instanton Homology. Springer Lecture Notes in Physics 375 (1991), 219–221.

[24] Coherent orientations for periodic orbit problems in symplectic geometry (with H. Hofer). Math. Zeit. 212 (1993), 13–38.

[25] Symplectic homology I: Open sets in $\mathbf{C}^n$ (with H. Hofer). Math. Zeit. 215 (1993), 37–88.

[26] Symplectic homology II: The general construction (with K. Cieliebak and H. Hofer). Math. Zeit. 218 (1995), 103–122.

[27] Transversality in elliptic Morse theory for the symplectic action (with H. Hofer and D. Salamon). To appear in Duke 1995.

[28] Applications of symplectic homology I (with H. Hofer and K. Wysocki). Math. Zeit. 217 (1994), 577–606.

[29] Applications of symplectic homology II (with K. Cieliebak, H. Hofer and K. Wysocki). To appear in Math. Zeit.

[30] The configuration space of Yang-Mills-Higgs theory on asymptotically flat manifolds. In this volume.

[31] Monopoles on asymptotically flat manifolds. In this volume.

[32] Instanton homology and Dehn surgery (unfinished manuscript). In this volume.

Floer's work on monopoles

Ian McAllister and Peter Braam

The following two preprints of Andreas Floer have been circulating in the mathematical community since 1986. They contain much of the author's PhD research which he was doing at Berkeley. In these preprints we have tried to correct the most obvious typographical errors, but we have not changed anything in the exposition.

The subject of the papers is the construction of moduli spaces of magnetic monopoles on asymptotically Euclidean 3-manifolds. The paper on the configuration space builds a framework in which the analysis can be carried out efficiently. The second paper constructs monopoles on three manifolds and establishes basic properties of the moduli spaces.

After Taubes studied monopoles (see [11]) and their configuration spaces [13] on Euclidean spaces, Floer, in 1984, began to consider both subjects on asymptotically Euclidean 3-manifolds. This led to the preprints and the announcement. As can be expected, several authors have considered monopoles during the intervening years. In 1985 Braam [4], [2] started studying monopoles on manifolds with hyperbolic ends (see also Braam and Austin [5]); he soon abandoned the three dimensional approach and called on the theory of instantons. Atiyah and Hitchin [1] led the way into understanding the dynamics of monopoles. This work was continued by, among others, Dancer and Leese [7], and is full of detailed geometrical considerations, special, so far, for Euclidean space. Monopoles for arbitrary groups on Euclidean space were studied by Hurtubise and Murray [10], following the well known papers of Hitchin to which Floer refers. The topology of the configuration space was studied by Taubes [13] and that of the solution space by Cohen and Jones [6]. Topological invariants for more general three manifolds were constructed by Braam [3].

More recently we have seen considerable activity along the lines which Floer was pursuing. Ernst [8] took up precisely the issues raised by Floer's announcement [9]. McAllister [12] has been constructing monopoles with high masses on general three manifolds, something Floer thought about a lot.

From personal contact with Floer we know that he saw great potential in the study of monopoles. It is very sad that he did not have the time to finish these preprints and lead the way in this area.

References

[1] M.F. Aityah and N.J. Hitchin, *The Geometry and Dynamics of Magnetic Monopoles*. Princeton Univeristy Press, 1988.

[2] P.J. Braam, A Kaluza Klein Approach to Hyperbolic 3-manifolds. *Enseignement Mathématique*, 34(3-4):275–311, 1988.

[3] P.J. Braam, Magnetic monopoles and topology. In Mabuchi and Mukai, editors, *Einstein Metrics and Yang-Mills connections*, pages 51–57, Marcel Dekker, 1993.

[4] P.J. Braam, Magnetic monopoles on 3-manifolds. *J. Diff. Geom.*, 30:425–464, 1989.

[5] P.J. Braam and D.M. Austin, Boundary values of hyperbolic monopoles. *Nonlinearity*, 3:809–823, 1990.

[6] R.L. Cohen and J.D.S. Jones. Representations of braid groups and operators couples to monopoles. In Donaldson and Thomas, editors, *Geometry of Low Dimensional Manifolds, I*, pages 191–205, Cambridge University Press, 1990.

[7] A. Dancer and R. Leese, Dynamics of $su(3)$ monopoles. *Proc. of the Royal Society*, 440(1909):421–430, 1993.

[8] K. Ernst, The Ends of the Monopole Moduli Space over $\mathbb{R}^3$#(homology sphere) – Parts I and II. In this volume.

[9] A. Floer, Monopoles on asymptotically euclidean 3-manifolds. *Bull. AMS.*, 16:125–127, 1987.

[10] J. Hurtubise and M.K. Murray, On the construction of monopoles for the classical groups. *Comm. Math. Phys.*, 122:35–89, 1989.

[11] A. Jaffe and C.H. Taubes, Vortices and Monopoles. Birkhäuser, Boston, 1980.

[12] I.G. McAllister, High-mass monopoles on 3-manifolds. *Oxford preprint*, 1993.

[13] C.H. Taubes, Monopoles and Maps from S^2 to S^2; the topology of the configuration space. *Comm. Math. Phys.*, 95:345–391, 1984.

Ian McAllister and Peter Braam
Mathematical Institute
24-29 St. Giles
Oxford OX1 3LB, UK

Monopoles on asymptotically flat manifolds

A. Floer

Abstract. We consider the relation between the theory of monopoles on $\mathbb{R}^3$ and the analogous problem on a manifold M obtained from $\mathbb{R}^3$ by forming the connected sum with a compact manifold M. The main result is that the moduli space of monopoles whose action is concentrated away from M is a smooth manifold which can be described in terms of the homology of M and the part, we determine the asymptotics of these functions. As a result, we determine the «end» of the moduli space in some special cases.

1 Introduction

In recent years, there has been an intensive study of the solution set of the Bogomolny equations

$$d_A\Phi = \pm * F_A \tag{1}$$

on Euclidean $\mathbb{R}^3$, see for example [9].

Here, A is a connection on a principal G-bundle P over M. We will restrict ourselves to the case $G = SU_2$, although most of the results have a generalization to arbitrary simple Lie groups. Φ is a section of the adjoint bundle

$$\operatorname{ad}(P) = P \otimes_{\text{ad}} su_2 \,, \tag{2}$$

where su_2 is the Lie algebra of SU_2. F_A is the curvature of A, d_A the covariant derivative corresponding to A and $*$ the Hodge duality isomorphism. It is well known that equation (1) characterizes the absolute minima of the action functional

$$\mathbf{a}(c) = \int_{\mathbb{R}^3} \{ \, |F_A|^2 \; + \; |d_A\Phi|^2 \, \} d^3x \tag{3}$$

on each connected component of the space of finite action configurations. The action (3) is invariant under the action of the gauge group of smooth sections g of the bundle $\operatorname{aut}(P)$, c.f. [9], by

$$g(c) = (g\Phi g^{-1}, \, (dg)g^{-1} + gAg^{-1}) \,. \tag{4}$$

The most interesting feature of Yang-Mills-Higgs theory is that the set of minima of (3), i.e. the set of solutions of (1), modulo the gauge equivalence of (4) is a finite dimensional manifold. In this and a series of forthcoming papers[1]), we will investigate this solution space in the case where $\mathbb{R}^3$ is replaced by a more general

1) c.f. [5], [3], [4] — some of these papers never appeared. See the notes accompanying these papers for alternative references

manifold. To find a suitable class of manifolds, first note that it is not likely to find critical points of $\mathbf{a}$ for manifolds of dimension other than 3. In fact, one can exclude this for $M = \mathbb{R}^d$ and $d \neq 3$ by a scaling argument, see [9], Section II. Moreover, equation (1) has only trivial solutions if M is compact. This is related to the fact that the solutions of (1) on $\mathbb{R}^3$ are to some extent characterized by their behavior at infinity. We will therefore restrict ourselves to manifolds whose ends resemble the end of $\mathbb{R}^3$. To be more exact, we will use the following definition:

Definition 1. A 3-dimensional Riemannian manifold (M, g) without boundary is called asymptotically flat with n ends ($n \geq 1$), if there exists a positive real number r_M and injective smooth maps

$$\omega_i : \mathbb{R}^3 - B_r \longrightarrow M, \quad 1 \leq i \leq n, \tag{5}$$

so that $M - \bigcup_{i-1}^{n} \omega_i(\mathbb{R}^3 - B_r)$ is compact and $\omega_i^* g = g_0 + \gamma_i$, where g_0 is the Euclidean metric on $\mathbb{R}^3$ and

$$|x| \cdot |\gamma_i| + |x|^2 \cdot |\nabla\gamma_i| + |x|^3 \; |\nabla\nabla\gamma_i| \leq c \tag{6}$$

is bounded.

It is easy to verify that for asymptotically flat manifolds (M, g), the curvature $R = R_g$ satisfies

$$R \in L^p \quad \text{for} \quad p \in (1, \infty]. \tag{7}$$

The particular decay condition on the metric is justified by the following special case: If $\omega_i^* g$ is of the form

$$(\omega_i^* g)_{jk} = \left(1 + \frac{\mu_i}{2r}\right)^4 \delta_{jk} + O(r^{-2}), \tag{8}$$

then μ_i can be interpreted as the total mass of M at the i-th end (see [11]). On the other hand, Definition 1 is strong enough so that the main features of Yang-Mills-Higgs theory on $\mathbb{R}^3$ carry over to asymptotically flat manifolds. For example, every finite action configuration c can be assigned n constants $m_i, i = 1, \ldots, n$, as the asymptotic values of $|\Phi|$, the length of the Higgs field, at the i-th end. We will call $m_i(c)$ the mass of c at the i-th end. Moreover, if $m_i(c) > 0$, then the homotopy class of the maps

$$\omega_i^* \acute{\Phi} : \mathbb{R}^3 - B_r \longrightarrow SU_2 - 0 \tag{9}$$

defines a «magnetic charge» $k_i \in \mathbb{Z}$. Here, $\acute{\Phi}$ is a regularized Higgs field defined in [2], Lemma 3. 3. Since an integration by parts yields

$$\mathbf{a}(c) = \pm 4\pi \underline{m} \cdot \underline{k} + \|\nabla_A \Phi \pm F_A\| , \tag{10}$$

see [2], Proposition 6, the solutions of (1) are, as in the case $M = \mathbb{R}^3$, the minima of the action (3). Let us denote by $\mathcal{M}^{\underline{k},\underline{m}}(M)$ the moduli space of equivalence classes of solutions of (1) with masses $\underline{m} = (m_1 \dots, m_n)$ and charges $\underline{k} = (k_1 \dots k_n)$. Moreover, we choose arbitrary numbers $0 < m_0, m_\infty < \infty$ and define

$$\mathcal{M}^{\underline{k}}(M) = \{[c] \in \mathcal{M}^{\underline{k},\underline{m}}(M) \mid m_o \leq m_i \leq m_\infty \text{ for } 1 \leq i \leq n\}. \tag{11}$$

The special phenomena that are connected with very high and very low masses will be the subject of a forthcoming paper. In the present paper, we investigate the question of under which conditions can one construct elements of $\mathcal{M}^{\underline{k}}(M)$ from monopole configurations on $\mathbb{R}^3$. As is well-known, the space $\mathcal{M}^{k,1}(\mathbb{R}^3)$ is a nonempty manifold of dimension $4k-1$ (see [1] and [7] for a more detailed description). Moreover, one can construct monopoles of masses other than 1 by scaling, see [9], Section II. We want to transplant monopole configurations on $\mathbb{R}^3$ with masses in $[m_0, m_\infty]$ to the ends of M. Define, therefore,

$$\mathcal{M}^k_R(\mathbb{R}^3) = \{[c] \in \mathcal{M}^k(\mathbb{R}^3) \mid \mathbf{a}(c, B_R) \leq \frac{1}{2}\}. \tag{12}$$

Here,

$$\mathbf{a}(c, B_R) = \int_{|x| \leq R} \left(|d_A\Phi|^2 + |F_A|^2 \right) d^3x \tag{13}$$

is the Yang-Mills-Higgs action of c restricted to B_R. Our main result is

Theorem 2. *Let M be an asymptotically flat manifold with only one end so that $H_1(M,\mathbb{Z})$ is finite. Then for every $k > 0$ there exists a positive real number R and an injective local diffeomorphism*

$$\mathcal{H} : H_1(M,\mathbb{Z}) \times \mathcal{M}^k_R(\mathbb{R}^3) \longrightarrow \mathcal{M}^k(M),$$

so that $[c]$ and $\mathcal{H}([\gamma],[c])$ can be represented by configurations c and c_γ with

$$\|\omega^* c_\gamma - c\|_c \leq \epsilon_R, \tag{14}$$

for $[c] \in \mathcal{M}^k_R$. Here $\| \ \|_c$ is the norm defined in Definition 9 and $\epsilon_{R'} \longrightarrow 0$ for $R' \longrightarrow \infty$.

In a special case we have

Theorem 3. *If $H^2(M,\mathbb{R}) = 0$, then the Higgs field of each configuration $[c] \in \mathcal{M}^1_R(M)$ has a unique nondegenerate zero $\zeta[c] \in M$ which defines a local diffeomorphism*

$$\zeta : \mathcal{M}^1_R(M) \longrightarrow M, \tag{15}$$

which is surjective up to a compact set if $H^2(M,\mathbb{R}) = 0$.

Now let M be a general asymptotically flat manifold with n ends and let $\mathcal{M}_R^k$ be the set of self-dual configurations on $\mathbb{R}^3$ of Definition 1. The approximate solutions will be modelled on the following set.

Definition 4. For $n \in \mathbb{R}_+$, set

$$\tilde{\mathcal{M}}_R^{\underline{k}}(M) = H^1(M, S^1) \times \{[\underline{c}] = ([c_1], \dots, [c_n]) \mid [\underline{c}] \in \oplus_{i=1}^n \mathcal{M}_R^{k_i}(\mathbb{R}^3), \max_{i,j} |m(c_i) - m(c_j)| \leq R^{-0.1}\}. \tag{16}$$

Here, $H^p(M, S^1)$ is the cohomology of M with coefficients in the abelian group S^1. Roughly speaking, the first factor corresponds to a choice of a «vacuum configuration» on the interior of M. Now recall that the set

$$\mathfrak{Q}(M) = \mathcal{C}(M)/\mathcal{G}(M) \tag{17}$$

of gauge equivalence classes of finite action configuration on M has the structure of a Hilbert manifold, see [2], Theorem 4 . We have

Theorem 5. *Let M be an asymptotically flat manifold with n ends. There exists a real number R and a smooth map*

$$\mathcal{H} : \tilde{\mathcal{M}}_R^{\underline{k}}(M) \longrightarrow \mathfrak{Q}(M)\,, \tag{18}$$

and a smooth map

$$\tilde{b} : \tilde{\mathcal{M}}_R^{\underline{k}}(M) \longrightarrow H^2(M, \mathbb{R}) \tag{19}$$

so that $\mathcal{H}([\gamma], [\underline{c}])$ is self-dual if and only if $\tilde{b}([\gamma], [\underline{c}]) = 0$.

If the zero-set of the function $\tilde{b}$ is regular, then the dimension of the moduli space thus obtained coincides with the Fredholm index of the linearization of the Bogomolny equation. In the second part of this paper, we will investigate the asymptotics of the function $\tilde{b}$. To simplify the analysis, we restrict ourselves to a smaller class of asymptotically flat manifolds.

Definition 6. A 3-dimensional asymptotically flat manifold is called flat at infinity if one can choose the maps ω_i of Definition 1 so that $\omega_i^* g$ is the Euclidean metric on $\mathbb{R}^3 - B_r$.

By de Rham's theory for asymptotically flat manifolds, see Proposition 19, we have

$$k_M := H^2(M, \mathbb{R}) \cong H^1(M, \mathbb{R}) \oplus \mathbb{R}^{n-1} = \text{kern } (d + d^*) \cap L^2(\Lambda^1(r))\,, \tag{20}$$

i.e. each of the obstructions is represented by a harmonic 1-form λ on M. Then

$$\lambda_i = \omega_i^* \lambda\,, \tag{21}$$

are harmonic 1-forms on $\mathbb{R}^3 - B_r$, since $\omega_i^* g$ is the Euclidean metric on $\mathbb{R}^3 - B_r$. Let us continue each λ_i smoothly to $\mathbb{R}^3$, and define the «charge density»

$$\rho_i = d * \lambda_i \,. \tag{22}$$

Then we can consider the multipole expansion for ρ_i, see for example [8]. From proposition it follows that the monopole moments define the first factor in the space of harmonic L^2-forms (21). Now if c is a self-dual monopole on $\mathbb{R}^3$, and we interpret the gauge invariant field

$$\phi := |\Phi|$$

as an electric potential, then its energy with respect to the «field» λ_i is given by the expression

$$e([c], \lambda_i) = \int_{\mathbb{R}^3} \phi \rho_i \,. \tag{23}$$

We will define the λ-component of the reduced function as

$$b^\lambda([\gamma], [\underline{c}]) = \left\langle \lambda \hat{\Phi}, b(c + \xi) \right\rangle \,. \tag{24}$$

where c is the approximate solution corresponding to $([\gamma], [\underline{c}])$ and ξ is the small correction of Proposition 23.

Theorem 7 *Define for* $([\gamma], [\underline{c}]) \in \tilde{\mathcal{M}}_{\underline{R}}^{\underline{k},\underline{m}}(M)$*:*

$$e^\lambda([\gamma], [\underline{c}]) := \sum_{i=1}^{n} e([c_i], \lambda_i) \,. \tag{25}$$

Let $p = 0$ if λ contains a monopole moment. Otherwise, let p be the minimum of the orders of the first nonvanishing multipole moment at those ends where c_i is nontrivial. Then there exists a constant K depending continuously on $\underline{m}$ and $\underline{k}$ so that

$$\left| b^\lambda - e^\lambda \right| \le K R^{-p-1.1} \,, \tag{26}$$

and for any vector $\xi \in T_{([\gamma],[\underline{c}])} \tilde{\mathcal{M}}_{\underline{R}}^{\underline{k},\underline{m}}(M)$,

$$\left| D_\xi (b^\lambda - e^\lambda) \right| \le K R^{-p-2.1} \,. \tag{27}$$

In (27), we have used the metric on $\tilde{\mathcal{M}}_{\underline{R}}^{\underline{k}}(M)$ defined by (179). We now consider two examples to show that the estimates of Theorem 5 are strong enough to draw conclusions about the structures of the moduli space. First assume that M has only one end and that $H^1(M, \mathbb{R}) = \mathbb{R}$. Then there exists a harmonic 1-form λ representing a generator of $H^1(M, \mathbb{R})$. In this case, Theorem 5 implies that the structure of the moduli space depends on the multipole expansion of λ.

In fact, assume that all multipoles of λ vanish up to order p, and that the leading order multipole is represented by the harmonic homogeneous polynomial P of degree p. Note that $p \ge 1$, since by Proposition 19, the sum of all monopole moments of λ over all ends must vanish. We call λ nondegenerate if P_λ has a nondegenerate zero-set $z(P_\lambda)$ in $\mathbb{R}^3 - 0$.

Theorem 8. *If the generator of $H^1(M,\mathbb{R})$ is nondegenerate, then the moduli space of charge 1 monopoles is nondegenerate outside a compact set and is diffeomorphic to*

$$H^1(M,S^1) \times (z(P_\lambda) - B_R) \times \mathbb{R}\,. \tag{28}$$

Proof. We have

$$\tilde{\mathcal{M}}^1_R(M) \approx H^1(M,S^1) \times (\mathbb{R}^3 - B_R) \times \mathbb{R}\,,$$

where the first factor corresponds to the choice of a flat $O(2)$-bundle over M and the other factors determine the position and the mass of a monopole on $\mathbb{R}^3$. Now Theorem 7 implies that the rescaled function

$$b^\lambda_R([\gamma],x) := R^{p+1} b^\lambda([\gamma], c_{R\cdot x}) \tag{29}$$

converges to

$$e^\lambda_R([\gamma],x) = R^{p+1} e^\lambda([\gamma], c_{R\cdot x}) \tag{30}$$

uniformly in C^1 on $\mathbb{R}^3 - B_1$. □

There are very similar statements for higher monopole numbers and higher dimensions of $H^1(M,\mathbb{R})$, since monopole configurations behave much like point changes for «long distances». However, they are formally more complicated and we will therefore not deal with the general case.

The second example concerns manifolds with n ends and with $H^1(M,\mathbb{R}) = 0$. In this case, only the monopole moments $g(\lambda_i)$ of the forms λ_i are important, and Theorem 5 implies that

$$\left| b^\lambda([\gamma],[\underline{c}]) - \textstyle\sum_i g(\lambda_i)\phi_i(0)\right| \le KR^{-1.1}\,. \tag{31}$$

In the same way as above, this yields a relation on the masses $m(c_i)$ and on the distances of the monopoles c_i from the origin.

2 Notation and preliminary results

In this section, we recall some of the notations conventions and results of [2].

The bundle of exterior forms of degree p on M with values in the adjoint bundle $\mathrm{ad}(P)$ will be denoted by Λ^p_{ad}. For example, any connection on P can be represented by a section A of Λ^1_{ad}, and the Higgs field Φ is a section of Λ^0_{ad}. For this reason, we use the abbreviation $\Omega = \Lambda^0 \oplus \Lambda^1$ and $\Omega_{\mathrm{ad}} = \Omega \otimes \mathrm{ad}(P)$. For ξ and ζ in $\mathrm{ad}(P)$, we denote by $[\xi,\zeta]$ the Lie bracket and by $\xi\cdot\zeta$ the negative of the Killing inner product on su_2. Moreover, we denote by $|\lambda|^2$ the square of $\lambda \in \Lambda^p_{\mathrm{ad}}$ with respect to «,» and the metric on M.

Definition 9. Let E be any tensor bundle over M and λ a compactly supported smooth section. Define

$$\|\lambda\|_p = \left(\int_M |\lambda|^p \, \mu_g\right)^{\frac{1}{p}} \tag{32}$$

$$\|\lambda\|_{V^0} = \|\lambda\|_2 \tag{33}$$

$$\|\lambda\|_{V^k} = \|\nabla\lambda\|_{V^{k-1}} \tag{34}$$

$$\|\lambda\|_{W^k} = \|\lambda\|_2 + \|\lambda\|_{V^k} \tag{35}$$

Here, μ_g is the Riemannian volume form on M. Moreover, for any $c = (\Phi, A) \in W^1_{\text{loc}}(\Omega_{\text{ad}})$, and any compactly supported smooth section ξ of E_{ad}, define

$$\|\xi\|_{Z_c} = \|\xi\|_2 + \|\Phi \cdot \xi\|_{\frac{6}{5}} \tag{36}$$

$$\|\xi\|_c^2 = \|\nabla_A \xi\|_2^2 + \|[\Phi, \xi]\|_2^2 \tag{37}$$

$$\|\xi\|_{X_c} = \|\nabla_A \xi\|_c + \|[\Phi, \xi]\|_c + \left\|\Phi \cdot \nabla_A^2 \xi\right\|_{\frac{6}{5}} \tag{38}$$

We denote by L^p, V^k, W^k, Z_c, H_c, and X_c the corresponding completions. Moreover, we define $\acute{H}_c$ as the space of all $\xi \in W^1_{\text{loc}}(E_{\text{ad}})$ with $\|\xi\|_c < \infty$. Similarly, we define $\acute{X}_c$ as the Banach space of all $\xi \in W^2_{\text{loc}}(E_{\text{ad}})$ so that $\|\xi\|_{X_c} < \infty$.

We now state some properties of these Banach spaces. Define the seminorms

$$\|\xi\|_{p,<r} = \left(\int_{M_r} |\xi|^p \, d\mu_g\right)^{\frac{1}{p}} \tag{39}$$

$$\|\xi\|_{p,>r} = \|\xi\|_p - \|\xi\|_{p,<r} \tag{40}$$

and similarly for the other norms of Definition 9. Throughout this paper, we will denote by C_i constants which depend only on the manifold M and its metric.

Lemma 10. *Let E be any metric bundle and ∇ any metric preserving connection. There exists a constant C_0 depending only on M and g so that for every $\lambda \in W_{1,\text{loc}}$ there exists $(\lambda_{(1)} \dots \lambda_{(n)}) \in \mathbb{R}^n$ so that for $1 \le i \le n$,*

$$\left\| \, |\omega_i^* \lambda| - \lambda_{(i)} \right\|_{6,>r} \le C_0 \, \|\nabla_A \lambda\|_{2,>r} \, . \tag{41}$$

Moreover, for each $c \in W^1_{\text{loc}}(\Omega_{\text{ad}})$ there exist linear maps

$$\acute{X}_c(E_{\text{ad}}) \subset \acute{H}_c(E_{\text{ad}}) \longrightarrow \mathbb{R}^n : \xi \longrightarrow \underline{\xi} := (\xi_{(1)} \cdots \xi_{(n)}) \tag{42}$$

with the property that

$$\left\|\omega_i^*(\xi - \Phi\xi_{(i)})\right\|_6 \le C_0 \, \|\xi\|_c \, .$$

Lemma 10 follows immediately from [2], Lemma 3. 1. Together with the definition of $\mathbf{a}(c)$, it gives rise to the following definition:

Definition 11. For $c \in W^1_{\text{loc}}(\Omega_{\text{ad}})$ with $\mathbf{a}(c) < \infty$, define $m_i(c)$ as the unique non-negative number with

$$|\omega_i^* \Phi| - m_i(c) \in L^6(\mathbb{R}^3 - B_r, \mathbb{R}) . \tag{43}$$

Consider now the configuration space

$$\mathcal{C}(M) = \{c = (\Phi, A) \in W^1_{\text{loc}}(\Omega_{\text{ad}}) \mid \mathbf{a}(c) < \infty, \quad m_i(c) > 0 \quad \text{for} \quad 1 \leq i \leq n\} . \tag{44}$$

It is acted upon by a gauge group

$$\mathcal{G} = \mathcal{G}(M) = W^2_{\text{loc}}(\text{aut}(P)) , \tag{45}$$

where we can interpret $\text{aut}(P)$ as a closed subset of the vector bundle over M with fibre the 2 by 2 complex matrices. The operation is given by

$$g(c) = g(\Phi, A) = (g\Phi g^{-1}, (dg)g^{-1} + gAg^{-1}) . \tag{46}$$

In [2], Theorem 4, we showed that the quotient space

$$\mathcal{Q}(M) = \mathcal{C}(M)/\mathcal{G}(M) \tag{47}$$

is in a natural way a Hilbert manifold. Note that the norms (36)–(38) are are gauge invariant, i.e. we have for example,

$$\|\xi\|_c = \|g\xi g^{-1}\|_{g(c)} \ . \tag{48}$$

We even have

Lemma 12. *The spaces defined by the norms (32) and (36)–(38) give rise to smooth Banach space bundles $\mathcal{L}^p, \mathfrak{X}, \mathcal{H}$, and $\mathcal{X}$, respectively, over $\mathcal{Q}(M)$. Moreover, there exist natural covariant derivatives, which are flat, i.e. derivatives in different directions commute.*

Proof. In order to exhibit local trivializations for the bundles $\mathfrak{X}, \mathcal{H}$, and $\mathcal{X}$, note that for $\xi \in H_c(\Lambda^1_{\text{ad}}) \oplus \acute{H}_c(\Lambda^0_{\text{ad}})$, we have

$$\|\zeta\|_{c+\xi} = K(\ \|\xi\|_c\ )\ \|\zeta\|_c \tag{49}$$

$$\|\zeta\|_{c+\xi} = K(\ \|\xi\|_c\ )\ \|\zeta\|_{X_c} \tag{50}$$

$$\|\zeta\|_{c+\xi} = K(\ \|\xi\|_c\ )\ \|\zeta\|_{Z_{c+\xi}} \ , \tag{51}$$

where the function $K : \mathbb{R}_+ \longrightarrow \mathbb{R}_+$ is increasing and depends only on $m_i(c)$ and $\mathbf{a}(c)$. The proof of (49) is given in Proposition B6.2 of [13]; the proof of the other two assertions is similar to the proof of Lemma 6. 4 of [2]. Using the local charts of [2], Theorem 4 of $\mathcal{C}(M)/\mathcal{G}$, we can define smooth trivializations

$$\begin{aligned} \mathcal{U}_c \times H_c &\longrightarrow \mathcal{H} : \\ (\xi, \zeta) &\longrightarrow [(c + \xi, \zeta)] \end{aligned} \tag{52}$$

of $\mathcal{H}$, and similarly of $\mathfrak{X}$ and $\mathcal{X}$. The natural flat covariant derivative, ∇ is now given by the standard connection in the charts (52). □

We will from now on denote by $K_i = K_i(c)$ constants which depend continuously on the quantities $\mathbf{a}(c)$ and m_i, $i = 1, \dots, n$.

Lemma 13. *We have for $\xi \in X_c$ and $\acute{\xi} \in \acute{X}_c$:*

$$\|\xi\|_\infty \leq K_0 \, \|\xi\|_{X_c} \tag{53}$$

$$\left\|[\Phi, \acute{\xi}]\right\|_c + \left\|\nabla_A \acute{\xi}\right\|_c \leq K_0 \left\|\acute{\xi}\right\|_{X_c} . \tag{54}$$

(53) was proved in Lemma 3. 4 of [2], for a proof of (54) see [2], Lemma 6. 1.

Let us now consider bilinear operations between the bundles defined above.

Lemma 14. *Let $\natural_i : E \times E' \longrightarrow E''$ be any bilinear operation between tensor bundles satisfying*

$$\nabla(\lambda \natural_i \mu) = (\nabla\lambda)\natural_i \mu + \lambda \natural_i (\nabla_\mu) . \tag{55}$$

Let $\natural_i : E_{\text{ad}} \times E'_{\text{ad}} \longrightarrow E''_{\text{ad}}$ be the extension of $\natural_i$ by the Lie bracket. Then $\natural_i$ induces continuous bilinear maps

$$\acute{X}_c(E_{\text{ad}}) \times \acute{X}_c(E'_{\text{ad}}) \longrightarrow \acute{X}_c(E''_{\text{ad}}) \tag{56}$$

$$\acute{H}_c(E_{\text{ad}}) \times \acute{X}_c(E'_{\text{ad}}) \longrightarrow \acute{H}_c(E''_{\text{ad}}) \tag{57}$$

$$\acute{Z}_c(E_{\text{ad}}) \times \acute{X}_c(E'_{\text{ad}}) \longrightarrow \acute{Z}_c(E''_{\text{ad}}) \tag{58}$$

$$\acute{H}_c(E_{\text{ad}}) \times \acute{H}_c(E'_{\text{ad}}) \longrightarrow \acute{Z}_c(E''_{\text{ad}}) \tag{59}$$

bounded by a constant K_0 which depends only on $\mathbf{a}(c)$ and $m_i(c)$. Moreover, we have for $\xi, \zeta \in H_c$:

$$\langle \xi, \zeta \rangle \leq K_0 \, \|\xi\|_c \, \|\zeta\|_{Z_c} . \tag{60}$$

Proof. (59) and (60) are Lemma 4. 2 of [2]. The proof of (56)–(58) can be found in the proof of Lemma 6. 2 of [2]. □

3 Vacuum configurations on M

Let us define the space

$$\mathcal{V}(M) = \{[c] \in \mathcal{C}(M)/\mathcal{G}(M) \mid \mathbf{a}([c]) = 0\} . \tag{61}$$

Note that for $[(\Phi, A)] \in \mathcal{V}(M)$, we have

$$\nabla_A \Phi = 0 . \tag{62}$$

Since the Higgs field of a configuration in $\mathscr{C}(M)$ cannot vanish completely, Φ is a covariantly constant section of the adjoint bundle. Hence the SU_2-bundle P is reduced to a bundle with structure group

$$G_\Phi = \{g \in G \mid g\Phi g^{-1} = \Phi\} \cong S^1 . \tag{63}$$

The condition $F_A = 0$ now implies that A reduces to a flat connection on the resulting principal S^1-bundle. It is well known that such flat connections are in 1-1-correspondence with the group

$$G\,\mathrm{Hom}(\pi_1(M), S^1) \cong H^1(M, S^1), \tag{64}$$

where $G\,\mathrm{Hom}$ denotes the set of group homomorphisms and the correspondence is given by the holonomy of the connection (see, for example [6]). We now show that the vacuum configurations are determined uniquely by the mass and the flat connection.

Proposition 15. *Let $[S^2]$ be a generator of $H^2(su_2 - 0, \mathbb{Z})$. Then there is a diffeomorphism*

$$\begin{aligned} \mathscr{V}(M) &\longrightarrow \mathbb{R}_+ \times H^1(M, S^1) \\ &\cong \mathbb{R}_+ \times Tor(H_1(M,\mathbb{Z})) \times H^1(M,\mathbb{R})/H^1(M,\mathbb{Z}) \end{aligned} \tag{65}$$

given by

$$[\Phi, A] \longrightarrow \Big(|\Phi| \,, (\hat{\Phi}^*[S^2])^v, [A \cdot \Phi] \Big)\,, \tag{66}$$

where v denotes Poincaré duality on M and $\hat{\Phi} = |\Phi|^{-1}\,\Phi$.

Proof. First consider the long exact sequence

$$\begin{aligned} 0 \longrightarrow H^1(M,\mathbb{Z}) &\longrightarrow H^1(M,\mathbb{R}) \longrightarrow H^1(M,S^1) \\ \longrightarrow H^2(M,\mathbb{Z}) &\longrightarrow H^2(M,\mathbb{R}) \longrightarrow H^2(M,S^1) \longrightarrow 0\,, \end{aligned}$$

induced by the exact sequence $0 \longrightarrow \mathbb{Z} \longrightarrow \mathbb{R} \longrightarrow S^1 \longrightarrow 0$ of abelian groups. This implies the second isomorphism in (65). Moreover, injectivity of the map (66) is obvious. In order to show that it is also surjective, note that every class in $H^2(M,\mathbb{Z})$ can be induced by a map $\hat{\Phi} : M \longrightarrow S^2$, since M is homotopy equivalent to its 2-skeleton. Then

$$\hat{A}_0 := [\hat{\Phi}, d\hat{\Phi}] \tag{67}$$

is a connection form which makes $\hat{\Phi}$ covariantly constant. Hence $[F_{\hat{A}_0}, \hat{\Phi}] = 0$ and $\hat{\Phi} \cdot F_{\hat{A}} = \hat{a}$ is closed. In fact, it is easy to see that it represents the reduction of the class $[\hat{\Phi}] = \hat{\Phi}^*[S^2]$ to the real numbers. Now we note that gauge transformations of the form

$$h = \exp(f\hat{\Phi}) \tag{68}$$

have the effect that

$$\hat{\Phi} \cdot F_{h(\hat{A})} = \hat{\Phi} \cdot F_{\hat{A}} + df\,. \tag{69}$$

We conclude that $\hat{\Phi}$ admits a flat connection if and only if the class $[\hat{\Phi}]$ is torsion. In this case, we can replace $\hat{A}$ by $A = \hat{A} + a\hat{\Phi}$ for any closed 1-form on M. This completes the proof of Proposition 15. □

4 The approximate solutions

In this section, we will define a procedure for gluing together vacuum configurations as defined in the preceding section on the interior of M and elements of the moduli space $\mathcal{M}_R^k(\mathbb{R}^3)$ on the Euclidean $\mathbb{R}^3$ transplanted to the ends of M. Recall therefore Definition 4. For $([\gamma], [\underline{c}])$ consider the vacuum configuration class $[v]$ corresponding to the mass

$$m = \frac{1}{n} \sum_i m(c_i) \tag{70}$$

and to the cohomology class $[\gamma]$ in Proposition 15. The aim of this section is to prove

Proposition 16. *For every $r > r_M$ and $\epsilon > 0$ there exists an $R > r$ and a smooth map*

$$\mu : \tilde{\mathcal{M}}_R^k \longrightarrow \mathfrak{Q}(M) \tag{71}$$

so that

$$\mu([\gamma], [\underline{c}]) = [v] \in \mathcal{V}(M) \text{ on } M_r \tag{72}$$

$$\omega_i{}^*(\mu([\gamma], [\underline{c}])) = [c_i] \qquad \text{on } \mathbb{R}^3 - B_{2r} \tag{73}$$

$$\|b(\mu([\gamma], [\underline{c}])\|_{Z_c} \leq K \cdot R^{-0.1}\,. \tag{74}$$

Proof. There exists a representative v of $[v]$ so that

$$v_i = \omega_i^* v = (m\tau, 0) \tag{75}$$

on $\mathbb{R}^3 - B_r$ by Proposition 15. Here τ is a constant element of $S^2 \subset su_2$. The next step is to find representatives c_i for $[c_i]$ which are close to $(m\tau, 0)$ on M_{2r}. To this end we use the fact that by Proposition 28, we can choose R large enough so that $|\Phi_i(x)| \geq \frac{m_0}{2}$ on B_{4r}. Hence we can choose a representative with

$$\Phi_i = \phi_i \tau \qquad \text{on } B_{4r}\,, \tag{76}$$

where $\phi_i = \Phi_i \cdot \hat{\Phi}_i$ is a real function. This immediately implies an estimate on the transversal component of A, since

$$\begin{aligned} A_T &= [\tau, [A, \tau]] = [\tau, \nabla_A \tau] = [\tau, (\nabla(\,|\Phi|^{-1}\,)\Phi + \,|\Phi|^{-1}\, \nabla_A \Phi] \\ &= \,|\Phi|^{-1}\, [\tau, \nabla_A \Phi]\,. \end{aligned} \tag{77}$$

This also implies an estimate on all terms of the form $A \wedge A$, since $A_L \wedge A_L = 0$. By Proposition 29 we have

$$\|(A_T, A \wedge A)\|_{W_2(B_{4r})} \leq K_2 \exp(-R/K_3)\,. \tag{78}$$

The gauge transformations on B_{4r} which preserve (76) are of the form $h(x) = \exp(f(x)\tau)$ with $f \in W^2_{\mathrm{loc}}(B_{4r}, \mathbb{R})$. They act on the longitudinal component of the connection according to the formula

$$\tau \cdot h(A_i) = \tau \cdot A_i + df\,, \tag{79}$$

see (4). In order to fix a gauge, we consider the continuous linear map

$$\Xi : W^1(\Lambda^1(B_{4r})) \longrightarrow W^2(\Lambda^0(B_{4r})) \tag{80}$$

which assigns to each 1-form α the unique $\Xi(\alpha)$ so that $\tilde{\alpha} = \alpha + d(\Xi(\alpha))$ satisfies $d^*\alpha = 0$ on B_{4r} and the normal component of $\tilde{\alpha}$ vanishes on the boundary of B_{4r}. The existence of this map follows from an abelian version of Lemmas 2. 6 and 2. 7 of [14]. We conclude that we can represent $[c_i]$ on B_{4r} by a configuration which in addition to (76) satisfies

$$d^* a_i = 0 \quad \text{on} \quad B_{4r} \tag{81}$$

$$x \cdot a_i = 0 \quad \text{on} \quad \partial B_{4r} \tag{82}$$

with $a_i := \tau \cdot A_i$. By Lemma 2. 5 of [14], these two equations imply that

$$\|a_i\|_{W_3(B_{4r})} \leq K_r\, \|da_i\|_{W_2(B_{4r})}\,. \tag{83}$$

Since

$$\tau \cdot F_{A_i} = \tau \cdot \left(dA_i + \frac{1}{2} A_i \wedge A_i \right) = da_i + \frac{1}{2}\,(A_i \wedge A_i) \cdot \tau\,, \tag{84}$$

we can use (78) and Proposition 29 to estimate

$$\begin{aligned} \|a_i\|_{W_3(B_{4r})} &\leq K_6 \left(\,\|F_{A_i}\|_{W_2(B_{4r})} + \,\|A_i \wedge A_i\|_{W_2(B_{4r})} \right) \\ &\leq K_7 R^{-2}\,. \end{aligned} \tag{85}$$

We now define the configuration $c = v\#\underline{c}$ by

$$\begin{aligned} c &= v \quad \text{on} \quad M_{2r} \\ \omega_i^* c &= \beta_r v_i + (1-\beta_r) c_i \quad \text{on} \quad \mathbb{R}^3 - B_r \end{aligned} \tag{86}$$

where, c.f. (46) in [2]

$$\begin{aligned} \beta_r &= 1 \quad \text{on} \quad M_r \\ \beta_r &= 0 \quad \text{on} \quad \mathbb{R}^3 - B_r \end{aligned} . \tag{87}$$

These definitions are consistent, since the Higgs field of the vacuum is parallel to τ on $B_{2r} - B_r$. Moreover, the gauge equivalence class of $v\#\underline{c}$ does not depend on the choice of the representatives v and c_i, as long as on B_{2r} they coincide with the representations defined above. It also does not depend of the choice of v outside M_{2r}. We can therefore define the map μ as

$$\mu([\gamma], [\underline{c}]) = [v\#\underline{c}] . \tag{88}$$

Let us now estimate $b(c)$ with $c = v\#\underline{c}$. Since c is self-dual on M_r, it suffices to estimate the terms

$$b_i(c) = \omega_i^*(\nabla_A \Phi - *F_A) \tag{89}$$

for $1 \le i \le n$. On $\mathbb{R}^3 - B_r$, we have

$$\begin{aligned} \omega_i^* \nabla_A \Phi &= (d\beta_r)(m\tau - \Phi_i) \cdot + \beta_r(1-\beta_r)[A_i, m\tau - \Phi_i] \\ &\quad + (1-\beta_r)\nabla_{A_i}\Phi_i \end{aligned} \tag{90}$$

$$\omega_i^* F_A = -(d\beta_r) \wedge A_i + \frac{1}{2}\beta_r(1-\beta_r) A_i \wedge A_i + (1-\beta_r) F_{A_i} \tag{91}$$

$$\begin{aligned} \omega_i^* \mu(c) &= \tau(d\beta_r(\phi_i - m) - *_i(d\beta_r \wedge a_i)) \\ &\quad + (1-\beta_r)(d_{A_i}\Phi_i - *_i F_{A_i}) + J_i \end{aligned} \tag{92}$$

where by (78),

$$\|J_i(x)\|_{W_2} \le K_{10} \exp(-R/K_{10}) . \tag{94}$$

Now it follows from (17) that for $p \ge \frac{6}{5}$,

$$\begin{aligned} \|d\beta_r(\phi_i - m) - *_i d\beta_r \wedge a_i\|_p &\le K_{11} \left(\|\phi_i - m(c_i)\|_{6<2r} + |m - m(c_i)| \right. \\ &\qquad \left. + \|a_i\|_{6<2r} \right) \\ &\le K_{11} R^{-0.1} \end{aligned} \tag{95}$$

by Definition 4. This completes the estimates for those parts of (57) which originate from the cutoff procedure. To estimate the metric dependent term, note that c_i is self-dual with respect to the Euclidean metric on $\mathbb{R}^3$. Using the formula

$$(*F)_k = (4 \det(g))^{-\frac{1}{2}} \epsilon_{klm} F_{lm} \tag{96}$$

for the Hodge dual with respect to a metric q and the property (6) of $\omega_i^* g$, we obtain the pointwise estimate

$$\left|\nabla_{A_i}\Phi_i - *_i F_{A_i}\right| \leq K_{12}\, |x|^{-1}\, \left|F_{A_i}\right| \,. \tag{97}$$

Now recall the pointwise estimate on the field strength in Proposition 28. A simple integration yields

$$\left\|\nabla_{A_i}\Phi_i - *_i F_{A_i}\right\|_{p,>r} \leq K_{13} R^{-0.2} \tag{98}$$

for $p \geq \frac{6}{5}$. This completes the proof of (74).

In order to show that μ is smooth, consider $[c] \in \mathcal{M}_R^k(\mathbb{R}^3)$ and a representative c satisfying (76), (81), and (82) on B_{4r}. In [2], Theorem 4, the structure of the Hilbert manifold $\mathcal{Q}(\mathbb{R}^3) = \mathcal{C}(\mathbb{R}^3)/\mathcal{G}$ is described by means of smooth charts s_c. Restricted to a neighborhood $\mathcal{U}_{[c]} \subset \mathcal{M}^k(\mathbb{R}^3) \subset \mathcal{Q}(\mathbb{R}^3)$, they define smooth maps

$$s_c : \mathcal{U}_{[c]} \longrightarrow Q_c \subset H_c(\Omega_{\mathrm{ad}}) \,. \tag{99}$$

In fact, by the results of [9], Section IV.2, this map takes values in $\tilde{Q}_c : Q_c \cap V_{2,A}(\Omega_{\mathrm{ad}})$ and is therefore a smooth map $\tilde{s}_c : \mathcal{U}_{[c]} \longrightarrow \tilde{\mathcal{U}} \subset \tilde{Q}_c$. Now, use [2], Definition 7.1, to construct a smooth map $\xi_T : \tilde{\mathcal{U}} \longrightarrow W^2(\Lambda^0_{\mathrm{ad}}(B_{4r}))$ so that

$$\exp(\xi_T)(c + (a,\phi)) =: c + (a',\phi') \tag{100}$$

satisfies (76) on B_{4r}. Then the map

$$\begin{aligned} g_c &: \tilde{\mathcal{U}} \longrightarrow W^2(\mathbb{R}^3, SU_2) \\ g_c(a,\phi) &= \exp(\beta_{2r}\Xi(a'\cdot\tau)\tau)\cdot\exp(\beta_{2r}\xi_T) \end{aligned} \tag{101}$$

with Ξ the continuous linear map of (80) is smooth and

$$c'(a,\phi) := g_c(a,\phi)(c + (a,\phi)) \tag{102}$$

coincides on B_{2r} with the unique representative of $c+(a,\phi)$ on B_{4r} satisfying (76), (81), and (82). We can now compose g_c with $\tilde{s}_c$. If we carry out this construction for each c_i in $\underline{c} = (c_1 \dots c_n)$, we obtain for each $([\gamma'],[\underline{c}'])$ close enough to $([\gamma],[\underline{c}])$ in $\tilde{\mathcal{M}}_R^k(M)$ a vector of configurations $\underline{c}'$. If we also define a smooth map

$$H^1(M,\mathbb{R}) \longrightarrow W^1(\Lambda^1) : [\gamma'] \longrightarrow v', \quad [\gamma] \longrightarrow v \tag{103}$$

so that v is supported in M_r and represents γ, then we can locally define a smooth map

$$\begin{aligned} \bar{\mu} &: \tilde{\mathcal{M}}_R^k \longrightarrow H_{v\#\underline{c}} \\ ([\gamma'],[\underline{c}']) &\longmapsto v'\#\underline{c}' - v\#\underline{c} \end{aligned} \tag{104}$$

so that $c + \bar{\mu}([\gamma'],[\underline{c}'])$ represents $\mu([\gamma],[\underline{c}])$. This completes the proof of Proposition 16. □

5 The linearized equation

The aim of this section is to show that for configurations which are flat on a large enough compact subset of M and which are approximately self-dual, the linearization

$$Db(c) : H_c(\Omega_{\mathrm{ad}}) \longrightarrow L^2(\Lambda^1_{\mathrm{ad}}) \tag{105}$$

has a finite dimensional cokernel which can be described in terms of the cohomology of M. This result will be necessary to perform the reduction in Section 6.

One readily obtains

$$Db(c)(a, \phi) = -\nabla_A \phi - [a, \Phi] + *d_A a\,. \tag{106}$$

$Db(c)$ has an infinite dimensional kernel due to the fact that b is equivariant under the operation of the automorphism group $C^\infty(\mathrm{Aut}\, P)$. One can combine $Db(c)$ with a linear «gauge fixing» term

$$f_c(a, \phi) = *d_A * a + [\Phi, \phi] = -d_A^* a + (\mathrm{ad}\,\Phi)\phi \tag{107}$$

to obtain an elliptic operator

$$\begin{aligned} \mathfrak{D}_c(a, \phi) &= (Db(c)(a, \phi),\ f_c(a, \phi)) \\ &= (*d_A a - d_A \phi, -d_A^* a) + ([\Phi, a], [\Phi, \phi]) \\ &= (\mathfrak{D}_A + \mathrm{ad}\,\Phi)(a, \phi)\,. \end{aligned} \tag{108}$$

Note that the operator $\mathfrak{D}_A$ is formally self-adjoint, i.e. for (a, ϕ) and (b, ψ) in $C_0^\infty(\Omega_{\mathrm{ad}})$, we have

$$\langle \mathfrak{D}_A(a, \phi), (b, \psi) \rangle = \langle (a, \phi), \mathfrak{D}_A(b, \psi) \rangle\ . \tag{109}$$

On the other hand, the operator $\mathrm{ad}\Phi$ is anti-self-adjoint since the Killing metric satisfies $((\mathrm{ad}\,\Phi)\alpha) \cdot \beta + \alpha \cdot (\mathrm{ad}\,\Phi)\beta = 0$. Therefore, the L^2 adjoint of $\mathfrak{D}_c$ is

$$\mathfrak{D}_c^+ = \mathfrak{D}_A - \mathrm{ad}\,\Phi\,. \tag{110}$$

To describe the symbol of $\mathfrak{D}_A$ as an elliptic operator, we use a Clifford module structure on the bundle Ω. For $\xi \in \Lambda^1$ and $(a, \phi) \in \Omega$ or Ω_{ad}, define

$$\xi \vee (a, \phi) = (\xi \surd a, -\xi\phi + *(\xi \wedge a))\,. \tag{111}$$

This defines a linear map $\Lambda^1 \longrightarrow \mathrm{End}\ (\Omega_{\mathrm{ad}})$ which is the symbol of $\mathfrak{D}_A$, i.e. in local terms, we have for a frame (ξ^i) in T^*M,

$$\mathfrak{D}_A = \sum_i \xi^i (\nabla_A)_i\,. \tag{112}$$

Lemma 17. *The operation* $\vee$ *defines the structure of a Clifford module on* Ω *and* Ω_{ad}, *i.e. for* $\xi,\ \nu \in \Lambda^1$, *we have*

$$\xi \vee \nu \vee + \nu \vee \xi \vee = -2(\xi \mid \nu)1\,. \tag{113}$$

Moreover, it defines a quaternionic structure on Ω *and* Ω_{ad} *in the sense that*

$$\xi \vee \nu \vee - \nu \vee \xi \vee = 2(\xi \times \nu)\vee \tag{114}$$

with $\xi \times \nu = *(\xi \wedge \nu)$.

This is proved by direct calculation. Lemma 17 can be used to derive the following well-known «Weizenböck formulas».

Lemma 18. *On* $C^\infty(\Omega_{\text{ad}})$, *we have*

$$\mathfrak{D}_c\mathfrak{D}_c^+ = -\nabla_A^2 - (\text{ad}\,\Phi)^2 + (*F_A - d_A\Phi)\vee + \text{Ric} \tag{115}$$

$$\mathfrak{D}_c^+\mathfrak{D}_c = -\nabla_A^2 - (\text{ad}\,\Phi)^2 + (*F_A + d_A\Phi)\vee + \text{Ric}\,, \tag{116}$$

where $\text{Ric}(a,\phi) = (Ric(a),0)$ *and* $\text{Ric}(a)_k = \text{Ric}^i_k\, a_i$ *with* $\text{Ric}^i_k = R^i_{lkm}g^{lm}$ *is the Ricci tensor.*

Now consider the special case where c is a vacuum configuration, see Proposition 15. Since Φ is covariantly constant, $\mathfrak{D}_c$ leaves the transversal and logitudinal subspaces invariant. On the subspace of longitudinal sections, it coincides with the operator

$$\mathfrak{D}(a,\phi) = (*da - d\phi, d^*a)\,. \tag{117}$$

The kernel of $\mathfrak{D}$ can be described in terms of the cohomology of M. Denote by $H^p_{DR}(M,\mathbb{R})$ the p-th de Rham cohomology group of M with real coefficients (see, for example [6]), and define

$$\mathbb{R}^n_0 = \{(x_1 \ldots x_n) \in \mathbb{R}^n \mid \sum_{i=1}^n x_i = 0\} \cong \mathbb{R}^{n-1}\,. \tag{118}$$

Proposition 19. *The operator* $\mathfrak{D}$ *defined in (117) is a self-adjoint operator on* $L^2(\Omega)$ *with domain* $W^1(\Omega)$ *and a Fredholm operator*

$$\mathfrak{D} : V^1(\Omega) \longrightarrow L^2(\Omega)$$

with

$$\text{kern}\,\mathfrak{D} = \text{cok}\,\mathfrak{D} =: k_M \cong \mathbb{R}^n_0 + H^1_{DR}(M,\mathbb{R})\,. \tag{119}$$

The isomorphism is given by

$$\lambda \longrightarrow \left(\int_{S_{i,r}} *\lambda\right)_{i=1,\ldots,n} + [\lambda]\,, \tag{120}$$

where $S_{i,r}$ *is the component of* ∂M_r, $r > r_M$, *at the i-th end of* M *and* $[\lambda]$ *is the class of the* Λ^1*-component of* λ *in de Rham cohomology.*

Proof. Clearly, $\mathfrak{D}$ is symmetric on $W^1(\Omega)$, since it is symmetric on $C_0^\infty(\Omega)$ and $C_0^\infty(\Omega)$ is dense in $W^1(\Omega)$. It remains to show that the domain of the adjoint operator

$$D(\mathfrak{D}^+) := \{\alpha \in L^2(\Omega) \mid \text{the distribution } \mathfrak{D}\alpha \in (C_0^\infty(\Omega))^*,$$
$$\text{can be represented by an element of } L^2(\Omega)\}$$

coincides with $W^1(\Omega)$. To see this, note that for every $\omega \in C_0^\infty(\Omega)$ we have with Lemma 14

$$\begin{aligned} \langle \mathfrak{D}\alpha, \mathfrak{D}\omega\rangle &= \left\langle \alpha, \mathfrak{D}^2\omega \right\rangle \\ &= \left\langle \alpha, (-\nabla^2 + \mathrm{Ric})\omega \right\rangle \\ &= \langle \nabla\alpha, \nabla\omega\rangle + \langle \alpha, \mathrm{Ric}\,\omega\rangle \ . \end{aligned} \tag{121}$$

Hence

$$\begin{aligned} |\langle \nabla\alpha, \nabla\omega\rangle| \ &\le \ |\langle \mathfrak{D}\alpha, \mathfrak{D}\omega\rangle| \ + \ |\langle \alpha, \mathrm{Ric}\,\omega\rangle| \\ &\le \ \|\mathfrak{D}\alpha\|_2 \ \|\mathfrak{D}\omega\|_2 \ + \ \|\alpha\|_2 \ \|\mathrm{Ric}\|_3 \ \|\omega\|_6 \\ &\le c \ \|\mathfrak{D}\alpha\|_2 \ \|\nabla\omega\|_2 \end{aligned} \tag{122}$$

by Proposition 16. Hence the map $\omega \longrightarrow \langle \alpha, \omega\rangle_{V^1}$ is continuous in ω in the V^1-norm. Applying the Riesz-representation theorem again yields $\alpha \in V^1$. Now we show that there exists $\gamma > 0$ so that for all $v \in V^1$ with $v \perp \ker \mathfrak{D}$ we have

$$\|\mathfrak{D}v\|_2 \ \ge \gamma \ \|v\|_{V^1} \ . \tag{123}$$

Assume therefore that for some sequence $v_i \in V^1(\Omega)$ we have

$$\text{(a)} \quad v_i \perp \mathrm{kern}\ \mathfrak{D} \quad \text{in the Hilbert space metric on } V^1(\Omega) \tag{124}$$

$$\text{(b)} \quad \|v_i\|_{V^1} \ \equiv 1 \tag{125}$$

$$\text{(c)} \quad \|\mathfrak{D}v_i\|_2 \ \longrightarrow 0 \, . \tag{126}$$

By (b), there exists a weakly convergent subsequence $v_i \rightharpoonup v_\infty$ in $V^1(\Omega)$. By (c) $\mathfrak{D}v_\infty = 0$, since the function $v \longrightarrow \|\mathfrak{D}v\|_2^2$ is weakly lower semi-continuous as a continuous convex function on $V^1(\Omega)$. By (a), $v_\infty = 0$. Since on the other hand,

$$\|\nabla v_i\|_2^2 \ \le \ \|\mathfrak{D}v_i\|_2^2 \ + \langle v_i, \mathrm{Ric}\, v_i\rangle \ ,$$

we find that for every $R > r_M$,

$$\begin{aligned} \|\nabla v_i\|_2^2 \ \le \ \|\mathfrak{D}v_i\|_2^2 \ &+ \ \|\mathrm{Ric}\|_\infty \ \cdot \ \|v_i\|_{2,<R}^2 \\ &+ \ \|\mathrm{Ric}\|_{\frac{3}{2},>R} \ \|v_i\|_6^2 \\ &\longrightarrow \ \|\mathrm{Ric}\|_{\frac{3}{2},>R} \ \|\nabla v_i\|_2^2 \ . \end{aligned}$$

For R large enough, this is a contradiction. This proves (123). In particular, we have shown that $\mathfrak{D}$ has a closed range. It remains to prove that the map in (120) is an isomorphism. Note therefore that for any closed form $\lambda \in C^\infty(\Lambda^1)$ the restrictions to the ends are exact since $H^1_{DR}(\mathbb{R}^3 - B_{r_M}) = 0$. Therefore, there exists a $\bar{\lambda} \in C_0^\infty$ so that $[\bar{\lambda}] = [\lambda]$. Now note that the functional

$$S_{\bar{\lambda}}[\phi] := \|\bar{\lambda} + d\phi\|_2^2$$

on $V^1(\Lambda^0)$ has a unique minimum ϕ_0 which satisfies $d^*(\bar{\lambda}+d\phi_0) = 0$. This shows that the map (120) is surjective. Now assume that λ is a harmonic 1-form in V^1 and that $[\lambda] = 0$, i.e. there exists a 1-form ϕ so that $\lambda = d\phi$. Since λ is also in the cokernel of $d + d^*$, $\lambda \in L^2$, and by Lemma 10, there exist numbers $m_1 \dots m_n$ so that $\omega_i^*\phi - m_i \in L^6(\mathbb{R}^3 - B_{r_0})$. Since ϕ is well defined only up to addition of constants, we can assume that $\Sigma_{i=1} m_i = 0$. The numbers then depend only on λ. Moreover, $\vec{m}\,(\lambda) := (m_1 \dots m_n)$ maps $\{\lambda \in \text{kern } (d + d^*) \mid [\lambda] = 0\}$ into $\mathbb{R}^n_0$. This map is injective, since if $m_i = 0$ for all i, then $\phi \in V^1(\Lambda^0)$, by Lemma 10, which contradicts the fact that λ is in the cokernel. To see that it is also surjective, consider the completion $V^1_{\underline{m}}$ of

$$C^\infty_{\underline{m}} = \{\phi \in C^\infty(M) \mid \omega_i^*\phi - m_i \quad \text{has compact support}\}$$

in the norm $\|\phi\|_{V^1} = \|\nabla\phi\|_2$. The functional $\|\phi\|^2_{V^1}$ has a unique minimum ϕ_0 and $\lambda := d\phi_0$ is a harmonic element of $L^2(\Lambda^1)$. By standard regularity theory, $\lambda \in V^1$ as well. □

We now derive some consequences of Proposition 19 for the operators

$$\mathfrak{D}_c : \hat{H}_c(\Omega_{\text{ad}}) \longrightarrow L^2(\Omega_{\text{ad}}),$$

with $c = v\#\underline{c}$ as defined in Section 4, and their restriction to $H_c(\Omega_{\text{ad}})$. Recall that we defined

$$\hat{H}_c(\Omega_{\text{ad}}) = H_c(\Omega_{\text{ad}}) \otimes \mathbb{R}^n,$$

where the additional n dimensions are generated by elements $(\Phi_{(i)}, 0)$, and $\Phi_{(i)}$, for $1 \le i \le n$, are certain generators of the gauge group associated to the configuration c, see [2], Definition 3.

Definition 20. Define

$$k_v = \{\beta_{r_M}\lambda\tau \mid \lambda \in k_M\}, \tag{127}$$

where we assume that r_M is large enough so that the bilinear form $(\mu, \lambda) \longrightarrow \langle\beta_{r_M}\lambda, \mu\rangle$ is nondegenerate on k_M. Moreover, define

$$\hat{k}_c = \{\lambda\#\underline{\xi} \mid \lambda \in k_M, \quad \xi_i \in \text{kern } \mathfrak{D}_{c_i}\}. \tag{128}$$

Here, $\lambda\#\underline{\xi}$ is defined in a similar way as $v\#\underline{c}$. Let μ_i, for $1 \leq i \leq n$, denote the asymptotic values of $|\Phi|^{-1}\ \Phi.\xi_0$, where $\xi = (\xi_0, \xi_1) \in \Lambda^0_{\text{ad}} \oplus \Lambda^1_{\text{ad}}$, and define

$$\mu = \frac{1}{n}\sum_{i=1}^{n} \mu_i .$$

Then set

$$\begin{aligned} \lambda\#\underline{\xi} &= (\mu, \lambda)\tau \quad \text{on} \quad M_r \\ \omega_i^*(\lambda\#\underline{\xi}) &= \beta_r(\mu, \lambda)\hat{\Phi} + (1-\beta_r)\xi_i \quad \text{on} \quad B_r . \end{aligned} \tag{129}$$

We also define

$$k_c = \hat{k}_c \cap H_c(\Omega_{\text{ad}}) .$$

Clearly, both k_v and k_c are finite dimensional subspaces of $L^2(\Omega_{\text{ad}}) \cap H_c(\Omega_{\text{ad}})$. According to [15], we have $\dim \mathcal{D}_{c_i} = 4k_i$. Hence

$$\dim k_c = 4\,|k|\ + \dim H^2(M, \mathbb{R}) .$$

Let us denote by π_v the L^2-orthogonal projection onto k_v and by π_c the H_c-orthogonal projection onto k_c. Moreover, define $\pi_v^\perp = 1 - \pi_v$ and $\pi_c^\perp = 1 - \pi_c$.

Proposition 21. *For every $\underline{k} \in \mathbb{N}^n$, there exist positive real constants R and C so that for $([\gamma], [\underline{c}]) \in \mathcal{M}_R^{\underline{k}}$ and $c = v\#\underline{c}$, the operator*

$$\mathcal{D}_c^\perp = \pi_v^\perp \mathcal{D}_c \mid_{k_c^\perp} : H_c(\Omega_{\text{ad}}) \supset k_c^\perp \longrightarrow k_v^\perp \subset L^2(\Omega_{\text{ad}}) \tag{130}$$

is an isomorphism. Its inverse $G_{v,\underline{c}}$ satisfies

$$\left\| G_{v,\underline{c}}\xi \right\|_c \leq C \, \|\xi\|_2 . \tag{131}$$

Proof. We first show that for R large enough, $\mathcal{D}_c^\perp$ is injective and has closed range, i.e. satisfies

$$\left\| \mathcal{D}_c^\perp \xi \right\|_2 \geq \frac{1}{C}\, \|\xi\|_c .$$

Assume therefore the contrary, i.e. assume that there exist sequences $c_{i\alpha} \in \mathcal{M}_\alpha^{k_i}(\mathbb{R}^3)$, $\alpha \in \mathbb{N}$, and $v_\alpha \in \mathcal{V}(M)$ and $\xi_\alpha \in H_{c_\alpha}(\Omega_{\text{ad}})$ with $c_\alpha = v_\alpha\#\underline{c}$ so that

$$\text{(a)}\, \xi_\alpha \perp k_{c_\alpha} \tag{132}$$

$$\text{(b)}\, \|\xi_\alpha\|_c = 1 \tag{133}$$

$$\text{(c)}\, \left\| \pi^\perp \mathcal{D}_{c_\alpha} \xi_\alpha \right\|_{2 \longrightarrow 0} . \tag{134}$$

To derive a contradiction, we first show that for every $r \geq r_M$

$$\xi_\alpha \longrightarrow 0 \text{ in } L^2(\Omega_{\mathrm{ad}}(B_r))\,. \tag{135}$$

Assuming that (135) holds, note that $\mathcal{D}_{c_\alpha} = \mathcal{D} + (\mathrm{ad}\,\Phi)^2$ on B_r for every r and for α large enough. Hence by Proposition 19,

$$\begin{aligned} \|\beta_r\xi_\alpha\|_{c_\alpha} &\leq C\,\left\|\pi^\perp\mathcal{D}_{c_\alpha}\beta_r\xi_\alpha\right\| \\ &\leq C\left(\,\left\|\pi_v^\perp\mathcal{D}_{c_\alpha}\xi_\alpha\right\|_2 + \|\xi_\alpha\|_{2,<2r}\,\|\nabla\beta_r\|_\infty\right), \end{aligned} \tag{136}$$

which converges to zero as $\alpha \longrightarrow \infty$. Now define

$$\xi_{i\alpha} = (1-\beta_n)\,\omega_i^*\xi_\alpha\,. \tag{137}$$

If for the self-dual configuration $c_{i\alpha}$, the operator $\mathcal{D}_{c_{i\alpha}}$ is defined with respect to the Euclidean metric on $\mathbb{R}^3$, then

$$\omega_i^*\mathcal{D}_{c_\alpha}(1-\beta_n)\xi_\alpha = \mathcal{D}_{c_{i\alpha}}\xi_{i\alpha} + \gamma_i(\xi_{i\alpha})\,, \tag{138}$$

where

$$|\gamma(\xi_{i\alpha})(x)| \;\leq\; |x|^{-1}\;|\nabla_A\xi_{i\alpha}(x)|\;. \tag{139}$$

This term is due to the decay properties of the metric ω_i^*g, see Definition 1. It is well known that $\mathcal{D}_{c_{i\alpha}}$ is Fredholm, i.e., invertible away from its kernel, see for example [1]. It follows from the definition of k_c and from the decay properties of the kernel elements of $\mathcal{D}_{c_{i\alpha}}$ that if $\pi_{i\alpha}$ denotes the $H_{c_{i\alpha}}$-orthogonal projection onto kern $\mathcal{D}_{c_{i\alpha}}$, then (132) implies that

$$\|\pi_{i\alpha}\xi_{i\alpha}\|_{c_{i\alpha}} \;\longrightarrow 0 \tag{140}$$

for $\alpha \longrightarrow \infty$. On the other hand,

$$\begin{aligned} \left\|\pi_{i\alpha}^\perp\xi_{i\alpha}\right\|_{c_\alpha} &\leq C\,\left\|\mathcal{D}_{c_{i\alpha}}\pi_{i\alpha}^\perp\xi_{i\alpha}\right\|_2 \\ &\leq C\left(\,\|\mathcal{D}_{c_\alpha}(1-\beta_r)\xi_\alpha\|_2 + \|\pi_{i\alpha}\xi_{i\alpha}\|_{c_{i\alpha}} + \|\gamma(\xi_{i\alpha})\|_2\right) \\ &\leq C\left(\,\left\|\pi_v^\perp\mathcal{D}_{c_\alpha}\xi_\alpha\right\|_2 + \|\xi_\alpha\|_{2.<2r}\,\|\nabla\beta_r\|_\infty \right. \\ &\quad \left. +\frac{1}{r}\,\|\xi_{i\alpha}\|_{c_{i\alpha}} + \|\pi_{i\alpha}\xi_{i\alpha}\|_{c_{i\alpha}}\right), \end{aligned} \tag{141}$$

which converges to C/r for $\alpha \longrightarrow \infty$. Here we have used the fact that $\pi_v\zeta = 0$ where ζ vanishes on M_{2r_M}, see the definition of k_v in (127). Since C does not depend on α, (140), (141), and (136) together contradict (134).

To prove (135), note that for every $R > r_M$, the decomposition

$$\xi_\alpha = \Phi_\alpha(\Phi_\alpha \cdot \xi_\alpha) + \nu_\alpha \tag{142}$$

is well defined on M_{2R} for α large enough. Let us consider the transversal part ν_α first. We have together with (117),

$$\begin{aligned} \|\mathfrak{D}_{C_\alpha}\beta_R\nu_\alpha\|_2^2 &= \|\mathfrak{D}_{A_\alpha}\beta_R\nu_\alpha]\|_2^2 + \|[\Phi_\alpha, \beta_R\nu_\alpha\|_2^2 + \langle\mathfrak{D}_{A_\alpha}\beta_R\nu_\alpha, [\Phi_\alpha, \beta_R\nu_\alpha]\rangle \\ &\quad + \langle[\Phi_\alpha, \beta_R\nu_\alpha], \mathfrak{D}_{A_\alpha}\beta_R\nu_\alpha\rangle \\ &= \|\mathfrak{D}_{A_\alpha}\beta_R\nu_\alpha\|_2^2 + \|\beta_R\nu_\alpha\|_2^2 \; |\Phi_\alpha|^2 \; . \end{aligned} \tag{143}$$

Here we have used the fact that $\mathfrak{D}_{A_\alpha}$ and $\mathrm{ad}(\Phi_\alpha)$ commute on M_{2R} for α large enough, since the Higgs field is covariantly constant, and that $\mathfrak{D}_{A_\alpha}$ is symmetric, whereas $\mathrm{ad}(\Phi_\alpha)$ is antisymmetric with respect to $< . >$. Estimating the left hand side, we obtain

$$\begin{aligned} \|\nu_\alpha\|_{2,<R} &\leq \|\beta_R\nu_\alpha\|_2 \\ &\leq m_i^{-1} \left(\|\nabla\beta_R \vee \nu_\alpha\|_2 + \|\beta_R\mathfrak{D}_{C_\alpha}\nu_\alpha\|_2 \right) \\ &\leq m_0^{-1} \left(\|\nabla\beta_R \vee \nu_\alpha\|_2 + \|\mathfrak{D}_{C_\alpha}\vee_\alpha\|_2 \right) . \end{aligned} \tag{144}$$

Since ν_α is bounded in L^6, we can choose a subsequence so that $|\nu_\alpha|$ converges weakly in $L^6(M, R)$. Let us denote the limit by f. Since $\|\nabla\beta_R \vee \nu_\alpha\|_2 \leq |\nabla\beta_R|_3 \; \|\nu_\alpha\|_6$ is continuous and convex, it is weakly lower semi-continuous on L^6, and we have with (138),

$$\limsup_{k\to\infty} \|\nu_k\|_{2<R} \leq m_0^{-1} \; \|\nabla\beta_R \vee f\|_2 \; =: \epsilon_R \, . \tag{145}$$

Since $\lim_{R\to\infty}\epsilon_R = 0$, this proves that $\nu_\alpha \longrightarrow 0$ in $L^6_{\mathrm{loc}}(\Omega_{\mathrm{ad}})$.

Now consider the sequence $\lambda_\alpha := \hat{\Phi}\cdot\xi_\alpha$. It is bounded in $L^6(\Omega)$ and therefore can be assumed to converge weakly to some $\lambda_\infty \in L^6(\Omega)$. Moreover, for any $R > r$, we have for α large enough,

$$\|\nabla\lambda_\alpha\|_{2,<R} = \left\|\hat{\Phi}\nabla_{A_\alpha}\xi_\alpha\right\|_{2,<R} \leq \|\xi_\alpha\|_{C_\alpha} \, , \tag{146}$$

so that λ_α is also bounded in $W^1(\Omega(B_R))$. By uniqueness of the weak limit we conclude that every weakly converging subsequence and therefore the whole sequence converges weakly to $\lambda_\infty \in W^1(\Omega(B_R))$. Together with (132), the estimate (140) also implies that $\|\nabla\lambda_\infty\|_2 \leq 1$, so that $\lambda_\infty \in V^1$. In a similar way we show that $\mathfrak{D}\lambda_\infty = 0$. In fact, for every $R > 0$, we have that

$$\|\mathfrak{D}\lambda_\alpha\|_{2,<R} = \left\|\hat{\Phi}.\mathfrak{D}_{C_\alpha}\xi_\alpha\right\|_{2,<R} \leq \|\mathfrak{D}_{C_\alpha}\xi_\alpha\|_2$$

converges to zero. Hence $\lambda_\infty \in k_M$. Now it follows from (129) and (133) that $\lambda_\infty = 0$. Hence the longitudinal part of ξ_α converges weakly to zero in V^1 and therefore converges to zero in L^2_{loc}. This completes the proof of Proposition 19. □

It follows similarly that Proposition 21 holds equally for $\mathfrak{D}_c$ replaced by $\mathfrak{D}_c^+$ and k_c replaced by $\hat{k}_c$. Moreover, Proposition 21 gives us the following:

Corollary 22. *For all $c \in \mathfrak{Q}^{\underline{k}}(M)$, the operator*

$$\mathfrak{D}_c : H_c(\Omega_{\text{ad}}) \longrightarrow L^2(\Omega_{\text{ad}})$$

has Fredholm index $4\,|k|$.

6 Newton's method

It is the aim of this section to prove

Proposition 23. *For R large enough and for $([\gamma], [\underline{c}]) \in \tilde{\mathcal{M}}_R^{\underline{k}}$, there exists $\epsilon > 0$ and a unique*

$$\xi(v, \underline{c}) \in H_{v\#\underline{c}}$$

satisfying

$$\|\xi(v, \underline{c})\|_c \leq \epsilon \tag{147}$$

$$b(v\#\underline{c} + \xi(v, \underline{c})) \in k_v \,. \tag{148}$$

Moreover, we have the estimate

$$\|\xi(v, \underline{c})\|_c \leq C\, \|b(v\#\underline{c})\|_2 \,. \tag{149}$$

By [2], Theorem 4, the tangent space of the quotient $\mathfrak{Q}(M) = \mathscr{C}(M)/\mathscr{G}$ is given by

$$\hat{\mathfrak{Q}}_c = \{\xi \in \hat{H}_c(\Omega_{\text{ad}}) \mid \langle \xi, \nabla_A \Phi_{(i)} \rangle_c = 0 \;\text{ for }\; 1 \leq i \leq n \;\text{ and }\; f_c\xi = 0\} \,. \tag{150}$$

Here, the $\Phi_{(i)}$ are the additional generators of $\hat{H}_c$ already used in the preceding section. The «gauge fixing» operator f_c is the L^2-adjoint of the linearization of the gauge group action, see (107). Expanding the section $\pi_v^{\perp} b$ around c, we obtain a smooth map

$$\begin{aligned} F : \hat{Q}_c &\longrightarrow k_v^{\perp} \subset L^2(\Omega_{ad}) \\ F(\xi) &= \pi_v^{\perp} b(c + \xi) \\ &= \pi_v^{\perp}(b(c) + Db(c)\xi + \xi\natural\xi) \,. \end{aligned} \tag{151}$$

By the definition (108) of $\mathfrak{D}_c$ and by Proposition 21, there exists an inverse

$$G : L^2(\Omega_{\text{ad}}) \supset k_v^{\perp} \longrightarrow \hat{Q}_c \cap k_c^{\perp} \,, \tag{152}$$

which is bounded independently of c as long as R is large enough. Moreover, Lemma 14 implies

$$\|\xi\natural\xi\|_2 \leq C\, \|\xi\|_c^2 \,. \tag{153}$$

Hence Proposition 23 follows from the following general result.

Proposition 24. *Assume that a smooth map*

$$f : E \longrightarrow F \tag{154}$$

between Banach spaces $(E, \| \quad \|)$ *and* F *has a Taylor expansion*

$$f(\xi) = f(0) + Df(0)\xi + N(\xi) \tag{155}$$

so that $Df(0)$ *has a right inverse G satisfying*

$$\|GN(\xi) - GN(\zeta\| \leq C_N (\|\xi\| + \|\zeta\|) \|\xi - \zeta\| \tag{156}$$

for some constant C_N. *If*

$$\|Gf(0)\| \leq \frac{1}{8C_N}, \tag{157}$$

then the zero-set of f *in* $B_\epsilon = \{\xi \in E \mid \|\xi\| < \epsilon\}$ *with* $\epsilon = \frac{1}{4C_N}$ *is a smooth manifold of dimension equal to the dimension of* kern $Df(0)$. *In fact, if*

$$K_\epsilon = \{\xi \epsilon \text{ kern } Df(0) \mid \|\xi\| < \epsilon\} \tag{158}$$

and

$$K_\epsilon^\perp = \{\xi \in GF \mid \|\xi\| < \epsilon\}, \tag{159}$$

then there exists a smooth function $\phi : K_\epsilon \longrightarrow K_\epsilon^\perp$ *so that* $f(\xi + \phi(\xi)) = 0$, *and all zeroes of* f *in* B_ϵ *are of the form* $\xi + \phi(\xi)$. *Moreover, we have the estimates*

$$\|\phi(\xi)\| \leq 2 \|Gf(0)\| \tag{160}$$

$$\|D\phi(\xi))\| \leq 8C_N \|Gf(0)\| . \tag{161}$$

Proof. Consider the decomposition

$$E = K \oplus K^\perp,$$

where $K =$ kern $Df(0)$ and $K^\perp = GF$. Then replacing f by Gf, we can restrict ourselves to the case $F = K^\perp$ and

$$f(\xi) = f(0) + \pi^\perp \xi + N(\xi), \tag{162}$$

where $\pi^\perp : E \longrightarrow K^\perp$ is the projection along K.

We first show that there exists a unique zero of f in $K_\epsilon^\perp$ for $\epsilon = (4C_N)^{-1}$. The other assertions then follows by the implicit function theorem.

The zeroes of f on $K^\perp$ are fixed points of the map

$$F(\xi) = -f(0) - N(\xi). \tag{163}$$

Note that F maps $K^{\perp}$ into itself since

$$\begin{aligned} \|F(\xi)\| &\leq \|f(0)\| + \|N(\xi)\| \\ &\leq \|f(0)\| + C_N \|\xi\|^2 \\ &\leq \epsilon . \end{aligned} \tag{164}$$

Moreover, F is contracting since, for $\xi, \zeta \epsilon K_{\epsilon}^{\perp}$,

$$\begin{aligned} \|F(\xi) - F(\xi)\| &= \|N(\xi) - N(\zeta)\| \\ &\leq 2C_N \cdot \epsilon \cdot \|\xi - \zeta\| \\ &\leq \frac{1}{2} \|\xi - \zeta\| . \end{aligned} \tag{165}$$

By the Banach fixed point theorem, there exists a unique solution of (162) in $K_{\epsilon}^{\perp}$. Moreover, we have the estimate

$$\begin{aligned} \|\xi\| &\leq \frac{1}{1-K} \|F(0) - 0\| \\ &\leq 2 \|f(0)\| . \end{aligned} \tag{166}$$

In order to prove the rest of the assertion, we show that $Df(\xi)$ restricted to $K^{\perp}$ is an isomorphism. In fact, we have

$$Df(\xi) |_{K^{\perp}} = \mathrm{id}(K^{\perp}) + DN(\xi) |_{K^{\perp}} , \tag{167}$$

and by the definition of the Frechet derivative

$$\|DN(\xi)\zeta\| \leq \|N(\xi) - N(\xi + \zeta)\| + \psi(\zeta) \|\xi\| , \tag{168}$$

where $\psi(0) = 0$ and ψ is continuous. Hence it follows from (165) that in the operator norm on E,

$$\begin{aligned} \|DN(\xi)\| &= \|Df(\xi) - Df(0)\| \\ &\leq 2C_N \|\xi\| . \end{aligned} \tag{169}$$

This proves that (167) is isomorphic for $\|\xi\| \leq \epsilon$, with bounded inverse. Hence the zero-set of f in B_{ϵ} is a smooth tangent space that projects isomorphically onto $K = \text{kern } Df(0)$. To prove (170), we use the identity

$$D\phi(0) = -\left(Df(\phi(0)) |_{K^{\perp}}\right)^{-1} \left(Df(\phi(0)) |_{K}\right) . \tag{170}$$

The operator norm of the first operator on the right hand side is bounded by the number 2. In order to estimate the second operator, note that $Df(0) |_{K} = 0$ and apply (169). □

7 Proof of the theorems

By Propositions 16 and 23, there exists for R large enough a smooth map

$$\mathcal{H} : \tilde{\mathcal{M}}_R^{\underline{k}}(M) \longrightarrow \mathcal{Q}^{\underline{k}}(M) : \mathcal{H}([\gamma], [\underline{c}]) = \mu([\gamma], [\underline{c}]) + \xi([\gamma], [\underline{c}]) \tag{171}$$

which satisfies conditions (14). Moreover, by Proposition 23, we have for every representative c of $\mathcal{H}([\gamma], [\underline{c}])$,

$$b(c) \in k_c\,, \tag{172}$$

where k_c is the finite dimensional space defined in (128). Hence $\tilde{b}$ has been reduced to

$$\tilde{b} : \tilde{\mathcal{M}}_R^{\underline{k}}(M) \longrightarrow \mathrm{Hom}(k_M, \mathbb{R}); \quad \tilde{b}([\gamma], [\underline{c}])(\lambda) := b^{\lambda}([\gamma], [\underline{c}])\,. \tag{173}$$

Now we assume that M satisfies the hypothesis of Theorem 2. We then have

$$\tilde{\mathcal{M}}_R^{\underline{k}}(M) = H_1(M, \mathbb{Z}) \times \mathcal{M}_R^{\underline{k}}(\mathbb{R}^3)\,, \tag{174}$$

see Definition 1. Moreover, since $k_M = 0$, the map $\mathcal{H}$ now takes values in the moduli space $\mathcal{M}^{\underline{k}}(M)$. The other assertions of Theorems 2 and 3 are now immediate consequences of the following result:

Proposition 25. *Suppose that* $\alpha_i \in T_{c_i}\mathcal{M}_R^{k_i}(\mathbb{R}^3)$, $1 \leq i \leq n$, *and* $u \in H^1(M, \mathbb{R})$. *Then, for R large enough,* $(u, \underline{\alpha})$ *represents an element of* $T_{([\gamma],[\underline{c}])}\tilde{\mathcal{M}}_R^{\underline{k}}$.

Since $H^1(M - M_r, \mathbb{R}) = 0$, we can represent u by a smooth closed 1-form with support in M_r. We now apply the procedure for the proof of Proposition 16 to the path

$$([v_t], [\underline{c}_t]) = ([v] + t[u], [\underline{c} + t\underline{\alpha}]) \tag{175}$$

in $\tilde{\mathcal{M}}_R^{\underline{k}}$, which is tangent to $(v, \underline{\alpha})$ at $t = 0$, where we have made the obvious identification between γ_t and v_t. Let $c := v\#\underline{c}$ be the representative of $\mu(\gamma, [\underline{c}_t])$ defined by (86)–(88). With the map $\bar{\mu}$ defined in (104), we construct the family

$$c_t = c + \bar{\mu}(tv, t\underline{\alpha}) \tag{176}$$

of representatives of $\mu([\gamma_t], [\underline{c}_t])$. If $\alpha_i = (a_i, \phi_i)$ then we define

$$\lambda_i := \hat{\Phi}.\phi_i \tag{177}$$

$$\lambda = \frac{1}{n}\sum_i \lambda_i\,. \tag{178}$$

By equation (101), we obtain

$$\dot{c} = (\lambda, u)\hat{\Phi} \quad \text{on} \quad M_r \tag{179}$$

$$\omega_i^* \dot{c} = \beta_r(\lambda_i, u)\hat{\Phi} + (1 - \beta_r)(\alpha_i + c_i X_i) \tag{180}$$

$$\text{on} \quad \mathbb{R}^3 - B_r \quad \text{for} \quad 1 \leq i \leq n\,. \tag{181}$$

where, with $\alpha_i = (a_i, \phi_i)$, the term X_i has the following form:

$$X_i = \frac{d}{dt} g_c(ta_i, t\phi_i) \\ = \beta_{2r}\left([\tau, \phi_i] + \Xi(\tau \cdot (a_i + d_{A_i}[\tau, \phi_i]))\right) , \tag{182}$$

see equation (101). We claim that

$$\|\mathfrak{D}_c \dot{c}\|_2 + \|\mathfrak{D}_c \dot{c}\|_{Z_c} \leq K_2 R^{-0.2} . \tag{183}$$

In fact, we have $\mathfrak{D}_c \dot{c} = 0$ on M_r, and on $\mathbb{R}^3 - B_{4r}$,

$$|\omega_i^* \mathfrak{D}_c \dot{c}| = |\mathfrak{D}^g_{c_i} \alpha_i| \\ \leq C\, |x|^{-1}\, |\nabla_{A_i} \alpha_i| ,$$

since $\mathfrak{D}_{c_i} \alpha_i = 0$ with respect to the Euclidean metric α_i, and by (6) of Definition 1. With (183) and (191), we can therefore estimate

$$\|\mathfrak{D}_c \dot{c}\| \leq K_4 \sum_i \left(\|\alpha_i\|_{W_1(B_{4r})} + \|\rho^{-1} \nabla_{A_i} \alpha_i\|_{2,>r} \left(+ \|\rho^{-1} \nabla_{A_i} \alpha_i\|_{\frac{6}{5},>r} \right) \right) . \tag{184}$$

On the other hand, α_i is concentrated at the zero-set of c_i, which lies outside B_R for $([\gamma], [\underline{c}]) \epsilon \tilde{\mathcal{M}}_R$. In fact, (211) of Proposition 28 completes the proof of (183).

Now Lemma follows immediately from

$$\|\pi^\perp \dot{c}\|_c \leq C\, \|D_c \pi^\perp \dot{c}\| \\ = C\, \|D_c \dot{c}\| \\ \leq C R^{-0.2}$$

and

$$\|\pi^\perp \dot{c}\| \leq C\, \|D_c \pi^\perp \dot{c}\|_{Z_c} \\ \leq C\, \|D_c \dot{c}\|_{Z_c} \\ \leq C R^{-0.2} .$$

This completes the proof of Proposition. 25

We will later need the following:

Lemma 26. *With*

$$\dot{\xi} = \left(\frac{d}{dt} \xi(c_t) \right)_{t=0} ,$$

we have

$$\|\dot{\xi}\|_c \leq K R^{-0.2} .$$

Proof. This is a consequence of (161) of Proposition 24.

8 The reduced function

In this section, we prove Theorem 7. Recall therefore the expansion (151).

$$\begin{aligned} b^{\lambda}([\gamma],[\underline{c}]) &= \left\langle \lambda\hat{\Phi}, b(c+\xi)\right\rangle \\ &= \left\langle \lambda\hat{\Phi}, b(c)\right\rangle + \left\langle \lambda\hat{\Phi}, \mathfrak{D}_c\xi\right\rangle + \left\langle \lambda\hat{\Phi}, \xi\natural\xi\right\rangle . \end{aligned} \tag{185}$$

Note that $b(c)$ is supported outside M_r, since c is a flat configuration on M_r. Moreover, since the metric g is assumed to be flat outside M_r, the only contribution to $b(c)$ comes from the cut-off function in (87). Let us set

$$x_i = \langle \lambda_i, d\beta_r \vee (\phi_i, a_i)\rangle + J_1 , \tag{186}$$

where we have used the abbreviations $\phi_i = |\Phi_i|$ and $a_i = A_i \cdot \tau$. Here and in the following, we will denote by J_i and $J_i(x)$ terms which satisfy estimates of the type

$$|J_i(x)| \le K e^{-R/K} . \tag{187}$$

We now use the equations

$$\tau \cdot F_{A_i} = da_i + \frac{1}{2}(A_i \wedge A_i)\cdot \tau , \tag{188}$$

see (84), and

$$\tau \cdot d_{A_i}\Phi_i = d\phi_i \tag{189}$$

to conclude that

$$d\phi_i = *da_i + J_2 . \tag{190}$$

Now we have the following lemma:

Lemma 27. *For any section (ϕ, a) of $\Omega(\mathbb{R}^3)$ we have with $e([c],\lambda)$ defined in* (23),

$$|\langle \lambda_i, d\beta_r \vee (\phi, a)\rangle - e([c],\lambda_i)| \le C\, \|d\phi - *da\|_{W_1(B_{2r})} . \tag{191}$$

Proof. Consider the 3-form

$$w = (\lambda \mid d\beta_r(m-\phi) + *(d\beta_r \wedge a))\mu_{g_i} , \tag{192}$$

on $\mathbb{R}^3$, where μ_{g_i} is the Riemannian volume form on the end which we are considering. Now the first term in (191) is the integral of w over $\mathbb{R}^3$. We can transform w using some of the relations listed in Section 2:

$$\begin{aligned} w &= (\lambda \mid d\beta_r(m-\phi)) + (\lambda \mid *(d\beta_r \wedge a))\,\mu_g \\ &= (\lambda \mid d\beta_r(m-\phi))\,\mu_g + (*(d\beta_r\wedge a)\mid\lambda)\,\mu_g \\ &= *\lambda \wedge d\beta_r(m-\phi) + **\,(d\beta_r \wedge a)\wedge\lambda \\ &= d\beta_r \wedge (*\lambda)(m-\phi) + d\beta_r \wedge a \wedge \lambda \\ &= d\beta_r \wedge (a\wedge\lambda + (*\lambda)(m-\phi)) \\ &=: d\beta_r \wedge \gamma . \end{aligned} \tag{193}$$

We can eliminate the cutoff function β_r in (193) in the following way. Let ψ be a function with $d\psi = \lambda$ on $\mathbb{R}^3 - B_{r_0}$. Then

$$\begin{aligned}\gamma &= a \wedge \lambda - *\lambda(\phi - m) \\ &= a \wedge d\psi - *d\psi(\phi - m) \\ &= d(a\psi) + da \wedge \psi - (*d\psi)(\phi\, m) \\ &= d(a\psi) + *(\psi d\phi - \phi d\psi) + J_1 .\end{aligned} \tag{194}$$

We conclude that for any $\rho \in [r, 2r]$,

$$\int_{S^2_\rho} \gamma = \int_{S^2_\rho} \{*(\psi d\phi - (\phi - m)d\psi) + J_1\} . \tag{195}$$

Hence by Green's formula,

$$\int_{S^2_\rho} *(\psi d\phi - \phi d\psi) = \int_{B_\rho} (\psi \Delta\phi - \phi\Delta\psi) . \tag{196}$$

Since

$$\Delta\phi = *d * d\phi = *d\ *(d\phi) - *da) , \tag{197}$$

this proves (191). □

Now Theorem 7 is proved as soon as we obtain proper estimates for the second and third term in (185). Since λ is harmonic on M, an integration by parts yields

$$\begin{aligned}\left|\left\langle \lambda\hat{\Phi}, \mathfrak{D}_c\xi\right\rangle\right| &= \left|\left\langle \mathfrak{D}_c(\lambda\hat{\Phi}), \xi\right\rangle\right| = \left|\left\langle \nabla_A\hat{\Phi}\natural\lambda, \xi\right\rangle\right| \\ &= \left(\left\|\nabla_A\hat{\Phi}\right\|_{\infty,<\frac{R}{3}} \|\lambda\|_2 + \left\|\nabla_A\hat{\Phi}\right\|_2 \|\lambda\|_{\infty,>\frac{R}{3}} \right) \|\xi\|_2 \\ &\le C\left(e^{-\mu R} + R^{-p+2}\right) R^{-0.1} .\end{aligned} \tag{198}$$

Moreover, by Proposition 29, we have

$$\begin{aligned}\left|\left\langle \lambda\hat{\Phi}, \xi\natural\xi\right\rangle\right| &= \left\langle \lambda, [\hat{\Phi}, \xi]\natural\xi\right\rangle \\ &\le \left(\|\lambda\|_2 \left\|[\hat{\Phi}, \xi]\right\|_{\infty,>\frac{R}{3}} + \|\lambda\|_{\infty,<\frac{R}{3}} \left\|[\hat{\Phi}, \xi]\right\|_2 \right) \|\xi\|_2 \\ &\le C(e^{-\mu R} + R^{-p+2})R^{-0.1} .\end{aligned} \tag{199}$$

This proves Theorem 7.

In order to prove the second assertion of Theorem 5, first recall the description of the tangent space of $\tilde{\mathcal{M}}_R^k$ in (176)–(181). We calculate in the notation of (176)–(184) that

$$D_\nu b^\lambda(v,[\underline{c}]) = \frac{d}{dt}\left\langle \lambda\hat{\Phi}_t, b(c_t+\xi_t)\right\rangle |_{t=0} \\ = \left\langle \lambda\frac{\partial}{\partial t}\hat{\Phi}_t |_{t=0}, b(c+\xi)\right\rangle + \left\langle \lambda\hat{\Phi}, \frac{d}{dt}(c_t+\xi_t) |_{t=0}\right\rangle . \tag{200}$$

The first term vanishes, since $b(c+\xi)\subset K_c$ has support in B_{2r} and is purely longitudinal there, whereas $\frac{\partial}{\partial t}\hat{\Phi}_t |_{t=0}$ is purely transversal on B_{2r}. Hence

$$D_\nu b^\lambda(v,[c]) = \left\langle \lambda\hat{\Phi}, Db(c)\dot{c}\right\rangle + \left\langle \lambda\hat{\Phi}, \mathfrak{D}_c\dot{\xi}\right\rangle + \left\langle \lambda\hat{\Phi}, \xi\natural(\dot{c}+\dot{\xi})\right\rangle . \tag{201}$$

We have by (183),

$$Db(c)\dot{c} = Db(c)(\beta_r(\lambda,\omega)\hat{\Phi} + (1-\beta_r)(\alpha_i + d_{c_i}X_i)) \\ = d\beta_r \vee ((\lambda,\omega)\tau - (\alpha_i + d_{c_i}X_i)\,. \tag{202}$$

Here we have used the fact that $Db(c)(\alpha_i + d_{c_i}X_i) = 0$ since the metric is flat on $M - M_r$. Hence

$$X_i = \langle \lambda_i, d\beta_r \vee (a,b)\rangle + J \tag{203}$$

with

$$(a,b) = (\lambda,\omega) - \tau\cdot\alpha_i + (o, d\tau\dot{X}_i) \tag{204}$$

and J exponentially small in R. It is now easy to see that (a,b) satisfies the hypothesis of Lemma 27. Moreover, by Lemma 26, we have for $\dot{\xi}$ the same estimates as for ξ, so that we can estimate the second and third term of (201) in a similar way as we did for (37). This completes the proof of Lemma 27.

9 Estimates

In this section, we want to give estimates on the fields of Yang-Mills configurations and on the correction terms ξ of Section 6 away from the zero-set of the Higgs field of c. For any $c\in C$, define therefore

$$d = d_c : M \longrightarrow \mathbb{R} : d_c(x) = \mathrm{dist}(\mathrm{zero}\ (\Phi), x) \tag{205}$$

and the set

$$U_r = U_{r,c} = \{x\in M \mid d_c(x) \geq r\}\,. \tag{206}$$

Proposition 28. *Let M be an asymptotically flat manifold and let $c = (A, \Phi)$ be a self-dual configuration in $\mathcal{C}(M)$. Then we have*

$$\|[\Phi, (B, E)]\|_{2,U_r} + \|\nabla_A \nabla_A [\Phi, (B, E)]\|_{2,U_r} \leq K_0 \exp(-r/K_0). \tag{207}$$

Moreover, if $\alpha \in \acute{H}_c(\Lambda^0) + H_c(\Lambda^1_{\mathrm{ad}})$ satisfies

$$\mathcal{D}_c \alpha = \mu E \tag{208}$$

for $\mu \in \mathbb{R}$, then

$$\|[\Phi, \alpha]\|_{2,U_r} + \|\nabla_A \nabla_A [\Phi, \alpha]\|_{2,U_r} \leq K_0 (\|\alpha\|_c + |\mu|) \exp(-r/K_0). \tag{209}$$

If $(M, g) = (\mathbb{R}^3, g_0)$, then there exists a continuous function $K : [0, \frac{1}{2}) \longrightarrow \mathbb{R}$ depending only on the masses and the action of c so that for $\epsilon \in [0, \frac{1}{2})$,

$$\|\delta^\epsilon (B, E)\|_{2,U_r} + \|\delta^\epsilon \nabla_A \nabla_A (B, E)\|_{2,U_r} \leq K(\epsilon) r^{-1} \tag{210}$$

$$\|\delta^\epsilon \alpha\|_{2,U_r} + \|\delta^\epsilon \nabla_A \nabla_A \alpha\|_{2,U_r} \leq K_0 (\|\alpha\|_c + |\mu|) \exp(\frac{-r}{K_0}). \tag{211}$$

In Proposition 28, K_0 can be assumed to coincide with the constant of Section 2. We remind the reader of the convention that K_i, $i \in \mathbb{N}$, denote constants depending only on the masses and the action of the configuration c in question. Our next result concerns the correction term ξ of Section 6.

Proposition 29. *Let $c = v \neq \underline{c}$ be as in Proposition 16 and $\xi = \xi(c)$ as in Proposition 23. Let us further denote $\dot{\xi} = D_\alpha \xi(c)$ for some $\alpha \in T_{[w,c]} \tilde{\mathcal{M}}^k_R$. Then we have for $R \geq r$ large enough and for $U_r = U_r(c)$,*

$$\|[\Phi, \xi]\|_{X,U_r} + \|[\Phi, \dot{\xi}]\|_{X,U_r} \leq K_0 \exp(-\frac{r}{K_0}) \tag{212}$$

$$\|\delta^\epsilon (\xi, \dot{\xi})\|_{2,U_r} + \|\delta^\epsilon \nabla_A \nabla_A (\xi, \dot{\xi})\|_{2,U_r} \leq K(\epsilon) r^{-1}. \tag{213}$$

To prove these two proposition, we will first show how in Propositions 28 and 29 the estimates on the full field follow from the estimates on the transversal part. We will then give a proof of (212) of Proposition 29, assuming that (210) and (211) hold. Finally, we will point out how the same method can be used to prove the transversal estimates of Proposition 28, without using the results of 29.

In order to derive (210) and (211) from (207) and (208), it suffices to consider the longitudinal components of the fields E, B and α. Therefore, define the cutoff function

$$\beta_r = \beta_{r,c} = 1 - \beta(\delta_c(x)/r) \tag{214}$$

and the normalized Higgs field

$$\hat{\Phi}_r = \beta_{r,c}\Phi\,|\Phi|^{-1}\;, \tag{215}$$

which for r large enough is well defined by Lemma C.14 of [12]. Now note that the self-duality equation $E = B$, together with the identities $d_A E = [F_A, \Phi] = *[B, \Phi]$ and $d_A^* B = *d_A F_A = 0$, implies that $\mathfrak{D}_c E = 0$. For the longitudinal component $e := E \cdot \hat{\Phi}$, this implies

$$\mathfrak{D}e = E \vee \nabla_A \hat{\Phi}\,. \tag{216}$$

Similarly, it follows from (209) that for $a := \alpha \cdot \hat{\Phi}$,

$$\mathfrak{D} = a \vee \nabla_A \hat{\Phi}\,. \tag{217}$$

Now (210) and (211) follow from (207) and (208) through [10], Theorem 1. We now come to the estimates (212) on the transversal part of the correction term. The method of the proof is the one of [9], V. We first derive a pointwise estimate on the fields and their covariant derivatives by using a second order elliptic equation. Then we consider the transversal part of this equation and apply the comparison principle.

Note that by Proposition 23 we already know that

$$\|\xi\|_c \;\leq K_1 R^{-0.1}\,. \tag{218}$$

In order to derive a second order equation for ξ, apply the operator $\mathfrak{D}_c^+$ to the defining equation for ξ:

$$0 = \mathfrak{D}_c^+ \pi_c^\perp b(c+\xi) = \mathfrak{D}_c^+(1-\pi_c)(b(c) + \mathfrak{D}_c\xi + \xi\natural\xi)\,. \tag{219}$$

By Lemma 18, we have

$$\mathfrak{D}_c^+\mathfrak{D}_c = L_c + \mathrm{Ric} + (E+B)\vee\,, \tag{220}$$

with $L_c = -\nabla_A^2 - (\mathrm{ad}\,\Phi)^2$. Hence

$$L_c\xi = -\mathfrak{D}_c^+\{b(c) + \xi\natural\xi\} + \mathfrak{D}_c^+\pi_c b(c+\xi) + \frac{1}{2}\,\mathrm{Ric}(\xi) + (E+B)\vee\xi\,. \tag{221}$$

Now note that

$$\|\mathfrak{D}_c^+ b(c)\|_Z \;\leq K_2 R^{-0.1}. \tag{222}$$

$$\xi\;\|\mathfrak{D}_c^+\xi\natural\|_Z \;\leq K_2\;\|\xi\|_C\;\;\|\xi\|_X \;\leq K_3 R^{-0.1}\;\|\xi\|_X \tag{223}$$

$$\|\mathfrak{D}_c^+\pi_c b(c+\xi)\|_Z \;\leq K_2\;\|b(c+\xi)\|_2 \;\leq K_3 R^{-0.1} \tag{224}$$

$$\|\mathrm{Ric}(\xi)\|_Z \;\leq (\;\|\mathrm{Ric}\|_\infty \;+\; \|\mathrm{Ric}\|_3\,)\;\|\xi\|_2 \;\leq K_3 R^{-0.1} \tag{225}$$

$$\|(E+B)\vee\xi\|_Z \;\leq K_2\;\|\xi\|_c \;\leq K_3 R^{-0.1}\,. \tag{226}$$

Inequality (222) follows easily from the formula (92) for $b(c)$. In fact, the cutoff-dependent part of $b(c)$, which has support in M_{2r}, can be estimated with the help of (78). The contribution from the metric can be estimated using Proposition 21 and the properties of $\gamma_i = \omega_i^* g - g_0$, see (6) of Definition 1. For later convenience, we note here already that by the above estimates, we even have in addition to (222)

$$\|\nabla_A \mathfrak{D}_c^+ b(c)\|_Z \le K_4 R^{-0.1}\,. \tag{227}$$

Putting (222)–(226 into (221), we obtain the estimate

$$\begin{aligned} \|\xi\|_X &\le K_5 \|L_c \xi\|_Z \\ &\le K_6 R^{-0.1} (1 + \|\xi\|_X)\,. \end{aligned} \tag{228}$$

Hence

$$\|\xi\|_X \le K_7 R^{-0.1}\,. \tag{229}$$

In a similar way, we obtain an equation for $\dot\xi$. In the notation of Section 7, we have

$$\begin{aligned} 0 &= \mathfrak{D}_c^+ \frac{d}{dt} \pi_{c_t}^\perp (b(c_t + \xi_t), f_c \xi_t) \\ &= \mathfrak{D}_c^+ \pi_c^\perp \left(Db(c+\xi)\dot\xi, f_c \dot\xi + \dot c \natural \xi\right) \\ &= \mathfrak{D}_c^+ \pi_c^\perp \{Db(c)\dot c + \mathfrak{D}_c \dot\xi + \xi \natural \dot c\}\,. \end{aligned} \tag{230}$$

Hence

$$\begin{aligned} L_c \dot c = & -\mathfrak{D}_c^+ \{\xi \natural \dot\xi + Db(c)\dot c + \xi \natural \dot c\} \\ & + \mathfrak{D}_c^+ \pi_c \{\xi \natural \dot\xi + db(c)\dot c + \xi \natural \dot c + \mathfrak{D}_c^+ \mathfrak{D}_c \dot\xi\} \\ & + \frac{1}{2}\operatorname{Ric}(\dot\xi) + (E+B) \vee \dot\xi\,. \end{aligned} \tag{231}$$

The term $Db(c)\dot c$ has already been estimated in (183). Using the estimates (211) of Proposition 28 on $\alpha \in \operatorname{kern} \mathfrak{D}_{c_i}$, we also obtain

$$\|\mathfrak{D}_c^+ \mathfrak{D}_c \dot c\| \le \|\mathfrak{D}_c \dot c\|_Z + \|\nabla_A \mathfrak{D}_c \dot c\|_Z \le K_8 R^{-0.1}\,. \tag{232}$$

Since the other terms of (231) can be estimated in a similar way as in (223)–(226), we obtain

$$\|\dot\xi\|_X \le K_9 R^{-0.1}\,. \tag{233}$$

Since we will also need a pointwise estimate on $\nabla_A \xi$ and $\nabla_A \dot\xi$, we have to push this method one step further. Let us therefore take the covariant derivatives of equations (221) and (228):

$$L_c \nabla_A(\xi, \dot\xi) = (L_c \nabla_A - \nabla_A L_c)(\xi, \dot\xi) + \nabla_A L_c(\xi, \dot\xi)\,. \tag{234}$$

By [4], Lemma 4.3, we have

$$
\begin{aligned}
(\nabla_A L_c - L_c \nabla_A)\xi &= (\nabla_A \nabla_A^2 - \nabla_A^2 \nabla_A)\xi \\
&\quad + (\nabla_A \operatorname{ad}\Phi - \operatorname{ad}\Phi \nabla_A)\operatorname{ad}\Phi\xi \\
&\quad + \operatorname{ad}\Phi(\nabla_A \operatorname{ad}\Phi - \operatorname{ad}\Phi\nabla_A)\xi \\
&= d_A^*\,[\xi, F(\nabla_A)] + (d_A\xi)\surd F(\nabla_A) \\
&\quad + \operatorname{ric}(\nabla_A\xi) \\
&\quad + E \vee [\Phi,\xi] + [\Phi, E\vee\xi]\,.
\end{aligned}
\tag{235}
$$

Here, $F(\nabla_A)$ denotes a certain combined operation of the metric curvature R_g and the curvature F_A of the connection. We can now use the fact that $R_g(x) \leq c\rho^{-3}(x)$ by Definition 1 and (207) of Proposition 28 to obtain

$$
\|(\nabla_A L_c - L_c\nabla_A)\xi\|_Z \leq K_{10}\,\|\xi\|_X\,. \tag{236}
$$

We can also estimate the covariant derivatives of the terms in (222)–(226) in a similar way. Moreover, we treat the $\dot{\xi}$-component of equation (234) to obtain

$$
\left\|\nabla_A(\xi,\dot{\xi})\right\|_X \leq K_{11}R^{-0.1}\,. \tag{237}
$$

Let us now consider the fields $E_i = B_i$ of the configurations c_i. Equation (116) together with the invertibility of $\mathfrak{D}_c$ on $\mathbb{R}^3$ see [9], Lemma IV 4. 3, imply that with β_r as in (214) and with equation (116)

$$
\begin{aligned}
\|E\|_{c,U_r} &\leq K_{12}\,\|\mathfrak{D}_c^+(\beta_r E)\|_2 \\
&\leq K_{13}\,(\,\|\nabla\beta_r\|_\infty\,\|E\|_2 + \|\beta_r E\|_2\,) \\
&\leq K_{14}\cdot r^{-0.5}\,.
\end{aligned}
\tag{238}
$$

Here, we dropped the index i and used the estimates of [12], Lemma C. 1.4 on the fields of self-dual configurations on $\mathbb{R}^3$. Similarly, we obtain a second order equation for E by

$$
\mathfrak{D}_c\mathfrak{D}_c^+ E = L_c E = E\natural E\,. \tag{239}
$$

The first equality is due to the Weizenböck formula (see Lemma 18) for $\mathfrak{D}_c\mathfrak{D}_c^+$ in the case of self-dual configurations on $\mathbb{R}^3$. We now obtain a second order estimate by

$$
\begin{aligned}
\|E\|_{X,U_{2r}} &\leq K_{15}\,\|L_c(\beta_r E)\|_Z \\
&\leq K_{16}\left\{\,\|E\|_{c,U_r}^2 + \alpha(c,U_r)^{\frac{1}{2}}\right\} \\
&\leq K_{17} r^{-0.5}\,,
\end{aligned}
\tag{240}
$$

where we have used the fact that $\|\nabla\beta_r\|_p$ and $\|\Delta\beta_r\|_p$ are bounded independently of r for $p \in [3, \infty]$. Again, we use (206) together with (236) and (250) to estimate the covariant derivative of the field

$$\|\nabla_A E\|_{X,\mathcal{U}_r} \leq K_{18} \,. \tag{241}$$

Finally, consider the section α on $\mathbb{R}^3$ with

$$\mathfrak{D}_c \alpha = \mu_E \,. \tag{242}$$

In a similar way as for the fields (E_i, B_i) we obtain a second order equation

$$\begin{aligned} L_c\alpha &= \mathfrak{D}_c\mathfrak{D}_c^+\alpha = E \vee \alpha + \mu\mathfrak{D}_c E \\ &= E \vee \alpha \,, \end{aligned} \tag{243}$$

and an estimate

$$\|\alpha\|_{X,\mathcal{U}_r} + \|\nabla_A\alpha\|_{X,\mathcal{U}_r} \leq K_{19}\, |\alpha|\, r^{-0.5} \tag{244}$$

with $|\alpha|^2 = \|\alpha\|_c^2 + |\mu|^2$, as in (179).

For $1 \leq i \leq n$, define now the quantities

$$\begin{aligned} \theta_i &= \left(\omega_i^*\xi, \omega_i^*\dot{\xi}\right) \\ \phi_i &= \left(E_i, \alpha_i\right) \,. \end{aligned} \tag{245}$$

We want to obtain exponential estimates for $[\Phi_i, \theta_i]$ and $[\Phi_i, \theta_i]$ on the set $\mathring{\mathcal{U}}_r = \mathcal{U}_r(c_i) - B_r$. Here we assume that r is large enough so that

$$(\omega_i^* g)_{\alpha\beta} = \delta_{\alpha\beta} \tag{246}$$

$$\omega_i^* = c_i \tag{247}$$

$$\omega_i^* \dot{c} = \alpha_i \tag{248}$$

$$\omega_i^* b(c) = \gamma_i \natural B_i \tag{249}$$

$$\omega_i^* D_c \dot{c} = \gamma_i \natural \nabla_A \alpha_i \tag{250}$$

$$|\Phi_i| \geq \frac{1}{2} m(c_i) \tag{251}$$

$$\lambda = 0 \quad \text{for} \quad \lambda \in k_c \,. \tag{252}$$

In the following, we drop the index i and set $\theta' = [\Phi, \theta]$. Following Taubes [9], we will prove that the real function on $\mathbb{R}^6$,

$$u(x_1, x_2) = |\Phi'(x_1)|^2 + |\theta'(x_2)|^2$$

is subharmonic on $\overset{\circ}{U}{}_r^2 := \overset{\circ}{U}{}_r^2 \times \overset{\circ}{U}{}_r^2$ in the sense that

$$-\Delta u + K_{20}^{-1} u \leq 0 \tag{253}$$

on this set. Here, Δ is the Laplacian on $\mathbb{R}^6$ with respect to the metric $g_0 \times \omega_i^* g$. Before proving (253), we will show that it implies exponential estimates. Note therefore that by (228), (233), (230) and (244), we have

$$|\phi'|^2 + |\theta'|_2 \leq K_{21} \quad \text{on} \quad \partial \overset{\circ}{U}{}_r^2 \,. \tag{254}$$

We want to use the comparison principle to prove that on $\overset{\circ}{U}{}_r^2$,

$$\begin{aligned} u = & \; |\Phi'(x)|^2 + |\theta'(x|^2 \leq v(x) \\ := & \; K_{22} \exp(-|x| / K_{22}) + \sum_{\alpha=1}^{k(c)} \exp\left(-|x - x_\alpha(c)| / K_{22}\right) , \end{aligned} \tag{255}$$

where $x_\alpha(c)$ are the «centers» of the monopole c as in Lemma C. 1.2 of [12]. One readily calculates that for r large enough, we have $\Delta v \leq -K_{23}^{-1} v$. Moreover, we have $v(x) \geq u(x)$ on $\partial \mathbb{R}^3_{r,c}$. Hence the difference function $w = u - v$ is negative on $\partial \mathbb{R}^3_{r,c}$ and is subharmonic in the sense that

$$\left(\Delta + K_{24}^{-1}\right) w \leq 0 \tag{256}$$

on $\mathbb{R}^3_{r,c}$. It is well known (see, for example, Taubes [9], VI. 3), that such a function cannot obtain a positive maximum on $\mathbb{R}^3_{r,c}$. Since w converges to zero for large x and is negative at the boundary, we conclude therefore that it is everywhere negative. This shows that (253) implies (255).

To prove (253), we calculate

$$\begin{aligned} \frac{1}{2} \Delta \, |\theta'|^2 & = |\nabla_A \theta'|^2 + \theta' \cdot (\nabla_A^2 \theta') \\ & = |\nabla_A \theta'|^2 + \theta' \cdot \{2E\nabla_A \theta + [\Phi, L_c \theta] - \theta'''\} . \end{aligned} \tag{257}$$

Note that since $|\Phi(x)| \geq \frac{m(c)}{2}$ on $\overset{\circ}{U}_r$, we have

$$\begin{aligned} |E\nabla_A \theta| & \leq 2 \, |[\Phi, E\nabla_A \theta]| \\ & \leq 2 \, |(E\nabla_A \theta)'| \\ & \leq 2 \left(|E'| \, |\nabla_A \theta| + |E| \, |(\nabla_A \theta)'| \right) \\ & \leq 2 \left\{ |\phi'|_u \, |\nabla_A \theta| + |\phi| \, (\nabla_A \theta)' \right\} \end{aligned} \tag{258}$$

and

$$
\begin{aligned}
|(\nabla_A \theta)'| &\leq |\nabla_A \theta'| + |E\theta| \\
&\leq |\nabla_A \theta'| + 2(\, |E'|\ |\theta| + |E|\ |\theta'|\, \} \\
&\leq |\nabla_a \theta'| + 4\, |\phi|\ |\theta'|_u \, .
\end{aligned} \tag{259}
$$

Moreover, we can combine equations (221), (231), (239), and (243) to an equation

$$
L_c \theta' = (\phi, \theta) \natural \theta + \gamma \phi \tag{260}
$$

with $|\mu| \leq |\alpha|$. In the same way as above, it follows that

$$
\begin{aligned}
\Delta\, |\theta'|^2 \geq\ & |\nabla_A \theta'|^2 + \frac{1}{4}\, |\theta'|^2 \\
& - \Big(2\, |\nabla_A \theta| + 8\, |\theta|^2 + 4\, |\theta| \Big)\, |\theta'|_\mu \\
& - 2\, |\theta|\ |\nabla_A \theta'| \, .
\end{aligned} \tag{261}
$$

In the same way, we use the equation

$$
L_c \phi = \phi \natural \phi + \mu \phi \, , \tag{262}
$$

where μ is the real number in (209), to conclude that

$$
\begin{aligned}
\Delta\, |\phi'|^2 \geq\ & |\nabla_A \phi'|^2 + \frac{1}{4}\, |\phi'|^2 \\
& - K_{25} \Big(|\nabla_A \phi| + |\phi|^2 + |\phi| + |\alpha| \Big) \\
& - 2\, |\phi|\ |\Delta_A \phi| \, .
\end{aligned} \tag{263}
$$

By (228), (224), (237), (243), and (244), we can choose r and R large enough and $|\alpha|$ small enough so that γ, $|\nabla_A(\theta, \phi)|$, and $|(\theta, \phi)|$ are so small that (258) and (263) imply (253). This completes the proof of (255).

In order to prove (212), we recall the definition (214) of the cutoff function β_r and estimate

$$
\begin{aligned}
\Big\| \beta_r [\hat{\Phi}, \theta] \Big\|_c^2 &\leq c\, \Big\| \mathfrak{D}_c^+ \beta_r [\hat{\Phi}, \theta] \Big\|_2^2 \\
&= \Big\langle \beta_r [\hat{\Phi}, \theta], \mathfrak{D}_c \mathfrak{D}_c^+ \beta_r [\hat{\Phi}, \theta] \Big\rangle \\
&\leq \Big\langle \beta_r [\hat{\Phi}, \theta], \beta_r \mathfrak{D}_c \mathfrak{D}_c^+ [\hat{\Phi}, \theta] \Big\rangle \\
&\quad + J_1
\end{aligned} \tag{264}
$$

with

$$J_1 = \langle [\Phi_r, \theta], \Delta\beta_r[\Phi, \theta] + d\beta_r \vee \mathcal{D}_c^+[\Phi, \theta\rangle \\ \leq K_{26} \, \|[\Phi, \theta\|_{c,A} \, . \tag{265}$$

Here, $A = U_{2r} - U_r$. Now consider the term

$$\begin{aligned} \mathcal{D}_c\mathcal{D}_c^+[\hat{\Phi}, \theta] &= \mathcal{D}_c^+\mathcal{D}_c[\hat{\Phi}, \theta] - 2B\, V[\hat{\Phi}, \theta] \\ &= \left(\mathcal{D}_c^+\mathcal{D}_c(\operatorname{ad}\hat{\Phi}) - (\operatorname{ad}\hat{\Phi})\mathcal{D}_c^+\mathcal{D}_c\right)\theta \\ &\quad + [\hat{\Phi}, \mathcal{D}_c^+\mathcal{D}_c\theta] - 2B\, v[\hat{\Phi}, \theta] \, . \end{aligned} \tag{266}$$

We have

$$\begin{aligned} &|(\mathcal{D}_c^+\mathcal{D}_c(\operatorname{ad}\Phi) - (\operatorname{ad}\Phi)\mathcal{D}_c^+\mathcal{D}_c)\theta| \\ &\leq c\left(\left|\nabla_A\nabla_A\hat{\Phi}\natural\theta\right| + \left|\nabla_A\hat{\Phi}\natural\nabla_A\Phi\right| \right) \\ &\leq K_{27}\exp(-R/K_{27}) \, , \end{aligned} \tag{267}$$

since θ and $\nabla_A\theta$ are bounded on M. Moreover,

$$\begin{aligned} \left|[\hat{\Phi}, \mathcal{D}_c^+\mathcal{D}_c\xi]\right| \;&\leq\; \left|[\hat{\Phi}, \mathcal{D}_c^+(b(c) + \xi\natural\xi)]\right| + K_{26}\left|[\hat{\Phi}, \mathcal{D}_c\hat{\Phi}]\right| \\ &\leq\; \left|\nabla_A\hat{\Phi}\right| \left(|b(c)| + |\xi|^2 + K \right) \\ &\quad + \left|\mathcal{D}_c^+[\hat{\Phi}, b(c)]\right| + |\nabla_A(\xi\natural[\Phi, \xi]| \\ &\leq K_{27}\exp(-r/K_{28}) \\ &\quad + K_{28}R^{-0.1}\left(|\nabla_A[\Phi, \xi]| + |[\Phi, \xi]| \right) . \end{aligned} \tag{268}$$

Similarly by (230),

$$\begin{aligned} \left|[\hat{\Phi}, \mathcal{D}_c^+\mathcal{D}_c\dot{\xi}]\right| \;&\leq\; \left|[\hat{\Phi}, \mathcal{D}_c^+\{Db(c)\dot{c} + \xi\natural\dot{c}\}]\right| + K_{29}\left|[\hat{\Phi}, \mathcal{D}_c^+\hat{\Phi}]\right| \\ &\leq\; \left|\nabla_A\hat{\Phi}\right| \left(|Db(c)\dot{c}| + |\xi\natural\dot{c}| + K_{30} \right) \\ &\quad + |\nabla_A\xi| \, |[\Phi, \dot{c}]| + |\nabla_A(\xi\natural[\Phi, \dot{c}] + [\Phi, \xi]\natural\dot{c})| \\ |B \vee [\Phi, \xi]| \;&\leq K_{31}R^{-2}\, |[\Phi, \theta]| \, . \end{aligned} \tag{269}$$

Hence (264) yields

$$\begin{aligned} \|\beta_r[\Phi, \theta]\|_c \;&\leq c\, \|[\Phi, \theta]\|_{c,A} \\ &\quad + K_{32}\exp(-R/K_{32}) \\ &\quad + K_{33}R^{-0.1}\, \|\beta_r[\Phi, \theta]\|_c \, . \end{aligned} \tag{270}$$

Therefore, Proposition 29 is proved as soon as we find an exponential estimate for $\|[\Phi,\theta]\|_{c,A}$. This is done as follows: Let $\beta_A = \beta_{2r}(1-\beta_r)$.

$$\begin{aligned} \left\|[\hat{\Phi},\theta]\right\|_{c,A} &\leq \left\|\beta_A[\hat{\Phi},\theta]\right\| \\ &\leq c\,K_{34}\left(\left\|d\beta_A[\hat{\Phi},\theta]\right\|_2\right) \\ &\quad + \left\|\beta_A\nabla_A\hat{\Phi}V\theta\right\|_2 \\ &\quad + \left\|\beta_A[\hat{\Phi},\mathfrak{D}_c^+\theta\right\|_2 \\ &\leq K_{35}e^{-R/K_{35}}\,. \end{aligned} \tag{271}$$

Here we have used the pointwise estimate

$$|E\vee\theta| \leq |E_T|\ |\theta| + |E|\ |\theta_T| \tag{273}$$

and the identity

$$\begin{aligned} [\Phi,\mathfrak{D}_c^+\xi] &= [\Phi,\mathfrak{D}_c\xi] - 2[\Phi,[\Phi,\xi]] \\ &= [\Phi,\xi\natural\xi] - 2[\Phi,[\Phi,\xi]] \\ &\leq c(\,|\xi|_T\ |\xi|\ +\ |\xi_T|\,) \end{aligned} \tag{273}$$

together with the corresponding identity for $\dot{\xi}$. Inequality (271) completes the proof of (212) of Proposition 29.

References

[1] S.K. Donaldson, Nahm's equation and the classification of monopoles. *Comm. Math. Phys*, 96:387–407, 1984.

[2] A. Floer. The configuration space of Yang-Mills-Higgs theory on asymptotically flat manifolds. In this volume.

[3] A. Floer. Monopole moduli spaces for asymptotically flat manifolds, to appear.

[4] A. Floer. Monopoles on asymptotically euclidean 3-manifolds. *Bull. AMS.*, 16:125–127, 1987.

[5] A. Floer. Monopoles on asymptotically flat manifolds. In this volume.

[6] W. Greub, S. Halperin, and R. Vanston. *Connections, curvature, and cohomology I.* Academic Press, New York and London, 1978.

[7] N. Hitchin. Monopoles and Geodesics. *Comm. Math. Phys.*, 86(299), 1982.

[8] J.D. Jackson. *Classical Electrodynamics*. Wiley, 2nd edition, 1975.

[9] A. Jaffe and C. Taubes. *Vortices and Monopoles*. Birkhäuser, Boston, 1980.

[10] L. Nirenberg and H. Walker. Null spaces of elliptic partial differential operators on $\mathbb{R}^n$. *J. Math. Anal. Appl.*, 42(27), 1973.

[11] T. Parker and C. Taubes. On Witten's proof of the positive energy theorem. *Comm. Math. Phys.*, 84, 1982.

[12] C. Taubes. Min-max theory for the Yang-Mills-Higgs Equation. *Comm. Math. Phys.*, 97:473–540, 1985.

[13] C. Taubes. Monopoles and maps from S^2 to S^2; the topology of the configuration space. *Comm. Math. Phys.*, 95:345–391, 1984.

[14] K.L. Uhlenbeck. Connections with L^p bounds on the curvature. *Comm. Math. Phys.*, 83(31), 1981.

[15] E. Weinberg. Parameter counting for multimonopole solutions. *Phys. Rev.*, D 20:936–944, 1979.

The configuration space of Yang-Mills-Higgs theory on asymptotically flat manifolds

A. Floer

Abstract. We show that the quotient space of the space of finite action Yang-Mills-Higgs configurations by the group of gauge transformations has a natural Hilbert manifold structure. Moreover, the solutions of the Bogomolny (self-duality) equations are the zero set of a Fredholm section of a smooth Hilbert space bundle over this manifold.

1 Introduction

Yang-Mills-Higgs theory is the variational theory of the functional

$$\mathbf{a}(A, \Phi) = \|F_A\|_{L^2}^2 + \|\nabla_A \Phi\|_{L^2}^2 , \tag{1}$$

where F_A is the curvature of a connection A on a principal bundle P over $\mathbb{R}^3$ and Φ is a section of the adjoint bundle. We will restrict ourselves to the case where the fibre of P is the group SU_2. For an introduction to Yang-Mills-Higgs theory, see the book [11], which also serves as a general background for this paper.

We want to consider the functional (1) for general manifolds. Since the theory does not seem to yield interesting results for dimensions other than 3, or for compact manifolds, we consider the following class of manifolds, compare [14]:

Definition 1. A 3-dimensional Riemannian manifold (M, g) without boundary is called asymptotically flat with n ends ($n \geq 1$), if there exists a positive real number r_M and injective smooth maps

$$\omega_i : \mathbb{R}^3 - B_{r_M} \longrightarrow M, \quad 1 \leq i \leq n, \tag{2}$$

such that $M - \bigcup_{i=1}^n \omega_i(\mathbb{R}^3 - B_{r_M})$ is compact and $\omega_i^* g = g_0 + \gamma_i$, where g_0 is the Euclidean metric on $\mathbb{R}^3$ and

$$|x| \cdot |\gamma_i| + |x|^2 \, |\nabla \gamma_i| + |x|^3 \, |\nabla\nabla \gamma_i| \leq C \tag{3}$$

is bounded.

A notable feature of Yang-Mills-Higgs theory on Euclidean $\mathbb{R}^3$ is that the set of minima of the action (1) is a nontrivial finite dimensional manifold after dividing out by the group of gauge transformations, i.e. the group of sections of the automorphism bundle of P. In fact, this so-called moduli space of (anti) self-dual monopoles has been described by means of algebraic geometry, see [10] and [3]. For $\mathbb{R}^3$ replaced by an asymptotically flat manifold, this moduli space is still expected to be a finite dimensional manifold, although it is not possible anymore to *construct* solutions. In this paper, we will therefore provide the setup for the analytic treatment of the theory, which will be carried out in [5], [4], [7] [1]). In particular, we will describe the quotient space of finite action configurations by the gauge group.

Choosing a smooth reference connection ∇ on P, we can identify the space of connections with the space of 1-forms with values in $\mathrm{ad}(P)$. Define therefore the bundles $\Omega = \Lambda^0 \oplus \Lambda^1$ and $\Omega_{\mathrm{ad}} = \Omega \otimes \mathrm{ad}(P)$. We will consider a configuration $c = (\Phi, A)$ as an element of $W^1_{\mathrm{loc}}(\Omega_{\mathrm{ad}})$, i.e. as a section with locally square integrable derivatives. Then, the fields F_A and $\nabla_A \Phi$ are locally square integrable. If we restrict ourselves to finite action configurations, that is, we assume that F_A and $\nabla_A \Phi$ are globally square integrable, then it follows from Lemma 13 below that there exists a collection of nonnegative numbers $\underline{m}(c) = (m_1 \dots m_n)$ such that

$$|\omega_i^*(\Phi)| - m_i \in L^6\left(\mathbb{R}^3 - B_{r_M}\right) . \tag{4}$$

Definition 2. We call $m_i = m_i(c)$ the mass of c at the i-th end. Moreover, we define for $m = (m_1 \dots m_n)$, $m_i > 0$ the configuration space

$$\mathscr{C}_{\underline{m}} = \{c \mid \mathbf{a}(c) < \infty \text{ and } \underline{m}(c) = \underline{m}\} , \tag{5}$$

and the total configuration space

$$\mathscr{C} = \bigcup_{\underline{m}>0} \mathscr{C}_{\underline{m}} . \tag{6}$$

The topologies on $\mathscr{C}_{\underline{m}}, \mathscr{C}$ are defined by the $W^1_{\mathrm{loc}}(\Omega_{\mathrm{ad}})$ topology and by the map

$$\begin{aligned} \mathscr{C} &\longrightarrow L^2\left(\Lambda^1_{\mathrm{ad}} \oplus \Lambda^2_{\mathrm{ad}}\right) \\ c &\longmapsto (\nabla_A \Phi, F_A) . \end{aligned} \tag{7}$$

A suitable topology for the gauge group is now given by

$$\mathscr{G} = W^2_{\mathrm{loc}}(\mathrm{Aut}(P)) ,$$

---

1) Some of these papers never appeared. See the notes accompanying these papers for alternative references

which acts on $\mathscr{C}$ by

$$g(c) = \left(g\Phi g^{-1}, gAg^{-1} + (dg)g^{-1}\right). \tag{8}$$

Note that neither the configurations nor the elements of $\mathscr{G}$ satisfy any decay conditions at the ends of M. As a result, it is impossible to define Banach manifold structures on $\mathscr{C}$ and $\mathscr{G}$, as one does for configuration spaces on compact manifolds. However, we will see that the quotient admits a natural Hilbert manifold structure.

Definition 3. For $c \in \mathscr{C}$ define the Hilbert space $H_c(\Omega_{\mathrm{ad}})$ as the completion of $C_0^\infty(\Omega_{\mathrm{ad}})$ in the norm

$$\|\xi\|_c^2 = \|\nabla_A\xi\|_2^2 + \|[\Phi, \xi]\|_2^2 . \tag{9}$$

Let $\Phi_{(i)}, 1 \leq i \leq n$, be the special generators of $\mathscr{G}$ associated with c as constructed in Definition 23 below. Set

$$\vartheta_c(a, \phi) = d_A^* a + [\Phi, \phi] \tag{10}$$

and

$$\begin{aligned} \grave{Q}_c = \{(a, \phi) \in H_c(\Omega_{\mathrm{ad}}) \oplus \left\langle (\Phi_{(1)}, 0) \ldots (\Phi_{(n)}, 0) \right\rangle \mid \\ d_A^* a + [\Phi, \phi] = 0 \text{ and } \left\langle a, \nabla_A \Phi_{(i)} \right\rangle_c = 0 \quad \text{for} \quad 1 \leq i \leq n\}. \end{aligned} \tag{11}$$

There is a linear map

$$\underline{\mu} : \grave{Q}_c \longrightarrow \mathbb{R}^n$$

with the property that

$$\omega_i^* \, |\phi| - \mu_i \left(a, \phi\right) \in L^6(\mathbb{R}^3 - B_{r_0}) . \tag{12}$$

We denote the kernel of μ in $\grave{Q}_c$ by Q_c.

Theorem 4. *Let $\mathscr{C}^*_{\underline{m}}(M)$ denote the set of irreducible Yang-Mills-Higgs configurations, i.e. of configurations $c = (A, \Phi)$ with $\nabla_A \Phi \neq 0$ in $\mathscr{C}_{\underline{m}}(M)$. Then*

$$Z_{\underline{m}}(M) = \mathscr{C}^*_{\underline{m}}/\mathscr{G} \tag{13}$$

is a Hilbert manifold with

$$T_{[c]} Z_{\underline{m}}(M) \cong Q_c . \tag{14}$$

In fact, for every $c \in \mathscr{C}^*_{\underline{m}}$ there exists a zero neighborhood $\mathscr{U} \subset Q_c$ so that the slice

$$\mathscr{S}_c = c + \mathscr{U} \subset \mathscr{C} \tag{15}$$

intersects each orbit in $\mathscr{G}\mathscr{S}_c \subset \mathscr{C}$ exactly twice, and $\mathscr{G}\mathscr{S}_c$ is open in $\mathscr{C}$. These intersection points define charts

$$s_c : \mathscr{G}\mathscr{S}_c/\mathscr{G} \longrightarrow \mathscr{S}_c : [c'] \longmapsto \mathscr{G}_{c'} \cap \mathscr{S}_c \tag{16}$$

of a smooth atlas of $Z^*_{\underline{m}}(M)$.

In the case without specified masses we have

Theorem 5. *If we set* $\mathcal{C}^*(M) = \{c \in \mathcal{C}(M) \mid \nabla_A \Phi \neq 0\}$, *then*

$$Z(M) := \mathcal{C}^*(M)/\mathcal{G}$$

is a Hilbert manifold modelled on the Hilbert space $\grave{Q}_c \cong Q_c \oplus \mathbb{R}^n$.

The proof of Theorem 4 proceeds as follows. First we note (see Section 5) that the operation of $\mathcal{G}$ on $\mathcal{C}$ has a closed graph. If $\mathcal{G}$ and $\mathcal{C}$ were Hilbert manifolds, then this result would immediately imply that the quotient is a Hilbert manifold, too, see [8], by a simple application of the implicit function theorem. In our case, however, we have to define a preliminary gauge parallelizing the Higgs field at the ends, before we can apply the implicit function theorem. This is carried out in Section 6.

As in the Euclidean case (see [6]), the configuration space decomposes into components labelled by a charge. If there are several ends, then each end can be assigned an integer charge as follows (see Section 5).

Proposition 6. *There is a continuous map* $\underline{k} : Z(M) \longrightarrow \mathbb{Z}^n$ *so that with* $\underline{m} \cdot \underline{k} = \Sigma m_i k_i$,

$$\mathbf{a}(c) = \pm 4\pi \underline{m}(c) \cdot \underline{k}(c) + \|\nabla_A \Phi \pm *FA\|_{L^2}^2 \ . \tag{17}$$

This is an immediate corollary of Proposition 24 below. As a consequence of Proposition 6, the minima of the Yang-Mills-Higgs action in each component $Z_{\underline{k},\underline{m}}(M)$ of configurations with charges k_i and masses m_i are the solutions of the equation

$$\nabla_A \Phi = \pm * F_A \, , \tag{18}$$

as is the case for Euclidean $\mathbb{R}^3$. Equation (18) has the following Fredholm property:

Proposition 7. *The map*

$$b : \mathcal{C}(M) \longrightarrow L^2(\Omega_{\mathrm{ad}}) : c \longmapsto \nabla_A \Phi - *F_A \tag{19}$$

is equivariant and gives rise to a smooth Fredholm section of a smooth Hilbert space bundle over $Z_{\underline{k}}(M)$ *with index* $4k$. *It is surjective if and only if the Fredholm operator*

$$\begin{aligned} \mathcal{D}_c &: H_c(\Omega_{\mathrm{ad}}) \longrightarrow L^2(\Omega_{\mathrm{ad}}) \\ \mathcal{D}_c(a, \phi) &= (-d_A^* a, *d_A a - d_A \phi) + ([\Phi, \phi], [\Phi, a]) \end{aligned} \tag{20}$$

is surjective.

In [5] and [4] we will examine the solution space of (18).

2 Notation

Let M be a smooth 3-dimensional manifold and let $G \hookrightarrow P \longrightarrow M$ be a smooth principal bundle over M, where G is a simple Lie group. In this paper, we will alway assume that $G = SU_2$. We are concerned with the following two associated vector bundles (c.f. [8]):

$$\mathrm{ad}(P) = P \times_{\mathrm{ad}} su_2 \tag{21}$$

$$\mathrm{aut}(P) = P \times_{\mathrm{ad}} G\,, \tag{22}$$

where su_2 is the Lie algebra of G and ad denotes the adjoint action of G on su_2 and G.

For $\xi, \zeta \in su_2$, we denote the Lie bracket by $[\xi, \zeta]$ and the negative of the Killing inner product by $\xi \cdot \zeta$. The bundle of skew symmetric p-forms on TM is denoted by Λ^p. Moreover, we set $\Lambda^p_{\mathrm{ad}} = \Lambda^p \otimes \mathrm{ad}(P)$. The exterior product in Λ^p (see [2]) have the following extensions:

Definition 8. For $\lambda \in \Lambda^p_{\mathrm{ad}}, \mu \in \Lambda^q_{\mathrm{ad}}, f \in \Lambda^q$, we define

$$\lambda \wedge \mu \in \Lambda^{p+q}_{\mathrm{ad}} \tag{23}$$

$$\lambda \wedge f \in \Lambda^{p+q}_{\mathrm{ad}} \tag{24}$$

$$\lambda \dot{\wedge} \mu \in \Lambda^{p+q} \tag{25}$$

as the exterior product based on the operation of the Lie bracket, the multiplication by real numbers, and the negative of the Killing metric, respectively.

A metric g on M defines the Hodge duality isomorphism $* : \Lambda^p \longrightarrow \Lambda^{3-p}$. It can be extended to an isomorphism $* : \Lambda^p_{\mathrm{ad}} \longrightarrow \Lambda^{3-p}_{\mathrm{ad}}$. If λ, μ, and f are as in Definition 8 with $p > q$, then we abbreviate

$$\lambda \surd \mu = * \left((*\mu) \wedge \lambda \right) \tag{26}$$

and $f \surd \mu, \lambda \dot{\surd} \mu$ analogously. If $p = q$, we set

$$(\lambda \mid \mu) = \lambda \dot{\surd} \mu\,. \tag{27}$$

An L^2-inner product between two compactly supported Λ^p_{ad}-sections is given by

$$\langle \lambda, \mu \rangle = \int_M (\lambda \mid \mu)\, \mu_g\,, \tag{28}$$

where $\mu_g = *1$ is the Riemannian volume form.

Consider now the covariant derivatives defined by connections on P. Given a connection ∇, there is a 1-1 correspondence between smooth connections on P

and smooth sections of Λ^1_{ad}. In terms of covariant derivations on the bundle $\mathrm{ad}(P)$, it is given by

$$\nabla_A \xi = \nabla \xi + [A, \xi]\,. \tag{29}$$

For any tensor bundle T over M, we will extend ∇_A to $T \otimes \mathrm{ad}(P)$ using the Levi-Civita connection of the metric g. The action of the gauge group on the space of connections can now be described by

$$g(A) = (\nabla g)g^{-1} + gAg^{-1}\,. \tag{30}$$

For any connection ∇_A, we can extend the operation of exterior differentiation $d : \Lambda^p \longrightarrow \Lambda^{p+1}$ to

$$d_A : \Lambda^p_{\mathrm{ad}} \longrightarrow \Lambda^{p+1}_{\mathrm{ad}}\,. \tag{31}$$

The formal (L^2-) adjoint of d_A with respect to the inner product $\langle , \rangle$ can be expressed in terms of d_A and the Hodge star as

$$d_A^* = (-1)^p * d_A * : \Lambda^p_{\mathrm{ad}} \longrightarrow \Lambda^{p-1}_{\mathrm{ad}}\,. \tag{32}$$

In particular, for $\lambda \in \Lambda^1_{\mathrm{ad}}$ we have $d_A^* \lambda = -g^{ij}(\nabla_A)_i \lambda_j$. We now list some properties of the operations defined above.

$$d_A(\lambda \wedge \mu) = (d_A \lambda) \wedge \mu + (-1)^j \lambda \wedge d_A \mu \tag{33}$$

$$d(\lambda \dot{\wedge} \mu) = (d_A \lambda) \dot{\wedge} \mu + (-1)^j \lambda \dot{\wedge} d_A \mu \tag{34}$$

$$* * \lambda = \lambda \tag{35}$$

$$(*\lambda \mid *\mu) = (\lambda \mid \mu)\,. \tag{36}$$

If M is a manifold with boundary, we have

$$\langle d_A \lambda, \mu \rangle = \langle \lambda, d_A^* \mu \rangle + \int_{\partial M} \lambda \dot{\wedge} (*\mu)\,. \tag{37}$$

The formula $dd\xi = 0$, which holds for $\xi \in \Lambda^p$, does not hold for d_A. Instead, there exists a 2-form $F_A \in \Lambda^2_{\mathrm{ad}}$ so that

$$d_A d_A \lambda = F_A \wedge \lambda\,, \tag{38}$$

see, for example [9]. F_A is called the curvature of the connection ∇_A. The curvatures of two connections on P are related by

$$F_{A+a} = F_A + d_A a + \frac{1}{2} a \wedge a\,. \tag{39}$$

3 Functional analysis on asymptotically flat manifolds

In this section we generalize results of functional analysis on $\mathbb{R}^3$ to asymptotically flat manifolds. We introduce the following Banach spaces:

Definition 9. Let M be a Riemannian manifold and let E be a vector bundle over M with a metric $(.\mid.)$ and a metric preserving connection ∇. On $C_0^\infty(E)$ consider the norms

$$\|\lambda\|_p^p = \int_M \left((\lambda \mid \lambda)^{\frac{p}{2}}\right) \mu_g \tag{40}$$

$$\|\lambda\|_{W(k,p)} = \|\lambda\|_p + \left\|\nabla^k \lambda\right\|_p \tag{41}$$

$$\|\lambda\|_{V(k,p)} = \left\|\nabla^k \lambda\right\|_p \tag{42}$$

and the corresponding completions $L^p, W^{k,p}$ and $V^{k,p}$ of $C_0^\infty(E)$. In particular, set $W^{k,2} = W^k$ and $V^{k,2} = V^k$.

Definition 10. Given a configuration $c = (A, \Phi)$ in $\mathcal{C}$ and $\xi \in C_0^\infty(E_{\text{ad}})$, where E is some tensor bundle on M, we define the Hilbert space norm

$$\|\xi\|_c^2 = \|[\Phi, \xi]\|_2^2 + \|\nabla_A \xi\|_2^2 \,. \tag{43}$$

Denote by $H_c(E_{\text{ad}})$ the corresponding completion of $C_0^\infty(E_{\text{ad}})$.

Definition 11. For any measurable subset $K \subset M$ we define the seminorms

$$\|\xi\|_{p,K}^p = \int_K (\xi \mid \xi)^{\frac{p}{2}} \mu_g \,. \tag{44}$$

In particular, for asymptotically flat manifolds, define a smooth function $\rho : M \longrightarrow \mathbb{R}$ so that $\rho \geq 1$ and $\rho(\omega(x)) = |x|$ on $\mathbb{R}^3 - B_{r_M}$. For any $R \geq r_M$, set $M_R = \{x | \rho(\omega(x)) \leq R\}$ and

$$\begin{aligned} \| \quad \|_{p,>R} &= \| \quad \|_{p,M-M_R} \\ \| \quad \|_{p,<R} &= \| \quad \|_{p,M_R} \,. \end{aligned} \tag{45}$$

We extend this notation to all norms defined above.

The completion of a normal space always exists in an abstract sense. Of course, we are mainly interested in those cases where the closure can be identified with a subset of the set of measurable functions on M. This is certainly the case for the L^p-spaces and for all spaces whose norms can be estimated by L^p norms. Moreover, most embedding theorems which hold on $\mathbb{R}^3$ generalize to asymptotically flat manifolds M. Before we state some of these theorems, we have to introduce the cut-off function.

Definition 12. Let $\beta : \mathbb{R} \longrightarrow \mathbb{R}$ be a smooth function with

$$\begin{aligned} &1) \quad \beta(x) = 0 \quad \text{for} \quad x \geq 2 \\ &2) \quad \beta(x) = 1 \quad \text{for} \quad x \leq 1 \\ &3) \quad \beta(x) \in [0,1] \quad \text{for all} \quad x\,. \end{aligned} \tag{46}$$

Moreover, define for $R > r_M$

$$\beta_R : M \longrightarrow \mathbb{R} : x \longmapsto \beta\left(\frac{1}{R} \cdot \rho(x)\right) . \tag{47}$$

The function β_R defines in a unique way a partition of unity

$$1 = \beta_R + \sum_{i=1}^{n} \beta_R^i\,, \tag{48}$$

where β_r^i is smooth with support in the i'th component of $M - M_R$.

Lemma 13. *Let $E \longrightarrow M$ be a vector bundle with metric $|\cdot|$ over an asymptotically flat manifold. Define $M_i := M_{r_M} \cup \omega_i\left(\mathbb{R}^3 - B_{r_M}\right)$. There exists a constant C_s so that for any metric preserving connection ∇ on E and for all $\xi \in C^\infty(E)$ with $\nabla\xi \in L^p$ there exist constants $m_i \geq 0$, $1 \leq i \leq n$, so that with $q = \frac{3p}{p-3}$*

$$\|\, |\xi| \;- m_i\|_{q,M_i} \;\leq C_s \,\|\nabla\xi\|_p \;. \tag{49}$$

We have $m_i = 0$ for all $1 \leq i \leq n$ if and only if $\xi \in W^{1,p}(E)$. Moreover, if $\nabla\xi \in W^1$, then for every end

$$\sup_{\rho(x)\geq R} \;|\, |\xi| \;- m_i| \;\longrightarrow 0 \quad \text{for} \quad R \longrightarrow \infty\,. \tag{50}$$

Proof. Since Kato's inequality (see [11],VI)

$$\|\nabla\, |\xi|\, \|_p \;\leq\; \|\nabla\xi\|_p \tag{51}$$

holds for any metric preserving connection and for any Riemannian manifold, it suffices to prove the assertion for $\xi \in C^\infty(M,\mathbb{R})$. The assertions are well known for $(M,g) = (\mathbb{R}^3, g_0)$. From the proof of [11], VI, Lemma 3.2, it even follows that on $\mathbb{R}^3$ with $q = \frac{3p}{p-3}$

$$\|\xi\|_{q,>r} \;\leq C\; \|\nabla\xi\|_{p,>r} \tag{52}$$

for some constant C and for any $r > 1$. We can apply the diffeomorphism ω_i to prove that (52) holds for each end of M. With a cutoff function β_r^i of Definition 11 we obtain

$$\begin{aligned}\|\beta_R^i\xi\|_q &\le C_1 \|d(\beta_R^i\xi)\|_p \\ &\le C_2 \left(\|(d\beta_R^i)\xi\|_p + \|(\beta_R^i)d\xi\|_p \right) \\ &\le C_3 \left(\|d\beta_R^i\|_3 \; \|\xi\|_{q,M_i} + \|d\xi\|_{p,M_i} \right) \\ &\le C_4 \|d\xi\|_{p,M_i} \; .\end{aligned}$$

By a usual Sobolev inequality, there exists a constant C_r which may depend on r with

$$\begin{aligned}\|\beta_r\xi\|_q &\le C_r \|d(\beta_r\xi)\|_p \\ &\le C_r \left(\|(d\beta_r)\xi\|_p + \|\beta_r d\xi\|_p \right) \\ &\le C_r \left(\|d\beta_r\|_3 \; \|\xi\|_{q,M_i} + \|d\xi\|_p \right) \\ &\le C_5 \|d\xi\|_p \; .\end{aligned}$$

This proves the first assertion if $\xi \in V_p^i$. Now assume only that $\|d\xi\|_p < \infty$. It is known by [16] that for every end there exists a constant m_i so that $\omega^* \, |\xi| - m_i \in L^q(\mathbb{R}^3 - B_{r_M})$. Consider the functions

$$\xi_r := \beta_r \, |\xi| + \sum_{i=1}^{n} \beta_r^i m_i \, .$$

We have

$$\|d\xi_r\|_p \le C \sum_{i=1}^{n} \|d\xi_r^i\|_3 \; \|w^* \, |\xi| - m_i\|_{q>r} + \|d \, |\xi| \, \|_{p>r} \; . \tag{53}$$

We can now approximate each ξ_r in $V^{1,p}$ by smooth functions (see, for example, [1]). (53) also proves Lemma 18.

In order to prove the last assertion, observe that $W_\nabla^1 \subset L^2 \cap L^6$ by the above and apply the following Lemma 14

Lemma 14. *For $f \in C_0^\infty(M, \mathbb{R})$ we have*

$$\|f\|_\infty \le \|\nabla f\|_2 + \|\nabla f\|_4 \; . \tag{54}$$

Proof. By (49) and a simple interpolation formula, we have

$$\|f\|_p \leq C_1 \left(\|\nabla f\|_2 + \|\nabla f\|_4 \right)$$

for all $p \in [6, \infty)$. Moreover, a Sobolev inequality yields

$$\|v\|_\infty \leq C_2 \left(\|\nabla v\|_{\frac{7}{2}} + \|v\|_{\frac{7}{2}} \right).$$

Applying this inequality to $v = |f|^2$, we obtain

$$\begin{aligned} \left\| |f|^2 \right\|_\infty &\leq C_2 \left(2 \, \|f \cdot \nabla f\|_{\frac{7}{2}} + \|f^2\|_{\frac{7}{2}} \right) \\ &\leq C_2 \left(2 \, \|f\|_{28} \, \|\nabla f\|_4 + \|f\|_7^2 \right) \\ &\leq C_3 \left(\|\nabla f\|_2 + \|\nabla f\|_4 \right)^2 . \end{aligned}$$

□

In addition to the configuration space $\mathscr{C}$ of Definition 2, we will use the subspace

$$\tilde{\mathscr{C}} = \left\{ (A, \Phi) \in \mathscr{C} \mid \nabla_A \nabla_A \Phi \in L^2 \left(\Lambda^1 \otimes \Lambda^1_{\mathrm{ad}} \right) \right\}, \tag{55}$$

which is topologized by the topology of $\mathscr{C}$ and by the map $c \longmapsto \nabla_A \nabla_A \Phi \in (\Lambda^1 \otimes \Lambda^1_{\mathrm{ad}})$. There are natural retractions of the space $\mathscr{C}$ to $\tilde{\mathscr{C}}$. For any $c \in \mathscr{C}$, consider the affine space

$$\Xi_c := \{ \Phi + \phi \mid \phi \in H_c(\Lambda^0_{\mathrm{ad}}) \} \tag{56}$$

and thereon the functional

$$S_c[\Psi] = \|\nabla_A \Psi\|_2^2 + \|[\Phi, \Psi]\|_2^2 = \|\Psi\|_c . \tag{57}$$

Lemma 15. *S_c has a unique critical point $\acute{\Phi} = \Phi + \phi_c$ in Ξ_c with*

$$\left(-\nabla_A^2 - (\mathrm{ad}\,\Phi)^2 \right) \phi_c = 0 \tag{58}$$

$$\left\| \nabla_A \nabla_A \acute{\Phi} \right\|_2 \leq K \tag{59}$$

$$\|\phi_c\|_c + \left\| \acute{\Phi} \right\|_c \leq 5 \, \|\nabla_A \Phi\|_2 \tag{60}$$

$$\left\| \acute{\Phi} \right\|_\infty \leq m(c), \tag{61}$$

where K depends only on $\mathbf{a}(c)$ and $m(c) = \max_{1 \leq i \leq n} m_i(c)$. Moreover, the assignment

$$c \longmapsto \acute{c} := (A, \acute{\Phi}) \tag{62}$$

defines a continuous map $\mathscr{C} \longrightarrow \tilde{\mathscr{C}}$.

Proof. Assertions (58)–(60) are easy generalizations of [16], Proposition B3.2 and B3.3 to asymptotically flat manifolds. Inequality (61) follows from (58) through

$$\begin{aligned} -\nabla^2 |\Phi|^2 &= -\left|\nabla_A \acute{\Phi}\right|^2 - \left\langle \Phi, \nabla_A^2 \Phi \right\rangle \\ &= -\left|\nabla_A \acute{\Phi}\right|^2 - \left|[\Phi, \acute{\Phi}]\right|^2 \\ &< 0 \end{aligned} \tag{63}$$

and the maximum principle, see for example [11].

4 Inverting the Laplacian

One of the technical problems in Yang-Mills-Higgs theory is to invert the linear operator

$$L_c : \vartheta_c \varphi_c = -\nabla_A^2 - (\operatorname{ad} \Phi)^2, \quad \nabla_A^2 = \nabla_A^* \nabla_A \tag{64}$$

where ϑ_c is the *gauge fixing* operator introduced in (10) and φ_c is the linearization of the operation (8) of the gauge group on the configuration c. Note that $\operatorname{ad} \Phi$is not an isomorphism on su_2. For sections ξ of $\operatorname{ad} P$ with $[\Phi, \xi] = 0$ the problem is similar to the problem of finding Banach norms so that the Laplacian

$$-\nabla^2 = d^* d : C_0^\infty(M, \mathbb{R}) \longrightarrow C_0^\infty(M, \mathbb{R}) \tag{65}$$

extends to a linear homeomorphism. For example, if we use the L^2-norm in the target space, then the domain would have to contain unbounded functions. If we want to avoid this, we have to use stronger norms in the target space, for example weighted norms (see Nirenberg and Walker [13]). We will instead use auxiliary L^p-norms.

Definition 16. Let E be a metric vector bundle over M with connection. The Banach spaces X_c and Z_c are the completions of $C_0^\infty(\Lambda^0_{\mathrm{ad}})$ in the norms

$$\|\xi\|_{X_c} = \|\xi\|_c + \|\nabla_A \nabla_A \xi\|_2 + \left\|(\nabla_A^2 \xi) \cdot \Phi\right\|_{\frac{6}{5}} \tag{66}$$

$$\|\xi\|_{Z_c} = \|\xi\|_2 + \|\xi \cdot \Phi\|_{\frac{6}{5}} \,. \tag{67}$$

We now list some properties of X_c and Z_c.

Lemma 17. *For $\xi \in X_c$, we have with K depending only on $\mathbf{a}(c)$ and $\underline{m}(c)$:*

$$\|\xi\|_\infty + \|\nabla_A \xi\|_c + \|[\Phi, \xi]\|_c \leq K \, \|\xi\|_{X_c} \,. \tag{68}$$

Proof. The estimate on $\|\xi\|_\infty$ follows from Lemma 14. Moreover, using the decomposition $\Phi = \grave{\Phi} - \phi_c \in L^\infty + H_c$ of Lemma 15, we find

$$\begin{aligned}\|\nabla_A\xi\|_c \le \|\nabla_A\nabla_A\xi\|_2 + m(c)\,\|\nabla_A\xi\|_2 \\ + \|\phi_c\|_6\,\|\nabla_A\xi\|_3 \le K\,\|\xi\|_{X_c}\,.\end{aligned}$$

By the same method we estimate $\|[\Phi,\xi]\|_c$. □

Lemma 18. *For $\xi,\zeta \in H_c$ and $\eta \in Z_c$, we have with K depending only on $\mathbf{a}(c)$ and $\underline{m}(c)$*

$$\langle\xi,\eta\rangle_2 \le K\,\|\xi\|_c\,\|\eta\|_{Z_c} \tag{69}$$

$$\|[\xi,\zeta]\|_{Z_c} \le K\,\|\xi\|_c\,\|\zeta\|_c\,. \tag{70}$$

Proof. Lemma 17 applied to Φ implies a bound on the volume of the set

$$K_c = \left\{x \in M \mid |\Phi| \le \frac{m_0}{2}\right\}$$

which depends only on the action of c and on $m_0(c) = \min_{1\le i\le n} m(c)$.

Hence

$$\begin{aligned}\langle\xi,\zeta\rangle &\le \|\xi\|_{2,K_c}\,\|\xi\|_2 + \frac{2}{m_0}\left(\|[\Phi,\xi]\|_2\,\|\zeta\|_2 + \|\xi\|_6\,\|\Phi\cdot\zeta\|_{\frac{6}{5}}\right)\\ &\le \left(((\mathrm{vol}(K_c))^{\frac{1}{3}} + \frac{2}{m_0}\right)\|\xi\|_c\,\|\zeta\|_{Z_c}\,.\end{aligned}$$

After these preparations, we can come to the main result of this section

Proposition 19. *Let M be an asymptotically flat manifold. Let E be a metric bundle over M with connection ∇, whose curvature is square integrable. For $c = (A,\Phi) \in \mathscr{C}$, we have a linear homeomorphism*

$$L_c = -\nabla_A^2 - (\mathrm{ad}\,\Phi)^2 : X_c(E_{\mathrm{ad}}) \longrightarrow Z_c(E_{\mathrm{ad}})\,. \tag{71}$$

Moreover, the operator norms of L_c and L_c^{-1} are bounded in terms of $\|F(\nabla)\|_2$, $\mathbf{a}(c)$ and $\underline{m}(c)$.

Proof. All constants K_i below will be assumed to depend only on $a(c)$ and $\underline{m}(c)$. To see that L_c is continuous, we estimate by Lemma 17

$$\|L_c\xi\|_{Z_c} \leq \|\nabla_A^2\xi\|_2 + \|[\Phi,[\Phi,\xi]]\|_2 + \|\Phi\cdot\nabla_A^2\xi\|_{\frac{6}{5}} \leq K_1 \|\xi\|_{X_c} .$$

To construct an inverse for L_c, we first show that for every $\xi \in Z$ there exists a $u_\xi \in W^2_{\mathrm{loc}}$ so that

$$L_c u_\xi = \xi \tag{72}$$

almost everywhere on M and

$$\|u_\xi\|_{X_c} \leq \|\xi\|_{Z_c} . \tag{73}$$

Consider therefore the functional

$$S_\xi(u) = \frac{1}{2}\|\xi\|_c^2 + \langle u,\xi\rangle \tag{74}$$

on H_c. By Lemma 18 we find a constant C_ξ so that

$$\frac{1}{4}\|u\|_c^2 - C_\xi \leq S_\xi(u) \leq \|u\|_c^2 + C_\xi . \tag{75}$$

Moreover, since S consists of a continuous linear term and of a continous positive definite quadratic term, it is smooth and strictly convex. Together with the coercivity condition, this implies (see [11], Proposition VI 7. 8 and Proposition VI 8. 5) that S_ξ has a unique minimum u_ξ, which is a weak solution of (72), i.e. for every $v \in H_c$ we have

$$\begin{aligned}\langle v,u_\xi\rangle_c &= \langle\nabla_A v,\nabla_A u_\xi\rangle + \langle[\Phi,v],[\Phi,u_\xi]\rangle \\ &= \langle v,\xi\rangle .\end{aligned} \tag{76}$$

Standard regularity theory for elliptic equations (see for example [12]) now implies that since $(A,\Phi) \in W^1_{\mathrm{loc}}(\Omega_{\mathrm{ad}})$, we have $u_\xi \in W^2_{\mathrm{loc}}$, and that (72) holds almost everywhere on M. Moreover, choosing $v = u_\xi$ in (76) yields (73). Now set $u := u_\xi$. Under the additional assumption that u has compact support, we now show that

$$\|\nabla_A\nabla_A u\| \leq K_2 \|\xi\|_{Z_c} . \tag{77}$$

Lemma 20. *For a bundle E_{ad} with connection ∇_A and curvature $F(\nabla_A)$, we have*

$$\begin{aligned}&\left(\nabla_A^2\nabla_A - \nabla_A\nabla_A^2\right)u \\ &= d_A^*\left(u\wedge F(\nabla_A)\right) + (d_A u)F(\nabla_A) + \mathrm{Ric}(d_A u),\end{aligned} \tag{78}$$

where Ric *denotes the operation of the Ricci-tensor of (M,g) on $\Lambda^1 \otimes E_{\mathrm{ad}}$.*

Proof. See the proof of [16], after Lemma B 3.8. The metric part comes from the commutator relation

$$(\nabla_i\nabla_j - \nabla_j\nabla_i)\nabla_j\xi = [F_{ij}, \nabla_j u] + \sum_k R_{ji}^{ik}\nabla_k u. \tag{79}$$

□

Now an integration by parts yields

$$\begin{aligned} \|\nabla_A\nabla_A u\|_2^2 &= -\left\langle \nabla_A^2\nabla_a u, \nabla_A u\right\rangle \\ &= \left\langle \nabla_A\left(\nabla_A^2 u\right), \nabla_A u\right\rangle - \langle d_A^*(F(\nabla_A)\wedge u) \\ &+\nabla_A u\surd F(\nabla_A) + \mathrm{Ric}(\nabla_A u), \nabla_A u\rangle \\ &= \left\langle \nabla_A^2 u, \nabla_A^2 u\right\rangle + \langle F(\nabla_A)\wedge u, d_A d_A u\rangle \\ &\quad - \langle \nabla_A u\surd F(\nabla_A) + \mathrm{Ric}(\nabla_A u), \nabla_A u\rangle\,. \end{aligned} \tag{80}$$

By the method already used in the proof of Lemma 18, we can estimate the first term as follows

$$\begin{aligned} \left\|\nabla_A^2 u\right\|_2 &= \left\|\left((\mathrm{ad}\Phi)^2 - L_c\right)u\right\|_2 \\ &\leq K\,\|u\|_c + \|\xi\|_{Z_c}\,. \end{aligned} \tag{81}$$

To estimate the second term, we use the defining relation for the curvature $d_A d_A u = F(\nabla_A)\wedge u$, see (38) and Lemma 18:

$$\begin{aligned} \|F(\nabla_A)\wedge u\|_2 &\leq \|F(\nabla_A)\|_2\ \|u\|_\infty \\ &\leq K_4\left(\|R_g\|_2 + \|F_A\|_2\right)\left(\|\nabla_A u\|_2 + \|\nabla_A u\|_4\right). \end{aligned} \tag{82}$$

Together with (80) we conclude that

$$\|\nabla_A\nabla_A u\|_2 \leq K_5\left(\|\xi\|_Z + \|\nabla_A u\|_4\right). \tag{83}$$

Now we apply Hölder's inequality to obtain

$$\begin{aligned} \|\nabla_A u\|_4 &\leq \|\nabla_A u\|_2^{\frac{1}{4}}\ \|\nabla_A u\|_6^{\frac{3}{4}} \\ &\leq K_6\,\|u\|_c^{\frac{1}{4}}\ \|\nabla_A\nabla_A u\|_2^{\frac{3}{4}} \\ &\leq K_7\,\|\xi\|_Z^{\frac{1}{4}}\ \|\nabla_A\nabla_A u\|_2^{\frac{3}{4}}\,. \end{aligned}$$

This together with (83) implies (77).

To complete the proof, we have to remove the condition of compact support on u. With β_r of Definition 12, we have $L_c\beta_R u \longrightarrow L_c u = \xi$ in Z for $R \longrightarrow \infty$. In fact

$$\begin{aligned}
&\|L_c u_R - L_c u\|_Z \\
&\le \|L_c\beta_R u - \beta_R L_c\|_Z + \|(1-\beta_R)Lu\|_Z \\
&\le 2\,\|\nabla\beta_R\sqrt{}\nabla_A u\|_Z \\
&\quad + \left\|\nabla^2\beta_R \cdot u\right\|_Z + \|(1-\beta_R)\xi\|_Z \\
&\le \left(\|\nabla\beta_R\|_\infty + \|\nabla\beta_R\|_3\right)\|\nabla_A u\|_{2>R} \\
&\quad + \left(\|\nabla\nabla\beta_R\|_3 + \|\nabla\nabla\beta_R\|_{\frac{3}{2}}\right)\|u\|_{6>R} \\
&\quad + \|(1-\beta_R)\xi\|_Z
\end{aligned} \tag{84}$$

converges to zero for $R \longrightarrow \infty$, since $u \in H_c$ and $\xi \in Z$ and since $\|\nabla\nabla\beta_R\|_p + \|\nabla\beta_R\|_q$ is bounded independently of R for $p \ge \frac{3}{2}$ and $q \ge 3$. Now we have by (77):

$$\begin{aligned}
\|\nabla_A\nabla_A u\|_2 &= \sup_R \|\nabla_A\nabla_A u\|_{2<R} \\
&\le \limsup_{R\to\infty} \|\nabla_A\nabla_A\beta_R u\|_2 \\
&\le K_8 \lim \|L\beta_r u\|_Z \le K_9\,\|\xi\|_Z\ .
\end{aligned} \tag{85}$$

This completes the proof of Proposition 19. □

If ∇ is a metric connection on a metric bundle E, define X as the completion of $C_0^\infty(E)$ in the norm

$$\|\xi\|_{X_\nabla} = \|\nabla\nabla\xi\|_2 + \|\nabla\xi\|_2 + \left\|\nabla^2\xi\right\|_{\frac{6}{5}}\ . \tag{86}$$

As an immediate corollary of Proposition 19, we then have

Proposition 21. *If the curvatures of ∇ and of g are square integrable, then the map*

$$\nabla^2 : X_\nabla(E) \longrightarrow L^2(E) \cap L^{\frac{6}{5}}(E) \tag{87}$$

is a linear homeomorphism.

5 The gauge operation

Recall the definitions of the gauge group $\mathcal{G}$ and of the configuration spaces $\mathcal{C}$ and $\mathcal{C}^*$ given in the introduction. The following proposition describes the properties of the operation of $\mathcal{G}$ on $\mathcal{C}$.

Proposition 22. *The map*

$$\mathcal{G} \times \mathcal{C} \longrightarrow \mathcal{C} : (g, c) \longmapsto g(c) = (g(A), g\Phi g^{-1}) \tag{88}$$

with $g(A)$ defined in (30) is continuous. Moreover, the operation of $\mathcal{G}$ on $\mathcal{C}^*$ is free up to the center $\mathbb{Z}_2 = \{\pm 1\} \subset \mathcal{G}$ and has a closed graph in the following sense: Let $(c_n) \subset \mathcal{C}^*$ and $(g_n) \subset \mathcal{G}$ be sequences so that $c_n \longrightarrow c$ and $g_n(c_n) \longrightarrow c'$ in $\mathcal{C}^*$. Then $g_n \longrightarrow g$ in $\mathcal{G}/_{\mathbb{Z}_2}$ with $g \in \mathcal{G}$ satisfying $g(c) = c'$.

Proof. Clearly, $g\Phi g^{-1} \in W^2_{\text{loc}}(\Lambda^0_{\text{ad}})$ is jointly continuous in $g \in \mathcal{G}$ and $\Phi \in W^2_{\text{loc}}(\Lambda^0_{\text{ad}})$. Moreover, $g\nabla g^{-1} + gAg^{-1} \in W^1_{\text{loc}}(\Lambda^1_{\text{ad}})$ is jointly continuous in $L^\infty(\text{Hom}(\text{ad}\, P, \text{ad}\, P))$. Hence $F_{g(A)} = gF_A g^{-1}$ is jointly continuous in $c \in \mathcal{C}$ and $g \in \mathcal{G}$.

We show now that $\mathcal{G}/_{\mathbb{Z}_2}$ operates freely on $\mathcal{C}^*$. In fact, if $r \geq r_M$ is large enough, then $g(c) = c'$ on M_r implies $g = \pm 1$ on M_r. To see this, consider the open set $M_g = \{x \in M_r \mid g(x) \neq -1\}$. There exists a unique continuous section ξ_r of $\text{ad}\, P$ over M_g so that $\exp \xi_r(x) = g(x)$ and $\|\xi_r\|_\infty \leq \pi$. By the definition (8) of the operation of g, we then have

$$\nabla_A \xi_r = 0 \quad \text{and} \quad [\Phi, \xi_r] = 0\,. \tag{89}$$

Hence $|\xi_r|$ is constant, which shows that either $g(x) \equiv -1$ or $M_g = M_r$.

Now assume that there exists such a $\xi_r \neq 0$ for all $r > M_r$. Since ξ is well defined by $\xi(x_0)$ for any $x_0 \in M_{r_M}$, we conclude that there exists a $\xi \neq 0$ on M satisfying (89).

Now consider the sequences $c_n = (A_n, \Phi_n)$. Let $P_{|U} = U \times SU_2 \subset U \times M_2(\mathbb{C})$ be a local trivialization on P, with $U \subset M$ open and $M_2(\mathbb{C})$ be linear space of 2×2-complex matrices. The equation $g_n(A_n) = A'_n$ can then locally be written as

$$dg_n = g_n A_n - A'_n g_n\,, \tag{90}$$

where d is the derivative of functions $M \longrightarrow M_2(\mathbb{C})$ and the product is the composition of matrices. Hence g_n is bounded in $W^1(U, M_2(\mathbb{C}))$. With $\nabla A'_n$ converging in $L^2(\Lambda^1(U) \times M_2(\mathbb{C}))$, we find that

$$\nabla\nabla g_n = (\nabla g_n)A_n - A'_n(\nabla g_n) + g_n(\nabla A_n) - (\nabla A'_n)g_n \tag{91}$$

implies that g_n is bounded in $W^2(U, M_2(\mathbb{C}))$. Hence there exist weakly convergent subsequences $g_n \longrightarrow g$ in $W^2(U, M_2(\mathbb{C}))$. Moreover, $g_n \longrightarrow g$ strongly in $L^\infty(U, M_2(\mathbb{C}))$ and $\nabla g_n \longrightarrow \nabla g$ strongly in $L^3(\Lambda^1(U) \otimes M_2(\mathbb{C}))$. Hence equation (91) implies that $g_n \longrightarrow g$ in $W^2(U, M_2(\mathbb{C}))$. Since $SU_2 \subset M_2(\mathbb{C})$ is closed, and since we can cover any compact set $M_r \subset M$ by finitely many open sets U with $P_{|U}$ trivial, we conclude that every subsequence of g_n contains a subsequence converging strongly in $W^2(\text{aut}\, P_{|M_r})$ to some g. Moreover, by continuity of the operation of M, we have $g(c') = c$ on M_r. By the above, we can conclude that for r large enough, g is uniquely determined by this property. Hence $g_n \longrightarrow g$ on every M_r with $r \geq r_M$. □

Consider now the following n infinitesimal generators of $\mathcal{G}$.

Definition 23. As in Lemma 15 we define $\acute{\Phi}_{(i)}$ for $1 \leq i \leq n$ as the unique critical point of the functional defined by (57) on the space $\beta_r^i \Phi + H_c(\Lambda^0_{\mathrm{ad}})$.

In analogy to $\acute{\Phi}$, they satisfy

$$\left\| L_c \acute{\Phi}_{(i)} \right\| = 0 \tag{92}$$

$$\left\| \acute{\Phi}_{(i)} \right\|_\infty \leq m_i(c) \tag{93}$$

$$\left\| \nabla_A \acute{\Phi}_{(i)} \right\|_2 + \left\| \nabla_A \nabla_A \acute{\Phi}_{(i)} \right\|_2 + \left\| [\Phi, \acute{\Phi}_{(i)} \right\|_2 \leq K, \tag{94}$$

where K depends only on $\mathbf{a}(c)$ and $\underline{m}(c)$. The corresponding n-dimensional subgroup of $\mathcal{G}$ will turn out to be responsible for the n additional gauge conditions in Definition 3.

Proposition 24. *The map*

$$\underline{k} : Q \longrightarrow \mathbb{R}^n : [c] \longmapsto \left(\frac{1}{4\pi m_i} \int_M (d_A \acute{\Phi}_{(i)}) \dot{\wedge} F_A \right) 1 \leq i \leq n \tag{95}$$

is continuous and takes values in $\mathbb{Z}^n \subset \mathbb{R}^n$. Moreover we have

$$\underline{m} \cdot \underline{k} = \Sigma m_i k_i = \frac{1}{4\pi} \int_M (d_A \Phi) \dot{\wedge} F_A = \frac{1}{4\pi} \int (d_A \Phi \mid * F_A) \mu_g . \tag{96}$$

Proof. Integrality of (95) is proved in the same way as Proposition 2 of [6]. The identity (96) follows from Lemma 2 of [6]. □

6 Proof of Theorem 4

The proof of Theorem 4 is carried out in two steps. First, we construct a gauge $g_1 : \mathcal{U}_c \longrightarrow \mathcal{G}$ on a neighborhood of c in $\mathcal{C}$ so that $g_1(c')(c') - c$ has certain decay properties. We then use the implicit function theorem to improve g_1 to a continuous gauge g so that $g(c')(c') - c \in Q_c$.

Define

$$\|(a, \phi)\|_{Y_c} = \|(a, \phi)\|_c + \left\| \Phi \cdot d_A^* a \right\|_{\frac{6}{5}} , \tag{97}$$

and let Y_c denote the subspace of H_c with finite $\| \quad \|_{Y_c}$. The proof of the following proposition, which essentially follows the proof of [16], Lemma B. 4.5 will be given in Section 7.

Proposition 25. *There exists a neighborhood $\mathcal{U}_1(c)$ of c in $\mathcal{C}$ and a map*

$$g_1 : \mathcal{U}_1(c) \longrightarrow \mathcal{G}; g_1(c) = 1 \tag{98}$$

so that the assignment

$$c' \longrightarrow g_1(c')(c') - c \tag{99}$$

is continuous at c' as a map $\mathcal{U}_1(c) \longrightarrow Y_c$.

□

Recall now the Banach spaces X_c and Z_c of Definition 16. We want to augment X_c by the fields $\acute{\Phi}_{(i)}$ of Section 5. For $\underline{t} \in \mathbb{R}^n$, set $\acute{\Phi}_{\underline{t}} = \sum_{i=1}^n t_i \acute{\Phi}_{(i)}$ and define

$$\acute{X}_c := \{\xi + \acute{\Phi}_{\underline{t}} \mid \xi \in X_c, \underline{t} \in \mathbb{R}^n\} \cong X_c \otimes \mathbb{R}^n \tag{100}$$

$$\acute{Z}_c := Z_c \otimes \mathbb{R}^n \tag{101}$$

with the product topologies. Note that in analogy to Lemma 17, we can derive from (70) the estimate

$$0 < \left\|\acute{\Phi}_{(i)}\right\|_{X_c} < \infty \tag{102}$$

for $1 \leq i \leq n$. Hence the norm $\left\|\xi_{\underline{t}}\right\|_X$ of Definition 16 is continuous on $\acute{X}$. We can extend the operator φ_c to an operator $\acute{\varphi}_c$ on $\acute{X}_c$ and define

$$\acute{\vartheta}_c : Y_c \longrightarrow \acute{Z}_c : (a, \phi) \longrightarrow (\vartheta_c(a, \phi), (\left\langle a, \nabla_A \acute{\Phi}_i \right\rangle_c)_{i=1\dots n}) . \tag{103}$$

The definitions of $\acute{X}_c, \acute{Y}_c, \acute{Z}_c$ are motivated by the following properties.

Lemma 26. *The linear maps*

$$\acute{X}_c \xrightarrow{\acute{\varphi}_c} Y_c \xrightarrow{\acute{\vartheta}_c} \acute{Z}_c \tag{104}$$

ar continuous and Kern $\acute{\vartheta}_c \subset Y_c$ *is equivalent to the Hilbert space Q_c of Definition 1. Moreover, $\acute{\vartheta}_c \acute{\varphi}_c : \acute{X}_c \longrightarrow \acute{Z}_c$ is an isomorphism.*

Proof. Continuity of $\varphi_c : X_c \longrightarrow H_c$ follows from Lemma 17. From (93) it follows that the extension $\acute{\varphi}_c$ is continuous, too. Continuity of ϑ_c follows immediately from the definitions of Y_c and Z_c. Moreover, since for $(a, \phi) \in H_c$ with $\vartheta_c(a, \phi) = 0$ we have $\Phi \cdot d_A^* a = 0$, it follows that $(a, \phi) \in Y_c$.

To prove the last assertion, we calculate

$$\acute{\vartheta}_c \acute{\varphi}_c(\xi + \acute{\Phi}_{\underline{t}}) = \left(L_c \xi, (\langle \nabla_A \xi + \nabla_A \Phi_i, \nabla_A \Phi_i \rangle_c)_{i=1\dots n}\right) . \tag{105}$$

Now injectivity of $\acute{\vartheta}_c \acute{\varphi}_c$ follows immediately from Proposition 19. Since $\acute{\vartheta}_c \acute{\varphi}_c$ is a Fredholm operator of index zero by Proposition 19 this also implies that $\acute{\vartheta}_c \acute{\varphi}_c$ is surjective. □

We want to modify the gauge g_1 of Proposition 25 by elements of the gauge subgroup

$$\mathcal{G}_c := \exp \acute{X}_c := \{\exp \xi \mid \xi \in \acute{X}_c\}. \tag{106}$$

Lemma 27. *$\mathcal{G}_c$ is a smooth Banach Lie group with Banach Lie algebra $\acute{X}_c$. It operates smoothly on Y_c by*

$$g[y] = g(c+y) - c. \tag{107}$$

Proof. We first show that $\acute{X}$ is a Banach algebra, i.e. that

$$\|[\zeta_s, \xi_t]\|_{X_c} \leq K \|\zeta_s\|_{\acute{X}_c} \|\xi_t\|_{\acute{X}_c} \tag{108}$$

for some constant K. In a similar way, we then show that the infinitesimal operation of $\xi \in C^\infty(\Lambda^0_{\mathrm{ad}})$ and $\acute{\Phi}_{\underline{t}}$ on $(a, \phi) \in Y_c$ extends to a continuous map $\Psi_c : \acute{X}_c \times Y_c \longrightarrow Y_c$

$$(\xi, (a, \phi)) \longrightarrow X_\xi(a, \phi) = (\nabla_A \xi, [\Phi, \xi]) + ([a, \xi], [\phi, \xi]). \tag{109}$$

The smoothness of the multiplication in $\mathcal{G}_c$ and of the map (107) will then follow from the theory of ordinary differential equations.

To prove (108), note that for $\xi, \zeta \in X_c$, we have

$$\|\nabla_A \nabla_A [\xi, \zeta]\|_2 \leq \|\nabla_A \nabla_A \xi\|_2 \|\xi\|_\infty + 2 \|\nabla_A \xi\|_4 \|\zeta_A\|_4 \tag{110}$$

$$+ \|\xi\|_\infty \|\nabla_A \nabla_A \zeta\|_2 \tag{111}$$

$$\|[\xi, \zeta]\|_c \leq \|\xi\|_\infty \|\zeta\|_c + \|\xi\|_\infty \|\zeta\|_c$$

$$\|\Phi \cdot \nabla_A^2 [\xi, \zeta]\|_{\frac{6}{5}} \leq \|\Phi \cdot [\nabla_A^2 \xi, \zeta]\|_{\frac{6}{5}} + 2 \|\Phi \cdot [\nabla_A \xi, \nabla_A \zeta]\|_{\frac{6}{5}} \tag{112}$$

$$+ \|\Phi \cdot [\xi, \nabla_A^2 \zeta]\|_{\frac{6}{5}} \tag{113}$$

$$\leq \|\nabla_A^2 \xi\|_2 \|[\Phi, \zeta]\|_3 + \|\nabla_A^2 \zeta\|_2 \|[\Phi, \zeta]\|_3 + 2 \|\nabla_A \xi\|_2 \|\nabla_A \zeta\|_3$$

so that $\|[\xi, \zeta]\|_{X_c} \leq K_1 \|\xi\|_{X_c} \|\zeta\|_{X_c}$ by Lemma 17. Here we have used the Jacobi identity and the equation $\Phi \cdot [\phi, \psi] = \phi \cdot [\psi, \Phi]$. By (102) it follows in the same way that $[\acute{\Phi}_{(i)}, \acute{\Phi}_{(j)}] \in X_c$ for $1 \leq i \leq j \leq n$ and that $\mathrm{ad}\, \acute{\Phi}_{(i)}$ operates continuously on X_c. This completes the proof of (108).

We show now that the vector field Ψ_c of (109) is continuous. The constant part $\acute{\vartheta}_c \xi$ depends continuously on $\xi \in \acute{X}_c$ by Lemma 26. The part which is bilinear in ξ and (a, ϕ) is estimated by

$$\|[a, \xi]\|_c \leq \|a\|_c \|\xi\|_\infty + \|a\|_6 \left(\|\nabla_A \xi\|_3 + \|[\Phi, \xi]\|_3 \right)$$

and

$$\left\|\Phi \cdot d_A^*[a,\xi]\right\|_{\frac{6}{5}} \leq \|\nabla_A a\|_2 \ \|[\Phi,\xi]\|_3 + \|\Phi \cdot (a\surd d_A \xi)\|_{\frac{6}{5}}$$

together with (102), and Lemma 17. Now note that

$$F_{\xi,y}(t) := \exp(t\xi)[y]$$

satisfies the ordinary differential equation

$$\frac{d}{dt} F_{\xi,y}(t) = \Psi(\xi, F_{\xi,y}(t)) \tag{114}$$

with initial condition $F_{\xi,y}(0) = y$. The solution depends smoothly on ξ and y. Moreover, since the vector field is the sum of a constant and a linear part, we have

$$\begin{aligned} \frac{d}{dt} \left\|F_{\xi,y}(t)\right\|_{Y_c} &\leq \left\|\tfrac{d}{dt} F_{\xi,y}(t)\right\|_{Y_c} \\ &\leq \left\|\Psi_\xi(F_{\xi,y}(t))\right\|_{Y_c} \\ &\leq \|\xi\|_{\acute{X}} \ \left\|F_{\xi,y}(t)\right\|_{Y_c} . \end{aligned} \tag{115}$$

Hence $\frac{d}{dt} log \left\|F_{\xi,y}(t)\right\|_{Y_c}$ is bounded independently of t. Consequently, the norm of $\left\|F_{\xi,y}(t)\right\|_{Y_c}$ cannot become infinite in finite time. In particular, the map

$$\begin{aligned} \acute{X}_c \times Y_c &\longrightarrow Y_c \\ (\xi, y) &\longrightarrow F_{\xi,y}(1) = \exp(\xi)[y] , \end{aligned}$$

exists and is smooth. In the same way we prove that multiplication in $\mathcal{G}_c$ is smooth. This completes the proof of Lemma 27. □

By the implicit function theorem, there exist neighborhoods $\mathcal{U}_{\acute{X}_c}$ of $1 \in \mathcal{G}_c$ and $\mathcal{U}_{Y_c}$ of $0 \in Y_c$ and a smooth map

$$g_2 : \mathcal{U}_{Y_c} \longrightarrow \mathcal{U}_{\acute{X}_c}; \Xi(g_2(y), y) := \acute{\vartheta}_c g_2(y)[y] = 0 . \tag{116}$$

Since the map $c' \longrightarrow g_1(c')(c') - c'$ of Proposition 25 is continuous at $c' = c$, we find a neighborhood $\mathcal{U}_c \subset \mathcal{U}_1(c)$ of c in $\mathcal{C}$ so that the assignment (99) maps $\mathcal{U}_c$ into $\mathcal{U}_{Y_c}$. Now we define the map

$$g : \mathcal{U}_c \longrightarrow \mathcal{G} : g(c') = g_2(g_1(c')(c') - c') . \tag{117}$$

and the induced map

$$\varsigma_c : \mathcal{U}_c \longrightarrow Q_c : c' \longrightarrow g(c')(c') - c , \tag{118}$$

where $Q_c = \operatorname{Kern} \acute{\vartheta}_c \subset Y_c$ by Lemma 26. Moreover, the map ς_c is continuous at $c' = c$ and satisfies $\varsigma_c(c) = 0$. This shows that for every zero neighborhood $\mathcal{U}_{Q_c}$ in Q_c there exists a neighborhood $\mathcal{U}_c$ of $c \in \mathcal{C}$ so that every orbit in the invariant neighborhood $\mathcal{G}\mathcal{U}_c \subset \mathcal{C}$ has at least one intersection with the set $c + \mathcal{U}_{Q_c}$.

It remains to prove that we can choose the neighborhoods small enough so that this intersection is unique. This is done in the following way. By the implicit function theorem, there exist neighborhoods $\mathcal{U}_{Y_c}$ of $0 \in Y$ and $\mathcal{U}_{\acute{X}_c}$ of $1 \in \mathcal{G}_c$ so that for each $y \in \mathcal{U}_{Y_c}$ there is a unique $g \in \mathcal{U}_{\acute{X}_c}$ with $g[y] \in Q_c$. In particular, if $y \in \mathcal{U}_{Y_c} \cap Q_c$, then the only $g \in \mathcal{U}_{\acute{X}_c}$ with $g[y] \in Q_c$ is the identity. Therefore, it suffices to show that if we choose $\mathcal{U}_{Q_c}$ small enough, then every $g \in \mathcal{G}$ with $g[y] \in \mathcal{U}_{Q_c}$ for some $y \in \mathcal{U}_{Q_c}$ has to be in $\mathcal{U}_{\acute{X}_c}$. This follows from

Proposition 28. *For each neighborhood $\mathcal{U}_{\acute{X}_c}$ of $1 \in \mathcal{G}_{\acute{X}_c}$ there exists a neighborhood $\mathcal{U}_{Y_c}$ of $0 \in Y_c$ so that if y and $g[y] \in \mathcal{U}_{Y_c}$, $g \in \mathcal{G}$ then $g \in \mathcal{U}_{\acute{X}_c} \subset \mathcal{G}_c$.*

The proof of Proposition 28 will be given in Section 8. We have finished the proof of assertion (1) of Theorem 5. Now the map ς_c of (118) defines a bijective map

$$\varsigma_c : \mathcal{G}\mathcal{U}_c/\mathcal{G} =: \mathcal{U}_\varsigma(c) \longrightarrow \mathcal{U}_c \tag{119}$$

by the unique intersection of the orbit with $c + \mathcal{U}_{Q_c}$. The proof of Theorem 5 is now completed by

Lemma 29. *For $c_0, c_1 \in \mathcal{G}$, define $Q(c_i, c_j) = \varsigma_{c_i}(\mathcal{U}_{Q_{c_i}} \cap \mathcal{U}_{Q_{c_j}}) \subset Q(c_i)$. Then we have a smooth map*

$$\tau_{c_0,c_1} := \varsigma_{c_0}\varsigma_{c_1}^{-1} : Q(c_0, c_1) \longrightarrow Q(c_1, c_0)\,. \tag{120}$$

Proof. We need the following lemma:

Lemma 30. *For $c, c' \in \mathcal{G}$ and $\xi \in Y_c$ we have*

$$\|\xi\|_{Y_{c'}} \leq K(\,\|c - c'\|_c\,)\,\|\xi\|_{Y_c}\,. \tag{121}$$

where $K : \mathbb{R}_+ \longrightarrow \mathbb{R}_+$ is an increasing function depending only on $\mathbf{a}(c)$.

Proof. We easily obtain (see also [16], Proposition B6. 2)

$$\|(a, \phi)\|_{c'} \leq (1 + K_1\,\|c - c'\|_c\,)\,\|(a, \phi)\|_c\,. \tag{122}$$

It therefore remains to estimate for $(\psi, b) = c - c'$:

$$\begin{aligned}\left\|(\Phi + \psi) \cdot d^*_{A+b}a\right\|_{\frac{6}{5}} \leq\ & \left\|\Phi \cdot d^*_A a\right\|_{\frac{6}{5}} + \|\Phi \cdot (b\surd a\|_{\frac{6}{5}} \\ & + \left\|\psi \cdot d^*_A a\right\|_{\frac{6}{5}} + \|\psi \cdot (b\surd a)\|_{\frac{6}{5}}\,,\end{aligned}$$

by $\|(a,\phi)\|_{Y_c}$ and $\|(b,\psi)\|_c$. The estimate of the first term is obvious from (97). By Lemma 18, we have $\|\Phi\cdot(b\surd a)\|_{\frac{6}{5}} \leq \|b\|_c \; \|a\|_c$. To estimate the third term, note that $d_A^*a \in Z_c$ and $\psi \in H_c$ and apply Lemmas 13 and 18. For the last term, we use the method of Lemma 18 and the fact that $b_L \surd a_L = 0$:

$$\begin{aligned}\|\psi(b\surd a)\|_{\frac{6}{5}} &\leq \|\psi\|_6 \left\{\left(\|b\|_{2,K_c} + \frac{2}{m_0}\|[\Phi,b]\|_2\right)\|a\|_6 \right.\\ &\quad \left. + \|b\|_6\left(\|a\|_{2,K_c} + \frac{2}{m_0}\|[\Phi,b]\|\right)\right\} \qquad (123)\\ &\leq \|\psi_c\| \; \|b\|_c \; \|a\|_c \,.\end{aligned}$$

□

To continue the proof of Lemma 29, set $c_0 := c$. If $c_1 = g(c)$ for $g \in \mathcal{G}$, then $\tau_{c,g(c)} : Q_{g(c)} \longrightarrow Q_c : q \longmapsto g^{-1}qg^{-1}$ is continuous linear and therefore smooth. To prove the general case, we can therefore assume that $c = c_1 + q$, where $q = \varsigma_{c_1}([c_0]) \in \mathcal{U}_{Q_{c_1}}$. By Lemma 30, q is also small in Q_c. We can therefore apply the implicit function theorem directly to obtain

$$\begin{aligned}\varsigma_c([c_1]) &= \varsigma_c([c-q]) = g_2(-q)\\ &= \varsigma_c\left(\overset{-1}{\underset{c_1}{}}(q)\right),\end{aligned}$$

where g_2 is the smooth function of 116. □

Now the proof of Theorem 4 is completed up to the proofs of Proposition 25 and Proposition 28, which will be given in the next two sections.

7 Proof of Proposition 25

We first construct a map $\tilde{g}_1$ satisfying the conditions of Proposition 25 on a neighborhood of c in $\tilde{\mathcal{C}}$. We then use Lemma 15 to extend $\tilde{g}_1$ to a neighborhood of c in $\mathcal{C}$.

The reason for working in $\tilde{\mathcal{C}}$ rather than $\mathcal{C}$ is that the second order condition $\|\nabla_A\nabla_A\Phi\|_2 < \infty$ implies by Lemma 13 that there exists $r \in \mathbb{R}_+$ so that $|\Phi| \geq \frac{m_0}{2}$ outside M_r. We can identify this number with the number r_M of Definition 1. Then the field

$$\hat{\Phi} = (1-\beta_{\frac{r}{2}})\frac{\Phi}{|\Phi|} \qquad (124)$$

is well defined and the map

$$\begin{aligned}\tilde{\mathcal{C}} &\longrightarrow L^2\left(\Lambda^1_{\mathrm{ad}} \oplus (\Lambda^1 \otimes \Lambda^1_{\mathrm{ad}})\right)\\ c &\longrightarrow \left(\nabla_A\hat{\Phi}, \nabla_A\nabla_A\hat{\Phi}\right)\end{aligned} \qquad (125)$$

is continuous. Following Taubes [16], we construct the map $\tilde{g}_1$ in two steps. First, we choose for each c' sufficiently close to c a gauge $g_0(c')$ which makes the Higgs field of c' parallel to Φ outside some compact set. This construction is well-defined up to the action of a certain Abelian subgroup of $\mathcal{G}$, the operation of which is similar to the operation of the gauge group in electro-dynamics. It can be fixed by a Lorentz gauge.

To carry out the first step, we use the following geometrical construction on $SU_2 \cong S^3$ and its Lie algebra:

Definition 31 Assume that $\psi, \psi' \in su_2$ with $|\psi| = |\psi'| = 1$ and $|\psi - \psi'| < 1$. Let $(\psi - \psi')_T$ be the component of $\psi - \psi'$ which is orthogonal to ψ. Let $\alpha : [0,1) \longrightarrow [0,\infty)$ be the unique function with $\alpha(0) = 0$ so that

$$\xi(\psi, \psi') := \alpha\left(|(\psi - \psi')_T| \right) (\psi - \psi')_T \tag{126}$$

has the property

$$\psi = \mathrm{ad}(\exp \xi(\psi, \psi'))\psi' .$$

Now consider the subset

$$\tilde{\mathcal{U}}_R(c) = \{c' = (A', \Phi') \mid \left|\hat{\Phi} - \hat{\Phi}'\right| < \frac{1}{3} \quad \text{on} \quad M_R\} \tag{127}$$

of $\tilde{\mathcal{C}}$. It is open in $\tilde{\mathcal{C}}$ by (125) and by the Sobolev embedding $W^2(M_R) \longrightarrow L^\infty(M_R)$. Now define

$$\tilde{\mathcal{C}}_\Phi = \{c' = (A', \Phi') \in \tilde{\mathcal{C}} \mid [\Phi', \Phi] = 0 \quad \text{on} \quad M - M_r\} . \tag{128}$$

Lemma 32. *For $R > 2r$, there exists a map $g_0 : \tilde{\mathcal{U}}_R(c) \longrightarrow \mathcal{G}$, so that $g_0(c')(c') \in \tilde{\mathcal{C}}_\Phi$ for all $c' \in \mathcal{C}$ and so that on M_R, we have*

$$g_0(c') = \exp\left\{(1 - \beta_r(x))\xi(\hat{\Phi}(x), \hat{\Phi}'(x))\right\} . \tag{129}$$

Proof. Clearly g_0 defined in (129) satisfies $\mathrm{ad}(g_0)\hat{\Phi}' = \hat{\Phi}$ on $M_R - M_r$. We want to extend this automorphism to all of M. Since $\hat{\Phi}$ and $\hat{\Phi}'$ are continuous on M, we can cover $M - M_{2r}$ by open balls $U_i, i \in \mathbb{N}$, so that for $x, y \in U_i$

$$\begin{aligned} \left|\hat{\Phi}(x) - \hat{\Phi}(y)\right| &< \frac{1}{3} \\ \left|\hat{\Phi}'(x) - \hat{\Phi}'(y)\right| &< \frac{1}{3} . \end{aligned} \tag{130}$$

We can assume that all intersections $U_i \cap U_j$ are contractible. Together with $U_0 := M_R$ this defines a covering

$$\mathcal{O} = (U_i)_{i \geq 0} \tag{131}$$

of M. For every $i \geq 0$, define an element $g_i \in SU_2$ by $g_i = 1$ if $U_i \cap U_0 \neq \emptyset$ and by $g_i \hat{\Phi}'(x_i) g_i^{-1} = \hat{\Phi}(x_i)$ for some $x_i \in U_i$ otherwise. Then by (130), we have for all $x \in U_i, i \geq 0$

$$\left| g_i \hat{\Phi}'(x) g_i^{-1} - \hat{\Phi}(x) \right| < 1 . \tag{132}$$

We can now apply Definition 31 to obtain automorphisms

$$\begin{aligned} g_i &\in W^2 \left(\mathrm{Aut} P \,|_{U_i} \right) \\ g_i(x) &= \xi(\hat{\Phi}, g_i \hat{\Phi}'(x) g_i^{-1}) g_i \end{aligned} \tag{133}$$

for $i \geq 1$. By construction, we have for all $i \geq 0$ and $x \in M - M_{2r}$

$$\mathrm{ad}(g_i(x)) \hat{\Phi}'(x) = \hat{\Phi}(x) . \tag{134}$$

Moreover, for $x \in U_i \cap U_j$ with $i, j \geq 0$, we have

$$\hat{\Phi}(x) = \mathrm{ad}(g_i(x) g_j^{-1}(x)) \hat{\Phi}(x) . \tag{135}$$

For any $g \in SU_2$ with $\mathrm{ad}(g)\hat{\Phi} = \Phi$ we have $g = \exp(\alpha \hat{\Phi})$ with $\alpha \in \mathbb{R}$. Moreover, α is well defined up to addition of integers. Since we assumed all intersections $U_i \cap U_j$ to be contractible, we can therefore define continuous maps $f_{ij} : U_i \cap U_j \longrightarrow \mathbb{R}$ so that $f_{ij} = 0$, if $i = 0$ or $j = 0$ and

$$\exp(2\pi (f_{ij}(x) \hat{\Phi}(x)) = g_i(x) g_j^{-1}(x) . \tag{136}$$

We now use the following construction of algebraic topology.

Definition 33. For $A = \mathbb{R}$ or $\mathbb{Z}$, define the Abelian groups

$$\begin{aligned} \mathscr{C}_A^k = \{ & \left(f_{i_0 \ldots i_k} \right)_{0 \leq i_0 < i_1 < \ldots < i_k} \mid \\ & f_{i_0 \ldots i_k} \in W^2 \left(U_{i_0} \cap \ldots \cap U_{i_k} \right) \\ & f_{i_0 \ldots i_k} = 0 \quad \text{if} \quad i_0 = 0 \} . \end{aligned} \tag{137}$$

Moreover, define the group homomorphisms

$$\begin{aligned} \delta : \mathscr{C}_A^k &\longrightarrow \mathscr{C}_A^{k+1} \\ (\delta f)_{i_0 \ldots i_k} &= \sum_{j=0}^{k} (-1)^j f_{i_0 \ldots \hat{i}_j \ldots i_k} , \end{aligned} \tag{138}$$

where the index $\hat{i}_j$ has to be omitted.

It is easy to see that the functions f_{ij} of (136) are in $W^2(U_i \cap U_j, \mathbb{R})$ and define therefore an element $f^{(1)}$ of $\mathcal{C}^1_{\mathbb{R}}$. Recall that, for $0 < i < j, f_{ij}$ is well defined by (136) up to addition of a constant integer. Since it is well defined for $i = 0$, we conclude that all $\tilde{f}^{(1)} = (\tilde{f}_{ij})_{0 \le i \le j} \in \mathcal{C}^1_{\mathbb{R}}$ satisfying (136) are of the form $\tilde{f}^{(1)} = f^{(1)} + z$, where $z \in \mathcal{C}^1_{\mathbb{Z}}$. We want to show that there exists such a z so that

$$\delta\tilde{f}^{(1)} = \delta f^{(1)} + \delta z = 0 . \tag{139}$$

To this end, note that $\delta\tilde{f}^{(1)} \in \mathcal{C}^2_{\mathbb{R}}$, as one easily derives from (136) and (135). Moreover, $\delta\delta f^{(1)}$ represents a nontrivial cohomology class in $H^2(M, M_R; \mathbb{Z})$. But since M_R is a deformation retract of M, this group vanishes, (see [15]). This proves the above claim.

Since the sheaf of W^2_{loc}-functions on M is fine, it follows from (139) that

$$\tilde{f}^{(1)} = \delta f^{(0)} \tag{140}$$

for some $f^{(0)} \in \mathcal{C}^0_{\mathbb{R}}$. In fact, since $\tilde{f}^{(1)}_{0j} = 0$ for all j, we can choose $f^{(0)} \equiv 0 : M_R = U_o \longrightarrow \mathbb{R}$. Then the gauge transformation

$$g_0(x) := \exp(f_i(x)\hat{\Phi}(x))g_i(x) \tag{141}$$

is well defined independently of i. This completes the proof of Lemma 32. □

Now define the gauge subgroup

$$\mathcal{G}_\Phi = \{\exp(f\hat{\Phi}) \,|\, f \in W^2_{\mathrm{loc}}(M, \mathbb{R}), f \equiv 0 \quad \mathrm{on} M_{2r}\} . \tag{142}$$

Clearly $\mathcal{G}_\Phi$ operates on the subset $\tilde{\mathcal{C}}_\Phi$ of (128). We want to modify the gauge transformation $g_0(c')$ by an element of $\mathcal{G}_\Phi$ to make it well defined. Therefore we use

Lemma 34. *There exists a continuous map*

$$h : \tilde{\mathcal{C}}_\Phi \longrightarrow \mathcal{G}_\Phi$$

so that the following map is continuous

$$\tilde{\mathcal{C}}_\Phi \longmapsto Y_c : \bar{c} \longrightarrow h(\bar{c})(\bar{c}) - c . \tag{143}$$

Proof. Set $\bar{c} = (\bar{A}, \bar{\Phi})$ and $(a, \phi) = \bar{c} - c$. Denote by a_T the pointwise projection of $a(x)$ to the orthogonal complement of $\hat{\Phi}(x)$ and define $\lambda(\bar{c}) = \hat{\Phi} \cdot a$. It is easy to see (see also [16]) that a gauge transformation

$$h = \exp(f\hat{\Phi}) \in \mathcal{G}_\Phi$$

satisfies

$$\lambda(h(\bar{c})(\bar{c})) = \lambda(\bar{c}) + df\,. \tag{144}$$

The map

$$\tilde{\mathcal{C}}_\Phi \longrightarrow L^2 \cap L^6(\Lambda^1_{\mathrm{ad}}) : \bar{c} \longrightarrow a_T \tag{145}$$

is continuous since

$$\begin{aligned} [a, \hat{\Phi}] &= \nabla_{\bar{A}}\hat{\Phi} - \nabla_A\hat{\Phi} \\ &= \left|\bar{\Phi}\right|^{-1} (\nabla_{\bar{A}}\Phi)_T - \left|\Phi\right|^{-1} (\nabla_A\Phi)_T \end{aligned}$$

is continuous in $L^6 \cap L^2$ for $c, \bar{c} \in \tilde{\mathcal{C}}$ by Lemma 13. Moreover, the map

$$\tilde{\mathcal{C}}_\Phi \longrightarrow L^2(\Lambda^2) : \bar{c} \longmapsto d\lambda(\bar{c})$$

is continuous, since we have on $M - M_r$:

$$\begin{aligned} d\lambda(\bar{c}) &= d(\hat{\Phi} \cdot a) = (d_A\hat{\Phi}) \cdot a + \hat{\Phi} \cdot d_A a \\ &= d_A\hat{\Phi} \cdot a_T + \hat{\Phi} \cdot (F_{\bar{A}} - F_A) + \frac{1}{2}\hat{\Phi} \cdot (a_T \wedge a_T)\,, \end{aligned}$$

and since $F_{\bar{A}} \in L^2$ and $a_T \in L^6 \cap L^2$ are continuous in $\bar{c}$. Here we have used the fact that $d_A\hat{\Phi}$ is purely transversal.

Now consider the linear operator

$$\begin{aligned} \mathcal{D} : V^1(\Omega) &\longrightarrow L^2(\Omega) \\ (\mu, \nu) &\longrightarrow (*d\mu - d\nu, d * \mu) \end{aligned} \tag{146}$$

of [5], (5.13). Note that $(*d\lambda, 0) \in (\operatorname{Kern}\mathcal{D})^\perp$, since for $(\mu, \nu) \in V^1(\Omega)$, we have $\langle(\mu, \nu), (*d\lambda, 0)\rangle = -\langle d^*\mu, 0\rangle = 0$.

By [5], Proposition 5.1, we have $(\operatorname{Kern}\mathcal{D})^\perp = \operatorname{Im}(\mathcal{D}) \subset L^2$ and there exists a continuous inverse $G : \operatorname{Im}\mathcal{D} \longrightarrow V^1(\Omega)$. With $G((*d\lambda, 0)) = (\mu, \nu)$, we have

$$d(\lambda - \mu) = *(*d\lambda - *d\mu) = *d\nu\,, \tag{147}$$

which by Hodge decomposition implies that $d\nu = 0$ and $d\lambda = d\mu$. Since $H^1_{DR}(M - M_r) = 0$, there exists by the Poincaré lemma a unique function $f \in W^2_{\mathrm{loc}}(M -$

$M_r, \mathbb{R})$ depending continuously on $d\lambda(\bar{c})$ and hence on $\bar{c}$ so that $f(x_0) = 0$ for some fixed $x_0 \in \partial M_r$ and

$$df = \mu - \nu \tag{148}$$

on $M - M_r$. We claim that

$$h(\bar{c}) := \exp((1 - \beta_r)\hat{\Phi} f) \tag{149}$$

makes the map (143) continuous. Define therefore $a' = h(\bar{c}) a h^{-1}(\bar{c})$. In view of continuity of (.) and h, it suffices to show that the following terms depend continuously on $\bar{c} \in \mathscr{C}_\Phi$:

$$\nabla_A \hat{\Phi} \cdot a' = \nabla(\lambda - (*d\lambda)) \subset L^2(\Lambda^1) \tag{150}$$

$$\nabla_A [a', \hat{\Phi}] = \nabla_A (\nabla_{\bar{A}} - \nabla_A)\hat{\Phi} \tag{151}$$

$$= \nabla_{\bar{A}} \nabla_{\bar{A}} \hat{\Phi} - [a, \nabla_{\bar{A}} \hat{\Phi}] - \nabla_A \nabla_A \hat{\Phi} \subset L^2(\Lambda^1 \otimes \Lambda^1_{\mathrm{ad}}) \tag{152}$$

$$\nabla_A \phi = \nabla_A \left|\bar{\Phi}\right| \hat{\Phi} = \left(\nabla \left|\bar{\Phi}\right| \right) \hat{\Phi} + \left|\bar{\Phi}\right| \nabla_A \hat{\Phi} \subset L^2(\Lambda^1_{\mathrm{ad}}) \tag{153}$$

$$\Phi \cdot d *_A a' = |\Phi| \cdot \left(d_A \hat{\Phi} \cdot a'_T + d * (\hat{\Phi} \cdot a') \right) \in L^{\frac{6}{5}}(\Lambda^0)\,. \tag{154}$$

Continuity of (150) follows from continuity of (145). In (152), we use the fact that $\hat{\Phi} = \hat{\hat{\Phi}}$ and continuity of (125). Continuity of (153) is obvious. In (154), we use continuity of (145) and the fact that, by construction of h,

$$\begin{aligned} d^*(\hat{\Phi} \cdot a') &= d^*(\hat{\Phi}, h(\bar{c})(a)) \\ &= d^*(\lambda + df) \end{aligned} \tag{155}$$

has support in M_{2r}. This completes the proof of Lemma 34. □

We now combine the gauges g_0 and h.

Lemma 35. *The map*

$$g_1 : \tilde{\mathscr{U}}_R(c) \longrightarrow \mathscr{G} : c' \longmapsto h(g_0(c')(c')) \tag{156}$$

is continuous at c *and satisfies* $g_1(c) = 1$.

Proof. We first show that $g_1(c')$ is independent of the particular choice of $g_0(c')$ as long as it satisfies (129) on M_{2r}. In fact, if g_0 and g_0' are two such automorphisms, then

$$g_0' = \exp(f\hat{\Phi}) g_0 \tag{157}$$

with $\operatorname{supp} f \subset M - M_{2r}$. Then it is easy to see that

$$h(g_0(c')(c')) = \exp(f\hat{\Phi}) h(g_0'(c')(c'))\,. \tag{158}$$

Now assume that $c_i \longrightarrow c$ in $\mathscr{C}$. Then we can choose a sequence $R_i \longrightarrow \infty$ so that

$$\left\|\hat{\Phi}_i - \hat{\Phi}\right\|_{W^2,<R_i} \longrightarrow 0 . \tag{159}$$

Then define $g_i = g_0(c_i)$ as in Lemma 32 replacing R by R_i. Since g_i satisfies (129) on M_{R_i}, we conclude that $g_i \longrightarrow 0$ in G. Since the operation of $\mathscr{G}$ on $\tilde{\mathscr{C}}$ is continuous, we conclude that $g_i(c_i) \longrightarrow 0$ and therefore $h(g_i(c_i')) \longrightarrow 0$ in $\mathscr{G}$ by continuity of h in Lemma 34. This completes the proof of Lemma 35.

□

Now define for c' close enough to c in $\mathscr{C}$:

$$g_1(c') := \tilde{g}_{\acute{c}}(\acute{c}') . \tag{160}$$

By Lemma 32 and Lemma 35, this map is continuous at $c' = c$ with $g_1(c) = 1$. Hence if c_i' converges to c in $\mathscr{C}$, so does $c_i := g_1(c_i')(c_i')$. In particular, the following sequence converges to zero in $L^2(\Lambda^1_{\mathrm{ad}})$:

$$\begin{aligned} \nabla_{A_i}\Phi_i - \nabla_A\Phi &= \nabla_{A_i}\acute{\Phi}_i - \nabla_A\acute{\Phi} + \nabla_{A_i}\phi_{c_i} - \nabla_A\phi_c \\ &= \left(\nabla_{A_i}\acute{\Phi}_i - \nabla_A\acute{\Phi}\right) + [a_i, \phi_{c_i}] + \nabla_A(\phi_{c_i} - \phi) . \end{aligned}$$

Since $\acute{c}_i \longrightarrow \acute{c}$ in $\tilde{\mathscr{C}}$, we have $\nabla_{A_i}\acute{\Phi} \longrightarrow \nabla_A\acute{\Phi}$ in L^2. Since $\|a_i\|_c \longrightarrow 0$ and $\|\phi_{c_i}\|_c$ is bounded, we have $\|[a_i, \phi_{c_i}]\|_2 \longrightarrow 0$. This proves that

$$\|\nabla_A(\phi_{c_i} - \phi)\|_2 \longrightarrow 0 \quad \text{for} \quad i \longrightarrow \infty .$$

Moreover, the sequence

$$\begin{aligned} [\Phi_i, [\Phi_i, \phi_{c_i}]] - [\Phi, [\Phi, \phi_c]] &= \left[\Phi_i[\Phi_i, \acute{\Phi}_i]\right] - \left[\Phi, [\Phi, \acute{\Phi}]\right] \\ &= \nabla^2_{A_i}\Phi - \nabla^2_i\Phi_i \end{aligned}$$

converges to zero in $L^2(\Lambda^0_{\mathrm{ad}})$ since $\acute{c}_i \longrightarrow \acute{c}_i$ in $\tilde{\mathscr{C}}$. Now Proposition 25 follows immediately.

8 Proof of Proposition 28

Let y_i be a sequence converging to zero in Y_c, and let (g_i) be any sequence in $\mathscr{G}$. The aim of this section is to prove that if $g_i[y_i] \longrightarrow 0$ in Y_c, then $g_i \in \mathscr{G}_c$ for i large enough and $g_i \longrightarrow 1$ in $\mathscr{G}_c$.

As a first step, we conclude that $g_i \longrightarrow 1$ in $\mathscr{G}$ by Proposition 22. Now define the norm

$$\|(a, \phi)\|_{\tilde{Y}_c} = \|(a, \phi)\|_{Y_c} + \|\nabla_A\nabla_A\phi\|_2$$

on the corresponding subspace $\tilde{Y}_c$ of y_c. With the map of Lemma 15, we have

Lemma 36.

$$g_i(c) - c \longrightarrow 0 \text{ in } Y_c \tag{161}$$

$$g_i(\acute{c}) - \acute{c} \longrightarrow 0 \text{ in } \tilde{Y}_{\acute{c}} \tag{162}$$

Proof. We use the identity

$$g[y] = g(c) - c + gyg^{-1}, \tag{163}$$

to estimate

$$\|g(c) - c\|_{Y_c} \leq \|g[y]\|_{Y_c} + \|gyg^{-1}\|_{Y_c} \, .$$

Since $g(c+y) = c + g[y]$, we have with Lemma 30:

$$\begin{aligned} \|gyg^{-1}\|_{Y_c} &\leq K\left(\|g[y]\|_{c+g[y]} \right) \|gyg^{-1}\|_{Y_{c+g[y]}} \\ &= K'\left(\|g[y]\|_{Y_c} \right) \|y\|_{Y_{c+y}} \\ &\leq K'\left(\|g[y]\|_{Y_c} \right) K\left(\|y\|_{Y_c} \right) \|y\|_{Y_c} \, , \end{aligned} \tag{164}$$

where K' is another continuous function $\mathbb{R}_+ \longrightarrow \mathbb{R}_+$. Together with (163), this proves (161). Since the map effects only the Higgs field and since Y_c and $Y_{\acute{c}}$ are equivalent by Lemma 30, it now suffices to show that $\phi_i := g_i \acute{\Phi} g_i^{-1} - \acute{\Phi}$ satisfies $\|\phi_i\|_c + \|\nabla_A \nabla_A \phi_i\|_2 \longrightarrow 0$. Therefore, we estimate

$$\begin{aligned} L_c(\acute{\Phi}_i - \acute{\Phi}) &= L_{c_i + (a_i, \phi_i)} \acute{\Phi} \\ &= 2a_i \surd \nabla_A \acute{\Phi}_i + [\nabla_A^* a, \acute{\Phi}_i] + a \surd [a, \acute{\Phi}_i] \\ &\quad + [\Phi_i, [\Phi_i, \acute{\Phi}_i]] + [\Phi_i, [\phi_i, \acute{\Phi}_i] - [\phi_i, [\phi_i, \acute{\Phi}_i]] \end{aligned}$$

in Z_c. We have

$$\begin{aligned} \left\| [\nabla_A^* a_i, \acute{\Phi}_i] \right\|_Z &\leq \|\nabla_A a_i\|_2 \left\| \acute{\Phi}_i \right\|_\infty \\ &\quad + \left\| \nabla_A^* a_i \cdot [\Phi, \Phi_i] \right\|_{\frac{6}{5}} \\ &\leq \|a\|_c \, , \end{aligned}$$

where we have used the identity $a \cdot [b,c] = b \cdot [c,a]$. A similar argument applies to

$$\left\| [\Phi_i, [\phi_i, \acute{\Phi}_i]] \right\|_Z \leq \|\phi_i\|_c + \left\| [\Phi_i, \Phi] \cdot [\phi_i, \acute{\Phi}_i] \right\|_{\frac{6}{5}} \, .$$

The other terms can be estimated by Lemma 13. Hence by Proposition 19:

$$\|\phi_i\|_{x_c} \leq K(\|L_c \phi_i\|_{Z_c}) \longrightarrow 0 \, .$$

□

Let us from now on drop the accent on c and assume that $c \in \tilde{\mathcal{C}}$ and that $(a_i, \phi_i) = (A_i, \Phi_i) - (A, \Phi) = g_i(c) - c \longrightarrow 0$ in $\tilde{Y}_c$. Moreover, define r and $\hat{\Phi}$ as in (124) and denote the transversal component of $\xi \in \mathrm{ad}(P)$ by ξ_T. We can now estimate the *transversal part* of g_i:

Lemma 37. *There is decomposition*

$$g_i = \exp(f_i \hat{\Phi}) \bar{g}_i \,, \tag{165}$$

so that $\bar{g}_i \longrightarrow 1$ in $\mathcal{G}_c$.

Proof. Since $\|\phi_i\| \leq \|(a_i, \phi)\|_{Y_c}$, we have for i large enough

$$\left\|\hat{\Phi}_i - \hat{\Phi}\right\|_\infty < 1 \,.$$

Now we use Definition 10. As in (129), we set with $r \geq r_M$ and β_r the cutoff function of Definition 12

$$g_{i\chi} = \exp\left((1 - \beta_r)\xi(\hat{\Phi}_i, \hat{\Phi})\right) . \tag{166}$$

Then,

$$g_{i\chi} \hat{\Phi}_i g_{i\chi}^{-1} = \hat{\Phi} \tag{167}$$

on $M - M_{2r}$. We claim that $g_{i\chi} \longrightarrow 0$ in $\mathcal{G}_c$. Since the exponential map $\hat{X} \longrightarrow \mathcal{G}_c$ is a diffeomorphism on a neighborhood of 1, it suffices to show that $\xi(\hat{\Phi}, \hat{\Phi}') \longrightarrow 0$ in $W^2_A(\Lambda^0_{\mathrm{ad}})$. Consider therefore (126). We have $|\Phi|^{-1} \in W^2(M - M_r)$ and

$$\left(\hat{\Phi}_i - \hat{\Phi}\right)_T = |\Phi_i|^{-1} (\phi_i)_T \tag{168}$$

on $M - M_r$. Since $(\phi_i)_T \longrightarrow 0$ in $W^2_A(\Lambda^0_{\mathrm{ad}})$, this implies that $(\Phi_i - \Phi)_T \longrightarrow 0$ in $W^2_A(\Lambda^0_{\mathrm{ad}})$. Moreover, the function $\alpha : [0, 1) \longrightarrow [0, \infty)$ is smooth on (0,1) with $\alpha'(0) < \infty$. Therefore, and by Kato's inequality, $\alpha(\left|\hat{\Phi}_i - \hat{\Phi})_T\right|) \longrightarrow 0$ in $W^2(M, \mathbb{R})$. This implies that $\xi(\hat{\Phi}_i) \longrightarrow 0$ in $W^2_A(\Lambda^0_{\mathrm{ad}})$.

On $M - M_{2r}$, we have

$$g_i = \exp(f_i \Phi) g_{i\chi} \tag{169}$$

for a function $f_i \in W^2_{\mathrm{loc}}(M - M_{2r}, \mathbb{R})$. (See Section 3 and note that $M - M_{2r}$ is simply connected.) Therefore, the automorphisms

$$g_{i\mu} := \exp\left(-(1 - \beta_{2r}) f_i \hat{\Phi}\right) g_i g_{i\chi}^{-1} \tag{170}$$

are supported in M_{4r}. Now note that $(g_i)\,|_{M_{4r}} \longrightarrow 0$ in $W^2(\mathrm{aut}\, P\,|_{M_{4r}})$ by Proposition 6. Moreover, we can choose f_i so that $f_i\,|_{M_{4r} - M_{2r}} \longrightarrow 0$ in $W^2(M_{4r} - M_{2r}, \mathbb{R})$. Hence $g_{i\mu} \longrightarrow 1$ in $W^2(\mathrm{aut}\, P\,|_{M_{4r}})$ and

$$\bar{g}_i := g_{i\mu} g_{i\chi} \longrightarrow 1 \text{ in } \mathcal{G}_c \,.$$

This completes the proof of Lemma 37. □

Let us now consider the sequence

$$(\bar{a}_i, \bar{\phi}_i) := \bar{g}_i[0] = \bar{g}_i(c) - c\,. \tag{171}$$

Since $\mathcal{G}_c$ operates continuously on Y_c by Lemma 6.2, Lemma 37 implies that $(\bar{a}_i, \bar{\phi}_i) \longrightarrow 0$ in Y_c. Now consider the real 1-forms

$$\begin{aligned}\lambda_i &= \hat{\Phi} \cdot (a_i - \bar{a}_i) \\ &= \hat{\Phi} \cdot (\exp(f\hat{\Phi})(A) - A) \\ &= df_i\,.\end{aligned}$$

Here we have used (144). We can estimate λ_i in the following way:

Lemma 38.

$$\begin{aligned}&\textbf{(1)} \quad \lambda_i \in V^1(\Lambda^1) \text{ for all } i \\ &\textbf{(2)} \quad \lim_{i \longrightarrow \infty} \left(\|d^*\lambda_i\|_2 + \|d^*\lambda_i\|_{\frac{6}{5}} \right) = 0\,.\end{aligned} \tag{172}$$

Proof. We have

$$\begin{aligned}\|\lambda_i\|_6 &\leq \|a_i\|_6 + \|\bar{a}_i\|_6 \\ &\leq \|a_i\|_c + \|\bar{a}_i\|_c\end{aligned}$$

by Lemma 13. Moreover,

$$\begin{aligned}\|\nabla\lambda_i\|_2 &\leq \left\|\nabla(\hat{\Phi} \cdot (a_i - \bar{a}_i))\right\| \\ &\leq \left\|\nabla_A\hat{\Phi}\right\|_3 \left(\|a_i\|_6 + \|\bar{a}_i\|_6 \right) + \|\nabla_A a_i\| + \|\nabla_a\| \\ &\leq C\, \|a_i\|_c + \|\bar{a}_i\|_c\,.\end{aligned}$$

This proves the first assertion since $\|a\|_c \leq \|(a, \phi)\|_{Y_c}$. It also shows that $\|d^*\lambda_i\|_2 \longrightarrow 0$. Hence it remains to estimate

$$\begin{aligned}\|d^*\lambda_i\|_{\frac{6}{5}} &= \left\|d^*\hat{\Phi} \cdot (a_i - \bar{a}_i)\right\|_{\frac{6}{5}} \\ &\leq \left\|d_A\hat{\Phi}\right\|_3 \|(a_i - \bar{a}_i)_T\|_2 + \left\|\hat{\Phi} \cdot d_A^* a_i\right\| + \left\|\hat{\Phi} \cdot d_A^* \bar{a}_i\right\|\,.\end{aligned} \tag{173}$$

Here we have used the fact that $d_A\hat{\Phi}$ has no longitudinal component outside M_r. Since outside M_r we have also $|\Phi| > \frac{m}{2}$, we can estimate (78) by $\|(a_i, \phi_i)\|_y$ and $\|(\bar{a}_i, \bar{\phi}_i)\|_y$. This completes the proof of Lemma 38. □

Now choose for $k = 1,\dots,n$ asymptotically flat metrics g_i on $\mathbb{R}^3$ with $g_k = \omega_k^* g$ outside B_{r_0}. By Lemma 38, we have

$$\omega_k \lambda_i \in V^1(\Lambda^1(\mathbb{R}^3), g_k) \tag{174}$$

and $d^*\omega_k^*\lambda_i \longrightarrow 0$ in $L^2 \cap L^{\frac{6}{5}}(\Lambda^1(\mathbb{R}^3), g_k)$. Hence by Proposition 21, there exist unique sequences

$$f_{ki} \longrightarrow 0 \text{ in } V^{2+}(\Lambda^0(\mathbb{R}^3), g_k) \tag{175}$$

with $i \longrightarrow \infty$ so that for all i and k:

$$d^* df_{ki} = d^* d\omega_k^* f_i = d\omega_k^* \lambda_i\,. \tag{176}$$

We show now that for each k and i, the difference

$$\gamma_{ki} := f_{ki} - \omega_k^* f_i \tag{177}$$

is a constant. Therefore, note that $d\gamma_{ki} \in V^1(\Lambda^1(\mathbb{R}^3), g_k)$, since this is the case for df_{ki} by (175) and for $d(\omega_k^* f_i) = \omega_k^* \lambda_i$ by Lemma 38. Moreover, $d\lambda_{ki}$ is harmonic by (176). By the preceding discussion this implies that $d\gamma_{ki} = 0$. From (175) one concludes using Definition 1 that

$$\lim_{i\longrightarrow\infty} \sum_{k=1}^{n} \left(f_{ik} \circ \omega_k^{-1}\right) \hat{\Phi}_k = 0 \text{ in } X_c\,.$$

Hence to complete the proof of 28, we need to show that

$$\lim_{i\longrightarrow\infty} \gamma_{ki} = 0$$

for $k = 1,\dots,n$. This follows from the fact $g_i \longrightarrow 1$ in $\mathcal{G}$ by Proposition 22.

References

[1] R.A. Adams. *Sobolev Spaces.* Academic Press, 1975.

[2] H. Cartan. *Formes différentielles.* Edition Scientifiques Hermann, Paris, 1967.

[3] S.K. Donaldson. Nahm's equation and the classification of monopoles. *Comm. Math. Phys*, 96:387–407, 1984.

[4] A. Floer. Monopole moduli spaces for asymptotically flat manifolds. To appear.

[5] A. Floer. Monopoles on asymptotically flat manifolds. In this volume.

[6] A. Floer. Integrality of the monopole number in $SU(2)$ Yang-Mills-Higgs theories on $\mathbb{R}^3$. *Comm. Math. Phys.*, 93:367–378, 1984.

[7] A. Floer. Monopoles on asymptotically euclidean 3-manifolds. *Bull. AMS.*, 16:125–127, 1987.

[8] Freed, D.S and Uhlenbeck, K.K. *Instantons and Four Manifolds*, volume No. 1. of *MSRI Publications*. Springer, 1984.

[9] W. Greub, S. Halperin, and R. Vanston. *Connections, curvature, and cohomology I*. Academic Press, New York and London, 1978.

[10] N. Hitchin. Monopoles and Geodesics. *Comm. Math. Phys.*, 86(299), 1982.

[11] A. Jaffe and C. Taubes. *Vortices and Monopoles*. Birkhäuser, Boston, 1980.

[12] C.B. Morrey. *Multiple integrals in the calculus of variations*. Springer, Berlin, Heidelberg, New York, 1966.

[13] L. Nirenberg and H. Walker. Null spaces of elliptic partial differential operators on $\mathbb{R}^n$. *J. Math. Anal. Appl.*, 42(27), 1973.

[14] T. Parker and C. Taubes. On Witten's proof of the positive energy theorem. *Comm. Math. Phys.*, 84, 1982.

[15] E. Spanier. *Algebraic Topology*. McGraw Hill, New York, 1966.

[16] C. Taubes. Monopoles and maps from S^2 to S^2; the topology of the configuration space. *Comm. Math. Phys.*, 95:345–391, 1984.

The following paper by Andreas Floer has not been published before and is of historical interest. The paper is not in its final form and was therefore only made available to a small group of mathematicians. We only corrected some of the obvious typographical mistakes but otherwise we left it in its original form. The paper by P. Braam and S.K. Donaldson entitled «Floer's work on instanton homology, knots and surgery» (in this volume) surveys and explains Floer's ideas in great detail.

Instanton homology and Dehn surgery

A. Floer

Abstract. We derive a long exact sequence relating the instanton homology of two homology 3-spheres which are obtained from each other by $\pm$ 1-surgery. The third term is a $\mathbb{Z}_4$-graded homology associated to knots in homology 3-spheres.

1 Introduction

Let M be a homology 3-sphere, i.e. an oriented closed 3-dimensional smooth or topological manifold whose first homology group $H_1(M, \mathbb{Z})$ vanishes. Poincaré, who first conjectured that M would have to be the standard 3-sphere, found as a counterexample the first nontrivial example, now known as the Poincaré sphere. Its fundamental group is of order 120. Since then, many other examples have been found, most (all) of which have an infinite fundamental group. For example,

$$M(p,q,r) = \{x \in C^3 \mid |x| = 1 \quad \text{and} \quad x_1^p + x_2^q + x_3^r = 0\}$$

is a homology 3-sphere if p, q, and r are relative prime. In this case, $M(p,q,r)$ is called a Brieskorn sphere. Properties of Brieskorn spheres were studied e.g. in [M]. Recently, Donaldson's theory of instantons on 4-manifolds applied successfully to the study of 3-manifolds. First, Fintushel and Stern [FS] proved that the Poincaré-sphere has infinite order in «integral cobordism». Pursuing the same approach Furuta [FU] proved that all manifolds $M(2,3,6k-1)$ are linearly independent for any $k \in \mathbb{N}$. There is a strong feeling that instantons have more to say about 3-manifolds, even though the above results rely very much on special properties of the Brieskorn spheres and of their fundamental groups. In [Fl] and in the present paper, we therefore approach the problem from the other side, by constructing instanton-invariants on 3-manifolds which can be defined rather generally, leaving computations and applications (some luck provided) to the future. The invariant, as defined in [Fl], takes on the form of a graded Abelian group $I_*(M)$ graded by $\mathbb{Z}_8$. This as well as the definition of I_* suggests that one should consider it as a homology theory, and we will in fact refer to it as instanton homology. They have been calculated for the Brieskorn spheres in [FS]. It is the purpose ot the present paper to expose further properties of I_* to justify this terminology.

Since we will need a slight extension of instanton homology, we briefly review the construction. Let M be a general SO_3-bundle over an oriented closed 3-manifold. Instanton homology is a result of applying methods of Morse theory to the following (infinite-dimensional) variational problem: Consider the space $A(M)$ of L_1^4-Sobolev connections on P. Choosing a reference connection (which we will always assume to be the product connection θ if M is a product bundle), we can identify $A(M)$ with the space $L_1^4(\Omega^1(M))$ of Sobolev 1-forms with values in the adjoint bundle $\mathrm{ad}(M) = (M \times so_3)/SO_3$. The Sobolev coefficients are fixed rather arbitrarily to ensure that each connection is actually continuous on M. $A(M)$ is acted upon by the gauge group

$$G = L_2^4(M \times_{Ad} SO_3).$$

We rather want to restrict ourselves to the subgroup

$$G_S = L_2^4(M \times_{Ad} SU_2)/\mathbb{Z}_2.$$

(Note that SO_3 is the group of inner automorphisms of SU_2 as well as of SO_3.) The double covering $SO_3 = SU_2/\mathbb{Z}_2$ defined an extension

$$1 \to G_S(M) \to G_0(M) \xrightarrow{\eta} H^1(M, Z_2) \to 0,$$

where $H^1(M, \mathbb{Z}_2)$ has the usual additive group structure. The homomorphism η can be described topologically as the obstruction to deforming g to the identity over the 1-skeleton of M. In fact, we can define G_S as the set of all gauge transformations in M which are homotopic to one of the «local» transformations which map the exterior of some 3-ball B^3 in M to the identity. We therefore have a natural isomorphism $\pi_0(G) \simeq \mathbb{Z}$, through the degree $\deg(g)$ of the map $M/(M - B^3) \to SU_2 \simeq S^3$. The quotient space $B(M) = A(M)/G$ is then a finite covering of the space of gauge equivalence classes of connections on M, with covering group $H_1(M, \mathbb{Z}_2)$.

To define the Chern-Simons function (see [CS]) note that $T_\alpha A(M) = L_1^4(\Omega^1(M))$, so that the integral of $\mathrm{tr}(F_a \wedge \alpha)$, for $(a, \alpha) \in TA(M)$ defines a canonical one-form on $A(M)$. It turns out to be closed; in fact, there exists a function $\mathbf{a}$ on $A(M)$ such that

$$d\mathbf{a}(a)\alpha = \int \mathrm{tr}(F_a \wedge \alpha). \tag{1}$$

It is almost gauge invariant in the sense that

$$\mathbf{a}(g(a)) = \mathbf{a}(a) + \deg(g). \tag{2}$$

Hence it defines, up to an additive constant, a function $\mathbf{a}: B(M) \to \mathbb{R}/\mathbb{Z}$. It can also be described as the «secondary Pontrjagin class». Recall that on a principal bundle X over a closed 4-manifold, the first Pontrjagin class is represented by the 4-form $p_1(A) = \operatorname{tr}(F_A \wedge F_A)$ for any connection A. Here, $F_A = dA + A \wedge A$ is the curvature 2-form. (For forms with values in $\operatorname{ad}(M)$, the exterior product is here extended by matrix composition in the adjoint representation of SO_3.) It follows that the integral $\int p_1(A)$ is an integer and is independent of A. If the boundary $\partial X = M$ is not empty, then this is generally not true anymore, but $\int p_1(A)$ modulo the integers depends only on the restriction of A to M and is given by $\mathbf{a}$. This can actually be used to define $\mathbf{a}$ since every SO_3-bundle M can be extended to some SO_3-bundle X as above.

By definition, the critical set of $\mathbf{a}$ is the set of flat connections on M, which we will denote by $R(M)$. It is well known that the holonomy yields an injective map

$$R(M) \to \operatorname{Hom}(\pi_1(M), SO_3)/\operatorname{ad}(SO_3).$$

Again, $\pi_1(M)$ refers only to the base manifold. Flat connections are therefore sometimes referred to as representations (of the fundamental group). Conversely, for each representation one can construct an SO_3-bundle with a flat connection whose monodromy is prescribed by the representation.

It is $R(M)$ which will become the set of simplices in instanton homology. To understand this, recall the following statement of (finite dimensional) Morse theory.

Theorem 1.1 (Thom-Smale-Witten). *Let f be a function on a closed manifold B with nondegenerate critical set $C(f)$. Let $C_p(f)$ denote the free abelian group over all $x \in C(f)$ with Morse index p. Then there exist homomorphisms*

$$\partial_p: C_p(f) \to C_{p-1}(f)$$

such that $\partial_p \partial_{p+1} = 0$ and

$$\ker \partial_p / \operatorname{im} \partial_{p+1} = H_p(B).$$

In fact, if g is a metric on B such that the gradient field on f induces a Morse Smale flow on B, then the matrix elements $\langle \partial a, b \rangle$ with respect to the natural basis can be defined as the intersection number of the unstable manifold of x and the stable manifold of y in arbitrarily level sets between $f(a)$ and $f(b)$.

In particular, if follows that $|C_p(f)| \geq \dim H_p$. It is the defining property of Morse-Smale flows that the intersections above are transverse. The manifolds involved can all be given natural orientations, so that the matrix elements are integers. They would therefore define homomorphism for any $\mathbb{Z}$-module (i.e. any abelian group) as coefficients, and the same is true for instanton homology. We will, however, restrict ourselves to coefficients in $\mathbb{Z}$ for the sake of brevity.

We want to apply a similar procedure to the Chern-Simons function. To define the gradient flow, note that the set $B^*(M)$ of irreducible (i.e. non-Abelian) connections is a smooth Banach manifold with tangent spaces

$$T_aB = \{\alpha \in L_1^p(M) \mid d_a^*\alpha = 0\}, \tag{3}$$

see e.g. [FU] or [Fl]. Note moreover that for any metric σ on the base manifold, the Hodge dual of F_a is $f_\sigma(a) := *F_a \in L^p(\Omega^1 \otimes su_2)$ which satisfies $d_a^* f_\sigma(a) = 0$ due to the «Bianchi identity» $d_aF_a = 0$. Finally, the gauge equivariance of F_a implies that $f_\sigma(g(a)) = gf_\sigma(a)g^{-1}$, so that f_σ is in fact a section of the bundle T_aB^* obtained by replacing L_1^4 by L^4 in (TB). Even though it is not a tangent field over B^* in the sense of Banach manifolds, it has properties similar to vector fields on finite dimensional manifolds. The reason is that the flow trajectories of f_σ, i.e. the solution of the «flow equation»

$$\frac{\partial a(\tau)}{\partial \tau} + f_\sigma(a(\tau)) = 0$$

are in 1-1 correspondence to self dual connections A on the infinite cylinder $M \times \mathbb{R}$ with vanishing τ-component and with $A\,|_{M\times\{\tau\}} = a(\tau)$. One can also show (see [Fl]) that they «connect» two critical points if and only if their Yang-Mills action $||F_A||_2^2$ is finite. Three problems arise if we try to fit this gradient flow into the framework of Theorem 1.1. First, we have mentioned above that the Chern-Simons function is well-defined only locally. Surprisingly, it turns out that we can simply ignore this point, since not only the function, but also the Morse index is ill-defined along nontrivial loops in $B(M)$. Second, since flat connections are not necessarily non-degenerate as critical points of **a**, we perturb **a** by a function of the form

$$h_\kappa : \mathcal{B}(A) \to \mathbb{R};\, h_\kappa(a) = \int_D h(\kappa_\theta(a)) \tag{4}$$

where h is a character of G, and $\kappa_\theta(a)$ the parallel transport along a thickened knot

$$\kappa : T \times D \times SO_3 \to A.$$

The measure $d\mu(\theta)$ can be assumed to be smooth and supported in the interior of D. Let us denote by s a triple $s = (\sigma, \lambda, h)$, where σ is a metric on A and (λ, h) a disjoint collection of knots labeled by characters of G. It defines a perturbation f_s of the Chern-Simons function, with L^2-gradients

$$f_s(a) = *F_a + s'(a)$$

where s' is a smooth section of $T\mathcal{B}(A)$ and in this sense a compact perturbation of $*F_a$. The critical points and trajectories of f_s are now given by

$$\begin{aligned} \mathcal{R}_s(A) &= \{\alpha \in \mathcal{B}(A) \mid f_s(a) = 0 \quad \text{and} \quad G_\alpha = 1\} \\ \mathcal{M}(A,s) &= \{a : \mathbb{R} \to \mathcal{A}(A) \mid \frac{\partial a(\tau)}{\partial \tau} + f_s(a(\tau)) = 0 \\ &\quad \text{and} \quad \lim_{r\to\pm\infty} a(\tau) \in \mathcal{R}_s(A)\}. \end{aligned} \tag{5}$$

Analytically, we consider $\mathcal{M}_s(A)$ as a perturbation of the space of self-dual connections on $\mathbb{R} \times A$. In fact, the temporal gauge defines a bijection

$$\mathcal{M}_s(A) \subset \{a \in \mathcal{A}(\mathbb{R} \times A) \mid F_a + *F_a = s'(a) \text{ and } \|F_a + s'(a)\|_2 < \infty\}/\mathcal{G}(\mathbb{R} \times A)$$

where

$$\mathcal{A}(\mathbb{R} \times A) = \cup_{\alpha,\beta \in \mathcal{R}_s(A)} \alpha_+ + \beta_- + L_1^p(\Omega'_{\mathrm{ad}}(A))$$

and for each $\alpha \in \mathcal{R}_s(A)$, $\alpha_\pm$ are chosen such that for $\pi: \mathbb{R} \times A \to A$,

$$\alpha_\pm = \pi^* \hat{\alpha} \quad \text{on} \quad \mathbb{R}_\pm \times A,$$

for some representative $\hat{\alpha}$ of α. The flow equation is then given by a non-linear $\mathcal{G}(\mathbb{R} \times A)$-equivariant map

$$f_s: \mathcal{A}(\mathbb{R} \times A) \to L^p(\Omega^-(\mathbb{R} \times A))$$
$$f_s(a) = \frac{1}{2}(F_a + *F_a) + s'(a).$$

We call s stable if the operators

$$f_s'(a) = d_{a,s}^- + s''(a): L_1^2(\Omega'_{\mathrm{ad}}(\mathbb{R} \times A) \to L^2(\Omega_+^2(\mathbb{R} \times A))$$

are surjective for all $a \in \mathcal{M}_s(A)$. This implies that $\mathcal{R}_s(A)$ is a non-degenerate as the critical set of S_s and contains no nontrivial reducible representations. By compactness, it is then also finite. The set of stable parameters plays the role of the Morse-Smale gradient flows, and is denoted by $\mathcal{S}(A)$. Then we proved in [Fl]:

Theorem 1.2 *$\mathcal{S}(A)$ is not empty and contains for any metric σ on A elements (σ, h) such that*

$$||h|| = \sum_{\kappa \in \lambda} |h_\kappa|^2$$

is arbitrarily small. Then $\mathcal{R}_s(A)$ is finite and $\mathcal{M}_s(A)$ decomposes into smooth manifolds of non-constant dimensions satisfying

$$\dim M_{\sigma\pi}(a,b) = \mu(a) - \mu(b) (\operatorname{mod} 8)$$

for some function $\mu = \mathcal{R}_s(M) \to \mathbb{Z}_8$. There exists a natural orientation on M_s which is well-defined up to a change of orientation of $a \in \mathcal{R}_s(M)$ (meaning a simultaneous change of the orientations on $M(a,b)$ and $M(b,a)$ for all b in $\mathbb{R}_8(M)$) and which has the following property: Denote by $R_p, p \in \mathbb{Z}_8$, the free abelian group over the elements a of $\mathcal{R}_{\sigma\pi}$ with $\mu(a) = p$. Define the homomorphism

$$\partial_p: R_p \to R_{p-1}$$
$$\partial_p(a) = \sum_{A \in \hat{M}(a,b)} o(A) b$$

with $o(A) = 0$ for $\dim_A M(a,b) \neq 0$. Then if A allows no nontrivial abelian connections, we have $\partial_p \partial_{p+1} = 0$, and the homology groups

$$I_p(M,s) := \ker \partial_P / \operatorname{im} \partial_{p+1}$$

are canonically isomorphic for any $s \in \mathcal{S}(A)$.

The only point in the proof of 1.2 that differs from the case of trivial SU_2-bundles is the question of orientations. This is not a property of the moduli space alone, but follows from a global property of the operator family D_A over $B(M)$. To be more precise, define the determinant line bundle of D as the real line bundle with fibres

$$\Lambda = \det D_A = \det(\ker D_A) \otimes_{\mathbb{R}} \det(\operatorname{cok} D_A)^*.$$

Local trivializations can be defined as follows. For any finite-dimensional subspace $E \subset L_A$ such that $\operatorname{cok} D_A$ maps injectively into $\operatorname{Hom}(E, \mathbb{R})$ define the projection $\pi_E^{\frac{1}{}}: L_A \to L_A/E$. The $\det\ker \pi_E^{\frac{1}{}}$ can be identified naturally with $\det(D_A)$, and can clearly be extended smoothly on a neighborhood of A and B. If now in addition $f_s(A) \in E$, then the set

$$\mathbf{M}_E = \{A + a \mid a \in T_A B \quad \text{and} \quad f_s(A+a) \in E\}$$

is a smooth manifold locally at A, and its orientation is determined by an orientation of Λ and E. In particular, any orientation on Λ defines an orientation on $\mathbf{M}$.

Lemma 1.3 *Let X be an SO_3-bundle over a compact 4-manifold X, and let $G_S(X)$ denote the group of associated SU_2-gauge transformations defined as in (G). Then Λ is orientable on $A(X)/G_S(X)$.*

Proof. Since G contains only the SU_2-gauge transformations, orientability can be proved in the same way as Proposition 3.20 and Corollary 3.22 of [D]. □

The canonical isomorphism of the instanton homology groups can be understood best in the following functorial framework.

Definition 1.4 *$SO_3[3]$ is the category whose objects are principal SO_3-bundles M, N over closed oriented 3-manifolds and whose morphisms $X: M \to N$ are principal G bundles over oriented smooth 4-manifolds together with an oriented bundle isomorphism $g_X: M \cup \bar{N} \to \partial X$.*

Note that every element g of $G(M) \ltimes D(M)$ defines an «endomorphism» $Z_g = (M \times [0,1], \bar{g})$ of M, where $g: M \cup \bar{M} \to M \times \{0,1\}$ is the union of $g_0(x) = (g(x), 0)$ and $g_1(x) = (x, 1)$. (The group $G_S \subset G(M)$ will act as identities in the following, just as we are really interested only in the equivalence class of the bundle X relative to ∂X.) We will restrict ourselves to the (full) subcategory $SO_3^*[3]$ consisting of bundles M which do not admit a nontrivial abelian connection. That is, the objects in $SO_3^*[3]$ are either trivial bundles over homology 3-sphere, or nontrivial bundles with no reducible flat connections. We define a functor I_* from $SO_3^*[3]$ into the $\mathbb{Z}_8$-graded Abelian groups by means of the following auxiliary structures on M. Consider conformal structures σ on

$$X_\infty = \mathbb{R}_- \times M_- \cup X \cup \mathbb{R}_+ \times M_+$$

which are constant equivalent to product metrics on the ends, with perturbation products which are constant on a collection $\lambda \times D \times \mathbb{R}$ of thickened cylinders in X_∞. Let h be a time-dependent character of G associated to the components of λ, and define

$$\mathcal{M}_s(X) = \{A \in \mathcal{A}(X) \mid F_a + *_\sigma F_a = s'(a) \quad \text{and} \quad ||F_a + s'(a)||_2 < \infty\}/\mathcal{G}(X)$$

as in (5). Denote by $\mathcal{S}(X)$ the set of all such (σ, λ, h) such that $\mathcal{M}_s(X)$ is stable, and which define stable limits $s_\pm \in \mathcal{S}(M_\pm)$.

Theorem 1.5 *$\mathcal{S}(X)$ is nonemtpy and for $\pi_X \in \mathcal{S}(X)$,*

$$v_{X,\pi_X}: R_*(M, \pi_M) \to R_*(N, \pi_N)$$
$$v_{X,\pi}(a) = \sum_{A \in M(a,b)} o(A) b$$

has the following properties

(1) $v_{X,\pi_X} \partial_{M,\pi_M} = \partial_{N,\pi_N} v_{X,\pi_X}$.

(2) *Let $(X, \pi_X): (M, \pi_M) \to (N, \pi_N)$ and $(Y, \pi_Y): (N, \pi_N) \to (L, \pi_L)$ be two stabilized cobordisms. For $\rho \in \mathbb{R}_+$ large enough define on the bundle*

$$X\#_\rho Y = (\mathbb{R}_- \times M) \cup X \cup ([-\rho, \rho] \times N) \cup Y \cup (\mathbb{R}_+ \times L): M \to L.$$

Then for ρ large enough, the obvious perturbed parameter $\pi_X \#_\rho \pi_Y$ on $X\#_\rho Y$ is regular, and induces the composite

$$v_\rho = v_{Y,\pi_Y} \circ v_{X,\pi_X}.$$

(3) *The homomorphism*

$$(X, \pi_X)_*: I_*(M, \pi_M) \to I_*(M, \pi_N)$$

does not depend on the choice of π_X on the compact bundle X.

These are three properties that allow us to consider I_* as a functor on the category $SO_3[3]$ rather than a functor on the category of pairs (A, s). (The same argument is used e.g. in the definition of algebraic homology theories through projective resolutions.) I_* has a cyclic $\mathbb{Z}_8$-grading, meaning that $\mathbb{Z}_8$ acts freely on I_* by increasing the grading. If M is a nontrivial bundle, then there usually does not exist a canonical identification of the grading label p with an element of $\mathbb{Z}_8$. However, if $H_1(M) = 0$ then the gauge equivalence class of the product connection θ_M is a nondegenerate (though reducible) element of $\mathbb{R}$. In this case, we can therefore define a canonical $\mathbb{Z}_8$-gradient of $I_*(M)$ through the convention that

$$\mu(a) = \dim M(a, \theta_M).$$

With these conventions, we can determine the degree of an induced homomorphism X_* through index calculations. Recall first the following facts about *closed* bundles:

Proposition 1.6 ([FS]) *SO_3-bundles X over closed 4-manifolds are uniquely determined by*

$$\begin{aligned} w(X) &= f_X^* w_2 \in H^2(X, \mathbb{Z}_2) \\ p(X) &= f_X^* p_1[X]. \end{aligned} \tag{6}$$

Here, $f_X: |X| \to BSO_3$ is the classifying map of X and $w_2 \in H^2(BSO_3, \mathbb{Z}_2)$ is the second Stiefel-Whitney class and $p_1 \in H^4(BSO_3, \mathbb{Z})$ is the first Pontrjagin class of the associated $\mathbb{R}^3$-bundle $\mathrm{ad}(X) = (X \times \mathbb{R}^3)/SO_3$. In fact, (wp) defines a bijection between the set of $G(X)$-equivalence classes of SO_3-bundles over X and the set

$$wp(X) = \{(w,p) \in H^2(X, \mathbb{Z}_2) \oplus \mathbb{Z} \mid w^2[X] = p (\mathrm{mod}\, 2)\}.$$

For any connection $A \in B(X)$, the Fredholm index of the elliptic complex

$$(d_A, d_A^+): 0 \to \Omega^0_{\mathrm{ad}}(X) \to \Omega^1_{\mathrm{ad}}(X) \to \Omega^-_{\mathrm{ad}}(X) \to 0$$

or equivalently of the elliptic operator $D_A = (d_A^+ \oplus d_A^*)$ is

$$\begin{aligned} \mu(X) &= 2p_1(X) - 3e_+(X), \\ e_+(X) &= \beta_0 - \beta_1 + \beta_2 \\ &= 1 - \dim(H^1(X, \mathbb{R})) + \dim(H^2_+(X, \mathbb{R})) \\ &= e(X) + \mathrm{sgn}(X). \end{aligned}$$

The classification of (wp) extends to morphisms X between (necessarily trivial) bundles over homology 3-spheres in $SO_3^*[3]$. Moreover, endomorphisms $X: M \to M$ in $SO_3[3]$ are classified by

$$\begin{aligned} \{(w, w^1, p) \in H^2(X, M) \oplus H^2(X, M^1) \oplus \mathbb{Z} \mid i_X^* w = i_X^* w^1 \\ \text{and} \quad (w, w^1)[X, g_X] = p \bmod 2\}, \end{aligned}$$

where $[X]$ is the fundamental class in $H^4(X, \partial X)$. With this notation, an index computation yields

Proposition 1.7 *If $M = N$ or $H_1(M) = H_1(N) = 0$, then a cobordism $X: M \to N$ in $SO_3^*[3]$ induces homomorphisms*

$$\begin{aligned} X_p: I_p M &\to I_{p+\mu(X)}(M) \\ \mu(X) &= 2p(X) + 3(\beta_2^-(X) - \beta_1(X)). \end{aligned}$$

2 Dehn surgery

For a knot κ in a homology 3-sphere M, denote by κM the homology 3-sphere obtained by $(+1)$-surgery on κ. We want to show that $I_*(M)$ and $I_*(M+\kappa)$ are related by an exact triangle (a long exact sequence) whose third term is an invariant of κ. In fact, it will turn out that it depends only on the knot complement K of κ, so that it is the same for all exact sequences relating $M+q\kappa$ and $M+(q+1)\kappa$. Note that κ is well-defined in K, by ν depends on the choice of an embedding $K \subset M$. We may therefore say that «integral closures» $K(\alpha)$ with $H_1(K(\alpha)) = 0$ is an affine $\mathbb{Z}$-family of manifolds. The only closure of K which is canonical is $\bar{K} := K(\kappa)$, which is homology-equivalent to $S^2 \times S^1$. Although there exist, consequently, reducible representations of $\pi_1(\bar{K})$, none of them can be represented by flat connections in the SO_3-bundle K over K with nonzero $w_2(P) \in H^2(\bar{K}, \mathbb{Z}_2)$. In fact, restriction yields a 1-1 correspondence

$$R(\bar{K}, P) \simeq \{\rho \in \mathrm{Hom}(\pi_1(K), SU_2) \mid \rho(\kappa) = -1\}.$$

Since κ is trivial in $H_1(K)$, all abelian representations of $\pi_1(K)$ would assign it the identity $1 \in SU_2$. It is easy to see that if $\eta(g) \neq 0$, then g^2 generates $\pi_0(G_s(\bar{K}))$. Our knot invariant now takes on the following form:

Definition 2.1 *The instanton homology of a knot in a homology 3-sphere (or of its complements) is defined as the $\mathbb{Z}_4$-graded abelian group*

$$I_*(K) = I_*(K \cup_\alpha T \times D \times SO_3),$$

where $\alpha(T \times \partial D \times 1)$ is nontrivial in the fibre but trivial in $H_1(\kappa)$. The surgery triangle of κ is the triple

$$I_*(K) \to I_*(M) \to I_*(\kappa M) \to I_*(K) \tag{7}$$

of surgery cobordisms, where w is trivial relative to the boundaries and with the ends identified in such a way that its total degree is -1.

The extensions of the framed surgery cobordisms to SO_3-bundles is unique except for a possible w-class between M and $M+K$, which is excluded. That the total degree of (ST) is equal to -1 modulo 4 follows from an index calculation. The construction becomes perfectly symmetric in the three maps of (7) if one formulates the surgery problem for general knots (K, M), which are principal SO_3-bundles M over a closed oriented 3-manifold $|M| = M/SO_3$, together with an equivariant embedding

$$\kappa \cong T \times D \times SO_3 \to M.$$

Viewed from the knot complement K, (κ, M) defines sections κ_0, κ_1 of the bundle ∂K over two simple closed curves in $|\partial K| = T^2$, through

$$\kappa_0 = \{1\} \times \partial D \times \{1\}$$
$$\kappa_1 = T \times \{1\} \times \{1\}.$$

Hence, any bundle K over an oriented 3-manifold together with an isomorphism

$$\iota: T^2 \times SO_3 \cong \partial K$$

defines a framed knot in the closed manifold $M = K \cup_2 T \times D \times SO_3$ with

$$\kappa_\nu := \iota(\{1\} \times T \times SO_3) = \partial D_\nu;\ D_\nu \cong D \times SO_3.$$

Note that by duality,

$$\ker[\iota_* H_1(\partial K) \to H_1(M)] = \mathbb{Z},$$

so that there exists an equivariant embedding

$$\iota_S(S_K, T \times SO_3) \subset (K, \partial K)$$

with $\iota_S(T) \subset \partial K$ generating $\ker(\iota_\partial)_*$. Hence for any knot complement, there exists a canonical simple closed curve. Moreover, if $H_1(M) = 0$, then α_K is complementary to ν_K. This defines a canonical embedding for any tame knot in a homology sphere M. In this sense, the following discussion extends the construction of the preceding section. This class of knots is acted upon by the group

$$\Gamma = \mathbb{Z}_2^2 \ltimes Sl_2\mathbb{Z} = \text{Diff}(T_2)/\underset{1}{\text{Diff}}(T_2) \rtimes H^1(T_2, \mathbb{Z})$$

of isotopy classes of equivariant diffeomorphisms of the bundle $\partial U = T^2 \times SO_3$. In fact, for $\Theta \in \Gamma$, we define

$$\Theta(K, M) = (U, U \cup_\Theta K).$$

or, in the complementary picture,

$$\Theta(\iota, K) = (\iota \circ \Theta^{-1}, K).$$

For $\Theta = \begin{pmatrix} n & p \\ m & q \end{pmatrix} \rtimes (a, b) \in Sl_2Z \rtimes \mathbb{Z}_2^2$ with $\mathbb{Z}_2 = \{0, 1\}$, we have

$$(\Theta_K)_\nu = \{(t^n, t^p, \rho(at) \mid t \in T\} \simeq a(n\kappa_\nu + p\kappa_\pi)$$
$$(\Theta_K)_\mu = \{(t^m, t^q, \rho(bt) \mid t \in T\} \simeq b(m\kappa_\nu + q\kappa_\pi).$$

Example 1: The restriction

$$\Gamma_0 = \text{Aut}(T \times D \times SO_3) = \mathbb{Z}_2 \ltimes \mathbb{Z} \to \text{Aut}(T^2 \times SO_3)$$

defines the subgroup of surgeries leaving $\kappa_\nu = \iota_\partial(\{1\} \times T \times \{1\})$ invariant. Hence it describes the subgroup of all surgeries on κ which do not change the topology of the bundle $M = K(\kappa_\nu)$. In fact, Γ_0 is generated by the operator ϵ changing the orientation of κ, the operator Φ changing the framing of κ and the operator Ψ of order 2 changing the trivialization of M over the core of κ:

$$\Gamma_0 = \left\{ \epsilon \begin{pmatrix} 1 & 0 \\ m & 1 \end{pmatrix} \rtimes (1, b) \,\middle|\, b, \epsilon \in \mathbb{Z}_2, \quad m \in \mathbb{Z} \right\}.$$

Example 2: The element

$$\Sigma = \begin{pmatrix} 0 & 1 \\ -1 & 0 \end{pmatrix}$$

of order 2 in $PSl(2, \mathbb{Z})$ corresponds to framed («honest») surgery on the framed (and lifted) knot κ.

Definition 2.2 *The surgery cobordism of* $(K,M) \cong (\iota, K)$ *is defined as*

$$X = M \times [0,1] \cup_\alpha D \times D \times SO_3,$$
$$M = \langle \iota, K\rangle \to \langle \Theta \circ \iota, K\rangle = \Theta_K M.$$

Σ together with Γ_0 spans Γ. However, there is a more canonical generating system

$$PSl_2\mathbb{Z} = \mathbb{Z}_2 * \mathbb{Z}_3$$

where $\mathbb{Z}_2$ is generated by Σ and $\mathbb{Z}^3$ by

$$\Xi = \Sigma\Phi = \begin{pmatrix} 0 & -1 \\ 1 & 0 \end{pmatrix} \begin{pmatrix} 1 & 0 \\ 1 & 1cr \end{pmatrix} = \begin{pmatrix} -1 & -1 \\ 1 & 0 \end{pmatrix}$$

satisfying

$$\Xi^2 = \begin{pmatrix} -1 & -1 \\ 1 & 0 \end{pmatrix} \begin{pmatrix} -1 & -1 \\ 1 & 0 \end{pmatrix} = \begin{pmatrix} 0 & 1 \\ -1 & -1 \end{pmatrix} = \Xi^{-1} \tag{8}$$

$$1 + \Xi + \Xi^2 = 1 + \Xi + \Xi^{-1} = 0. \tag{9}$$

Hence $(\Xi + \psi)^3 = \text{id}$ for any $\psi \in H^1(T^2, \mathbb{Z}_2)$. In particular, the surgery sequence

$$X_{\Xi^k(K)} \colon \Xi^k M \to \Xi_\kappa^{k+1} M$$

is periodic with period 3. The surgery triangle is defined as follows:

Definition 2.3 *Define*

$$\xi = \psi \circ \Sigma \circ \Phi.$$

Then for any knot K*, we define the surgery triangle* (ST) *of* K *as the 3-periodic sequence* $X_{\xi^k K}$*. We abbreviate*

$$\begin{array}{ccc} & \sigma_{4K} M = K_+ & \\ \bar{X}_{\Sigma,\psi K} = Y_K \swarrow & & \nwarrow Z_K = X_{\xi,\xi K} \\ M & \longrightarrow & \xi K = M + K \\ & X_K = X_{\xi K} & \end{array}$$

$$K_+ = (\psi\Sigma K)^1 = K \cup_{\tilde{\alpha}} (D^2 \times T \times SO_3)$$
$$M + K = K \cup_\alpha (D \times T \times SO_3),$$

where the attaching maps α *and* $\tilde{\alpha}$ *are determined by*

$$\tilde{\alpha}(\partial D \times \{1\}) = \psi K_1$$
$$\alpha(\partial D \times \{1\}) = K_0 + K_1.$$

The surgery triangle of K coincides with (ST) if $H_1(K) = \mathbb{Z}$ and κ_1 is canonical, i.e. $\kappa_1 = \partial S_K$ for some subbundle $(S_K, \partial S_K) \subset (K, \partial K)$. Therefore exactness of (2.1) follows from the following result

Theorem 2.4 *If all bundles M, $M + K$, and K_+ are in $SO_3^*[3]$, then $(X, Y, Z) := (X_K, Y_K, Z_K)$ induces an exact triangle.*

Because of the symmetry of (1.7) with respect to Ξ, we only need to prove exactness of the two homomorphisms

$$I_*K \overset{X_*}{\to} I_*M \overset{Y_*}{\to} I_*(M+K).$$
$$\operatorname{im} Y_* = \ker X_* \subset I_*(M).$$

The sequence property $X_* Y_* = 0$ follows from a connected sum decomposition. (This is similar to the vanishing theorem for Donaldson invariants.)

Lemma 2.5

$$XY = \bar{Z} \# P$$

where P is the SO_3-bundle over $\mathbb{C}P^2$ with $w_1(P) = 1$ and $p_1(P) = 0$.

Proof. We have

$$X, Y = (D \times D \times SO_3) \cup_{K_-} ([-1,1] \times M) \cup_{K_+} (D \times D \times SO_3)$$

where $K_+ = \Phi K$ and $K_- = \Psi K$. Hence

$$\overset{\circ}{P} := (D \times D^2 \times SO_3) \cup_{\Psi K} ([-1,1] \times K) \cup_{\Phi K} (D \times D \times SO_3)$$

defines the neighborhood of a nontrivial SO_3-bundle over S^2 embedded in XY, with self-intersection number 1. Hence $\partial \overset{\circ}{P} = S^3 \times SO_3$ and Lemma 2.5 follows. To prove (9), consider a conformal structure σ_Z on $\bar{Z}$ and σ_P on $\mathbb{C}P^2$ which are standard on a 4-ball B^4, and a family $\sigma_\rho = \sigma_Z \cup \sigma_0 \cup \sigma_P$ of conformal structures on

$$XY \cong (\bar{Z} \setminus B^4) \cup [-\rho, \rho] \times S^3 \cup (P \setminus B^4).$$

Here, σ_0 is standard on the neck $[-\rho, \rho] \times S^3$. Perturb M_{σ_Z} outside B^4 such that $M_{\sigma_Z}(Z)$ is regular and assume that the ball in P is chosen small enough such that

$$M_0(P) = \{L \oplus R\}$$
$$\dim M_k(P) = 8k - 1$$

are regular. Then for ρ large enough, we have

$$M^{(0)}_{\sigma_\rho}(XY) = \emptyset$$

where $M^{(k)}(X,Y)$ is the union over all k-dimensional components of $M(a|XY|b)$ for all irreducible $a,b \in R(M+K)\times R(K)$. In fact, for any sequence $a \in M_{\sigma_{\rho_n}}(XY)$ for $\rho_n \to \infty$, consider $A_n^- = A^n \mid Z_{\rho_n} \in M_\sigma(Z_\rho)$ with

$$Z_\rho = (Z \setminus B^4) \cup ([-\rho_n, 0] \times S^3) = (Z \setminus B_n^4) \subset Z.$$

Then A_n^- accumulates at some $A^- \in M_{\sigma_Z}(\bar{Z})$. Similarly, A_n^+ accumulates to some $A^+ \in M_{\sigma_P}(P)$. Hence

$$\mathrm{Index}(D_{A^+}) = \mathrm{Index}(D_{A^+}) + \mathrm{Index}\, D_{A^-} + 3 \geq 0 - 1 + 3 = 2,$$

since $\mathrm{Index}(D_{A^+}) \geq 0$ by regularity, and $\mathrm{Index}(\{L \oplus R\}) = -1$ is the minimal index of an instanton in P. This proves assertion (9). □

3 Hamiltonians of knots

Let (M,κ) be as in Theorem 2.4. The purpose of this section is to define certain perturbations of the Chern-Simons function which «simulate» the effect of surgery. To this end, fix a smooth compactly supported volume from ω on the interior of D_2 with $\int_{D^2} d\omega(\theta) = 1$. Let $\rho_{\kappa,\theta}$ denote the holonomy of a along $\hat{\kappa} \mid_{\theta \times S^1}$, and let $h\colon SU_2 \to R$ be any smooth function invariant with respect to the adjoint action. Then the Hamiltonian associated with h and κ is defined as the function

$$\begin{aligned} &h_\kappa\colon B(M) \to \mathbb{R} \\ &h_\kappa(a) = \int_{D^2} h\big(\rho_{\kappa,\theta}(a)\big)\, d\omega(\theta). \end{aligned} \tag{10}$$

It was shown in [Fl] that h_κ is smooth. Moreover, there exists a smooth section $\mathrm{grad}\, h_\kappa$ of $TB(M)$ such that for all $\xi \in T_a B(M)$,

$$\langle \mathrm{grad}\, h_\kappa(a), \xi \rangle = Dh_\kappa(a)\xi.$$

Note that it is not necessary that h defines an invariant function on SO_3. If we define $h^-(g) = h((-r\,\mathrm{id}) \circ g)$ and if κ^- is the same framed knot with the opposite lifting, then $h_{\kappa^-} = h_\kappa^-$. Since $\mathrm{grad}_\sigma \phi$ is a smooth section even of $TB(M)$, the linearizations of $F_{\sigma\phi}$ at $R(M,\phi)$ are Fredholm operators which can be extended to

$$\begin{aligned} D_a = D_{\sigma,\phi,a} L_k^p(\Omega_{\mathrm{ad}}(M)) &\to L_{k-1}^p(\Omega_{\mathrm{ad}}(M)) \\ D_a(\alpha_0, \alpha) &= (d_a^*\alpha, d_a\alpha_0 + Df_{\sigma,\pi}). \end{aligned}$$

One verifies that D_a is self-adjoint and on $L_2(\Omega_{\mathrm{ad}}(M))$ and hence has in Fredholm index zero. (Other than that, the construction of $f_{\sigma\phi}$ and hence of $R(M,\phi)$ extends to manifolds with boundary.) We call $R(M,\phi)$ nondegenerate if D_a is an isomorphism for each $a \in R(M,\phi)$. Our first aim is to show that this is the case for «generic» ϕ. Therefore, define the Hilbert space

$$H_0 = \left\{ h(\rho) = \sum_{n=0}^{\infty} e^{-n} a_n \operatorname{tr}(\rho^n) \,\middle|\, (a_n)_n \in l^2(\mathbb{N}_0) \right\}$$

of smooth invariant functions on $SU_2 \subset M_2(\mathbb{C})$. Then any framed and lifted link λ defines a Hilbert space

$$\otimes^{\lambda} H_0 \cong H_\lambda \subset C^\infty(B(M))$$

of smooth functions on $B(M)$ with smooth L_2-gradients.

Proposition 3.1 *For every irreducible bundle M there exists a framed and a lifted link λ such that*

$$H_\lambda(M) = \{\phi \in H_\lambda \mid R(M,\phi) \quad \text{is nondegenerate}\}$$

is of first category in H_λ. Moreover, for knot κ in M we can choose a link in $M \setminus \kappa$ such that

$$\begin{aligned} H_\lambda(M,\kappa) = \{\phi \in H_\lambda \mid \phi \in H_\lambda(M^1) &\text{ for each } M^1 \text{ obtained from } M \text{ by Dehn} \\ \text{surgery on } \kappa, \text{ and any nontrivial } a &\in R(M^1,\phi) \text{ is nontrivial on } \partial K.\} \end{aligned} \tag{11}$$

Proof. Consider a Hegaard splitting $M = B \cup_\Sigma B^1$ of M. Then $R(M)$ is the intersection of $L = \iota^* R(B)$ and $L^1 = \iota^* R(B^1)$ in $P := R(\Sigma)$. The condition that M is irreducible means that $L \cap L^1 \subset R^*(\Sigma) \cup \{\theta\}$. Now

$$\bigcup_n C^\infty(G^n/\operatorname{ad}(G)^n) \times SCC^n(\Sigma) \times L \to T^*(\delta,\gamma,x) \mapsto d\delta_\gamma(x)$$

is surjective at each point. Moreover, for any compact $K \subset B(M)$ one can find a link $M = B \cup_\Sigma B^1$ such that

$$\begin{array}{rccc} \operatorname{tr}\rho_\lambda\colon & K & \hookrightarrow & (-1,1)^n \subset \mathbb{R}^n \\ & \cup & & \cup \\ & B(M) & \to & [-1,1]^n \end{array}$$

restricts to an injection $K \mapsto (-1,1)^n$. To prove this, we only have to show that for any $a \neq b$ in $B(M)$ there exists a link with $\operatorname{tr}\rho_\lambda(a) \neq \operatorname{tr}\rho_\lambda(b)$. Note that if κ is contractible in M, then all $a \in R(M)$ have trivial restrictions to T^2. However, they can be «perturbed away» from the trivial connection on T^2, with the possible exception of the product connection. Note also that standard elliptic regularity theory implies that $R(M,\phi)$ is compact as a subset of the Banach manifold $B(M)$.

Hence, for $\phi \in H_\lambda(M)$, it is a finite set. The significance of $H_\lambda(M,T)$ now lies in the fact that the lifted restriction map $R(M,\phi)\setminus\{\theta\} \to R(\partial\kappa, SU_2)$ takes values outside an ϵ-neighborhood U_ϵ of the central set, which is also the fixed point set of the conjugation c with

$$R(\partial K, SU_2) = H^1(\partial K, S^1)/c.$$

Our next aim is to define a function on (the neighborhood of) κ itself to «simulate» the effect of surgery on κ. Note that the definition of $R(M,\phi)$ as the zero set of $(f\circ\pi)$ makes sense even for manifolds with boundary. For example, for $\kappa \cong S^1 \times D \times SU_2$,

$$R(\kappa) = H^1(S^1\times D, S^1)/c \cong \{a \in R(\partial K, SU_2) \mid a(\partial D) = 1\}. \qquad \square$$

The second description is obtained by restriction. Now we have

Proposition 3.2 *For any neighborhood U of $(-1,-1) \in R(\partial K, SU_2)$ there exists a function $h \in \mathcal{S}(K)$ such that restriction to ∂K defines a bijection*

$$R(\kappa,h)\setminus U = \{a \in R(\partial K, SU_2)\setminus U \mid a(\nu+\kappa) = 1 \text{ or } a(\kappa) = -1\} \tag{12}$$

for $\kappa = S^1\times\{1\}$ and $\nu = \{1\}\times\partial D^2$.

In fact, consider $\kappa \cong S^1\times D$, the knots $\kappa_0 \cong S^1 \times D^1_{1/5}(0)$ with framing 0 and the knot $\kappa_- \cong S^1\times D_{1/5}(1/2)$ with framing -1 and define

$$h = (h_0)_{\kappa_0} + (h_1)_{\kappa_1} \quad \text{with}$$
$$h_0^1(S) = \frac{\epsilon}{1+\epsilon} S \quad \text{for} \quad S \le \frac{1-\epsilon}{1+\epsilon}$$
$$H^1_-(S) = \begin{cases} S/\epsilon & \text{for } S \le \epsilon-\epsilon^2 \\ 1-s & \text{for } S \ge \epsilon \end{cases}$$

and $h_0^1([0,1]) \subset [0,\epsilon]$ and $h_1^1([\epsilon-\epsilon^2,\epsilon]) \subset [1-\epsilon,1]$. Then (12) holds with $U = U_{2\epsilon}(-1,-1)$.

Proof. Let us first consider the effect of one knot Hamiltonian. In T_κ, we have

$$Dh_\kappa(a)(\theta,\phi) = h'(\rho_{\kappa,\theta}(a))d^2\theta \tag{13}$$

where $h'(g) = \nabla h(g) g^{-1} \in su_2$. (We always identify $T^*_g SU_2$ with $\Omega^1 SU_2$ and su_2^* with su_2 by means of the invariant metric on SU_2. We also use here the canonical embedding of $C^\infty(\Omega^2(M))$ in the dual of $C^\infty(\Omega^1(M))$, which depends only on this bilinear form.) The expression simplifies crucially for $a \in R(M, h_\kappa)$, since then $\frac{\partial}{\partial\phi}\neg F_a = 0$ and

$$\frac{\partial\rho_{\kappa,\theta}}{\partial\theta_i} = \int_{S^1} F_a\left(\frac{\partial\hat{k}}{\partial\theta_i}, \frac{\partial\hat{k}}{\partial\phi}\right) d\phi = 0.$$

Hence the holonomy $\rho_{\kappa,\theta}(a)$ does not depend on θ and can be denoted by $\rho_\kappa(a)$. In particular, it remains in a constant abelian subgroup $A_\kappa \subset SU_2$, which is well-defined unless $\rho_\kappa(a)$ is central. (The latter case we can disregard here, since it would imply that a is actually flat.) Given an orientation of κ, it is canonically isomorphic to $S^1 = \mathbb{R}^2/\mathbb{Z}$

$$R_0 = \{(x,y,z) \in [0,1]^3 \mid x-z \in [\epsilon-\epsilon^2, \epsilon+\epsilon^2] \text{ and } z = g(x-z) \text{ and } y = f(x)\}$$

$$R_+ = \{(x,y,z) \in [0,1]^3 \mid y = f(x) \text{ and } x = 1 \text{ and } 1-z \geq \epsilon\}$$
$$= \{(1,0)\} \times [0,1-\epsilon]\}$$

$$R_- = \left\{ (x,y,z) \in [0,1]^3 \,\middle|\, y = f(x) \text{ and } z = \frac{x}{1+\epsilon} \text{ and } x \underbrace{\left(1 - \frac{1}{1+\epsilon}\right)}_{\epsilon/(1+\epsilon)} \leq (1-\epsilon)\epsilon \right\}$$

$$= \left\{ \left(x, f(x), \frac{x}{1+\epsilon}\right) \,\middle|\, x \leq \frac{1-\epsilon}{1+\epsilon} \right\}.$$

In R_+ we have used the constraint that $f(1) = 0$. If we parametrize $R_\pm$ by $(x, y+z) \in [0,1] \times \mathbb{R}$, then

$$R_+ = \{1\} \times [0, 1-\epsilon]$$
$$R_- = \left\{ \left(x, f(x) + \frac{x}{1+\epsilon}\right) \mid x \leq \frac{1-\epsilon}{1+\epsilon} \right\} = \left\{ (x,x) \mid x \leq \frac{1-\epsilon}{1+\epsilon} \right\}.$$

As to R_0, note that if g and f are continued smootly such that g

$$g([\epsilon - \epsilon^2, \epsilon]) \subset [1-\epsilon, 1]$$
$$f([0,1]) \subset [0, \epsilon],$$

then $z \in [1-\epsilon, 1]$, $x \in [1-2\epsilon, 1]$, and $y \in [0,\epsilon]$. Hence $(x, y+z) \in \mathcal{U}_{2\epsilon}((1,1))$, which corresponds to $\mathcal{U}_{2\epsilon} \subset R(\partial\kappa)$. This completes the proof of Proposition 3.2. □

Propositions 3.1 and 3.2 together with compactness of $R(M,\phi)$ now yield the following splitting:

Proposition 3.3 *For each framed and lifted knot K in an irreducible bundle M there exists $s \in \mathcal{S}(M+K)$ supported in K such that*

$$R(M + K_s) = R(M_s) \cup R(M_{K,s}).$$

To show that the cobordisms $[K]$ and $[K_+]$ induce the inclusion of the first and the projection onto the second component. Let us abbreviate $\Xi(\kappa, M) = (\lambda, L)$ and $K = M - \kappa$.

Identify ∂K with the 2-torus $T^2 = \partial D_0 \times \partial D_1 \subset \mathbb{R}^3 \subset S^3$ by $\partial D_0 = \nu$ and $\partial D_1 = \nu + \kappa$. (The orientation of this map is independent of the chosen direction.) Consequently, identify the «interior» and «exterior» of T^2 with the knot neighborhoods in L and M, respectively. (The orientations induced by the K-orientations on L and M are then opposite.) Choose a conformal structure γ on Z so as to restrict to the product structures on

$$(-\infty, 0] \times M \quad \text{and} \quad [0, \infty) \times L,$$

while being standard on B^4. On the punctured 4-ball $B^4 - \{0\} \simeq (-\infty, a] \times S^3$, it is then product, too. We may smooth out γ near the boundary, as long as the conformal structure is product in the set $R \times T_\kappa$ composed of $[0, \infty) \times T_\kappa \subset [0, \infty) \times L$ and the «cone» $(-\infty, 0] \times D_1 \times S^1$ on the interior of T^2 in S^3 is product. To construct connections which satisfy $f_s(a) = 0$ for $s = (h, \kappa, \gamma)$ as in Section 1, consider $\alpha \in R(M)$ and the unique $\bar{a} \in R(M, H)$ which coincides with α on K. Clearly, the constant connection $a_K(\tau) = (0, a)$ on $K \times R$ is H-flat, and so is the constant extension of $\bar{a}$ to $(M' \setminus K) \times \mathbb{R}$ and of a to $(M \setminus K) \times \mathbb{R}$. We denote the corresponding connection on Z by a_α.

Let us now have a closer look at the situation in S^3: If the interior of $T^2 \subset S^3$ is identified with $T_\kappa \subset M$, and embedded in 3-space such that $\nu - \kappa$ bounds a disc there, then the zero framed knot κ in $M + \kappa$ becomes the $(+1)$-framed trivial knot in S^3. Hence

$$\begin{aligned} R(S^3, H) = \big\{ & a \in R(D_- \times S^1) \mid \nu(a) = \kappa(a) \quad \text{and} \\ & 0 \le \kappa(a) \le 1 - \epsilon \quad \text{or} \quad \kappa(a) = 1 \quad \text{and} \quad 0 \le \nu(a) \le 1 - \epsilon \big\}. \end{aligned}$$

Since flatness in $D \times S^1$ implies that $\nu(a) = \kappa(a)$, and since $\kappa(a) = 1$ has no solution in the indicated domain, we have the following unreduced critical set

$$\tilde{R} = \tilde{R}(S^3, H_\epsilon) = \{(0, \rho_\kappa) \quad \text{for} \quad \rho_\kappa \in SU_2 - U_\epsilon(-id)\} \simeq SU_2 - U_\epsilon(-\mathrm{id}).$$

In other words, the perturbation H on S^3 deforms the nondegenerate critical point $\theta \in \tilde{B}(S^3)$ into a «critical plateau» $\bar{P}_\epsilon - U_\epsilon(-id)$, which is a closed 3-ball. Of course, this is a highly non-generic situation, but the interior P_ϵ of $\bar{P}_\epsilon$ is a nondegenerate critical manifold. Define therefore

$$A(\alpha, \beta) = \cup_{\mu \in \tilde{R}} \big(A_{a,b,\mu} + L^p_{1,\epsilon}(\Omega^1_{\mathrm{ad}}) \big)$$

with $A_{a,b,\mu}$ some smooth connection which is temporal outside a compact subset with constant restrictions $A(\tau) = a$, b, or μ, respectively. $L^p_{k;\epsilon}$ is the Sobolev space with exponential weight $e^{r\epsilon}$ on $\mathbb{R}_+ \times S^3$ for some $\epsilon < 1$. Then f_s defines a smooth function $A(a, b) \to L^p_{a;\epsilon}(\Omega^-(\tilde{Z}))$ which is equivariant under the action of $G = L^p_{2;\epsilon}(\mathrm{Aut}\, \tilde{Z})$.

Lemma 3.4 *The zero set of the section f_s on $A(\tilde{Z})$ is in 1-1 correspondence to the set*

$$M_s(\tilde{Z}) = \{a \in A(\tilde{Z}) \mid f_s(a) = 0 \quad \text{and} \quad \|f_s(A)\|_2 < \infty\}/G.$$

Moreover, it is a smooth submanifold of $B(\tilde{Z})$ with a smooth map

$$\begin{aligned} M_s(\tilde{Z}) &\to R(M) \times R(S^3, s) \times R(M', s) \\ A &\mapsto (a_0, a_\infty, a_1). \end{aligned}$$

Proof. This is Proposition 3b.1 of [Fl]. □

In this Fredholm setting, we have

Lemma 3.5 *$[a_\alpha]$ is the only element of $M_s^+(a, \bar{a})$, as well as of $M_s^{-1}(a, \bar{a})$, and it is nondegenerate in both spaces.*

Proof. Consider the restrictions $\bar{a}_0$, $\bar{a}_1$ and $\bar{a}_\infty$ of A to the manifolds $M \times 0$, $L \times 0$, and $S^3 \times 0$. Then in each of the three half cylinders,

$$\begin{aligned} \frac{d\mathbf{a}_s(a(\tau))}{dt} &= D\mathbf{a}_s(a(\tau))\frac{\partial a}{\partial \tau} \\ &= \left\langle \operatorname{grad} \mathbf{a}_s(a(\tau)) \frac{\partial a}{\partial \tau} \right\rangle \\ &= \frac{1}{4}\left\{ \left\| \frac{\partial a}{\partial \tau} + \operatorname{grad} \mathbf{a}_s(a(\tau)) \right\|_2^2 - \left\| \frac{\partial a}{\partial \tau} - \operatorname{grad} \mathbf{a}_s(a(\tau)) \right\|_2^2 \right\} \\ &= \frac{1}{4}\|f_s(a)\|_2^2 + \|\bar{f}_s(a)\|_2^2. \end{aligned}$$

Hence

$$\mathbf{a}_s(a_0) + \mathbf{a}_s(a_\infty) = \mathbf{a}_s(a_1) + \|f_s(a)\|_2^2 + \|\bar{f}_s(a)\|_2^2.$$

Since $\mathbf{a}_s(a_\infty) = 0$, we get a relation between a_0 and a_1 alone. Since $f_s(a) = \bar{f}_s(a) = 0$ for $a = a_\alpha$ as constructed above, it follows that $\mathbf{a}_s(a_0) = \mathbf{a}_s(a_1)$, and hence that $\bar{f}_s(a) = 0$ for all $A \in M(a_0, a_1)$. Nondegeneracy of a_α in $\mathcal{M}_s(\tilde{Z})$ follows from a linearized version of the same argument. □

The set $\mathcal{M}_s(\tilde{Z})$ defines a homomorphism $\tilde{Z}_\#$ inducing Z_* despite the additional end of $\tilde{Z}$. In fact, we can perturb the functional at S^3 such that $\mathcal{M}$ is not changed but fits the framework of Theorem 1.5, extended to the case of the boundary components. We can then apply a further deformation to the unperturbed functional on S^3, which has the ℓ product connection as its only critical point instantons on $\tilde{Z}$ whose limit on S^3 are the product connection are gauge equivalent to smooth instantons on Z.

It is now clear that the homomorphism $\tilde{Z}_{\#}$ is injective since it differs from the identity only by a homomorphism that strictly decreases a_s. In order to identify the quotient complex we consider the cobordism $X: M' \to \bar{K}$ corresponding to zero framed surgery on κ in M. As before we actually consider the pointed cobordism (that is we omit an arbitrary point) so that

$$X = M^1 \times \mathbb{R}_- - \cup \bar{K} \times \mathbb{R}_- + \cup S^3 \times \mathbb{R}_- + .$$

Here, S^3 is the union of the known neighborhoods in M^1 and $\bar{K}$ oriented with M^1 and against $\bar{K}$ (Note that in $\bar{T} \subset \bar{K}$ it is κ which bounds a disc $\bar{D}$ whereas the core $0 \times S^1$ is homologous to ν and generates $H^1(K)$. Since ν pushed inside K (i.e. outside $\bar{T}$) intersections Σ_κ positively ν pushed inside $\bar{T}$ will intersect $\bar{D}$ negatively i.e. ν and κ form a right hand pair in $\bar{T}$).

We now have an embedding

$$\mathbb{R} \times T_\kappa \subset (-\infty, 0] \times M^1 \cup [0, \infty) \times S^3$$

preserving the first coordinate. By construction the connections $\bar{b} \in R(M^1, H)$ for $b \in R(\tilde{K})$ i.e. the ones contained in the second component of $R(M^1, H)$ coincide with b on K so that they define flat connections on X outside $\mathbb{R} \times T_\kappa$. On $\mathbb{R} \times T_\kappa$ it can be extended continuously to a connection as in the case of $\tilde{Z}$. The crucial point is again to determine the critical set of $\mathbf{a}_s$ on S^3. Here s is given by (1.4) for the zero framed trivial knot in S^3. The lifting however is not the standard one on $S^3 \times su_2$ since that one would extend to a lifting over $\bar{D}$ and is therefore opposite to the one that extends to K and hence to M^1. Hence H on S^3 does not vanish on θ but is in fact maximal there. Switching off s now amounts to adding a function which has a minimum at θ in the critical plateau so that according to the above discussion the H-flat connections stabilize as instantons «from» $\bar{b}_\lambda \in \bar{R}(\tilde{K}) \subset R(M + \kappa H)$ «to» $b \in R(\tilde{K})$. By the same argument as before the moduli space $M(U, H_\lambda)$ for λ less than but close to 1 defines a surjective chain operator. We therefore have two homomorphisms

$$R_*(M) \xrightarrow{i} R(M^1; H) = (RM)_* \oplus (RK)_* \xrightarrow{j} R_*K$$

where i is an isomorphism after projecting into RM_* and j is an isomorphism when restricted to RK_*. The last problem in proving exactness of (ET) (see Theorem 2.4) at $M + \kappa$ is that we do not know that $ji = 0$. It is here that we have to link the analytical construction with the topological vanishing result of Section 3. By Proposition 3.1 there exists a chain homotopy $\gamma: R_*(M) \to R_*(\tilde{K})$ such that

$$ji = \gamma \partial_M + \partial_K \gamma.$$

Now let $j^{-1}: R_*(\tilde{K}) \to RK_* \subset R(M^1 H)$ denote the inverse of j restricted to RK_* and define $\bar{\gamma} = j^{-1}\gamma$.

Lemma 3.6 *We have a chain isomorphism*

$$\Phi = i - \partial_{M+K}\bar{\gamma} - \bar{\gamma}\partial_M : R_*(M) \to \ker j.$$

Proof. To show that Φ takes values in $\ker j$, note that $j\bar{\gamma} = \gamma$ and hence

$$\begin{aligned} j\Phi(a) &= jia - j\partial_{M+K}\bar{\gamma}a - j\bar{\gamma}\partial_M a \\ &= jia - \partial_K j\bar{\gamma}a - \gamma\partial_M a \\ &= jia - \partial_K \gamma a - \gamma\partial_M a \\ &= 0 \end{aligned}$$

by (CHt). To show that it is a chain homomorphism, we calculate for $\mu(a)$ even

$$\begin{aligned} \Phi(\partial_M a) &= i\partial_M a - \partial_{M+K}\bar{\gamma}\partial_M a - \bar{\gamma}\partial_M\partial_M a \\ &= \partial_{M^1}(ia - \bar{\gamma}\partial_M a) \\ &= \partial_{M^1}(ia - \bar{\gamma}\partial_M a - \partial_{M^1}\bar{\gamma}a) \\ &= \partial_{M^1}(\Phi a). \end{aligned}$$

Finally, to show Φ is an isomorphism we use the fact that γ is constructed from instantons, so that it decreases the Chern-Simons function. (The perturbation does not change this fact, since the Chern-Simons function on the S^3-asymptotics of any H_λ instanton vanishes for $0 \leq \lambda \leq 1$.) So is its restriction j to $R_*(\tilde{K})$, in fact we have $j = id + \alpha$, where α is *strictly* decreasing in the action. It then follows that $j^{-1} = \text{id} + \beta$, where β is also strictly decreasing. (Just order $R(\tilde{K})$) according to the action, such that α is strictly triangular.) It follows that $\bar{\gamma}$ is decreasing, such that $\Phi = \text{id} + \phi$ for a strictly decreasing ϕ. Hence Φ is injective. On the other hand, we know that $\ker j$ projects injectively to R_*M, since no pure element of R_*K can be annihilated by j. Now the composition of these two injective maps is again triangular and hence defines an isomorphism of R_*M. This proves that Φ is an isomorphism of abelian groups. □

References

[AM] Akbulut, S. and McCarthy, J., *Casson's invariant for oriented homology 3-spheres — a exposition*, Mathematical Notes, Princeton University Press.

[A] Atiyah, M. F., *New invariants of 3 and 4 dimensional manifolds*, in «Symp. on the Mathematical Heritage of Hermann Weyl», Univ. of North Carolina, May 1987 (eds. R. Wells et al.).

[C] Casson, A., *An invariant for homology 3-spheres*, Lectures at MSRI Berkeley, 1985.

[CS] Chern, S. S. and Simons, J., *Characteristic forms and geometric invariants*, Ann. Math. **99** (1974), 48–69.

[D] Donaldson, S. K., *The orientation of Yang-Mills moduli spaces and 4-manifolds topology*, J. Diff. Geom. **26** (1987), 397–428.

[FS] Fintushel, R. and Stern, R. J., *Pseudofree orbifolds*, Ann. Math. **122** (1985), 335–346.

[Fl] Floer, A., *An instanton-invariant for 3-manifolds*, Commun. Math. Phys. **118** (1988), 215–240.

[FU] Furuta, M., *Homology cobordism group of homology 3-spheres*, Invent. Math. **100** (1990), no. 2, 339–355.

[G1] Goldman, W. M., *The symplectic nature of the fundamental group of surfaces*, Adv. in Math. **54** (1984), 200–225.

[G2] Goldman, W. M., *Invariant functions on Lie groups and Hamiltonian flows of surface group representations*, MSRI Preprint (1985).

[H] Hempel, J., *3-manifolds*, Ann. of Math. Studies **86**, Princeton University Press, Princeton (1967).

[Ki] Kirby, R., *A calculus for framed links in S^3*, Invent. Math. **45** (1978), 35–56.

[Ko] Kondrat'ev, V. A., *Boundary value problems for elliptic equations in domains with conical or angular points*, Transact. Moscow Math. Soc. **16** (1967).

[M] Milnor, J., *On the 3-dimensional Brieskorn manifolds $M(p,q,r)$, knots, groups, and 3-manifolds*, Ann. of Math. Studies **84**, Princeton University Press (1975), 175–225.

[S] Smale, S., *An infinite dimensional version of Sard's theorem*, Amer. J. Math. **87** (1973), 213–221.

[T1] Taubes, C. H., *Self-dual Yang Mills connections on non-self-dual 4-manifolds*, J. Diff. Geom. **17** (1982), 139–170.

[T2] Taubes, C. H., *Gauge theory on asymptotically periodic 4-manifolds*, J. Diff. Geom. **245** (1987), 363–430.

[W] Witten, E., *Supersymmetry and Morse theory*, J. Diff. Geom. **17** (1982), 661–692.

Some remarks on symplectic monodromy of Milnor fibrations

V.I. Arnold

There exist interesting and unexplored relations between symplectic geometry and the theory of critical points of holomorphic functions. One of the manifestations of these relations is the Wahl-Neumann theorem, according to which the Euler characteristics of the Floer homology of the boundary of a Milnor fiber of a quasi-homogeneous function of 3 variables is proportional to the signature of the Milnor fiber, provided that the boundary is a homological 3-sphere [1].

The signature is well-defined even when the boundary is not a homological sphere of dimension three, namely for all boundaries of dimension 3, 7, 11 and so on. It is hence natural to ask whether some generalized Floer homology (or, at least, some Casson invariant) can be defined for higher dimensional boundaries. Of course, one expects that they will depend not only on the smooth structure of the boundary, but rather on some more delicate structure, related to the symplectic structure of the fiber, for instance, on the contact or on the CR structure of the boundary.

[The situation might be similar to that in Lie group theory. The natural generalization of SL(2) (or of the group SDiff(2) of area-preserving plane diffeomorphisms) is from many points of view not SL(n) (correspondingly not the group SDiff(n) of volume-preserving space diffeomorphisms) but rather the linear symplectomorphisms group Sp(n) of $2n$-space (correspondingly, rather the group SpDiff of all symplectomorphisms). Thus, the natural generalization of the «jacobian determinant problem» (the classification of the plane polynomial mappings with jacobian 1) might be rather the classification of the polynomial symplectomorphisms of 4-space than that of the polynomial volume preserving mappings of 3-space.]

Floer was the only symplectic geometry expert who had expressed some optimism in discussing the preceding conjectures with me (in January 1991 at the Bombay Tata Institute). A lot of preliminary work should probably have been done before these conjectures are made precise. First of all, one has to understand the symplectic geometry of the Milnor fibers, the contact geometry of their boundaries, the Lagrange geometry of the vanishing and covanishing cycles, and the Legendre geometry of their boundaries. One has to study the higher dimensional version of the Bennequin invariants of Legendre curves, the signature of infinite dimensional quadratic forms, generalizing the Bennequin self-linking form, constructed from the symplectic or contact manifolds and from their Lagrange or Legendre submanifolds or currents. One might speculate that the Milnor fiber should have the maximal signature among all the symplectic manifolds with the same contact boundary, but even this has not been proved (or disproved).

In this article two simple and elementary facts on the symplectic geometry of the Milnor fibrations are presented. They might be of no use for the goals described above (but may also have other applications).

1 Symplectic geometry of regular coadjoint orbits

We start from the standard complex symplectic structure Ω on a regular orbit of the coadjoint representation of the complex Lie algebra $A_n = \mathrm{SL}(n+1, \mathbb{C})$. The theorem that follows is perhaps known for a long time, while I was unable to find it explicitly in the literature:

Theorem 1 *A regular orbit, equipped with the real symplectic structure* $\mathrm{Im}\,\Omega$*, is isomorphic to the cotangent bundle space of the manifold of the complete complex flags in* $\mathbb{C}^{n+1}$*, equipped with the standard Liouville symplectic structure, if and only if all the eigenvallues of some (and hence any) matrix in the orbit are real.*

Comments. A regular orbit consists of all matrices with a fixed characteristic polynomial (having no multiple roots). A point of an orbit is fixed by the choice of the invariant lines, one for each eigenvalue. Hence each orbit is complex diffeomorphic to the configuration space of the sets of $n+1$ ordered independent lines in $\mathbb{C}^{n+1}$. The union of all regular orbits is fibered holomorphically over the manifold of polynomials with no multiple roots. This fibration possesses a holomorphic locally flat connection (one may move the eigenvalues, fixing the eigendirections). The monodromy group of this connection is the standard representation of the braid group onto the Weyl group (of the permutations of the eigendirections).

As a complex manifold, the orbit is holomorphically fibered over the manifold of complete complex flags in $\mathbb{C}^{n+1}$. Namely, a matrix is sent to its first eigendirection, to the plane, spanned by the first two 1's, and so on. This fibration depends on the ordering of the eigendirection. The number of such fibrations is equal to the order of the Weyl group, $(n+1)!$, and the monodromy group permutes these fibrations.

Each of these fibrations is a holomorphic Lagrange fibration with respect to the standard symplectic structure of the coadjoint orbit. Hence its fibers have a local affine structure, as any Lagrange fibration's fibers. In fact they are holomorphic affine spaces globally. Each fibration of the orbit over the flag manifold is real diffeomorphic to the (co)tangent bundle fibration of the flag manifold as real vector bundles. However, as holomorphic fibrations, these two fibrations (with affine complex fibers) are different: the (co)tangent bundle has a zero section while the orbit fibration has no holomorphic sections.

To fix a real algebraic equivalence of the orbit fibration to the cotangent bundle space, one fixes an hermitian metric in the space $\mathbb{C}^{n+1}$. Then one obtains a real section of the orbit fibration, associating to a flag the ordered set of hermitian orthogonal lines, generating the spaces of the flag. The natural complex symplectic structure of the orbit is transformed by this identification into the sum of the standard Liouville symplectic structure of the cotangent bundle space of the flag manifold with a two-form, lifted from the flag manifold. This unitary-invariant 2-form ω on the flag manifold can be computed explicitly. The 2-dimensional homology group of the flag manifold $F(n+1)$ is generated by n spheres CP^1 (consisting of the flag , whose spaces, with one exception, coincide with those of a fixed flag). The periods of the 2-form ω on these cycles are equal to $2\pi(\lambda_j - \lambda_{j+1})$, where $j = 1, \ldots, n$ (λ_j, $j = 1, \ldots, n+1$, are the eigenvalues, $\lambda_1 + \ldots + \lambda_{n+1} = 0$).

If there exists a nonreal eigenvalue, then the symplectic structure $\operatorname{Im}\Omega$ on the orbit is not exact. If they are all real, then the calculations show that the form $\operatorname{Im}\omega$ on the flag manifold vanishes (the vanishing of the periods follows from the above formula). Thus the fibration of the orbit with the symplectic structure $\operatorname{Im}\Omega$ over the flag manifold is in this case real equivalent to the cotangent bundle of the flag manifold with its real Liouville symplectic structure. Let us fix an orbit, corresponding to real eigenvalues, and the real symplectic structure $\operatorname{Im}\Omega$ on it.

Problem *Can one choose real symplectic structures (equivalent to that one) on the other orbits in such a smooth way, that the monodromy of the resulting fibration would be symplectic?*

Comments. More precisely, the symplectic monodromy problem for a fibration with symplectic fibers consists of the following questions:

1) Let $\{p(t)\}$, $0 \le t \le 1$, be a loop in the base space. Does there exist a family of symplectomorphisms $s(t) : X(0) \to X(t)$ of the fiber over $p(0)$ onto the fibers over $p(t)$, such that $s(0) =$ identity (one may also ask whether these symplectomorphisms verify some boundary conditions, say that the boundary or «the infinity» remains pointwise fixed).
2) Is it possible to choose the symplectomorphisms $s(t)$ in such a way, that the symplectomorphisms $s(1)$ of the initial fiber onto itself, corresponding to different loops of the base space, would form a representation of the fundamental group of the base space (in our case — of the braid group) into the symplectomorphisms group of the fiber.

Remark. Weaker variants of both questions are still interesting: one may drop the boundary conditions, or one may require the symplectic condition only for the diffeomorphisms of the initial fiber into itself, $s(1)$ (forgetting the symplectic structures of other fibers).

The symplectic monodromy problem is solved below for the simplest case $A_1 = \mathrm{SL}(2, \mathbb{C})$.

2 Symplectic monodromy of Morse singularity

The regular orbits fibration for the coadjoint representation of $A_1 = \mathrm{SL}(2)$ coincides with the fibration of the regular level surfaces of the Morse function

$$f(x, y, z) = x^2 + y^2 + z^2 : \mathbb{C}^3 \to \mathbb{C}$$

where (x, y, z) are suitable linear coordinates in the dual space of the Lie algebra of zero-trace 2×2 complex matrices. Each regular level surface is real diffeomorphic to the 4-manifold T^*S^2. We choose the initial fiber $x^2+y^2+z^2 = 1$. The symplectic structure of §1 is then $\mathrm{Im}\,\Omega$, where $\Omega = dx \wedge dy \wedge dz/df$.

Remark. The complex symplectic structure of the Milnor fiber of complex dimension 2, induced from the natural Poisson structure of the dual space of a simple complex Lie algebra via the Brieskorn mapping of the transversal to the subregular orbit, always has the form $dx \wedge dy \wedge dz/df$, where $f = const$ is the fiber equation (A.B. Givental, 1980). The imaginary part over the fiber corresponding to real eigenvalues is the right (exact) symplectic structure of the Milnor fiber.

The monodromy diffeomorphism lifts the loop $f = e^{2\pi it}$. It is a four-dimensional generalization of the Dehn twist on a 2-cylinder, and we shall see that it can be made symplectic.

We start from a «finitely supported» version — from the Milnor monodromy acting on the discs bundle B^*S^2. This four-manifold with boundary is formed by the unite discs ($p^2 \leq 1$) in the fibers of the cotangent bundle. The monodromy preserves the points of the boundary, the orientation and the zero section. The monodromy action on the zero section is the antipodal involution. The image of a disc-fiber, B, is a disc B' with the same boundary, such that $B' - B$ is a cycle, homologous to the zero section.

We shall construct a symplectomorphism with these properties. Let $I : S^2 \to S^2$ be the antipodal involution, and $I^* : B^*S^2 \to B^*S^2$ — its action on the (co)tangent vectors. The mapping I^* is a symplectic involution.

Let us consider the function $H = p^2/2 : B^*S^2 \to \mathbb{R}$ (the kinetic energy of a free particle on the sphere). The transformation g^t of the phase flow with Hamiltonian function H at time t is symplectic. It preserves the points of the zero section, and it moves the (co)tangent vectors of length l along their geodesics at distance lt. For $t = \pi$ the (co)vectors of length 1 are translated by the flow to the antipodal point the same way as they are translated by the involution I^*.

Now let us consider the product I^*g^π. This symplectomorphism leaves fixed the points of the boundary and acts on the zero section as the antipodal involution. The unit cotangent disc centered at the North pole is transformed in a (co)vector field with a singularity at that pole. Hence I^*g^π acts on the fibers as required.

Theorem 2 *The mapping I^*g^π provides a symplectic monodromy of the Milnor fibration of the Morse function (in the strong sence, explained at the end of §1).*

To extend the result to the coadjoint orbits (that is to the unbounded cotangent bundle space T^*S^2 instead of B^*S^2), one replaces $H = p^2/2$ by a monotonic function of $u = p^2$ such that $\sqrt{u}dH/du$ tends monotonically to $\pi/2$ for $u \to \infty$.

Remark. A similar construction works for Morse functions in any dimension.

To study the orbits of $A_n = \mathrm{SL}(n+1, \mathbb{C})$ one may replace the involution I by the set of n involutions of the flag manifold. The involution I_j moves only the j-dimensional subspace of the flag. The new j-subspace is defined by the following condition: the line, representing it in the quotient plane of the $j+1$-subspace over the $j-1$-subspace is replaced by its hermitian orthocomplement in this plane.

The Hamiltonian of the free particle on the sphere is replaced by the set of n unitary-equivariant Hamiltonian flows on the space of the cotangent bundle of the flag manifold. These flows might be described as the left-invariant geodesic flows on the Lie group $\mathrm{SU}(n+1)$. The corresponding energy functions are then naturally defined on the space of antihermitian matrices with zero diagonal entries. Each Hamiltonian function is the squared modulus of an element immediately over the main diagonal. I have not checked whether the symplectomorphisms of the cotangent bundle space of the flag manifold, generated by this construction, form a representation of the braid group into the symplectomorphism group of the orbit, corresponding to real eigenvalues equipped with the symplectic structure $\mathrm{Im}\,\Omega$. The question whether the Milnor fibrations of the simple singularities (different from A_1) of functions of 3 or more variables have symplectic monodromies or flat symplectic connections also remains open. A locally flat connection for the A_2 singularity of functions of two variables can be constructed using the real affine structures of the elliptic curves and probably exists also in the case of three variables.

References

1 W. Neumann and J. Wahl, *Casson invariant of links of singularities*, Comment. Math. Helvetici **65** (1990), pp. 58–78.

2 V.M. Kharlamov and Ja.M. Eliashberg, *On the number of complex points of a real surface in a complex surface*, Proceedings of the Leningrad Topology Conference August 1982, Nauka, Leningrad, 1983, pp. 143–148.

3 A. Douady, *Noeux et structures de contact en dimension 3 d'après D. Bennequin*, sem. N. Bourbaki, 1982–1983, Exp. 604, pp. 60401–60420

Steklov Mathematical Institute
42, Vavilova Street
117966 Moscow GSP-1
Russia

CEREMADE
Université Paris-Dauphine
75775 Paris, Cedex 16e
France

Floer homology

Michael Atiyah

1 Introduction

One of the more remarkable developments of recent times has been the discovery and gradual elucidation of deep links between classical geometry and quantum field theory. This process continues apace and on several different levels. It also involves mathematicians and physicists with very different backgrounds, each making their own distinctive contribution. It is, as a part of this intricate web, that Floer's own work is best understood.

His background and training was as an analyst with special interest in symplectic geometry, and this has always had a close relation with quantum mechanics. However, it was the more recent and speculative ideas on quantum field theory, as developed and expounded by Witten, that were to influence Floer and lead him to his key ideas. On the mathematical side, Floer was also influenced by, and made extensive use of, the ideas of Gromov on pseudo-holomorphic curves. Finally, in a somewhat related direction, Floer built on the pioneering work of Donaldson in 4-dimensional geometry.

Many strands therefore combined in Floer's work, and he skilfully wove them together to produce an elegant and integrated theory which occupies a key place in current developments. «Floer Homology» will remain as his lasting memorial.

Although interested in, and influenced by, unrigorous ideas in quantum field theory Floer's own contributions were completely rigorous and involved delicate analytical arguments. His inspiration, and much of his informal thinking, was however clearly geometrical.

2 Symplectic Geometry

A Hamiltonian vector field, on a symplectic manifold, has zeros at the critical points of the corresponding Hamiltonian function. This implies the usual Morse inequalities and Arnold suggested that similar inequalities should hold for the fixed points of global symplectic diffeomorphisms (in the identity component). Establishing this «Arnold conjecture» became a major challenge in symplectic geometry and considerable progress was made in various special cases. However, it was Floer who finally established it in great generality. Moreover, the ideas he introduced led far beyond the Arnold conjecture and opened up completely new territory.

If $f : X \to X$ is a symplectic diffeomorphism then the diagonal and the graph Γ of f are Lagrangian submanifolds of $X \times X^*$ (where X^* is X with the opposite symplectic structure). The fixed points of f then correspond to the intersection $\Delta \cap \Gamma$. Floer set out, more generally, to analyze the intersection of any two Lagrangian submanifolds L_+, L_- in a compact symplectic manifold M. For this purpose he considered the space P of paths on M which begin on L_+ and end on L_-. The classical action defines (under appropriate conditions) a functional S on P whose critical points correspond to the constant paths located at the intersections $L_+ \cap L_-$. More precisely, if $p_0, p_1 \in P$ are two nearby paths, connected by a family p_t $(0 < t < 1)$ of paths then p_t defines a surface π in M and

$$S(p_1) - S(p_0) = \int_\pi \omega$$

where ω is the 2-form on M defining the symplectic structure. If $\pi_1(M) = \pi_2(M) = 0$ then P is connected and simply-connected and S is a single-valued function. If $\pi_2(M) \neq 0$ then P is not simply-connected and S is multi-valued.

Floer's main idea was to define appropriate «homology groups» of P, related to the critical points of S as in usual Morse theory. There are two initial difficulties:

(1) the Morse indices of S are all infinite

(2) S is multi-valued.

The solution to (1) lies in the fact that one can define a finite difference between the Morse indices of any two critical points. The difference depends only on the homotopy class of a path joining the critical points. Because of (2) there is then a global ambiguity which means that the index difference is only well-defined modulo a suitable even integer. This means that the eventual Floer homology groups are graded not by the integers but by the integers modulo $2n$.

To construct the relevant homology from the critical points Floer follows the ideas of Witten [4] and uses the pseudo-holomorphic curves of Gromov [3]. These correspond to the trajectories of grad S connecting critical points.

There are of course many analytical details in carrying out this programme and showing, in particular, stability under deformation.

Returning to symplectic diffeomorphisms, the stability of the Floer homology groups implies that, in the identity component, the Floer groups coincide with the homology of the original manifold. This leads to the solution of the Arnold conjecture.

As Witten has heuristically shown, the Floer groups can be interpreted as the «Hilbert space» of a topological quantum field theory in $1 + 1$ dimensions.

3 Connections on 3-manifold

It is well known to theoretical physicists that there are strong analogies between 2-dimensional «non-linear sigma models» and 4-dimensional gauge theories. This is reflected at the classical level by the similarity between, for example, Gromov's theory of pseudo-holomorphic curves and Donaldson's theory of instantons on 4-manifolds. It also means that the Hamiltonian approaches to the two theories are similar.

Floer followed through this analogy by studying the Chern-Simons functional on the space of G-connections on a compact 3-manifold Y. Analytically this is very similar to the action functional S on the space of paths P described in Section 2 and Floer was able to define analogous «Floer homology groups». These are then invariants of Y and depend on the representations modulo cancellations due to instants on $Y \times R$ which interpolate between flat connections at the two ends.

Donaldson was able to show how these Floer homology groups of Y were related to the instanton invariants that he had introduced for 4-manifolds. Roughly speaking if $\partial X = Y$ then the Donalson invariants of X take their values in the Floer homology of Y. Moreover if a closed 4-manifold X is cut into two parts X_+, X_- along a 3-manifold Y then the (absolute) Donalson invariants of X can be calculated by suitably pairing the relative Donaldson invariants of X_+ and X_-.

Unfortunately both Donaldson invariants and Floer homology are in general extremely difficult to compute. However the relation between them is useful and may assist in computations.

Witten [5] has interpreted the relation between Donaldson invariants and Floer homology in terms of a topological quantum field theory in $3+1$ dimensions, in complete analogy with the $1+1$ dimensional theory alluded to in Section 2.

4 Flat connections on a surface

The analogy between the two Floer theories becomes much closer if we introduce the moduli space M of flat G-connections on a compact orientable surface Σ. This has a natural symplectic structure (with singularities). Moreover if $\Sigma = \partial Y$ then the flat connections extending to Y define a Lagrangian subspace L of M. If Y is a closed 3-manifold cut along Σ into two parts Y_+, Y_-, then we have two Lagragian subspaces L_+, L_- of M. Ignoring singularities, we are then in the situation of Section 2 so that we might expect to define Floer homology groups using the path space P. On the other hand, for the closed 3-manifold Y, we have directly the Floer homology groups of Section 3. Moreover, the critical points of the two functionals both correspond to flat connections on Y. It is therefore reasonable to expect that the two theories agree. Considerable progress in this direction has been made by Dostoglou and Salamon [1], using a limiting process which stretches Y in the direction orthogonal to Σ.

If this can be shown in general then one has a method for calculating the Floer homology groups of the 3-manifold in terms of a decomposition $Y = Y_+ \cap_\Sigma Y_-$. This will reduce the calculation to the presumably easier case of the (lower-dimensional) Floer homology groups of Section 2.

There are other techniques for calculation introduced by Floer involving various modifications of his homology groups. In particular the exact sequence of [2] should prove useful.

References

[1] S. Dostoglou and D. Salamon, *Instanton homology and symplectic fixed points*, in *Symplectic Geometry*, LMS Lecture Notes Series, Vol. **192** (1993), 57–93.

[2] A. Floer, *Instanton homology and Dehn surgery*, in this volume.

[3] M. Gromov, *Pseudo-holomorphic curves in symplectic manifolds*. Invent. Math. **82** (1985), 307–347.

[4] E. Witten, *Super-symmetry and Morse theory*. J. Diff. Geom. **17** (1982), 661–692.

[5] E. Witten, *Topological quantum field theory*. Commun. Math. Phys. **117** (1988), 335–386.

Michael Atiyah
Trinity College
Cambridge
England

Topologie des systèmes de Moser en dimension quatre

M. Audin

Abstract. Dans cet article, je vais étudier les tores de Liouville et leurs bifurcations pour deux groupes de systèmes intégrables discutés par Moser dans [14]: le flot géodésique sur certaines orbites coadjointes de $\mathfrak{so}(4)$ d'une part (cas (a)) et les géodésiques des quadriques ou le problème de Neumannn dans $\mathbf{R}^3$ (cas (b)).

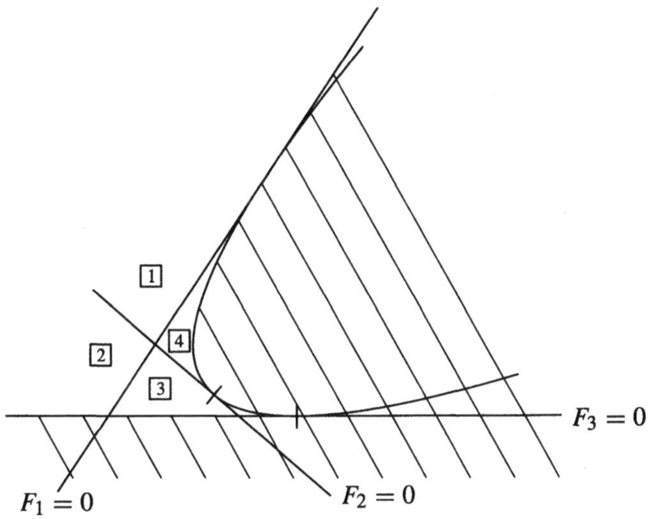

Fig. 1 application moment pour les quadriques de $\mathbf{R}^3$

Il est assez remarquable que, malgré la célébrité du problème des géodésiques des quadriques (un des plus vieux systèmes intégrables du monde) et l'abondance de la littérature qui lui a été consacrée depuis Jacobi, les aspects topologiques en aient été assez négligés: je n'ai trouvé d'énoncé pour les tores de Liouville couvrant *tous* les cas de quadriques que dans le très beau papier de Knörrer [13], et, malheureusement, ledit énoncé est erroné... même dans le cas des quadriques de $\mathbf{R}^3$!

De même, je n'ai pas été capable de trouver dans un travail antérieur un dessin du lieu critique et de l'image de l'application moment qu'on peut voir sur la figure 1, ni a fortiori une analyse des singularités de ce lieu critique, qui sont pourtant assez spéciales.

On sait que ces systèmes sont liés à des courbes hyperelliptiques. Je vais étudier ici assez naïvement la structure *réelle* de ces courbes pour en déduire des informations topologiques. Ces questions sont traitées avec des méthodes plus fines dans [3].

J'ai déjà signalé que la littérature consacrée à ce problème est énorme. Je choisis ici assez arbitrairement ce qui pourrait servir d'introduction. Tout d'abord, on peut contempler les beaux dessins et commentaires de Hilbert et Cohn-Vossen [9] et d'Arnold [2]. Klingenberg [12] traite en détail et de façon à la fois classique et lisible le cas de l'ellipsoïde (mais naturellement on peut consulter directement Jacobi [10] et Weierstrass [17]). Pour l'hyperboloïde à une nappe je n'ai trouvé qu'Hadamard [7]. Comme j'ai déjà dit, le point de vue le plus moderne est dans l'article [14] de Moser et il y a aussi celui de Knörrer [13]. Presque tous les articles consacrés à la méthode algébro-géométrique notamment [1] citent ces exemples — avec des commentaires plus ou moins heureux sur les bifurcations et/ou le cas réel.

Ce travail a été terminé pendant un séjour à l'Université du Québec à Montréal en 1992. Je remercie S. Boyer, F. Lalonde et M. Troyanov pour leur accueil.

Diverses personnes ont dû écouter ou lire diverses versions de ce travail. Que toutes soient remerciées, et plus particulièrement Horst Knörrer et Jean-Yves Mérindol pour leurs commentaires.

1 Description des systèmes et des coubes associées

1.1 Des variétés symplectiques

Fixons un entier n et deux vecteurs x et y unitaires orthogonaux de l'espace euclidien $\mathbf{R}^{n+2}$. Alors $x \wedge y \in \Lambda^2\mathbf{R}^{n+2} \cong \mathfrak{so}(n+2)$ est une matrice antisymétrique qui ne dépend que du plan orienté dont (x, y) est une base orthonormée directe. On a ainsi un plongement de la grassmannienne $\widetilde{G}_2(\mathbf{R}^{n+2})$ dans $\mathfrak{so}(n+2)$ dont l'image est l'ensemble des matrices antisymétriques de norme 1 et de rang 2. C'est aussi une orbite (co-)adjointe de $SO(n+2)$, en particulier une variété symplectique (de dimension $2n$).

On va aussi considérer l'espace $\mathcal{D}_n$ des droites affines orientées de $\mathbf{R}^{n+1}$. Une telle droite est bien définie par la donnée d'un vecteur unitaire p et de la projection orthogonale u de l'origine:

$$\mathcal{D}_n \cong \{(p, u) |\ \|p\| = 1 \text{ et } p \cdot u = 0\}$$

ce qui l'identifie à TS^n, une première façon de le décrire comme variété symplectique. Plus canoniquement, c'est aussi l'espace des caractéristiques[1]) du flot géodésique de la métrique euclidienne sur $\mathbf{R}^{n+1}$.

1.1.1. Remarque On peut facilement plonger $\mathcal{D}_n$ dans $\widetilde{G}_2(\mathbf{R}^{n+1})$ simplement en plongeant $\mathbf{R}^{n+1}$ comme l'hyperplan affine $x_0 = 1$ dans $\mathbf{R}^{n+2}$: à (p, u) on associe le plan

$$\begin{pmatrix} 0 \\ p \end{pmatrix} \wedge \frac{1}{\sqrt{1 + \|u\|^2}} \begin{pmatrix} 1 \\ u \end{pmatrix},$$

1) En fait les deux structures symplectiques ainsi définies ne coïncident qu'au signe près.

l'unique plan vectoriel contenant la droite des $u+tp$. Naturellement ce plongement n'est pas symplectique: le volume de $\mathcal{D}_n$ est infini.

1.2 Les systèmes intégrables

Appelons n la dimension de l'espace euclidien considéré: $m = n+2$ ou $n+1$. Les coordonnées sont nommées $(x_0, \dots, x_{n+1})$ dans le premier cas et $(x_1, \dots, x_{n+1})$ dans l'autre. Fixons une matrice diagonale inversible α dans End$\mathbf{R}^m$ et considérons les matrices

$$A_h = h\alpha + a \wedge b + h^{-1} B_{a,b}$$

où h est un paramètre formel destiné à vivre dans $\mathbf{P}^1(\mathbf{C})$, $(a, b) = (x, y)$ ou (p, u) selon le cas et $B_{a,b}$ est une matrice symétrique de rang ≤ 2. Concrètement, on va s'intéresser aux systèmes décrits dans l'article de Moser:

(a) Les équations d'Euler-Arnold pour le flot géodésique sur les orbites de matrices de rang 2 dans $\mathfrak{so}(n+2)$, cas où $m = n+2$, $(a, b) = (x, y)$, $B_{a,b} = 0$.

(b) Le flot géodésique sur une quadrique de $\mathbf{R}^{n+1}$ ou le problème de Neumann, cas où $m = n+1$, $(a, b) = (p, u)$ et $B_{a,b} = -p \otimes p$.

Il n'y a pas de difficulté à traiter de la même façon les autres exemples de l'article de Moser, en particulier le cas d'une force centrale appliquée à un point massique sur un ellipsoïde.

Les systèmes hamiltoniens du cas (a) et des deux cas de (b) peuvent se mettre sous forme de Lax en

$$\dot{A}_h = [A_h, \Gamma + \beta h]$$

où Γ est une matrice antisymétrique fonction de (a, b) (ce ne sont pas des systèmes linéaires!) et β une matrice diagonale fonction seulement de α. Les valeurs précises de ces deux matrices dépendent de celui des trois systèmes que l'on considère, voir [1] par exemple.

1.1.2. Remarque Ici je ne m'intéresse pas réellement au système, c'est à dire à l'équation différentielle ou au hamiltonien, mais en fait seulement à une application moment dont ce hamiltonien est une projection linéaire. Ceci explique pourquoi du point de vue topologique (sinon dynamique) les géodésiques de l'ellipsoïde et le problème de Neumann sont traités de façon identique! Du point de vue de l'équation de Lax, ceci explique pourquoi je ne m'intéresse ici qu'à A_h et pas à $\Gamma + \beta h$.

En calculant le polynôme caractéristique de A_h, on en déduit des intégrales premières, dont on montre qu'elles commutent (ces exemples sont justiciables du théorème d'Adler-Kostant-Symes), et qui sont en nombre suffisant (n indépendantes) pour que ces systèmes soient intégrables au sens d'Arnold-Liouville.

En effet, puisque A_h est une perturbation de rang 2 d'une matrice diagonale inversible, il est facile d'évaluer son polynôme caractéristique [14]:

$$\det(A_h - \mu I) = h^p [h^2 - \Psi_z(a, b)]$$

pour un entier p convenable, et pour $\mu = -hz$, et où $\Psi_z(a,b)$ est une fraction rationnelle de z dont les coefficients sont nos intégrales premières et dont les pôles sont les coefficients de la matrice α.

Une conséquence de la forme particulière des matrices A_h est donc de donner des courbes spectrales hyperelliptiques, les courbes C d'équation $h^2 = \Psi_z(a,b)$. Supposons les coefficients α_i de α distincts et, pour fixer les idées

$$0 < \alpha_1 < \cdots < \alpha_{n+1}, \quad \text{et} \quad \alpha_0 < \alpha_1 \quad \text{si} \quad m = n+2.$$

Considérons encore la forme bilinéaire symétrique

$$\varphi_z(a,b) = \sum_i \frac{a_i b_i}{\alpha_i - z}$$

sur $\mathbf{R}^m$, ainsi que la forme quadratique f_z associée. La fraction rationnelle $\Psi_z(a,b)$ est

- $\varphi_z(x,y)^2 - f_z(x) f_z(y)$ dans le cas (a)
- $\varphi_z(p,u)^2 - f_z(p)[f_z(u) - 1]$ dans le cas (b).

Par exemple, dans le cas (b), $\Psi_z(p,u) = 0$ exprime le fait que la droite des $u + tp$ est tangente à la quadrique Q_z d'équation $f_z = 1$. Quand z est fixé, le flot du champ hamiltonien de Ψ_z (au niveau $\Psi_z = 0$) décrit les tangentes le long des géodésiques de Q_z. De même dans le cas (a), $\Psi_z(x,y) = 0$ exprime le fait que le plan défini par (x,y) est isotrope pour la forme quadratique f_z.

Un calcul direct de résidu permet d'exprimer

$$\Psi_z(a,b) = \sum_i \frac{F_i(a,b)}{\alpha_i - z}$$

avec, dans le cas (a):

$$F_i(x,y) = \sum_{j \neq i} \frac{(x_i y_j - x_j y_i)^2}{\alpha_i - \alpha_j}$$

et dans le cas (b):

$$F_i(p,u) = p_i^2 + \sum_{j \neq i} \frac{(p_i u_j - p_j u_i)^2}{\alpha_i - \alpha_j}.$$

Ces intégrales premières sont dues à Uhlenbeck. Elles vérifient des relations plus ou moins évidentes:

- dans le cas (a), $\sum F_i = 0$ et $\sum \alpha_i F_i = 1$, qui dit que $\|x \wedge y\|^2 = 1$,
- dans le cas (b), $\sum F_i = 1$, qui dit que $\|p\|^2 = 1$.

Compte tenu de ces relations, on a le bon nombre (n) d'intégrales premières. Appelons E le sous-espace de dimension n de $\mathbf{R}^m$ décrit par ces relations, et $F = (F_i)_i$, considérée comme à valeurs dans E: «application moment».

On peut aussi écrire

$$\Psi_z(a,b) = \frac{\mathcal{P}_{a,b}(z)}{\prod(\alpha_i - z)}$$

où $\mathcal{P}_{a,b}$ est un polynôme de degré n dans tous les cas (à cause des relations entre les F_i). Ainsi la courbe C d'équation $h^2 = \Psi_z$ est revêtement ramifié de $\mathbf{P}^1(\mathbf{C})$ en les m valeurs des α_i, les n racines de $\mathcal{P}$ et le point à l'infini si $m = n+1$, soit $2n+2$ points, son genre est donc n.

2 Géométrie réelle de C et image de l'application moment

2.1 La courbe C

Il y a un certain nombre de points à coordonnées réelles évidents sur la courbe C. Par exemple, on remarque que l'équation (en z) $f_z(a) = 0$ s'écrit aussi, en réduisant au même dénominateur, $\mathcal{R}_a(z) = 0$, où $\mathcal{R}_a$ est un polynôme de degré $m-1$ dont toutes les racines sont réelles puisque les signes des $\mathcal{R}_a(\alpha_i)$ alternent[2]). Appelons les ζ_j en les numérotant de façon que

$$\alpha_0 < \zeta_0 < \alpha_1 < \zeta_1 < \cdots < \zeta_n < \alpha_{n+1}.$$

Posons $\eta_j = \varphi_{\zeta_j}(a,b)$ de sorte que le point R_j de coordonnées (η_j, ζ_j) est un point réel de la courbe C.

2.1.1. Proposition *La partie réelle de la courbe hyperelliptique C de genre n possède $n+1$ composantes connexes. Une et une seule de ces composantes, notée C_i, se projette à l'intérieur de l'intervalle $]\alpha_i, \alpha_{i+1}[$.*

La démonstation est directe et consiste à étudier le signe de Ψ_z pour $z \in \mathbf{R}$: on sait que Ψ_z ne peut changer de signe qu'en les α_i et en les racines de son numérateur $\mathcal{P}_{a,b}$. En plus, on sait qu'il est positif en chacun des points ζ_j. A cause de la position des ζ_j par rapport aux α_i, on voit que $\mathcal{P}_{a,b}$ doit avoir une racine réelle dans chaque intervalle $[\zeta_j, \zeta_{j+1}]$. Comme son degré est justement n, toutes ses racines sont réelles.

On en déduit l'assertion sur les composantes de la partie réelle de la courbe C. La composante C_i est celle qui contient le point R_i. □

Appelons $z_1 < \cdots < z_n$ les racines de $\mathcal{P}_{a,b}$, et $S_1, \ldots, S_n$ les points de C correspondants. Une façon équivalente mais utile de caractériser les positions des points S_j sur les composantes de C ou des z_j par rapport aux α_j est la suivante (due à Knörrer [13] au moins dans le cas (b)). Je montrerai plus bas que c'est elle qui donne l'image de l'application moment:

---

2) Faisons la convention que $\alpha_0 = -\infty$ (et de même ζ_0) si $m = n+1$.

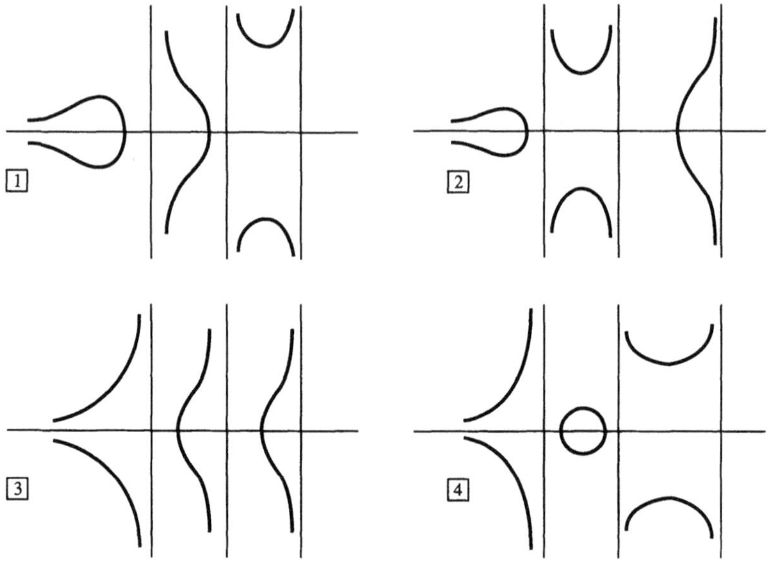

Fig. 2 la partie réelle de C pour $n = 2$

2.1.2. Corollaire *Ecrivons l'ensemble* $\{\alpha_0, \ldots, a_{n+1}, z_1, \ldots, z_n\}$ *sous la forme* $\{b_0, \ldots, b_{2n+1}\}$ *avec* $b_0 < b_1 < \cdots < b_{2n+1}$. *Alors* $z_j = b_{2j-1}$ *ou* b_{2j}. □

Supposons par exemple que $n = 2$ et, pour fixer les idées, que nous soyons dans le cas (b). Il y a $2^n = 4$ configurations possibles pour les positions des S_j et pour la forme de $C(\mathbf{R})$. Elles sont schématisées sur la figure 2.

Une autre application immédiate est:

2.1.3. Corollaire *Quand* a *et* b *sont réels, le polynôme* $\mathcal{P}_{a,b}$ *n'a jamais de racine d'ordre* ≥ 3. *Chacune de ses éventuelles racines doubles est une racine de* $f_z(a)$. □

2.1.4. Remarque Le fait que les z_j soient réels est bien classique et fait partie du folklore du problème dans le cas des quadriques (cas (b)): il exprime que les tangentes à une quadrique sont en fait tangentes à n des quadriques de la famille des quadriques homofocales Q_z. H. Knörrer m'en signale une autre démonstration, implicite dans [13] et basée sur la dualité projective (la famille des quadriques homofocales est duale à une famille *linéaire* de quadriques, d'où, d'ailleurs, une bonne partie de la géométrie algébrique classique liée au problème des géodésiques sur les quadriques).

2.1.5. Remarque Les points R_j (plus ou moins) ont été utilisés par Moser [14] dans son étude de la relation entre la jacobienne de C et la «variété isospectrale» (un quotient d'un niveau de F). Dans le contexte présent, où C est une courbe spectrale, on peut montrer que $R_1 + \cdots + R_n + S_1 + \cdots + S_n$ (+ un autre point si $m = n + 2$) est un diviseur qui représente le fibré des vecteurs propres de A_h dans $\operatorname{Pic} C$ (voir [3]).

2.2 Lieu critique de l'application moment

Je laisse la démonstration de la

2.2.1. Proposition *La courbe C est lisse si et seulement si elle correspond à une valeur régulière de l'application moment.*

pour la fin de ce paragraphe. On en déduit que le lieu critique de l'application moment F est le discriminant de la famille des courbes C (les valeurs de F sont les paramètres qui figurent dans l'équation de C) que j'appellerai, en abusant des notations, discriminant de C. Celui-ci est très facile à comprendre: comme l'équation de C est

$$h^2 = \sum \frac{F_i}{\alpha_i - z} = \frac{\mathcal{P}_{a,b}(z)}{\prod(\alpha_i - z)}$$

il est bien clair que son discriminant est formé

- du discriminant du polynôme $\mathcal{P}_{a,b}$, enveloppe de la famille d'hyperplans $P_z : \sum F_i/(\alpha_i - z) = 0$ dans l'espace E (points où $\mathcal{P}_{a,b}$ a une racine multiple).
- des hyperplans $P_{\alpha_i} : F_i = 0$ (c'est P_z quand $z \to \alpha_i$) qui lui sont tangents.

Par exemple, pour $n = 2$ (cas où la variété symplectique est de dimension 4), dans le plan E, on a une famille de droites enveloppant une parabole. Le discriminant est formé de 4 ou 3 de ces droites et de la parabole. Voir les figures 3 et 1 pour les cas (a) et (b) respectivement.

Démonstration. La proposition 2.2.1. est vraie sur $\mathbf{R}$ comme sur $\mathbf{C}$. Montrons la sur $\mathbf{R}$. Fixons une valeur F telle que C soit lisse. Appelons $\mathcal{T}_F$ le niveau correspondant. Comme C est lisse, on a une applivation

$$\mathcal{T}_F \xrightarrow{\Phi_F} \operatorname{Pic}^d C$$

qui associe à un point (a, b) de $\mathcal{T}_F$ le fibré sur C dont la fibre en (h, z) est la droite propre de A_h pour la valeur propre $\mu = -hz$ correspondant (pour avoir un degré d positif, ici égal à $2n + 1$ ou $2n$, on prend en fait le dual de ce fibré).

Je vais montrer que les images par l'application tangente $T\Phi_F$ des champs hamiltoniens des fonctions F_i engendrent tout l'espace tangent à $\operatorname{Pic}^d C$: ainsi ces champs eux-mêmes engendreront un espace de dimension $\geq n$ et F sera bien une valeur régulière.

Calculons donc l'application tangente $T\Phi_F$. Ce calcul fait partie du folklore de la méthode algébro-géométrique (voir [16] par exemple): si X_H est le gradient symplectique d'un hamiltonien du type

$$H(a, b) = f(A_h) = \operatorname{Rés}\left(g(A_h) h^{-k-1} dh\right)$$

où g est une fonction invariante sur l'algèbre de Lie des matrices, alors $df(A_h)$ commute avec A_h, ainsi elle a les mêmes vecteurs propres, pour une certaine valeur propre $\psi(h,\mu)$. Cette fonction ψ définit un 1-cocycle du recouvrement de C par les ouverts $\mathcal{U}_+$ où $h \neq \infty$ et $\mathcal{U}_-$ où $h \neq 0$ et en fin de compte un élément de $H^1(C;\mathcal{O}_C)$, l'espace tangent à $\operatorname{Pic}^d C$, qui est l'image du vecteur X_H par l'application tangente à Φ_F.

Comme $\Psi_z = F_i/(\alpha_i - z)$, les fonctions $\Psi_{z_1},\ldots,\Psi_{z_n}$ sont des combinaisons linéaires des F_i. Il suffit naturellement de montrer que les images de leurs gradients symplectiques engendrent l'espace vectoriel $H^1(C;\mathcal{O}_C)$ de dimension n. Le calcul est simplifié: il suffit de calculer l'image du gradient symplectique de Ψ_0 et de remplacer ensuite α par $\alpha - z_j$ $(1 \leq j \leq n)$. On considère donc Ψ_0 comme un hamiltonien sur la variété symplectique $W^{2n} = \widetilde{G}_2(\mathbf{R}^{n+2})$ ou $\mathcal{D}_n$ et on commence par l'exprimer à l'aide d'une fonction invariante de A_h.

Par définition, Ψ_0 est le coefficient de h^p dans $\det A_h$, de sorte que la fonction invariante est $g = \det$, $g'(A_h)$ s'identifie à $\det(A_h)A_h^{-1}$ et que $\psi(h,z) = -hz^{-1}$.

On veut donc montrer maintenent que les classes représentées par les n cocycles $h(z-z_j)^{-1}$ pour $1 \leq j \leq n$ dans $H^1(C;\mathcal{O}_C)$ sont indépendantes. C'est un calcul de résidu: l'appariement $f \otimes \omega \mapsto \operatorname{Rés}_{h=0}(f\omega)$ induit un isomorphisme $H^1(C;\mathcal{O}_C) \to H^0(C;\Omega^1_C)^\star$.

Soit donc

$$\omega_i = \frac{z^{i-1}dz}{h\prod_s(\alpha_s - z)} \quad \text{pour} \quad 1 \leq i \leq n.$$

Ces formes constituent une base de $H^0(C;\Omega^1_C)$ et on calcule aisément:

$$a_{i,j} = \operatorname{Rés}_{h=0} h(z-z_j)^{-1}\omega_i = \operatorname{Rés}_{S_j} \frac{h\omega_i}{z-z_j} = \frac{2z_j^{i-1}}{\prod_s(\alpha_s - z_j)}$$

et la matrice des $a_{i,j}$ est de rang n dès que les z_j sont distincts, ce qui est le cas puisque C est lisse.

Nous avons donc montré que, si C est lisse, F est non critique. Inversement, supposons C singulière. Alors la valeur de F considérée est sur le discriminant.

Si elle est sur un des hyperplans $F_i = 0$, en supposant qu'elle soit effectivement dans l'image, elle est l'image d'un point (a,b) avec $a_i = b_i = 0$ et ce point est critique (par une vérification directe).

Si elle est sur le discriminant de la famille des polynômes $\mathcal{P}$, alors le polynôme a une racine multiple ζ. Comme on l'a déjà remarqué dans 2.1.3., on a $\zeta = z_i = \zeta_{i+1} = z_{i+1}$ et la racine ζ est aussi une racine de $f_z(a) = 0$. A cause de la forme de Ψ_z, on a aussi $\varphi_\zeta(a,b) = 0$. Alors $f_\zeta(b+ta)$ est constant égal à $f_\zeta(b)$. On sait donc

– dans le cas (a) que le plan (x,y) est dans le cône isotrope de f_ζ et donc la constante est nulle: $f_\zeta(y) = 0$,

– dans le cas (b) que la droite $u + tp$ est tangente à la quadrique Q_ζ et donc la constante vaut 1: $f_\zeta(u) = 1$.

C'est exactement ce qu'il faut pour conclure que (a, b) est un point critique de Ψ_ζ. Il ne reste alors plus assez de Ψ_{z_j} indépendants: (a, b) est bien un point critique de F. □

2.2.2. Remarque Dans le cas (b), les valeurs de F situées sur le discriminant du polynôme $\mathcal{P}$ donnent des niveaux $\mathcal{T}_F$ qui sont constitués des droites de la quadrique correspondant à la racine multiple. Bien entendu, ce sont des points singuliers pour le flot géodésique. Le niveau complexifié, quant à lui, contient plus (il contient des points réguliers): l'interprétation de la racine multiple comme une racine de $f_z(a)$ est basée sur la réalité de a.

2.3 Image de l'application moment

Le corollaire 2.1.2. décrit en fait l'image de l'application moment:

2.3.1. Proposition *Un point de E est dans l'image de F si et seulement si les n racines de*

$$\mathcal{P}_{a,b}(z) = \sum_i F_i \prod_{j \neq i} (\alpha_j - z)$$

sont réelles et telles que si l'ensemble des α_i et des z_j est ordonné en

$$b_0 < b_1 < \cdots < b_{2n+1}$$

alors $z_j = b_{2j-1}$ ou b_{2j}.

On a vu que la condition était nécessaire. Pour montrer qu'elle est suffisante, il suffit de montrer qu'au moins un point de chaque zône est atteint, ou encore que chacune des configurations possibles de la partie réelle de C est effectivement réalisée pour des vecteurs (a, b) réels. C'est une vérification simple que je laisse au lecteur.

2.3.2. Remarque L'ensemble des valeurs régulières de F a 2^n composantes connexes.

Les figures 1 et 3 représentent les images des applications moments pour $n = 2$ et dans les cas (b) et (a) respectivement.

Remarquons que la partie utile (atteinte) du discriminant (de la famille des courbes C ou de F) est assez peu singulière: ses singularités sont des points multiples (intersections des hyperplans $F_i = 0$, auto-intersections du discriminant de $\mathcal{P}$) et des points où deux composantes (un des hyperplans et le discriminant de $\mathcal{P}$) sont tangentes entre elles. Par exemple, pour $n = 2$ (voir la figure 1), il n'y a dans le lieu critique ni point de rebroussement, comme on en trouve notamment dans le cas de Kowalevski, ni de point isolé comme par exemple dans le cas de la toupie symétrique (voir [5] ou [4] par exemple). En effet, un corollaire immédiat de 2.1.3. est

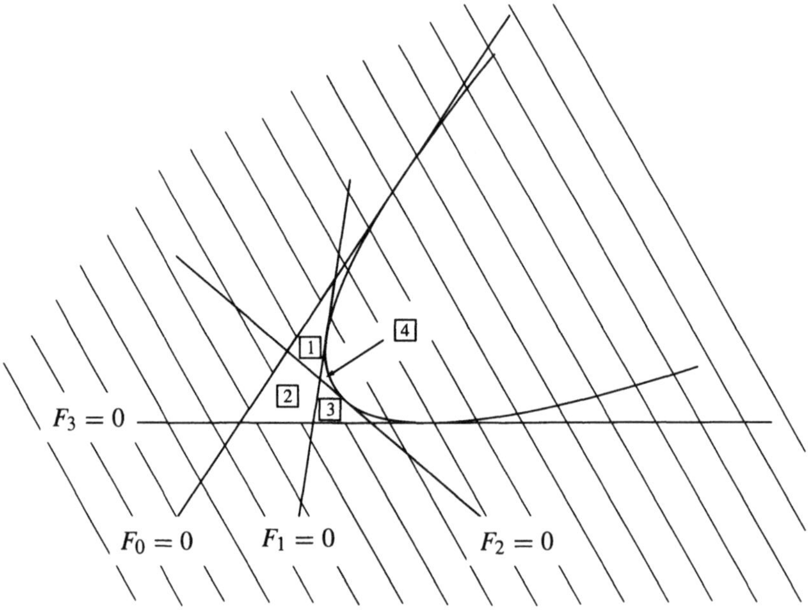

Fig. 3 application moment pour le flot géodésique de $\mathfrak{so}(4)$

2.3.3. Corollaire *L'image de l'application moment F ne peut contenir aucun point singulier du discriminant du polynôme $\mathcal{P}$ (mais peut contenir des points multiples de ce discriminant dès que $n \geq 4$).* □

2.3.4. Remarque L'annonce [15] contient trois schémas représentant les valeurs atteintes par le moment pour le flot géodésique sur $\mathfrak{so}(4)$. La figure 3 n'y figure pas, mais le troisième des schémas correspond à des orbites coadjointes de $SO(4)$ à petit pfaffien et est assez proche de celui obtenu ici où l'on a considéré une orbite à pfaffien nul (matrices de rang 2). Les bifurcations que nous trouverons sont identiques à celles annoncées dans ce schéma. (Voir aussi [4] pour le flot géodésique sur $SO(4)$.)

3 Les tores de Liouville et leurs bifurcations en dimension 4

Dans tout ce §, je suppose que $n = 2$.

3.1 Les tores de Liouville

Considérons à nouveau, sur les figures 3 et 1, l'image de l'application moment. Les parties hachurées ne sont pas atteintes: à l'intérieur de la parabole, $\mathcal{P}$ n'a pas ses racines réelles, les autres zônes hachurées correspondent à des racines réelles qui ne vérifient pas les inégalités imposées par 2.3.1. Les droites P_z sont les tangentes à la parabole. Les numéros $\boxed{1}$, $\boxed{2}$, $\boxed{3}$, $\boxed{4}$ des quatre zônes atteintes correspondent aux positions des z_i définies par les mêmes chiffres sur la figure 2. Par exemple dans le cas (b), ces zônes vont correspondre à des tangentes communes

- à un ellipsoïde et à un hyperboloïde à une nappe pour la zône $\boxed{1}$,
- à un ellipsoïde et à un hyperboloïde à deux nappes pour la zône $\boxed{2}$,
- à un hyperboloïde à une nappe et à un hyperboloïde à deux nappes pour la zône $\boxed{3}$,
- à deux hyperboloïdes à une nappe pour la zône $\boxed{1}$.

3.1.1. Remarque Le lecteur sera sans doute satisfait de constater que l'intersection d'une droite P_z avec l'image de l'application moment est un segment (compact) si et seulement si cette intersection décrit bien les géodésiques d'un authentique ellipsoïde. il remarquera aussi que, pour des raisons de convexité évidentes, il n'y a aucune tangente commune à deux ellipsoïdes ou à deux hyperboloïdes à deux nappes de la famille. Autrement dit deux droites P_z pour deux valeurs distinctes de z soit toutes deux $< \alpha_1$, soit toutes deux dans $]\alpha_2, \alpha_3[$ doivent se couper *en dehors* de l'image de F: c'est encore une façon de trouver cette image dans le cas $n = 2$.

3.1.2. Proposition *Le niveau $\mathcal{T}_F$ a 2 composantes connexes quand F est dans une des zônes $\boxed{1}$, $\boxed{2}$, $\boxed{3}$, et 4 quand F est dans la zône $\boxed{4}$.*

Pour démontrer la proposition 3.1.2., on utilise d'abord un argument permettant de ne traiter qu'un seul des deux cas (a) et (b), par exemple en montrant que (b) s'obtient à partir de (a) en faisant tendre convenablement α_0 vers $-\infty$. Il suffit alors[3]) de traiter le cas (b) que l'on peut reformuler en:

3.1.3. Proposition *$\mathcal{T}_F$ a 2 composantes si au plus une des deux quadriques qu'il définit est un hyperboloïde à une nappe (zônes $\boxed{1}$, $\boxed{2}$, $\boxed{3}$) et 4 si toutes les deux sont des hyperboloïdes à une nappe.* □

Cette différence s'explique par la différence entre l'hyperboloïde à une nappe et les deux autres types de quadriques de la famille:

- soit — directement — pour la métrique: après tout, on s'intéresse aux géodésiques, et il n'est pas indifférent d'être en courbure négative!
- soit — de façon équivalente — pour la topologie: une fois complété, l'hyperboloïde à une nappe est un tore, alors que les autres sont des sphères.

On peut alors se reposer sur les résultats classiques sur les géodésiques des quadriques: le cas où l'une des quadriques est un ellipsoïde est exprimé en termes

3) Le lecteur peu convaincu par cet argument trouvera une démonstration directe pour le cas (a) dans 3.1.6..

de tores de Liouville par exemple dans le livre de Klingenberg [12]. Le cas où l'une des deux est un hyperboloïde à deux nappes n'est pas vraiment différent. Les géodésiques de l'hyperboloïde à une nappe ont été étudiées par Hadamard [7]. La projection du niveau $\mathcal{T}_F$ sur l'une des quadriques (complétée à l'infini) qu'il définit est un anneau si F est dans l'une des zônes numérotées de 1 à 3 (voir les beaux et classiques dessins de [9]), et l'hyperboloïde tout entier (tore) dans la zône $\boxed{4}$. Les points génériques de ces images sont images de 4 points du niveau. On peut alors conclure. □

3.1.4. Remarque Le point méritant réellement une démonstration, puisque dans le cas considéré, c'est celui qui est faux dans [13], est celui de $\boxed{4}$. Voilà une autre démonstration: les points de la frontière (parabole), on l'a vu, correspondent aux droites de l'hyperboloïde à une nappe. Comme on considère ici des droites orientées, les deux familles de droites donnent lieu à un niveau (singulier) formé de 4 cercles. Un niveau régulier proche est donc formé de 4 tores!

3.1.5. Remarque Il est assez facile de corriger l'erreur dans [13] et d'en déduire, ici, que le quotient de $\mathcal{T}_F$ par le groupe $\mathbf{Z}/2$ qui change l'orientation de la géodésique a 1 ou 2 composantes selon les cas. Il faut ensuite faire encore un petit effort pour en déduire 3.1.3.

3.1.6. Remarque On peut aussi traiter directement le cas (a) en appliquant les résultats de Haine [8] dans ce cas particulier: le niveau $\mathcal{T}_F$ s'identifie à la surface abélienne $\operatorname{Prym}(Y|E)$ où E est la courbe elliptique $y^2 = \prod(\alpha_i - z)$ et Y la courbe de genre 3, normalisée de $h^2 = \mathcal{P}_{a,b}/y^2$, qui est à la fois un revêtement double ramifié en 4 points de E et un revêtement double non ramifié de C. Presque par définition, $\operatorname{Prym}(Y|E)$ est le quotient de la jacobienne de C par le sous-groupe engendré par l'élément d'ordre 2 représenté par le diviseur $S_1 + S_2 - 2P_0$. Connaissant les positions de S_1 et S_2 sur les C_i, on en déduit facilement comment ce sous-groupe opère sur $\pi_0(\operatorname{Jac} C(\mathbf{R}))$ et donc les nombres de composantes de $\operatorname{Prym}(Y|E)(\mathbf{R})$ suivant les valeurs de F.

3.1.7. Remarque Une autre option serait d'utiliser l'application de vecteurs propres que j'ai déjà mentionnée. Ici utilisée brutalement, elle donne assez peu de renseignements: très précisément, l'application $\Phi_F : \mathcal{T}_F \to \operatorname{Pic} C$ induit un difféomorphisme de $\mathcal{T}_F/G$ sur *une* des composantes de la partie réelle de la jacobienne, où G est le groupe $(\mathbf{Z}/2)^{n+1}$ engendré par les symétries autour des axes de coordonnées dans $\mathbf{R}^{n+1}$. Même quand $n = 2$, ce résultat est bien loin de donner 3.1.3. La méthode est affinée dans [3].

3.1.8. Remarque Le fait que Moser montre que $\mathcal{T}_F/G$ est la jacobienne de C n'est pas plus utile... et surtout n'est en rien contradictoire avec 3.1.7: son résultat est précisément que $(\mathcal{T}_F(\mathbf{C})/G)$ est isomorphe à la jacobienne de C. Même si cet isomorphisme préserve les structures réelles, ce qui est le cas, il permet tout au plus

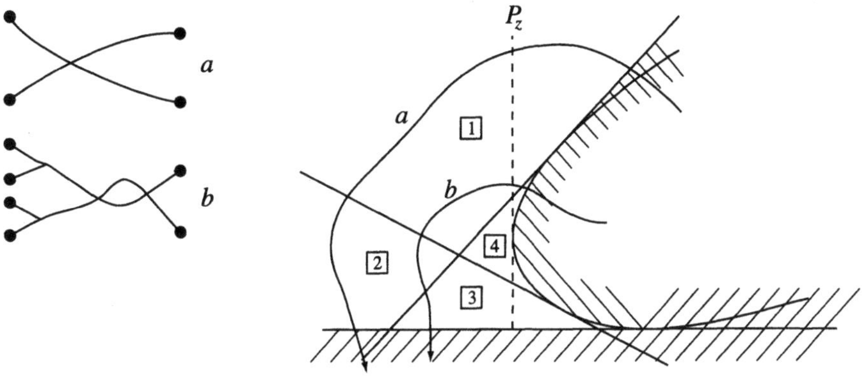

Fig. 4 bifurcations des tores de Liouville

d'identifier la partie réelle de $(\mathcal{T}_F(\mathbf{C})/G)$ à la partie réelle de la jacobienne. Le quotient par G crée des composantes connexes et cette variété est assez différente de $\mathcal{T}_F(\mathbf{R})/G$.

3.2 Les bifurcations

Toutes les méthodes proposées ci-dessus permettent aussi de démontrer

3.2.1. Proposition *Les bifurcations des tores de Liouville sont celles indiquées sur la figure* 4. □

Les bifurcations sont indiquées ici par des diagrammes à la Kharlamov, ou à la Fomenko (voir [11] et [6] par exemple). Il faut remarquer que le problème strict des géodésiques des quadriques, c'est à dire le hamiltonien Ψ_z sur $\mathcal{D}_2$ est précisément un exemple qui ne vérifie pas les conditions de Fomenko dès que $z \in]\alpha_1, \alpha_2[$ (cas des géodésiques d'un hyperboloïde à une nappe): le niveau $\Psi_z = 0$ n'est *pas* régulier. On a vu en effet que les droites de la quadrique en sont des points critiques, ce qui se traduit graphiquement par le fait que la droite P_z qui décrit ce niveau est *tangente* au discriminant. Les chemins le long desquels j'ai représenté les bifurcations sont, comme il se doit, transverses au discriminant.

Références

[1] M. Adler, P. vanMoerbeke, *Completely integrable systems, euclidean Lie algebras and curves*, et *Linearization of hamiltonian systems, Jacobi varieties and representation theory*, Advances in Math. **38** (1980), 267–317 et 318–379.

[2] V.I. Arnold, *Méthodes mathématiques de la mécanique classique*, Mir, Moscou, 1974.

[3] M. Audin, *Courbes algébriques et systèmes intégrables: géodésiques des quadriques*, Expo. Math. **12** (1194), 193–226.

[4] M. Audin, *Toupies, un cours sur les systèmes intégrables*, prépublication IRMA, Strasbourg (1994).

[5] M. Audin, R. Silhol, *Variétés abéliennes réelles et toupie de Kowalevski*, Compositio Math. **87** (1993), 153–229.

[6] A.T. Fomenko, *Integrability and nonintegrability in geometry and mechanics*, Math. and its Applications, Kluwer, 1988.

[7] J. Hadamard, *Sur la forme des lignes géodésiques à l'infini et sur les géodésiques des surfaces réglées du second ordre*, Bull. Soc. Math. France **26** (1898), 195–216.

[8] L. Haine, *Geodesic flow on $SO(4)$ and abelian surfaces*, Math. Ann. **263** (1983), 435–472.

[9] D. Hilbert, S. Cohn-Vossen, *Anschauliche Geometrie*, Springer Verlag, 1932.

[10] C. Jacobi, *Vorlesungen über Dynamik*, Gesammelte Werke, Supplementband, Berlin, 1884.

[11] M.P. Kharlamov, *Bifurcation of common levels of first integrals of the Kowalevskaya problem*, PMN U.S.S.R. **47** (1983), 737–743.

[12] W. Klingenberg, *Riemannian Geometry*, de Gruyter, 1982.

[13] H. Knörrer, *Geodesics on the ellipsoid*, Invent. Math. **39** (1980), 119–143

[14] J. Moser, *Geometry of quadrics and spectral theory*, in *The Chern Symposium*, Springer (1980), 147–188.

[15] A.A. Oshemkov, *The topology of surfaces of constant energy and bifurcation diagrams for integrable cases of the dynamics of a rigid body on* $so(4)$, Russian Math. Surveys **42** (1987), 241–243.

[16] A. Reiman, *Integrable hamiltonian systems connected with graded Lie algebras*, J. Soviet Math. **19** (1982), 1507–1545.

[17] K. Weierstrass, *Über die geodätischen Linien auf dem dreiachsigen Ellipsoid*, in *Math. Werke*, I (1861), 257–266.

Michèle Audin,
Institut de Recherche Mathématique Avancée
Université Louis Pasteur,
7, rue René Descartes,
F-67084 Strasbourg Cedex, France

Morse-Bott theory and equivariant cohomology

D.M. Austin and P.J. Braam

1 Introduction

Critical points of functions and gradient lines between them form a cornerstone of physical thinking. In Morse theory the topology of a manifold is investigated in terms of these notions with equally profound success: Smale proved the h-cobordism and generalized Poincaré conjectures using surgery cobordisms, see [Mi2].

Morse's original application (see [Mo] and [Mi1]) illustrates an important theme. He chose to investigate the energy functional on the infinite dimensional space of loops in a manifold (see [Mo] and [Mi1]). Not only is this physically relevant, for it describes geodesic motion, but it also led to the Bott periodicity theorem, see Bott [B4] and Bott and Samuelson [BS]. The condition of non-degenerate critical points was relaxed by Bott to allow for non-degenerate manifolds of critical points. This was more or less necessary to deal with the natural group actions on the loop space of a symmetric space.

This theme is borne out in a later application of Morse theory by Atiyah and Bott [AB2]. They consider the Yang-Mills functional as a function on the space of connections over a Riemann surface which is invariant under the group of gauge transformations. Morse-theoretic ideas applied in the setting of equivariant cohomology enable the computation of cohomological information about the space of holomorphic bundles on the surface, which coincide with the minima of the Yang-Mills functional. The same ideas were subsequently used by Frances Kirwan [K] to enable the computation of the cohomology of geometric quotients in a very general setting.

Acting from the perspective of quantum field theory, Witten [Wi] renewed interest in Morse's complex. Morse's complex is the free module generated by the critical points and graded by their index. A differential on this complex is given by counting the number of gradient lines between critical points differing in index by 1. Inspired by this, Floer made his dramatic contributions by building a Morse complex both for the symplectic action functional on the loop space of a symplectic manifold (Floer [F2]) and for the Chern-Simons functional on the space of connections over a homology 3-sphere, [F1]. Not only is the configuration

---

1) The first author gratefully acknowledges support from a National Science Foundation Postdoctoral Research Fellowship.

space infinite dimensional, so is the index of each critical point. These theories had important applications: they solved (part of) the Arnold conjecture and established an important gluing relation for Donaldson polynomial invariants.

Floer's homology groups for homology 3-spheres asked for many generalizations to describe more general gluing laws governing Donaldson's polynomial invariants, both through incorporating other auxiliary information (see Fukaya [Fu] and Braam and Donaldson [BD]) as well as by considering general 3-manifolds [AuB]. In the course of this work, new techniques were developed in these infinite dimensional cases relating to equivariant cohomology, cup products and various alternative approaches to problems of classical Morse theory. The purpose of this paper is to give a self-contained finite dimensional description of these new aspects.

When a function has nondegenerate critical manifolds, we say it is Morse-Bott. A complex which computes the homology of the manifold in terms of the critical submanifolds and gradient flows was missing. In this paper, we build such a Morse-like complex which computes the deRham cohomology of the manifold. The chains of the complex are differential forms on the critical submanifolds and the boundary map is defined in terms of integration over the flow lines of the gradient vector field. In the case when the function is truly Morse, our complex reduces to the standard Morse complex. The existence of our complex easily implies the Morse inequalities of Bott in the same way that the standard Morse inequalities follow from the existence of a complex computing homology. Furthermore, in the spirit of Atiyah and Bott, we show how to compute the equivariant cohomology of a manifold with group action in terms of an invariant function on the manifold. We do so in various ways, using a Morse-Bott function on approximations of the homotopy quotient as well as through equivariant differential forms.

The isomorphism between the homologies of the deRham complex and ours is obtained at the chain level by integrating differential forms over the unstable manifolds of critical submanifolds. This clearly illuminates the fundamental role played by the Thom isomorphism in Morse theory. Furthermore, it leads to a detailed investigation of the structure of the unstable manifolds near their frontiers. With this in hand, additional algebraic topological invariants, such as the cup product structure and Poincaré duality pairing, can be easily obtained from the Morse complex. When studying the cup product, we come particularly close to Witten's idea of deforming the exterior derivative using the Morse function; instead we use a classical analogue and pull back differential forms under the gradient flow for large time. Using these tools we also find isomorphisms at the chain level between the Morse and Morse-Bott complexes. These considerations should be intimately connected with the results of Cohen, Jones and Segal, [CJS].

This paper is organized as follows: §2 contains a review of finite dimensional Morse theory. Since so many of these ideas are applicable to the later Morse-Bott case, we have attempted to be as thorough as possible. A comprehensive discussion of the asymptotics of the gradient line spaces is found in an appendix. A new feature included within this section is an explanation of the cup product in terms of the Morse complex. Next we revise standard Morse theory for Morse-Bott functions in §3 by developing a complex to compute deRham cohomology in the case when the function has nondegenerate critical submanifolds. Cup products are again explained. §4 is a brief introduction to some tools from equivariant cohomology which we apply in §5 where it is shown how to compute the equivariant cohomology of a manifold with group action in terms of an invariant function. This is an extension of the Morse-Bott ideas in §3. Finally, §6 contains a more detailed look at the module structure for equivariant cohomology. This will be useful in [AuB] to explain the «u»-map on Floer homology.

Acknowledgements
Discussions of this work with Andy Keck, Graeme Segal and Ron Stern have been very useful. The first author would like to thank the Institute for Advanced Study for their hospitality, and both authors gratefully acknowledge NSF support.

2 Review of classical Morse theory

This section presents a review of classical finite dimensional Morse theory. The fundamental observation is that there is an intimate relationship between the critical points of a function on a manifold and the topology of the manifold. Succinctly stated, this says that critical points correspond to handles added to the manifold. As such, one can build a complex (C_*, ∂) whose homology equals the singular homology of the manifold — a purely topological invariant. We will make this relationship clear in this review. For additional background, we refer the reader to the following: Smale [Sm1], Milnor [Mi1], Milnor [Mi2], Witten [Wi], and Floer [F3]. Bott [B1] gives an interesting historical perspective.

While this section is primarily a review, it is also designed to motivate our techniques in the more general Morse-Bott case. For this reason, the discussion is self-contained and many details are given. Several of the more technical results are proven in Appendix A. We also give a proof here that the homology of this complex is independent of the Morse function and metric on the manifold. This is useful for infinite dimensional applications, such as Floer homology, where the complex is not computing any known homology of the space. Finally in §2.4, a new feature of Morse theory is described: the cup product structure on the cohomology ring is accessible from a Morse function in terms of critical points and gradient flow information.

2.1 The complex (C_*, ∂)

Suppose that B is a smooth closed, oriented Riemannian manifold of dimension n and $f : B \to \mathbb{R}$ is a smooth function. A point $\alpha \in B$ is called a critical point of f if $df_\alpha = 0$. At a critical point, the Hessian $d^2 f_\alpha$ is well defined as a quadratic form on $T_\alpha B$, and we call a critical point non-degenerate if its Hessian is a non-degenerate symmetric bilinear form. We will henceforth assume that f is a Morse function meaning that all the critical points of f are non-degenerate.

For a critical point α, let H_α be a maximal negative definite subspace of $T_\alpha B$ with respect to $d^2 f_\alpha$. The index is defined by

$$i(\alpha) = \dim H_\alpha.$$

Non-degenerate critical points are necessarily isolated, and in suitable local coordinates the function is a quadratic form. To such a critical point α, we can assign an unstable manifold $\mathcal{U}_\alpha$ defined by

$$\mathcal{U}_\alpha = \{b \in B; \phi_t(b) \to \alpha \text{ as } t \to -\infty\}$$

where ϕ_t is the gradient flow for the vector field $-\operatorname{grad} f$ (see Smale [Sm2]). It is well known that $\mathcal{U}_\alpha$ is an embedded $\mathbb{R}^{i(\alpha)}$. Similarly, one defines a stable manifold $\mathcal{S}_\alpha$ for α using the flow as $t \to \infty$. Again, $\mathcal{S}_\alpha$ is an embedded $\mathbb{R}^{n-i(\alpha)}$. The *moduli spaces of gradient lines* between critical points α and β are of central interest in our discussion. These spaces have an $\mathbb{R}$-action so we define:

$$\mathcal{M}(\alpha, \beta) = \mathcal{S}_\beta \cap \mathcal{U}_\alpha$$
$$\widetilde{\mathcal{M}}(\alpha, \beta) = (\mathcal{S}_\beta \cap \mathcal{U}_\alpha)/\mathbb{R}\ .$$

We begin by stating that «most» functions are Morse:

Proposition 2.1 *For a Baire set of functions $f \in C^\infty(B)$, all critical points are nondegenerate and $\mathcal{U}_\alpha$ and $\mathcal{S}_\beta$ intersect transversely for all α and β. Such an f is called a generic Morse function.*

Proof. See Appendix B.

To begin to build a complex, we arbitrarily choose an orientation of all unstable manifolds. At a critical point α, the tangent space splits into the stable and unstable parts so that $T_\alpha B = T_\alpha \mathcal{U}_\alpha \oplus T_\alpha \mathcal{S}_\alpha$. Since B is oriented, the orientations of the $\mathcal{U}_\alpha$ orient all stable manifolds. Because the intersections of stable and unstable manifolds are assumed to be transverse. All moduli spaces of gradient lines are now oriented and in particular if $i(\alpha) = i$ and $i(\beta) = i - 1$, then $\mathcal{U}_\alpha \cap \mathcal{S}_\beta$ is smooth, oriented and of dimension 1. Moreover it is invariant under the flow ϕ_t so that

$$(\mathcal{U}_\alpha \cap \mathcal{S}_\beta)/\mathbb{R}$$

is simply a collection of oriented points.

Now form a complex with chain groups

$$C_i = \text{ free } \mathbb{Z}\text{-module generated by critical points of index } i.$$

The boundary operator $\partial : C_i \to C_{i-1}$ is described by

$$\partial(\alpha) = \sum m_k \beta_k,$$

where α is a critical point of index i, the β_k are the critical points of index $i-1$ and

$$m_k = \#(\mathcal{S}_{\beta_k} \cap \mathcal{U}_\alpha)/\mathbb{R} \in \mathbb{Z},$$

where # denotes the algebraic count of the oriented points. The finiteness of the intersection is guaranteed by the compactness of B.

It is also useful to consider the dual complex (C^*, ∂^*). Here

$$C^i = \text{ free } \mathbb{Z}\text{-module generated by critical points of index } i$$

and the coboundary operator $\partial^* : C^i \to C^{i+1}$ is given by

$$\partial^*(\alpha) = \sum \#\widetilde{\mathcal{M}}(\gamma_j, \alpha)\gamma_j$$

where the sum is taken over the critical points γ_j of index $i+1$.

The fact that (C_*, ∂) and (C^*, ∂^*) form complexes will be discussed in some detail and proven as Proposition 2.8 since the ideas are relevant for our discussion in later sections. Assuming this fact for now, we state the fundamental theorem concerning this complex:

Theorem 2.2 *The homology of the complex* (C_*, ∂) $((C^*, \partial))$ *is naturally isomorphic to the singular (co)homology of* B*; that is,*

$$H_j(B, \mathbb{Z}) \cong H_j(C_*, \partial) \qquad H^j(B, \mathbb{Z}) \cong H^j(C^*, \partial^*).$$

Of course, this theorem holds for general coefficient groups. It certainly was known to Milnor [Mi2], although he did not define the boundary operator equally explicitly. For a self contained proof see, for example, Floer [F3]; in Section 3, a new proof will be given over the real numbers.

There are two interesting corollaries of this theorem. That the critical points form a basis of a complex which computes the cohomology of B immediately implies the well-known Morse inequalities. Let $c_i = \text{ rank } C_i$, $h_i = \text{ rank } H_i(B)$ and $z_i = \text{ rank ker } (\partial : C_i \to C_{i-1}) \leq c_i$. Defining $k_i = c_i - z_i \geq 0$ gives

Corollary 2.3 [Morse inequalities] *There are* $k_i \geq 0$ *such that*

$$\sum_i (c_i - h_i)t^i = (1+t)\sum_i k_i t^i.$$

Secondly, we may consider the Morse function $-f$ which gives a new complex $(\overline{C}_*, \overline{\partial})$; by Theorem 2.2, the homology of $(\overline{C}_*, \overline{\partial})$ also computes the homology of B. The critical points of f and $-f$ agree; however, a critical point of f of index i is a critical point of $-f$ of index $n-i$. Hence

$$\overline{C}_i = \text{free } \mathbb{Z}\text{-module generated by critical points of } f \text{ of index } n-i.$$

Moreover, $\overline{\partial}(\alpha) = \sum \#\widetilde{\mathcal{M}}(\gamma_j, \alpha)\gamma_j$. This shows that the two complexes, (C^*, ∂^*) and $(\overline{C}_*, \overline{\partial})$, are isomorphic by the natural map taking

$$C^i \to \overline{C}_{n-i}.$$

This gives the well known isomorphism

Corollary 2.4 [Poincaré Duality] $H^*(B;\mathbb{Z}) \cong H_{n-*}(B;\mathbb{Z})$.

Next we will discuss why C_* is a complex and why its cohomology is independent of the generic Morse function f. These are the key ideas in our paper.

2.2 (C_*, ∂) is a complex

The basic idea is quite simple. If $\partial \circ \partial(\alpha) \neq 0$ then there must be gradient lines from α to some β and from β to some γ with $i(\alpha) = i(\beta) + 1 = i(\gamma) + 2$. This means that the space parametrizing gradient lines from α to γ (which is of dimension 1) has a boundary component equal to the factorization through β. But boundary points of 1-dimensional manifolds come in pairs, so there must be another factorization $\alpha \to \beta' \to \gamma$ (as in figure 1).

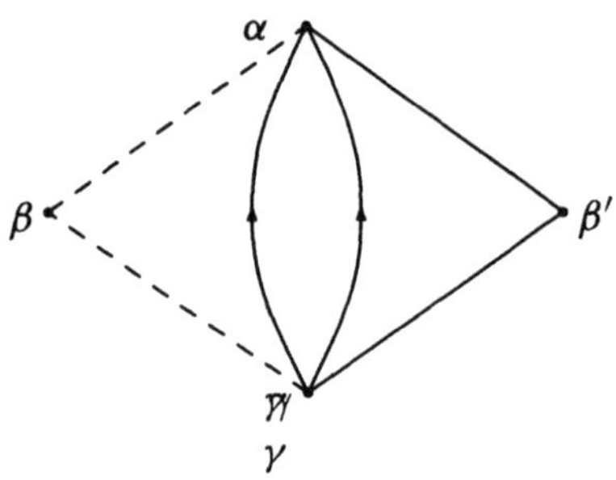

Fig. 1

The proof consists of showing that this is indeed so and that the orientations cause these two factorizations to cancel out. For this reason, we begin a careful analysis of the asymptotic structure of the space of gradient lines referring to Appendix A for the more technical proofs. Our attention to detail here will ease the more general Morse-Bott case in later sections. The discussion here owes much to Floer [F1].

The assumption that f is a generic Morse function — that is, a Morse function whose stable and unstable manifolds intersect transversely — implies that for $\alpha \neq \beta$, $\widetilde{\mathcal{M}}(\alpha,\beta)$ is a smooth manifold of dimension $i(\alpha) - i(\beta) - 1$. In particular, this says that if $i(\alpha) \leq i(\beta)$, then $\widetilde{\mathcal{M}}(\alpha,\beta) = \emptyset$. Notice that, as a set, $\widetilde{\mathcal{M}}(\alpha,\beta)$ can also be regarded as a submanifold of $f^{-1}(c)$ for any $c \in (f(\beta), f(\alpha))$. However, the topology of pointwise convergence on the space of gradient trajectories can be much stronger than the induced topology from B. This will be important for our study of the asymptotic structure of the $\widetilde{\mathcal{M}}(\alpha,\beta)$.

First of all, we have a weak compactness property which states that limits of a sequence of gradient lines are broken gradient lines; the breaking points are intermediate critical points.

Lemma 2.5 *If α, β are two critical points and $\{\gamma_i\}$ is a sequence of gradient lines in $\widetilde{\mathcal{M}}(\alpha, \beta)$, then there are*

1) *a subsequence $\{\gamma_j\}$*
2) *a finite set of critical points $\alpha = \alpha_1, \alpha_2, \ldots, \alpha_k, \alpha_{k+1} = \beta$*
3) *a finite set of strictly decreasing real numbers $r_1, \ldots, r_k$ with the property that $r_i \in (f(\alpha_{i+1}), f(\alpha_i))$*

such that

1) *the points $x_{i,j} = \gamma_j(s)$ such that $f(x_{i,j}) = r_i$ converge to a regular point on B which lies on a gradient line in $\widetilde{\mathcal{M}}(\alpha_i, \alpha_{i+1})$*
2) *the indices of the α_j are strictly decreasing with j; that is,*

$$i(\alpha_1) > i(\alpha_2) > \cdots i(\alpha_{k+1}).$$

Pictorially, this may be seen as in figure 2.

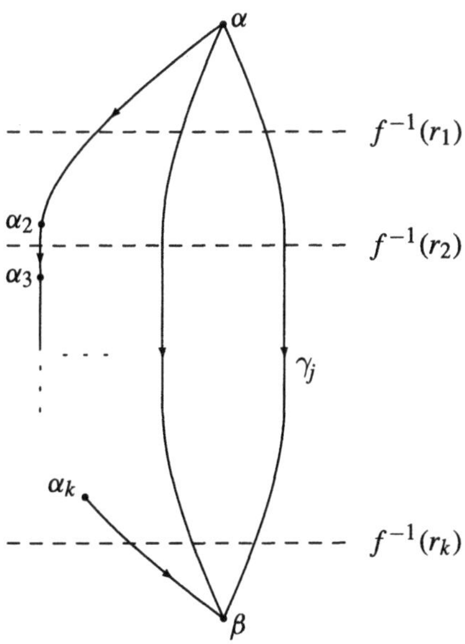

Fig. 2

Proof. The proof is in Appendix A (A.1.5).

Notice that for $i(\alpha) - i(\beta) = 1$ the lemma states that $\widetilde{\mathcal{M}}(\alpha, \beta)$ is compact. We can now describe a local model of the boundary of the moduli of gradients.

Lemma 2.6 *Let $\alpha = \alpha_0, \alpha_1, \ldots, \alpha_{k-1}, \alpha_k = \beta$ be a finite set of critical points, the indices of which are strictly decreasing with j. For sufficiently small ϵ, there is a diffeomorphism, called the gluing map,*

$$G : \prod_{j=0}^{k} \widetilde{\mathcal{M}}(\alpha_j, \alpha_{j+1}) \times \prod_{j=1}^{k} (0, \epsilon) \to \widetilde{\mathcal{M}}(\alpha, \beta)$$

mapping onto an open set in $\widetilde{\mathcal{M}}(\alpha, \beta)$. Moreover, if γ_i is a sequence of gradient lines as in Lemma 2.5, then $\gamma_i \in$ image G for all large i. The compactification in the sense of Lemma 2.5 proceeds by letting the real parameters in $(0, \epsilon)$ approach zero.

Proof. See A.2.2. Figure 3 indicates geometrically what the parameters

$$(\{\gamma_i\}_{j=0}^{k}, \{t_i\}_{j=1}^{k}) \in \prod \widetilde{\mathcal{M}}(\alpha_i, \alpha_{i+1}) \times \prod [0, \epsilon)$$

signify. □

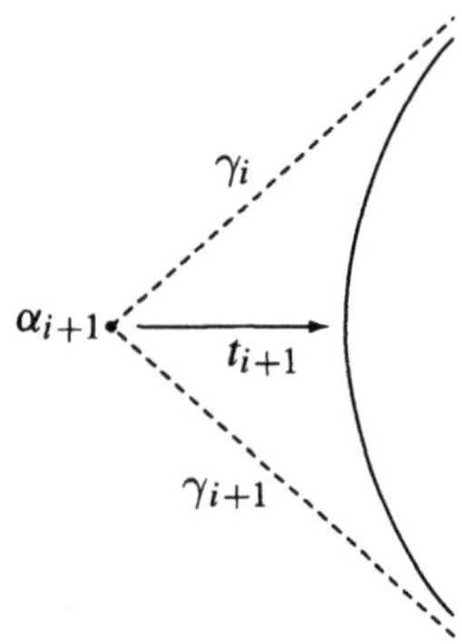

Fig. 3

Together the two lemmas prove that $\widetilde{\mathcal{M}}(\alpha, \beta)$ has a compactification

$$\overline{\widetilde{\mathcal{M}}}(\alpha, \beta) = \widetilde{\mathcal{M}}(\alpha, \beta) \cup \bigcup_{k \in \mathbb{Z}} \bigcup_{\alpha = \alpha_0, \ldots, \alpha_k = \beta} \left(\prod_{j=0}^{k} \widetilde{\mathcal{M}}(\alpha_j, \alpha_{j+1}) \times \prod_{j=1}^{k} [0, \epsilon)\right)$$

which is a smooth manifold with corners.

In showing that (C_*, ∂) is a complex, we need to understand how the orientations of the various moduli spaces of gradient lines relate to one another through the gluing map. We now develop a calculus for discussing orientations of submanifolds and their intersections.

Recall that an oriented manifold with boundary induces an orientation on its boundary. This orientation is defined by the rule that the outward pointing normal followed by the boundary orientation must add up to the orientation of the full manifold; with this convention, Stokes' theorem is true. If R is an oriented submanifold of B then the normal bundle is oriented by the rule $[R][\mathcal{N}_R] = [B]$, where the notation means juxtaposition of oriented frames. Transverse intersections of oriented submanifolds R and T are oriented by the rule that the sum of the oriented normal bundles $\mathcal{N}_R \oplus \mathcal{N}_T$ is the oriented normal bundle of $R \cap T$. These rules are more conveniently expressed by noting that the juxtaposition of frames can be expressed as Clifford multiplication in $C(T^*B)$, which in terms of an orthonormal framing e_i is defined by $e_i^2 = 1$ and $e_i e_j = -e_j e_i$ if $i \neq j$. At different points in a submanifold of B we can compare the two framings by using parallel transport.

Starting from a choice of orientations of stable manifolds $\mathcal{S}_\alpha$ for critical points α we orient the unstable ones by

$$[\mathcal{S}_\alpha][\mathcal{U}_\alpha] = [B].$$

For oriented intersections $\mathcal{M}(\alpha, \beta) = \mathcal{S}_\beta \cap \mathcal{U}_\alpha$, we find

$$[\mathcal{M}(\alpha, \beta)] = [\mathcal{U}_\alpha][B]^{-1}[\mathcal{S}_\beta].$$

Orient the gradient line space $\widetilde{\mathcal{M}}(\alpha, \beta)$ in a level set by $[\mathcal{M}(\alpha, \beta)] = [\widetilde{\mathcal{M}}(\alpha, \beta)][\nabla f]$.

Proposition 2.7 *The gluing map*

$$G : \widetilde{\mathcal{M}}(\alpha, \beta) \times (0, \epsilon) \times \widetilde{\mathcal{M}}(\beta, \gamma) \to \widetilde{\mathcal{M}}(\alpha, \gamma)$$

is orientation reversing.

Proof. Consider the image of G in the level set $f^{-1}(\delta)$ where δ is in $(f(\beta), f(\alpha))$ (as in the figure 4).

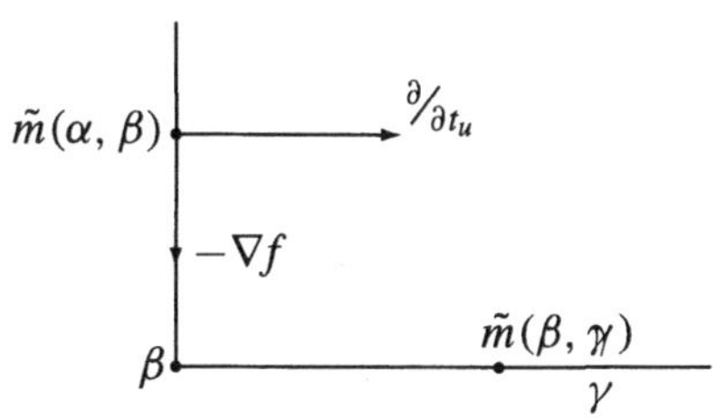

Fig. 4

The rules above give:

$$[\widetilde{\mathcal{M}}(\alpha,\gamma)] = -[\widetilde{\mathcal{M}}(\alpha,\beta)][t_u][\widetilde{\mathcal{M}}(\beta,\gamma)]. \qquad \square$$

To complete the discussion that (C_*, ∂) is a complex, notice that

$$\begin{aligned}\partial\circ\partial(\alpha) &= \sum_{i(\gamma)=i(\alpha)-2}\ \sum_{i(\beta)=i(\alpha)-1} \#\widetilde{\mathcal{M}}(\alpha,\beta)\#\widetilde{\mathcal{M}}(\beta,\gamma)\gamma \\ &= \sum_{\gamma} \partial\widetilde{\mathcal{M}}(\alpha,\gamma)\gamma \\ &= 0\end{aligned}$$

since the boundary of a 1-manifold has zero points counted with orientation. This gives our result

Proposition 2.8 (Complex) $\partial\circ\partial = 0$.

The only fact used in the proof is that the oriented boundary of a 1-manifold has zero points when counted with sign. These arguments, and the analogous ones in the following sections, call on a detailed knowledge of the asymptotics: limits of a sequence of trajectories should be understood in terms of gluing trajectories which connect intermediary critical points.

2.3 Independence of function and metric

The construction above is at the heart of the definition of Floer homology. To show that Floer homology is a topological invariant, one cannot proceed by comparing it to a known topological invariant. We therefore present the usual alternative proof in the finite dimensional setting which shows that the homology groups computed are independent of the choices. This proceeds along the lines of Floer [F1] and Fukaya [Fu]. It makes use of time dependent gradient flows on $B \times \mathbb{R}$.

The construction of the complex (C_*, ∂) depended on a Morse function and metric on the manifold. Of course, since the cohomology of this complex is just the singular cohomology of the manifold, the cohomology is completely independent of the choice of Morse function and metric. However, in many modern applications, the complex is defined for an infinite dimensional manifold and there is no known cohomology theory with which to compare. For this reason, we will give an alternative proof, in finite dimensions, that the homology groups computed from the complex (C_*, ∂) are independent of the Morse function and metric, following Floer [F1] and Fukaya [Fu]. This will also be useful in the definition of equivariant Floer cohomology (see [AuB]).

If two Morse functions f_i and metrics g_i on B are given, we obtain two complexes, (C^1_*, ∂^1) and (C^2_*, ∂^2), as above. We now follow a three step procedure. First we construct chain homomorphisms from C^1 to C^2, and vice versa, using a time dependent flow. Secondly we show that their composition is a chain homomorphism which also arises from a time dependent flow. Finally we construct a homotopy operator showing that the composition induces the same map as the identity (which arises from a constant family). The three together show that the homology is canonical up to isomorphism.

To construct a map of complexes we use a family of functions f_t and metrics g_t on B parametrized by $\mathbb{R}$ and equal to f_1 (resp. g_1) for $t < -1$ and to f_2 (resp. g_2) for $t > 1$. This pair is conveniently denoted by $F = (f_t, g_t)$, and defines a vector field

$$-\operatorname{grad}_{g_t} f_t + \frac{\partial}{\partial t}$$

on $B \times \mathbb{R}$ equal to $-\operatorname{grad} f_1 + \frac{\partial}{\partial t}$ for $t < -1$ and to $-\operatorname{grad} f_2 + \frac{\partial}{\partial t}$ when $t > 1$. By abuse of notation we will also denote the vector field by F or F_{21} if we want to indicate that it interpolates between the $\operatorname{grad} f_i$, $i = 1, 2$.

For such vector fields F with flow ϕ_s, one can easily define submanifolds

$$\mathcal{U}_\alpha = \{(x,t) \in B \times \mathbb{R}; \phi_s(x,t) \to (\alpha, -\infty) \text{ as } s \to -\infty\},$$

of dimension $i(\alpha)+1$ when α is a critical point of f_1, and similarly manifolds $\mathcal{S}_\beta$ for critical points of f_2. They will inherit orientations from the stable and unstable manifolds in B and intersect transversely for a Baire set of vector fields F on $B \times \mathbb{R}$ equal to the $\operatorname{grad} f_i$ in the prescribed areas (see the transversality discussion in Appendix B). Assume that we have chosen such an F on $B \times \mathbb{R}$ and let

$$\widetilde{\mathcal{M}}_F(\alpha, \beta) = (\mathcal{S}_\beta \cap \mathcal{U}_\alpha)/\mathbb{R}$$

of dimension $i(\alpha) - i(\beta)$ denote the oriented intersection.

A chain map $F_{21,*} : C_*^1 \to C_*^2$ can now easily be defined: for basis elements $\alpha \in C_i^1, \beta \in C_i^2$, we define

$$F_{21,*}(\alpha) = \sum_\beta \#\widetilde{\mathcal{M}}_F(\alpha, \beta)\beta$$

using an algebraic count of oriented points.

Before explaining why this is a chain map, we observe that if F_{21} interpolates between f_1 and f_2, and F_{32} between f_2 and a third Morse function f_3 on B, then

Lemma 2.9 $(F_{31})_* = (F_{32})_*(F_{21})_*$ *where*

$$F_{31,t} = F_{32,t-T}, \quad t > \frac{1}{2}, \text{ and } F_{31,t} = F_{21,t+T}, \quad t < -\frac{1}{2},$$

for large T.

Proof. This is a minor variant of the proof of Proposition 2.6 describing the gluing of gradient lines and it can be found in the appendix (see A.2.4). □

Lemma 2.10 $F_* : (C^1_*; \partial^1) \to (C^2_*; \partial^2)$ *is a chain map; that is,* $F_*\partial^1 + \partial^2 F_* = 0$.

Proof. Consider α a critical point of f_1 of index i and β a critical point of f_2 of index $i-1$. This space of gradient lines, $\widetilde{\mathcal{M}}_F(\alpha,\beta)$, has dimension 1 and boundary consisting of factorizations:

$$(\cup_\gamma \widetilde{\mathcal{M}}_{f_1}(\alpha,\gamma) \times \widetilde{\mathcal{M}}_F(\gamma,\beta)) \bigcup (\cup_\delta \widetilde{\mathcal{M}}_F(\alpha,\delta) \times \widetilde{\mathcal{M}}_{f_2}(\delta,\beta))$$

where γ runs over critical points of f_1 of index $i-1$ and δ runs over critical points of f_2 of index i (see A.2.3). This is shown as before keeping in mind that due to the presence of the time factor in the flow, only one of the factorizing trajectories can go through the time dependent area. The gluing and orientation proceeds as before. Then $F_*\partial^1(\alpha) = \sum_{\beta,\gamma} \#\widetilde{\mathcal{M}}_{f_1}(\alpha,\gamma)\#\widetilde{\mathcal{M}}_F(\gamma,\beta)\beta$ and $\partial_2 F_*(\alpha) = \sum_{\beta,\delta} \#\widetilde{\mathcal{M}}_F(\alpha,\delta)\#\widetilde{\mathcal{M}}_{f_2}(\delta,\beta)\beta$. Thus the coefficient of β in $F_*\partial^1 + \partial^2 F_*$ is precisely the number of oriented boundary points of $\widetilde{\mathcal{M}}_F(\alpha,\beta)$ and hence zero. □

Let F^i, $i = 1, 2$, be two families interpolating between f_1, g_1 and f_2, g_2. To complete the proof that the Morse complex defines a topological invariant, we construct a chain homotopy to show that the induced maps F^i_* on homology are the same. To do so, we use an appropriate family of interpolating functions and metrics F_u on $B \times \mathbb{R}$, with $u \in [0,1]$. We require that $F_u = F^1$ when $u < \frac{1}{4}$ and $F_u = F^2$ when $u > \frac{3}{4}$. Furthermore for $t > T$ we have $F_u(t) = (f_1, g_1)$ for all $u \in [0,1]$, and similarly for $t < -T$ we want (f_2, g_2).

We shall show on the level of complexes:

Lemma 2.11 *The maps* $(F_1)_*$ *and* $(F_2)_*$ *are chain homotopic; that is, there is a map* $H : C^1_i \to C^2_{i+1}$ *such that* $(F_1)_* - (F_2)_* = -H\partial^1 - \partial^2 H$.

Proof. To find the homotopy and the relation we once more consider the boundary of a space of trajectories: this time trajectories for the flow of the vector field G on $B \times \mathbb{R} \times [0,1]$ which is tangent to the $B \times \mathbb{R}$ for every $u \in [0,1]$, and agrees with the vector field F_u considered above on $B \times \mathbb{R} \times \{u\}$. We pick critical points α of f_1 of and β of f_2 of index j. After perturbing our vector field slightly, we can assume that

$$\widetilde{\mathcal{M}}_G(\alpha \times [0,1], \beta \times [0,1])$$

is a smooth manifold of dimension $i - j + 1$ (see A.3.3). There are endpoint maps defined on this space mapping it to $\alpha \times [0,1]$ and $\beta \times [0,1]$.

When the index of α is i and that of β is $i+1$, then $\widetilde{\mathcal{M}}_G(\alpha \times [0,1], \beta \times [0,1])$ is a compact zero-dimensional manifold. The fibers of the endpoint maps are non-empty for only finitely many $u \in [0,1]$. Now define the chain homotopy $H : C^1_i \to C^2_{i+1}$ as

$$H(\alpha) = \sum_\beta \#\widetilde{\mathcal{M}}_G(\alpha \times [0,1], \beta \times [0,1])\beta.$$

For δ a critical point of f_2 of index i, $\widetilde{\mathcal{M}}_G(\alpha \times [0,1], \delta \times [0,1])$ is a 1-dimensional manifold whose boundary, when studied as above, equals:

$$
\begin{aligned}
&- \widetilde{\mathcal{M}}_{F_0}(\alpha,\delta) \cup \widetilde{\mathcal{M}}_{F_1}(\alpha,\delta) \bigcup \\
&- \{\{\cup_\gamma(\widetilde{\mathcal{M}}_{f_1}(\alpha,\gamma) \times \widetilde{\mathcal{M}}_G(\gamma \times [0,1], \delta \times [0,1]))\} \bigcup \\
&- \{\cup_\delta(\widetilde{\mathcal{M}}_G(\alpha \times [0,1], \beta \times [0,1]) \times \widetilde{\mathcal{M}}_{f_2}(\beta,\delta))\}\}.
\end{aligned}
$$

Here γ, β range over the critical points of f_1, f_2 of index $i-1$, $i+1$ respectively. Counting boundary points gives the desired relation. □

Observe that the homotopy measures which gradient lines cease or begin to exist under variation of the parameter u. Putting everything together we obtain:

Proposition 2.12 *The assignment $(M,f,g) \to H_*(C_*,\partial)$ defines topological invariants of M up to canonical isomorphism, i.e. changing the metric or Morse function induces a canonical isomorphism between the corresponding homology groups.*

Proof. If (f_1,g_1), (f_2,g_2) are as above, then the crucial observation to be made is that the composition $F_{12} \circ F_{21}$ can be realized by the constant family $F_t = (f_1,g_1)$. By 2.11, this induces the identity on homology. This shows that $F_{12} = F_{21}^{-1}$. □

Observe that the compactness assumption on B can be relaxed in the following way. Let $\mathcal{C}$ be a class of functions and metrics such that:

1) for any $(f,g) \in \mathcal{C}$ the critical point set of f is compact, and families of gradient lines connecting critical points of at most index +2 apart can only degenerate into factorizations into gradient lines connecting intermediary critical points;
2) the family of critical points and the moduli of gradients between critical points of index at most two apart in $\mathcal{C} \times B$ are cut out transversally;
3) for generic families $F_t = (f_t,g_t) : \mathbb{R} \to \mathcal{C}$ connecting two elements in $\mathcal{C}$, the stable and unstable manifolds intersect transversally, and the boundary of the $\widetilde{\mathcal{M}}_F(\alpha,\beta)$ for $i(\alpha) = i(\beta)$ or $i(\alpha) = i(\beta)+1$ is as described in the proof of Lemma 2.11;
4) between any two F_1, F_2 there exist families $G = F_u$ with smooth moduli space $\widetilde{\mathcal{M}}_G(\alpha \times [0,1], \beta \times [0,1])$ which has a boundary as in the proof of Lemma 2.11 if $i(\alpha) = i(\beta)$.

Observe that the last two conditions exclude the possibility of critical points walking off to infinity in families. The homology is now an invariant of the manifold and the class of functions $\mathcal{C}$.

Example

Let $f : B \to \mathbb{R}$ be a function such that its critical points lie in a compact subset equal to $K \equiv f^{-1}([-N,N])$ of B. Let $O \subset B$ be a relatively compact neighborhood of K, and let $\mathcal{C}$ be the affine space of functions consisting of functions $f+g$ where g has support in O. The metric may be arbitrary. Apart from the generic intersection properties, which are proved in Appendix B, it is obvious that all conditions are met.

The importance of existence, orientation and asymptotics of the stable and unstable manifolds must be stressed. All this found its origin in the fundamental work of Smale in dynamical systems and in the transversality theory of R. Thom.

2.4 Cup products

The complex (C_*, ∂) derived from a Morse function computes the homology of B using only information about the one-dimensional gradient line moduli spaces. It is natural to ask if additional information is obtained by studying the higher dimensional moduli spaces. In this section, we indicate that these spaces determine the cap and cup product structures on homology and cohomology. As we will cover this in more generality in §3, we delay some of the proofs until then. Braam and Donaldson use this construction in a gauge-theoretic application (see [BD]).

Let (C_*, ∂) denote the Morse complex defined by f with coefficient group $\mathbb{R}$. Given a differential form $\omega \in \Omega^k(B)$, define

$$c(\omega) : C_i \to C_{i-k} : \alpha \mapsto \sum \Big(\int_{\mathcal{M}(\alpha,\beta_l)} \omega \Big) \beta_l \tag{2.1}$$

where the sum runs over the critical points β_l of index $i-k$.

A careful study of the asymptotics of the gradient line spaces (as is becoming routine) shows that

$$(c(d\omega)(\alpha)) = c(\omega)(\partial\alpha) + \partial(c(\omega)(\alpha)). \tag{2.2}$$

This follows from

$$\begin{aligned} c(d\omega)(\alpha) &= \sum \Big(\int_{\mathcal{M}(\alpha,\beta)} d\omega \Big) \beta \\ &= \sum \Big(\int_{\partial\mathcal{M}(\alpha,\beta)} \omega \Big) \beta. \end{aligned}$$

By Proposition 2.6, it follows that the relevant part of $\partial\mathcal{M}(\alpha,\beta)$ is

$$\cup_{i(\gamma)=i-1} \widetilde{\mathcal{M}}(\alpha,\gamma) \times \mathcal{M}(\gamma,\beta) \bigcup \cup_{i(\delta)=i-k} \mathcal{M}(\alpha,\delta) \times \widetilde{\mathcal{M}}(\delta,\beta).$$

Then

$$\begin{aligned} c(d\omega)(\alpha) &= \sum_\beta \Big(\sum_\gamma \# \widetilde{\mathcal{M}}(\alpha,\gamma) \Big(\int_{\mathcal{M}(\gamma,\beta)} \omega \Big) \beta + \sum_\delta \# \widetilde{\mathcal{M}}(\delta,\beta) \Big(\int_{\mathcal{M}(\alpha,\delta)} \omega \Big) \beta \Big) \\ &= \partial(c(\omega)(\alpha)) + c(\omega)(\partial\alpha) \end{aligned}$$

which proves (2.2).

There are induced maps:

$$\begin{aligned} H^k_{dR}(B) \otimes H_i(C_*; \mathbb{R}) &\to H_{i-k}(C_*; \mathbb{R}) \\ H^k_{dR}(B) \otimes H^i(C^*; \mathbb{R}) &\to H^{i+k}(C^*; \mathbb{R}) . \end{aligned} \tag{2.3}$$

Theorem 2.13 *The maps 2.3 induced by 2.1 agree with the usual cap and cup products, under the isomorphism between the (co)homology of the Morse complex with singular (co)homology.*

A simple consequence of this theorem is the following. We may define a pairing

$$C^i \otimes \overline{C}^{n-i} \to \mathbb{R}$$

by $\sum m_\alpha \alpha \otimes \sum n_\alpha \alpha \mapsto \sum m_\alpha n_\alpha$.

Corollary 2.14 *This induces the Poincaré duality pairing*

$$H^i(B; \mathbb{R}) \otimes H^{n-i}(B; \mathbb{R}) \to \mathbb{R}$$

on cohomology.

Perhaps the most important point we encountered here is that the spaces $\mathcal{M}(\alpha, \beta)$ contain definite homological information. To obtain a product structure on $H^*(C^*, f)$ without invoking differential forms on B, one would have to relate the cycles defined using these moduli spaces to Morse cycles. This is done by taking intersection numbers with stable (or unstable) manifolds of complementary dimension, and warrants further study.

We will see how to generalize this to equivariant cohomology in §5.

3 Morse-Bott theory

The theory outlined in §2 has two deficiencies. Proposition 2.1 tells us that «most» functions on a manifold are Morse. However, often one is presented with a natural function which is not Morse. Moreover, this theory is ill-suited for the study of equivariant functions, for in this case, the critical point set will necessarily contain orbits of the group action and so the critical points are generally not isolated.

In this section, we will rectify these deficiencies by studying so-called Morse-Bott functions. Here we allow the critical point sets to occur as positive dimensional submanifolds while requiring that the Hessian be non-degenerate on the normal bundle. Additionally, the equivariant case suggests some natural assumptions on the function which will amount to assuming that the stable and unstable manifolds intersect transversely — this is not the generic case for equivariant functions. Our methods give a complex, defined in terms of the deRham complex of the critical point sets and gradient line spaces, which computes the deRham cohomology of the manifold. As such, this theory will generalize the usual Morse theory of §2. A lengthy overview will describe the complex with the technical details supplied afterwards. Later, we show how cup products may be computed and demonstrate the theory with a few examples. For the sake of exposition, we will often refer to the proofs of the special cases given in §2 when no new ideas are involved.

3.1 Overview

Let $f : B \to \mathbb{R}$ be a function on the closed, oriented Riemannian manifold B, non-degenerate in the sense of Bott; that is, the critical points are parametrized by a submanifold, S, and on S, the Hessian of f defines a fiberwise non-degenerate pairing on the normal bundle. For a component of S, S_α, we use the metric to decompose the normal bundle into the positive and negative eigenspaces of the Hessian, viz.

$$\nu(S_\alpha) = \nu_\alpha^+ \oplus \nu_\alpha^- .$$

As in the special case of §2, we define an index

$$i(\alpha) = \text{ rank } \nu_\alpha^- .$$

Grouping the components of equal index defines $S_i = \cup_{i(\alpha)=i} S_\alpha$. Denote by $\mathcal{M}(S_i, S_j)$ the collection of points in B which are connected by gradient lines to S_i and S_j. The flow ϕ_t induces an $\mathbb{R}$ action on this space whose quotient, denoted $\widetilde{\mathcal{M}}(S_i, S_j)$, parametrizes the space of gradient lines beginning at S_i and ending at S_j. Following the flow as $t \to \pm\infty$ gives lower and upper endpoint maps

$$\begin{aligned} l_j^i &: \widetilde{\mathcal{M}}(S_i, S_j) \to S_j \\ u_j^i &: \widetilde{\mathcal{M}}(S_i, S_j) \to S_i . \end{aligned} \tag{3.1}$$

We will now state our assumptions on f and the metric g. First of all, we require that f be «weakly» self-indexing so that $\widetilde{\mathcal{M}}(S_i, S_j) = \emptyset$ if $j \geq i$ (this is in contrast to f being «strictly» self-indexing meaning $f(S_i) = i$). Secondly, a transversality assumption on the stable and unstable manifolds defined by f enables us to conclude that $\widetilde{\mathcal{M}}(S_i, S_j)$ is a smooth manifold and that the endpoint maps are locally trivial fiber bundles with oriented fibers. We will be more precise momentarily.

An example to keep in mind is the following: Consider S^2 as the unit sphere in $\mathbb{R}^3$ and let $f : S^2 \to \mathbb{R}$ be given by the function $f(x, y, z) = z^2$. The critical points of index 2 are the north and south pole — $S_2 = \{(0, 0, 1), (0, 0, -1)\}$ — while the minima of f are the points on the equator $\{z = 0\} = S_0$. Then the moduli space of gradient lines $\widetilde{\mathcal{M}}(S_2, S_0)$ is diffeomorphic to the disjoint union of two copies of S^1.

We now define our filtered complex. Let $C^{i,j} = \Omega^j(S_i)$ with the operators

$$\partial_r : C^{i,j} \to C^{i+r, j-r+1} : \omega \mapsto \begin{cases} d\omega & \text{when } r = 0 \\ (-1)^j (u_i^{i+r})_* (l_i^{i+r})^* (\omega) & \text{otherwise} \end{cases}$$

where the map $(u_i^{i+r})_*$ is integration along the fiber of the bundle in (3.1). Let

$$C^p = \oplus_{i+j=p} C^{i,j} = \oplus_{i+j=p} \Omega^j(S_i)$$

with boundary operator $\partial = \sum \partial_i$. The complex is filtered by

$$0 \subset C_p^p \subset C_{p-1}^p \subset \cdots \subset C_0^p = C^p$$

where

$$C_k^p = \oplus_{i+j=p, i \geq k} C^{i,j}.$$

The main theorem of this section gives

Theorem 3.1 *The cohomology of the complex* (C^*, ∂) *is isomorphic to the de Rham cohomology of* B*; that is,*

$$H^*(C^*; \partial) = H^*_{\mathrm{dR}}(B).$$

Notice that the theory of §2 is contained within this more general framework. In the case when f is a generic Morse function, the critical points are isolated. Hence $C^i = \Omega^0(S_i)$ is the vector space generated by the critical points of index i. Furthermore, $\partial_r = 0$ for all $r \neq 1$ for dimensional reasons. As before, $\widetilde{\mathcal{M}}(S_{i+1}, S_i)$ is an oriented 0-manifold — that is, a collection of signed points. Integrating over the upper endpoint map simply counts these points with orientations. In this way, we recover the ordinary Morse complex.

3.2 The Morse-Bott complex

We will now make this more precise. First, we introduce the stable and unstable manifolds of the gradient flow and describe the assumptions needed to insure that $\mathcal{M}(S_i, S_j)$ and $\widetilde{\mathcal{M}}(S_i, S_j)$ are smooth and that the endpoint maps are fibrations. As our definition of the boundary operator ∂ involves integration along the fiber of the upper endpoint maps $u_j^i : \widetilde{\mathcal{M}}(S_i, S_j) \to S_i$, we will carefully study the structure of the boundary of $\widetilde{\mathcal{M}}(S_i, S_j)$ (Lemmata 3.3 and 3.4). Stokes' Theorem leads naturally to the fact that $\partial \circ \partial = 0$ (Proposition 3.5). Next we set up a chain map between the ordinary de Rham complex and (C^*, ∂) (Lemma 3.6). Knowledge of the boundary structure of the unstable manifolds enables us to show that this induces a map of filtered complexes inducing an isomorphism of filtered cohomology groups (Theorem 3.8). This proves Theorem 3.1.

Proposition 3.2 *The unstable manifold,* $\mathcal{U}_i$*, of* S_i *which is defined as*

$$\mathcal{U}_i = \{b \in B \mid \lim_{t \to -\infty} \phi_t(b) \in S_i\}.$$

is the image of a one-to-one immersion of ν_i^- *into* B*. There is a smooth endpoint map* $u^i : \mathcal{U}_i \to S_i$ *given by* $u^i(b) = \lim_{t \to -\infty} \phi_t(b)$ *which when restricted to a neighborhood of* S_i *has the structure of a locally trivial fiber bundle.*

Proof. This is shown in the appendix (A.9). □

Define $\mathcal{U}_i^s = (u^i)^{-1}(s)$ for $s \in S_i$. Similarly, the stable manifold $\mathcal{S}_i$ is defined using the flow as $t \to \infty$ and leads to the map $l_i : \mathcal{S}_i \to S_i$. Of course, the moduli space of gradient lines is $\widetilde{\mathcal{M}}(S_i, S_j) = (\mathcal{U}_i \cap \mathcal{S}_j)/\mathbb{R}$. We will make the following assumptions in the sequel:

Assumptions:

1) $\widetilde{\mathcal{M}}(S_i, S_j) = \emptyset$ if $i \leq j$. That is, f is weakly self-indexing.
2) For all i, j and all $s \in S_i$, $\mathcal{U}_i^s$ intersects $\mathcal{S}_j$ transversally.
3) Both the critical submanifolds S_i and their negative normal bundle ν_i^- are orientable for all i.

By the transversality assumption (2) above, the following spaces are manifolds:

$$\mathcal{M}(S_i, S_j) = \mathcal{S}_j \cap \mathcal{U}_i$$
$$\widetilde{\mathcal{M}}(S_i, S_j) = \mathcal{M}(S_i, S_j)/\mathbb{R}$$

where the $\mathbb{R}$ action is given by translation along the flow lines. We have smooth endpoint maps

$$\begin{aligned} l_j^i &: \widetilde{\mathcal{M}}(S_i, S_j) \to S_j \\ u_j^i &: \widetilde{\mathcal{M}}(S_i, S_j) \to S_i. \end{aligned} \tag{3.2}$$

Furthermore, the transversality assumption (2) implies that $u_j^i : \widetilde{\mathcal{M}}(S_i, S_j) \to S_i$ has the structure of a locally trivial fiber bundle.

To define and study our filtered complex, we must first describe the structure of the boundary of the spaces $\widetilde{\mathcal{M}}(S_i, S_j)$ and $\mathcal{U}_i$. This is entirely analogous to §2.

For a sequence $j = i_0 < i_1 < \cdots < i_m = i$, using the fact that the u_j^i induce fibrations, define

$$X_{i_0,i_1,\ldots,i_m} = (l_{i_{m-1}}^{i_m})^* \ldots (l_{i_1}^{i_2})^* \widetilde{\mathcal{M}}(S_{i_1}, S_{i_0})$$

and

$$Y_{i_0,i_1,\ldots,i_m} = (l_{i_{m-1}}^{i_m})^* \ldots (l_{i_0}^{i_1})^* \mathcal{U}_{i_0}.$$

Using the usual notation for fibered products, we may more explicitly write,

$$X_{i_0,i_1,\ldots,i_m} = \widetilde{\mathcal{M}}(S_{i_m}, S_{i_{m-1}}) \times_{S_{i_{m-1}}} \widetilde{\mathcal{M}}(S_{i_{m-1}}, S_{i_{m-2}}) \times_{S_{i_{m-2}}} \cdots \times_{S_{i_1}} \widetilde{\mathcal{M}}(S_{i_1}, S_{i_0})$$
$$Y_{i_0,i_1,\ldots,i_m} = \widetilde{\mathcal{M}}(S_{i_m}, S_{i_{m-1}}) \times_{S_{i_{m-1}}} \widetilde{\mathcal{M}}(S_{i_{m-1}}, S_{i_{m-2}}) \times_{S_{i_{m-2}}} \cdots \times_{S_{i_0}} \mathcal{U}_{i_0}\,.$$

One thinks of the $X_{i_0,i_1,\ldots,i_m}$ as describing the space of «broken» gradient lines from S_i to S_j factorizing through intermediate critical submanifolds as in Lemma 2.5. We may topologize the disjoint unions

$$\widetilde{\mathcal{M}}(S_i, S_j) \qquad \bigcup_{j=i_0<i_1<\cdots<i_m=i} \qquad X_{i_0,i_1,\ldots,i_m}$$

and

$$\mathcal{U}_i \qquad \bigcup_{i_0<i_1<\cdots<i_m=i} Y_{i_0,i_1,\ldots,i_m}$$

by requiring that sequences converge as in Lemma 2.5. As in §2, the following lemma holds.

Lemma 3.3 (Compactification) *The spaces of gradient lines* $\widetilde{\mathcal{M}}(S_i, S_j)$ *and the unstable manifolds* $\mathcal{U}_i$ *may be compactified so that*

$$\partial\overline{\widetilde{\mathcal{M}}}(S_i, S_j) = \bigcup_{j=i_0<i_1<\cdots<i_m=i} X_{i_0,i_1,\ldots,i_m}$$

$$\text{and } \partial\overline{\mathcal{U}_i} = \bigcup_{i_0<i_1<\cdots<i_m=i} Y_{i_0,i_1,\ldots,i_m}.$$

These compactified spaces both have the structure of a manifold with corners — that is, they are locally diffeomorphic to an open set in $U^N = \{(x_i \in R^N | x_i \geq 0 \text{ for all } i\}$. *There is a natural injective immersion of the* $\bar{\mathcal{U}}_i$ *into* B, *which is not usually proper.*

Proof. See Appendix §A.3. It should be noted that the transversality assumption (2) above is necessary. □

The maps $u^i_j : \widetilde{\mathcal{M}}(S_i, S_j) \to S_i$ extend smoothly to $X_{i_0,i_1,\ldots,i_m}$ and agree with $u^{i_m}_{i_{m-1}}$. Similarly for l^i_j. Furthermore, $u^i_j : \partial\widetilde{\mathcal{M}}(S_i, S_j) = \bigcup X_{i_0,i_1,\ldots,i_m} \to S_i$ and has the structure of a bundle map which we denote $u^i_{j,\partial}$ and similarly for $u^i : \partial\mathcal{U}_i \to S_i$.

Before defining our filtered complex we need to choose orientations of the S_i and their normal bundles. This orients all the spaces of gradient lines. Let $(\widetilde{\mathcal{M}}(S_{i+l}, S_i))^s = (u^{i+l}_l)^{-1}(s)$ for $s \in S_{i+l}$.

Lemma 3.4 (Boundary orientation) *The codimension 1 stratum*

$$(l^{i+k}_{i+l})^* \widetilde{\mathcal{M}}(S_{i+l}, S_i) \to (\widetilde{\mathcal{M}}(S_{i+k}, S_{i+l}))^x$$

of $(\widetilde{\mathcal{M}}(S_{i+k}, S_i))^x$ *has boundary orientation*

$$(-1)^{k-l-1}[(\widetilde{\mathcal{M}}(S_{i+k}, S_{i+l}))^x][(\widetilde{\mathcal{M}}(S_{i+l}, S_i))^y].$$

The codimension one stratum

$$(l^k_l)^* \mathcal{U}_l \to (\widetilde{\mathcal{M}}(S_k, S_l))^x$$

of $\bar{\mathcal{U}}^x_k$ *has boundary orientation:*

$$(-1)^{k-l-1}[(\widetilde{\mathcal{M}}(S_k, S_l)^x][\mathcal{U}^y_l].$$

Proof. As in Lemma 2.7. □

We digress briefly to discuss integration along the fiber of a bundle $\pi : E \to C$ with fiber F. We furthermore assume that C is a closed manifold and that the typical fiber is an oriented compact d-dimensional manifold with corners so that $\pi_\partial : \partial E \to C$ is also a bundle with fiber ∂F. Take coordinates x on C and t on F. A form on the total space E may locally be written as a linear combination of forms $\pi^*(\phi)f(x,t)dt_{i_1} \wedge dt_{i_2} \wedge \cdots \wedge dt_{i_r}$ where ϕ is a form on C. Then integration along the fiber, $\pi_* : \Omega^j(E) \to \Omega^{j-d}$ is defined by

$$\pi_*(\pi^*(\phi)f(x,t)dvol_F) = \phi \int f(x,t)dvol_F$$
$$\pi_*(\pi^*(\phi)f(x,t)dt_{i_1} \ldots dt_{i_r}) = 0 \text{ for } r < d.$$

It is important to consider how exterior differentiation is related to π_*. Since Stokes' theorem holds for manifolds with corners, we have for $\omega \in \Omega^j(C)$

$$\pi_*(d\omega) = d\pi_*(\omega) + (-1)^{j-d+1}(\pi_\partial)_*(\omega|_{\partial E}). \tag{3.3}$$

To see that we have defined a complex, we must first show that $\partial^2 = (\sum_r \partial_r)^2 = 0$. This is equivalent to the following proposition.

Proposition 3.5 (Complex) *For each* k, $\sum_{l=0}^{l=k} \partial_{k-l}\partial_l = 0$.

Proof. The fact that $d^2 = 0$ on $\Omega^j(S_i)$ proves the case $k = 0$.

For $k > 0$, we are interested in studying $\partial_k \partial_0 \omega$ for a form $\omega \in \Omega^j(S_i)$. Using (3.3) above and the study of the asymptotics of the space of gradient lines, we arrive at

$$\begin{aligned} \partial_k\partial_0\omega &= (-1)^{j+1}(u_i^{i+k})_*(l_i^{i+k})^* d\omega \\ &= (-1)^{j+1}(u_i^{i+k})_* d(l_i^{i+k})^*\omega \\ &= (-1)^{j+1} d(u_i^{i+k})_*(l_i^{i+k})^*\omega + (-1)^{j+1}(-1)^{j-k}(u_{i,\partial}^{i+k})_*(l_i^{i+k})^*\omega \\ &= -\partial_0\partial_k\omega + (-1)^{k-1}(u_{i,\partial}^{i+k})_*(l_i^{i+k})^*\omega. \end{aligned} \tag{3.4}$$

We now study the boundary integral by considering the codimension one strata of the boundary. These strata fit into the following commutative diagram:

$$\begin{array}{ccccc} & (l_{i+l}^{i+k})^*\widetilde{\mathcal{M}}(S_{i+l},S_i) & \to \widetilde{\mathcal{M}}(S_{i+l},S_i) & \xrightarrow{l_i^{i+l}} & S_i \\ \swarrow & & \downarrow & & \downarrow u_i^{i+l} \\ S_{i+k} & \xleftarrow{u_{i+l}^{i+k}} & \widetilde{\mathcal{M}}(S_{i+k},S_{i+l}) & \xrightarrow{l_{i+l}^{i+k}} & S_{i+l}. \end{array}$$

Integration over the fiber commutes with the pullback of bundles; hence,

$$\begin{aligned}(u_{i,\partial}^{i+k})_*(l_i^{i+k})^*\omega &= \sum_{0<l<k}(-1)^{k-l-1}(u_{i+l}^{i+k})_*(l_{i+l}^{i+k})^*(u_i^{i+l})_*(l_i^{i+l})^*\omega \\ &= \sum_{0<l<k}(-1)^k\partial_{k-l}\partial_l.\end{aligned} \tag{3.5}$$

These two expressions, (3.4) and (3.5), together prove the lemma. □

3.3 Comparison with the de Rham complex

We now claim that this complex computes the deRham cohomology of the manifold B. To do this, we will assume that f is strictly self-indexing ($f(S_i) = i$) and define the sets $B_k = f^{-1}((k - \frac{1}{2}, \infty))$. The essential feature of B_k is that it contains all critical submanifolds of index less than k and their unstable manifolds while containing no critical submanifolds of higher index. When f is not strictly self-indexing, it is possible, using the transversality assumption, to define sets B_k also having this property. This is a simple extension of an idea of Smale [Sm2]. Thus, the self-indexing assumption only simplifies our notation without restricting the generality of the conclusion.

Consider the standard deRham complex $(\Omega^j(B), d)$ filtered by

$$0 \subset \Omega_c^j(B_n) \subset \Omega_c^j(B_{n-1}) \subset \cdots \subset \Omega_c^j(B_{-1}) = \Omega^j(B)$$

where $B_k = f^{-1}((k - \frac{1}{2}, \infty))$ and the subscript denotes compactly supported forms. With $C'^j = \Omega^j(B)$ and $C_k'^j = \Omega_c^j(B_k)$, we will describe a map of filtered complexes

$$\Phi : C'^j \to C^j$$

inducing a map on the filtered cohomology groups. To define $\Phi_k : C'^j \to C^{k,j-k}$, recall that $u^k : \mathcal{U}_k \to S_k$ has the structure of a locally trivial bundle. For $\omega \in C'^j = \Omega^j(B)$, let

$$\Phi_k(\omega) = (u^k)_*(\omega|_{\mathcal{U}_k}).$$

Then $\Phi = \oplus\Phi_k$.

Lemma 3.6 (Chain map) *Φ is a map of filtered complexes — that is, $\Phi \circ d = \partial \circ \Phi$.*

Proof. The statement of the lemma is equivalent to

$$\Phi_k(d\omega) = \sum_{l=0}^{l=k} \partial_{k-l}\Phi_l(\omega).$$

The codimension one stratum of the boundary $\partial\mathcal{U}_k$ is given by $Y_{l,k}$ with $l < k$. We rewrite $Y_{l,k} = (l_l^k)^*\mathcal{U}_l$. Then

$$\begin{aligned}\Phi_k(d\omega) &= (u^k)_*(d\omega)\\ &= d(u^k)_*(\omega) + (-1)^{j-k-1}(u_\partial^k)_*(\omega)\\ &= d(u^k)_*(\omega) + (-1)^{j-k-1}\sum_{l<k}(-1)^{k-l-1}(u_l^k)_*(l_l^k)^*(u^l)_*(\omega)\\ &= \partial_0\Phi_k(\omega) + (-1)^{j-k-1}\sum_{l<k}(-1)^{k-l-1}(-1)^{j-l}\partial_{k-l}\Phi_l(\omega).\end{aligned}$$

Furthermore $\Phi_{k-1}(\omega) = 0$ if $\omega \in \Omega_c^j(B_k)$ so that $\Phi : C_k'^j \to C_k^j$. This shows that Φ defines a map of filtered complexes. □

The previous lemma enables us to say that Φ induces a map on the filtered cohomology groups of the complexes. In fact, this map is an isomorphism as will be proven in the next theorem. First, we review how a filtered complex induces a spectral sequence (see [BT]). Let K be a complex filtered by

$$\cdots \subset K_n \subset \cdots \subset K_1 \subset K_0 = K = K_{-1} = \dots$$

and let $GK_n = K_n/K_{n+1}$ be the associated graded complex. The short exact sequence

$$0 \to K_{n+1} \to K_n \to GK_n \to 0$$

induces the long exact sequence in cohomology

$$\cdots \to H^p(K_{n+1}) \to H^p(K_n) \to H^p(GK_n) \to \cdots$$

This exact couple results in a spectral sequence whose E_1 term is $E_1^{n,p} = H^p(GK_n)$. The naturality of this construction implies that a map between filtered complexes induces a map on the spectral sequences. The following algebraic lemma will be useful in what follows.

Lemma 3.7 *Let $f : K^1 \to K^2$ be a chain map of filtered complexes. If f induces an isomorphism of the E_1 term of the associated spectral sequences, then f induces an isomorphism on homology.*

Proof. Since f induces an isomorphism of the E_1 terms of the spectral sequences, it must also induces an isomorphism of the E_∞ terms. However, the E_∞ term is the associated graded group of the homology of the complex. The result follows by noting that a homomorphism of filtered groups which is an isomorphism on the associated graded groups must be an isomorphism. □

We are now ready to prove our main theorem (Theorem 3.1). For the remainder of this section, we will use real coefficients without explicit notation. Let $N_k = f^{-1}((-\infty, k - \frac{1}{2}])$ so that the cohomology of the complex $H^p(\Omega_c^*(B_k), d) = H^p(B, N_k)$. Thus the filtration on the deRham complex induces a filtration on $H^p(B)$ given by the image subgroups $(H^p(B, N_k) \to H^p(B))$. The spectral sequence obtained by filtering forms on B by level was studied by Fary [Fa] and is described in [Bo].

Theorem 3.8 *Φ induces an isomorphism of filtered cohomology groups; hence*

$$\Phi : H^p(C_k^*, \partial) \simeq \text{ image } (H^p(B, N_k) \to H^p(B)).$$

Theorem 3.1 appears as a special case.

Proof. We begin by showing that Φ induces an isomorphism on the E_1 term of the spectral sequences. The associated graded complexes of the two sequences are given by

$$GC_k^p = \Omega^{p-k}(S_k)$$
$$\text{and} \qquad GC_k'^p = \Omega_c^p(B_k)/\Omega_c^p(B_{k+1}).$$

On the E_1 term, Φ induces $\Phi : H^p(GC_k') \to H^p(GC_k) = H^{p-k}(S_k)$. Recall that this map is defined by $\Phi(\omega) = (u_k)_*(\omega|_{\mathcal{U}_k})$ for $\omega \in \Omega_c^p(B_k)$.

Let $F_k = B_k \cap \mathcal{U}_k$ be a neighborhood of S_k in $\mathcal{U}_k$. By the stable manifold theorem A.9, F_k may be identified with the disk bundle D_k^- in the negative normal bundle ν_k^-. The map Φ factors as

$$H^p(GC_k') \to H_c^p(F_k) \to H^{p-k}(S_k)$$

where the first map is induced by restriction of forms $\Omega_c^p(B_k) \to \Omega_c^p(F_k)$ and the second map is the Thom isomorphism given by integrating over the fibration $F_k \to S_k$. To show that Φ is an isomorphism, is suffices to show that restriction of forms induces an isomorphism $H^p(GC_k') \to H_c^p(F_k)$ as follows.

The short exact sequence

$$0 \to \Omega_c^p(B_{k+1}) \to \Omega_c^p(B_k) \to GC_k'^p \to 0$$

induces a long exact sequence in cohomology related to the exact sequence of the triple (B, N_{k+1}, N_k) as

$$\begin{array}{ccccccc}
\dots H_c^p(B_{k+1}) & \to & H_c^p(B_k) & \to & H^p(GC_k') & \to \dots \\
\downarrow & & \downarrow & & \downarrow & \downarrow \\
\dots H^p(B, N_{k+1}) & \to & H^p(B, N_k) & \to & H^p(N_{k+1}, N_k) & \to \dots
\end{array}$$

By the Five Lemma, $H^p(GC_k') \to H^p(N_{k+1}, N_k)$ is an isomorphism. Furthermore, the gradient flow defines a deformation retract of the pair (N_{k+1}, N_k)

onto $(F_k \cup N_k, N_k)$ (as in [Mil] for the case of nondegenerate critical points). Hence, $H^p(N_{k+1}, N_k) \to H^p(F_k, \partial F_k) = H_c^p(F_k)$ is an isomorphism by excision. Putting these together implies that restriction of forms gives an isomorphism $H^p(GC'_k) \to H_c^p(F_k)$ as claimed.

Now that Φ induces an isomorphism on the E_1 term of the spectral sequences, it must induce an isomorphism on cohomology by Lemma 3.7. □

Without the transversality and self-indexing assumptions, the unstable manifolds $\{\mathcal{U}_i\}$ still form a stratification of B. From this, we could compute the cohomology of B resulting in a theory over the integers. However, as this method does not generalize to infinite dimensions — the stable manifolds need not form a stratification — we will not pursue this course.

As in §2, the existence of the Morse-Bott complex implies the more general Morse inequalities of [B3] and [Bi]:

Corollary 3.9 *There is a polynomial $Q(t)$ with non-negative coefficients such that*

$$\sum_{i,j} \dim H^j(S_i; \mathbb{R}) t^{i+j} = \sum_k \dim H^k(B; \mathbb{R}) t^k + (1+t)Q(t).$$

A Morse-Bott function f is called *perfect* if $Q(t) = 0$. If either the higher differentials vanish in the spectral sequence or the components of the boundary operator ∂_k for $k > 0$ are zero, then f is perfect. This is the case in Bott [B4].

3.4 Comparison with the Morse complex

Remember that Proposition 2.1 states that a Morse function necessarily lies arbitrarily close to our given Morse-Bott function f. It is instructive to prove how the two complexes are related. This is in fact not much different from the comparison with the de Rham complex.

Choose tubular neighborhoods T_i of the critical submanifolds S_i and generic Morse functions f_i on S_i. Let ρ_i be bump functions which are identically 1 near S_i and identically zero outside of T_i. Define

$$f_\epsilon = f + \epsilon \sum_i \rho_i f_i.$$

For sufficiently small ϵ, f_ϵ is a nondegenerate Morse function. We denote the Morse complex that it defines by $(C_\epsilon^*, \partial_\epsilon^*)$. We will describe a chain map between (C^*, ∂) and $(C_\epsilon^*, \partial_\epsilon^*)$ which induces an isomorphism on cohomology. Fintushel and Stern [FS] have used this idea in computing Floer homology groups for Seifert fibered homology 3-spheres.

The critical points of f_ϵ are exactly the critical points of f_i; moreover, if α is a critical point of index j for f_i, it is critical of index $i+j$ for f_ϵ. Denote the unstable manifold of α in S_i by $\mathcal{U}_\alpha^i$.

Proposition 3.10 *For small ϵ, the map $F : (C^*, \partial) \to (C^*_\epsilon, \partial^*_\epsilon)$ defined by*

$$F(\omega) = (\int_{\mathcal{U}^i_\alpha} \omega)\alpha,$$

where $\omega \in \Omega^j(S_i)$, is a chain morphism inducing an isomorphism on cohomology.

Proof. Given critical points α and β of f_ϵ, denote the moduli space of gradient lines of f_ϵ by $\widetilde{\mathcal{M}}_\epsilon(\alpha, \beta)$. Showing that F is a chain map is equivalent to checking

$$F(\partial_k \omega) = \sum \#\widetilde{\mathcal{M}}_\epsilon(\beta, \alpha)\beta$$

where the sum is taken over the critical points β of f_{i+k} of index $j - k + 1$. For $k = 0$, this is just Proposition 3.6. The proof for $k > 0$ involves the asymptotics of the gradient line moduli spaces. Let p denote the minima of f_i so that

$$\mathcal{M}_\epsilon(\beta, p) = u^{-1}_{i+j}(\mathcal{U}^{i+j}_\beta) \cap l^{-1}_i(\mathcal{S}^i_p).$$

Then the boundary of $\mathcal{M}_\epsilon(\beta, p)$ contains factorizations

$$u^{-1}_{i+j}(\mathcal{U}^{i+j}_\beta) \cap \widetilde{\mathcal{M}}(S_{i+j}, S_i) \bigcup \widetilde{\mathcal{M}}_\epsilon(\beta, \alpha) \times \mathcal{U}^i_\alpha.$$

Moreover the integral $\int_{\mathcal{M}_\epsilon(\beta,p)} l^*_i \, d\omega$ vanishes since the form has no component along the flow lines. Putting this together implies that

$$\begin{aligned}
0 &= \int_{\mathcal{M}_\epsilon(\beta,p)} l^*_i \, d\omega \\
&= \int_{\partial\mathcal{M}_\epsilon(\beta,p)} \omega \\
&= \int_{\mathcal{U}^{i+j}_\beta} ((u_{i+j})_* l^*_i \omega)|_{\mathcal{U}^{i+j}_\beta} - \#\widetilde{\mathcal{M}}_\epsilon(\beta, \alpha) \int_{\mathcal{U}^i_\alpha} \omega|_{\mathcal{U}^i_\alpha} \\
&= F(\partial_k \omega) - \partial^*(F(\omega)).
\end{aligned}$$

This implies that F is a chain map.

Moreover, both complexes are filtered by the index of the critical submanifolds of the original Morse-Bott function f and F is a chain map of filtered complexes. It follows that the E_1 of both spectral sequences is $H^j(S_i)$ and by Proposition 3.1, F induces an isomorphism on E_1 and hence on the homology of the complexes (Lemma 3.7). □

3.5 Cup products

Here we will generalize the discussion on cup products in §2 to include the Morse-Bott case and a proof will be given for Theorem 2.13. The idea is that the cup product can be realized by integrating over moduli spaces of gradient lines.

For $\omega \in \Omega^k(B)$, define

$$c(\omega) : C^l \to C^{l+k}$$

in terms of its decomposition $c(\omega) = \oplus_{m \geq 0} c(\omega)_m$ where

$$c(\omega)_m : \Omega^j(S_i) \to \Omega^{j-m+k}(S_{i+m}).$$

For $m = 0$, $c(\omega)_0(\alpha) = \omega|_{S_i} \wedge \alpha$. For $m > 0$, define

$$c(\omega)_m(\alpha) = (u^{i+m})_*(\omega|_{\mathcal{M}(S_{i+m},S_i)} \wedge l_i^* \alpha).$$

Studying the gradient line spaces, exactly as in §2.4, gives the chain relation

$$c(d\omega)(\alpha) = \partial(c(\omega)(\alpha)) + c(\omega)(\partial\alpha).$$

Then there is an induced map on cohomology such that the generalization of Theorem 2.13 holds.

Theorem 3.11 *The map*

$$c : H^k_{dR}(B) \otimes H^l(B;\mathbb{R}) \to H^{l+k}(B;\mathbb{R})$$

is the cup product map.

Proof. We will prove this in the case in which f is a nondegenerate Morse function as in §2. The general Morse-Bott case is no more complicated.

The result is proven once we construct a commutative diagram

$$\begin{array}{ccc} H^k_{dR}(B) \otimes H^l_{dR}(B) & \xrightarrow{\wedge} & H^{k+l}_{dR}(B) \\ \downarrow & & \downarrow \\ H^k_{dR}(B) \otimes H^l(C^*,\partial) & \xrightarrow{c} & H^{k+l}(C^*,\partial) \end{array}$$

where the vertical arrows are induced by the isomorphism in Theorem 3.1 — that is, by integration over the unstable manifolds.

Let $\omega \in \Omega^k(B), \eta \in \Omega^l(B)$ and α, β be critical points of f of index l and $l+k$ respectively. To demonstrate the commutative diagram above, we wish to show that the chains $\int_{\mathcal{U}_\beta} \omega \wedge \eta$ and $\int_{\mathcal{M}(\beta,\alpha)} \omega \int_{\mathcal{U}_\alpha} \eta$ define the same cohomology class.

In the spirit of Witten's deformation of the deRham complex [Wi], we consider the integral $\int_{\mathcal{U}_\beta} \phi_{-t}^* \omega \wedge \phi_t^* \eta$. Since we have pulled the forms back by a diffeomorphism isotopic to the identity, they still define the same cohomology classes and hence the integral over the unstable manifold $\mathcal{U}_\beta$ must be cohomologous to $\int_{\mathcal{U}_\beta} \omega \wedge \eta$. In the limit that $t \to \infty$, the support of the forms becomes concentrated around the critical points. Let γ be a critical point of index i and denote the eigenvalues of the Hessian of f at γ by λ_j where $\lambda_j > 0$ for $j \leq i$ and $\lambda_j < 0$ for $j > i$. Set $\rho = \min |\lambda_j|$ and $R = \max |\lambda_j|$. For convenience, assume that around γ there is a coordinate chart $(x_1, \ldots, x_i, y_{i+1}, \ldots, y_n)$ such that the vector field $-\operatorname{grad} f$ is linear so that the gradient flow may be written

$$\phi_t(x,y) = (e^{\lambda_1 t} x_1, \ldots, e^{\lambda_i t} x_i, e^{\lambda_{i+1} t} y_{i+1}, \ldots, e^{\lambda_n t} y_n).$$

Near γ, we write $\omega(x,y) = \omega_{I,J} dx^I \wedge dy^J$ and similarly for η. Then using multi-index notation,

$$\phi_{-t}^* \omega(x,y) = \omega_{I,J}(\phi_{-t}(x,y)) e^{-\lambda_I t} e^{\lambda_J t} dx^I dy^J$$

where $\lambda_K = \sum_{k \in K} \lambda_k$. The form $\phi_t^* \eta$ admits a similar expression. Near the critical point γ, the unstable manifold $\mathcal{U}_\beta$ factorizes as $\mathcal{M}(\beta, \gamma) \times \mathcal{U}_\gamma$. We can then apply Fubini's theorem to estimate

$$\begin{aligned} \left| \int_{\mathcal{U}_\beta} \phi_{-t}^* \omega \wedge \phi_t^* \eta \right| &\leq \int_{\mathcal{M}(\beta,\gamma)} \int_{\mathcal{U}_\gamma} |\omega_{IJ}(\phi_{-t}(x,y)) \eta_{I'J'}(\phi_t(x,y))| e^{(\lambda_{I'} - \lambda_I)t} dx^I dx^{I'} \\ &\qquad e^{(\lambda_J - \lambda_{J'})t} dy^J dy^{J'} \\ &\leq \int_{\mathcal{M}(\beta,\gamma)} \int_{\mathcal{U}_\gamma} |\omega_{IJ}(\phi_{-t}(x,y)) \eta_{I'J'}(\phi_t(x,y))| e^{(R|I'| - \rho|I|)t} dx^I dx^{I'} \\ &\qquad e^{(\lambda_J - \lambda_{J'})t} dy^J dy^{J'}. \end{aligned}$$

Counting dimensions shows that any non-zero terms must satisfy

$$\begin{aligned} |I| + |I'| &= i \\ |J| + |J'| &= l + k - i \\ |I| + |J| &= k \\ |I'| + |J'| &= l. \end{aligned}$$

Making the coordinate change

$$g(x,y) = (e^{\lambda_1 t} x_1, \ldots, e^{\lambda_i t} x_i, y_{i+1}, \ldots, y_n),$$

we may rewrite the inner integral over the unstable manifold $\mathcal{U}_\gamma$. The contribution to each term from the inner integral then becomes bounded by

$$\int_{\mathcal{U}_\gamma} |\omega_{IJ}(g(\phi_{-t}(x,y)))\eta_{I'J'}(g(\phi_t(x,y)))| e^{(-(\rho+R)|I|)t} dx^I dx^{I'}.$$

As t grows, the contributions from terms with $|I| > 0$ become increasingly small and can hence be neglected. For this reason, we can suppose that $|I| = 0$. A similar argument shows that the only significant contribution is from the term with $|J'| = 0$. Putting together the equalities above, we conclude that the index of γ must be l and the integral becomes

$$\int_{\mathcal{M}(3,\gamma)} \omega \int_{\mathcal{U}_\gamma} \eta.$$

This proves the result when the vector field is linear about a critical point. In general, this is not so; however, we may use the estimates in Lemma A.3 to bound the corrections arising from nonlinearity of the vector field. The same argument then holds. □

Furthermore, the Poincaré pairing can be understood by reversing the gradient flow as in Corollary 2.14. First suppose that $-f$ is a Morse-Bott function satisfying the assumptions above. Then $-f$ defines a Morse-Bott complex $(\overline{C}^*, \overline{\partial})$ which again computes the cohomology of B. Then we have

Corollary 3.12 *The Poincaré duality pairing $H^k(B;\mathbb{R}) \otimes H^{n-k}(B;\mathbb{R}) \to \mathbb{R}$ can be expressed in terms of the Morse-Bott complex*

$$C^k \otimes \overline{C}^{n-k} \to \mathbb{R}$$

by $\omega_i \otimes \eta_j \to \int_{S_i} \omega_i \wedge \eta_i$.

Proof. The key observation is that the unstable manifold, $\overline{\mathcal{U}_i}$, of the critical submanifold S_i under the function $-f$ is (up to factorizations) just the gradient line space $\mathcal{M}(S_k, S_i)$ where S_k runs over all the local maxima for f. Then, arguing as in Theorem 3.11 shows that

$$\begin{aligned}\int_B \omega \wedge \eta &= \int_{\mathcal{M}(S_k,S_i)} \omega \int_{\mathcal{U}_i} \eta \\ &= \int_{\overline{\mathcal{U}_i}} \omega \int_{\mathcal{U}_i} \eta.\end{aligned}$$

□

3.6 Examples

Though the Morse-Bott complex may appear somewhat daunting, it is often quite easy to compute. In this section, we will present some examples to bolster this claim.

1. Consider the two-sphere $S^2 = \{(x,y,z) \in \mathbb{R}^3 | x^2+y^2+z^2 = 1\}$ with the Morse-Bott function $f(x,y,z) = z^2$. There are two critical points, p and q, of index 2, while the equatorial S^1 is critical of index 0. Hence the E_1 term of the spectral sequence is

$$\begin{array}{cccc} 0 & 0 & 0 & 0 \\ \mathbb{R} & 0 & 0 & 0 \\ \mathbb{R} & 0 & \mathbb{R}^2 & 0 \end{array} .$$

The differential $d_2 : \mathbb{R} \to \mathbb{R}^2$ is given by integrating the volume form on S^1 along the moduli spaces $\widetilde{\mathcal{M}}(p, S^1)$ and $\widetilde{\mathcal{M}}(q, S^1)$ which are both circles. Then $d_1(\omega) = (1, -1)$ and the E_∞ term of the spectral sequence looks like

$$\begin{array}{cccc} 0 & 0 & 0 & 0 \\ 0 & 0 & 0 & 0 \\ \mathbb{R} & 0 & \mathbb{R} & 0 \end{array} .$$

2. Consider $\mathbb{C}\mathbb{P}^n = S^{2n+1}/S^1$ with coordinates $[z_0 : z_1 : \ldots : z_n]$. Define the Morse-Bott function

$$f([z_0 : z_1 : \ldots : z_n]) = |z_n|^2.$$

There is a critical point $p = [0 : 0 : \ldots : 1]$ of index $2n$ and a critical submanifold $\mathbb{C}\mathbb{P}^{n-1} = [z_0 : z_1 : \ldots : z_{n-1} : 0]$ of index 0. Then the E_1 term of the spectral sequence looks like

$$\begin{array}{ccccc} 2n-2 & \mathbb{R} & 0 & \ldots & \\ 2n-3 & 0 & 0 & \ldots & \\ 2n-4 & \mathbb{R} & 0 & \ldots & \\ \vdots & \vdots & & & \\ 0 & \mathbb{R} & 0 & \ldots & \mathbb{R} \\ & 0 & 1 & \ldots & 2n \end{array} .$$

Notice that all higher differentials vanish for dimension reasons. This gives a simple way to inductively compute the cohomology of complex projective spaces. By changing f to $-f$, one can easily obtain the cup product structure.

3. Let $\pi : E \to B$ be a fiber bundle with fiber F and let f be a nondegenerate Morse function on B. Pulling back to E gives a Morse-Bott function $\tilde{f} = \pi^* f$ on E. The critical submanifolds of $\tilde{f}$ are the fibers over the critical points of f and hence all diffeomorphic to F. Moreover, the index of a fiber is exactly the index of the critical point over which it lies. The Morse-Bott complex then gives a spectral sequence whose E_2 term is the cohomology $H^*(F) \otimes H^*(B)$. The resulting spectral sequence is the Leray spectral sequence for the fibration (see [BT] and [Bo]).

As an example, consider the Hopf fibration $S^3 \to S^2$ with fibers S^1. Let f be the height function on S^2 with two critical points, one at the north pole of index 2 and one at the south pole of index 0. Now the E_2 term of the spectral sequence given by $\tilde{f}$ is

$$\begin{matrix} \mathbb{R} & 0 & \mathbb{R} \\ \mathbb{R} & 0 & \mathbb{R} \end{matrix}.$$

The differential $d_2 : \mathbb{R} \to \mathbb{R}$ precisely measures the Euler class of the bundle. This is a general phenomena; for k-sphere bundles, the differential d_{k+1} will measure the Euler class.

4. There is one example of a fiber bundle which is relevant to our study of equivariant cohomology in the next sections. Let X be a manifold with G-action and $EG \to BG$ the classifying bundle for G. As we will see in the next section, the homotopy quotient, defined as $X_G = EG \times_G X$, forms a fiber bundle $\pi : X_G \to BG$ with fiber X. The cohomology of X_G is called the equivariant cohomology of X. If f is a Morse function on BG, then $\pi^* f$ is a Morse-Bott function on X_G as in example 3 whose critical submanifolds are all diffeomorphic to X. The Morse-Bott complex then computes the equivariant cohomology of X.

For example, suppose that $G = SU(2)$. In this case, the classifying bundle is $S^\infty \to \mathbb{HP}^\infty$. On $\mathbb{HP}^\infty$, we may take a perfect Morse function whose critical points all have index divisible by 4. The only non-zero components of the boundary operator are ∂_0 and ∂_4. Of course, ∂_0 is just the usual exterior derivative on $\Omega^*(X)$. The component $\partial_4 : \Omega^{j+3}(X) \to \Omega^j(X)$ is easily described. Given critical submanifolds X_{i+4} and X_i, the moduli space of gradient lines is

$$\widetilde{\mathcal{M}}(X_{i+4}, X_i) = X \times SU(2).$$

The upper endpoint map is simply projection onto X while the lower endpoint map is given by the group action: $l_i(x,g) = g \cdot x$. Then if $\omega \in \Omega^{j+3}(X)$ and M is a submanifold of X of dimension j,

$$\int_M \partial_4(\omega) = \int_{G\cdot M} \omega$$

where $G \cdot M = \{g \cdot m | g \in G, m \in M\}$. In particular, if $\omega \in \Omega^3(X)$, then $\partial_4(\omega)(x) = \int_{G\cdot x} \omega$.

Many properties of equivariant cohomology can be understood geometrically in terms of this picture. In the next sections, we will build a different complex to compute equivariant cohomology.

4 Equivariant cohomology

Here we review the basic definitions and standard results about equivariant cohomology. This will be useful in the next section where we discuss the computation of equivariant cohomology using critical submanifolds and gradient flows as in the previous sections.

Let G a compact Lie group and denote the universal G-fibration by $EG \to BG$. If X is a space with a G-action, the space $X_G \equiv EG \times_G X$, called the homotopy quotient, is a fiber bundle over BG with fiber X. The G-equivariant cohomology of X is defined by

$$H_G(X;\Gamma) = H^*(X_G;\Gamma).$$

Moreover, pullback by the fibration $X_G \to BG$ gives $H_G(X;\Gamma)$ the structure of a module over the ring $H^*(BG;\Gamma)$. This module structure will be investigated more fully in §6.

The following examples illustrate some properties of equivariant cohomology. First of all $H_G(\{point\};\mathbb{Z}) = H^*(BG;\mathbb{Z})$. If the G-action on X is free, then $H_G(X;\mathbb{Z}) = H^*(X/G;\mathbb{Z})$ and the equivariant cohomology of an orbit is given by $H_G(G/K;\mathbb{Z}) = H^*(BK;\mathbb{Z})$ for a closed subgroup K.

The equivariant cohomology of a space intertwines cohomological information from the quotient with the subgroups appearing as stablizers of points. When G acts freely on X, the module structure is easily understood.

Proposition 4.1 *Suppose G acts freely on X, then the $H^*(BG;\mathbb{Z})$ module structure of $H_G(X;\mathbb{Z}) = H^*(X/G)$ is described as follows: if $\phi \in H^*(BG;\mathbb{Z})$ and $\omega \in H^*(X/G)$, then*

$$\phi \cdot \omega \mapsto c^*(\phi) \cup \omega,$$

where $c : X/G \to BG$ is the classifying map for the principal G-bundle $X \to X/G$.

Observe that the image of $c^* \subset H^*(X/G;\mathbb{Z})$ consists of the characteristic classes of the G-bundle $X \to X/G$.

There is a convenient model with which we will consider equivariant cohomology: the equivariant cohomology with real coefficients can be computed from a deRham like complex of differential forms. The version which we discuss is due to Cartan [C] and we refer to the excellent book by Berline, Getzler and Vergne [BGV]. Denote the Lie algebra of G by $\mathfrak{g}$. Define the so-called equivariant differential forms

$$\Omega_G(X) = (\Omega^*(X) \otimes S^*(\mathfrak{g}^*))^G,$$

where G acts on its Lie algebra by conjugation. Linear polynomials on $\mathfrak{g}$ have degree two, and this turns $\Omega_G(X)$ into a bigraded complex. Pick a basis X_α of $\mathfrak{g}$ with dual basis ϕ_α. The differential is a graded derivation defined on the generators by

$$d_G\omega = d\omega - \sum_\alpha \phi^\alpha i_{\tilde{X}_\alpha}\omega, \qquad d_G\phi^\alpha = 0$$

for $\omega \in \Omega^*(X)$ and $\tilde{X}_\alpha$ the vector field on X induced by $X_\alpha \in \mathfrak{g}$. Cartan shows that this defines a complex and:

Proposition 4.2 $H_G(X;\mathbb{R}) = H^*(\Omega_G(X), d_G)$.

Proof. The isomorphism is induced by the Chern-Weil homomorphism which we now describe. Let $\pi : P \to M$ be a principal G-bundle. The pullback map π^* identifies differential forms on M, $\Omega^*(M)$, with *basic* differential forms on P,

$$\Omega^*_{\text{basic}}(P) = \{\eta \in \Omega^*(P) | \eta \text{ is horizontal and } G\text{-invariant}\}.$$

Now consider the classifying G-bundle $\pi : EG \to BG$ with a connection whose curvature is $\Omega \in \Omega^2(EG; \mathfrak{g})$. For a G-manifold X, define a map

$$\chi : \Omega^*_G(X) \to \Omega^*_{\text{basic}}(EG \times X) : \omega \otimes \phi) \mapsto \phi(\Omega) \otimes \omega.$$

Using the identification of basic forms on $EG \times X$ with forms on X_G, we have an induced map $\chi : \Omega^*_G(X) \to \Omega^* X_G$. The differential d_G was defined so that χ is a chain map. The induced map on cohomology is the isomorphism which gives the proposition. □

When X is a point, we see the well known fact $H^*(BG;\mathbb{R}) = S^*(\mathfrak{g}^*)^G$, which elegantly demonstrates the module structure of $H^*_G(X;\mathbb{R})$.

It is important to notice that the Chern-Weil homomorphism is natural with respect to pullback under equivariant maps and integration over equivariant fibrations. Suppose $f : X \to Y$ is a G-equivariant map. Then there is an induced map $f^* : \Omega^*_G(Y) \to \Omega^*_G(X)$ by $f^*(\omega \otimes \phi) = f^*(\omega) \otimes \phi$ which commutes with the Chern-Weil homomorphism

$$\begin{array}{ccc} \Omega^*_G(Y) & \to & \Omega^*(Y_G) \\ \downarrow & & \downarrow \\ \Omega^*_G(X) & \to & \Omega^*(X_G). \end{array}$$

We digress briefly to discuss integration along the fiber of an equivariant fiber bundle. Let $E \to B$ be a G-equivariant fiber bundle with fiber F. Furthermore, let $\widetilde{X}$ denote the vector fields on E and B induced by an element of the Lie algebra $\mathfrak{g}$. We claim that integration along the fiber commutes with contraction by $\widetilde{X}$; that is,

$$\pi_*(i_{\widetilde{X}}\omega) = i_{\widetilde{X}}\pi_*(\omega). \tag{4.1}$$

This follows since if $\omega = \pi^*\phi f(x,t) dvol_F$, then

$$i_{\widetilde{X}}(\pi^*(\phi) f(x,t) dvol_F) = \pi^*(i_{\widetilde{X}}\phi) f(x,t) dvol_F + (-1)^{\deg \phi}\pi^*(\phi) f(x,t) i_{\widetilde{X}} dvol_F.$$

Integrating along the fiber kills the second term while the first term integrates to $i_{\widetilde{X}}\pi_*(\omega)$. There is a map $h_* : \Omega^*_G(E) \to \Omega^{*-n}_G(B)$, where n is the dimension of the fiber F, induced by

$$h_*(\omega \otimes \phi) = h_*(\omega) \otimes \phi.$$

By (4.1), h_* induces a map, integration over the fibers, on equivariant cohomology. Notice that $\chi(h_*(\omega \otimes \phi)) = \phi(\Omega) \otimes h_*(\omega) = h_*(\chi(\omega \otimes \phi))$ so that h_* commutes with the Chern-Weil homomorphism.

In the case when $\pi : E \to B$ is an oriented vector bundle over a compact manifold of rank k, the Thom isomorphism $H_c^{*+k}(E;\mathbb{R}) \to H^*(B;\mathbb{R})$ can be realized by integration over the fiber — that is, $\omega \to \pi_*(\omega)$. An inverse is constructed using the Thom class $u \in H_c^k(E)$ to define $\omega \to u \wedge \pi^*\omega$. The class u is characterized by the fact that it is closed and has integral 1 over each fiber of π. Beautiful representatives for the Thom class were found by Matthai and Quillen [MQ] (see also [BGV]) in which the Thom form is acquired by substitution of the curvature of a connection on E into a standard formula.

The Thom isomorphism also holds in the more general setting of equivariant cohomology. Let $E \to B$ be an oriented G-equivariant vector bundle of rank k. We define $\Omega^*_{G,c}(E) = (\Omega^*_c(E) \otimes S^*(\mathfrak{g}^*))^G$ and a map $\pi_{G,*} : \Omega^{*+k}_{G,c}(E) \to \Omega^*_G(B)$ by $\pi_{G,*}(\omega \otimes \phi) = \pi_*(\omega) \otimes \phi$. As before, this induces a map on equivariant cohomology $H^{*+k}_{G,c}(E) \to H^*_G(B)$. Constructing an equivariant Thom form will give the inverse map as above. Let $P \to B$ be the G-equivariant principal K-bundle associated to E and choose a G-invariant connection A. The equivariant curvature $F^A_G \in \Omega^2_G(P, Lie(K))$ is defined as follows (see [BGV]). Fixing a point $p \in P$, an element $\psi_p \in \mathfrak{g}^* \otimes Lie(K)$ can be produced by assigning to a vector $X \in \mathfrak{g}$ the vector

$$A_p(\tilde{X}) \in Lie(K)$$

where $\tilde{X}$ is the infinitesimal action of X on P. The equivariant curvature is now:

$$F^A_G = F^A - \psi.$$

The usual substitution into the formula of [MQ] produces a Thom form and proves:

Proposition 4.3 (The Thom Isomorphism) *Given a G-equivariant vector bundle $E \to B$ of rank k, integration over the fiber defines an isomorphism $H^{*+k}_{G,c}(E;\mathbb{R}) \to H^*_G(B;\mathbb{R})$.*

5 Equivariant cohomology and Morse-Bott theory

Many examples of Morse-Bott functions arise as invariant functions on a manifold with group action. Here the critical point sets are invariant under the group action and so must necessarily contain orbits. The theory in §3 tells us how to build a complex which computes the deRham cohomology of the manifold. It is but a slight generalization, using equivariant differential forms rather than ordinary differential forms, to compute the equivariant cohomology of the manifold. We take up this question in this section whose outline closely follows that of §3.

5.1 The equivariant Morse complex

Let X be a compact G-manifold. An interesting theorem of Wasserman [Wa] asserts that the G-invariant functions on X with isolated non-degenerate critical orbits form a Baire set. Let $f : X \to \mathbb{R}$ be a G-invariant function with nondegenerate critical orbits. The critical point set of index i of such a function is of the form

$$S_i = \cup_\alpha G/K^i_\alpha$$

for certain subgroups K^i_α which are the stabilizers of these critical orbits. Following the notation of §3, $\mathcal{M}(S_i, S_j)$ denotes the set of points connected by a gradient line to S_i and S_j, $\widetilde{\mathcal{M}}(S_i, S_j) = \mathcal{M}(S_i, S_j)/\mathbb{R}$, and the lower (upper) endpoint maps are l^i_j $(u^i_j) : \mathcal{M}(S_i, S_j) \to S_j$ (S_i). Notice that these maps are equivariant with respect to the G-action.

We shall now assume that the stable and unstable manifolds intersect transversally. This is not an assumption which will be satisfied generically for equivariant non-degenerate functions (see [P]), but it is often satisfied by interesting group actions. For instance, if the complement of the critical orbits contains only one orbit type, then the usual obstructions to equivariant transversality disappear and we can make this assumption. Observe that due to equivariance, $\mathcal{M}(S_i, S_j) = \emptyset$ for $j > i$, i.e. our weakly self-indexing assumption is implied by the transversality assumption. Similarly the assumption that the endpoint maps induce fibrations is an immediate consequence of the presence of a transitive G-action on the components of the critical point set.

We will form a complex which computes the equivariant cohomology of X, again using equivariant forms on the critical submanifolds. Using the Cartan model for equivariant cohomology, we define a filtered complex (C^*, ∂_G) with

$$C^p = \oplus_{i+l=p}\Omega^l_G(S_i) = \oplus_{i+j+2k=p}(\Omega^j(S_i) \otimes S^k(\mathfrak{g}^*))^G.$$

The filtration is given by

$$C^p_n = \oplus_{i+l=p, i\geq n}\Omega^l_G(S_i)$$

so that

$$C^p = C^p_0 \subset C^p_1 \subset \cdots \subset C^p_n \subset \cdots$$

As before, the boundary operator $\partial_G : C^p_n \to C^{p+1}_n$ splits into $\partial_G = \sum_k(\partial_G)_k$ where $(\partial_G)_k : C^p_n \to C^{p-k+1}_{n+k}$ is a map of $S^*(\mathfrak{g}^*)^G$ modules as follows: for $\omega \otimes \phi \in (\Omega^j(S_i) \otimes S^j(\mathfrak{g}^*))^G$, let $(\partial_G)_0(\omega \otimes \phi) = d_G(\omega \otimes \phi)$ and for $k > 0$, let

$$(\partial_G)_k(\omega \otimes \phi) = \partial_k\omega \otimes \phi.$$

The following theorem shows that the filtered complex introduced above is in fact a complex.

Theorem 5.1 $\partial_G \circ \partial_G = 0$.

Proof. The proof closely follows that of Theorem 3.5. Again we must show that

$$\sum_l (\partial_G)_l (\partial_G)_{k-l} = 0$$

for all fixed k. The case that $k = 0$ is trivial so we concentrate on $k > 0$. From our definitions,

$$\sum (\partial_G)_l (\partial_G)_{k-l}(\omega \otimes \phi) = \sum (\partial_l \partial_{k-l}\omega) \otimes \phi \\ - \sum i_{\tilde{X}_\alpha} \partial_k \omega \otimes \phi\phi^\alpha - \sum \partial_k i_{\tilde{X}_\alpha} \omega \otimes \phi\phi^\alpha.$$

As the first sum vanishes by Proposition 3.5, our investigation focuses on the remaining terms. The theorem is proven if we have $i_{\tilde{X}_\alpha} \partial_k \omega + \partial_k i_{\tilde{X}_\alpha} \omega = 0$ which is equivalent to

$$i_{\tilde{X}_\alpha} (u_{i+k})_* l_i^*(\omega) = (u_{i+k})_* l_i^* i_{\tilde{X}_\alpha}(\omega).$$

This is true precisely because integration along the fiber and pullback commute with contraction by $\tilde{X}$ for equivariant maps. □

Remark 5.2 *If $\mathcal{I}$ is an ideal in $S^*(\mathfrak{g})^G$, then we can also consider the complex $(C^*/\mathcal{I}, \partial_G)$. This has a variety of interesting applications, one of which we shall discuss below.*

Still following the outline of §3, we now show that the cohomology of our filtered complex is the equivariant cohomology. Again we filter the equivariant forms on X. Because the transversality assumption is satisfied, there is no loss of generality in assuming the invariant function to be strictly self-indexing (that is, $f(S_i = i)$), and so let $X_n = f^{-1}((n - \frac{1}{2}, \infty))$. Notice that X_n is a G-space since f is G-invariant. Define

$$C'^* = \Omega_G^*(X) = (\Omega^*(X) \otimes S^*(\mathfrak{g}^*))^G$$

with the filtration

$$C_n'^* = \Omega_{G,c}^*(X_n).$$

With the usual boundary operator d_G, (C'^*, d_G) forms a filtered complex.

We define a map of filtered complexes

$$\Psi : (C'^*, d_G) \to (C^*, \partial_G) : \Psi(\omega \otimes \phi) \mapsto \Phi(\omega) \otimes \phi.$$

Recall the chain morphism $\Phi = \oplus \Phi_k$ where $\Phi_k : \Omega^p(X) \to \Omega^{p-k}(S_k) : \omega \mapsto (u_k)_*(\omega|_{\mathcal{U}_k})$, integration along the unstable manifold of S_k. To show that Ψ is a map of complexes, we must show $\Psi \circ d_G = \partial_G \circ \Psi$. Applying Lemma 3.6, this is equivalent to

$$\Phi_k(i_{\tilde{X}}\omega) = i_{\tilde{X}} \Phi_k(\omega)$$

which follows easily from the fact that contraction by $\widetilde{X}$ commutes with integration along the fiber of an equivariant bundle. Notice that Ψ is a map of $S^*(\mathfrak{g}^*)$ modules.

In analogy with Theorem 3.1, the main theorem of this section is

Theorem 5.3 *The map Ψ induces an isomorphism of filtered complexes. In particular, $\Psi : H^*_G(X;\mathbb{C}) \to H^*(C^*, \partial_G)$ is an isomorphism of $S^*(\mathfrak{g}^*)$ modules.*

Proof. The proof closely follows that of Theorem 3.8. We first consider the E_1 terms of the spectral sequences induced by the two filtered complexes. As before, the associated graded complexes are

$$GC^p_k = \Omega^{p-k}_G(S_k) \qquad \text{and} \qquad GC'^p_k = \Omega^p_{G,c}(X_k)/\Omega^p_{G,c}(X_{k+1})$$

and the E_1 terms give the cohomology of the associated graded complexes. Hence Ψ induces a map on the E_1 term, $H^p(GC'_k) \to H^p(GC_k) = H^p_G(S_k)$.

Recalling the notation of §3, we define $F_k = X_k \cap \mathcal{U}_k$ as a neighborhood of S_k in the unstable manifold $\mathcal{U}_k$. Again, Ψ factors through the restriction of forms and integration over the fibration $F_k \to S_k$. The same argument as in the proof of Theorem 3.8 shows that restriction of equivariant forms gives an isomorphism $H^p(GC'_k) = H^p_{G,c}(F_k)$. Furthermore, the Thom isomorphism holds for equivariant cohomology so that $\Psi : H^p(GC'_k) \to H^p(GC_k) = H^{p-k}_G(S_k)$ is an isomorphism of real vector spaces. Then Ψ induces an isomorphism on the E_1 terms of the spectral sequences and hence by Lemma 3.7, Ψ induces an isomorphism of $S^*(\mathfrak{g}^*)$-modules in cohomology. □

As in §3, the filtered complex gives rise to a spectral sequence whose E_1 term equals

$$H^*_G(S_i;\mathbb{C}) = \oplus_\alpha H^*(BK^i_\alpha;\mathbb{C}).$$

Since we are assuming that the manifold X is finite dimensional, convergence of the spectral sequence is guaranteed. However, in infinite dimensional applications, this may not hold (see [Au] and [AuB]).

As always, the existence of the complex leads to equivariant Morse inequalities and a notion of perfection.

Corollary 5.4 *There is a polynomial $Q(t)$ with non-negative coefficients such that*

$$\sum_{i,j} \dim H^j_G(S_i;\mathbb{R})t^{i+j} = \sum_k \dim H^k_G(B;\mathbb{R})t^k + (1+t)Q(t).$$

The function is called *equivariantly perfect* if $Q(t) = 0$. The Yang-Mills function on the space of connections over a Riemann surface is equivariantly perfect (see [AB2]), and more generally, the norm squared of the moment map for a algebraic group action is perfect, Kirwan [K].

At this point it would be logical to insert a discussion which shows that the cohomology of the equivariant Morse complex is independent of metric and function. This issue is of importance in equivariant Floer cohomology and will be taken up in [AuB].

5.2 Comparison with the Morse-Bott complex

The invariant function f on X defines a Morse-Bott function on the homotopy quotient. Thus, we have two complexes which compute equivariant cohomology: the equivariant complex above and the Morse-Bott complex on the homotopy quotient. We will use the Chern-Weil homomorphism to construct a chain map between the two complexes which induces an isomorphism on cohomology.

Without too much work, it would be possible to apply our Morse-Bott theory directly to the infinite dimensional homotopy quotient. Instead, we choose to work with finite dimensional approximations as follows. Let $\{EG_n \to BG_n\}$ be a sequence of finite dimensional principal G-bundles approximating $EG \to BG$; that is, the homotopy groups $\pi_i(EG_n) = 0$ for $i < n$ and EG_{n+1} is constructed from EG_n by adding cells of sufficiently large dimension. Then we may form the approximate homotopy quotients to X, $X_{G,n} = EG_n \times_G X$. It is a standard fact that for a fixed j, $H^j_G(X) = H^j(X_{G,n})$ for all $n >> 0$. For example, when $G = SU(2)$, we may take $S^{4n+3} \to \mathbb{HP}^n$ as an approximating sequence. Define the ideal $\mathcal{I}_n = \ker\{H^*(BG;\mathbb{C}) \to H^*(BG_n;\mathbb{C})\}$.

Now a G-invariant function on X with non-degenerate critical orbits gives a Morse-Bott function on $X_{G,n}$. Hence the method of §3 is applicable to the computation of $H^*(X_{G,n};\mathbb{R})$ and thus $H^*_G(X)$. This produces a Morse-Bott complex which we denote by (C'^*_n, ∂'). We will show that there is a natural chain map $\chi : (C^*, \partial_G) \to (C'^*_n, \partial')$.

Notice that the critical submanifolds of index i on $X_{G,n}$ are precisely $(S_i)_{G,n}$ while the gradient line moduli spaces are $\widetilde{\mathcal{M}}((S_i)_{G,n}, (S_j)_{G,n}) = \widetilde{\mathcal{M}}(S_i, S_j)_{G,n}$. The Chern-Weil homomorphism (see §4) defines

$$\chi : \Omega^*_G(S_i) \to \Omega^*((S_i)_{G,n}).$$

To check that this defines a chain map, one must show that

$$\chi((\partial_G)_k(\omega \otimes \phi)) = \partial'_k \chi(\omega \otimes \phi).$$

For $k = 0$, this is just the fact that the Chern-Weil homomorphism induces a chain map between equivariant differential forms and forms on the homotopy quotient. For $k > 0$, this follows from the fact that the Chern-Weil homomorphism commutes with pullback and integration over equivariant fiber bundles.

Now note that we also have a chain morphism:

$$\chi : C^*/\mathcal{I}_n \to C'^*_n/\chi(\mathcal{I}_n).$$

and that the cohomology of $C'^*_n/\chi(\mathcal{I}_n)$ equals that of C'^*_n, because $\chi(\mathcal{I}_n) \subset \operatorname{im}\partial'$. This gives:

Proposition 5.5 *The cohomology groups $H^*(C^*/\mathcal{I}_n, \partial_G)$ and $H^*(X_{G,n};\mathbb{C})$ are isomorphic through χ.*

Proof. As always, both complexes are filtered by index and the chain map χ is a map of filtered complexes. By Proposition 4.2, χ induces an isomorphism on the E_1 terms and hence by Lemma 3.7, χ is an isomorphism on homology. □

Notice that $\mathcal{I}_n$ is zero in degrees small compared to n. It is now quite clear that using equivariant differential forms is an effective way to deal only with low dimensional moduli spaces of gradient lines. This is important in gauge theoretic applications where higher dimensional spaces may exhibit new forms of non-compactness (bubbling-off).

5.3 Cup products

The cup product structure on $H^*_G(X)$ can be recovered as in §3. Given $\omega \otimes \phi \in \Omega^*_G(X)$, define

$$c_G(\omega \otimes \phi) : C^l \to C^{l+k} : (\eta_i \otimes \psi_i) \mapsto c(\omega)(\eta_i) \otimes \phi\psi_i.$$

Again, we have

Theorem 5.6 *The map c_G induces a map on cohomology*

$$c_G : H^*_G(X) \otimes H^*(C^*, \partial_G) \to H^*(C^*, \partial_G)$$

which agrees with the cup product map.

The analog to the Poincaré duality pairing in equivariant cohomology is integration over the fiber $X_G \to BG$, assuming that X is oriented and closed:

$$H^k_G(X) \otimes H^l_G(X) \to H^{l+k-n}(BG),$$

where n is the dimension of X. In terms of the equivariant complex, this is given by reversing the flow (or considering the invariant function $-f$) to obtain an equivariant complex $(\overline{C}^*, \overline{\partial}_G)$.

Corollary 5.7 *The pairing $H^k_G(X) \otimes H^l_G(X) \to H^{l+k-n}(BG)$ is described by the equivariant complexes as a map*

$$C^* \otimes \overline{C}^* \to S^*(\mathfrak{g}^*)$$

given by $(\omega_i \otimes \phi_i) \otimes (\eta_j \otimes \psi_j) \mapsto (\int_{S_i} \omega_i \wedge \eta_i)\phi_i\psi_i$.

5.4 Examples

1. The simplest non-trivial example of equivariant cohomology is $X = S^2$ where S^1 acts by equatorial rotations. It is well known that

$$H^*_{S^1}(S^2;\mathbb{R}) = \mathbb{R}[u] \oplus \mathbb{R}[v]/(c,-c)$$

where $(c,-c)$ is the subgroup of constant polynomials. Consider S^2 as the unit sphere in $\mathbb{R}^3$ with coordinates (x,y,z). We will study the following three S^1-invariant functions:

1) $f(x,y,z) = z$ with critical points of index 0 at the south pole and of index 2 at the north pole. Then $(\partial_G)_l = 0$ for $l > 0$. Then $E_\infty^{i,j} = E_1^{i,j} = \mathbb{R}$ if $i = 0, 2$ and j is even and is 0 otherwise.
2) $f(x,y,z) = 1 - z^2$ takes on minima at the poles and a maximum at the equator. Then
$$E_1^{i,j} = \begin{cases} \mathbb{R} & \text{if } i = 1,\ j = 0 \\ \mathbb{R}^2 & \text{if } i = 0,\ j \text{ even} \\ 0 & \text{otherwise.} \end{cases}$$
The differential $d_1(a,b) = (\partial_G)_1(a,b) = a - b$ by counting gradient lines.
3) $(x,y,z) = z^2$ takes on maxima at the poles and a minimum at the equator. Then
$$E_1^{i,j} = E_\infty^{i,j} = \begin{cases} \mathbb{R} & \text{if } i = 0,\ j = 0 \\ \mathbb{R}^2 & \text{if } i = 2 \\ 0 & \text{otherwise.} \end{cases}$$

2. Let (B,ω) be a symplectic manifold with an S^1 action preserving the symplectic form. A moment map is a function $\mu : B \to \mathbb{R}$ such that

$$d\mu = i_{\frac{\partial}{\partial\theta}}\omega$$

where $\frac{\partial}{\partial\theta}$ is the vector field on B generating the S^1 action. In this example, we will consider μ as a Morse function. Since ω is a nondegenerate 2-form, the critical point set of μ equals the fixed point set of the S^1 action; that is

$$\{b \in B | d\mu_b = 0\} = B^{S^1}.$$

It is well known that the critical point set is nondegenerate in the sense of Bott. The fibers of normal bundle of the fixed point set are S^1 representations with no trivial factors; hence the critical point sets are of even index. Denote the components of index i by $B_i^{S^1}$. We will furthermore suppose that the gradient flow is Morse-Smale so that our theory applies.

For this example, $(\partial_G)_k = 0$ for $k > 0$ for the following reason. The moduli spaces $\widetilde{\mathcal{M}}(B_{j+k}^{S^1}, B_j^{S^1})$ have an S^1 action commuting with the endpoint maps. However, the S^1 action fixes the endpoints. Then for $\eta \in \Omega^*(B_j^{S^1})$,

$$i_{\frac{\partial}{\partial\theta}}(l_j^{j+k})^*\eta = (l_j^{j+k})^* i_{\frac{\partial}{\partial\theta}}\eta = 0.$$

However, $\frac{\partial}{\partial\theta}$ is tangent of the fiber of $u_j^{j+k} : \widetilde{\mathcal{M}}(B_{j+k}^{S^1}, B_j^{S^1}) \to B_i^{S^1}$ and hence

$$\partial_k \eta = (u_j^{j+i})_*(l_j^{i+j})^*\eta = 0.$$

Then $\partial_G = (\partial_G)_0$ which implies that μ is equivariantly perfect; that is,

$$H_{S^1}^*(B; \mathbb{R}) = \oplus_i H_{S^1}^{*-i}(B_i^{S^1}; \mathbb{R}).$$

This is a special case of a general theory developed by Atiyah-Bott [AB2] and Kirwan [K] in which it is seen that for reductive groups acting algebraically and Kählerian on projective varieties, the length of the moment map squared is a perfect 'Morse' function; it is known in general if the function is non-degenerate, but techniques with stratifications establish perfection. Ginzburg [G] has similar results for a torus action on a Kähler manifold.

3. We will consider the case when $G = SU(2)$. Additionally, we impose the constraint that any stabilizer K is the centralizer of a subgroup of $SU(2)$ and that the action is of a single orbit type away from the critical points, equal to $SU(2)/(\mathbb{Z}/2)$. In the sequel [AuB], we are interested in precisely this situation when studying the $SU(2)$-action on the space of based $SU(2)$-connections over a 3-manifold. We will see that the possible orbit types and the possible nonzero boundary operators are quite restricted.

There are only three orbit types possible:

1) For $K = \mathbb{Z}/2$, the orbits are $SO(3)$'s and called irreducible.
2) For $K = U(1)$, the orbits are S^2's and called $U(1)$-orbits.
3) For $K = SU(2)$, the orbits are points and called $\mathbb{Z}/2$-orbits.

The terminology is borrowed from the situation with connections on principal $SU(2)$-bundles. Notice that any irreducible critical orbit contributes $H^*(B\mathbb{Z}_2; \mathbb{R}) = H^*(pt; \mathbb{R})$ to the E_1-term of the spectral sequence, a critical $U(1)$-orbit contributes a $H^*(\mathbb{C}P^\infty; \mathbb{R})$ and a critical $\mathbb{Z}/2$-orbit a $H^*(BSU(2); \mathbb{R}) = H^*(\mathbb{HP}^\infty; \mathbb{R})$.

For dimension reasons, many differentials vanish. For example, the differential d_1 is a count of gradient lines. The second differential d_2 always vanishes since all of the cohomology is in even vertical dimensions. For $k = 3$, only $d_3 : H^2(\mathbb{C}P^\infty; \mathbb{R}) \to H^0(\mathbb{R}P^\infty; \mathbb{R})$ is nonzero. Furthermore, $d_4 \equiv 0$.

In the next section, we will consider this example further in relation to the $H^*(BG)$-module structure.

6 The $H^*(BG)$ module structure

The G-equivariant cohomology of a G-space X has a rich algebraic structure: $H_G^*(X)$ appears as an $H^*(BG)$-module. In this section, we will investigate this module structure in terms of the complex introduced in §5 paying special attention to the case $G = SU(2)$. This material explains the «u»-map on Floer homology and is relevant to the equivariant Floer cohomology developed in [AuB].

First, we say a few words about the origins of this structure and its expression in the Cartan model. Recall that the homotopy quotient X_G is a fibration over the classifying space BG, $\pi : X_G \to BG$. Given an element of $u \in H^*(BG)$, define $u : H^*_G(X) \to H^*_G(X)$ by cupping with $\pi^*(u)$. Using the Cartan model for equivariant cohomology and identifying $H^*(BG) = (S^*(\mathfrak{g}^*))^G$, the map $u : H^*_G(X) \to H^*_G(X)$ is described by multiplying a G-equivariant form on X with an invariant symmetric polynomial; that is,

$$u(\omega \otimes \phi) = \omega \otimes u\phi.$$

In the case that X admits a G-invariant Morse-Bott function with nondegenerate isolated critical orbits, we have the complex described in §5; namely,

$$C^p = \oplus_{i+l=p} \Omega^l_G(S_i) = \oplus_{i+j+2k=p} (\Omega^j(S_i) \otimes S^k(\mathfrak{g}^*))^G.$$

Furthermore, we have the chain map $\Psi : \Omega^*_G(X) \to C^*$, given by integrating forms along the unstable manifolds, which induces an isomorphism on cohomology. As Ψ does not act on the polynomial part of an equivariant form, we see that the module structure on $H^*_G(X)$ in terms of the equivariant complex is simply

$$u(\omega \otimes \phi) = \omega \otimes u\phi$$

for $\omega \otimes \phi \in \Omega^*_G(S_i)$. That is, in the array of cochains $\Omega^j_G(S_i)$, multiplying by elements of $H^*(BG)$ translates an element vertically.

It is interesting to investigate the situation when $G = SU(2)$ as in the final example from §5; that is, the stabilizers which occur are always the centralizer of some subgroup and the stablizer away from the critical orbits is exactly $\mathbb{Z}/2$. For instance, let $S^k_i = SO(3)$ be a critical orbit of index i and with $\mathbb{Z}/2$ stabilizer. Choose a cohomology class represented by $\alpha \in \Omega^0_G(S^k_i)$ so that $\partial_G(\alpha) = 0$. Let $u \in H^4(BSU(2))$ be the universal second Chern class. We wish to study the cohomology class $u(\alpha)$. Since $H^*_G(S^k_i) = H^*(pt)$, it follows that $u(\alpha) = (\partial_G)_0(\beta) = d_G(\beta)$ for some $\beta \in \Omega^3_G(S^k_i)$. Then in cohomology, $u(\alpha)$ is represented by $\sum_{k=1}^4 (\partial_G)_k(\beta)$. We will compute these components of the boundary operator and see that they contain interesting information about the configuration of gradient lines. Notice that this discussion only applies to orbits with trivial stabilizer.

Choosing an orthonormal basis $\{\phi_i\}$ for $\mathfrak{g}^*$, u is represented as an element of $S^*(\mathfrak{g}^*)$ by $\sum \phi_i^2$. Let X_i be the left-invariant vector fields on S^k_i generated by ϕ^*_i, and ω_i left-invariant 1-forms dual to X_i. Denoting the G-equivariant volume form on S^k_i by $dvol_{SO(3)}$, one checks that defining $\gamma = dvol_{SO(3)} + \sum \omega_i \otimes \phi_i$ leads to $d_G(\gamma) = \sum \phi_i^2$. Hence, $\beta = \alpha \wedge \gamma$. We will discuss the components of

$\partial_G(\beta)$. As α is simply a constant function, it is insignificant in the computations that follow.

First, we study the component $(\partial_G)_2(\gamma) \in \Omega^2_G(S_{i+2})$. Let S^l_{i+2} be a component of S_{i+2}. Notice that if the stabilizer of S^l_{i+2} is not $U(1)$, the $(\partial_G)_0$ cohomology vanishes in dimension 2. We then assume that S^l_{i+2} is stabilized by $U(1)$ and hence $S^l_{i+2} = S^2$. For convenience, assume that $\widetilde{\mathcal{M}}(S^l_{i+2}, S^k_i)$ is compact. Then $\widetilde{\mathcal{M}}(S^l_{i+2}, S^k_i) = \bigcup SO(3)$ and $\widetilde{\mathcal{M}}(S^l_{i+2}, S^k_i)/SO(3)$ is a collection of oriented points. Denote the algebraic count of these points by n. From the Gysin sequence, we see that integrating over the fiber in $SO(3) \to S^2$ takes the G-equivariant volume form on $SO(3)$ to the G-equivariant volume form on S^2; that is $\partial_2(dvol_{SO(3)}) = n\ dvol_{S^2}$. Furthermore, notice that the infinitesimal action of ϕ^*_i on $S^l_{i+2} = S^2$ defines two poles and hence height functions h_i taking the values ± 1 on the poles. One checks that $\partial_2(\omega_i) = nh_i$. Then the component of $(\partial_G)_2(\gamma)$ in $\Omega^2_G(S^l_{i+2})$ is $n(dvol_{S^2} + \sum h_i \otimes \phi_i)$. Notice that $dvol_{S^2} + \sum h_i \otimes \phi_i$ generates $H^2_G(S^l_{i+2})$.

Consider now the component $(\partial_G)_4(\gamma) \in \Omega^0_G(S_{i+4})$. By counting dimensions, we see that $(\partial_G)_4(\gamma) = \partial_4(dvol_{SO(3)})$. For convenience, we will assume that $\widetilde{\mathcal{M}}(S^l_{i+k}, S^k_i)$ is compact in the sequel. Suppose the stabilizer of S^l_{i+k} is:

1) $SU(2)$. As above, $\widetilde{\mathcal{M}}(S^l_{i+4}, S^k_i) = \bigcup SO(3)$ so again let n denote the algebraic count of points in the quotient $\widetilde{\mathcal{M}}(S^l_{i+4}, S^k_i)/SO(3)$. The component of $(\partial_G)_4(\gamma)$ in $\Omega^0(S^l_{i+4})$ is n.

2) $U(1)$. Then $\widetilde{\mathcal{M}}(S^l_{i+4}, S^k_i)$ is a principal $SO(3)$ bundle over a compact surface Σ. In fact, a fiber of the lower endpoint map trivializes this bundle so that $\widetilde{\mathcal{M}}(S^l_{i+4}, S^k_i) = \Sigma \times SO(3)$. Meanwhile, a fiber of the upper endpoint map has the form, $u^{-1}_{i+4}(pt) = \Sigma \times S^1$ and as such the component of $\partial_G(\gamma)$ in $\Omega^0_G(S^l_{i+4})$ must vanish.

3) $\mathbb{Z}/2$. Thinking of $\widetilde{\mathcal{M}}(S^l_{i+4}, S^k_i)$ as a 6-dimensional submanifold of X, form

$$P = \bigcup_{-\infty \leq t \leq \infty} \phi_t(\widetilde{\mathcal{M}}(S^l_{i+4}, S^k_i))$$

where ϕ_t is the flow of the gradient vector field of our Bott-Morse function. Then $P \to P/SO(3)$ is a principal $SO(3)$ fibration. We claim that the component of $(\partial_G)_4(\gamma)$ in $\Omega^0_G(S^l_{i+4})$ is related to the first Pontrjagin number $p_1(P)[P/SO(3)]$. Over $\cup_{-\infty \leq t \leq 0} \phi_t(\widetilde{\mathcal{M}}(S^l_{i+4}, S^k_i))/SO(3)$, a fiber of the upper endpoint map $u^{-1}_{i+4}(pt)$ provides a trivialization. Likewise a fiber of the lower endpoint map gives a trivialization over the rest. The lower endpoint map on $u^{-1}_{i+4}(pt)$ gives the transition function and hence $\partial_4(dvol_{SO(3)})$ measures its degree. The Pontrjagin number is 4 times this degree. In [AuB], we will show that this explains «u»-map on Floer homology as arising from the

$H^*(BSO(3))$-module structure on the equivariant cohomology of the space of based connections.

To summarize,

Proposition 6.1 *When $\alpha \in \Omega^*_G(S^k_i)$ with S^k_i is an irreducible orbit, $u(\alpha)$ measures the number of gradient lines to $U(1)$-orbits of index $i+2$ and $\mathbb{Z}/2$-orbits of index $i+4$ as well as the Pontrjagin number of the framed moduli space connecting irreducibles of index $i+4$. For a cohomology class represented by $\alpha \in \Omega^*_G(S^k_i)$ with $S^k_i = pt$ or S^2, the module structure simply carries α vertically.*

Appendix A
Asymptotics of gradient line moduli spaces

The methods of this paper depend in a crucial way on a detailed understanding of the ends of the moduli spaces of gradient lines. In this appendix, we will discuss the techniques necessary for a careful examination of these spaces as required in the body of the paper. We begin by assuming that the critical points of the flow are isolated before attacking the more general case of manifolds of critical points. Furthermore, these methods enable us to prove a local stable manifold theorem for the flow of the Chern-Simons functional about a nondegenerate critical point.

In this introduction, we explain a special case of the motivating boundary value problem. Our first concern is the structure of solution curves near nondegenerate critical points of vector fields. To begin, consider the special case of the linear vector field on $\mathbb{R}^k \oplus \mathbb{R}^l$ given by $v(x,y) = (-x,y)$. Trajectories of this vector field are solutions to the differential equation

$$\frac{d}{dt}\begin{pmatrix} x^s(t) \\ x^u(t) \end{pmatrix}(t) = \begin{pmatrix} -x^s(t) \\ x^u(t) \end{pmatrix}.$$

For all $T > 0, p \in \mathbb{R}^k, q \in \mathbb{R}^l$ with $\|p\| = \|q\| = 1$, we have a unique solution curve

$$[-T,T] \to \mathbb{R}^k \oplus \mathbb{R}^l : t \to (x^s(t), x^u(t))$$

such that

$$\begin{aligned} x^s(-T) &= p \\ x^u(+T) &= q. \end{aligned}$$

This solution curve is explicitly given by $t \to (e^{-(T+t)}p, e^{-(T-t)}q)$.

We call the set of points $(x,0)$ the stable manifold since the trajectories through these points are bounded. Likewise, the points $(0,y)$ comprise the unstable manifold. Notice that as $T \to \infty$, the trajectory described above approaches a trajectory in the stable manifold on $[-T,0]$ and a trajectory in the unstable manifold on $[0,T]$.

To change our perspective, we can consider this family of trajectories as the result of *gluing* trajectories in the stable and unstable manifolds in the following sense. Let $a(t)$ be a trajectory in the stable manifold, $b(t)$ in the unstable manifold. As $t \to \pm\infty$ we get the graphs in figure 5 for the length of the vector field $\|v(a(t))\|$ and $\|v(b(t))\|$.

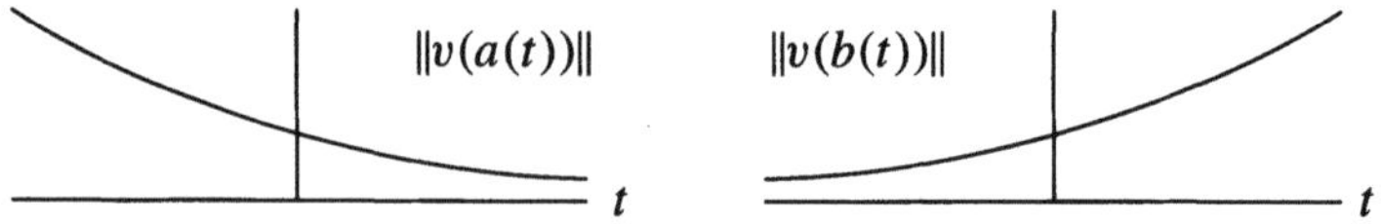

Fig. 5

We now assert that if we chop off the trajectories at $t_o > 0, t_1 < 0$, respectively, then we can find a unique trajectory $c(t)$ such that figure 6 results.

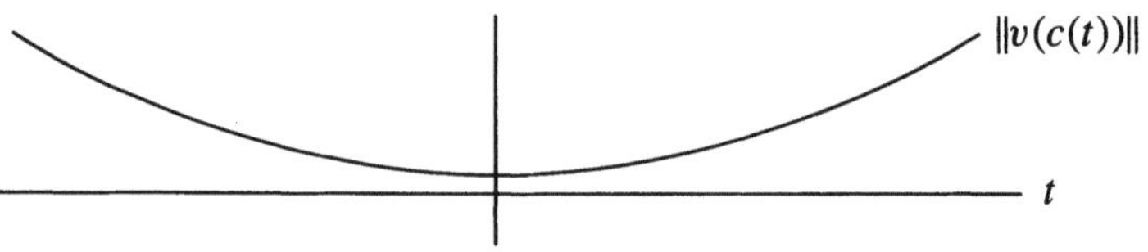

$$\|c(t) - a(t)\| = O(e^{+t}), \text{ for } t < t_1$$
$$\|c(t) - b(t)\| = O(e^{-t}) \text{ for } t > t_0$$

Fig. 6

We have glued the trajectory $a(t)$ to $b(t)$. It is this sort of graphical picture that appears in Floer's work [F1] on trajectories for the anti-self-duality equation. This property was recognized in the context of gradient like systems under the name of transitivity (Smale [Sm2]). In fact it holds in considerable generality as we shall see.

A.1 Gluing trajectories near isolated critical points

Consider a bundle of Banach spaces

$$\pi : E \to \mathbb{R}$$

with coordinates x in the fiber and t along the base and a «linear» vector field $(x, t) \to v(x, t)$ such that

$$\pi_*(v(x,t)) = \frac{\partial}{\partial t}$$
$$v(0,t) = 0$$

for $t \in \mathbb{R}$, $x \in E_t = \pi^{-1}(t)$. We assume that the flow of v is defined by linear maps $\Phi(t,\tau) : E_\tau \to E_t$ satisfying the following hyperbolicity assumptions. The bundle E is a direct sum:

$$E = E^s \oplus E^u$$

of Banach subbundles and $\Phi(t,\tau)$ preserves the splitting. Furthermore, there are constants $C > 0, \rho > 0$ such that

$$\begin{aligned} \|\Phi(t,\tau)x^s\| &\leq Ce^{-\rho(t-\tau)}\|x^s\| \quad \text{for } t \geq \tau \text{ and } x^s \in E^s_\tau \\ \|\Phi(t,\tau)x^u\| &\leq Ce^{\rho(t-\tau)}\|x^u\| \quad \text{for } t \leq \tau \text{ and } x^u \in E^u_\tau. \end{aligned}$$

We also assume that the projections

$$\pi^i_t : E_t \to E^i_t \text{ for } i = s, u$$

along E^u_t and E^s_t have bounded norm

$$\|\pi^i_t\| \leq A$$

uniformly in t.

This is a general situation: it is easy to check that the flow of the linear part of a vector field in the neighborhood of a hyperbolic critical point can easily be cast in this form. Furthermore, nondegenerate periodic orbits and more generally orbits in a hyperbolic set satisfy such estimates.

Let g be a vector field on E, to be considered as a vertical perturbation of v, such that

$$\pi_*(g(x,t)) = 0.$$

We shall be interested in the flow of

$$w(x,t) = v(x,t) + g(x,t).$$

Before stating our gluing theorem, let us first recall how the stable manifold theory applies in this situation (see [Du]).

Theorem A.1 *(i) Assume that*

$$\begin{aligned} \|g(0,t)\| &\leq \alpha, \\ \|D_x g(x,t)\| &\leq \beta \end{aligned}$$

for all $(x,t) \in E$. *If* $\gamma\beta < 1$ *with* $\gamma = \frac{A}{\rho}(C + \frac{1}{C})$, *then for each*

$$a \in E^s_0,$$

there is a unique $\xi(a) \in E_0$ *with,* $\pi^s(\xi(a)) = a$*, such that the trajectory* $(x(t), t)$ *satisfying*

$$\begin{cases} \dfrac{d(x(t),t)}{dt} = w(x,t) \\ x(0) = \xi(a) \end{cases}$$

is bounded for all t*. It satisfies the a priori bound:*

$$\|x(t)\|_t \leq (1-\gamma\beta)^{-1}(\|a\| + \gamma\alpha)$$

for all $t \geq 0$*. Further the assignment* $a \to \xi(a)$ *is as smooth as* $(x,t) \to g(x,t)$*.*

(ii) Assume that

$$\|g(0,t)\| = 0 \text{ and } \|D_x g(x,t)\| < \beta$$

and that for ρ' *such that* $0 < \rho' < \rho$*, we have*

$$\beta\gamma' < 1 \text{ with } \gamma' = \frac{A}{(\rho-\rho')}(C + \frac{1}{C}).$$

Then the bounded solution above satisfies:

$$\|x(t)\|_t \leq (1-\gamma\beta)^{-1} C e^{-\rho' t}\|a\|.$$

This theorem can be applied *locally* to vector fields $w(x,t) = v(x,t) + h(x,t)$ when $h(x,t)$ is quadratic in x and satisfies an estimate

$$\|D_x h(x,t)\| \leq B\|x\|$$

for all t. One shows easily that if $\psi_\delta(x,t)$ is a cutoff function equal to one on a ball of radius δ and zero outside radius 2δ, then for δ sufficiently small

$$g(x,t) = \psi_\delta(x,t)h(x,t)$$

is such that the estimates of Proposition 2.1 apply.

Henceforth we assume that

$$w(x,t) = v(x,t) + g(x,t)$$

satisfies

$$g(0,t) = 0 \text{ for all } t$$
$$\|D_x g(x,t)\| \leq B\|x\| \text{ for all } (x,t) \in E$$

for fixed $B > 0$. In this case, E_0^s is tangent to the stable manifold at O. Now we restrict ourselves to a δ neighborhood of zero section in E. Using Theorem A.1 we may assume that E is the fibred product of E^u, E^s the unstable and stable manifold for w; that is, we have used a coordinate transformation $(x,t) \to (\xi_t(x), t)$ for $x \in E^s$ and similarly for $x \in E^u$.

In analogy with the situation described in the introduction, we can now state the gluing theorem:

Theorem A.2 *Let $S^s_{\epsilon,t}$ be the sphere of radius ϵ in E^s_t and $S^u_{\epsilon,t}$ and sphere of radius ϵ in E^u_t. For $T_1 < T_2$, ϵ sufficiently small and $p \in S^s_{\epsilon,T_1}$, $q \in S^u_{\epsilon,T_2}$, there is a unique solution $x(t)$ of*

$$\frac{dx}{dt} = w(x(t),t)$$

such that

$$\begin{aligned}\pi^s(x(T_1)) &= p\\ \pi^u(x(T_2)) &= q\\ \|x(t)\| &\leq \epsilon.\end{aligned}$$

The solution depends smoothly on the parameters p, q, T_1, and T_2 and will be denoted $x(t, p, q, T_1, T_2)$.

Proof. Denote by $x^s(t), x^u(t), w^s(x,t), w^u(x,t)$ the images of $x(t)$, $w(x,t)$ under the projections π^s, π^u. A solution $x(t)$ to the initial value problem for

$$\frac{dx}{dt} = v(x(t),t) + g(x(t),t)$$

is generally described by the integral equation:

$$x(t) = \Phi(t,t_0)x(t_0) + \int_{t_0}^{t} \Phi(t,s)g(x(s),s)ds.$$

This is the Lagrange formula for solutions to inhomogeneous linear equations.

More generally we can write flow lines of w as

$$\begin{aligned}x(t) = \Phi(t,T_1)p &+ \int_{T_1}^{t} \Phi(t,s)g^s(x(s),s)ds\\ &+ \Phi(t,T_2)q - \int_{t}^{T_2} \Phi(t,s)g^u(x(s),s)ds\end{aligned} \tag{A.1}$$

Notice that a solution to (A.1) satisfies the 'initial' conditions

$$\begin{aligned}x^s(T_1) &= p\\ x^u(T_2) &= q.\end{aligned}$$

We can write (A.1) as a fixed point equation:

$$x = F(x,p,q,T_1,T_2)$$

where F is the right hand side of (A.1) and x lies in the Banach space

$$C^0([T_1,T_2],E)$$

equipped with the supremum norm. Now the set of $t \to x(t)$ with $\|x\| < \eta$ is mapped into itself since

$$\begin{aligned} \|F(x,p,q,T_1,T_2)\| &\leq (C+\frac{1}{C})\epsilon + \frac{A}{\rho}(C+\frac{1}{C}) \sup_{\|x\|<\delta} \|g(x,t)\| \\ &\leq (C+\frac{1}{C})\epsilon + \frac{A}{\rho}(C+\frac{1}{C})B\eta^2 \leq \eta \end{aligned}$$

provided that ϵ and η are sufficiently small. In the same way, one sees that F is a contraction when η is sufficiently small.

The theorem now follows from the contraction principle. Applying the implicit function theorem shows that solutions depend smoothly on s, u, T_1, T_2 since

$$\|\frac{\partial F}{\partial x}\| \leq \frac{A}{\rho}(C+\frac{1}{C}) sup_{\|x\|<\eta} \|\frac{\partial g}{\partial x}\| < 1$$

for η sufficiently small. □

Lemma A.3 *For all $0 < \rho' < \rho$, we may choose ϵ sufficiently small and D sufficiently large so that the following a priori estimates hold:*

$$\begin{aligned} \|x^s(t,p,q,T_1,T_2)\| &\leq \epsilon D e^{-\rho'(t-T_1)} \\ \|\frac{\partial x^s}{\partial p}(t,p,q,T_1,T_2)\| &\leq \epsilon D e^{-\rho'(t-T_1)} \\ \|\frac{\partial x^s}{\partial q}(t,p,q,T_1,T_2)\| &\leq \epsilon D e^{-\rho'(t-T_1)}. \end{aligned}$$

Similarly, for x^u we have

$$\begin{aligned} \|x^u(t,p,q,T_1,T_2)\| &\leq \epsilon D e^{-\rho'(T_2-t)} \\ \|\frac{\partial x^u}{\partial p}(t,p,q,T_1,T_2)\| &\leq \epsilon D e^{-\rho'(T_2-t)} \\ \|\frac{\partial x^u}{\partial q}(t,p,q,T_1,T_2)\| &\leq \epsilon D e^{-\rho'(T_2-t)}. \end{aligned}$$

All estimates hold uniformly in p, q, and T_1, T_2.

Proof. The estimates hold by considering the integral equation (A.1) and applying the method of continuity. We will demonstrate the ideas by showing

$$\|x^s(t,p,q,T_1,T_2)\| \leq \epsilon D e^{-\rho'(t-T_1)}.$$

Define the open set $\mathscr{E} = \{t \in [T_1,T_2] : \|x^s(t,p,q,T_1,T_2)\| < \epsilon D e^{-\rho'(t-T_1)}\}$. If $D > 1$, then $T_1 \in \mathscr{E}$. So suppose that $[T_1, t) \subset \mathscr{E}$ where

$$x^s(t) = \Phi(t,T_1)p + \int_{T_1}^{t} \Phi(t,\sigma) g^s(x(\sigma),\sigma) d\sigma.$$

Estimating the norm

$$\begin{aligned}
\|x^s(t)\| &\leq Ce^{-\rho(t-T_1)}\|p\| + C\int_{T_1}^{t} e^{-\rho(t-\sigma)}\|g^s(x(\sigma),\sigma)\|d\sigma \\
&< Ce^{-\rho(t-T_1)}\epsilon + ABCD\epsilon^2\int_{T_1}^{t} e^{-\rho(t-\sigma)}e^{-\rho'(\sigma-T_1)} \\
&< (C + \frac{ABC\epsilon}{\rho-\rho'}D)\epsilon e^{-r\rho'(t+T_1)}.
\end{aligned}$$

Choose ϵ to be small so that the quantity

$$\frac{ABC\epsilon}{\rho-\rho'} < 1.$$

Then require D to satisfy

$$A + \frac{ABC\epsilon}{\rho-\rho'}D < D.$$

Now $t \in \mathscr{E}$ showing that $\mathscr{E}$ is closed and hence $\mathscr{E} = [T_1, T_2]$. □

The previous lemma conveniently leads to the following compactness theorem. For a large value of T, let $T_1 = -T$ and $T_2 = T$. Then denote $x(t,p,q,T) = x(t,p,q,-T,T)$.

Theorem A.4 *Let $t \to x(t,p,q,T)$ be as in Theorem A.2. For $T \to \infty$, the trajectories*

$$t \to x(t-T,p,q,T) : [0,T] \to E$$

approach the solution to

$$\frac{dx(t)}{dt} = w(x(t),t), \quad x(0) = (s,0)$$

uniformly in T and on compact sets parameters p and q. Similarly, the trajectories

$$t \to x(T-t,p,q,T) : [0,T] \to E$$

approach the unstable manifold solutions.

Proof. The initial value for $x(t-T,p,q,T)$ is the point $(s, x^u(-T,p,q,T))$. But by Lemma A.3,

$$\|s^u(-T,p,q,T)\| < \epsilon De^{-2\rho' T}$$

so that $x(-T,p,q,T) \to (s,0)$ as $T \to \infty$. □

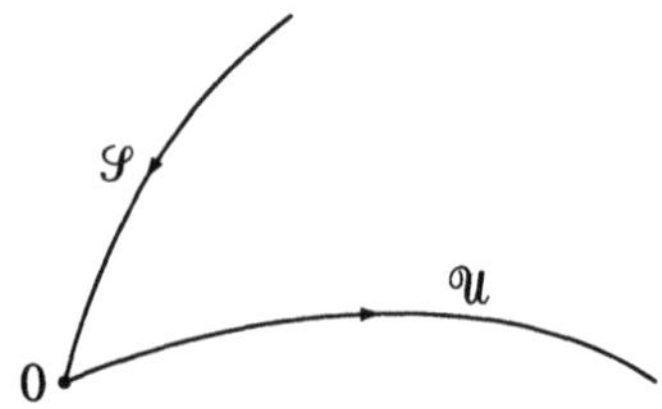

Fig. 7

The implication of this theorem is that the limit of the trajectories is really the «broken» trajectory where the first piece lies in the stable manifold and the second in the unstable (see figure 7).

These preliminaries allow us to proceed to our main gluing theorem.

A.2 Gluing theorems

Let v be a vector field on a manifold X and suppose that α, β, γ are nondegenerate critical points of v such that the intersections $\mathcal{U}_\alpha \cap \mathcal{S}_\beta$ and $\mathcal{U}_\beta \cap \mathcal{S}_\gamma$ are transverse. Denote the space of trajectories from α to β by

$$\widetilde{\mathcal{M}}(\alpha, \beta) = \mathcal{U}_\alpha \cap \mathcal{S}_\beta / \mathbb{R}$$

where the $\mathbb{R}$ action is given by the flow. Similarly, define $\widetilde{\mathcal{M}}(\beta, \gamma)$ and $\widetilde{\mathcal{M}}(\alpha, \gamma)$. We will study the relationship between the ends of $\widetilde{\mathcal{M}}(\alpha, \gamma)$ and the spaces $\widetilde{\mathcal{M}}(\alpha, \beta)$ and $\widetilde{\mathcal{M}}(\beta, \gamma)$.

We use the framework of §A.1. Consider a small ball $B \subset X$ around the critical point β and let $E \to \mathbb{R}$ be a bundle of Banach spaces such that the unit disk subbundle $E_1 = B \times \mathbb{R}$. There is a vector field w on E induced by v which has the form $v + \frac{\partial}{\partial t}$. Consider $\widetilde{\mathcal{M}}(\alpha, \beta) \subset S^s_{\epsilon, T_1}$ and $\widetilde{\mathcal{M}}(\beta, \gamma) \subset S^u_{\epsilon, T_2}$. Due to the assumption on the transversality of the intersections, there are maps

$$u : \widetilde{\mathcal{M}}(\alpha, \beta) \times B^u_{\epsilon, T_1} \to B^s_{\epsilon, T_1}$$
$$s : \widetilde{\mathcal{M}}(\beta, \gamma) \times B^s_{\epsilon, T_2} \to B^u_{\epsilon, T_2}$$

with

$$u(\phi, 0) = \phi$$
$$s(\psi, 0) = \psi$$

such that a slice to the $\mathbb{R}$-action on $\mathcal{U}_\alpha$ is locally described by the graph $(u(\phi, \sigma), \sigma)$. Likewise, a slice to the $\mathbb{R}$-action on $\mathcal{S}_\gamma$ is given by $(\tau, s(\psi, \tau))$.

Theorem A.5 *For $T = T_2 - T_1$ large, there is a smooth injection*

$$G_T : \widetilde{\mathcal{M}}(\alpha, \beta) \times \widetilde{\mathcal{M}}(\beta, \gamma) \to \widetilde{\mathcal{M}}(\alpha, \gamma)$$

depending smoothly on T.

Technically, the following proof works for compact subsets $K_{\alpha,\beta} \subset \widetilde{\mathcal{M}}(\alpha,\beta)$ and $K_{\beta,\gamma} \subset \widetilde{\mathcal{M}}(\beta,\gamma)$. The generalization to the statement given above is simple. In the gauge theory case this generalization cannot be made so easily.

Proof. First we describe the boundary value problem. For trajectories $\phi \in \widetilde{\mathcal{M}}(\alpha,\beta)$ and $\psi \in \widetilde{\mathcal{M}}(\beta,\gamma)$, we wish to perturb the broken trajectory slightly to produce a trajectory from α to γ. This is done by finding small $\sigma \in B^u_{\epsilon,T_1}$ and $\tau \in B^s_{\epsilon,T_2}$ and a trajectory $x(t)$ so that

$$\begin{aligned} x(T_1) &= (u(\phi,\sigma),\sigma) \in \mathcal{U}_\alpha \\ x(T_2) &= (\tau, s(\psi,\tau)) \in \mathcal{S}_\gamma. \end{aligned} \tag{A.2}$$

Notice that this amounts to specifying the two components at the two endpoints T_1 and T_2.

Our principal tool for accomplishing this perturbation is Theorem A.2. This says that for gradient lines $\gamma \in \widetilde{\mathcal{M}}(\alpha,\beta)$ and $\psi \in \widetilde{\mathcal{M}}(\beta,\gamma)$ and $\sigma \in B^u_{\epsilon,T_1}, \tau \in B^s_{\epsilon,T_2}$, there is a gradient line $x(t)$ so that $x^s(T_1) = u(\gamma,\sigma)$ and $x^u(T_2) = s(\psi,\tau)$. Moreover, $x(t)$ depends smoothly on all the parameters. This gives one of the components at each endpoint. The other is given by considering the map

$$F_{(\gamma,\psi,T)} : B^u_{\epsilon,T_1} \times B^s_{\epsilon,T_2} \to B^u_{\epsilon,T_1} \times B^s_{\epsilon,T_2}$$

by

$$F_{(\gamma,\psi,T)}(\sigma,\tau) = (x^u(T_1), x^s(T_2)).$$

Notice that a fixed point (σ,τ) will satisfy (A.2) and hence the accompanying trajectory $x(t)$ is a gradient line from α to γ. Fixed points exist for large T since F is a contraction as we demonstrate now.

We have

$$|F(v_1,w_1) - F(v_2,w_2)| < \eta_v|v_1 - v_2| + \eta_w|w_1 - w_2|$$

where

$$\begin{aligned} \eta_v &< (|\frac{\partial x^u(T_1)}{\partial p}| + |\frac{\partial x^s(T_2)}{\partial p}|)|\frac{\partial u}{\partial \tau}| \\ \eta_w &< (|\frac{\partial x^u(T_1)}{\partial q}| + |\frac{\partial x^s(T_2)}{\partial q}|)|\frac{\partial s}{\partial \sigma}|. \end{aligned}$$

The partials $|\frac{\partial u}{\partial \tau}|$ and $|\frac{\partial s}{\partial \sigma}|$ are bounded and Lemma A.3 shows that

$$|\frac{\partial x^u(T_1)}{\partial p}|, |\frac{\partial x^u(T_1)}{\partial p}|, |\frac{\partial x^s(T_2)}{\partial q}|, |\frac{\partial x^s(T_2)}{\partial q}|$$

decay exponentially with T. Hence for large T, F is a contraction and has a unique fixed point denoted

$$(\sigma(\gamma,\psi,T), \tau(\gamma,\psi,T)).$$

In the same way, Lemma A.3 shows that the derivative $\frac{dF}{dz}$ is small for large T. The implicit function theorem implies that the gluing map

$$G_T : \widetilde{\mathcal{M}}(\alpha,\beta) \times \widetilde{\mathcal{M}}(\beta,\gamma) \to \widetilde{\mathcal{M}}(\alpha,\gamma)$$

is smooth. □

This theorem and its proof have several important consequences for us.

Theorem A.6 *For large T, there is an injective local diffeomorphism*

$$G : \widetilde{\mathcal{M}}(\alpha,\beta) \times \widetilde{\mathcal{M}}(\beta,\gamma) \times [T,\infty) \to \widetilde{\mathcal{M}}(\alpha,\gamma)$$

onto an end of $\widetilde{\mathcal{M}}(\alpha,\gamma)$. For $A_1 \in \mathcal{M}(\alpha,\beta), A_2 \in \mathcal{M}(\beta,\gamma)$, the paths $G(A_1, A_2, T)$ converge to the broken trajectory formed by A_1 and A_2 at the rate $\epsilon De^{-\rho' T}$.

Proof. The proof of the first part mimics the proof of Theorem A.5 while the second part follows from the estimates of Lemma A.3. □

Given two Morse functions f_1 and f_2, consider the manifold $X \times \mathbb{R}$ with a vector field F equal to

$$-\nabla f_1 + \frac{\partial}{\partial t} \text{ for } t < -1$$
$$-\nabla f_2 + \frac{\partial}{\partial t} \text{ for } t > 1.$$

For critical points α, β of f_1 and γ of f_2, we have the following

Theorem A.7 *For large T, there is an injective local diffeomorphism*

$$G : \widetilde{\mathcal{M}}_{f_1}(\alpha,\beta) \times \widetilde{\mathcal{M}}_F(\beta,\gamma) \times (T,\infty) \to \widetilde{\mathcal{M}}_F(\alpha,\gamma)$$

onto an end of $\widetilde{\mathcal{M}}_F(\alpha,\gamma)$.

Proof. Proceed as in Theorem A.5 considering the stable manifold of γ, $\mathcal{S}_\gamma \subset B_{\epsilon,-1}$. □

Consider two vector fields F_1, F_2 on $X \times \mathbb{R}$ with

$$F_1 = \begin{cases} -\nabla f_1 + \frac{\partial}{\partial t} & \text{for } t < -1 \\ -\nabla f_2 + \frac{\partial}{\partial t} & \text{for } t > 1 \end{cases}$$

and

$$F_2 = \begin{cases} -\nabla f_2 + \frac{\partial}{\partial t} & \text{for } t < -1 \\ -\nabla f_3 + \frac{\partial}{\partial t} & \text{for } t > 1 \end{cases}$$

Define $F_{3,T}$ to be

$$F_{3,T}(t) = \begin{cases} F_1(t+T) & \text{for } t < 0 \\ F_2(t-T) & \text{for } t > 0. \end{cases}$$

Let α, β, γ be critical points for f_1, f_2, f_3 respectively.

Theorem A.8 *For large T, there is an injective local diffeomorphism*

$$G_T : \widetilde{\mathcal{M}}_{F_1}(\alpha,\beta) \times \widetilde{\mathcal{M}}_{F_2}(\beta,\gamma) \to \widetilde{\mathcal{M}}_{F_3}(\alpha,\gamma)$$

onto an end of $\widetilde{\mathcal{M}}_{F_3}(\alpha,\gamma)$.

Proof. Consider the flow $-\nabla f_2 + \frac{\partial}{\partial t}$ on $X \times [-T, T]$ and apply Theorem A.5 where we regard $\mathcal{U}_\alpha \subset E_{-T}$ and $\mathcal{S}_\gamma \subset E_T$. □

A.3 Manifolds of critical points

We now consider the case in which the critical points are no longer isolated. Let w be a vector field on a manifold M and assume that S is a submanifold consisting of zeroes of w. Denote the normal bundle of S by $N_S \equiv TM|_S/TS$. Suppose that for all $s \in S$,

$$dw_s : N_s \to N_s$$

has no eigenvalues with real part 0. When $w = \nabla f$, we say that f is a Morse-Bott function.

In this situation, N splits smoothly into subbundles

$$N = N^+ \oplus N^-$$

such that the real parts of the eigenvalues of dw_s on $N_s^\pm$ have sign $\pm$. Around a base point $s \in S$, there are coordinates in a neighborhood $U \subset M$ such that

$$U \to S \times N_0^- \times N_0^+ : m \mapsto (m^0, m^-, m^+).$$

In these coordinates, the spaces $\{0\} \times N_0^- \times \{0\}$ and $\{0\} \times \{0\} \times N_0^+$ are tangent to the strictly stable and strictly unstable manifolds of $(0,0,0)$ for w. We shall first show that this family $\mathcal{S}_s$ of stable manifolds of $(s,0,0)$ is a smooth fibration isomorphic to $N^- \to S$, when intersected with a small neighborhood of S. This is the analogue of Theorem A.1.

Choose a small coordinate neighborhood U in which to work and denote the coordinates $z = (s,x,y)$. Write $L = dw_{(0,0,0)}$ for the linear part of the vector field. Locally $w(z) = Lz + g(z)$ where $g(z)$ may be decomposed into its components $g(z) = (g^0(z), g^-(z), g^+(z))$ which are at most quadratic in z. More specifically, notice that $g(s,0,0) = 0$ and that $g^-(s,0,y) = 0$ and $g^+(s,x,0) = 0$. This implies that the leading order terms of g are

$$\begin{aligned} g^0(s,x,y) &\sim xy \\ g^-(s,x,y) &\sim xy + xs + x^2 \\ g^+(s,x,y) &\sim xy + ys + y^2. \end{aligned} \tag{A.3}$$

With this in mind, we prove the local stable manifold theorem for families of critical points.

Theorem A.9 *Given $s \in S$, $n^- \in N^-$, there is a unique bounded trajectory of w, $x : [0,\infty) \to U$ such that $x^-(0) = n^-$ and $\lim_{t\to\infty} x(t) = (s,0,0)$. Moreover, for varying (s,n^-) the map $s \mapsto x^0(0)$ defines a diffeomorphism between a neighborhood of the zero section of N^- and a neighborhood of S in $\mathcal{S}_S$, intertwining the projection and endpoint map.*

This theorem is not new: it can be seen as a consequence of results in [HPS].

Proof. In the same way, we set up the boundary value problem. Define the complete metric space

$$\begin{aligned}\mathcal{F} = \{y : [0,\infty) \to U| \quad & \|x^-(t)\| \le \epsilon D e^{-\rho' t} \\ & \|x^+(t)\| \le \epsilon \\ & \|x^0(t)\| \le \epsilon\}\end{aligned}$$

for constants D and small ϵ and ρ'. Let $\Phi(t,\sigma)$ denote the flow of L from time σ to t and $g^{+,0}(z) = g^+(z) + g^0(z)$. Then define the operator

$$\begin{aligned}F(y)(t) = \phi(t,0)n^- &+ \int_0^t \Phi(t,\sigma) g^-(y(\sigma))d\sigma \\ &+ s - \int_t^\infty \Phi(t,\sigma) g^{+,0}(y(\sigma))d\sigma\end{aligned}$$

A fixed point of F describes a bounded trajectory with the appropriate boundary data: $x^-(0) = n^-$ and $x^0(t) \to s$ as $t \to \infty$. Using the same reasoning as before, we conclude that $F : \mathcal{F} \to \mathcal{F}$ and that F is a contraction.

The fact that $s \mapsto x^0(0) = s - \int_0^\infty g^0(x(\sigma))d\sigma$ is a submersion follows from estimating

$$\frac{\partial}{\partial s}\int_0^\infty g^0(x(\sigma))d\sigma.$$

Using the information about the leading order terms in (A.3), it follows that

$$|\frac{\partial}{\partial s}\int_0^\infty g^0(x(\sigma))d\sigma| < C\int_0^\infty |x^+(\sigma)\frac{\partial x^-(\sigma)}{\partial s} + x^-(\sigma)\frac{\partial x^+(\sigma)}{\partial s})|d\sigma.$$

The integrand decays exponentially with σ and hence the variation of the integral with s can be made sufficiently small. □

We continue by demonstrating the generalization of Theorem A.2: incoming and outgoing gradient lines can be glued provided they converge to the same point on S.

Theorem A.10 *Let $(s,n^-,n^+) \in U$ and T be sufficiently large. There is a trajectory $x : [-T,T] \to U$ so that $x^-(-T) = n^-$, $x^+(T) = n^+$, and $x^0(0) = s$. Furthermore, as $T \to \infty$ the trajectories converge to a broken trajectory consisting of a solution in the stable manifold followed by one in the unstable.*

Proof. We define the complete metric space

$$\begin{aligned}\mathcal{F} = \{y : [-T,T] \to U \mid &\|y^-(t)\| \leq \epsilon D e^{-\rho(t+T)}\\ &\|y^+(t)\| \leq \epsilon D e^{-\rho(T-t)}\\ &\|y^0(t)\| \leq \epsilon\}\end{aligned}$$

for suitable constants D, ϵ, ρ. Define the operator

$$\begin{aligned}F(y)(t) = &\Phi(t,-T)n^- + \int_{-T}^{t} \Phi(t,\sigma)g^-(y(\sigma))d\sigma\\ &+ \Phi(t,T)n^+ - \int_{t}^{T} \Phi(t,\sigma)g^+(y(\sigma))d\sigma\\ &+ s - \int_{t}^{0} g^0(y(\sigma))d\sigma.\end{aligned}$$

As before, $F : \mathcal{F} \to \mathcal{F}$ and is a contraction. The fixed point is a trajectory satisfying the hypotheses of the theorem. One sees that as $T \to \infty$, the integral equation breaks into the integral equations for the stable and unstable manifold solutions. □

Suppose that $S_\alpha, S_\beta, S_\gamma$ are critical submanifolds so that the intersections of stable and unstable manifolds are transverse. Suppose, in addition, that $\mathcal{S}_\gamma$ intersects $(\mathcal{U}_\beta)_s$ transversely for all s. Applying the reasoning of Theorem A.5 leads to

Theorem A.11 *For T large, there is an injective local diffeomorphism*

$$G : \widetilde{\mathcal{M}}(S_\alpha, S_\beta) \times_{S_\beta} \widetilde{\mathcal{M}}(S_\beta, S_\gamma) \times (T,\infty) \to \widetilde{\mathcal{M}}(S_\alpha, S_\gamma)$$

onto an end of $\widetilde{\mathcal{M}}(S_\alpha, S_\gamma)$ where $\widetilde{\mathcal{M}}(S_\alpha, S_\beta) \times_{S_\beta} \widetilde{\mathcal{M}}(S_\beta, S_\gamma)$ denotes the fibered product over the endpoint maps.

A.4 The stable manifold theorem for the Chern-Simons function

The discussion in section A.1 lends itself to the study of the gradient flow of the Chern-Simons function. In particular, it follows that the local stable manifold theorem, as in Theorem A.1, holds in this setting.

Let A be a nondegenerate critical point of the Chern-Simons function on $\mathcal{B}$, the space of gauge equivalence classes of connections on the compact 3-manifold M. A neighborhood of A in $\mathcal{B}$, and also the tangent space $T_A\mathcal{B}$, may be identified with $\ker d_A^* : \Omega^1(M, su(2)) \to \Omega^0(M, su(2))$. We will work with the Sobolev norm L_1^2 without explicitly denoting it. The gradient flow of the Chern-Simons function, at a connection $\nabla = A + A'$, is $- * F^\nabla$, which decomposes into linear and quadratic pieces:

$$- * F^\nabla = - * d_A A' - *[A', A'].$$

The Hessian at the critical point A is given by

$$*d_A : \ker d_A^* \to \ker d_A^*.$$

This is a self-adjoint elliptic operator and as such has a discrete real spectrum. Decompose $\ker d_A^* = \oplus_\lambda H^\lambda$ into its eigenspaces and define $H^\pm = \oplus_{\lambda<(>)0} H^\lambda$. The nondegeneracy of A implies that $\ker *d_A = 0$ so that $\ker *d_A = H^- \oplus H^+$.

Notice that the flow of the linear part of the vector field, $-*d_A$, is defined by

$$\Phi(t,\tau)\psi = e^{(t-\tau)*d_A}\psi.$$

Therefore, the flow is continuous on H^- only for positive time, where it is a smoothing operator, and similarly the flow in continuous on H^+ only for negative time. Remember that the vector field is not continuous on L_1^2 connections.

Reflecting on the methods of §A.1, we see that this is sufficient to prove the stable manifold theorem for the Chern-Simons flow. Consider a ball B of radius ϵ in $L_1^2(M, su(2))$. For $a \in H^-$ and $x \in C^0([0,\infty), B)$, define

$$F(x,a)(t) = \Phi(t,0)a + \int_0^t \Phi(t,\sigma)g^-(x(\sigma))d\sigma - \int_t^\infty \Phi(t,\sigma)g^+(x(\sigma))d\sigma$$

where $g^\pm(A) = [A,A]^\pm$, the decomposition of the quadratic term given by $H^\pm$. Notice that this is well defined since we use the forward flow on H^- and the backwards flow on H^+. Also $\Phi(t,\sigma)$ satisfies the estimates of section A.1 so that the techniques used in the proof of the stable manifold theorem apply. This leads to:

Theorem A.12 *Let $a \in H^-$ be sufficiently small in L_1^2. There is $\xi(a) \in H^+$ such that the trajectory of the flow $-*F^\nabla$ through $(a, \xi(a))$ is bounded. Moreover, $a \mapsto \xi(a)$ is a smooth map.*

The fact that $a \mapsto \xi(a)$ follows from the implicit function theorem for Banach spaces. It is important here to use L_1^2 connections since ξ is as smooth as g is. Using L_1^2 implies that g is smooth.

Appendix B
Transversality

In the interest of completeness, we present in this appendix a discussion of transversality theory, a principal tool throughout this paper, and a proof of Proposition 2.1 This presentation proceeds as in Abraham and Robbin [AR].

We recall the basic definitions of transversality. Let X, Y be manifolds, $W \subset Y$ a submanifold and $f : X \to Y$ a smooth map. We say that f is *transverse* to W at $x \in X$ if either $f(x) \notin W$ or $f(x) \in W$ and

$$T_{f(x)}W + (f_*)(T_xX) = T_{f(x)}Y.$$

Gobally, f is transverse to W if it is transverse for all $x \in X$. When f is transverse to W, the implicit function theorem implies that $f^{-1}(W)$ is a submanifold of X.

An important application comes from considering a family of maps parametrized by a manifold $\mathscr{C}$, or equivalently a map

$$F : X \times \mathscr{C} \to Y.$$

The following theorem shows that transverse maps are «generic».

Theorem B.1 *Suppose that $F : X \times \mathscr{C} \to Y$ is transverse to W. Then there is a Baire set $\mathscr{C}_g \subset \mathscr{C}$ so that for all $c \in \mathscr{C}_g$, $f_c : X \to Y$ is transverse to W.*

The importance of this theorem, as we shall see, is that it can sometimes be quite easy to show that a family of maps is transverse. It is easy to see that if

$$F : X \times \mathscr{C} \to Y$$

is transverse to W, then the canonically defined one-parameter version of F,

$$\mathscr{F} : (X \times [0,1]) \times \mathrm{Maps}([0,1], \mathscr{C}) \to Y,$$

is also transverse to W. Theorem B.1 implies that if $F : X \times \mathscr{C} \to Y$ is transverse to W, then for a «generic» one parameter family, $\gamma : [0,1] \to \mathscr{C}$, the map $\mathscr{F}_\gamma$ is transverse to W and hence the space

$$\mathscr{F}_\gamma^{-1}(W) \subset X \times [0,1]$$

is a submanifold of $X \times [0,1]$. The set of $s \in [0,1]$ for which $F_{\gamma(s)}^{-1}(W)$ is not cut out transversely is the bifurcation locus of the path. In the case that W is compact and F is proper then the bifurcation points are an isolated closed subset of [0,1].

For submanifolds $R, S \subset Y$, we say that R intersects S transversally at $y \in R \cap S$ if

$$T_yR + T_yS = T_yY.$$

Furthermore, R intersects S transversally if this is true at every point in the intersection. There are equivalent definitions that are useful. Let i_R(resp i_S) : R(resp S) $\to Y$ be the inclusions of R and S. Then R intersects S transversally if i_R is transverse to the submanifold S. Again, this is equivalent to

$$i_R \times i_S : R \times S \to Y \times Y$$

being transverse to the diagonal submanifold. A theory for families of submanifolds parametrized by $\mathscr{C}$ holds as above.

Now we are ready to discuss the genericity properties of the gradient flow of Morse functions. Recall that a function is Morse if its critical points are nondegenerate and that the gradient flow of a Morse function is Morse-Smale if all stable and unstable manifolds intersect transversely. The following is then a variant of the celebrated Kupka-Smale theorem and we shall just outline the interesting elements. For details we refer to Abraham and Robbin [AR].

Proposition B.2 *Given a metric g on B, there is a Baire set $\mathcal{C}_g \subset C^\infty(B)$ so that $f \in \mathcal{C}_g$ is Morse with Morse-Smale gradient flow.*

Proof. The fact that the set $\mathcal{C}_M$ of Morse functions in $C^\infty(B)$ is a Baire set follows from the discussion above as follows. Notice that a function f is Morse precisely when the differential form df, considered as a section of T^*B, is transverse to the 0-section. It is easy to verify that the map

$$B \times C^\infty(B) \to T^*B$$

given by $(b, f) \mapsto df_b$ is transverse to the 0-section. Theorem B.1 then implies that the set of Morse functions $\mathcal{C}_M$ is a Baire set. Notice that $\mathcal{C}_M$ is independent of the metric. When B is compact, we can also say that $\mathcal{C}_M$ is open in $C^\infty(B)$. However, this open set is not connected since critical points can appear and disappear as we vary the function.

When B is compact, there is a neighborhood $\mathcal{O}_f$ of $f \in \mathcal{C}_M$ in $C^\infty(B)$ consisting entirely of Morse functions and such that the number of critical points of a given index on this neighborhood is constant. This gives a a universal parametrization of the stable and unstable manifolds for the critical points α_j of all functions in $\mathcal{O}_f$:

$$u : \cup_j \mathbb{R}^{i(\alpha_j)} \times \mathcal{O}_f \to B$$
$$s : \cup_j \mathbb{R}^{n-i(\alpha_j)} \times \mathcal{O}_f \to B$$

In fact one can cover $\mathcal{C}_g$ with a countable collection of such $\mathcal{O}_f$'s, such that the resulting cover is locally finite ([§33.2][AR]).

The important step is to show that

$$u \times s : \cup_{j,k} \mathbb{R}^{i(\alpha_j)} \times \mathbb{R}^{n-i(\alpha_k)} \times \mathcal{O}_f \to B$$

is transversal to the diagonal. For this, it is necessary to know how the gradient flow is perturbed by varying the function. As a first step, consider perturbing the gradient flow by an arbitrary vector field ([§32][AR]): let $X + sY$ be a one parameter family of vector fields and let Φ^s denote the flow of $X + sY$. Choose an arbitrary point $b \in B$ and let $t \to \gamma(t) = \Phi^0_t(b)$ be the trajectory of X through b. Then we compute

$$\frac{d\Phi^s_{t_0}}{ds}(b)|_{s=0} = \int_0^{t_0} (\Phi^0_t)_*(Y_{\gamma(t_0-t)})dt. \tag{B.1}$$

It is important to notice that this derivative only depends on the perturbing vector field Y along the trajectory $\gamma(t)$.

The heart of the proof of the Kupka-Smale theorem is then to show that the vector fields (in our case, the gradient vector fields) are rich enough to make the universal stable and unstable manifolds transversal. To this end, suppose that there is $(x,y,f) \in \mathbb{R}^i \times \mathbb{R}^{n-i} \times \mathcal{O}_f$ so that $(u \times s)(x,y,f) = (b,b)$. Denote the derivative of u and s along $\mathcal{O}_f$ by $\frac{\partial u}{\partial X}$ and $\frac{\partial s}{\partial X}$ respectively. Then $u \times s$ is transverse to the diagonal if for every $v \in T_b B$, there is a vector field Y supported away from the critical points, such that either

$$\begin{aligned} \frac{\partial u}{\partial X}(x,y,f)(0,0,Y) &= v \\ \frac{\partial s}{\partial X}(x,y,f)(0,0,Y) &= 0, \end{aligned} \tag{B.2}$$

or

$$\begin{aligned} \frac{\partial s}{\partial X}(x,y,f)(0,0,Y) &= v \\ \frac{\partial u}{\partial X}(x,y,f)(0,0,Y) &= 0, \end{aligned} \tag{B.4}$$

a relation which in fact is considerably stronger than transversality. We need to perturb by vector fields whose support is disjoint from an open set containing the critical points. Suppose that γ connects critical points α and β. When b lies on γ and is close to β, we choose Y to satisfy (B.2). Otherwise when b is close to α, we find Y satisfying (B.4).

We construct the vector field Y as follows. Choose $h \in C^\infty(\mathbb{R})$ with support in $[-t_0, 0]$ so that $\int_{\mathbb{R}} h\,dt = 1$. Let $\gamma(t)$ be the gradient line through b and define Y along $\gamma(t)$ by

$$Y(\gamma(t)) = h(t)(\Phi_t)_*(v).$$

Using (B.1), it is clear that

$$\begin{aligned} \frac{\partial u}{\partial X}(x,y,f)(0,0,Y) &= v \\ \frac{\partial s}{\partial X}(x,y,f)(0,0,Y) &= 0. \end{aligned}$$

It is elementary to further verify that Y may be extended to B with support in a neighborhood of $\gamma(t)$ so that $Y = -\nabla\tilde{f}$ for some function $\tilde{f}$. □

The proof for genericity of time dependent flows is slightly easier, as one may now use time dependent functions to perturb the given vector field. The property of the flow used in the homotopy equivalence of complexes is the one parameter version of the transversality of time dependent gradient flows.

References

[AR] R. Abraham and J. Robbin, «Transversal Mappings and Flows», W.A. Benjamin, 1967.

[AB1] M.F. Atiyah & R. Bott, *The moment map and equivariant cohomology*, Topology **23** (1984), 1–28.

[AB2] M.F. Atiyah & R. Bott, *The Yang-Mills equations over Riemann Surfaces*, Phil. Trans. Roy. Soc. London A **308** (1982), 523–615.

[Au] D.M. Austin, *Equivariant Floer cohomology for spherical space forms*, in preparation.

[AuB] D.M. Austin & P.J. Braam, *Equivariant Floer Cohomology*, in preparation.

[BGV] N. Berline, E. Getzler & M. Vergne, «Heat kernels and Dirac operators», Springer-Verlag, 1992.

[Bi] J.M. Bismut, *The Witten Complex and the degenerate Morse inequalities*, J. Diff. Geom. **23** (1986), 207–240.

[Bo] A. Borel, «Seminar on Transformation Groups», Annals of Math. Studies 46, Princeton University Press, 1960.

[B1] R. Bott, *Morse theory indomitable*, Publ. Math. IHES **68** (1988), 99–114.

[B2] R. Bott, *An application of the Morse theory to the topology of Lie groups*, Bull. Soc. Math. France **84** (1956), 251–281.

[B3] R. Bott, *Nondegenerate critical manifolds*, Ann. Math. **60** (1954), 248–261.

[B4] R. Bott, *The stable homotopy of the classical groups*, Annals of Mathematics **70** (1959), 313–337.

[BS] R. Bott & H. Samuelson, *Applications of the theory of Morse to symmetric spaces*, Am. J. Math. **80**, 964–1029.

[BT] R. Bott & L. Tu, «Differential Forms in Algebraic Topology», Springer-Verlag, 1982.

[Br1] P.J. Braam, *Floer homology groups for homology 3-spheres*, Adv. in Math. **88** (1991), 131–144.

[BD] P.J. Braam & S.K. Donaldson, *Gluing $SO(3)$-polynomial invariants*, in this volume.

[C] H. Cartan, *La transgression dans un groupe de Lie et dans un espace fibré principal*, Colloque de Topologie (Espaces Fibrés), 57–71, C.B.R.M. Bruxells.

[CJS] R.L. Cohen, J.D.S. Jones & G.B. Segal, *Morse Theory and Classifying Spaces*, preprint.

[Du] J.J. Duistermaat, *Stable Manifolds*, preprint (1976).

[Fa] I. Fary, *Valeurs critiques et algèbres spectrales d'une application*, Annals of Math. **63** (1956) 437–490.

[FS] R. Fintushel & R. Stern, *Instanton Homology of Seifert Fibered Homology Three-Spheres*, Proc. London Math. Soc (3) **61** (1990), 109–137.

[F1] A. Floer, *An Instanton Invariant for 3-Manifolds*, Comm. Math. Phys. **118**, (1988), 215–240

[F2] A. Floer, *Morse theory for Lagrangian intersections*, J. Diff. Geom. **29** (1988) 513–547.

[F3] A. Floer, *Witten's Complex and infinite dimensional Morse theory*, J. Diff. Geom. **30** (1989) 207–221.

[Fu] K. Fukaya, *Instanton homology for oriented 3-Manifolds*, preprint (1989).

[G] V.A. Ginzburg, *Equivariant cohomology and Kähler geometry*, Funktsional. Anal. i Prilozhen. **21** 19–34.

[HPS] M. Hirsch, C. Pugh & M. Shub, «Invariant Manifolds», Springer-Verlag, 1977.

[K] F. Kirwan, *Cohomology of Quotients in Symplectic and Algebraic Geometry*, Mathematical Notes 31, Princeton University Press, 1986.

[MQ] V. Matthai & D. Quillen, *Superconnections, Thom Classes and equivariant differential forms*, Topology **25** (1986), 85–110.

[Mi1] J. Milnor, «Morse Theory», Annals of Math. Studies 51, Princeton University Press, 1973.

[Mi2] J. Milnor, «Lectures on the h-cobordism Theorem», Mathematical Notes, Princeton University Press, 1965.

[Mo] M. Morse, *The calculus of variation in the large*, Amer. Math. Soc. Colloquium Publications, Vol. XVIII. (1934).

[P] T. Petrie, *Obstructions to transversality for compact Lie groups*, Bull. Amer. Math. Soc. **80** (1974), 1133–1136.

[Sm1] S. Smale, *Morse Inequalities for a dynamical system*, Bull AMS (1960), 43–69.

[Sm2] S. Smale, *On gradient dynamical systems*, Annals of Math **74** (1961), 199–206.

[Wa] A.G. Wasserman, *Equivariant differential topology*, Topology **8** (1969), 127–150.

[Wi] E. Witten, *Supersymmetry and Morse theory*, J. Diff. Geom. **17** (1982), 661–692.

D.M. Austin
Institute for Advanced Study
Princeton, NJ
University of British Columbia
Vancouver, BC Canada V6T 1Y4
E-mail address daustin@unixg.ubc.ca

P.J. Braam
The Mathematical Institute
23–29 St. Giles
Oxford OX1 3LB,UK

Some simple continuity properties of symplectic capacities

S.M. Bates

Introduction

According to the Eliashberg-Gromov C^0-symplectic rigidity theorem (see [Gro2]), a smooth diffeomorphism which is C^0-approximable by smooth symplectomorphisms must itself be symplectic. Motivated by this fact, recent developments in symplectic geometry have led to the concept of symplectic capacity, a «two-dimensional» invariant c for subsets of $\mathbb{R}^{2n}$ satisfying the following axioms:

C1. (Monotonicity) If $U \subset V$, then $c(U) \leq c(V)$.
C2. (Conformality) If $f^*\omega = \alpha \cdot \omega$, then $c(f(U)) = |\alpha| \cdot c(U)$ for all $U \subset \mathbb{R}^{2n}$.
C3. (Non-triviality) $c(B^{2n}(1)) > 0$ and $c(Z^{2n}(1)) < \infty$.

Here $f\colon \mathbb{R}^{2n} \to \mathbb{R}^{2n}$ is assumed to be a smooth diffeomorphism, $B^{2n}(1)$ denotes the unit ball in $\mathbb{R}^{2n}$, and $Z^{2n}(1) = B^2(1) \times \mathbb{R}^{2n-2}$.

Early examples of symplectic capacities were provided by Gromov [Gro1], followed by Ekeland and Hofer [EH1], [EH2]; further examples and extensions of capacities to subsets of more general symplectic manifolds are described in [Hof1], [Hof2], [HZ], [V1], [V2] and notably in the joint work of Floer with Hofer [FH] and Hofer-Wysocki [FHW]. As recently pointed out in [Hof3], the existence of symplectic capacities and their relation to the C^0-symplectic rigidity theorem motivate in turn the definition of a continuous symplectomorphism of $\mathbb{R}^{2n}$ as any homeomorphism $h\colon \mathbb{R}^{2n} \to \mathbb{R}^{2n}$ such that h and $h \times \mathrm{Id}_{\mathbb{R}^2}\colon \mathbb{R}^{2n+2} \to \mathbb{R}^{2n+2}$ preserve every capacity of all open subsets of their domains. In particular, these conditions guarantee that $Dh \in Sp(\mathbb{R}^{2n})$ wherever the derivative Dh of such a map exists.

The main goal of this note is to present a version of the symplectic rigidity theorem for C^0-limits of continuous symplectomorphisms of $\mathbb{R}^{2n}$. To this end, we consider the closely related topic of (dis)continuity properties of symplectic capacities on spaces of subsets of $\mathbb{R}^{2n}$ endowed with Hausdorff-like topologies. This leads in particular to the concept of symplectically regular subsets of $\mathbb{R}^{2n}$, and our main result states that a (bijective) C^0-limit of continuous symplectomorphisms preserves the capacity of all regular open sets.

I am happy to thank H.Hofer and my advisor A.Weinstein for helpful discussions and criticisms of the manuscript.

---

*) Supported by an NSF Graduate Fellowship in Mathematics.

1 Preliminaries

Consider any euclidean space (X,d) with its usual metric, and let $\mathcal{B}_X \subset 2^X$ denote the space of bounded open subsets of X. Each subset U of X gives rise to a *distance function* $d_U \colon X \to \mathbb{R}$ defined by

$$d_U(x) = \inf\{d(x,y) : y \in U\}.$$

The triangle inequality for the metric d implies that this function is continuous. If $\overline{U}$ denotes the topological closure of U, then clearly $d_U = d_{\overline{U}}$. The *Hausdorff distance* between two subsets A, B of X is given by the formula

$$d_{\mathcal{H}}(A,B) = \sup_{x \in B} d_A(x) + \sup_{y \in A} d_B(y).$$

The Hausdorff distance defines a pseudometric on $\mathcal{B}_X$ which for our purposes gives too coarse a picture of the differences between certain subsets (by Example 2.1. below, the function $B \mapsto \mathrm{vol}(B)$ is everywhere discontinuous with respect to the Hausdorff topology on the space of embedded balls in X), and we will instead equip $\mathcal{B}_X$ with the much finer topology induced by the pseudometric

$$d_S(A,B) = d_{\mathcal{H}}(A,B) + d_{\mathcal{H}}(\partial A, \partial B).$$

Here, $\partial A = \overline{A} \setminus A^\circ$ is the topological boundary of A.

The goal of this section is to describe d_S-continuity points of functions $c \colon \mathcal{B}_X \to \mathbb{R}$ which are monotone with respect to inclusions, i.e., $U \subset V$ implies $c(U) \leq c(V)$ for all $U, V \in \mathcal{B}_X$. For this purpose, it will be useful to consider those $U \in \mathcal{B}_X$ at which the value of c equals the limit of its values on topologically significant sets containing or contained in U.

Definition 1.1. *For a monotone function $c \colon \mathcal{B}_X \to \mathbb{R}$, define $c_i, c_o \colon \mathcal{B}_X \to \mathbb{R}$ by*

$$c_i(U) = \sup_{\substack{K \text{ open} \\ \overline{K} \subset U}} c(K) \qquad c_o(U) = \inf_{\substack{V \text{ open} \\ \overline{U} \subset V}} c(V).$$

From this definition it follows easily that $c_i \leq c \leq c_o$ and $(c_i)_i = c_i$, $(c_o)_o = c_o$. If $c = c_i$ ($c = c_o$), then c will be called inner (outer) regular. A set $U \in \mathcal{B}_X$ is said to be *c-regular* if $c_i(U) = c_o(U)$.

Lemma *If $c \colon \mathcal{B}_X \to \mathbb{R}$ be a monotone function, then:*

(1) *c_i, c_o are lower- and upper-semicontinuous, respectively.*
(2) *Every c-regular open set is a d_S-continuity point of c.*

Proof If $U \in \mathcal{B}_X$ and $K \subset U$ is compact, then there exists $\varepsilon > 0$ such that $K^\varepsilon = d_K^{-1}([0,\varepsilon))$ satisfies $d_{\mathcal{H}}(K^\varepsilon, \partial U) > \varepsilon$. If $U' \in \mathcal{B}_X$ does not contain K, then either U' is disjoint from some component of K^ε, or $\partial U' \cap K^\varepsilon \neq \emptyset$. In both cases, $d_S(U,U') > \varepsilon$, and so $K \subset U'$ whenever U' is d_S-close to U. This implies lower-semicontinuity of c_i, and the upper-semicontinuity of c_o is checked similarly. Statement (2) follows from (1) and the inequality $c_i \leq c \leq c_o$. □

The converse of statement (2) of this lemma is generally false, as illustrated by the punctured open disk $D_0 \subset \mathbb{R}^2$ and the function whose value at $U \in \mathcal{B}_{\mathbb{R}^2}$ is 1 if $\overline{D}_0 \subset U$ and zero otherwise. Specifically, since any set U containing $\overline{D}_0$ must satisfy $d_S(U, D_0) \geq 1$, it follows that D_0 is a continuity point of c, although it fails to be c-regular. Despite such examples, continuity and regularity can be further related via *regular chains*.

Definition 1.3. *A regular chain is a family $\{U_t\}_{t\in\mathbb{R}}$ of bounded open sets in X such that*

(1) $\overline{U}_{t'} \subset U_t$ *for* $t' < t$.
(2) *If V is any open set containing $\overline{U}_t$, then there exists $t_V > t$ such that $\overline{U}_{t_V} \subset V$.*
(3) *If $K \subset U_t$ is compact, then there exists $t_K < t$ such that $K \subset U_{t_K}$.*

Given an open set $U \in \mathcal{B}_X$, a regular chain $\{U_t\}$ for which $U = U_0$ is easily constructed by setting

$$U_t = \begin{cases} U^t & \text{for } t > 0 \\ U & \text{for } t = 0 \\ U \setminus \overline{(\partial U)^{|t|}} & \text{for } t < 0 \end{cases}$$

where U^ε denotes the open ε-neighborhood of U in X. The main remark of this section is the following.

Theorem 1.4. *Let U_t be a regular chain and let $c : \mathcal{B}_X \to \mathbb{R}$ be monotone. If $c_U : \mathbb{R} \to \mathbb{R}$ is defined by $c_U(t) = c(U_t)$, then:*

(1) *The function c_U is continuous at all but at most countably many points of $\mathbb{R}$.*
(2) *The set U_t is c-regular if and only if c_U is continuous at t.*

Proof Statement (1) is an immediate consequence of the fact that $c_U : \mathbb{R} \to \mathbb{R}$ is monotone: Since the value of a monotone function must increase by a positive amount at each point of discontinuity, there can be only countably many such points within any interval on which c_U is defined (the sum of any uncountable collection of positive numbers is necessarily infinite).

The definition of the regular chain U_t implies that for any compact subset $K \subset U$ and any open set V containing $\overline{U}$, there exists $\varepsilon > 0$ such that $K \subset U_t$ and $\overline{U}_t \subset V$ for any $t \in (-\varepsilon, \varepsilon)$. Statement (2) thus follows from the definition of c-regular sets. □

This implies a converse of Lemma 1.2. for open sets with smooth boundary.

Corollary 1.5. *If c is continuous at an open set $U \in \mathcal{B}_X$ bounded by a smooth hypersurface, then U is c-regular.*

Proof Using a tubular neighborhood of ∂U, it is easy to construct a regular chain U_t such that $U = U_0$ and which is *continuous* as a map $t \mapsto U_t$ into $\mathcal{B}_X$. Continuity of c at U implies that c_U is continuous at 0, and the assertion follows from Theorem 1.4.(2). □

Next we consider homeomorphisms which preserve the value of c on c-regular open sets.

Lemma 1.6. *If $f: X \to X$ is a homeomorphism, then the following are equivalent:*

(1) $c_0(f(U)) = c_0(U)$ *and* $c_i(f(U)) = c_i(U)$ *for all* $U \in \mathcal{B}_X$.
(2) $c(f(U)) = c(U)$ *for all c-regular* $U \in \mathcal{B}_X$.

Proof The assertion $(1 \Rightarrow 2)$ follows from the definition of c-regular sets. To prove the converse, let U_t be a regular chain with $U = U_0$. By Theorem 1.4., there exists a positive sequence $t_k \to 0$ such that U_{t_k} is c-regular for all k. Thus,

$$c_o(U) = \lim_{k\to\infty} c(U_{t_k}) = \lim_{k\to\infty} c(f(U_{t_k})) = c_o(f(U)).$$

Similarly for c_i. □

We will call a map $f : X \to X$ a *c-homeomorphism* if it satisfies the hypotheses of Lemma 1.6.; from the lemma, it follows easily that these mappings comprise a subgroup of the homeomorphism group of X.

Theorem 1.7. *The set of c-homeomorphisms is a closed subgroup of the homeomorphism group of X in the C^0-topology.*

Proof If f_k is a sequence of homeomorphisms of X which C^0-converge to a homeomorphism f, then $f_k(U)$ must d_S-converge to $f(U)$ for any $U \in \mathcal{B}_X$. From the upper-semicontinuity of c_0, it therefore follows that any homeomorphism $f: X \to X$ which is C^0-approximable by c-homeomorphisms must satisfy $c_0(f(U)) \geq c_0(U)$ for any $U \in \mathcal{B}_X$. Since the homeomorphism group of X is a topological group with the C^0 topology (see [A]), f^{-1} is also C^0-approximable by c-homeomorphisms, hence $c_0(f^{-1}(U)) \geq c_0(U)$. These inequalities imply that f *preserves* c_0, and a similar argument for c_i completes the proof. □

2 Regularity properties of symplectic capacities

Using the discussion of the preceding section, we now study some properties of volume and symplectic capacity of subsets of $\mathbb{R}^{2n}$ equipped with its standard metric, symplectic, and volume structures. In answer to a basic problem mentioned by Viterbo in the survey [V1p.353], we begin with the following simple construction, which shows in particular that volume and each symplectic capacity are everywhere discontinuous with respect to the Hausdorff metric on the space of compact sets in $\mathbb{R}^{2n}$ bounded by smooth hypersurfaces.

Example 2.1. Let $U \subset \mathbb{R}^{2n}$ be any connected bounded open set. Given $\varepsilon > 0$, we choose a finite subset A of U which approximates U in the Hausdorff metric. If $r > 0$ and $p \in U$ are chosen so that the ball $B^{2n}(p, r)$ of radius r centered at p is contained in U, then the standard proof (see [Bo]) of the k-fold transitivity of the symplectomorphism group of $\mathbb{R}^{2n}$ produces a hamiltonian isotopy of $\mathbb{R}^{2n}$ under which $B^{2n}(p, r)$ maps to a subset B of U containing A. $\triangle$

On the other hand, capacities are better behaved on the space $\mathcal{B}_{\mathbb{R}^{2n}}$ with respect to the pseudometric d_S introduced in the preceding section. For example:

Proposition 2.2. *Any countable family of symplectic capacities is d_S-continuous at a dense subset of the space of open sets in $\mathbb{R}^{2n}$ with smooth compact boundary.*

Proof Open sets $U \in \mathcal{B}_{\mathbb{R}^{2n}}$ with smooth boundary belong to *continuous* regular chains U_t as seen in the proof of Corollary 1.5., and the continuity points of any countable family of capacities must include U_t for all but countably many $t \in \mathbb{R}$ by Theorem 1.4.. $\square$

To contrast this result, it would be interesting to find a non-regular open set with *smooth* boundary, in the spirit of Viterbo's question mentioned above. Presently we know of no such example and instead turn to some specific regular capacities and regular sets. Denote by $\mathcal{D}$ the group of smooth symplectomorphisms of $\mathbb{R}^{2n}$, and set

$$\bar{c}(U) = \inf\{\pi r^2 : \exists \psi \in \mathcal{D} \text{ with } \psi(U) \subset Z^{2n}(r)\},$$

$$\underline{c}(U) = \sup\{\pi r^2 : \exists \psi \in \mathcal{D} \text{ with } \psi(B^{2r}(r)) \subset U\}$$

for $U \in \mathcal{B}_{\mathbb{R}^{2n}}$. As indicated in [EH1], $\underline{c}$ and $\bar{c}$ are extremal symplectic capacities in the sense that $\underline{c} \leq c \leq \bar{c}$ for any c satisfying the «normalization» condition $c(B^{2n}(1)) = c(Z^{2n}(1)) = \pi$. Evidently $\underline{c}$ is inner-regular and $\bar{c}$ is outer-regular; the preceding inequality implies furthermore that $\bar{c}_i$ and $\underline{c}_o$ are maximal inner- and minimal outer-regular members for the class of normalized capacities. Further examples include the inner-regular Hofer-Zehnder capacity [HZ] and the outer-regular Ekeland-Hofer capacities [EH2].

An open set which is c-regular for every capacity c will be referred to as (symplectically) regular. A familiar collection of such sets are those with «symplectically convex» boundaries.

Proposition 2.3. *If the boundary of an open set $U \in \mathcal{B}_{\mathbb{R}^{2n}}$ is a smooth connected hypersurface of restricted contact type, then U is symplectically regular.*

Proof Since ∂U is of restricted contact type, there exists an isotopy $\psi_t : \mathbb{R}^{2n} \to \mathbb{R}^{2n}$ such that $\psi_t^* \omega = e^t \omega$ and $U_t = \psi_t(U)$ is a regular chain with $U = U_0$. The conformality axiom C2 implies that every capacity c must satisfy $c(U_t) = e^t\, c(U)$, and so the assertion follows from Theorem 1.4.. $\square$

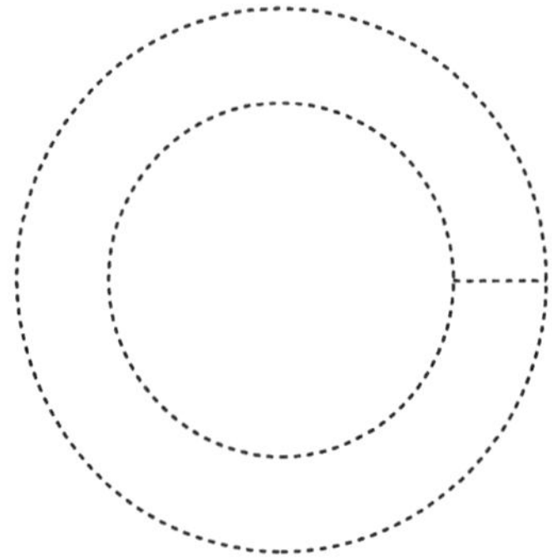

Fig. 1

To contrast this proposition, we now describe a non-regular open set. By a general theorem of Greene and Shiohama [GS], there exists an area-preserving embedding of the unit disk $B^2(1)$ into $\mathbb{R}^2$ whose image A consists of an open annulus of inner radius $R > 1$ with a single radial segment removed (see Fig.1). By the «extension after restriction» principle (see [EH1p.371]), it is clear that A can be exhausted by ambient symplectic embeddings of disks of radius < 1, thus $c_i(A) = c(B^2(1))$ for any capacity c. On the other hand, $c_o(A) = c_o(\overline{A}) \geq c_o(\partial B^2(R))$; since $\partial B^2(R)$ is of restricted contact type, this inequality combined with [EH1Thm.1] implies in particular that $c_{EH}(A) > R^2 \cdot c_{EH}(B^2(1))$, where c_{EH} denotes the outer-regular capacity constructed in [EH1], and so A is not regular.

One interesting consequence of this example is that not every subset of $\mathbb{R}^2$ which is symplectomorphic to the unit disk must be symplectically regular (in fact, the latter inequality also implies that $\underline{c}(A) < \overline{c}(A)$; compare [EH1p.372]). To produce a similar example with connected boundary in dimension 4, we recall that by the Floer-Hofer product formula (see [V1]), the lagrangian torus $\partial B^2(R) \times \partial B^2(R) \subset \mathbb{R}^4$ has c_{EH}-capacity equal to πR^2, and by monotonicity we see therefore that $c_{EH}(\overline{A \times A}) \geq \pi R^2$. On the other hand, $\overline{A \times A}$ can be $d_{\mathcal{H}}$-approximated by symplectic embeddings of polydisks $B^2(r) \times B^2(r) \to A \times A$ for $r < 1$. Since the c_{EH}-capacity of these polydisks is bounded by π, it follows from similar reasoning as before that $A \times A$ is non-regular.

A diffeomorphism φ of $\mathbb{R}^{2n}$ is said to *displace* a subset $B \subset \mathbb{R}^{2n}$ provided that the intersection $\varphi(B) \cap B$ is empty. From [Hof2] we recall that the displacement energy capacity κ (resp. $\hat{\kappa}$) of a bounded subset $B \subset \mathbb{R}^{2n}$ equals the infimum of the energies of all hamiltonian diffeomorphisms which displace B (resp. $\overline{B}$). Using the fact that a map which displaces $\overline{B}$ must also displace a neighborhood of $\overline{B}$, it is easily verified that $\hat{\kappa}$ is outer-regular; examples such as the set A described above illustrate on the other hand that κ is not inner-regular. (For related observations on the continuity properties of κ, we refer to [P].)

A further check of definitions shows that $\kappa_i = \hat{\kappa}_i$, $\kappa_o = \hat{\kappa}$, and thus $\kappa(S) = \hat{\kappa}(S)$ for every symplectically regular set S. In order to see that this equality does

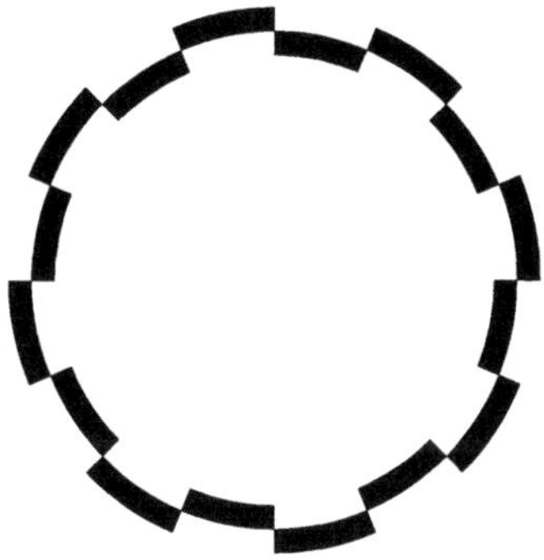

Fig. 2

not hold for more general $S \subset \mathbb{R}^{2n}$, we need a lemma. For $B \in \mathcal{B}_{\mathbb{R}^{2n}}$, let $\underline{B}$ denote the union of B and its bounded complementary components.

Lemma 2.4. *For any bounded, connected set $B \subset \mathbb{R}^{2n}$,*

$$\kappa(B) = \kappa(\underline{B}).$$

Proof Note that $\underline{B}$ cannot be mapped symplectically into any bounded component U of $\mathbb{R}^{2n} \setminus B$: If $f : \underline{B} \to \mathbb{R}^{2n}$ is any continuous map such that $\overline{f(\underline{B})} \subset U$, the obvious inclusions $U \subset \operatorname{int}(\underline{B})$, $\overline{U} \subset \underline{B}$ imply that $\operatorname{int}(\underline{B}) \setminus \overline{f(\operatorname{int}(\underline{B}))} \neq \emptyset$, and so f fails to preserve volume. □

Fix $\varepsilon > 0$ and let S_ε consist of disjoint open pieces of two concentric annuli such that each piece has angular length ε, and the closure $\overline{S}_\varepsilon$ contains the circle of radius 1 (see Fig.2). To bound $\kappa(S_\varepsilon)$ from above, we note that the time-1 map of the hamiltonian flow associated to the «harmonic oscillator» $H_\varepsilon(q,p) = \frac{\varepsilon}{2} \cdot (q^2 + p^2)$ displaces S_ε from itself. By altering H_ε appropriately away from S_ε, we can conclude that $\kappa(S_\varepsilon) \leq 2\varepsilon$. Meanwhile, Lemma 2.4. implies that $\hat{\kappa}(S_\varepsilon) \geq \kappa(B^2(1)) = \pi$, and so the strict inequality $\hat{\kappa}(S_\varepsilon) > \kappa(S_\varepsilon)$ holds for small ε. Combined with our earlier remarks, this example proves

Corollary 2.5. *$\hat{\kappa}$ is not inner-regular, while κ is neither inner- nor outer-regular.*

Finally we note that the preceding example illustrates the sharpness of the inequality $c_{EH} \leq \hat{\kappa}$ proven in [Hof2Thm.1.6(a)]: Although $\kappa(S_\varepsilon) \leq 2\varepsilon$, the fact that $\overline{S}_\varepsilon$ contains the unit circle implies that $c_{EH}(S_\varepsilon) \geq \pi$. In other words, a given set with large c_{EH}-capacity may be displaced from itself with relatively little energy, although this is never true of the set's closure.

3 C^0 symplectic maps

With the «symplectic size» of subsets of $\mathbb{R}^{2n}$ described in concrete terms by symplectic capacities, it is natural to define a C^0 symplectomorphism as any (compactly supported) homeomorphism of $\mathbb{R}^{2n}$ which preserves the capacity of every subset of $\mathbb{R}^{2n}$. This definition is supported by the following result from [EH1].

Theorem (Ekeland-Hofer). *If c is a symplectic capacity, and if $f\colon \mathbb{R}^{2n} \to \mathbb{R}^{2n}$ and $f \times \mathrm{Id}_{\mathbb{R}^2}$ preserve the c-capacity of every ball in their respective domains, then the derivative Df is symplectic where it exists.*

We will consider the following apparently weaker definition of C^0 symplectic map.

Definition 3.1. *If c is a symplectic capacity, then a homeomorphism $f\colon \mathbb{R}^{2n} \to \mathbb{R}^{2n}$ is called a c-symplectomorphism provided that f and $f \times \mathrm{Id}_{\mathbb{R}^2}$ preserve the c-capacity of every c-regular open set of their respective domains.*

By Proposition 2.3. and the preceding theorem, the derivative of a c-symplectomorphism is symplectic wherever it exists. Lemma 1.6. and Theorem 1.7. imply

Theorem 3.2. *If c is a symplectic capacity, then the set of c-symplectomorphisms is a closed subgroup of the homeomorphism group of $\mathbb{R}^{2n}$ in the C^0 topology. If c is inner-regular or outer-regular, then $c(U) = c(f(U))$ for every open set U and c-symplectomorphism f.* □

Intersecting these subgroups for all symplectic capacities proves the following C^0 version of the Eliashberg-Gromov rigidity theorem:

Corollary 3.3. *The set of homeomorphisms of $\mathbb{R}^{2n}$ which preserve all capacities of each regular open set is a C^0 closed subgroup of the homeomorphism group of $\mathbb{R}^{2n}$.* □

Nonsmooth elements of this subgroup are provided, for example, by the hamiltonian flow of any compactly supported C^1 function on $\mathbb{R}^{2n}$ whose hamiltonian vector field is *uniquely integrable* (see [Ba]).

References

[A] Arens, R., Topologies for homeomorphism groups, *Am. J. of Math.* **68** (1946) 593–610.

[Ba] Bates, S.M., *Symplectic End Invariants and C^0 Symplectic Topology*, Ph.d. Thesis, U.C.Berkeley, 1994.

[Bo] Boothby, W.M., Transitivity of the automorphisms of certain geometric structures, *Trans. Amer. Math. Soc.* **137** (1969) 93–100.

[EH1] Ekeland, I. and Hofer, H., Symplectic topology and hamiltonian dynamics, *Math. Zeit.* **200** (1989) 355–378.

[EH2] —, Symplectic topology and hamiltonian dynamics II., *Math. Zeit.* **203** (1990) 553–567.

[FH] Floer, A. and Hofer, H., Symplectic homology I., *Math. Zeit.* **215** (1994) 37–88.

[FHW] Floer, A., Hofer, H., and Wysocki, K., Applications of symplectic homology, preprint.

[GS] Greene, R.E., and Shiohama, K., Diffeomorphisms and volume-preserving embeddings of noncompact manifolds, *Trans. Amer. Math. Soc.* **255** (1979) 403–414.

[Gro1] Gromov, M., Pseudoholomorphic curves in symplectic manifolds, *Invent. Math.* **82** (1985) 307–347.

[Gro2] —, *Partial Differential Relations,* Springer-Verlag, Berlin, 1986.

[Hof1] Hofer, H., Estimates for the energy of a symplectic map, *Comment. Math. Helv.* **68** (1993) 48–72.

[Hof2] —, On the topological properties of symplectic maps, *Proc. of the Royal Soc. Edinburgh*, **115A** (1990) 25–38.

[Hof3] —, Symplectic invariants, *Proc. of the ICM*, Kyoto, 1990, vol.I, 521–528.

[HZ] Hofer, H. and Zehnder, E., A new capacity for symplectic manifolds, in *Analysis, et cetera,* P.Rabinowitz and E.Zehnder, eds., Academic Press, 1990, 405–428.

[P] Polterovich, L., An obstacle to non-lagrangian intersections, *this volume.*

[V1] Viterbo, C., Capacités symplectiques et applications, *Astérisque* 177–178, Sém. Bourbaki, exp.714, (1988–89) 345–362.

[V2] —, Symplectic topology as the geometry of generating functions, *Math. Ann.* **292** (1992) 685–710.

S.M. Bates
Department of Mathematics
University of California
Berkeley, CA. 94720

Floer's work on instanton homology, knots and surgery

P.J. Braam and S.K. Donaldson

Introduction

This paper is an exposition of Floer's work which was completed *circa* 1989 and distributed in the shape of the two preprints [F1] (which is the preceding paper in this volume), [F2] (which was distributed as a «Preliminary version»). A description of the results was published in the Durham Proceedings [F3]. In this first part of the paper we deal with the «gauge theory» content of this work of Floer: that is, the proofs of the exact triangle in [F1] and the «excision axiom» of [F2]. This part is written with the aim of coming quickly to grips with the main geometrical points involved. The second part of our paper will take the topics further, introducing a more general framework for the results and describing the calculation scheme Floer developed in [F2]. The salient points here are in Sections II.1.3 where automorphisms of Floer homology are explained and in II.1.4, where Floer homology is computed for some important special manifolds. In section II.2.2 the exact triangle is discussed in a general setting, and in II.3.1, where it is explained how Kirby calculus gives rise to a set of exact sequences.

Acknowledgments
We would like to thank S. Akbulut and W.B.R. Lickorish for assisting us in understanding the Kirby calculus, and David Austin for proofreading our manuscript. The first author would like to thank the University of British Columbia for its hospitality.

Part I
Floer's exact triangle and excision axiom

1 Background

We begin by reviewing some fundamental points about Floer's homology groups for 3-manifolds (the «instanton homology groups»). These fundamentals are now quite widely known: more details and background can be found in Floer's original paper [F], in the exposition of Braam [B], the broad survey of Atiyah [A1], the papers of Fintushel and Stern [FS1,2], Fukaya [Fuk] and Dostoglou and Salamon [DS1,2] and the forthcoming book [DFK]. Let Y be an oriented homology 3-sphere: the Floer homology groups $HF_i(Y), i = 0, \dots, 7$ are abelian groups intrinsically associated to Y. To define the groups one considers the Chern-Simons functional **cs** on the space of connections on a (trivial) $SU(2)$ bundle over Y. The critical points of **cs** are the flat connections, and if we fix a Riemannian metric on Y the solutions of the anti-self-dual instanton equation on the tube $Y \times \mathbf{R}$ can be viewed as integral curves of the gradient vector field of **cs**. The critical points of **cs** may be very degenerate but in that case one can choose a small perturbation $\mathbf{cs}'$ with a finite set of non-degenerate critical points (modulo gauge equivalence). If the perturbation is small the critical points are all irreducible except for the trivial connection: there are no flat S^1 connections since Y is a homology sphere. Let $\mathfrak{R}$ be the finite set of irreducible critical points, modulo gauge equivalence. One defines a grading function $\nu : \mathfrak{R} \to \mathbb{Z}/8$ using the spectral flow of the Hessian of $\mathbf{cs}'$ along 1-parameter families of connections over Y, or equivalently, by the index of the operator $d^* + d^+$ coupled to connections over the tube. We write $\mathfrak{R}_i$ for $\nu^{-1}(i)$, with $i \in \mathbb{Z}/8$. The grading function is normalised by considering paths to the trivial connection. For any flat connections $\rho, \sigma \in \mathfrak{R}$, the connections over the tube which are asymptotic to ρ and σ respectively at $-\infty$, $+\infty$ are divided into classes labelled by a relative second Chern class. Within one of these classes one can set up the usual Fredholm machinery to study the solutions of the instanton equation (or the relevant deformed equation in the case that one has chosen a perturbation of the Chern-Simons functional). The virtual dimension of the moduli space $M(\rho, \sigma)$ of such solutions is equal to $\nu(\rho) - \nu(\sigma)$ modulo 8. And, for generic choices of perturbation and metric on Y, the moduli space is a smooth manifold of this dimension. If ρ is not equal to σ, the translations of the tube act freely on the moduli space so we get a reduced space $M_0(\rho, \sigma) = M(\rho, \sigma)/\mathbb{R}$. For our purposes we will only need low dimensional moduli spaces, corresponding to a particular class of connections over the tube with the given limits, so we will not bother to distinguish explicitly between the different components of moduli spaces with same limits.

To define the groups HF(Y) one forms the $\mathbb{Z}/8$-graded chain complex with groups

$$C_i(Y) = \bigoplus_{\rho \in \mathfrak{R}_i} \mathbf{Z}. < \rho >$$

generated by the critical points $\mathfrak{R}$. To be more precise one should really choose certain data attached to each critical point which will permit the orientation of all moduli spaces $M(\rho, \sigma)$, much as in the definition of the homology of a simplicial complex. Then the Floer differential

$$\partial : C_i(Y) \to C_{i-1}(Y)$$

is defined by

$$\partial < \rho > = \sum n_{\rho\sigma} < \sigma >,$$

where $n_{\rho\sigma}$ is the number of points in the 0-dimensional space $M_0(\rho, \sigma)$, counted with appropriate signs. The main result of [F] is that $\partial^2 = 0$ and the homology groups HF$_*(Y)$ of $(C_*(Y), \partial)$ are independent of the choices of perturbation and metric, up to canonical isomorphism. It follows immediately from the definition that changing the orientation of Y changes the homology groups to the corresponding cohomology groups so there is a dual pairing:

$$\mathrm{HF}_*(Y) \otimes \mathrm{HF}_*(\overline{Y}) \to \mathbf{Z}. \tag{1}$$

These Floer homology groups allow one to extend the invariants for closed manifolds defined by Yang-Mills instantons to manifolds with boundary. Consider an oriented 4-manifold X with boundary a homology sphere Y. There is a number $\nu(X) \in \mathbf{Z}/8$ such that if ρ is a critical point with $\nu(\rho) = \nu(X)$ there is a moduli space of instantons (or, if appropriate, solutions of a deformation of the instanton equation) $M(X, \rho)$, over the 4-manifold obtained by adjoining a semi-infinite tube $Y \times [0, \infty)$ to X, with limit ρ, having dim $M(X, \rho) = 0$. Once again one needs to build in extra data (an orientation of the homology of X) which will allow the orientation of the moduli spaces, but when this is done we get a number $n(X, \rho)$ of points in $M(X, \rho)$. One can then consider the element

$$\psi_X = \sum n(X, \rho) < \rho >$$

in $C_{\nu(X)}(Y)$. The numbers $n(X, \rho)$, and hence the class ψ_X, can alter as we change the various choices made (e.g. the Riemannian metric on X); the basic idea is that the corresponding Floer homology class will *not* change. Some further discussion is necessary for a complete treatment here, since in general one may run into trouble from the trivial (reducible) connection, much as one runs into trouble in defining the invariants for closed manifolds in some cases. For a full discussion see [DFK]: we will not go into this here since it will turn out that these difficulties do not arise in the cases we will be concerned with. Under suitable hypotheses on X then, one shows that $\partial\psi_X = 0$ and that the homology class $\Psi_X = [\psi_X] \in \mathrm{HF}_{\nu(X)}(Y)$ is an invariant of X.

There is a corresponding discussion for cobordisms. If W is an oriented cobordism between homology spheres Y_0, Y_1, there are (again under some restrictions on the homology of W) induced maps

$$\Psi_W : \mathrm{HF}_i(Y_1) \to \mathrm{HF}_{i+\nu}(Y_2),$$

where the dimension shift ν is determined by W, in fact $\nu = 3(b_2^+(W) - b_1(W))$. This is defined by «counting instantons» over W. We define a chain map with matrix entries $n(\rho_1, \rho_2)$ for $\rho_1 \in C_i(Y_1)$, $\rho_2 \in C_{i+\nu}(Y_2)$, equal to the number of solutions — counted with signs — of a deformed instanton equation over the manifold obtained by adjoining tubes to W. Different choices give chain-homotopic maps, which have the same action on homology. One can generalise further to cobordisms with many boundary components: for example a cobordism V from a disjoint union $Y_0 \cup Y_1$ to Y_2 induces a map Ψ_V from $\mathrm{HF}_*(Y_0) \otimes \mathrm{HF}_*(Y_1)$ to $\mathrm{HF}_*(Y_2)$.

These invariants for manifolds with boundary enjoy functorial properties, which are manifestations of the «gluing» of instanton solutions. For example if W' is a cobordism from Y_1 to Y_2 and W'' is a cobordism from Y_2 to Y_3 then

$$\Psi_{W''} \circ \Psi_{W'} = \Psi_W : \mathrm{HF}_*(Y_1) \to \mathrm{HF}_*(Y_3) \tag{2}$$

where W is the obvious composite cobordism. This reflects the geometrical principle that if the metric on W'' is deformed to have a very long tube $Y_2 \times (-T, T)$ in the middle, then instantons over W'' are obtained by gluing instantons over W and W' with the same flat limit over Y_2: indeed Floer's whole theory of instanton homology is a way of systematically encoding this principle. The same idea applies to other cases, for example if the two boundary components Y_0, Y_1 of the cobordism V above are identified by an orientation-reversing diffeomorphism and if X is the 4-manifold with one boundary component Y_2 obtained from V by pasting Y_0 to Y_1, the invariant $\Psi_X \in \mathrm{HF}_*(Y_2)$ is obtained by «contracting» Ψ_V using the dual pairing (1). (Strictly, if we want to work over the integers we should refine this discussion by taking account of the «Tor» terms in the homology of a tensor product. We shall encounter this point again in our discussion of the «excision axiom».)

To complete these preliminaries we will discuss the Floer homology groups of a more general oriented 3-manifold Z: the case we have in mind most is where Z has the homology of $S^1 \times S^2$. In any case we we need to assume that there is an *admissible* bundle over Z, in the terminology of Part II below. The theory proceeds much as before, but using connections on such a (non-trivial) bundle. The theory is simpler in some respects; one does not run into problems with the reducible connection, but a little more complicated in others, and there are slightly different ways of setting it up. The way we prefer in this Part I is to work with a $U(2)$ bundle $\tilde{P}$ over Z such that $c_1(\tilde{P})$ is an «odd» element of $H^2(Z)/\mathrm{Torsion}$. We consider *projectively flat* connections on $\tilde{P}$, whose curvature is some arbitrary fixed 2-form (multiplied by the identity in the adjoint bundle), and with a fixed «determinant» $U(1)$ connection. This is equivalent to working with flat connections on P modulo a restricted gauge group — the gauge transformations which are homotopic to the identity over the 1-skeleton of Z. We can equally work with flat $SU(2)$ connections over the complement in Z of a loop representing $c_1(\tilde{P})$, and with holonomy -1 around a small linking circle. One gets a $\mathbb{Z}/8$-graded homology $\mathrm{HF}_*(Z)$ (which depends on the choice of bundle $\tilde{P}$). If we choose a class $[\Sigma]$ in $H^1(Z) = H_2(Z)$ which has an odd pairing with $c_1(\tilde{P})$ then we get an involution $g = g_\Sigma$ of degree 4

$$g : \mathrm{HF}_*(Y'') \to \mathrm{HF}_{*+4}(Y''),$$

so one can form the quotient $\mathrm{HF}'_*(Z) = \mathrm{HF}^*_Z$ which is $\mathbb{Z}/4$- graded. In simple cases the Floer homology HF'_* can be obtained directly by working with $SO(3)$ connections. From the point of view of $U(2)$ connections, the involution g is induced by the tensor product with a flat complex line bundle having holonomy ± 1. The instanton theory goes through much as before, with maps induced by $U(2)$

bundles over 4-dimensional cobordisms defined by projectively ASD connections. It is harder to normalise the grading in $\mathrm{HF}_*(Z)$. If Z is a homology $S^1 \times S^2$ one can make this normalisation by requiring that if X is a 4-manifold with boundary Z, such that the pair (X, Z) has the homology of $(D^2 \times S^2, S^1 \times S^2)$ and we work with the $U(2)$ bundle over X whose first Chern class corresponds to the dual of $D^2 \times$ point, then the index $\nu(X)$ is 0.

To simplify our language, later in this paper, we will sometimes just refer to «flat» connections, and «instantons», when we really mean the projectively flat connections, and projectively ASD connections on non-trivial $U(2)$ bundles, as discussed above.

2 Statement of results

Suppose that, as above, Y is an oriented homology 3-sphere and let $K \subset Y$ be a knot: a smoothly embedded circle in Y. We introduce two other manifolds:

1) The homology 3-sphere Y' obtained from Y by $+1$ surgery on K,
2) The homology $S^1 \times S^2$ Y'' obtained by 0-surgery on K.

The definitions here are standard fare in 3-manifold topology. Let N be a tubular neighbourhood of K in Y and let

$$\chi : \partial N \to S^1 \times S^1$$

be the natural identification fixed so that $\chi(S^1 \times \{\text{pt.}\})$ maps to zero in $H_1(N)$ and $\chi(\{\text{pt.}\} \times S^1)$ maps to zero in $H_1(Y \setminus N)$. (Thus the first generator is the «meridian» and the second is the «longitude» of the knot.) These conditions fix χ up to isotopy once we chose an oriention of the knot (and compatible orientation of the normal bundle), and in fact the orientation of K will not affect our constructions. The manifold Y'' is obtained by cutting out N and gluing it back using the diffeomorphism of $S^1 \times S^1$ which interchanges the two factors: in homology the situation is the same as going from S^3 to $S^1 \times S^2$ by regluing two $S^1 \times S^2$'s. The manifold Y' is obtained similarly as

$$Y' = N \cup_\xi (Y \setminus N)$$

where $\xi : \partial N \to \partial Y \setminus N$ is the conjugate by χ of the diffeomeorphism

$$(\theta, \phi) \mapsto (\theta - \phi, \phi)$$

of the torus. There are standard, oriented, surgery cobordisms W_1 from Y' to Y and W_2 from Y to Y'' which are illustrated in Diagram 1. In this diagram dimensions have been reduced by two; so we see 2-manifolds with 1-dimensional boundaries, but it provides a fairly accurate guide to the higher dimensional case. In particular, if we let W be the composite cobordism from Y' to Y'', there is an embedded

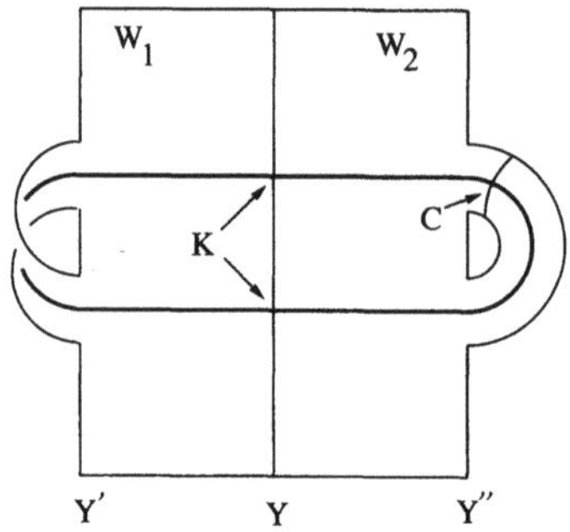

Diagram 1

2-sphere Σ in W with self-intersection -1, as illustrated in the diagram. It is the union of discs $D_i \subset W_i$ whose common boundary is the original knot K. This self-intersection number fixes the signs ($+1$ surgery) in the set-up, and it is crucial that we consider this sign rather than the oppposite: $+1$ surgery on Y yields in general a different manifold to -1 surgery.

We now consider the Floer homology groups of the three manifolds Y', Y, Y''. The cobordisms W_1, W_2 induce maps on the Floer homology, where in the second case we use a non-trivial bundle over W_2 extending that on Y''. (With our grading convention we should use the bundle over W_2 whose first Chern class is dual to the small disc C in the handle, as illustrated.) We will abbreviate Ψ_{W_1}, Ψ_{W_2} to a_*, b_* respectively, so we have

$$\mathrm{HF}_*(Y') \xrightarrow{a_*} \mathrm{HF}_*(Y) \xrightarrow{b_*} \mathrm{HF}_*(Y'') \tag{3}$$

and the result we wish to prove is:

Theorem 1 *The sequence (3) is exact at* $\mathrm{HF}_*(Y)$*: the kernel of* b_* *is the same as the image of* a_*.

A few remarks are in order. First the maps a_*, b_* have degree 0, with respect to the grading we have defined: this is a matter of checking the definitions, using the addition property for indices. For example, the map a_* has degree zero because the cobordism W_1 has $b_1(W_1) = b_2^+(W_1) = 0$ (whereas $b_2^-(W_1) = 1$, with a class represented by the disc D_1 capped off by a surface in Y). Second, the maps a_*, b_* are part of a long exact sequence, or exact triangle,

$$\ldots \mathrm{HF}_{i+1}(Y'') \xrightarrow{c_*} \mathrm{HF}_i(Y') \xrightarrow{a_*} \mathrm{HF}_i(Y) \xrightarrow{b_*} \mathrm{HF}_i(Y'') \xrightarrow{c_*} \mathrm{HF}_{i-1}(Y') \ldots \tag{4}$$

where the map c is defined by another cobordism W_3 from Y'' to Y'. (In fact W_3 is the manifold obtained from $\overline{W}$ by collapsing or «blowing down» the 2-sphere Σ.) In this part of the paper we wish to focus on the single segment (3) in the exact sequence, taking up the long sequence again in Part II. We will see there that there is a certain symmetry between the three terms involved, so the proof of

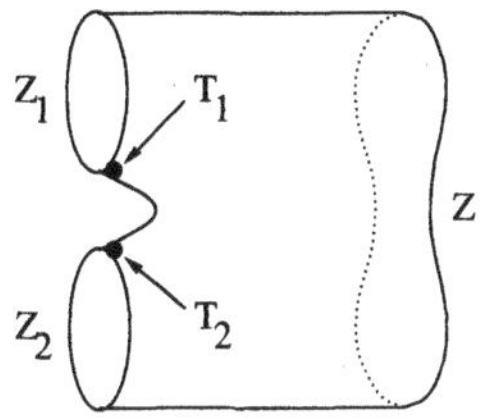

Diagram 2

exactness at each stage of (4) is essentially the same. Note in particular that, since $+1$ and -1 surgery are inverse operations, we have an exact sequence

$$\mathrm{HF}_*(Y'') \to \mathrm{HF}_*(Y) \to \mathrm{HF}_*(Y^{(-1)})$$

where $Y^{(-1)}$ is the result of -1 surgery on K.

We now turn to Floer's «excision property», a substantially simpler result. This deals with the Floer homology groups in the second case of connections on a non-trivial bundle. To motivate the result, consider first a very simple observation. Let $\tilde{P}$ be a non-trivial $U(2)$ bundle over a 3-manifold U which defines Floer groups $\mathrm{HF}_*(U)$. Suppose there is an embedded 2-sphere S in U over which $\tilde{P}$ has odd degree. Then there are no projectively flat connections on $\tilde{P}$, since the same is already true over S. Thus in this situation $\mathrm{HF}_*(U) = 0$. The excision property deals with the next case, of embedded tori, and as we shall see the proof hinges on the fact that there is up to isomorphism, a *unique* projectively flat connection on the bundle of odd degree over the torus. (Or equivalently, flat connection on a non-trivial $SO(3)$ bundle.) To state Floer's result, we consider a pair of 3-manifolds Z_1, Z_2 and embedded tori $T_1 \subset Z_1, T_2 \subset Z_2$ which define non-zero classes in homology. We consider Floer groups defined by bundles $\tilde{P}_i \to Z_i$ which have odd degree over T_i. Fix a diffeomorphism between T_1 and T_2 and let Z be the 3-manifold obtained by cutting open Z_1 and Z_2 along the tori and gluing together the resulting boundary components in pairs, in a way determined by the orientations of Z_1, Z_2. There is a standard cobordism V from the disjoint union $Z_1 \cup Z_2$ to Z, illustrated in Diagram 2.

Again this diagram is a fairly accurate guide, since near the surgery region the cobordism V is a product of a portion of the 2-dimensional cobordism illustrated in the diagram and a 2-torus. There is a bundle over V which extends $\tilde{P}_i$ over the boundary components Z_i, and restricting this to the other end Z of the cobordism we get Floer groups $\mathrm{HF}_*(Z)$. We obtain then an induced map

$$\Psi_V : \mathrm{HF}_*(Z_1) \otimes \mathrm{HF}_*(Z_2) \to \mathrm{HF}_*(Z).$$

Now consider the involution

$$h = g_1 \otimes g_2 : \mathrm{HF}_*(Z_1) \otimes \mathrm{HF}_*(Z_2) \to \mathrm{HF}_*(Z_1) \otimes \mathrm{HF}_*(Z_2),$$

where the involution g_i is defined by the class of T_i in Z_i. There is a class in $H^1(V)$ which restricts to these classes on the boundary components Z_i and to 0 on the component Z, and it then follows that $\Psi_V \circ h = \Psi_V$. Thus we get an induced map from the quotient $\mathrm{HF}_*(\mathrm{Z}_1) \otimes \mathrm{HF}_*(\mathrm{Z}_2)/h$ to the Floer homology of Z. Now to state the result in a simple form we use Floer homology groups with coefficients in the field of rational numbers $\mathbb{Q}$.

Theorem 2 *For the Floer homology groups with coefficients in* $\mathbb{Q}$, *the cobordism* V *induces an isomorphism from*

$$\mathrm{HF}_*(\mathrm{Z}_1) \otimes \mathrm{HF}_*(\mathrm{Z}_2)/h$$

to $\mathrm{HF}_*(\mathrm{Z})$.

Again, a few remarks are in order. First, note that the result is the same for any diffeomorphism between the tori used to construct Z. Going in the other direction, another way of saying this is that if we are given a 3-manifold Z which contains a region N bounded by a pair of tori, we can express the Floer homology as a tensor product of two contributions obtained by «excising» N. Second, if we worked with integral coefficients we would have a more complicated result, involving Tor, using the Kunneth theorem for the homology of a tensor product. Third, the quotient of $\mathrm{HF}_*(\mathrm{Z}_1) \otimes \mathrm{HF}_*(\mathrm{Z}_2)$ by the involution h can also be written, as in Floer's paper, as the tensor product over the group ring $\mathbb{Q}[h]$ of $\mathbb{Z}/2$. Finally, the result becomes tidier in the $\mathbb{Z}/4$-graded set-up in which it just becomes the assertion that, with rational coefficients,

$$\mathrm{HF}'_*(\mathrm{Z}) = \mathrm{HF}'_*(\mathrm{Z}_1) \otimes \mathrm{HF}'_*(\mathrm{Z}_2).$$

3 Techniques

Floer's proof of exactness, Theorem 1, involves the delicate application of three important ideas:

1) the «monotonicity» property of the cobordism functor with respect to filtrations of the Floer groups;
2) deformation of the flatness and instanton equations;
3) analysis of moduli spaces with degenerate limiting data.

The proof of the excision result, Theorem 2, involves only the first of these. In this subsection we will explain these ideas in turn, and give the proof of Theorem 2.

3.1 Monotonicity and filtrations

It is well-known that an $SU(2)$ bundle over a closed 4-manifold which admits an instanton connection must have a positive second Chern class. This follows from that fact that the Chern-Weil integrand Tr F^2 is equal to $|F|^2$ for an anti-self-dual connection. In this subsection we want to discuss analogous results for manifolds with boundary in the setting of Floer's theory. Suppose W is a cobordism between 3-manifolds Y_1, Y_2 and $\mathbf{A}$ is a connection (with group $SU(2)$ say) over W with boundary values A_i over Y_i. Then the basic formula for the secondary Chern-Simons invariant is:

$$\mathbf{cs}(A_1) - \mathbf{cs}(A_2) = \frac{1}{8\pi^2} \int_W \text{Tr } F(\mathbf{A})^2 \mod \mathbb{Z} \tag{5}$$

Now we know that if A is an anti-self-dual connection then Tr $F^2 = |F|^2 d\mu$ so the integral appearing on the right hand side of (5) is positive. One is tempted to conclude that this imposes restrictions on the Chern-Simons invariants of the boundary values; but one is stopped by the fact that the formula (5), and the Chern-Simons invariants themselves only take values in $\mathbb{R}/\mathbb{Z}$ where the ordering is lost. However a closer analysis, which we shall now explain, shows that that the formula (5) does indeed give constraints on the map on Floer homology induced by the cobordism. Similar ideas have been used by Fintushel and Stern [FS2] and Furuta [Fu].

To begin, suppose for simplicity that the Chern-Simons functions on a homology sphere Y has non-degenerate critical points (flat connections), so the Floer homology groups are computed from a complex $C_*(Y)$ generated by the irreducible flat connections. Let ρ, σ be flat connections which define elements in the same chain group $C_\nu(Y)$, i.e. $\nu(\rho) = \nu(\sigma) = \nu$. This means that the spectral flow along any path from ρ to σ (in the space $\mathcal{B}_Y$ of gauge equivalence classes of connections over Y) is equal to zero modulo 8. The fundamental group of $\mathcal{B}_Y$ is $\mathbb{Z}$ and there is just one homotopy class of paths from ρ to σ along which the spectral flow is 0. We can lift the Chern-Simons function along this path to the real numbers and we conclude that there is a preferred integral lift $\tilde{\mathbf{cs}}(\rho, \sigma) \in \mathbb{R}$ of $\mathbf{cs}(\rho) - \mathbf{cs}(\sigma) \in \mathbb{R}/\mathbb{Z}$. This means that there is an intrinsic (partial) ordering of the generators of the chain group $C_\nu(Y)$ defined by $\rho \geq \sigma \Leftrightarrow \tilde{\mathbf{cs}}(\rho, \sigma) \geq 0$.

Now let us return to our cobordism W, and suppose the ends Y_1, Y_2 are homology spheres satisfying the non-degeneracy condition above. For clarity we will not distinguish between connections over W, with boundary values on the Y_i, and connections over the manifold obtained from W by adding on cylindrical ends, with corresponding asymptotic limits. Also, for simplicity let us supppose that the chain map induced by W has degree zero, so maps each $C_i(Y_1)$ to $C_i(Y_2)$. We can now repeat the discussion above, but with connections over W in place of paths in $\mathcal{B}_Y$ (which correspond to connections over the tube). Given ρ_1, ρ_2 over Y_1, Y_2 with $\nu(\rho_i) = \nu$ there is a unique topological class of connections, with the given flat structures over the ends, such that the index of the coupled $d^* + d^+$

operator over W, which gives the dimension of the instanton moduli space, is 0. Within this class the Chern-Weil integral appearing on the right hand side of (5) is constant, so we get a real number

$$I(\rho_1, \rho_2) = \int_W \operatorname{Tr} F_A^2,$$

for any connection A of index 0, depending only on the limits. The additivity property of the index implies that if we have another pair σ_1, σ_2 then:

$$I(\rho_1, \rho_2) + \tilde{\mathbf{cs}}(\rho_2, \sigma_2) = \tilde{\mathbf{cs}}(\rho_1, \sigma_1) + I(\sigma_1, \sigma_2).$$

This implies that there are functions J_1, J_2 such that

$$\begin{aligned} I(\rho_1, \rho_2) &= J_1(\rho_1) - J_2(\rho_2), \\ \tilde{\mathbf{cs}}(\rho_1, \sigma_1) &= J_1(\rho_1) - J_1(\sigma_1), \\ \tilde{\mathbf{cs}}(\rho_2, \sigma_2) &= J_2(\rho_2) - J_2(\sigma_2). \end{aligned}$$

On the other hand, the basic principle linking the instanton equation with the Chern-Weil integrand tells us that the instanton moduli space is empty, and so *a fortiori* the matrix element of the chain map ψ_W is 0, if $I(\rho_1, \rho_2) < 0$, and if $I = 0$ the only instantons are flat connections. To spell out what this means, if we choose ordered bases for the chain groups:

$$\rho_1^1 \geq \rho_1^2 \geq \rho_1^3 \geq \dots,$$

of $C_\nu(Y_1)$ and similarly ρ_2^j for $C_\nu(Y_2)$ then the matrix of ψ_W has zero entries in the terms corresponding to basis elements ρ_1^i, ρ_2^j with $J_1(\rho_1^i) < J_2(\rho_2^j)$, and for basis elements where equality holds the entries are determined by the flat connections over W (or, if these are not regular solutions of the instanton equation, by the flat connections plus the infinitesimal data which determines how they deform under generic perturbations of the equation). This is what Floer means by saying that the map ψ_W «decreases the Chern-Simons functional».

Let us now pass from the chain groups to the Floer homology. The situation is harder to explain if one works with all the groups HF_i simultaneously (for this one has to go to the $\mathbb{Z}$ cover, with $\mathbb{Z}$-graded groups). However we shall only need to consider homology groups in adjacent dimensions, say $C_{\nu-1}, C_\nu, C_{\nu+1}$. We can compare the Chern-Simons function of connections in $C_{\nu-1}(Y), C_\nu(Y)$ by using paths with spectral flow 1. One concludes that one can define a map

$$J : \mathcal{R}_{\nu-1}(Y) \cup \mathcal{R}_\nu(Y) \cup \mathcal{R}_{\nu+1}(Y) \to \mathbb{R},$$

whose differences give real valued lifts of the differences of the Chern-Simons functional, and such that the Floer differential ∂ is decreasing with respect to J.

Now define, for $\alpha > 0$, a subgroup $C_i^{(\alpha)}(Y) \subset C_i(Y)$ for $i = \nu, \nu \pm 1$ by using the basis elements $< \rho >$ with $J(\rho) \leq \alpha$. Clearly there are only a finite number of different subgroups involved, and as α increases, these give a filtration of the chain groups. The decreasing property of the differential means that we have a filtered complex, so we get a corresponding intrinsic filtration of the Floer group $\mathrm{HF}_\nu(Y)$, with a spectral sequence converging to the quotients of the filtration. The only confusing point is the lack of a convenient way to label the filtration, since the function J is only unique up to the adddition of a constant.

Moving back now to the cobordism W, we define functions J_1, J_2 on the three chain groups of each end as before. The function I fixes these up to a single overall constant, so the filtrations at the two ends are compatible and our assertions above about the matrix of the chain map can now be put into the form:

1) the map ψ_W is a filtered chain map between filtered complexes;
2) the induced map on the E_1 term of the spectral sequences is defined by the flat connections over W.

To complete the story, we consider variations in the metric on W. The instantons defined by two different metrics g_1, g_2 may give different chain maps, say ψ_1, ψ_2: $C_*(Y_1) \to C_*(Y_2)$ but, as we have stated in Section 1, these are chain homotopic:

$$\psi_1 - \psi_2 = H\partial + \partial H.$$

Recall how the chain homotopy H is constructed. One considers a generic path of metrics between g_1 and g_2 and the instantons on bundles where the virtual dimension of the moduli space is -1. Thus for $\rho \in \mathcal{R}_i(Y_1)$, $\sigma \in \mathcal{R}_{i-1}(Y_2)$, we let $h_{\rho\sigma}$ be the number of instantons in the parametrised moduli space, with limits ρ and σ (i.e. the number of points along the path when the moduli space is non-empty, counted with suitable signs). These numbers $h_{\rho\sigma}$ are the matrix entries of H. The same principle then shows that H is a strictly decreasing map with respect to the filtrations. So in sum the maps ψ_1, ψ_2 differ by a chain homotopy which respects the filtrations. (This implies that the induced maps, defined using the two metrics, on all stages of the spectral sequences are the same.) In a similar vein one can consider the case when the Chern-Simons function is degenerate, so to define the Floer homology one has to go to a perturbed function $\mathbf{cs}'$. The perturbed functional induces a filtration as before, and one can show that this filtration is independent of the perturbation, provided the perturbation is small enough. So we can sum up by saying that the Floer homology theory has a filtration which is natural with respect to the cobordism functor.

Of course the same principles apply in other situations. For example we can consider a 4-manifold X with a single boundary component Y. Then the Floer homology class Ψ_X is constrained to lie in a given level of the filtration of $\mathrm{HF}_*(Y)$, and the leading component is determined by the flat connections over X.

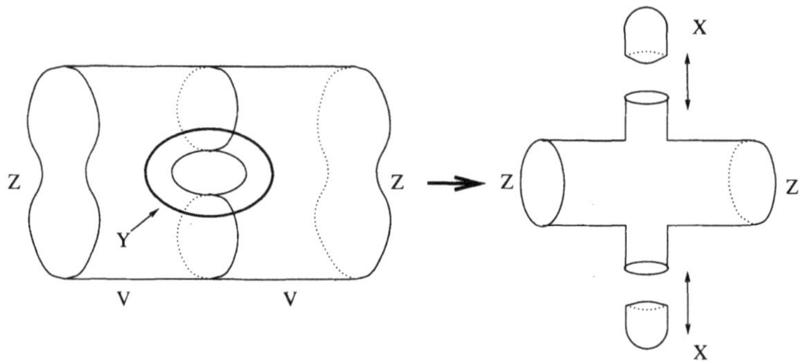

Diagram 3

3.2 The excision theorem

As an illustration of the ideas in 3.1 above, we will now explain the proof of Theorem 2. Let $\overline{V}$ be the manifold V with reversed orientation: i.e. an oriented cobordism from Z to $Z_1 \cup Z_2$. We will show that $\Psi_V \circ \Psi_{\overline{V}} : \mathrm{HF}_*(Z) \to \mathrm{HF}_*(Z)$ is the identity map and $\Psi_{\overline{V}} \circ \Psi_V : \mathrm{HF}_*(Z_1) \otimes \mathrm{HF}_*(Z_2) \to \mathrm{HF}_*(Z_1) \otimes \mathrm{HF}_*(Z_2)$ is the projection $1 + h$ onto the index two subgroup representing the quotient by the involution h. This clearly implies the desired result. We can use the functorial property to compare the composites $\Psi_V \Psi_{\overline{V}}$, $\Psi_{\overline{V}} \Psi_V$ with the maps induced by the composite cobordisms $V \circ \overline{V}, \overline{V} \circ V$. The theory sketched in Section 1 above needs to be amended, in a rather trivial way, in the first case because the intermediate manifold $Z_1 \cup Z_2$ has two components, and there is a class in $H^1(V \circ \overline{V})$ which restricts trivially on each boundary component (the two copies of Z), and is also trivial on the individual cobordisms $V, \overline{V}$. The tensor product with the corresponding flat line bundle effectively doubles the number of projectively flat line bundles, and this means that our gluing formula is $\Psi_{V \circ \overline{V}} = 2\Psi_V \circ \Psi_{\overline{V}}$. (The point is that one gets an additional discrete gluing parameter, multiplying by -1 on *one* of the components of the manifold $Z_1 \cup Z_2$.)

The idea now is to compare the composite cobordisms with products, via surgeries. Let γ be the loop in the cobordism $V \circ \overline{V}$ illustrated in Diagram 3. We get the product $Z \times I$ by cutting $V \circ \overline{V}$ along a 3-torus $S^1 \times T^2 = \gamma \times T^2$ and gluing in two copies of $X = D^2 \times T^2$, with γ regarded as the boundary of the disc. Thus the essential thing is to understand the Floer homology of the 3-torus, and the vector $\Psi_X \in \mathrm{HF}_*(S^1 \times T^2)$.

On an odd degree $U(2)$ bundle over the torus, there is an essentially unique (projectively) flat connection, up to equivalence, as we have mentioned above, and as the reader may readily check. This gives two (projectively) flat connections over $S^1 \times T^2$, α^+, α^- say with holonomy ± 1 respectively around the S^1 factor. One deduces then that there are just two non-zero Floer homology groups, say $\mathrm{HF}_0 = \mathbb{Z}\cdot\alpha^+, \mathrm{HF}_4 = \mathbb{Z}\cdot\alpha^-$ of $S^1 \times T^2$ and the involution g interchanges $\alpha^\pm$. Changing the orientation of the circle gives an orientation-reversing diffeomorphism of $S^1 \times T^2$, so we need not distinguish between this manifold and its oppposite: thus (1) becomes an «intersection pairing» on $\mathrm{HF}_*(S^1 \times T^2)$ and it is clear that α^+, α^- give an orthonormal basis for this pairing.

Now consider the manifold $X = D^2 \times T^2$ with boundary $S^1 \times T^2$. The vector Ψ_X in the Floer homology of the boundary is determined by the instantons over X, and in a general situation this could be very hard to compute. However in this particular case the calculation is easy, and gives a simple illustration of the ideas from 3.1. The flat connection α^+ plainly extends uniquely to a flat connection A^+ over X and it is easy to check that this has deformation index 0 (and is a regular point of the moduli space). Thus if A is any connection over X with index 0 and boundary value α^+, then

$$\int_X \mathrm{Tr}\, F^2 = 0,$$

(where F is the curvature of the corresponding $SO(3)$ connection) and we deduce that, if A is an instanton, it must be flat. So we conclude that A^+ is the only relevant solution over X. The same argument shows that there are no instantons in the relevant moduli space with boundary value α^- (the indices and curvature integrals only involve the $SO(3)$ connections where the distinction between $\alpha^\pm$ disappears), since α^- does not bound a flat connection. Thus we see that

$$\Psi_X = \alpha^+.$$

Now consider the manifold U, with four boundary components, obtained from $V \circ \overline{V}$ by cutting along γ. It defines a Floer class

$$\Psi_U \in \mathrm{Hom}\,(\mathrm{HF}_*(\mathrm{Z}), \mathrm{HF}_*(\mathrm{Z})) \otimes \mathrm{HF}_*(S^1 \times T^2) \otimes \mathrm{HF}_*(S^1 \times T^2),$$

which we can regard as four endomorphisms $U(\alpha^\pm, \alpha^\pm)$ of $\mathrm{HF}_*(Z)$. The map $\Psi_{V\circ\overline{V}}$ is obtained by contracting with the intersection pairing on $\mathrm{HF}_*(S^1 \times T^2)$, i.e.

$$\Psi_{V\circ\overline{V}} = U(\alpha^+, \alpha^+) + U(\alpha^-, \alpha^-).$$

On the other hand the gluing rule tells us that when we attach two copies of $X = D^2 \times T^2$ to get the product, we have

$$\psi_{Z\times I} = U(\alpha^+, \alpha^+).$$

There is a cohomology class in $H^1(U; \mathbb{Z}/2)$ which is zero on the two Z boundary components and 1 on the two copies of γ. Taking the tensor product with the corresponding flat connection gives a correspondence between connections with boundary values α^+, α^-; so we see that

$$U(\alpha^+, \alpha^+) = U(\alpha^-, \alpha^-),$$

and hence

$$\Psi_{V\circ\overline{V}} = 2\ \mathrm{Id}_{\mathrm{HF}_*(Z)}.$$

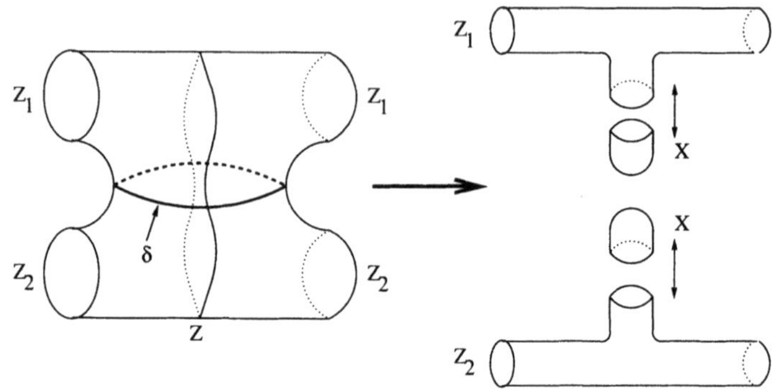

Diagram 4

This means, in view of our remarks above, that $\Psi_{\overline{V}}$ is a right inverse to Ψ_V. Now consider the composite $\overline{V} \circ V$. We recover the product $(Z_1 \cup Z_2) \times I$ by cutting along $\delta \times T^2$ where δ is the loop indicated in Diagram 4, and gluing in two copies of X.

Thus if we let W be the intermediate manifold with four boundary components we again get endomorphisms

$$W(\alpha^{\pm}, \alpha^{\pm}) : \mathrm{HF}_*(Z_1) \otimes \mathrm{HF}_*(Z_2) \to \mathrm{HF}_*(Z_1) \otimes \mathrm{HF}_*(Z_2).$$

Our gluing rules tell us that

$$\Psi_{\overline{V} \circ V} = W(\alpha^+, \alpha^+) + W(\alpha^-, \alpha^-), \;\; 1 = \Psi_{(Z_1 \cup Z_2) \times I} = W(\alpha^+, \alpha^+).$$

On the other hand the effect of tensoring with a flat line bundle shows that

$$W(\alpha^-, \alpha^-) = h \circ W(\alpha^-, \alpha^-) = h,$$

so $\Psi_{\overline{V} \circ V} = 1 + h$ as required.

Remarks

1) The proof here has been made more complicated by the fact that we are working with the $\mathbb{Z}/8$-graded groups. In the $\mathbb{Z}/4$-graded set-up we just have a single generator of $\mathrm{HF}'_*(S^1 \times T^2)$ and the argument becomes a good deal clearer.
2) The assertion about the Floer homology groups can be seen as a refinement of the more elementary fact that the relevant representation space $\mathcal{R}_Z$ is the quotient of $\mathcal{R}_{Z_1} \times \mathcal{R}_{Z_2}$ by an involution.

3.3 Deformations and the geometric triangle

We now move on to the second main technique, the perturbation of the Chern-Simons functional. Of course this has already entered into the discussion, in the very definition of Floer homology, but in that case the perturbations that are required were rather arbitrary — any small sufficiently generic perturbation will do. Now in contrast we need to make specific, large, perturbations which, in Floer's words, «simulate the effect of surgery».

Let γ be an embedded, oriented, circle in a 3-manifold Y. We consider perturbations defined by the holonomy of connections around γ. Let

$$\phi : SU(2) \to \mathbb{R}$$

be a smooth function which is invariant under the adjoint action of $SU(2)$.Of course such a function is determined by a real valued function f on the unit interval by

$$\phi(\begin{pmatrix} e^{i\pi t} & 0 \\ 0 & e^{-i\pi t} \end{pmatrix}) = f(t),$$

and smoothness requires that the derivative of f vanishes at the end points $0, 1$. Now let P be an $SU(2)$ bundle over Y and let $T(A, \gamma)$ be the holonomy of A around γ, which can be viewed as an automorphism of the fibre of P over any point of γ. The holonomy yields a conjugacy class in $SU(2)$, so it makes sense to define a real number

$$\Phi_0(A) = \phi(T_{A,\gamma}).$$

Let us now consider the derivative of Φ_0, regarded as a function on the space of connections. Let a be in $\Omega^1_Y(\mathfrak{g}_P)$, where $\mathfrak{g}_P$ is the vector bundle associated to the adjoint representation, so $A + sa$ is a one parameter family of connections, for $s \in \mathbb{R}$. The derivative $\delta\Phi_0$ is the derivative of $\Phi_0(A+sa)$ at $s = 0$. To write down a formula for this, observe that, if we identify the Lie algebra with its dual by the trace, the derivative of ϕ is a Lie algebra valued 1-form over $SU(2)$ and that if we apply this fibrewise to the holonomy, regarded as a section of the bundle Aut P over γ, we get a section $\phi'(T_A)$, say, of $\mathfrak{g}_P$ over γ. Then we have

Lemma 3 *The derivative of Φ is*

$$\delta\Phi_0 = -\int_\gamma \mathrm{Tr}\ (a\ \phi'(T_A)\ ).$$

To express this more concretely we parametrise γ by the unit interval, and choose a covariant constant trivialisation of the bundle along the interval with respect to which the holonomy is the diagonal matrix

$$\Lambda = \begin{pmatrix} e^{\pi i\tau} & 0 \\ 0 & e^{-i\pi\tau} \end{pmatrix},$$

for a constant $\tau \in [0, 1]$. The variation a over γ is represented by a matrix-valued function $a(x)$, and the formula to be proved is:

$$\delta\Phi_0 = -\int_0^1 \mathrm{Tr}\ (a(x)\phi'(\Lambda))dx,$$

where $\phi'(\Lambda) = \begin{pmatrix} if'(\tau) & 0 \\ 0 & -if'(\tau) \end{pmatrix}$. To prove this formula we consider a covariant constant section ξ_s of the bundle over the interval, with respect to the connection $A + sa$ where s is small, in other words a solution of the differential equation:

$$(\frac{d}{dx} + sa)\xi_s(x) = 0.$$

Differentiating with respect to s and interchanging the order of the derivatives gives:

$$\xi_s(1) = \xi_s(0)(1 - s\int_0^1 a(x)dx) + O(s^2),$$

so the holonomy Λ_s of $A + sa$ is $\Lambda_s = \Lambda(1 - s\int_0^1 a(x)dx) + O(s^2))$ and the result follows since

$$\frac{d}{ds}\phi(\Lambda_s) = \text{Tr } (\Lambda^{-1}\frac{d\Lambda_s}{ds}\phi'(\Lambda)).$$

From the point of view of analysis on Y, the derivative of Φ_0 is a rather singular operator, defined by a distribution supported on the loop. So the next step is to mollify this function by averaging over nearby loops. Fix coordinates to identify a neighbourhood of γ with $\gamma \times D^2$, and write γ_z for the obvious loop, parallel to γ, labelled by a point $z \in D^2$. We fix a compactly supported 2-form μ on D^2, of integral 1, and also denote by μ the corresponding 2-form on Y (an approximation to the current defined by γ). Now for each fixed z we can construct a function Φ_z on the space of connections, in the manner above, using the same function ϕ but the loop γ_z in place of $\gamma = \gamma_0$. Then we put

$$\Phi(A) = \int_{D^2} \Phi_z \mu(z).$$

We can read off the derivative of Φ from Lemma 3 by linearity: the integrations over the loops and the disc combine into a single three-dimensional integral and we get

$$\delta\Phi = -\int_Y \text{Tr } (a\phi'(T_A)) \wedge \mu.$$

Here we have abbreviated our notation somewhat so that T_A is the section of Aut P over the support of μ defined using the holonomy around γ_z at a point on γ_z.

Recall now the basic fact that the derivative of the Chern-Simons functional is given by the curvature:

$$\delta\mathbf{cs} = \int_Y \text{Tr } (a \wedge F_A).$$

Thus the derivative of the ($\mathbb{R}/\mathbb{Z}$-valued) function $\mathbf{cs} + \Phi$ on the space of connections is

$$\delta(\mathbf{cs} + \Phi) = \int_Y \mathrm{Tr}\ (a \wedge (F_A - \phi'(T_A)\mu))$$

and the critical points of $\mathbf{cs} + \Phi$ — the analogues of the flat connections in the deformed theory are the solutions A of the equation

$$F_A = \phi'(T_A)\mu. \tag{6}$$

Clearly the solutions of this equation are flat outside the support of μ. It is a striking fact that the solutions can be described precisely. To do this we consider the torus boundary of our tubular neighbourhood. Any flat $SU(2)$ connection over this torus reduces to S^1 so is determined by two parameters $\alpha, \beta \in (\mathbb{R}/2\mathbb{Z})$ where, in a suitable trivialisation, the holonomy around the meridian (the boundary of our disc D^2) is $\begin{pmatrix} e^{i\pi\alpha} & 0 \\ 0 & e^{-i\pi\alpha} \end{pmatrix}$ and around the longitude (parallel to γ) is $\begin{pmatrix} e^{i\pi\beta} & 0 \\ 0 & e^{-i\pi\beta} \end{pmatrix}$. These give a complete description of the equivalence classes of flat connections except for an ambiguity of sign: the points $(\alpha, \beta), (-\alpha, -\beta)$ correspond to the same connection. In other words, we have the well-known fact that the moduli space of flat $SU(2)$ connections over T^2 is the quotient $\mathcal{V}/\pm 1$ of a dual torus $\mathcal{V}$ (with coordinates α, β) by the group ± 1, which is intrinsically a 2-sphere. This action of ± 1 can be rather confusing, and it is useful to have in mind both the picture in the quotient space S^2 and in the dual torus. This is indicated in Figure 5.

Lemma 4 *Restriction to the complement of the tubular neighbourhood defines a one-to-one correspondence between*

1) *the gauge equivalence classes of solutions of the equation (6) over Y,*
2) *the gauge equivalence classes of flat connections over the complement such that the holonomies $\alpha, \beta \in \mathbb{R}/2\mathbb{Z}$ over the boundary satisfy $\alpha = f'(\beta)$ (where f is extended to a periodic function with period 2 by putting $f(2-b) = f(b)$, for $0 < b < 1$.)*

Notice that, with the given extension of f, the condition in (2) is preserved by $(\alpha, \beta) \mapsto (-\alpha, -\beta)$, so the condition is fully gauge invariant.

To prove this we first show that a solution A of (6) is reducible inside the solid torus N. The essential observation is that the holonomy is constant around the different loops γ_z. Suppose that in general B is a connection on a bundle Q over a manifold X and v is a vector field on X, generating a 1-parameter family $g_t : X \to X$. We use the connection B to lift these diffeomorphisms to the bundle, so we get a family of connections $g_t(B)$, on the same bundle Q, and we can differentiate this family at time 0. The formula for the derivative is

$$\frac{d}{dt} g_t(B) = i_v(F_B),$$

where i_v is the contraction by v. In our case we consider a vector field v transverse to γ, so the diffeomorphisms g_t take γ to nearby loops in the family γ_z. The equation (6) implies that the contraction of the curvature with v has no component along γ and this implies that the holonomy is constant in the family. This is true at the level of conjugacy classes — so that $\Phi_z(A)$ is independent of z — and also in the sense that the covariant derivative $\nabla_A T_A$ vanishes along v. Since it is always true that the covariant derivative of the holonomy is zero along the loop which defines it, we deduce that $\nabla_A T_A$ is zero throughout N. But then T_A defines a reduction of the connection over N. (Of course one has to treat separately the case when T_A is zero, but in this case the connection is flat and the results follow even more easily.)

Once we know that the connection is reducible, we can easily deduce Lemma 4. We choose an eigenspace of the holonomy around γ to define an S^1 connection over N: this is specified up to gauge equivalence by the holonomy $e^{\pi i\beta}$ around γ and by the curvature 2- form. But this curvature form is constrained by the equation (6) to be $f'(\beta)\mu$. The holonomy e^{α} of the S^1 connection around the meridian is given by integrating the curvature over a small spanning disc; thus, since μ has integral 1 and the holonomy around the γ_z is independent of z,

$$\alpha = \int_{D^2} f'(\beta)\mu = f'(\beta).$$

Conversely if we have a pair of holonomies α, β which satisfy this constraint, then we can construct an S^1 connection over the solid torus N with curvature $f'(\beta)\mu$ and holonomy α around each γ_z, and this gives a solution of our deformed equation (6) in N.

We see then that by making these deformations we can deform the axis $\alpha = 0$, which is the condition for extending a flat connection from the boundary of N to the interior into any graph $\alpha = f'(\beta)$, where f is an even function of period 2. We will now see what this has to do with surgery. We can regard the knot complement $Y_0 = Y \setminus N$ as a subset of each of the other manifolds Y', Y'' and so flat connections on these manifolds are given by flat connections over Y_0 which satisfy appropriate conditions on the boundary torus. In the case of Y' it is clear from the definition of the surgery that these condition is $\alpha = \beta$. For the other manifold Y'' the condition would be $\beta = 0$ if we wanted to work with $SU(2)$ connections, but the move to a non-trivial bundle means that we take instead the translated axis $\beta = 1$. (That is, we can represent the relevant projectively flat connections over Y'' by connections over the knot complement with holonomy -1 around a small linking circle.) The situation is summarised in Diagram 5 in which we have drawn the three lines, $L_Y, L_{Y''}, L_{Y'}$ say, in the dual torus $\mathcal{V}$ which correspond to the three manifolds. The corresponding picture in the quotient space, under the involution ± 1, is a triangle in the 2- sphere, as indicated. As we shall see, this «geometric triangle» lies at the root of the «exact triangle» of maps of the Floer homology groups, the basic point being that any side can be deformed into the union of the other two.

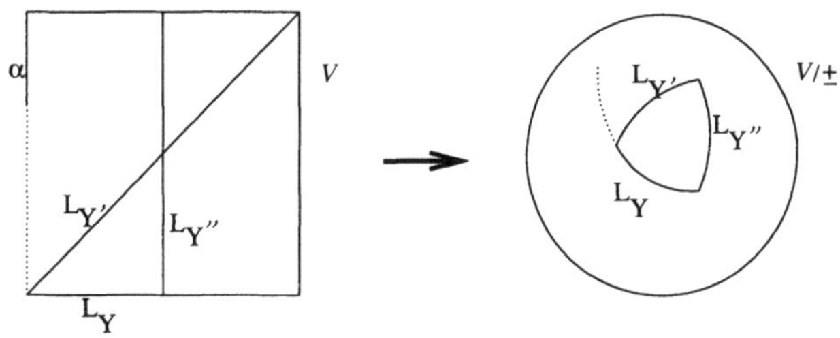

Diagram 5

By choosing a suitable function f, we can deform L_Y to a graph which lies very close to the other sides of the triangle. It would be possible to carry out the proof with such a deformation but the main argument is made a little easier by using a slightly more complicated deformation, which allows one to deform L_Y more precisely into the other two sides. To understand this it is convenient to think of the deformation of the boundary conditions in the following way. Choose the 2-form μ to vanish in a small neighbourhood of $0 \in D^2$. Thus, on the 3-manifold, μ is supported in a region $N \setminus N_0$, where N_0 is a smaller solid torus. Let θ_f be the «shearing» diffeomorphism

$$\theta_f(\beta, \alpha) = (\beta, \alpha + f'(\beta)),$$

of the representation torus $\mathcal{V}$. Then Lemma 4 says that the boundary condition over the outer torus boundary ∂N is obtained from that over the inner torus boundary ∂N_0 (i.e. the condition $\alpha = 0$ for a flat extension) by applying the diffeomorphism θ_f in the representation space. Suppose now that we modify the construction by using a foliation of the thickened torus $N \setminus N_0$ by circles which wind k times around the «core», i.e. which represent the class longitude $+ k \times$ meridian in homology in place of the circles which run parallel to the meridian that were used before (the case $k = 0$). The previous discussion extends without change: indeed it is just a matter of changing the trivialisation of the normal bundle of the loop. Thus the solutions of the resulting deformed equation, starting with a real function g, correspond to pairs (β, α) with

$$\alpha = g'(\beta + k\alpha).$$

In other words we now shift the boundary condition $\alpha = 0$ by applying a diffeomorphism ϕ_g to the representation space, which is a shear parallel to the line $\beta + k\alpha = 0$.

Now suppose that we combine these two cases, by using a deformation $\Phi_1 + \Phi_2$, where Φ_1 is defined by a function g and a foliation of $N \setminus N_0$, as above, with $k = -1$; and Φ_2 is defined by the original construction, using a function f and a foliation of the inner solid torus N_0 by circles with winding number 0. It is clear then that the solutions of the critical point equation for $\mathbf{cs} + \Phi_1 + \Phi_2$ correspond to boundary conditions (on the outer torus) obtained by applying the

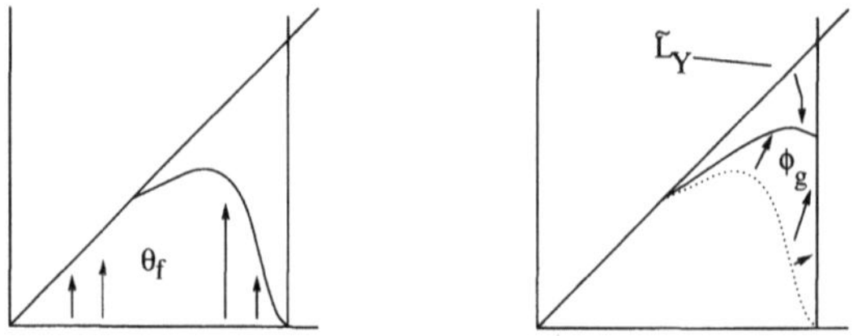

Diagram 6

composite diffeomorphism $\phi_g \circ \theta_f$ to $L_Y = \{\alpha = 0\}$. Now we can choose f and g such that this composite maps L_Y to a curve $\tilde{L}_Y$ formed by «rounding off» the vertex in $L_{Y'} \cup L_{Y''}$ by an arbitrarily small amount. It is straightforward to write down explicit formulae, but the idea is perhaps better conveyed by Diagram 6.

We now want to go on to the 4-dimensional picture and consider the analogues of the instantons for these deformed equations. It will be important to distinguish between two different notions, which we might call «time dependent» and «time independent» deformations.

We begin with the case of the tube $Y \times \mathbb{R}$, with a fixed Riemannian metric on Y. If Φ is a perturbation of the kind considered above, supported on the neighbourhood of a loop $\gamma \subset Y$, the corresponding gradient-line equation for a 1-parameter family A_t is

$$\frac{\partial A(t)}{\partial t} = *_3(F_{A(t)} - \mu\phi'(T_{A(t)})\ ),$$

where $*_3$ is the Hodge $*$-operator on Y. (With a suitable interpretation of the notation, this formula includes also the case of a «composite» deformation, as discussed above.) The solutions of this correspond to connections $\mathbf{A}$ over $Y \times \mathbb{R}$ which satisfy an equation

$$F^+_{\mathbf{A}} = \phi'(T_{\mathbf{A}})\mu^+. \tag{7}$$

The notation here means that we regard μ as a form on $Y \times \mathbb{R}$ (pulling back from Y under the projection map), and μ^+ is the self-dual component. Similarly, we define $T_{\mathbf{A}}$ over a point (y, t) in the support of μ^+ by using the holonomy of a circle $\gamma_z \times \{t\}$, for $y \in \gamma_z$. This translation between the 3 and 4 dimensional points of view runs just as in the familiar, undeformed, case.

Now let X be a 4-manifold with tubular ends, say with two ends $Y_1 \times (-\infty, 0], Y_2 \times [0, \infty)$ and a compatible metric. Suppose there are loops $\gamma_1 \subset Y_1, \gamma_2 \subset Y_2$ and that we can find a «product region» $S^1 \times D^2 \times \mathbb{R} \subset X$ compatible with product structures $\gamma_i \times D^2$ in Y_i. (Again, this extends in the obvious way to the case of a composite deformation.) We let μ be a 2-form on X suported in this product region and pulled back from a form on the disc, just as before. In this situation, with a fixed function ϕ, we can write down a *«time independent» deformation of the instanton equation on* X: $F^+(\mathbf{A}) = \phi'(T_{\mathbf{A}})\mu^+$, where the right hand side is interpreted as before using the product structure over the support of μ^+. This kind of deformation is a particular case of a more general «time dependent» class, consisting roughly speaking of any equation $F^+_{\mathbf{A}} + \epsilon(\mathbf{A}) = 0$ which restricts to the right form over the ends of X. For example, we can make a time dependent deformation

$$F^+_{\mathbf{A}} = \beta(t)\phi'(T_{\mathbf{A}})\mu^+,$$

where β is a cut-off function equal to 1 on the ends, and zero in the interior of X. The advantage of these time-dependent deformations is that they always exist: one does not need the product region running through X but merely over the ends. If the functional $\mathbf{cs} + \Phi$ has nondegenerate critical points on the boundary we can define homology groups using the gradient flow, and the solutions of any generic time-dependent deformation of the instanton equation define a chain map between the chain groups, as in the unperturbed situation. As we shall now see, the advantage of the special time-independent deformations, when they exist, is that these give maps which are compatible with filtrations, as in 3.1 above, a feature which does not hold for the general time-dependent case.

Suppose that there is a product region in X, with a 2-form μ, as above. Consider any connection over X, asymptotic to constants at the ends, and the deformed Chern-Weil integral

$$\tilde{I}(\mathbf{A}) = \frac{1}{8\pi^2}\int_X \mathrm{Tr}\ (F_{\mathbf{A}} + \mu T_{\mathbf{A}})^2.$$

The integrand is $\mathrm{Tr}\ F^2_{\mathbf{A}} + 2\mathrm{Tr}\ (F_{\mathbf{A}}T_{\mathbf{A}}) \wedge \mu$, since $\mu \wedge \mu = 0$. The second term in this is suported in the product region, where we can represent $\mathbf{A}$ as a 1-parameter family A_t. Then $F_{\mathbf{A}} \wedge \mu = \frac{d}{dt}A_t \wedge \mu$, so

$$\int_{D^2\times S^1} \mathrm{Tr}\ (F_{\mathbf{A}}T_{\mathbf{A}}) \wedge \mu = \frac{d}{dt}\Phi(A_t),$$

and the right-hand side of this formula makes sense since Φ is supported in the product region. Now, performing the integral over the remaining «time» variable, we get

$$\tilde{I}(\mathbf{A}) = \int_X \mathrm{Tr}\ (F_{\mathbf{A}} + \mu T_{\mathbf{A}})^2 = \int_X \mathrm{Tr}\ (F^2_{\mathbf{A}}) + \Phi(A_\infty) - \Phi(A_{-\infty}).$$

We deduce, applying the usual Chern-Weil theory to the first term, that $\tilde{I}(\mathbf{A})$ is constant on all connections $\mathbf{A}$ with the same limits, and the same relative Chern class. On the other hand we can write

$$\mathrm{Tr}\ (F_{\mathbf{A}} + \mu T_{\mathbf{A}})^2 = |F^+_{\mathbf{A}} + \mu^+ T_{\mathbf{A}}|^2 + |F^-_{\mathbf{A}} + \mu^- T_{\mathbf{A}}|^2,$$

so in the familiar way we see that $\tilde{I}(\mathbf{A}) \geq 0$ if $\mathbf{A}$ is a solution of the deformed instanton equations. The case where equality holds will be important: we call a connection *deformed-flat* if

$$F_{\mathbf{A}} = \mu T_{\mathbf{A}}; \tag{8}$$

so a deformed-instanton is deformed-flat if and only if $\tilde{I} = 0$.

It is now clear that the whole discussion of 3.1 goes over to the case of a time independent deformation of the instanton equations: the Floer chain groups for the deformed functionals $\mathbf{cs} + \Phi$ at each end have filtrations and the chain map defined by the solutions over the cobordism is decreasing with respect to these filtrations.

3.4 Analytical results for instantons with degenerate boundary conditions
A third plank in the main proof is a circle of analytical ideas involving index, moduli and gluing theorems for the instanton equations in a degenerate situation. Of course in the nondegenerate case these ideas are built in to the very definition of the Floer homology: the extension to the degenerate case is important in a number of situations and has been discussed by various authors. From the point of view of differential topology this extension corresponds to the passage from Morse theory to Morse-Bott theory. We shall now summarise the relevant results, without proof.

Suppose that X is a 4-manifold with a tubular end modelled on a 3-manifold Z, and that σ is a flat connection over Z. We can associate to σ two integers $h^0(\sigma), h^1(\sigma)$, the dimension of the cohomology groups of Y in the twisted coefficient system ad σ. These have the geometric significance that h^0 is the dimension of the stabiliser Γ_σ of σ in the gauge group, while $h^1(\sigma)$ is the dimension of the Zariski tangent space to the moduli space of flat connections at the point σ. Now let A be any connection over X which is asymptotic to σ in a suitable sense. One considers the operator $d_A^+ + d_A^*$, familiar in the deformation theory of instantons, acting on weighted Sobolev spaces with a weight function $e^{-\epsilon t}$ on the end, where $t > 0$ is standard coordinate on the tube. Informally, if ϵ is negative we force the elements of our function spaces to decay exponentially along the tube, while if ϵ is positive we allow exponential growth. If ϵ is either small and positive or small and negative, the operator is Fredholm on these weighted spaces, so we get Fredholm indices $\text{ind}^+(A), \text{ind}^-(A)$ in the two cases. These indices are different and are related by

$$\text{ind}^+(A) = \text{ind}^-(A) + h^0(\sigma) + h^1(\sigma). \tag{9}$$

There is a nonlinear moduli theory for instantons based on the weighted spaces with $\epsilon < 0$. This describes a moduli space M_σ of gauge equivalence classes of instantons which differ from A by exponentially decaying terms along the end, and in particular have the same limit σ. For *irreducible* points in this moduli space, after generic perturbation, we have

$$\dim M_\sigma = \text{ind}^-(A). \tag{10}$$

If all points in the moduli space are reducible, so A has a non-trivial isotropy group Γ_A with dimension $h^0(A)$, this formula should be modified to

$$\dim M_\sigma - h^0(A) = \text{ind}^-(A). \tag{11}$$

Now suppose that σ is contained in a moduli space V of flat connections over Y (which may be an open set in the full moduli space) along which the Chern-Simons function is «nondegenerate in the sense of Bott». This means that V is a smooth manifold of dimension $h^1(\sigma')$, for any $\sigma' \in V$. We also suppose that $h^0(\sigma')$, and hence the conjugacy class of $\Gamma_{\sigma'}$, is constant over V. Then we can construct a moduli space M_V which, as a set, is the union of the $M_{\sigma'}$. Thus there is an obvious map

$$r_X : M_V \to V.$$

Analytically, we can set this up as a smoothly varying family of Fredhom problems. The dimension of M_V, after generic perturbation is

$$\dim M_V(X) - h^0(A) = \text{ind}^-(A) + h^1(\sigma). \tag{12}$$

Here, again, the term $h^0(A)$ is only of relevance in the reducible case. The formulae (11),(12) should be understood in the sense that if the predicted dimension $\dim M_\sigma$ is negative but greater than $-h^1(\sigma)$, so the predicted dimension d of M_V is positive, then generically M_V has dimension d and the *generic* fibre of the map r_X is empty (but, of course, not necessarily all the fibres).

Now let X' be another manifold having the same end but with reversed orientation, so we can make a generalised connected sum $X\sharp_T X'$ in the usual way, with a positive parameter T measuring the length of the neck. Suppose A' is a connection over X' which is asymptotic to σ. We can form a connected sum connection $A\sharp A'$ over the closed manifold $X\sharp X'$ and this has an index, which is given by the gluing formula,

$$\text{ind}\,(A\sharp A') = \text{ind}^+(A) + \text{ind}^-(A') = \text{ind}^-(A) + \text{ind}^-(A') + h^0(\sigma) + h^1(\sigma). \tag{13}$$

Rearranging these index formulae we see that the dimension of the moduli space $M_{X\sharp X'}$, say, of instantons which are approximately connected sums satisfies:

$$\begin{aligned}&\dim M(X\sharp X') - h^0(A\sharp A') \\ &\quad = (\dim M_X(V) - h^0(A)) + (\dim M_{X'}(V) - h^0(A')) - h^1(\sigma) + h^0(\sigma),\end{aligned} \tag{14}$$

or if A and A' are irreducible:

$$\dim M_{X\sharp X'} = \dim M_X(V) + \dim M_{X'}(V) - h^1(\sigma) + h^0(\sigma). \tag{15}$$

It is natural to interpret (15) as saying that the instantons on the connected sum are made up of instantons over X and X' which have the same limit in V, and with a «gluing parameter» in a copy of Γ_σ. (Then the extra terms in (14) just keep track of the extent to which this gluing parameter can be cancelled by the automorphisms of A, A'.) A precise gluing theorem of this kind was proved by Mrowka [Mr], [MMR], and a proof by Fukaya is available in [Fuk]. A result sufficient for our needs is as follows:

Proposition 5 *Suppose that $r_X, r_{X'}$ are proper maps and all points in $M_V(X)$ and $M_V(X')$ are regular. Let N be the product over V,*

$$N = M_V(X) \times_V M_V(X') \equiv \{(A, A') \in M_X(V) \times M_{X'}(V) | r_X(A) = r_{X'}(A')\},$$

and suppose that the maps $r_X, r_{X'}$ are transverse along N. There is a fibre bundle E over N, with fibre the quotient $\frac{\Gamma_\sigma}{\Gamma_A \times \Gamma_{A'}}$ and for any compact subset E′ of E and large enough parameter T a diffeomorphism G from E′ to an open subset in the moduli space $M(X \sharp X')$.

Here a few remarks are in order: first the isotropy groups $\Gamma_A, \Gamma_{A'}, \Gamma_\sigma$ are compared in the obvious way, by evaluation at infinity. The statement of the Proposition is not completely precise, since different gluing parameters may lead to different products $\Gamma_A \times \Gamma_{A'} \subset \Gamma_\sigma$, but this will not be relevant for our application and we will leave the interested reader to formulate a more precise result. In the same vein, one can easily extend the result to the case where the isotropy groups change in the various moduli spaces. (This is best done by working with moduli spaces of «based» connections.) In another direction this gluing theorem should go hand-in-hand with completeness results, which assert that the image of the diffeomorphism G contains all connections that are close to connections in $M_{X'}, M_X$ on the two halves of the connected sum. It would take up too much space to formulate this here, so we will just refer to the many similar statements in the literature.

The final analytical point we shall need concerns the comparison of the «degenerate» situation considered here with a nondegenerate one. Thus we suppose that we introduce a perturbation ϵ and study the gradient lines of $\mathbf{cs} + \epsilon$. The main point is that these can be modelled on a *finite dimensional* problem: the gradient lines of the restriction of ϵ to the finite dimensional manifold V. This principle is important in a number of parts of Floer's theory, for example in computing the Floer homology when the Chern-Simons functional has critical manifolds (see [Fuk]). The result we shall use involves a version of the principle for a time-dependent equation:

$$\frac{\partial A}{\partial t} = \text{grad } (\mathbf{cs} + \epsilon_t)(A). \tag{16}$$

The functions ϵ_t define, by restricion, functions $\epsilon_t|_V$ on the manifold V and these have a gradient vector field $\text{grad}_V \epsilon_t|_V$. The idea is that when ϵ is small the solutions of (16) are well-approximated by the integral curves of $\text{grad}_V \epsilon_t|_V$ on V. To formulate a precise result, we introduce a small parameter δ, and let $\mathbf{cs} + \eta_s$ be a family of variations, for $s \in \mathbb{R}$, $\eta_0 = 0$. Fix a function $\chi(t)$ equal to 0 negative t and 1 for $t \geq 1$. Then consider a variation $\mathbf{cs} + \epsilon_t = \mathbf{cs} + \eta_{\delta\chi(t)}$. Let $F : V \to \mathbb{R}$ be the partial derivative $F = \frac{\partial \eta_s}{\partial s}$, so that $\epsilon_t|_V$ is approximately $\delta\chi(t)F$. Suppose $w(\tau), 0 \leq \tau < \infty$ is a path in V which is an integral curve of the gradient of F: $\frac{dw}{dt} = \text{grad}F$; and $w(\tau)$ has a limit v_∞ which is a critical point of F. Then the

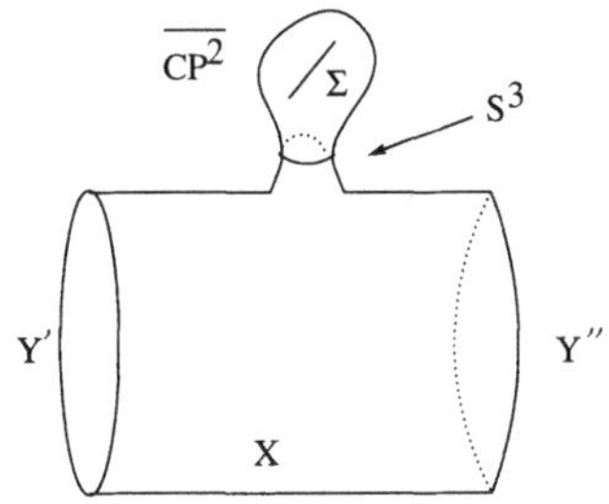

Diagram 7

reparametrised curve $v_\delta(t) = w_{\delta\Xi(t)}$, where Ξ is the integral of χ from 0, satisfies

$$\frac{dv_\delta}{dt} = \delta\chi(t)\mathrm{grad}F.$$

We can extend $v_\delta(t)$ to all $t \in \mathbb{R}$ by making $v_\delta(t) = w(0)$ for $t < 0$. The result we want is:

Proposition 6 *If the critical point v_∞ of F on V is non-degenerate with Morse index λ, then:*

1) *For small δ there is a single nearby critical point σ^*_∞ of $\mathbf{cs}+\eta_\delta$, in the space of all connections, and σ^*_∞ is non-degenerate.*
2) *For small δ there is a regular solution $A(t)$ to the deformed-instanton equation (16), for $-\infty < t < \infty$, with limits $\sigma_{w(0)}$ and σ^*_∞, which is near to the curve v_δ in V. The index of the corresponding connection $\mathbf{A}$ over $Y \times \mathbb{R}$ (with respect to the deformed instanton equation) is* $\mathrm{ind}^-(\mathbf{A}) = -(h^0(\sigma_{w(0)}) + \lambda)$.

The first part of Proposition 6 is straightforward differential topology. The second part can be proved using the Fredholm theory in weighted spaces and a continuity argument. (One may compare this with with the deeper approximation results in [DS2].)

4 The proof of exactness

With these preliminaries in place we can now go on to explain Floer's proof of the exactness property (Theorem 1). We begin with the easier part, the fact that the composite $b_* a_* = 0$. This is well-understood and, as Floer mentions, is a counterpart of similar «vanishing theorems» for closed 4-manifolds [DK]. Recall that there is an embedded 2-sphere Σ in the composite cobordism W from Y' to Y'', with self-intersection -1. The boundary of a tubular neighbourhood of Σ is a 3-sphere, and we can write

$$W = X \sharp \overline{\mathbf{CP}^2},$$

where X is another cobordism from Y' to Y''.

Next we remember that by the gluing rule (naturality property) the composite $b_* a_*$ is the same as the map defined by the instantons over W (extended by cylindrical ends) for any metric on W. More precisely we want to use projectively ASD connections on a $U(2)$ bundle with $c_1 = 1$ on Σ, which give instantons on a *non-trivial* $SO(3)$ bundle. We consider a family of metrics $g_\lambda, \lambda \to 0$ on W in which the neck of the connected sum is made very small, or what is conformally the same, the manifolds are joined by a very long tube. Then we claim that, for small λ, the relevant moduli space, of instantons with index 0, over W is *empty*. For, to state the argument somewhat informally, such instantons are made up by gluing an $SO(3)$-instanton A_X over X to an $SO(3)$ instanton, with non-zero w_2, $A_{\overline{\mathbf{CP}^2}}$ over $\overline{\mathbf{CP}^2}$. The instanton A_X is irreducible, since it is already irreducible on the boundary, hence its deformation index, ind (A_X), is non-negative. The instanton over $\overline{\mathbf{CP}^2}$ might *a priori* be reducible, but only to S^1, since w_2 is non-zero. The deformation index ind $(A_{\overline{\mathbf{CP}^2}})$ is therefore not less than -1 (i.e. minus the dimension of the isotropy group in the gauge group). The addition rule for the index tells us that a connection $A_W = A_X \sharp A_{\overline{\mathbf{CP}^2}}$ made by gluing these together has index

$$\text{ind}\,(A_W) = \text{ind}\,(A_X) + \text{ind}\,(A_{\overline{\mathbf{CP}^2}}) + 3 \geq 2,$$

so in particular there are no instantons with index 0 over X.

Before really coming to grips with the harder part of Floer's proof, we need some discussion of transversality and general position. For this we return to the picture of the representation torus $\mathcal{V}$ of the boundary ∂N of the knot-neighbourhood. In fact we shall gain our understanding by studying the situation in its double cover $\mathcal{V}$, compare diagram 5. We write Y_0 for the complement $Y \setminus N$ and $R = R_{Y_0}$ for the space of equivalence classes of representations of the knot complement. There is a restriction map $r : R \to \mathcal{V}$, such that the flat connections over Y are in $1-1$ correspondence with points of $r^{-1}(L_Y) \subset R$. There is a good deal of technology available for studying this set-up. First, there is a deformation theory for flat connections over Y_0 which gives a local model for a neighbourhood of a point $\rho \in R_{Y_0}$ in terms of the cohomology groups of the base, Y_0, in the twisted co-efficient system ad ρ. The Zariski tangent space of R_{Y_0} is $H^1(Y_0; \text{ad}\ \rho)$ and the moduli space is locally smooth if ρ is irreducible and the «obstruction space» $H^2(Y_0; \text{ad}\ \rho)$ is zero. For irreducible connections an Euler characteristic calculation tells us that

$$\dim H^1(Y_0; \text{ad}\ \rho) - \dim H^2(Y_0; \text{ad}\ \rho) = 1,$$

so R_{Y_0} has «virtual dimension» 1. By this we mean the appropriate Fredholm index when the deformation theory is carried out in the framework of Fredholm manifolds and maps, compare [Tau]. In this framework one can show that for generic deformations of the flatness equation, supported in the interior of Y_0, the corresponding deformed space is a 1-manifold at irreducible points.

We need a simple Lemma which relates the deformation theory on the complement with that on the closed manifold Y.

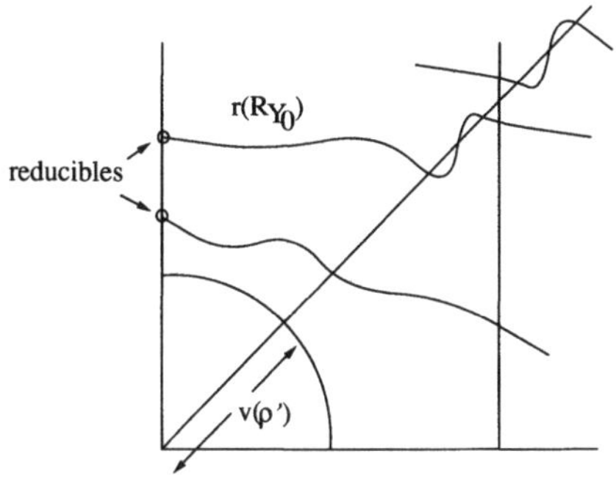

Diagram 8

Lemma 7 *Let ρ be an irreducible flat connection over Y_0 with $r(\rho) \in \text{int}\, L_Y$, and let ρ_Y be the corresponding flat connection over Y. Then ρ_Y is a non-degenerate critical point (i.e $H^1(Y;\rho_Y) = 0$) if and only if the two conditions*

1) $H^2(Y_0;\rho) = 0$, *so R_{Y_0} is a smooth 1-manifold near ρ;*
2) *the map r is transverse to the submanifold $L_{Y_0} \subset \mathcal{V}$ at ρ.*

The proof of this lemma is just the Mayer-Vietoris sequence:

$$\ldots H^1(Y) \to H^1(N) \oplus H^1(Y_0) \to H^1(\partial N) \equiv T\mathcal{V} \to H^2(Y)\ldots \tag{17}$$

with all cohomology in the appropriate twisted coefficients. The same ideas go over to give a similar result when we perturb the flatness equations in the interior of Y_0. One can show that for generic perturbations of this kind the two conditions analogous to those of Lemma 7 hold, so the perturbed moduli space $\tilde{R}_{Y_0}$ is smooth away from reducibles and the map r is transverse to L_Y. Then all irreducible solutions of the perturbed problem over Y are nondegenerate.

Let us now consider the *reducible* flat connections on the knot complement. These are given by representations

$$\pi_1(Y_0) \to H_1(Y_0) \to S^1.$$

By Alexander duality the homology of the complement is one-dimensional, generated by the meridian of the knot. So in our picture of $\mathcal{V}$ the reducible connections make up the other axis $\beta = 0$. One can extend the deformation machinery to study these points (although we do not need to do this here), and one finds that after small generic perturbation the irreducible part of the moduli space is a 1-manifold, whose closure meets the reducible set in a finite number of «bifurcation points». The corresponding picture in $\mathcal{V}$ is illustrated in Diagram 8.

Our hypotheses mean that, apart from the trivial connection, these reducible points do not enter when we study flat connections over Y, since their image under r does not meet L_Y. Of course we can study the flat connections over Y' and Y'' in the same way, replacing L_Y by $L_{Y''}$ or $L_{Y'}$, and again we do not encounter reducibles — in our Diagram 5 the reducibles make up the dotted line, which only intersects the basic geometric triangle at the point corresponding to the trivial connection.

We need to discuss one more genericity condition. The trivial connection θ_{Y_0} over the complement is a solution of any deformed equation. One can show that, for generic deformations, the image $r(\tilde{R}_{Y_0} \setminus \{\theta_{Y_0}\})$ avoids the vertices of the triangle: i.e. the other points map into the interiors of $L_Y, L_{Y'}, L_{Y''}$. With all this said, however, we will simplify our exposition of the proof by assuming that all these generic properties already hold for the undeformed set-up over the knot complement, so we assume:

Hypothesis 8 *For each irreducible flat connection ρ over the knot-complement Y_0:*

1) *the restriction of ρ to the torus ∂N does not have holonomy contained in the centre ± 1 of $SU(2)$;*
2) *the cohomology group $H^2(Y_0, \text{ad}\,\rho)$ is zero;*
3) *the map r is transverse to any of the three submanifolds* $\text{int}\, L_Y$, $\text{int}\, L_{Y'}$, $\text{int}\, L_{Y''}$ *at ρ.*

If these conditions do not hold then one makes a generic auxiliary perturbation in the interior of Y_0 and modifies the remainder of the proof appropriately.

We have now cleared the way to reach the main part of Floer's argument. Assuming Hypothesis (8) we can choose a pertubation Φ supported in the neighbourhood N, as in 3.3, such that the set $\tilde{R}$ of irreducible critical points of $\mathbf{cs} + \Phi$ is identified, under restriction to the common knot complement Y_0, with the union

$$\tilde{\mathcal{R}} = \mathcal{R}_{Y'} \cup \mathcal{R}_{Y''}.$$

The only reducible critical point of $\mathbf{cs} + \Phi$ is the trivial connection, and moreover the functional $\mathbf{cs}$ can be deformed to $\mathbf{cs} + \Phi$ through a family $\mathbf{cs} + \Phi_s$ having, for each s, no reducible critical points except for the trivial connection. For this we just take a family corresponding to a homotopy of L_Y to $\tilde{L}_Y$ across the basic geometric triangle in $\mathcal{V}$, avoiding all other reducibles. Now the restriction map r is transverse to the perturbed boundary-value set $\tilde{L}_Y$ and, by a simple extension of Lemma 7 to the deformed case, this means that the critical points of $\mathbf{cs} + \Phi$ are nondegenerate. Putting these facts together we conclude that there is a chain complex $(\tilde{C}_*, \tilde{\partial})$ with chain groups generated by the set $\tilde{\mathcal{R}}$, and differential defined by the gradient curves of $\mathbf{cs} + \Phi$ — the deformed-instantons over the tube — and the homology of this chain complex is canonically isomorphic to the Floer homology $\text{HF}_*(Y)$. By construction the chain group $\tilde{C}_*$ is identified *as a group* with the direct sum

$$\tilde{C}_* = C_*(Y'') \oplus C_*(Y'), \tag{18}$$

(and we shall see presently that this is compatible with the gradings). If we choose any time-dependent perturbations of the instanton equation over the cobordisms W_1, W_2, compatible with the given data on the boundaries, we get chain maps

$$C_*(Y') \xrightarrow{a} \tilde{C}_* \xrightarrow{b} C_*(Y''),$$

and these induce the same maps a_*, b_* on the Floer homology.

Now if it happened that the isomorphism (18) was an isomorphism of chain complexes, and that the chain maps a, b could be taken to be the inclusion $i : C_*(Y') \to C_*(Y') \oplus C_*(Y'')$, and projection $\pi : C_*(Y') \oplus C_*(Y'') \to C_*(Y'')$, then the exactness would be immediate: indeed we would get a long exact sequence in homology induced by the short exact sequence (i, π) of chain complexes. The real situation is rather more complicated, but the basic idea is that the situation above holds «modulo filtrations». We will now extract the basic algebraic lemma to this effect, used in Floer's proof. For this we consider a pair of chain maps

$$a : C'_* \to \tilde{C}_* \ , \ b : \tilde{C}_* \to C''_*,$$

between *filtered chain complexes* $C'_*, \tilde{C}_*, C''_*$. By this we mean that there are subcomplexes

$$\dots \tilde{C}_*^{(\mu)} \subset \tilde{C}_*^{(\mu-1)} \dots \subset \tilde{C}_*$$

$$\dots C_*'^{(\mu)} \subset C_*'^{(\mu-1)} \dots \subset C'_*,$$

$$\dots C_*''^{(\mu)} \subset C_*''^{(\mu-1)} \dots \subset C''_*,$$

and a and b are compatible with these.

Lemma 9 *Let a, b be chain maps between filtered complexes, as above. Suppose that*

1) $a = i + \alpha, b = \pi + \beta$, *where α, β are strictly decreasing with respect to the filtration (so, for example, α maps $C_*'^{\mu}$ to $\tilde{C}_*^{(\mu+1)}$), and i, π form a short exact sequence of group homomorphisms.*
2) *The composite ba is chain homotopic to zero by a homotopy $H : C'_* \to C''_*$,(i.e. $\partial H + H\partial = ba$), where H is strictly decreasing with respect to the filtration.*

Then there is a long exact homology sequence,

$$\dots \to H_*(C') \xrightarrow{a_*} H_*(\tilde{C}) \xrightarrow{b_*} H_*(C'') \to H_*(C') \to \dots$$

and in particular the kernel of b_* is equal to the image of a_*.

This lemma is elementary (and is very likely a standard result in some circles). Floer shows that a is chain homotopic to another map $a_1 = a - (\partial K + K\partial)$, such that the sequence

$$0 \to C'_* \xrightarrow{a_1} \tilde{C}_* \xrightarrow{b} C''_* \to 0 \tag{19}$$

is a short exact sequence of chain complexes. This implies the result, since a_1 induces the same map a_* on homology. First note that the hypothesis (1) implies that a is injective, like i, and b is surjective, like π. (One way of saying this is that an upper triangular matrix with 1's on the diagonal is invertible.) Furthermore there is a right inverse, R, to b, which is non-increasing with respect to the filtrations, and we put $K = R \circ H$. Then

$$b \circ (\partial K + K\partial) = \partial bRH + bRH\partial = \partial H + H\partial = ba,$$

so $b \circ (a - (\partial K + K\partial) = 0$. We define a_1 to be $a - (\partial K + K\partial)$. Then the sequence (19) is exact since

1) b is surjective, as established above;
2) a_1 is injective since it differs from a, and hence also from i, by maps which are strictly decreasing with respect to the filtration;
3) the composite ba_1 is zero. One way of seeing this is to regard the pair of maps a_1, b as a chain complex (ignoring the other differential). These are compatible with the filtrations so we get a spectral sequence converging to the homology (i.e. ker $b/\text{im } a_1$) and with E_1-term the homology of the quotients up the filtration. Now the hypothesis that $a_1 = i + \text{lower order}, b = \pi + \text{lower order}$ means that the E_1 term is zero, whence the result. Of course there are other ways of expressing the proof. If one works with coefficients in a field there is an obvious argument using dimensions.

The goal then, in the proof of exactness, is to show that there are deformed instanton equations over the cobordisms W_1, W_2 whose associated chain maps satisfy the conditions of Lemma 9, with respect to filtrations of the complexes $C'_* = C_*(Y'), \tilde{C} = \tilde{C}_*(Y), C''_* = C_*(Y'')$. These filtrations will be the ones defined by the Chern-Simons functional (over Y', Y'') and the perturbed functional $\mathbf{cs} + \Phi$ (over Y). We know that time independent deformations of the instanton equations give decreasing maps, with respect to the filtration, but it is clear that we cannot use such deformations directly, since on each of W_1, W_2 the problem is only perturbed at one end. The master stroke in Floer's proof is to *puncture* the manifolds W_1, W_2 to obtain manifolds $\hat{W}_1, \hat{W}_2$ with an additional end, and then to compare time-dependent deformations on the W_i with time-independent deformations on $\hat{W}_i$, as in Diagram 9.

To explain this in more detail: let $\hat{W}_i$ be the manifold obtained from W_i by removing an interior point, which we may take to lie on the disc D_i, and with the metric giving the manifold an additional tubular end modeled on $S^3 \times \mathbb{R}$. We can find a local product region in $\hat{W}_i$ which interpolates between the neighbourhood

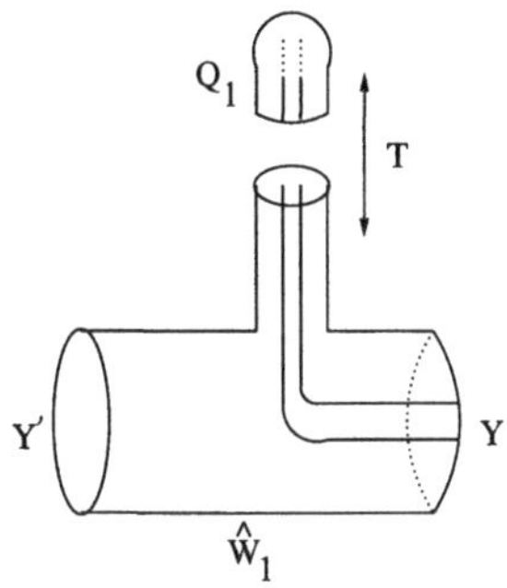

Diagram 9

of $K \subset Y$ and neighbourhoods of the standard circle in S^3. To find this product region we identify the punctured disc $D_i \setminus$ Point , with $S^1 \times \mathbb{R}$ and then take a tubular neighbourhood in the 4-manifold. Although the standard circle in S^3 is the same in the two cases, the neighbourhoods are essentially different, due to the different framings created by the relative chern class of the normal bundles to the discs D_i. That is, if N is the neighbourhood of K on which Φ is supported, we obtain different maps λ_1, λ_2 from N to the standard solid torus in S^3. We can thus regard Φ as giving two functionals $\Phi^{(1)}, \Phi^{(2)}$ on the space of connections over S^3, using these two identifications. In either case we get a time-independent deformed equation $F_A^+ + \mu^+\psi'(T_A) = 0$ over $\hat{W}_i$, with two deformed ends (Y with $\mathbf{cs} + \Phi$ and S^3 with $\mathbf{cs} + \Phi^{(1)}$ and $\mathbf{cs} + \Phi^{(2)}$) and one undeformed end (Y' or Y'').

We can easily read off the critical points of the deformed functional $\mathbf{cs} + \Phi^{(1)}, \mathbf{cs} + \Phi^{(2)}$ over S^3. Let l, m be the longitude and meridian of the standard torus in S^3. Then λ_1 maps the meridian of K to m and the longitude of K to $l+m$. The constraint that the holonomy around the longitude and meridian of K be equal goes over to the constraint that the holonomy around l be trivial. Similarly, λ_2 maps the meridian of K to m and the longitude to l. When we take account of the non-trivial bundle involved over W_2, the constraint that the holonomy around the meridian of K be -1 goes over to the constraint that the holonomy around m be the identity. (To see this, represent the connections used over W_2 by $SU(2)$ connections with a singularity along a surface C, as in Diagram 1, and let C and D_2 intersect at the puncture point.) So in each case the critical-point equation just asserts that we have a flat connection over the complement of the solid torus in S^3, with holonomy 1 around the longitude, and with the holonomy around the meridian curve lying outside a small neighbourhood of -1 (due to the « rounding off» of the triangle). The first constraint is vacuous, since the longitude bounds a disc in the complement. Thus the solutions are essentially the same in each case: they are all reducible over S^3, and are just extensions of the flat connections over $S^3 \setminus S^1$. They can be parametrised by a «moduli space» consisting of the interval $[0, 1]$ where 0 corresponds to the trivial connection. For v in $[0, 1]$ we write σ_v for the corresponding solution of the deformed equation. (Strictly we should distinguish between the two cases, but we will not bother to do this.)

Let us now fix our attention on W_1, the case of W_2 will be essentially the same. Let Q_1 be the 4-ball with a metric having one tubular end, so that we can regard W_1 as the connected sum of $\hat{W}_1$ and Q_1 across the S^3 ends, and introduce a parameter T defining the metric on the connected sum, in the familiar way. (See Diagram 9.) If we choose any (necessarily time-dependent) deformation of the instanton equation over Q_1, compatible with $\mathbf{cs} + \Phi$ on the end, we can glue this to the time-independent deformation over $\hat{W}_1$ to get a deformed instanton equation over W_1, defining a chain map $a_T : C_*(Y') \to \tilde{C}_*(Y)$. Similarly, we cap off $\hat{W}_2$ by a ball Q_2, carrying a deformed instanton equation, and we get a chain map $b_T : \tilde{C}_*(Y) \to C_*(Y'')$. We want to show that, with suitable deformations over Q_1, Q_2, and when the parameter T is sufficiently *large*, the chain maps a_T, b_T satisfy the conditions of Lemma 9.

The complement $W_1 \setminus D_1$ retracts onto the boundary component Y', so any flat connection ρ' over Y' automatically extends to $W_1 \setminus D_1$, that is, to the complement of the product region in $\hat{W}_1$. Furthermore, by the construction of Φ, this connection extends over the product region to a deformed-flat solution, $A_{\rho'}$ say, of the deformed equations. Just extend the given deformed flat solution over the end Y in a time-independent way throughout the product region. These connections $A_{\rho'}$ will be the essential ingredients in the proof. The first use we can make of them is to compare the functionals at the ends Y, Y' of the cobordism. If ρ' is a flat connection over Y', as above, write $\tilde{\rho}'$ for the corresponding solution of the deformed-flatness equation over Y, that is, the restriction of $A_{\rho'}$ to Y. Then our modified Chern-Weil theory of 3.3 (extended to three boundary components) gives:

$$\mathbf{cs}(\rho') = (\mathbf{cs} + \Phi)(\tilde{\rho}') - (\mathbf{cs} + \Phi^{(1)})(\sigma_{v(\rho')}), \operatorname{mod} \mathbb{Z} \tag{20}$$

where $v(\rho')$ is the parameter defining the restriction of $A_{\rho'}$ to the S^3 boundary. (This is the distance indicated on Diagram 8.) Now since the σ_v are all critical points the function $(\mathbf{cs} + \Phi^{(1)})(\sigma_v)$ is constant, and the constant is zero since this is the value at the trivial connection. So $\mathbf{cs}(\rho') = (\mathbf{cs} + \Phi)(\tilde{\rho}')$ modulo integers.

We now move to the final stage of the proof where we appeal to the gluing theory of 3.4, extended to the deformed equations in the obvious way. Along the interior of our interval $[0, 1]$, the perturbed functional is nondegenerate in the sense of Bott, so fits the framework we introduced above with $h^0(\sigma) = h^1(\sigma) = 1$. (Recall that by construction the limits $v(\rho')$ for the flat connections ρ' lie in this interior.)

Our first technical proposition describes the situation in $\hat{W}_1$.

Proposition 10

1) *The index of the solution $A_{\rho'}$ to the deformed equations over $\hat{W}_1$ is -1, thus the virtual dimension of the moduli space $M_V(\hat{W}_1)$ containing $A_{\rho'}$ is zero.*
2) *$A_{\rho'}$ is a regular point of this moduli space.*

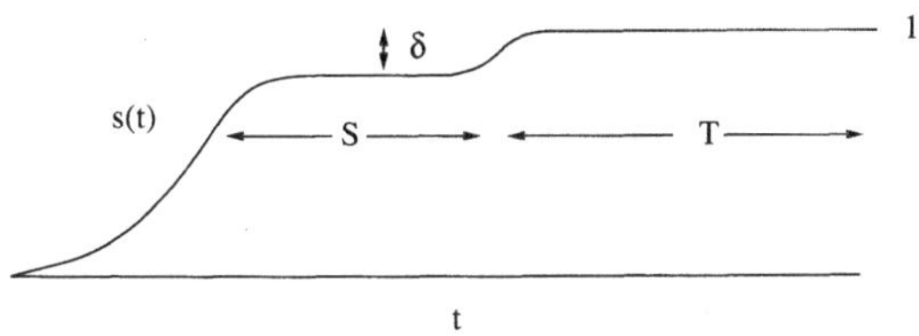

Diagram 10

We postpone the proof of Proposition 10 until the end of this section, and we go on now to the other piece Q_1. We choose any generic deformed equation over Q_1; then we have:

Proposition 11 *For a generic deformed equation over Q_1 and for each $v \in (0, 1)$ there is a solution B_v to the deformed equation, asymptotic to σ_v and with the properties:*

1) *the index* $\mathrm{ind}^-(B_v)$ *is* -1;
2) *the connection B_v is reducible, with $\Gamma_{B_v} = S^1$;*
3) *the connection B_v is a regular point of its moduli space. Moreover the connection B_v with property (1) is unique up to gauge equivalence.*

There are a number of ways to see the existence of these solutions B_v. One way is to deform from the case $v = 0$ (which is not actually covered by our proposition) where the trivial connection provides a solution, and then show by computing the relevant index that this extends to a family B_v. The solutions are all abelian — indeed because the index is negative there are no irreducible solutions — thus the analysis only involves *linear* equations, and one gets a strong uniqueness result. A more vivid way, which is used by Floer, is to apply the «gradient approximation» results of 3.4, adapted to the deformed equations. In this approach it is convenient to choose a more specific kind of deformation. Let $\mathbf{cs} + \Phi_s l$, $0 \leq s \leq 1$ be a homotopy from the undeformed functional ($s = 0$) to the deformed functional $s = 1$. These give corresponding functionals $\mathbf{cs} + \Phi_s^{(1)}$ over the 3-sphere. We may suppose that for $0 \leq s < 1$ the only critical point of $\mathbf{cs} + \Phi_s^{(1)}$ is the trivial connection. This just corresponds to chooosing a suitable isotopy on the the representation variety $\mathcal{V}$, as illustrated in Diagram 10.

Now for small positive δ we consider a time-dependent equation over the tube $S^3 \times \mathbb{R}$ of the form:

$$\frac{\partial A}{\partial t} = \mathrm{grad}\ \ (\mathbf{cs} + \Phi^{(1)}_{s(t)})(A)\,, \tag{21}$$

where $s(t)$ is a monotone function, equal to 0 for $t << 0$ and $1 - \delta$ for $t >> 0$. The trivial connection is a solution to this equation, and it is a regular point since there is no spectral flow in the Hessian of $\mathbf{cs} + \Phi_{s(t)}$. The equation is not deformed over the end $t \to -\infty$ so we may regard it as an equation for connections over Q_1, using Uhlenbeck's removal of singularities theorem. Now we consider a second time-dependent equation over $S^3 \times \mathbb{R}$ of the same form as above, but with limits corresponding to the functionals $\mathbf{cs} + \Phi^{(1)}$ as $t \to \infty$, $\mathbf{cs} + \Phi^{(1)}_{1-\delta}$ as $t \to -\infty$. We choose the time variation to have the shape considered in Proposition 6, which will tell us that solutions are modelled on the gradient flow lines of the induced function on the interval V. The S^3 end of $\hat{W}_1$ is regarded as an «ingoing» end, so the signs work out that the instanton equation corresponds to the *ascending* gradient flow, with our parametrisation. Thus the essential thing is to see that the perturbed function $\mathbf{cs} + \Phi^{(1)}_{1-\delta}$ is *negative* on the interval, with a maximum point at $v = 0$. For this tells us that there is a gradient line $w(\tau)$ of the finite-dimensional problem which runs from $v = v_0$ up to $v = 0$. Then the analogue of Proposition 6, for the case where where the initial problem is defined using a deformation, shows that there is a corresponding solution $\hat{B}(v_0)$ to the equation over the tube. (One has to modify Proposition 6 in two other ways: first, trivially, the «time» parameter on the end has been reversed and second because we need to include the end point $v = 0$ where the isotropy changes. This makes no real difference to the approximation of the gradient equation.)

To check this sign of $\mathbf{cs} + \Phi^{(1)}_{1-\delta}$, one has to work through the various definitions we have made. The sign of $\mathbf{cs} + \Phi^{(1)}_{1-\delta}$ on the σ_v is the same as the sign of $f_{1-\delta} - f$, where f is the function used to construct the deformation as in 3.3, and $f_{1-\delta}$ is the similar function appearing in the isotopy from the identity (the case $f_0 = 0$). It is easy to see then that this sign is the same as the sign of $-f$ and we get the desired answer since f is positive. Notice that this is the place where the sign in the surgery is used: if we tried to do the proof with the -1 surgered manifold in place of Y', we would need to deform to the anti-diagonal $\alpha = -\beta$ in $\mathcal{V}$ and this would need a negative function f.

It is easy to see that $\hat{B}(v_0)$ is reducible; indeed we could have carried out the whole discussion for S^1 connections. Finally then we define our deformed equation over Q_0 by making the connected sum of these two time-dependent deformations, with another parameter $S > 0$, as indicated on Diagram 11. For large S we get the desired solutions $B(v_0)$ by applying the gluing theory to the connection $\hat{B}(v_0)$ and the trivial connection.

We can now finally complete the argument. For each flat connection ρ' over Y', let $A^*(\rho') = A(\rho')\sharp B(v(\rho'))$ be the connection formed by gluing together the deformed-flat solution $A(\rho')$ and the reducible solution $B(v_0)$, with $v_0 = v(\rho')$. (We choose the parameter T large enough for the gluing results to apply for each of the finitely many ρ'.) Since the isotropy groups of $B(v)$ and σ_v are both S^1, there is no gluing parameter involved, and we get a solution with the properties

1) $A^*(\rho')$ is irreducible (since $A(\rho')$ is);
2) $\operatorname{ind} A^*(\rho') = 0$;
3) $A^*(\rho')$ is a regular solution.

It follows first, from the index calculation, that the Floer-grading of the flat connection ρ' over Y' and of the deformed solution $\tilde{\rho}'$ over Y are the same. Second, by the comparison of the functionals in (18) above, we see that the orderings of the flat connections ρ' over Y' induced by the Chern-Simons functional and of the deformed solutions $\tilde{\rho}'$ by the deformed functional over Y are the same. Consider now the moduli space $M(\rho', \tilde{\rho}')$ of index 0 solutions over W_1 interpolating between ρ' and $\tilde{\rho}'$. We claim that for suitable choices of our parameters, the connection $A^*(\rho')$ is the *only* point in this moduli space. Here we will use, for definiteness, a deformed equation constructed by the two-stage procedure outlined above, so we have parameters T, S, δ. This is essentially just the monotonicity property of 3.1, but the argument is slightly delicate because of the singular point $v = 1$ in the variety $\overline{V}$ of solutions to the critical point equation over S^3: we do not want to consider a moduli or gluing theory for connections with this singular limit. The reader may prefer to think through the proof first, ignoring this singular point, in which case the argument is a good deal cleaner. However, to deal with the singular point, we begin by considering any deformed-instanton B' over Q_1, with some limit σ_v. We observe that there is a relative Chern class $\kappa(B') \in \mathbb{Z}$, comparing B' with our known solutions $B(v_0)$, for $0 < v_0 < 1$ (If $v = 1$ we just compose $B(v_0)$ with a short path in $\overline{V}$.) The essential point is to choose our deformed problem so that there is no solution B' with $\kappa(B') < 0$. We claim that this is true for the deformations considered above if the parameter S is large and the parameter δ is small. For, arguing by contradiction, if we first let $S \to \infty$, we see by standard arguments using compactness results that such a solution would be made up of a sum of a connection θ' with limit the trivial connection σ_0 and a solution $\hat{B}'$, which runs from σ_0 to σ_v. We have relative Chern classes $\kappa(\hat{B}'), \kappa(\theta')$ comparing with our known $\hat{B}_{v_0}$ and with the trivial connection θ, and we must have $\kappa(\hat{B}') + \kappa(\theta') = \kappa(B')$. The equation governing the connection θ' is time dependent, but it has non-degenerate limits so we have a straightforward index theory for the solutions. If the Chern class $\kappa(\theta')$ is strictly negative then the solution θ' has negative index (in fact ≤ -11) and for generic choices of our deformation these will not appear. Thus we must have $\kappa(\hat{B}') < 0$, where $\hat{B}'$ is a solution of an equation of the form (21), with $s(t)$ running from $1 - \delta$ to 1. Now let δ tend to zero, and we see that we would get a limit of the $\hat{B}'$ which would be a solution of the *time-independent* equation

$$\frac{\partial A}{\partial t} = \text{grad}(\mathbf{cs} + \Phi^{(1)})(A)$$

but with $\kappa < 0$, and this would be the usual contradiction to the fact that κ can be expressed as the integral of the $\frac{d}{dt}(\mathbf{cs} + \Phi)dt$, which is positive by the gradient condition. The essential point of this argument is to control the existence of solutions in the region where the «large» time-dependent deformation occurs by the index theory.

Now, to resume the discusssion of the space $M(\rho', \tilde{\rho}')$, when T is large we see that any solution breaks up into a solution A' over $\hat{W}_1$ and a solution B' over

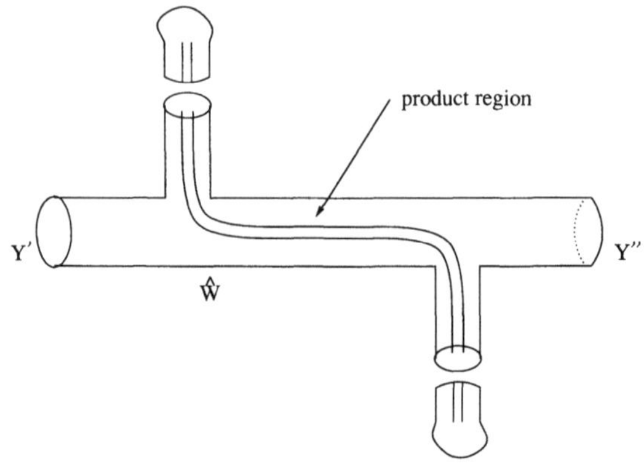

Diagram 11

Q_1, with the same limit in $\overline{V}$. (Standard arguments rule out more complicated limits.) Again, we define relative Chern classes $\kappa(A'), \kappa(B') \in \mathbb{Z}$ by comparing with our known solutions. Thus

$$\kappa(A') + \kappa(B') = 0.$$

The Chern-Weil theory for the time-independent deformed equations over $\hat{W}_1$ tells us that $\kappa(A') \geq 0$. If $\kappa(A') = 0$ then A' is a deformed-flat connection, and thus is equal to one of our connections $A(\rho')$ which correspond to the known solutions $A^*(\rho')$. If $\kappa(A') > 0$ then we would have to have $\kappa(B') < 0$ and we have seen that this does not occur, for a suitable deformation.

We conclude then that the moduli space $M(\rho', \tilde{\rho}')$ has just one point, which is regular. This gives the contribution i to the chain map a (one might want to discuss signs here, to say whether the matrix entries should be $+1$ or -1, but the conclusions of Lemma 9 hold in either case). The same argument as above, combined with the general discussion of 3.1, shows that all the other contributions are decreasing with respect to the filtration. The argument for the other map b, comparing with the projection π is just the same, building up deformed solutions from flat connections over Y'' and abelian solutions over Q_2, so we have established the first hypothesis of Lemma 9.

We now want to establish the second hypothesis of Lemma 9: that ba is chain-homotopic to zero by a chain homotopy $H : C_*(Y') \to C_*(Y'')$ which is decreasing with respect to the filtration. Once again, this involves the combination of time-dependent and time-independent deformations, and again one needs to take care because of the singular point in $\overline{V}$. We know that the instanton equation of a suitable metric g_W on $W = W_2 \circ W_1$ gives a zero chain map. Let $\hat{W}$ be the manifold made by identifying the Y boundary components of $\hat{W}_1, \hat{W}_2$, with a long neck, as in Diagram 11.

Adding on the caps Q_1, Q_2 we get a deformed equation over W which induces the chain map $b \circ a$. On the other hand the homotopy Φ_s gives a 1-parameter family of time-independent deformed equations over $\hat{W}$, with two undeformed ends (Y', Y'') and two deformed ends (the two copies of S^3). When $s = 0$ we get an undeformed problem, whose solutions correspond to the ordinary instantons for a suitable metric h_W on W (by removal of singularities). For any $s < 1$ the only critical points for the deformed equations over S^3 are the trivial connection. There is thus a straightforward moduli theory for solutions of the deformed equations over $\hat{W}$, which define chain maps $c(s)$, for generic parameter values s in $(0, 1)$, from $C_*(Y')$ to $C_*(Y'')$.

We will divide the construction of the chain homotopy into three stages. First we compare the chain maps defined by the instanton equations for the two metrics g_W, h_W on W. This is the standard case mentioned in 3.1. One considers a 1-parameter family of metrics interpolating between g_W and h_W. The two chain maps are chain-homotopic by a homomorphism which «counts» the instantons in the family of moduli spaces of index -1. The decreasing property with respect to the filtrations just reflects the fact that the Chern-Weil integral is positive for all these instantons. The homomorphism is strictly decreasing because there are no non-trivial flat connections on W, by our hypothesis 8.

Next we compare the chain maps $c(s)$ for $s < 1$. The same discussion applies since these maps are defined by a family of time-independent deformations, with nondegenerate limits. Two of these maps differ by a chain homotopy which counts index -1 solutions of the deformed equation, and this is decreasing by the extended Chern-Weil theory.

We see then that, since $c(0)$ is the same as the map induced by the metric h_W, it suffices to show that $c(s)$ differs from $b \circ a$ by a decreasing chain homotopy when s is close to 1. To do this recall that our deformed equations over W_i, and hence over W are defined by parameters S, T, δ. Introduce another parameter $u \in [0, \delta]$ and for each u consider a deformed equation over Q_1 defined as the connected sum of a family

$$\frac{\partial A}{\partial t} = \text{grad } (\mathbf{cs} + \Phi^{(1)}_{s(t)}),$$

where $s(t)$ runs from 0 to $1 - \delta$ and a similar family with $s(t)$ running from $1 - \delta$ to $1 - u$. We use a neck-length parameter S joining these, as before. The same argument as before shows that when δ is small and S is large there is no solution of this equation, for any $u \in [0, \delta]$ with $\kappa < 0$. Now fix δ and S and make the connected sum of this family over Q_1, and the similar one over Q_2 with the family defined by $\mathbf{cs} + \Phi_{1-\delta}$ over $\hat{W}$, with the neck parameter T. The deformed instantons in this family give chain maps between $C_*(Y'), C_*(Y'')$. If T is large then when $u = 0$ we get the chain map $b \circ a$ (by definition), whereas when $u = \delta$, we get the chain map $c_{1-\delta}$ (applying our gluing theory across the trivial connection σ_0 — the only critical point of $\mathbf{cs} + \Phi^{(1)}_{1-\delta}$ and $\mathbf{cs} + \Phi^{(2)}_{1-\delta}$). As before, these maps are chain-homotopic by a chain homotopy which counts solutions of index -1. To see that these are decreasing we argue again by contradiction and make T large. The important thing is again the fact that there are no solutions $\hat{B}_i$ over Q_i with negative κ. From a sequence of solutions over W, with $T \to \infty$, we extract a subsequence converging to a limit A over $\hat{W}$ and solutions $\hat{B}_i$ over Q_i. The limit A satisfies a time-independent equation corresponding to Φ_s for some s and since $\kappa(\hat{B}_i) \geq 0$, we must have $\kappa(A) < 0$. Then the modified Chern-Weil theory, for the time-independent deformation over $\hat{W}$, contradicts the supposition that all the solutions increase the level in the filtration. (One neeeds to make a separate argument for the case of equality, to rule out the possibility of deformed-flat solutions over $\hat{W}$.)

The only task remaining is to go back to explain the proof of the technical Proposition 10, about the deformed-flat solutions $A_{\rho'}$. The calulation will use two special devices. Recall that the index $\text{ind}^{\,-}(A_{\rho'})$ is defined as the difference:

$$\text{ind}^{\,-}(A_{\rho'}) = \dim \ker^{-} D_\mu - \dim \operatorname{coker}^{-} D_\mu,$$

where $D_\mu = d^* + d^+_\mu$ is a deformation of the coupled $d^* + d^+$ operator (appropriate to the deformed instanton equations) and the minus subscript refers to the fact that we take the kernel and cokernel in weighted spaces with an exponentially increasing weight function. Now, because $A_{\rho'}$ is a deformed-flat connection we have another operator d^-_μ, such that the equation $(d^+_\mu + d^-_\mu)a = 0$ is the linearisation of the deformed flat condition about $A_{\rho'}$. More explicitly, d^+_μ, d^-_μ are the derivatives with respect to a of the self-dual and anti-self-dual parts of $F_{A+a} + \mu\phi'(T_{A+a})$. If a decays rapidly at infinity then we can apply the deformed Chern-Weil theory and differentiate with respect to the connection to see that

$$\int_{\hat{W}_1} |d^+_\mu a|^2 - |d^-_\mu a|^2 = 0. \tag{22}$$

(As usual, this would be a familiar identity in the undeformed case.) A few moments consideration of the relevant boundary term in the integral shows that this equality holds if a is bounded at infinity and $d^\pm_\mu a$ decay exponentially. In particular any such a with $d^+_\mu a = 0$ also satisfy $d^-_\mu a = 0$; i.e they are infinitesimal variations of A_ρ as a deformed flat connection. Now it is fairly clear, and can be proved by the same kind of techniques as Lemma 7, that our solution $A_{\rho'}$ is «infinitesimally rigid», i.e. that any variation a as above with $d^\pm_\mu a = 0$ can be written as $a = d_A u$ for some infinitesimal gauge transformation u. Suppose a also satisfies the Coulomb gauge condition $d^*_{A_{\rho'}} a = 0$, and is bounded at infinity. Then a straightforward argument shows that we must have $d_{A_{\rho'}} u = a = 0$. We conclude then that *there is no non-trivial solution to the equation* $D_\mu a = 0$ *with a bounded limit at infinity and* $d^-_\mu a$ *decaying exponentially at infinity.*

This tells us, first of all, that $\ker^- D_\mu = 0$, so that $\text{ind}^{\,-} A_{\rho'} \leq 0$. However we can do more by constructing a map

$$\Delta : H^1_\sigma \to \operatorname{coker}^- D_\mu.$$

Here H^1_σ denotes the tangent space of the variety V of solutions to the deformed equations at the S^3 boundary of $\hat{W}_1$, which is 1-dimensional. (The analogue of the map Δ in the ordinary, undeformed case, of a flat connection τ over a manifold X with boundary, would be the connecting homomorphism $\Delta : H^1(\partial X) \to H^2(X, \partial X)$, where the cohomology is taken with coefficients in the flat bundle ad τ.) To define Δ, observe that an element h of H^1_σ gives a «constant» section $\tilde{h}$ over the end $S^3 \times \mathbb{R}^+$ with $d^*_{A_{\rho'}}\tilde{h}, d^+_\mu \tilde{h}, d^-_\mu \tilde{h}$ all zero. We let β be a cut-off function supported on the end, equal to 1 in a neighbourhood of infinity, and let $\Delta(h)$ be the class represented by $D_\mu(\beta\tilde{h})$ in the cokernel.

We claim now that Δ is injective. For suppose that $\Delta\tilde{h} = 0$, so that $D_\mu(\beta\tilde{h}) = D_\mu b$ for some exponentially decaying b. Then $a = \beta\tilde{h} - b$ satisfies $D_\mu a = 0$, and a is bounded and $d^-_\mu a$ decays exponentially. But then our conclusion above tells us that $a = 0$, and this clearly implies that $h = 0$. We conclude then that the cokernel is non-trivial, so ind $^-(A_{\rho'}) \leq -1$.

To complete the calculation we now use a doubling argument. Let $\overline{\hat{W}_1}$ be $\hat{W}_1$ with the reversed orientation, and Z be the double of $\hat{W}_1$ obtained by attaching $\overline{\hat{W}_1}$ to $\hat{W}_1$ along the three boundary components. The deformed equations and the solution $A_{\rho'}$ over $\hat{W}_1$ go over to similar equations and solution $\overline{A_{\rho'}}$ over $\overline{\hat{W}_1}$, which match up on the boundary to give a deformed-flat connection A_Z on a bundle P_Z over Z. Our gluing theory tells us that

$$\text{ind } A_Z = \text{ind }^-A_{\rho'} + \text{ind }^-\overline{A_{\rho'}} + 2,$$

where the contribution 2 is $h^0(\sigma) + h^1(\sigma)$. Now the index for the connection A_Z over the closed manifold Z is easy to calculate, because it depends only on the topological data

$$\text{ind } A_Z = 8c_2(P_Z) - \frac{3}{2}(\chi(Z) - \tau(Z)).$$

The manifold Z admits an orientation-reversing symmetry so it has signature zero; by the same token the Chern class $c_2(P_Z)$ is zero. Then a simple calculation shows that the Euler characteristic is $\chi(Z) = 2\chi(\hat{W}_1) = 0$. So we see that

$$\text{ind }^-A_{\rho'} + \text{ind }^-\overline{A_{\rho'}} = -2. \tag{23}$$

Now the same arguments applied to $A_{\rho'}$ also work for $\overline{A_{\rho'}}$; indeed this just corresponds to working with the operator d^-_μ over $\overline{\hat{W}_1}$. Thus we see that ind $^-A_{\rho'}$ and ind $^-\overline{A_{\rho'}}$ are both ≤ -1, and hence we deduce from (21) that equality must hold:

$$\text{ind }^-A_{\rho'} = \text{ind }^-\overline{A_{\rho'}} = -1.$$

This completes the proof of Proposition 10, since one easily sees that $A_{\rho'}$ is a regular point in the moduli space M_V if the map Δ is surjective. (Let us note in passing that one can define a map like Δ in the general setting of the moduli theory of 3.4, and it fits into a diagram of maps which reflects, at the linear level, the relationship between the various moduli spaces.)

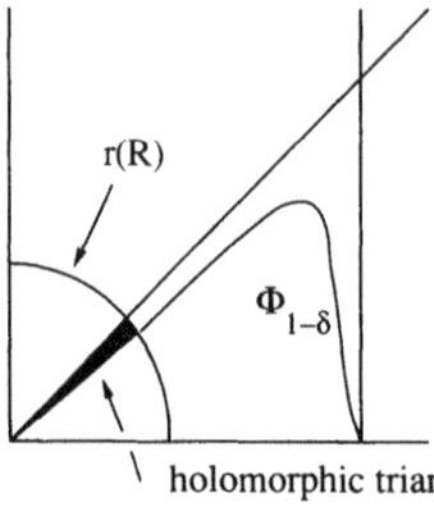

Diagram 12

Concluding remarks

There are many ways in which one might try to modify and extend the ideas in Floer's exact sequence. We will mention briefly three points which seem interesting. First, observe a slightly strange feature of the proof: the argument of Lemma 9 gives the existence of a long exact sequence, via a short exact sequence of chain complexes, but it does not seem obvious whether this is the same as the sequence defined by the triangle of cobordisms. Second, there are some similarities between the use of the deformed equations, which are an essential part of Floer's argument, and instanton solutions with codimension 2 singularities, as developed by Kronheimer and Mrowka [KM1],[KM2]. It would be interesting to develop a theory which combined these connections with Floer homology, and which would perhaps give a different way of proving the exactness. Third, there is a useful way of thinking–at least heuristically–about some of the arguments in Floer's exactness proof using the holomorphic geometry of the representation space $\mathcal{V}$. One would expect that, following the ideas of Dostoglou and Salamon [DS1], a partial model for the instantons on the cobordism W_1 is furnished by «holomorphic triangles», i.e holomorphic maps from a standard triangle in $\mathbf{C}$ to $\mathcal{V}$ which map the three sides to the subsets L_Y, $L_{Y'}$, $r(R_{Y_0})$, as indicated by the shaded region on Diagram 12.

Similarly, one can think of the solutions of the deformed problem as corresponding roughly to triangles with one deformed boundary condition. This way of thinking makes it easier to understand the somewhat involved gluing argument used in the proof. The essential point is to see that the intersection points ρ' give rise to «thin» holomorphic triangles when we deform away from the nongeneric situation. One also sees the role of orientations more clearly: one needs the triangles to have the correct orientation, with the sides taken in a given order.

Part II
Floer homology and knots

1 Floer homology and bundles

1.1 Flat connections and gauge transformations

In the sequel we will extensively work with $SO(3)$-bundles on 3-manifolds. It is important to understand precisely when such bundles carry flat, reducible connections, and what are the homotopy classes of gauge transformations.

Lemma 1.1 *Let $P \to Y$ be a principal $SO(3)$-bundle over an orientable, not necessarily closed, 3-manifold. The following are equivalent:*

1) *P carries no flat connection with holonomy group contained in an S^1 subgroup of $SO(3)$.*
2) *$w_2(P) \in H^2(Y;\mathbb{Z}_2)$ does not lift to a torsion class in $H^2(Y;\mathbb{Z})$ under the coefficient map:*
$$H^2(Y;\mathbb{Z}) \to H^2(Y;\mathbb{Z}_2).$$
3) *$w_2(P)$ is not zero as a map*
$$H_2(Y;\mathbb{Z}) \to \mathbb{Z}_2.$$
4) *$w_2(P)$ is not zero when evaluated on the fundamental class of an oriented closed surface $\Sigma \subset Y$.*

The proof is elementary.

Example 1.2

1) *The fundamental group of $Y = RP^3 = S^3/\mathbb{Z}_2$ has two representations in $SO(3)$, up to conjugation. One of these is the trivial representation, the other has monodromy* $\mathrm{diag}(1,-1,-1) \in SO(3)$ *along a generator of $\pi_1(SO(3))$, and gives a flat connection on the non-trivial $SO(3)$-bundle over Y. Yet it is reducible, in line with the above lemma.*
2) *On a surface Σ_g of genus g there is precisely one isomorphism class of non trivial $SO(3)$-bundles. On this bundle there are no S^1-reducible flat connections.*

The lemma above selects a good class of bundles on which the reducible flat connections do not play an important role:

Definition 1.3 *An $SO(3)$-bundle $P \to Y$ is called free of flat S^1 reductions if any of the conditions of lemma 1.1 holds. P is called admissible if Y is an oriented homology sphere or if P is free of flat S^1 reductions.*

When considering bundles the group of bundle automorphisms or gauge transformations enters picture. This group is defined as

$$\mathcal{G}(P_Y) = \Gamma(Y, P \times_{\mathrm{Ad}} SO(3)),$$

in other words as the space of sections of a fibration over Y with fibre $SO(3)$. We get an exact sequence:

$$\mathcal{G}_S(P_Y) \to \mathcal{G}(P_Y) \xrightarrow{\eta} H^1(Y;\mathbb{Z}_2),$$

where η measures the obstruction to deforming a gauge transformation to the identity over the one-skeleton of Y and $\mathcal{G}_S(P_Y)$ is defined as $\ker \eta$.

Definition 1.4 *The group $\mathcal{G}_S(P_Y)$ is called the restricted gauge group.*

From elementary obstruction theory one proves easily:

Lemma 1.5 *η is onto and the group $\mathcal{G}_S(P_Y)$ consists of the gauge transformations which lift to $SU(2)$-gauge transformations, i.e it equals the space of sections of*

$$P_Y \times_{\mathrm{Ad}} SU(2)$$

modulo $\mathbb{Z}_2$. Moreover,

$$\pi_0(\mathcal{G}_S(P_Y)) = \mathbb{Z}.$$

Notice that the component group of $\mathcal{G}(P_Y)$ now appears as an extension:

$$0 \to \mathbb{Z} \to \pi_0(\mathcal{G}(P_Y)) \to H^1(Y;\mathbb{Z}_2) \to 0.$$

The gauge group acts on the space of flat connections on P_Y. The centralizer $Z(H) \subset SO(3)$ of the holonomy group of a connection is isomorphic to the the stabilizer of the connection in $\mathcal{G}(P_Y)$. Recall that every non-identity element in $SO(3)$ lies in a unique one-parameter subgroup $S^1 \subset SO(3)$. This S^1 subgroup consists of rotations around a fixed axis in $\mathbb{R}^3$, and associated to it is an $O(2)$ subgroup. An $O(2)$ subgroup contains many Kleinian subgroups $K_0 \simeq \mathbb{Z}_2 \times \mathbb{Z}_2$ consisting of the identity, the rotation over π and two reflections.

Some simple computations classify the sorts of reduction one can encounter for $SO(3)$-connections:

- if a subgroup $H \subset SO(3)$ is not contained in an $O(2)$ it has trivial centralizer in $SO(3)$;
- the centralizer of $O(2)$ equals $\mathbb{Z}_2 \subset S^1 \subset O(2)$;
- if $H = K_0$ for a Kleinian subgroup K_0 then $Z(H) = K_0$;
- if $H \subset O(2)$, $H \not\subset S^1$ and H has more than 4 elements then $Z(H) = \mathbb{Z}_2$;
- if $H = \mathbb{Z}_2 \subset S^1$ then $Z(H) = O(2)$;
- if $\mathbb{Z}_2 \subsetneq H \subset S^1$ then $Z(H) = S^1$.

We see that stabilizers in the gauge group can occur even for admissible bundles, but they are always finite groups. As a result, for example, the flat isomorphism classes of flat connections on a non-trivial bundle over a surface form an *orbifold*, not a smooth manifold. However, as we shall see next, the restricted gauge group contains no such finite stabilizers. If $\mathcal{A}(P_Y)$ denotes the space of connections, then there is an action of $H^1(Y;\mathbb{Z}_2)$ on $\mathcal{A}(P_Y)/\mathcal{G}_S(P_Y)$ with qoutient the orbit space of connections: $\mathcal{A}(P_Y)/\mathcal{G}(P_Y)$.

At this point it is convenient to point out that after choosing an integral lift of $w_2(P_Y)$ and a 2 form representing this lift, we can fix a $U(2)$ bundle $E \to Y$ and uniquely lift connections from P_Y to E, where the central part is prescribed. The gauge transformations fixing this central part are the same as elements of $\mathcal{G}_S(P_Y)$ and the action of tensoring E with flat line bundles with monodromy in $\mathbb{Z}_2$ coincides with the action of $H^1(Y;\mathbb{Z}_2)$. From this one sees easily:

Proposition 1.6 *Let P_Y be admissible and let A be a non-trivial flat connection on P. Then the stabiliser $\mathcal{G}_S(P_Y)^A$ of A in $\mathcal{G}_S(P_Y)$ is trivial.*

From the previous proposition we have:

Corollary 1.7 *If an $SO(3)$-connection has non-trivial stabilizer in $\mathcal{G}(P_Y)$ and along a loop γ the monodromy does not lie in the S^1-subgroup determined by an element g in the stabilizer, then*

$$\eta(g)(\gamma) \neq 0.$$

If $P \to Y$ is *any* $SO(3)$-bundle then it is possible that flat connections exist which have nontrivial monodromy in an S^1 subgroup of $SO(3)$. Such connections always have stabilizer S^1 in $\mathcal{G}_S(P_Y)$. In $\mathcal{G}(P_Y)$ the stabilizer is S^1 if the monodromy is not contained inside $\mathbb{Z}_2 \subset S^1$ and $O(2)$ otherwise.

A very useful concept of a gauge transformation is its degree. Since the fundamental class of Y may not lift to a class in P_Y one has to be carful in defining the homological degree. We follow a geometric definition:

Definition 1.8 *The degree $d(g) \in \mathbb{Z}$ of a gauge transformation $g \in \mathcal{G}(P_Y)$ is the intersection number of its graph in*

$$P_Y \times_{\mathrm{Ad}} SO(3)$$

with the graph of the identity. For $g \in \mathcal{G}_S(P_Y)$ we define $d_S(g) \in \mathbb{Z}$ by taking the intersection number in

$$P_Y \times_{\mathrm{Ad}} SU(2).$$

If $g \in \mathcal{G}(P_Y)$ lifts to $\tilde{g} \in \mathcal{G}_S(P_Y)$ then

$$d(g) = 2d_S(\tilde{g}).$$

Proposition 1.9 *Let $g \in \mathcal{G}(P_Y)$ be such that $\eta(g)$ lifts to an integral class. Then:*

$$d(g) = (\eta(g) \cup w_2(P))[Y] \pmod 2.$$

In general:

$$d(g) = (\eta(g) \cup w_2(P) + \eta(g)^3)[Y] \pmod 2.$$

Proof. The first formula is proved by Dostoglou and Salamon [DS1] and the second formula can be proved along the same lines. □

Some examples will illustrate the situation:

Example 1.10

1) *Consider the trivial $SO(3)$-bundle over $SO(3)$. Here a gauge transformation exists of homological degree one, and the general formula in proposition 1.9 agrees with this.*
2) *Consider the non-trivial $SO(3)$ bundle over $SO(3)$. Here the proposition predicts that the degree is always even. This can also be seen directly.*

Finally we shall discuss the action of diffeomorphisms. Let $P_Y \to Y$ be a bundle as before. A orientation preserving diffeomorphism $\phi \in \mathrm{Diff}^+(Y)$ can be lifted to a bundle automorphism of P_Y precisely when $\phi^*(P_Y) \simeq P_Y$. Let $\mathrm{Diff}(P_Y)$ be this set of diffeomorphisms, then we define the group $\mathrm{Aut}(P_Y)$ through the exact sequence:

$$e \longrightarrow \mathcal{G}(P_Y) \longrightarrow \mathrm{Aut}(P_Y) \longrightarrow \mathrm{Diff}(P_Y) \longrightarrow e$$

This group acts naturally on the space of connections on P_Y. If the bundle is trivial $\mathrm{Aut}(P_Y)$ is the semidirect product:

$$\mathcal{G}(P_Y)cx \rtimes \mathrm{Diff}^+(Y)$$

1.2 Floer's category and functor

In this section we begin Floer's functorial approach to his instanton homology groups. This approach is that of a *topological field theory*, supplemented with additional information. We shall work throughout with the Lie group $SO(3)$. Denote by $\mathbb{C}$ or $SO(3)[3]$ the category with objects:

$$\mathrm{Obj}(\mathbb{C}) = \begin{array}{l}\{\text{principal SO(3)-bundles } P \to Y \\ \text{over oriented closed 3-manifolds } Y\}.\end{array}$$

We will abbreviate an object by P_Y and denote by $\overline{P_Y}$ the object consisting of the same bundle and three manifold, but with orientation reversed. A morphism between $(P_1)_{Y_1}$ and $(P_2)_{Y_2}$ consists of the following data:

- An $SO(3)$-bundle $Q \to W$ over an oriented 4-manifold W, together with an orientation preserving bundle morphism

$$Q_{|\partial W} = \overline{(P_1)_{Y_1}} \sqcup (P_2)_{Y_2}$$

- a lift of $w_2(Q)$ to $H^2(W; \mathbb{Z})$
- a choice of orientations of instanton moduli spaces for Q.

Definition 1.11 *The category $\mathbb{C}^*$ is the full subcategory consisting of admissible objects.*

The item of orientability can be tricky in the $SO(3)$-setup. The best way around this is to use $U(2)$-connections with fixed central part; the projectively anti-self dual connections on such bundles always form an orientable moduli space.

Recall that an $SO(3)$-bundle P over an oriented 3-manifold Y is determined, up to isomorphism, by its second Stiefel-Whitney class $w_2(P) \in H^2(Y;\mathbb{Z}_2)$. An $SO(3)$-bundle is the projectivization of a $U(2)$-bundle E with first Chern class c_1 when $w_2(P)$ lifts to an integral class $c_1 \in H^2(Y;\mathbb{Z})$. Generally, the obstruction to lifting $w_2(P)$ is the image of $w_2(P)$ under the Bockstein map

$$H^2(Y;\mathbb{Z}_2) \to H^3(Y;\mathbb{Z}),$$

which is zero as $H^3(Y;\mathbb{Z}) = \mathbb{Z}$. The $SO(3)$-bundle lifts to an $SU(2)$-bundle precisely when $w_2(P) = 0$.

On closed 4-manifolds W, the isomorphism classes of $SO(3)$-bundles Q are determined by

$$w_2(Q) \in H^2(Y;\mathbb{Z}_2) \text{ and } p_1(Q) \in H^4(W;\mathbb{Z}),$$

which satisfy the relation: $w_2(Q)^2 = p_1(Q) \bmod 4$ if $w_2(Q)$ lifts to the integers. On 4-manifolds there may be an obstruction to lifting the structure group of an $SO(3)$-bundle to $U(2)$.

If P_Y is an admissible bundle then we can consider the Chern-Simons function **cs** on the restricted orbit space

$$\mathcal{B}_S(P_Y) = \mathcal{A}(P_Y)/\mathcal{G}_S(P_Y)$$

of connections on P_Y. On $\mathcal{B}_S(P_Y)$ we consider a slightly perturbed **cs** as a Morse function, with Morse-Smale flow. The finitely many critical points have an index in an *affine* $\mathbb{Z}_8$ space, and we have:

Definition 1.12 *The Floer functor assigns to an object* $P_Y \in \mathrm{Obj}(\mathbb{C}^*)$ *the module* $\mathrm{HF}_*(Y)$ *which is affinely* $\mathbb{Z}_8$ *graded.*

To a morphism $Q_W \in \mathrm{Mor}((P_1)_{Y_1}, (P_2)_{Y_2})$ we need to assign a morphism of graded modules

$$\mathrm{HF}_*((P_1)_{Y_1}) \to \mathrm{HF}_*((P_2)_{Y_2}),$$

which is done by considering the instantons on Q. Here we equip W with a complete cylindrical metric near its boundary components, and consider once again only restricted gauge groups for $Q \to W$.

1.3 Further structure

The group $H^1(Y;\mathbb{Z}_2)$ may not act on the graded module $\mathrm{HF}_*(P_Y)$ due to equivariant transversality problems for the moduli spaces. It does act on Floer's complex, and if an element acts freely we will get an induced isomorphism of $\mathrm{HF}_*(P_Y)$. The degree of this isomorphism can be computed as follows. Let $g \in \mathcal{G}(P_Y)$ be a lift of a specific element in $H^1(Y;\mathbb{Z}_2)$. Then we can form a bundle $Q \to Y \times S^1$ using this gauge transformation, and the degree μ of the map

$$g_* : \mathrm{HF}_*(P) \to \mathrm{HF}_{*+\mu}(P)$$

equals $2p_1(Q) \pmod 8$, the dimension of the instanton moduli on this mapping torus. (It may be necessary here to assume that $\eta(g)$ lifts to the integers to ensure orientability.)

Proposition 1.13

$$p_1(Q) = 2d(g)$$

In particular no gauge transformation can act with a degree different from $0 \pmod 4$.

Proof. The Pontrjagin number $p_1(Q)$ is easily seen to equal $2d(g)$. □

Note that proposition 1.9 gives a formula for the degree. It should prove interesting to investigate if the full group $H^1(Y;\mathbb{Z}_2)$ acts on the Floer homology, and to investigate the nature of the action.

1.4 Examples

We shall now pay attention to a few special bundles and compute their Floer homologies. This is definitely not a comprehensive list of what is known, it merely paves the way for later sections.

First consider torus bundles over a circle. These form a special case of work by Dostoglou and Salamon [DS1] on the Floer homology of surface bundles. Let T^2 denote a two dimensional torus. The non-trivial $SO(3)$-bundle $P_{T^2} \to T^2$ carries a unique flat connection, up to gauge transformations, with holonomy a Kleinian group K_0. For any automorphism of P_{T^2} we can form a torus bundle over S^1 for which this still is the unique flat connection. If γ is a section of the torus bundle then a gauge transformation g with $\eta(g)(\gamma) \neq 0$ acts non-trivially on this connection considered as an element of $\mathcal{B}_S(P_Y)$, see proposition 1.13 above. This shows:

Proposition 1.14 *Let $Y \to S^1$ be a T^2-bundle, P_Y admissible and nontrivial on a fibre T, then $\mathrm{HF}_j(P_Y)$ is zero in all but exactly two dimensions differing in degree by 4, in which it is isomorphic to $\mathbb{Z}$.*

We shall see later that the independence of the Floer homology on the gluing map for P_{T^2} is a general phenomenon.

Let $V = S^1 \times (T^2 - D^2)$. If (α, β) is a basis of $H^1(T^2 - D^2)$ let P_V denote the non-trivial $SO(3)$bundle with w_2 equal to $\beta \wedge \delta$ where δ is dual to S^1. Up to automorphism this is the unique non-trivial bundle on V. We can glue two copies of P_V along their boundaries, mapping S^1 to S^1 and identifying the boundaries of the missing dics in the torus and obtain a unique bundle:

$$P_{U_2} \to U_2 \equiv S^1 \times \Sigma_2,$$

where Σ_2 is a surface of genus 2. The bundle is characterized by the fact that it is trivial on every $\theta \times \Sigma_2$, for $\theta \in S^1$.

Proposition 1.15

$$\mathrm{HF}_*(P_{\mathcal{U}_2}) \simeq H_*(T^2;\mathbb{Z}) \otimes \mathbb{Z}^2$$

Proof. We first compute the space of representations of $\pi_1(S^1 \times \Sigma_2)$ in $SO(3)$. Let α_i, β_i be two sets of generators of $\pi_1(T^2 - D^2)$, and let δ be a generator of $\pi_1(S^1)$. Then

$$\pi_1(S^1 \times \Sigma_2) =< \delta,\, \alpha_1,\, \alpha_2,\, \beta_1,\, \beta_2 | [\alpha_1, \beta_1] = [\alpha_2, \beta_2]^{-1},\, [\delta, \alpha_i] = [\delta, \beta_i] = 1 >$$

and we can assume that $w_2(\mathcal{U}_2)$ is dual to $(\beta_1 + \beta_2) \wedge \delta$.

Up to conjugation by K_0 we find that a representation has to map

$$\delta \to \mathrm{diag}(1,-1,-1) \qquad \beta_1 \to \mathrm{diag}(-1,1,-1).$$

To keep the Stiefel-Whitney class as prescribed and satisfy the commutation relations α_i must map to elements of the torus determined by $\mathrm{diag}(1,-1,-1)$, that is they can be written as

$$\begin{pmatrix} 1 & 0 \\ 0 & r_i \end{pmatrix}$$

for a 2×2 rotation matrix r_i. Similarly β_2 must map to

$$\begin{pmatrix} 1 & 0 \\ 0 & s_2 \end{pmatrix}$$

for a 2×2 reflection matrix s_2. The final relation now reads:

$$r_1^2 = r_2^{-2}$$

so that up to conjugation by K_0 the representation variety equals

$$S^1 \times S^1 \times \mathbb{Z}_2,$$

and this space is also the space of flat connections in $\mathcal{A}/\mathcal{G}_S$.

To see that it is cut out transversely we note that it is in terms of generators and relations. But then it is transverse in the space of connections as well, see e.g. Taubes [Tau].

Now we call on the Morse-Bott spectral sequence, see Furuta-Steer [FuSt]. It has E_1 term equal to two copies of $H^*(T^2)$ in dimensions differing by 4. It follows that

$$\mathrm{HF}_*(P_{\mathcal{U}_2}) \simeq H_*(S^1 \times S^1;\mathbb{Z}) \times \mathbb{Z}^2. \qquad \square$$

2 The surgery triangle and excision

2.1 Knots and Dehn surgery

Let D denote the two dimensional disc. In the sequel we will continually use knots and it is convenient to denote by

$$P_L \to L$$

the product bundle $L \times SO(3) \to L$ over $L = D \times S^1$. We shall make use of $P_L \to L$ as a *trivialised* $SO(3)$-bundle. Note that the boundary orientation of L is $\partial D \wedge S^1$.

Definition 2.1

1) *A framed knot in Y is an orientation preserving embedding*

$$L \to Y$$

or in $\mathbb{C}$ an equivariant embedding

$$\kappa : P_L \to P_Y$$

The oriented simple closed curve $\mu = \kappa(\theta \times \partial D)$ is called a meridian for the knot and $\lambda = \kappa(S^1 \times 1)$ a longitude; here $\theta \in S^1$, $1 \in \partial D$.

2) *The knot complement is defined as $K = Y - L$ and $P_K = P - im\,\kappa$; note the boundary orientation $\lambda \wedge \mu$, of K.*

3) *A Dehn surgery on a knot κ is an object in $\mathbb{C}$ equal to:*

$$P_K \cup_{\kappa \circ \Phi} P_L$$

for an orientation preversing bundle isomorphism:

$$\Phi : \partial P_L \to \partial P_L$$

A framed knot is determined by two homotopy classes of loops λ, μ in ∂K, together with a homotopy class of a trivialization of $(P_K)_{|\partial K}$. The loops form an oriented basis of $\pi_1(\partial K)$. In terms of these loops a Dehn surgery is determined by two coprime integers p/q, defined by the fact that $\partial D \to p\mu + q\lambda$, and a trivialisation of $(P_K)_{|p\mu+q\lambda}$.

On the set of Dehn surgeries there is a right action of:

$$\Gamma = \pi_0(\mathrm{Aut}(\partial P_L)) \simeq (\mathbb{Z}_2)^2 \rtimes \mathrm{SL}(2, \mathbb{Z})$$

The bundle automorphisms of ∂P_L which extend to P_L form the group

$$\Gamma_0 = \pi_0(\mathrm{Aut}(P_L)) \simeq (\mathbb{Z}_2 \times 0) \rtimes \Lambda_0$$

and act trivially on the diffeomorphism type of the surgeries (they may act non-trivially on the spaces of connections and Floer homology groups); here:

$$\Lambda_0 = \{\pm \begin{pmatrix} 1 & 0 \\ m & 1 \end{pmatrix}; m \in \mathbb{Z}\}$$

Note that Γ/Γ_0 maps to a quotient of $PSL(2,\mathbb{Z})$.

Now $PSL(2,\mathbb{Z}) \simeq \mathbb{Z}_2 * \mathbb{Z}_3$ is generated by $\Sigma = \begin{pmatrix} 0 & 1 \\ -1 & 0 \end{pmatrix}$ of order 2 and $\Xi = \begin{pmatrix} -1 & 1 \\ -1 & 0 \end{pmatrix}$ of order 3. Moreover, Ξ satisifes:

$$1 + \Xi + \Xi^2 = 0.$$

Therefore the element

$$\xi = ((0,0),\Xi) \cdot ((0,1), Id) \in \mathrm{Aut}(P_L)$$

has order three. Its precise action is described by

1) change the trivialisation over ∂D before attaching the bundles
2) $\partial D \to S^1$
3) $S^1 \to -\partial D - S^1$

This element is of profound importance in knot theory:

Example 2.2 *Let κ be an oriented knot in a homology sphere Y, and assume that P_Y is trivialised. Then in ∂K we have a simple loop, λ, oriented as κ, which generates the kernel of $H_1(\partial K;\mathbb{Z}) \to H_1(K;\mathbb{Z})$. It is determined solely by the knot complement. The canonical closure, or 0-surgery is the surgery where we attach ∂D to this loop l, S^1 to $-\mu$, and change the trivialization of $(P_K)_l$. This admissible object is denoted by $P_{\overline{K}} \to \overline{K}$; it is homologically equivalent to a non-trivial bundle over $S^2 \times S^1$.*

The closure can be defined through the map:

$$\partial D \to \lambda \qquad S^1 \to -\mu.$$

If we let ξ act on this once we get:

$$Y: \quad \begin{cases} \partial D \to -\mu \\ S^1 \to -\lambda + \mu \end{cases} \overset{\Lambda_0}{\simeq} \begin{cases} \partial D \to \mu \\ S^1 \to \lambda \end{cases}$$

Clearly this is the original bundle. Applying ξ twice we obtain:

$$Y_{-1}: \quad \begin{cases} \partial D \to \mu - \lambda \\ S^1 \to \lambda \end{cases}$$

This is a homology sphere, called the -1 surgery on κ in Y; it depends on the framing.

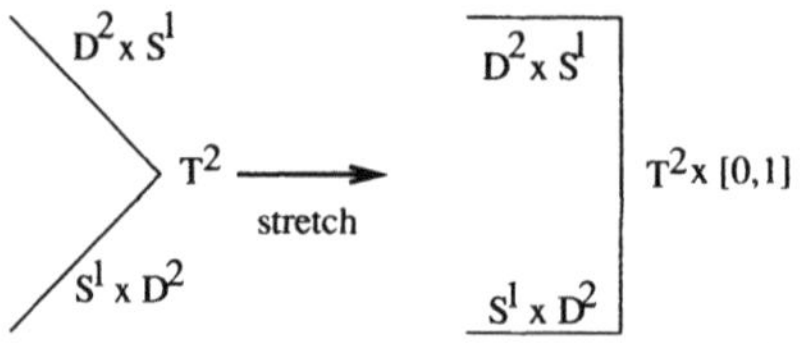

Diagram 13

A word of warning is in place. Let $\kappa : P_L \to P_Y$ be a framed knot, and P_Y admissible. It is not necessarily true that $\kappa \circ \xi$ defines an admissible object.

2.2 Surgery cobordisms and the exact triangle

We shall now see that certain surgeries define cobordisms, this is all well known material, see eg Rolfsen [Ro] 9.G. Recall that a surgery on a knot in S^3 is the effect on the boundary of attaching a 2-handle to D^4 precisely when the surgery coefficient p/q is an integer. The following simple proposition generalises this idea:

Proposition 2.3 *Let $\Phi_{1,2} : \partial(P_L) \to \partial(P_K)$ be two surgeries on κ, such that $\{\Phi_1(\partial D), \Phi_2(\partial D)\}$ is an integral basis of the group $\pi_1(\partial K)$. Then there is a $\Phi \in \mathrm{Mor}(\partial P_L, \partial P_K)$ such that*

$$(\Phi(1,0), \Phi(0,1)) = (\Phi_1(\partial D), \Phi_2(\partial D)),$$

and the bundle gluings induced by Φ over the loops $(0,1)$ and $(1,0)$ agree with those of Φ_1 and Φ_2 over ∂D.

Proof. Trivial □

Notice that Φ realises two surgeries in one blow, depending on which circle we «fill» in.

Definition 2.4 *Let the situation described in proposition 2.3 apply. The surgery cobordism is the morphism defined by the 4-manifold:*

$$Q_\Phi = (SO(3) \times D^2 \times D^2) \cup_\Phi ([0,1] \times (P_K))$$

where $D \times D$ is stretched so as to have boundary components

$$(D \times S^1) \cup ([0,1] \times S^1 \times S^1) \cup (S^1 \times D)$$

If Φ_1 sends ∂D to μ and S^1 to λ then the condition on Φ_2 is that $\Phi_2(\partial D) = \lambda + p\mu$, so that we can still say that integral surgeries give rise to cobordisms.

We now return to the exact triangle. First of all observe that applying ξ gives rise to a surgery cobordism. Therefore, if κ is a framed knot in P_Y and we denote by $P_{Y\xi}$ and $P_{Y\xi^2}$ the objects acquired by applying ξ we get cobordisms between each succesive pair. The composition of the three cobordisms is not the identity but induces a map of degree -1.

As long as all objects occuring in the surgery triangle are admissible we have a triangle of morphisms between Floer homology groups:

$$\begin{array}{ccc} & \mathrm{HF}_*(P_Y) & \\ \swarrow & & \nwarrow \\ \mathrm{HF}_*(P_{Y\xi}) & \longrightarrow & \mathrm{HF}_*(P_{Y\xi^2}) \end{array} \tag{24}$$

We can now state Floer's fundamental theorem:

Theorem 2.5 *The surgery triangle (24) obtained from ξ is exact, and of total degree -1, provided all bundles are objects of $\mathbb{C}^*$.*

Proof. Clearly all three objects should be admissible in order for the triangle to make sense. Due to the cyclic nature of this triangle we only need to consider two consecutive maps, a, b considering instantons on two cobordisms, and ascertain that the considerations of part I are still valid.

The first key condition in the proof is that the flat connections extending over the three manifolds correspond to the intersection with the three edges of a triangle instead of the intersection with three lines. To understand why this is so when applying ξ first observe that an $SO(3)$-bundle over the circle possesses two distinct flat connections with trivial monodromy. The homotopy classes of horizontal lifts of the circle for these two connections are different. However, if such a bundle over S^1 is the boundary of a bundle over the disc, only one of the flat connections will extend to the disc. For this triangle to appear the trivialisations over ∂D used in closing up the three 3-manifolds must not all be trivial, and any two must add up to the third. This is exactly realised by ξ.

The key condition in the proof is that in the representation space of ∂K the triangle formed by monodromy conditions for the three objects does not contain in its interior the restriction of a $U(1)$-reducible flat connection on P_K. The $U(1)$ reducible connections on P_K are parametrized by a torsion Chern class (modulo $\pm$) and a point in an affine space of dimension $\dim H^1(K;\mathbb{R})$. It follows easily that the three objects are admissible only if P_K is is free of flat S^1 reductions or if it is the complement of a knot in a homology sphere, thereby proving the theorem. □

Note that when P_K is free of flat S^1 reductions one can build many triangles from different framings of the knot.

2.3 Excision

Let

$$P_{T^2} \longrightarrow T^2$$

denote the nontrivial bundle over the torus; we shall refer to $[0,1] \times P_{T^2}$ as a *cylinder*.

Let P_{Y_1} and P_{Y_2} be connected admissible objects, each containing a cylinder C_i, $(i = 1, 2)$. Define a connected object

$$P_{Y_1} \#_{P_{T^2}} P_{Y_2} \equiv P_{Y_1} \backslash C_1 \cup_{P_{T^2}} P_{Y_2} \backslash C_2$$

for a choice of attaching maps. Then:

Theorem 2.6 *For any choice of attaching maps we have over the rational numbers:*

$$\mathrm{HF}_*(P_{Y_1} \#_{P_{T^2}} P_{Y_2}) = (\mathrm{HF}_*(P_{Y_1}) \otimes \mathrm{HF}_*(P_{Y_2}))/\mathbb{Z}_2.$$

If either group on the right hand side is torsion free the statement holds over the integers.

As we saw in Part I, the isomorphism is induced by a cobordism. This leads to another important:

Example 2.7

1) *Any P_{T^2} fibration over S^1 has the Floer homology of $S^1 \times P_{T^2} \simeq \mathbb{Z} \oplus \mathbb{Z}$ in dimensions differing by 4.*
2) *Recall the bundle $P_{\mathcal{U}_2}$ over $\mathcal{U}_2 = S^1 \times \Sigma_2$ introduced in II.1.4. Notice that $P_{\mathcal{U}_2}$ contains obvious cylinders in its handles. We can cut open and reglue along such cylinders, to get a large class of objects including bundles free of flat S^1 reductions over all circle bundles. Let $P_{\mathcal{U}_g}$ be an object where $g-1$ copies of $P_{\mathcal{U}_2}$ have been glued together. From $\mathrm{HF}_*(P_{\mathcal{U}_2}) = H_*(T^2) \otimes \mathbb{Z}^2$ we obtain that:*
$$\mathrm{HF}_*(\mathcal{U}_g) = H_*((T^2)^{g-1}) \otimes \mathbb{Z}^2.$$
This holds over the integers.
3) *Let $\mathcal{L} \in S^3$ be a fibered knot of genus one, i.e. $S^3 - \mathcal{L}$ is a fibration over S^1 with fibre $T^2 - D^2$. Let $\tilde{\mathcal{L}}$ denote the -1 surgery on $\mathcal{L}$, a homology sphere. Then*
$$\mathrm{HF}_*(\mathcal{L}) \simeq \mathrm{HF}_*(S^1 \times P_{T^2}) \simeq \mathbb{Z} \oplus \mathbb{Z}.$$

We are now in a position to put this basic knowledge to use.

3 Kirby calculus and the exact triangle

3.1 Introduction

In this section we wish to discuss the following:

Conjecture 3.1 *(Floer [F3], Theorem 7) Let J_* be a functor on $\mathbb{C}^*$ for which the surgery triangle is exact, the Künneth formula is true, and which gives the same module as* HF_* *for S^3 and $S^1 \times P_{T^2}$. Then $J_* = \mathrm{HF}_*$.*

Floer states a slightly different version of this in his Preliminary preprint 'Instanton Homology for Knots' and gives a large part of a proof. We give it here as a conjecture because it seems to us that the proof is not quite complete, although the result is very likely true. Our aim is now to present some of the key ideas in Floer's approach, although we would not like to pretend that we have reached the same understnading of these ideas as Floer had.

The proof Floer suggests is very interesting because it proceeds by simplifying the 3-manifold while keeping track of the Floer groups. One can present (in many different ways) any 3-manifold by (integral) surgery on an oriented link $\mathcal{L} = (\mathcal{L}_i)_{i=1,\dots,N}$ in S^3. By abuse of notation we shall write $\mathcal{L}$ for the manifold and $P_{\mathcal{L}}$ for the object in $\mathbb{C}$ obtained from the surgery on $\mathcal{L}$. In integral surgery each component is labeled by an integer, but we shall have a need for some other labelings as well.

The simplification of the link proceeds by changing the crossings of $\mathcal{L}_i$ over $\mathcal{L}_j$ for $i < j$ and by unknotting each individual component. In finitely many steps one reaches a collection of N unknots.

Changing a crossing in $\mathcal{L}$ is done through -1 surgery on an auxiliary loop γ around the crossing, as in diagram 14. Denote the object we acquire through this by $\mathcal{L}^-$. In order to invoke the exact triangle, applied to surgery on γ, we also need to understand 0 surgery on γ, which leads to an object $\mathcal{L}^0$. We shall show that $\mathcal{L}^0$ can also be described in terms of surgery on the link $\mathcal{L}$ where now the crossing has been changed to a «0-crossing», see diagram 14, provided the labelings are suitably adapted. The point is that both for $\mathcal{L}^-$ and for $\mathcal{L}^0$ fewer crossings need to be undone to reach unlinks than for $\mathcal{L}$.

Under the assumption that all objects are admissible we get the following exact triangle when changing a crossing:

$$\begin{array}{ccc} & \mathrm{HF}_*(P_{\mathcal{L}^0}) & \\ \swarrow & & \nwarrow \\ \mathrm{HF}_*(P_{\mathcal{L}}) & \longrightarrow & \mathrm{HF}_*(P_{\mathcal{L}^-}) \end{array} \tag{25}$$

The idea is then to proceed by induction on the number of crossings.

The problems one encounters with Floer's scheme have to do with admissibility; when changing a crossing between 2 components of a link neither $\mathcal{L}^-$ nor $\mathcal{L}^0$ may be admissible, thereby precluding an application of (25). Another, possibly less serious problem is that the final configurations have unknown Floer homology. This can likely be be overcome using the theory for connected sums, see Fukaya [Fuk]. We believe that Floer's approach to his conjecture is promising and clearly of independent interest.

Informally, one can think of the statement in 3.1 as saying that there is a method for computing all Floer homology groups «modulo exact sequences»: i.e. if we imagine that any time we know two out of three terms in a long exact sequence we can deduce the third, then we would be able to compute all the Floer groups.

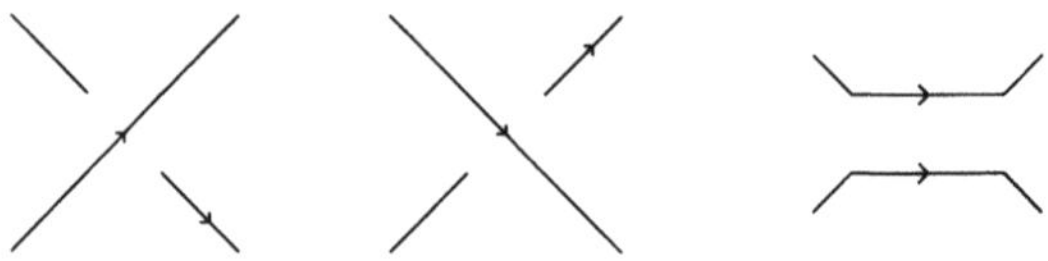

a) crossings

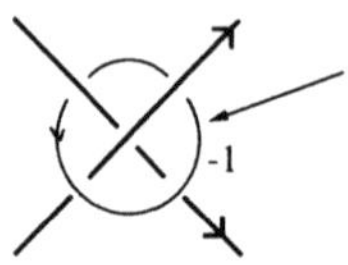

Diagram 14 b) surgery and crossings

3.2 Link diagrams and Kirby calculus

Every oriented three manifold can be obtained by doing integral surgery on the components $\mathcal{L}_i$ of a link $\mathcal{L} = (\mathcal{L}_i) \in S^3$. Every component $\mathcal{L}_i$ is given an orientation and the canonical longitude λ_i, which vanishes in $H_1(S^3 - \mathcal{L}_i; \mathbb{Z})$, and meridian μ_i give the framing. In integral surgery $D^2 \times S^1$ is reattached by sending ∂D^2 to $n_i\mu_i + \lambda_i$, for an integer n_i; we can fix the framing by sending S^1 to $-\mu$.

Mostly to keep track of our bundles we digress for a moment on the homology of the three manifold $\mathcal{L}$ which we obtain in this way. The meridians μ_i can be taken as generators of $H_1(\mathcal{L}; \mathbb{Z})$ and the relations come from

$$n_i\mu_i + \sum_{j \neq i} \mathrm{lk}(\mathcal{L}_i, \mathcal{L}_j)\mu_j = 0. \tag{26}$$

We refer to the symmetric matrices with linking number off the diagonal and the integers n_i on the diagonal as the linking matrix. It is a presentation matrix for $H_1(Y; \mathbb{Z})$. To understand the relations just note that $\lambda_i = \sum_{j \neq i} \mathrm{lk}(\mathcal{L}_i, \mathcal{L}_j)\mu_j$ in the link complement and attaching the handles gives: $n_i\mu_i + \lambda_i = 0$. We can specify a bundle over the surgered 3-manifold by giving its Stiefel-Whitney class:

$$w_2 \in H^2(\mathcal{L}; \mathbb{Z}_2) \simeq H_1(\mathcal{L}; \mathbb{Z}_2).$$

Thus this Stiefel Whitney class w_2 is merely a $\mathbb{Z}_2$-vector in the kernel of the linking matrix. The Poincaré duality ismorphism holds over the integers and reduces the verification of admissibility of the bundle to a simple exercise in matrix algebra.

Let $\mathcal{L} = (\mathcal{L}_i)$ be an oriented link in S^3, each component labeled with an integer n_i, which indicates the surgery that is performed, together with a $\mathbb{Z}_2$-vector $(w_2)_i$ in the kernel of the linking matrix. We indicate the object obtained from $\mathcal{L}$ by $P_{\mathcal{L}} \to \mathcal{L}$.

We have use for further labelings.

Definition 3.2

1) *The labeling $\mathcal{L}_1^D$ will be used to indicate that integral surgery is performed on $\mathcal{L}_1$ (according to an integer n_1 as above).*
2) *The labeling $(\mathcal{L}_1, \mathcal{L}_2)^C$ indicates that the boundary components corresponding to $\mathcal{L}_1$ and $\mathcal{L}_2$ on which the bundle is assumed to be nontrivial are attached (according to an element $c \in \mathrm{Aut}(P_{T^2})$). In other words the cylinder $[0, 1] \times P_{T^2}$ is attached.*
3) *The labeling $\mathcal{L}_1^T$ means that the nontrivial bundle P_V over $V = S^1 \times (T^2 - D^2)$ (see 1.4) is attached to $\mathcal{L}_1$ (according to an element in $\mathrm{Aut}(\partial P_L)$).*

We refer to the attachment of three manifolds different from L to the link complement as generalised surgery.

In the definition D, C, T stand for disc, cylinder and torus. This notation is a summary notation; the details are maybe best kept track of in a link diagram with only integral surgery.

We now return to a link diagram with integral labels. The Kirby moves, see Kirby [K1], [K2], describe which links give rise to the same three manifold. The following theorem is a trivial adaptation of this for objects of $\mathbb{C}$.

Theorem 3.3 *The object P_Y does not change if we remove any isolated, unknotted loops $\mathcal{L}_i$ with framing ± 1 from our labeled link and remove the i-th entry (which must equal 0) in the w_2 vector, or if we replace a component $\mathcal{L}_i$ with the band connected sum of $\mathcal{L}_i$ with a push off (according to the framing) of another component $\mathcal{L}_j$, where the new component is framed according to $n_i + n_j \pm 2\mathrm{lk}(\mathcal{L}_i, \mathcal{L}_j)$. The sign is plus exactly when the band sum is oriented. The i-th entry in the w_2 vector is not changed, and the j-th entry is replaced by $(w_2)_j - (w_2)_i$.*

The band sum operation is referred to as a *handle slide*. The link matrix changes by adding (or subtracting in the un-oriented case) the j-th row to the i-th row, and subsequently adding the j-th column to the i-th. It is important to note that if we do a generalised surgery on $\mathcal{L}_1$ and a normal integral surgery on $\mathcal{L}_2$ then we can still slide $\mathcal{L}_1$ over $\mathcal{L}_2$, that is apply the generalised surgery to the band connected sum of $\mathcal{L}_1$ with a push off of $\mathcal{L}_2$.

3.3 Knot crossings

We shall now use a little auxiliary loop γ around a crossing. We choose orientations as in diagram 14 b) and define a framing for γ by assigning it integer 0, that is we attach $\partial D \to \lambda_\gamma$ and $S^1 \to -\mu_\gamma$, and giving the w_2 vector an entry 1 for γ. This sets the scene for a correct application of the exact triangle. When γ goes around a crossing of two pieces of the link (which could either belong to a single or to two components of $\mathcal{L}$), as in diagram 14 b), the handle slide introduces an extra twist in the strands in the direction of γ. This enables one to change a + crossing into a $-$ crossing in a planar link diagram. We shall indicate the corresponding link with $\mathcal{L}^-$. This also holds for generalised crossings:

Proposition 3.4 *When we do a generalised surgery on $\mathcal{L}$ and a -1-surgery on γ the result is the generalised surgery applied to $\mathcal{L}^-$, and the labeling is unaffected as long as the orientations are chosen as in diagram 14 b).*

Note that one can also change crossings when the orientations are not as in diagram 14 b), but in that case one needs to look at labelings and orientations carefully, using Kirby's theorem.

Our next task is to investigate what effect 0-surgery on γ has on the manifold $\mathcal{L}$. As this appears to work out quite differently in various cases, we shall first deal with knot crossings. From this it will become apparent why generalised surgeries need to be taken into account. When working with a crossing γ of a single component $\mathcal{L}_1$ taking the connected sum with γ will split it into two components: $(\mathcal{L}_1^1, \mathcal{L}_1^2)$.

Proposition 3.5 *The result of 0-surgery on γ going around a crossing of $\mathcal{L}_1$ labeled D with integer n_1 is diffeomorphic to the object corresponding to $(\mathcal{L}_1^1, \mathcal{L}_1^2)^C$ and the remaining labelings unaffected. This object is always admissible and its Floer homology is independent of the attaching map for the cylinder.*

Proof. This is done with Kirby calculus. From the link matrix of the configuration given the components of $\mathcal{L}$ and γ, the admissibility of $\mathcal{L}^-$ and $\mathcal{L}^0$ is clear since $lk(\gamma, \mathcal{L}_1) = 0$.

0-surgery on γ corresponds to the effect of adding a 1-handle, see [K2]. The two strands of $\mathcal{L}_1$ run accross this one handle in different directions. After removing the solid torus corresponding to $\mathcal{L}_1$ adding the one handle means adding $(S^2 - 2 \text{ discs}) \times [0,1]$ to $\mathcal{L}$ minus two balls.

On the other hand adding a $[0,1] \times T^2$ to two components of a link corresponds to adding a one handle $[0,1] \times S^2$, where the two attaching S^2's each intersect a component in two circles, followed by attaching a two handle $D^2 \times S^1$, along a path obtained by drilling two $[0,1] \times D^2$ out of the one handle, connecting the circles of intersection, to connect the link components. This is clearly the same.

The resulting Floer homology is independent of the attaching map for the two torus by the excision theorem. □

This shows the need to introduce generalised labelings. Here we see that 0-surgery on the auxiliary loop transforms $\mathcal{L}_1^D$ into $(\mathcal{L}_1^1, \mathcal{L}_1^2)^C$. We can now apply the triangle:

Theorem 3.6 *If an auxiliary loop is used to change a crossing in a single component of $\mathcal{L}$ as in diagram 14 b), then the triangle (25) is exact.*

Example 3.7 *Consider a knot $\mathcal{L}$ labeled with D, $n_1 = -1$, which can be undone by changing a single crossing. Since $\mathcal{L}^-$ is S^3 we have*

$$\mathrm{HF}_*(\mathcal{L}) \simeq \mathrm{HF}_*((\mathcal{L}^1, \mathcal{L}^2)^C).$$

We have the following simple:

Lemma 3.8 *If a pair of components of an oriented link $\mathcal{L}$ carries label C and these components are separated by a 2-sphere in $\mathcal{L}$ then* $\mathrm{HF}_*(P_{\mathcal{L}}) = 0$.

Proof. There is no flat connection on the two sphere which has non-zero Stiefel-Whitney class. Consequently the Floer complex of $P_{\mathcal{L}}$ is 0. □

This lemma shows that it is unnecessary to unknot a single component of a C-pair, one merely needs to separate it from the other components. The lemma applies in particular to the final configurations obtained from the simplification scheme. If a component of the N component unlink is labelled with a C the Floer homology vanishes.

3.4 Crossings between two strands

The next proposition deals with a loop γ around two components $(\mathcal{L}_1, \mathcal{L}_2)$. The connected sum of $\mathcal{L}_1, \gamma$ and $\mathcal{L}_2$ gives rise to a one component link $\mathcal{L}_1\#\mathcal{L}_2$, described by the 0-crossing in a projection. The result of integral surgery on each, that is, each labeled with D, is fully described by the handle slide theorem, but we already have a need to consider crossings between a D and a C labeling and between the strands of a C pair.

Proposition 3.9 *Doing 0-surgery on γ going around a crossing of $\mathcal{L}_1$ and $\mathcal{L}_2$ as in diagram 14 b) affects labelings as follows:*

1) $\mathcal{L}_1^D, \mathcal{L}_2^D$	goes to	$(\mathcal{L}_1\#\mathcal{L}_2)^D$	$(DD \to D)$
2) $\mathcal{L}_1^D, (\mathcal{L}_2, \mathcal{L}_3)^C$	goes to	$((\mathcal{L}_1\#\mathcal{L}_2), \mathcal{L}_3)^C$	$(DC \to C)$
3) $(\mathcal{L}_1, \mathcal{L}_2)^C$	goes to	$(\mathcal{L}_1\#\mathcal{L}_2)^T$	$(CC \to T)$

The objects acquired by 0 surgery in case 2) and 3) are always admissible. In each case the exact triangle (25) holds if the three objects $P_{\mathcal{L}}$, $P_{\mathcal{L}^-}$, $P_{\mathcal{L}^0}$ are admissible.

The proof of the proposition is much the same as the one above. A few remarks are in order. Whereas the attaching map for the cylinder was irrelevant it definitely is not when attaching the punctured torus times the cirle. The admissibility of the objects $\mathcal{L}^-$ and $\mathcal{L}^0$ acquired by changing a crossing of two components on each of which we do an integral surgery is most easily read off from the matrix data, by including an extra row and column for γ. As a rule they are not admissible. Aside from admissibility questions we now have a powerful tool to compute Floer homology. The following example shows how to keep track of he grading:

Example 3.10 *Let H_{-n} for $n \geq 0$ be the generalised Hopf link $(\mathcal{L}_1, \mathcal{L}_2)^C$ with linking number $-n$, labelled with a C; compare diagram 15. The first two have easily understood Floer groups: H_0 contains an S^2 with non-trivial Stiefel Whitney class and H_{-1} is the object $S^1 \times P_{T^2}$. If we change a single crossing in H_{-n} we get the H_{-n+1}. On the other hand $\mathcal{L}_1\#\mathcal{L}_2$ is the unknot U labeled with a T for any $n \geq 0$, which is $S^1 \times P_{T^2}$. We can compute* $\mathrm{HF}(H_{-n})$ *by induction as follows.*

Assume that grading is given on $HF_*(H_{-1})$, such that it has non-zero homology in dimensions 0 and 4. In the triangle

$$\begin{array}{ccc} & \mathrm{HF}_*(U^T) & \\ \delta_n \swarrow & & \nwarrow \phi_n \\ \mathrm{HF}_*(H_{-n}) & \xrightarrow{\qquad} & \mathrm{HF}_*(H_{-n+1}) \\ & \psi_n & \end{array}$$

we want to show that $\phi_n = 0$. Choose the grading i_n on the group $\mathrm{HF}_*(U^T)$ and on $\mathrm{HF}_*(H_{-n+1})$ such that δ_n has degree -1 and ϕ_n and ψ_n have degree 0. We want to compute inductively in which dimension d_n a generator α of $\mathrm{HF}_*(U^T)$ hits $\mathrm{HF}_*(H_{-n})$, starting with $d_1 = 0$. Note that the moduli spaces on the cobordism between U^T and H_{-n} have dimension

$$0 = \dim \mathcal{M}(\alpha, \delta_n(\alpha)) = -d_n + i_n(\alpha) + 1$$

so that $d_n = i_n(\alpha) + 1$. On the cobordism between H_{-n} and U^T we find:

$$\dim \mathcal{M}(\delta_n(\alpha), \alpha) = -i_{n+1}(\alpha) + d_n.$$

Therefore the composition, which is a cobordism between U^T and itself, has moduli of dimension $i_n(\alpha) - 1 - i_{n+1}(\alpha)$, which we can alternatively compute from the dimension formula from a closed manifold by taking the mapping torus. The result equals $-3 \pmod 4$, so that $i_n(\alpha) - i_{n-1}(\alpha) = 2 \pmod 4$. Thus we find for the Betti numbers $\dim \mathrm{HF}_j(H_{-n})$:

j \ n	1	2	3	4	5
0	1	1	2	2	3
1	0	0	0	0	0
2	0	1	1	2	2
3	0	0	0	0	0
4	1	1	2	2	3
5	0	0	0	0	0
6	0	1	1	2	2
7	0	0	0	0	0

Note that once again we see that the grading is not absolute once the degrees of the maps in the triangle is fixed. We thank Ron Stern for informing us about the correct pattern.

It may seem a bit mysterious how to compute the groups of H_n for positive n. This is done by using a γ with opposite orientation from what is drawn in diagram 14 b). The -1 surgery transforms H_n into H_{n+1} and the zero surgery again gives the unlink. The computation of degrees is similar and we find that

$$\mathrm{HF}_*(H_n) \simeq \mathrm{HF}_*(H_{-n}).$$

This example can be continued considering knots K_n labeled with a D and integer -1 which reduce to H_n when a crossing is opened to a 0-crossing, and undo when a crossing is changed. A collection of such knots is drawn in diagram 15; it includes the figure 8 knot.

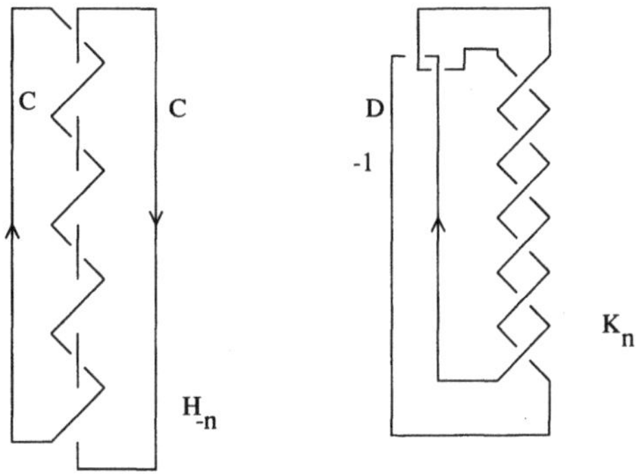

Diagram 15

In connection with label T it should be mentioned that we do not know the Floer homology of an unlink of N components some labeled with a T others with a D. This seems an excellent candidate for study with the connected sum techniques of Fukaya [Fuk].

A simple example where the in-admissibility of $\mathcal{L}^-$ and $\mathcal{L}^0$ appears is in a two component link as above with link number 5, but each component labeled D, one with integer 2, the other with 13. This configuration represents a Brieskorn sphere, and its Floer homology is known, see Fintushel and Stern [FS1]. Another one is three times +1 surgery on the components of the Borromean rings. It would be interesting to see if these links can be simplified with the surgery techniques of this section.

The problems outlined in the preceeding two paragraphs seem to be only obstructions to proving Floer's conjecture.

3.5 Further surgeries

It is quite surprising that the complexity of the scheme of simplifications does not increase further, although a new phenomenon appears when undoing, for example, knot crossings labelled with a T. Recall the object $P_{\mathcal{U}_2} \to \mathcal{U}_2$, from section 1.4. Let us first of all describe what happens when a crossing in a single T-labeled component is encountered:

Proposition 3.11 *Doing 0-surgery on γ going around a crossing of $\mathcal{L}_1$ as in diagram 14 b) affects labelings as follows:*

1) $\mathcal{L}_1^D$ *transforms to* $(\mathcal{L}_1^1, \mathcal{L}_1^2)^C$
2) $\mathcal{L}_1^T$ *transforms to the object which is the* P_{T^2}*-connected sum of* $P_{\mathcal{U}_2}$ *and the object obtained from the labeling* $(\mathcal{L}_1^1, \mathcal{L}_1^2)^C$
3) $(\mathcal{L}_1, \mathcal{L}_2)^C$ *transforms to* $(\mathcal{L}_1^1, \mathcal{L}_1^2, \mathcal{L}_2)^P$ *where label P indicates that*

$$(S^2 - 3 \text{ discs}) \times S^1$$

(«trousers» times a circle) is attached (the attaching map is relevant).

In all cases the objects obtained are admissible, and the exact triangle (25) holds.

The first case was already mentioned above. Recall that there is no need to undo a single C component. In the case of unknotting a T component the exact sequence takes the following form, by virtue of the Künneth theorem:

$$\begin{array}{ccc} & \mathrm{HF}_*(P_{\mathcal{L}0}) \otimes_{\mathbb{Z}} H_*(T^2) & \\ \swarrow & & \nwarrow \\ \mathrm{HF}_*(P_{\mathcal{L}}) & \longrightarrow & \mathrm{HF}_*(P_{\mathcal{L}-}) \end{array}$$

There is a similar proposition for crossings between two components:

Proposition 3.12 *The effect of 0 surgery on a loop around a crossing of $\mathcal{L}_1$ and $\mathcal{L}_2$ in a labeled link diagram is as follows:*

1)	$\mathcal{L}_1^D, \mathcal{L}_2^D$	goes to	$(\mathcal{L}_1\#\mathcal{L}_2)^D$	$(DD \to D)$
2)	$\mathcal{L}_1^D, \mathcal{L}_2^T$	goes to	$(\mathcal{L}_1\#\mathcal{L}_2)^T$	$(DT \to T)$
3)	$\mathcal{L}_1^D, (\mathcal{L}_2, \mathcal{L}_3)^C$	goes to	$((\mathcal{L}_1\#\mathcal{L}_2), \mathcal{L}_3)^C$	$(DC \to C)$
4)	$\mathcal{L}_1^T, (\mathcal{L}_2, \mathcal{L}_3)^C$	goes to	$((\mathcal{L}_1\#\mathcal{L}_2), \mathcal{L}_3)^{C\#}$	$(TC \to C\#)$
5)	$\mathcal{L}_1^T, \mathcal{L}_2^T$	goes to	$(\mathcal{L}_1\#\mathcal{L}_2)^{T\#}$	$(TT \to T\#)$
6)	$(\mathcal{L}_1, \mathcal{L}_2)^C$	goes to	$(\mathcal{L}_1\#\mathcal{L}_2)^T$	$(C_1, C_1 \to C)$
7)	$(\mathcal{L}_1, \mathcal{L}_4)^{C_1}, (\mathcal{L}_2, \mathcal{L}_3)^{C_2}$	goes to	$(\mathcal{L}_1\#\mathcal{L}_2, \mathcal{L}_3, \mathcal{L}_4)^P$	$(C_1, C_2 \to P)$

Here the labeling C# and T# mean that the object is formed from the result of a C or T label followed by taking the connected sum along P_{T^2} with P_{U_2}. The label P in (7) was explained in the preceeding proposition. In all but the first case the objects in the sequence will be admissible without question, and the triangle (25) is exact precisely when all objects are admissible.

Observe that case 7) is not necessary: by applying the excision argument one can reattach the cylinders and reduce to case 6). Armed with these surgery diffeomorphisms one can do another

Example 3.13 *Let $\mathcal{L} = (\mathcal{L}_1, \mathcal{L}_2)^C$ be a two component link. We shall prove that the Euler characteristic*

$$|\chi(\mathrm{HF}_*(\mathcal{L}))| = 2|lk(\mathcal{L}_1, \mathcal{L}_2)|.$$

We already saw that this is true for the Hopf links H_n. To get it in general, note that upon changing a crossing between the components $\mathcal{L}_1$ and $\mathcal{L}_2$ the Euler characteristic changes by $\chi(\mathrm{HF}_*((\mathcal{L}_1\#\mathcal{L}_2)^T))$, while the linking number changes by 1. For any knot κ in S^3 we have that $\chi(\mathrm{HF}_*(\kappa^T)) = 2$. This follows from applying the exact sequence belonging to proposition 3.11 (2), where the top term has Euler characteristic 0, allowing one to change κ to the unknot without affecting the Euler characteristic. The signs work out as for the Hopf link and the statement follows.

References

[A1] M.F. Atiyah, *New Invariants for 3 and 4-dimensional manifolds*, Proc. Sympos. Pure Math. 48(1988).

[B] P. Braam, *Floer Homology Groups for Homology 3-spheres*, Adv. Math. 88, no. 2(1991), pp. 131–144.

[DFK] S.K. Donaldson, M. Furuta & D. Kotschick, *On Floer Homology*, in preparation.

[DK] S.K. Donaldson & P.B.K. Kronheimer, *The Geometry of 4-Manifolds*, Oxford University Press, Oxford, 1990.

[DS1] S. Dostoglou & D. Salamon, *Instanton Homology and Symplectic Fixed Points*, preprint.

[DS2] S. Dostoglou & D. Salamon, *Self-Dual Instantons and Holomorphic Curves*, preprint.

[FS1] R. Fintushel & R. Stern, *Instanton homology of Seifert fibred homology three spheres*, Proc. LMS (3), 61(1990), 109–137.

[FS2] R. Fintushel & R. Stern, *Invariants for Homology 3-spheres*, in «Geometry of Low-Dimensional Manifolds 1», Ed. S.K. Donaldson and C.B. Thomas, LMS Lecture Note Series 150, CUP 1990.

[F] A. Floer, *An instanton invariant for 3-manifolds*, Comm. Math. Phys. 118(1988), 215–240.

[F1] A. Floer, *Instanton Homology and Dehn surgery*, in this volume.

[F2] A. Floer, *Instanton Homology for Knots*, preliminary preprint.

[F3] A. Floer, *Instanton Homology, Surgery and Knots*, in «Geometry of Low-Dimensional Manifolds 1», Ed. S.K. Donaldson and C.B. Thomas, LMS Lecture Note Series 150, CUP 1990.

[Fuk] K. Fukaya, *Floer homology of connected sums of homology 3-spheres*, preprint.

[Fu] M. Furuta, *The Homology Cobordism Group of Homology 3-Spheres*, preprint.

[FuSt] M. Furuta & B. Steer, *Seifert fibred homology 3-spheres and the Yang-Mills equations on Riemann surfaces with marked points*, to appear Adv. Math.

[K1] R. Kirby, *A calculus for framed links in* S^3, Inv. Math. 45(1978), 35–56.

[K2] R. Kirby, *The topology of 4-manifolds*, Lecture Notes in mathematics 1374, Springer Verlag 1989.

[KM1] P. Kronheimer & T. Mrowka, *Gauge Theory for Embedded Surfaces I*, preprint.

[KM2] P. Kronheimer & T. Mrowka, *Gauge Theory for Embedded Surfaces II*, preprint.

[Mr] T. Mrowka, *A Local Mayer-Vietoris Principle for Yang-Mills Moduli Spaces*, Harvard PhD thesis.

[MMR] J. Morgan, T. Mrowka & D. Ruberman, *L^2-moduli spaces for manifolds with tubular ends*, preprint.

[Ro] D. Rolfsen, *Knots and Links*, Mathematics Lecture Series 7, Publish or Perish Inc, 1976.

[Tau] C.H. Taubes, *Casson's Invariant and Gauge Theory*, Jour. Diff. Geom. 31(1990), 547–599.

P.J. Braam and S.K. Donaldson
The Mathematical Insitute
23–29 St. Giles
Oxford OX1 3LB, UK

Fukaya-Floer homology and gluing formulae for polynomial invariants

P.J. Braam and S.K. Donaldson

1 Introduction

The moduli spaces of instantons over a compact 4-manifold X carry a great deal of differential-topological information leading to many invariants of X. In some simple cases these invariants are just numbers, obtained by counting points in 0-dimensional spaces moduli spaces, but more generally one gets polynomial functions on the homology, particularly the 2-dimensional homology, of X. To any homology class $\Sigma \in H_2(X)$ one associates a cohomology class $\mu(\Sigma)$ over the moduli space and, assuming that this is even dimensional, one can then evaluate the top-dimensional power of $\mu(\Sigma)$ on the moduli space.

These invariants for closed manifolds lead naturally to the following problem: if a closed 4-manifold X is split into pieces $X = X_1 \cup_Y X_2$ by a 3-dimensional submanifold Y, how can we record differential-topological data about the pieces X_1, X_2 from which we can recover the invariants of X? In this problem, of course, we are following a standard theme in topology, which one can think of as embracing such things as the Mayer-Vietoris sequence for homology and addition formulae for Euler characteristics and signatures. A full solution to the problem would probably be important both for practical calculations for particular manifolds and also for a deeper understanding of the nature of the invariants (and perhaps of the role played by geometry and gauge theory in 4-dimensional topology).

While a good deal is already known about this question, many challenging problems remain. A case where one has a good theoretical understanding is when Y is a homology 3-sphere, and one knows that one can associate invariants to X_1, X_2 which take values in the Floer homology of their boundary, and that the invariants are obtained from these by a bilinear pairing. These ideas have been one of the motivations for the introduction of the notion of a «Topological Field Theory» [At2], which abstracts this simplest kind of «gluing formula'. However, to understand more general splittings one needs to overcome significant further problems. These problems have, in the main, to do with two things:

1) The effect of *reducible* connections.
2) The decomposition of homology classes in X by the hypersurface Y.

By (2) we mean the following. If Y is a homology sphere then any homology class $\Sigma \in H_2(X)$ can be decomposed into a sum of a contribution in $H_2(X_1)$ and a contribution in $H_2(X_2)$, and these can be carried through the discussion of relative invariants in a very straightforward way. When Y has non-trivial homology, by contrast, one may encounter the situation that a class $\Sigma \in H_2(X)$ can not be realised as combinations of classes in X_1, X_2, but rather arises from relative classes in $H_2(X_1, Y), H_2(X_2, Y)$ with the same boundary γ in $H_1(Y)$.

In many realistic cases one will encounter these two problems at the same time, but nevertheless it is possible to separate them out to a large degree, and this is the policy we follow in this paper. We shall study a situation in which reducible connections do not occur and we shall show that a construction of Fukaya, adapted very slightly to our hypotheses, furnishes a complete answer to the second difficulty. We shall see that there is a «Fukaya-Floer' group associated to Y and γ and that the invariants of X can be recovered from relative invariants associated to X_1, X_2, which have values in the Fukaya-Floer groups. In essence our conclusion will be that the «Topological Field Theory» formalism can be extended to take acount of the problem (2).

This paper is a part of a more general programme in which we hope to combine the ideas here, associated to (2) above, with extensions of ideas developed in [Do2], on the one hand, and [AuBr1] on the other, associated with the problems (1) of reducible connections. The goal of this programme is very similar to that of Taubes, described in his paper [Tau], and it should be interesting to compare the different approaches. One difference of our approach from that described by Taubes is that we work with generic deformations of the instanton equations over tubes, as in Floer's definition of his instanton homology groups, rather than the unperturbed equations considered by Taubes. This means that we avoid most analytical difficulties, and the problems become much more algebro-topological in nature.

This paper is to a large degree a work of exposition, since the geometrical results we need are mostly small variants of those proved in detail by Fukaya. Another starting point for us, particularly with regard to Section 4 below, was unpublished work of Furuta [Fur1] which we found enormously helpful in understanding Fukaya's work. More recently, a paper of Furuta [Fur2] describes some related ideas.

We begin the body of our paper, in Section 2, by discussing a simple case of the «gluing problem' which contains in a prototype form most of the ideas arising in the general case. In Section 3 we define the Fukaya-Floer groups and state the main results about them, with outline proofs. In Section 4 we explain how to fit the ideas into the framework of standard algebrtaic topology, explaining to some extent the detailed Morse-theoretic definitions of Section 3. In Section 5 we make some simple observations about the polynomial invariants in the setting of Topological Field Theories: we find relations between the invariants arising from a splitting of a 4-manifold.

Acknowledgements: The first author would like to thank David Austin for discussions and the Mathematics Department of the University of British Columbia for its hospitality.

2 Examples and techniques

In this section we will illustrate the mechanisms underlying our gluing formulae in a case which is quite simple and yet contains most of the essential features of our construction. Let $P \to X$ be a $U(2)$ bundle over the Riemannian 4-manifold X such that $\mathcal{M}(P)$, the moduli space of projectively anti-self dual connections on P, with fixed central part of the curvature[1]), is 2-dimensional and consists solely of irreducible connections. Using the map

$$\mu : H_2(X;\mathbb{Z}) \to H^2(\mathcal{B}^*_X;\mathbb{Z}) : \Sigma \to p_1(\mathbb{P})/\Sigma,$$

where $\mathbb{P} \to \mathcal{B}^*_X \times X$ is the universal bundle over the irreducible connections times X, we obtain a linear function on $H_2(X;\mathbb{Z})$:

$$\Sigma \to \psi(X,\Sigma) =< \mathcal{M}(X), \mu(\Sigma) > . \tag{1}$$

Of course we need to orient $\mathcal{M}(X)$ through a homology orientation of X (cf. Donaldson [Do3]) and we shall moreover assume that $b^+(X) > 1$, since this will ensure that (1) is independent of the (generic) metric and defines a differential topological invariant $\Psi(X,\Sigma)$ of X.

We assume next that $X = X_1 \cup_Y X_2$, and that P_Y carries precisely two projectively flat connections ρ_1 and ρ_2, both irreducible and non-degenerate, such that the following situation applies. We consider the manifolds X_1 and X_2 and assume that their Riemannian metrics have cylindrical ends: $\mathbb{R}_{>0} \times Y$. Our assumption then is that the moduli spaces of instantons $\mathcal{M}(X_i, \rho_j)$, with asymptotic boundary value ρ_j and curvatures in L^2, have dimensions:

$$\begin{aligned} \dim \mathcal{M}(X_1,\rho_1) &= 2 \qquad & \dim \mathcal{M}(X_1,\rho_2) &= 0 \\ \dim \mathcal{M}(X_2,\rho_1) &= 0 \qquad & \dim \mathcal{M}(X_1,\rho_2) &= 2. \end{aligned}$$

These assumptions imply that the *reduced* moduli space $\mathcal{M}_0(\rho_1,\rho_2)$ on the cylinder $\mathbb{R} \times Y$, which is the quotient of $\mathcal{M}(\rho_1,\rho_2)$ by translations, is a compact one-manifold, without boundary.

Any homology class $\Sigma \in H_2(X;\mathbb{Z})$ can be written as the sum of two *relative* classes $\Sigma_i \in H_2(X_i, \partial X_i;\mathbb{Z})$ such that

$$\partial\Sigma_1 = -\partial\Sigma_2 = \gamma \in H_1(Y;\mathbb{Z}).$$

We shall identify these classes with surfaces in X (with boundary in the relative case) and a loop in Y. The μ-map extends directly to a map

$$\mu : H_2(X_i, \partial X_i;\mathbb{Z}) \to H^2(\mathcal{B}^*(X_i,\rho_j);\mathbb{Z})$$

where $\mathcal{B}^*(X_i,\rho_j)$ is the space of connections with a *fixed* boundary value ρ_j. In the case that $\gamma \neq 0$ the classes $\mu(\Sigma_j)$ do not naturally have compact support on the moduli spaces $\mathcal{M}(X_i,\rho_j)$ but we shall show below that they can be lifted suitably to classes $\hat{\mu}(\Sigma_i)$ with compact support. The example we want to discuss in this section is:

---

1) From here on, we shall ignore the central part of the curvature of connections. Of course this central part needs to be fixed to obtain finite dimensional moduli spaces, compact spaces of flat connections etc.

Example 1. Let

$$m_0 = \#\mathcal{M}(X_1, \rho_2), \qquad n_0 = \#\mathcal{M}(X_2, \rho_1)$$

where # denotes a count with sign, and let

$$m_1 =< \mathcal{M}(X_1, \rho_1), \hat{\mu}(\Sigma_1) >, \qquad n_1 =< \mathcal{M}(X_2, \rho_2), \hat{\mu}(\Sigma_2) > .$$

Then

$$\Psi(X, \Sigma) = n_0 m_1 + n_1 m_0.$$

Proof. The target of the proof is to write $\mu(\Sigma)$ as the sum of two compactly supported cohomology classes, μ_j, which are defined on open sets of $\mathcal{M}(X)$.

The fundamental technique involved in doing so is to «stretch the neck', that is, to consider the invariant (1) when X is equipped with a metric g_T for which it contains a long cylindrical part $[-T, T] \times Y$. Arguing as in [DK] Chapter 7, one sees that, for large T every connection in $\mathcal{M}(X)_{g_T}$ must be approximately flat on at least one copy $t_0 \times Y$ in the cylinder, and consequently lies in the image of at least one of the two gluing maps:

$$\mathbf{g}_j : \mathcal{M}(X_1, \rho_j) \times \mathcal{M}(X_2, \rho_j) \to \mathcal{M}(X)_{g_T} \qquad j = 1, 2.$$

The images of these maps define our open sets in $\mathcal{M}(X)$. Strictly speaking, these maps are only defined on compact subsets, but we shall ignore this here for the sake of exposition. The overlap between these areas can be thought of as the image of a double gluing:

$$\mathcal{M}(X_1, \rho_2) \times \mathcal{M}(\rho_2, \rho_1) \times \mathcal{M}(X_2, \rho_1) \to \mathcal{M}(X_1), \tag{2}$$

with image a collar on the circles: $\mathcal{M}(X_1, \rho_2) \times \mathcal{M}(\rho_2, \rho_1) \times \mathcal{M}(X_2, \rho_1)$. This type of «factorization» also exhibits $\mathcal{M}(X_1, \rho_1)$ as a manifold with boundary:

$$\partial \mathcal{M}(X_1, \rho_1) = \mathcal{M}(X_1, \rho_2) \times \mathcal{M}_0(\rho_2, \rho_1), \tag{3}$$

and similarly for X_2.

Next we need to see how the gluing process deals with the cohomology classes. From a precise form of the excision theorem one obtains that when restricting $\mu(\Sigma)$ to the image $\mathcal{M}(X_1, \rho_j) \times \mathcal{M}(X_2, \rho_j)$ it equals, under the Künneth map:

$$\mu(\Sigma)_{|\operatorname{Im} \mathbf{g}_j} = \mu(\Sigma_1) \otimes 1 + 1 \otimes \mu(\Sigma_2) \in H^2(\mathcal{M}(X_1, \rho_j) \times \mathcal{M}(X_2, \rho_j)). \tag{4}$$

This becomes clear through considering the classes $\mu(\Sigma)$ and $\mu(\Sigma_j)$ as first Chern classes of *determinant* line bundles $\mathcal{L}_\Sigma$ and $\mathcal{L}_{\Sigma_j}$ for the Dirac operators $\mathcal{D}_A$ on these surfaces coupled to connections A in the moduli spaces. In the case of the surfaces Σ_j, which have cylindrical ends, we use boundary conditions of the sort introduced by Atiyah-Patodi-Singer [APS]. Again, this is possible thanks to fixing the asymptotic boundary value. The determinants of the Dirac operators then become sections s and s_j of the line bundles, and vanish at a connection A when $\mathcal{D}_A$ has a zero mode. In Donaldson-Kronheimer [DK] it is explained that for large T an L^2-projection from the kernel of $\mathcal{D}_\Sigma$ to $\ker \mathcal{D}_{\Sigma_1} \otimes \ker \mathcal{D}_{\Sigma_2}$ is an isomorphism. This works for families of connections and gives an isomorphism between $\mathbf{g}_j^*(\mathcal{L}_\Sigma)$ and $\mathcal{L}_{\Sigma_1} \boxtimes \mathcal{L}_{\Sigma_2}$ which maps the section s to $s_1 \boxtimes s_2$, thereby proving (4). A similar statement holds for multiple gluing as in (2).

Finally we show how to decompose $\mu(\Sigma)$ into compactly supported classes. It is clear on dimensional grounds that $\mu(\Sigma_1)$ can be lifted to an element

$$\mu_1 \equiv \hat{\mu}(\Sigma_1) \in H^2(\mathcal{M}(X_1, \rho_j), \partial(\mathcal{M}(X_1, \rho_j))),$$

but such lifts are ambiguous up to the image of $H^1(\mathcal{M}(X_1, \rho_j))$. These compactly supported cohomology classes are dual to sections of $\mathcal{L}_{\Sigma_1}$ which do not vanish outside a compact set. We obtain a homotopy class of such sections by a homotopy class of trivialisations of $\mathcal{L}_{\Sigma_1}$ restricted to $\partial(\mathcal{M}(X_1, \rho_j))$. One similarly defines a class μ_2. We must be careful that the lifts are made consistently to ensure $\mu(\Sigma) = \mu_1 + \mu_2$.

For X_1 we need to trivialise over $\partial\mathcal{M}(X_1, \rho_1)$ and from (3) we now see that trivialisations are determined entirely by trivialisations of the line bundle $\mathcal{L}_{\gamma \times \mathbb{R}}$ over $\mathcal{M}_0(\rho_2, \rho_1)$, i.e. the choice can be made on Y, independently of X_1. This also matches the homotopy classes of trivialisations over $\partial(\mathcal{M}(X_2, \rho_2))$ which we need for $\mu(\Sigma) = \mu_1 + \mu_2$. This completes the proof of the example.

One further point should be made here. Assume that $\mathcal{M}_0(\rho_2, \rho_1)$ is a circle. If we change the trivialisation by a single unit, then m_1 changes by m_0 while n_1 changes by $-n_0$; notice that $n_1 m_0 + n_0 m_1$ remains unaffected. This explains that, without the choice of trivialisations, the numbers obtained for left and right half may appear ambiguous. We shall now cast this result in a general form.

3 The Fukaya construction and gluing

3.1 Statements of the results

Let $P_Y \to Y$ be a $U(2)$ bundle over an oriented 3-manifold Y, such that $c_1(P_Y)$ is odd in $H^2(Y;\mathbb{Z})/\text{torsion}$. This leads to a situation in which P_Y carries only *irreducible* (projectively) flat connections, and we say such bundles are *free of flat reductions*. Possibly after making a small perturbation, there will be finitely many flat connections, and they will all be non-degenerate. We denote them by ρ_j; these form the generators of the Floer complex $(C_*(Y), \partial)$, where the grading lies in an affine $\mathbb{Z}/8$-space. It will be convenient to realise that this is equivalent to considering an integer graded complex, with an automorphism shifting degrees by 8. Let $\gamma \in H_1(Y;\mathbb{Z})$. We shall make suitable choices, as in Section 2, so that through the μ map we can obtain a map[2]) :

$$c_\gamma^\alpha : C_j(Y) \to C_{j-1-2\alpha}(Y) : \rho_k \to \sum_{\substack{\rho_l \\ \mathrm{ind}(\rho_l)=j-1-2\alpha}} < \mu(\gamma \times \mathbb{R})^\alpha, \mathcal{M}_0(\rho_k, \rho_l) > \rho_l,$$

where $\mathcal{M}_0(\rho_k, \rho_l)$ denotes the moduli space of instantons on the cylinder, quotiented out by the translation action, which is of dimension $2\alpha = \mathrm{ind}(\rho_k) - \mathrm{ind}(\rho_l) - 1$; notice that $\partial = c_\gamma^0$. Finally, recall that the Floer complex of $\overline{Y}$ is related to the Floer *co-complex* of Y through $C_{-j-3}(\overline{Y}) = C^j(Y)$, and this gives the Floer pairing: $C_{-j-3}(\overline{Y}) \otimes C_j(Y) \to \mathbb{Z}$.

For an integer $m \geq 0$ define the Fukaya-Floer complex by

$$CFF_j^{(m)}(Y) = C_j(Y) \oplus C_{j-2}(Y) \ldots C_{j-2m}(Y).$$

We see that $CFF_*^{(m)}(Y) \simeq C_*(Y) \otimes H_*(\mathbf{CP}^m)$, and we introduce the usual double complex notation $CFF_{j-2\alpha,2\alpha}^{(m)}(Y)$ for the summand $C_{j-2\alpha}(Y)$ in $CFF_j^{(m)}(Y)$.

Theorem 1. *Define a boundary operator $\partial^m : CFF_j^{(m)}(Y) \to CFF_{j-1}^{(m)}(Y)$ by a lower triangular matrix with entries*

$$\partial_{\alpha\beta}^m = \binom{\alpha}{\beta} c_\gamma^{\alpha-\beta} : CFF_{j-2\beta,2\beta}(Y) \to CFF_{j-1-2\alpha,2\alpha}(Y),$$

where $0 \leq \beta \leq \alpha \leq m$. Then

1) $(\partial^m)^2 = 0$, *with cohomology group denoted by $HFF_*^{(m)}(Y,\gamma)$.*
2) *The group $HFF_*^{(m)}(Y,\gamma)$ is independent of the metric on Y and choices of perturbations and trivialisations (as specified in Lemma 3), up to canonical isomorphism. It transforms naturally under diffeomorphisms of Y acting on (Y, P_Y, γ).*
3) *There is a pairing*

$$\sigma_m : HFF_j^{(m)}(Y,\gamma) \otimes HFF_{-j-3+2m}^{(m)}(\overline{Y}, -\gamma) \to \mathbb{Z}$$

$$((\psi_0, \psi_1, \ldots, \psi_m), (\phi_0, \phi_1, \ldots, \phi_m)) \to \sum_k \binom{m}{\alpha} \psi_\alpha \cdot \phi_{m-\alpha}$$

where $\psi_\alpha \cdot \phi_{m-\alpha}$ is a Floer pairing of $\psi_\alpha \in CFF_{j-2\alpha,2\alpha}(Y)$ and $\phi_{m-\alpha} \in CFF_{-j-3+2\alpha,2m-2\alpha}(\overline{Y})$ for $\alpha = 0, 1, \ldots, m$.

---

2) To define this map we need to orient all moduli spaces on the cylinder. When working with projectively anti-self dual $U(2)$-connections there are no obstructions to doing this and we shall omit further mention of orientations.

Let $P \to X$ be a principal $U(2)$ bundle over a closed 4-manifold, such that for a generic Riemannian metric g on X the *moduli space of instantons* $\mathcal{M}(X)$ on P is of dimension $2m = -2p_1(P) - 3(1 + b^+(X) - b^1(X))$. Under some additional assumptions we have a polynomial invariant, Donaldson [Do1], Donaldson-Kronheimer [DK]:

$$\Psi(X, \Sigma^m) = < \mathcal{M}(X), \mu(\Sigma)^m > .$$

Let $X = X_1 \cup_Y X_2$, assume that $P_Y \equiv P_{|Y}$ is free of flat reductions and write $\Sigma = \Sigma_1 + \Sigma_2$ for relative classes $\Sigma_j \in H_2(X_j, \partial X_j; \mathbb{Z})$ with $\partial\Sigma_1 = \gamma$, $\partial\Sigma_2 = -\gamma$. As in Section 2 we equip X_j with cylindrical ends. This gives rise to instanton moduli spaces $\mathcal{M}(X_1, \rho_i)$, and $\mathcal{M}(X_2, \rho_i)$, and, without loss of generality, we may assume that they have dimensions $2m - \mathrm{ind}_Y(\rho_i) \pmod 8$ and $\mathrm{ind}_Y(\rho_i) \pmod 8$, respectively. Again we can evaluate the cohomology classes $< \mathcal{M}(X_1, \rho_j), \mu(\Sigma_1)^k >$, when the dimensions match.

We now define a Floer-Fukaya chain:

$$\psi(X_1, \Sigma_1) = (\psi_0, \psi_1, \dots, \psi_m) \in CFF^{(m)}_{2m}(Y) \tag{5}$$

where ψ_α is an ordinary Floer chain obtained by evaluating $\mu(\Sigma_1)^\alpha$ on moduli spaces of dimension 2α:

$$\psi_\alpha = \sum_{\mathrm{ind}(\rho_j) = 2m - 2\alpha} < \mathcal{M}(X_1, \rho_j), \mu(\Sigma_1)^\alpha > \rho_j \quad \in CFF_{2m-2\alpha, 2\alpha}(Y)$$

and similarly for X_2 a chain $\psi(X_2, \Sigma_2) \in CFF_0{}^{(m)}(\overline{Y})$.

Theorem 2.

1) *Let $P \to X = X_1$ be a $U(2)$-bundle and assume that $P_Y \equiv P_{|Y}$ is free of flat reductions. The chains $\psi(X_j, \Sigma_j)$ are closed, and their cohomology classes in $HFF^{(m)}_*(Y, \gamma)$ and $HFF^{(m)}_{2m-*}(\overline{Y}, -\gamma)$ are independent of choices of metrics and trivialisations (as specified in Lemma 4).*
2) *Let $P \to X = X_1 \cup_Y X_2$ be a $U(2)$-bundle over a 4-manifold with $b_1(X) > 1$ and assume that $P_Y \equiv P_{|Y}$ is free of flat reductions. Then $\Psi(X, \Sigma^m) = \sigma_m(\Psi(X_1, \Sigma_1), \Psi(X_2, \Sigma_2))$.*

A few remarks are in place. As usual, the condition $b^+(X) > 1$ implies that for generic one-parameter families of metrics the moduli space $\mathcal{M}(X)$ contains no reducibles. On the other hand the Fukaya-Floer cycles and the pairing make sense regardless of the value of $b^+(X)$ and one may wonder what it evaluates. The fact that P_Y is free of flat reductions forces $b^+(X) \geq 1$. when $b^+(X) = 1$, the polynomial invariant is a function on «chambers» in $\mathbf{P}(H^2(X; \mathbb{R}))$ defined by conformal structures on X for which no reducible connections exist in $\mathcal{M}(X)$, see Donaldson [Do4]. The metrics on X with long necks $[-T, T] \times Y$ select a chamber, and the expression $\sigma_m(\Psi(X_1, \Sigma_1), \Psi(X_2, \Sigma_2))$ computes the polynomial on this chamber.

3.2 Outline of the proofs

The proof of Theorem 3.1 is given in detail by Fukaya [Fuk], and Theorem 3.2 is proved in a similar fashion. We give some of the main ingredients here, starting with Theorem 3.1.

The key point in the proofs is to choose representative submanifolds for the cohomology classes $\mu(\gamma \times \mathbb{R})$ which are «compatible» at the boundary. Recall the line bundle $\mathcal{L}_{\rho,\sigma} \to \mathcal{B}_0(\rho,\sigma)$, over the reduced orbit space of connections on $Y \times \mathbb{R}$ with boundary values flat connections ρ, σ, with $c_1(\mathcal{L}_{\rho,\sigma}) = \mu(\gamma \times \mathbb{R})$. We begin by choosing a section $s_{\rho,\sigma}$ such that its zero set is transverse. Secondly recall that for compact sets $K_1 \subset \mathcal{M}(\rho_1,\rho_2),\ldots,K_{j-1} \subset \mathcal{M}_0(\rho_{j-1},\rho_j)$ we have a gluing map (Floer [Fl]):

$$\mathbf{g} : K_1 \times [T,\infty) \times K_2 \times [T,\infty) \times \ldots \times [T,\infty) \times K_{j-1} \to \mathcal{M}(\rho_1,\rho_j).$$

Using the gluing maps we can *partially* compactify the moduli space $\mathcal{M}(\rho_1,\rho_j)$ to a manifold with corners: we add strata of codimension $j-1$ of the form:

$$\mathcal{M}(\rho_1,\rho_2) \times \mathcal{M}_0(\rho_2,\rho_3) \times \ldots \times \mathcal{M}_0(\rho_{j-1},\rho_j). \tag{6}$$

More precisely, this compactifies the portion of the moduli space which consist of instantons which are not concentrated near any point.

Lemma 3. *There are choices of transverse sections s_{ρ_i,ρ_j} such that s_{ρ_1,ρ_j} extends smoothly over (6) and restricts to the exterior tensor product: $s_{\rho_1,\rho_2} \boxtimes \ldots \boxtimes s_{\rho_{j-1},\rho_j}$.*

The proof of this was indicated in Section 2 and is based on the geometric form of the excision theorem.

Next we consider m sections $s_{\rho_i,\rho_j,l}$, each a small perturbation of s_{ρ_i,ρ_j}, such that for each $l = 1,\ldots,m$ the compatibility condition of the lemma holds. This gives rise to codimension two submanifolds $V_{\rho_i,\rho_j,l}$ and we let

$$V_{\rho_i,\rho_j}(k) = \cap_{l=1,\ldots,k} V_{\rho_i,\rho_j,l}$$

denote the intersection of the first k of these. Then the matrix elements of the boundary operator become intersection numbers:

$$c_\gamma^\alpha(\rho_i) = \#(V_{\rho_i,\rho_j}(\alpha) \cap \mathcal{M}_0(\rho_k,\rho_j))\rho_j.$$

We refer to this evaluation procedure as *cutting down the moduli space*. It is important to note that since the sections were all small perturbations of a fixed one we could have replaced $V_{\rho_i,\rho_j}(k)$ with the intersection of any k of the $V_{\rho_i,\rho_j,l}$ without affecting the intersection products.

We are now in a position to explain why the square of the boundary operator is zero. Consider flat connections ρ_β, ρ_α of indices $j-2\beta$, $j-2-2\alpha$, with $\alpha \geq \beta$. We consider the one-dimensional moduli space:

$$V_{\rho_\beta,\rho_\alpha}(\alpha-\beta) \cap \mathcal{M}_0(\rho_\beta,\rho_\alpha)$$

Standard arguments ([Fl],[Fuk],[Tau]) exclude localized instantons in this manifold of dimension one and show that the boundary consists of the following number of points:

$$\bigcup_{\rho_\gamma} \binom{\alpha-\beta}{\gamma-\beta} \# \left(V_{\rho_\beta,\rho_\gamma}(\gamma-\beta) \cap \mathcal{M}_0(\rho_\beta,\rho_\gamma)\right) \# \left(V_{\rho_\gamma,\rho_\alpha}(\alpha-\gamma) \cap \mathcal{M}_0(\rho_\gamma,\rho_\alpha)\right),$$

where ρ_γ runs over flat connections with index γ between α and β. When approaching the boundary of the one-dimensional moduli space, the instanton «breaks'. The component in $\mathcal{M}_0(\rho_\beta,\rho_\gamma)$ can lie in any intersection of $\gamma-\beta$ of the $V_{\rho_\beta,\rho_\gamma,j}$'s, and this explains the appearance of the binomial coefficient. The boundary is exactly the ρ_α component of

$$\binom{\alpha}{\beta}^{-1} \sum_\gamma \partial^m_{\alpha\gamma}\partial^m_{\gamma\beta}(\rho_\beta)$$

which therefore vanishes.

One shows more easily that $\sigma_d(\partial^m\rho_1,\rho_2) = \pm\sigma_d(\rho_1,\partial^m\rho_2)$, from which it follows that the pairing decends. We omit the construction of canonical isomorphisms induced by changing choices (a similar argument is outlined below).

Let us now pay attention to theorem 3.2. For the closed manifold X we also pick m-submanifolds V_i representing $\mu(\Sigma)$, then the polynomial invariant can be regarded as the intersection number

$$\Psi(X,\Sigma^m) = \#\mathcal{M}(X) \cap V(d).$$

To understand the gluing formula for this invariant we consider a family of metrics g_T as in Section 2. When letting T tend to infinity the instantons break into instantons on X_1 and X_2, both of which lie in cut-down moduli spaces.

We can find m sections over $\mathcal{B}(X_1,\rho)$, all close to a single one, which define submanifolds $V_{\rho,l}$ of codimension 2 respresenting $\mu(\Sigma_1)$, and $V_\rho(k)$, the intersection of the first k, of codimension $2k$. For $\mathcal{B}(X_2,\rho)$ we denote these manifolds by $W_{\rho,l}$ and $W_\rho(k)$. For compact sets $K_1 \subset \mathcal{M}(X_1,\rho_1)$, $K_2 \subset \mathcal{M}(\rho_1,\rho_2),\ldots,K_j \subset \mathcal{M}_0(\rho_{j-1},\rho_j)$ we have a gluing map:

$$\mathbf{g}_{X_1} : K_1 \times [T,\infty) \times K_2 \times [T,\infty) \times \ldots \times [T,\infty) \times K_j \to \mathcal{M}(X_1,\rho_j).$$

and also one for $K_i \subset \mathcal{M}(X_i,\rho)$:

$$\mathbf{g}_{X_1,X_2,T} : K_1 \times K_2 \to \mathcal{M}(X)_{g_T},$$

where the subscript g_T indicates that the target is the moduli space of instantons for a metric g_T on X containing a long neck. The first of these two gluing maps defines a partial compactification of the moduli spaces $\mathcal{M}(X_i,\rho)$ to a manifold with corners. Products of these can be considered as the «simplices» building up $\mathcal{M}(X)$, *c.f.* Taubes [Tau]. We can now formulate the second compatibility criterion:

Lemma 4. *The manifolds $V_{\rho,i}$ (and $W_\rho(j)$) can be made compatible with the gluing maps $\mathbf{g}_{X_1}$, $\mathbf{g}_{X_1,X_2,T}$ (and $\mathbf{g}_{X_2}$).*

This lemma states, for example, that the boundary of the cut-down moduli space is the same as cutting down the boundary of the moduli space. Phrased differently one can say that it makes cutting down compatible with the «simplicial» decomposition of $\mathcal{M}(X)$. With this lemma at our disposal the gluing formula is clear. The fact that the Fukaya-Floer chains are closed is proved much like $(\partial^m)^2 = 0$.

Let us finally consider a one-parameter family of metrics g_s on X_1, with $s \in [0,1]$, and show that $\Psi(X_1, \Sigma_1)$ changes by a boundary. Consider a one-dimensional moduli space

$$\mathbb{M} = \cup_s V_\rho(\alpha) \cap \mathcal{M}(X_1, \rho)_{g_s} \subset [0,1] \times \mathcal{B}^*(X_1, \rho).$$

Again the boundary of $\mathbb{M}$ governs the situation. Of course, for $s = 0, 1$ we find the Fukaya-Floer chains for g_0 and g_1. Other boundary terms occur for finitely many $s_k \in (0,1)$ where a moduli space $V_\sigma(\beta) \cap \mathcal{M}(X_1, \sigma)_{g_{s_k}}$ of *virtual* dimension -1 is not empty; these points are best thought of as points where the path hits the boundary of a compactified moduli space. For a generic path g_s this moduli space consists of finitely many points, and the number of points in this part of the boundary of $\mathbb{M}$ equals:

$$\binom{\beta}{\alpha} \# \left(\left(V_\sigma(\beta) \cap \mathcal{M}(X_1, \sigma)_{g_{s_k}} \right) \times \left(V_{\sigma,\rho}(\beta - \alpha) \cap \mathcal{M}_0(\sigma, \rho) \right) \right).$$

When we consider this simultaneously for all ρ we see that the difference in the two chains equals the boundary:

$$\partial^m \left(\cup_{k,\rho} V_\sigma(\beta) \cap \mathcal{M}(X_1, \sigma)_{g_{s_k}} \rho \right)$$

where the union goes over moduli of virtual dimension -1.

There are several ways in which one may hope to generalise the results in this section. Straightforward extensions arise from studying the polynomial invariant arising from evaluating both four and two dimensional classes on the moduli space. When the condition is relaxed that P_Y be free of reducible flat connections, gluing factors arise, and it is much more delicate to see how the invariant breaks into pieces.

4 Algebraic topology and the Fukaya construction

The aim of this section is to compare the Fukaya construction with standard ideas in algebraic topology, and thereby to obtain more insight into the definitions and properties that we have stated in Section 3.

4.1 Filtrations and cohomology

We will begin by recasting the results of Section 3 slightly. We continue with our fixed oriented 3-manifold Y, a $U(2)$ bundle P over Y and a loop γ in Y, as before. In Section 3 we introduced, for each $m \geq 0$, groups

$$CFF_i^{(m)} = C_i \oplus C_{i-2} \oplus C_{i-4} \dots \oplus C_{i-2m},$$

with a differential $\partial : CFF_i^{(m)} \to CFF_{i-1}^{(m)}$, giving homology groups $HFF_*^{(m)}(Y)$. We now observe that there are obvious chain maps

$$C_*(Y) \leftarrow CFF_*^{(1)}(Y) \leftarrow CFF_*^{(2)}(Y) \leftarrow \dots,$$

and we can pass to the inverse limit. Thus we put

$$CFF_i(Y) = \prod\nolimits_{n \geq 0} C_{i-2n},$$

(so elements of CFF_* are «infinite vectors» with entries in the ordinary Floer chains). The differentials discussed in Section 3 obviously induce a differential $\partial : CFF_* \to CFF_{*-1}$, and we get homology groups $HFF_i(Y)$. From now on we will work with this group, to avoid the bother of the parameter m: there is however no real change involved, since working with $HFF_*(Y)$ is essentially the same as working with the «finite» versions $HFF_*^{(m)}(Y)$ for large m. A manifold X_1 with boundary Y, and a class $\Sigma_1 \in H_2(X_1, \partial X_1)$ with boundary γ define an element $\Psi(X_1, \Sigma_1) \in HFF_*(Y)$. This is the homology class of a cycle

$$\psi(X_1, \Sigma_1) = (\psi_0, \psi_1, \psi_2, \dots),$$

where ψ_n is defined by «evaluating» the product of n representatives of the cohomology class $\mu(\Sigma_1)$ on $2n$-dimensional moduli spaces, as explained in Section 3. Changing the orientation of Y we get another graded group $HFF_*(\overline{Y})$. There is a pairing

$$\sigma_m : HFF_*(Y) \otimes HFF_*(\overline{Y}) \to \mathbb{Z},$$

induced by a canonical pairing of chain groups, which pairs the term $C_{i-2n}(Y)$ in $CFF_i(Y)$ with the term $C_{i-2n}(Y)^*$ in $CFF_{2m-i-3}(\overline{Y})$ for $0 \leq n \leq m$.

The basic gluing rule is that if $X = X_1 \cup_Y X_2$ and Σ is the homology class formed by Σ_1, Σ_2 as before, then for a moduli space $\mathcal{M}(X)$ of dimension $2m$ we have

$$\Psi(X, \Sigma) \equiv < \mu(\Sigma)^m, \mathcal{M}(X) >= \sigma_m(\Psi(X_1, \Sigma_1), \Psi(X_2, \Sigma_2)).$$

Now let $K_*^{(m)} \subset CFF_*$ be the kernel of the projection map from CFF_* to $CFF_*^{(m)}$, *i.e* the vectors with zeros in the first m entries. These are subcomplexes which give a filtration of the chain complex CFF_*, so we have a homology spectral sequence

$$E^2_{p,q} = HF_p(Y) \otimes R_q \Rightarrow HFF_*(Y)$$

where $HF_p(Y)$ is the ordinary Floer group and R_q is defined by $R_q = \mathbb{Z}, q$ even, $R_q = 0$, q odd. (Recall that we are building in the $\mathbb{Z}/8$ grading implicitly by using $\mathbb{Z}$ grading throughout with periodicity $HF_{p+8}(Y) = HF_p(Y)$.)

It is sometimes convenient to work with cohomology in place of homology, and there is some scope for confusion here, because in the ordinary Floer groups the analogue of the Poincaré duality isomorphism is built into the definitions, and passing from homology to cohomology just corresponds to reversing the orientation of the manifold. We write

$$C^i(Y) = C_i(Y)^* \equiv \mathrm{Hom}(C_i(Y), \mathbb{Z}),$$

so that $C^i(Y) = C_{3-i}(\overline{Y})$. We get a coboundary $d = d_Y : C^*(Y) \to C^{*+1}(Y)$ which is identical with the boundary map $\partial_{\overline{Y}}$. Passing to Fukaya-Floer groups, we now define a dual cochain complex

$$CFF^i = \mathrm{Hom}(CFF_i, \mathbb{Z}) = C^i \oplus C^{i-2} \oplus C^{i-4} \oplus \dots.$$

with cohomology $HFF^*(Y)$. The filtration gives a spectral sequence

$$E_2^{pq} = HF^p(Y) \otimes R^q \Rightarrow HFF^*(Y),$$

where $R^q = \mathrm{Hom}(R_q, \mathbb{Z})$. If we adopt rational coefficients then $HFF^*(Y)$ is the dual of $HFF_*(Y)$, but this is not the same as $HFF_*(\overline{Y})$, although the pairing σ_m gives a map from $HFF_*(\overline{Y})$ to to $HFF^*(Y)$.

4.2 Floer homology and ordinary homology

We will now step back and recall why the Floer homology $HF_*(Y)$ groups can be viewed, in a formal way, as the «middle-dimensional» homology groups of the infinite-dimensional space $\mathcal{B}_Y$ of gauge equivalence classes of connections on our bundle P over Y. There are two different points of view one can take. One is to see that Floer's definition of $HF_*(Y)$, using the critical points and gradient curves of the Chern-Simons functional, would give in an analogous finite-dimensional situation, of a function on a manifold B, the ordinary Morse complex computing $H_*(B)$, see [Sa] for example. The other point of view, which is heuristic rather than rigorous, is based on the fact that a 4-manifold X_1 with boundary Y defines a class $\Psi(X_1) \in HF_*(Y)$, and is explained in [At1]. Fix a metric on the manifold-with-boundary X_1 and let $\Lambda(X_1) \subset \mathcal{B}_Y$ be the set of boundary values of instantons over X_1. We imagine that there is a way to define infinite dimensional homology $HF_*(Y)$ with infinite-dimensional chains, so that $\Lambda(X_1)$ is a cycle whose homology class $\Psi(X_1)$ is independent of the metric. We think of the intersection pairing on $HF_*(Y) \otimes HF_*(\overline{Y})$, as being induced by intersection of cycles, then the gluing formula $\Psi(X) = \Psi(X_1)\Psi(X_2)$ just expresses the fact that the instantons on X can be identified with the intersection in $\mathcal{B}_Y$, $\Lambda(X_1) \cap \Lambda(X_2)$. To relate this way of thinking to the Morse-complex point of view one varies the metrics by adding on long, but finite, tubes to the boundaries, and sees that $\Lambda(X_1), \Lambda(X_2)$ are deformed to the ascending and descending submanifolds associated to the different critical points of the Chern-Simons functional, and the intersection localises near these critical points (flat connections).

We want now to move on to discuss *cup products* in Floer homology. The formalism alone might suggest that there should be a product from $H^*(\mathcal{B}_Y) \otimes HF^*(Y)$ to $HF^*(Y)$; multiplying a finite-dimensional cohomology class by an infinite-dimensional class to get another infinite-dimensional class. Such a product does exist, at least for the part of $HF^*(\mathcal{B}_Y)$ generated by the image of the map

$$\mu : H_*(Y) \to H^*(\mathcal{B}_Y).$$

(If we work with rational coefficients this gives all the cohomology.) The class relevant to us here is the 3-dimensional class $\nu = \mu(\gamma) \in H^3(\mathcal{B}_Y)$. We want to see that we can define a multiplication map in the Morse complex, i.e. a cochain map

$$\hat{\nu} : CF^* \to CF^{*+3}$$

commuting with the differential $d = d_Y$. (We will sometimes denote $\hat{\nu}$ also by ν.) The definition of this map will involve the choice of a representative 3-cochain for ν, but different representatives give chain-homotopic maps and we get an intrinsic cup-product map $\hat{\nu} : HF^*(Y) \to HF^{*+3}(Y)$. Of course it would be just the same to work with Floer homology, and we then obtain a «contraction» map which decreases degree by 3.

To define a homomorphism $\hat{\nu}$ of the cochains we need to define matrix entries $\hat{\nu}_{\rho\sigma}$ where ρ and σ are flat connections whose Floer indices differ by 3. To do this one uses the 3-dimensional moduli space $\mathcal{M}(\rho, \sigma)$ of instantons over $Y \times \mathbb{R}$ with limits ρ and σ, which is of course a product $\mathcal{M}(\rho, \sigma) = \mathcal{M}_0(\rho, \sigma) \times \mathbb{R}$. We write $\mathcal{M}_0^c(\rho, \sigma)$ for the suspension $\Sigma M_0(\rho, \sigma)$ obtained from $M_0 \times [0, 1]$ by collapsing the ends; so we can regard $\mathcal{M}(\rho, \sigma)$ as a subset of $\mathcal{M}_0^c(\rho, \sigma)$. There are three equivalent approaches that one can take to the definition of the matrix entry $\hat{\nu}_{\rho\sigma}$.

1) *The three dimensional approach.* Evaluation on $Y \times \{0\}$ maps $\mathcal{M}(\rho, \sigma)$ into $\mathcal{B}_Y$, and the map extends to $e_Y : \mathcal{M}_0^c \to \mathcal{B}_Y$. We choose a particular representative for ν over $\mathcal{B}_Y$ and evaluate this on the 3-chain $e_Y(\mathcal{M}_0^c(\rho, \sigma)) \subset \mathcal{B}_Y$ to set

$$\hat{\nu}_{\rho\sigma} = < \nu, e_Y(\mathcal{M}_0^c(\rho, \sigma)) > .$$

If $\mathcal{M}_0(\rho, \sigma)$ is compact then $e_Y(\mathcal{M}_0^c(\rho, \sigma))$ is a genuine 3-cycle in singular homology, so $\hat{\nu}_{\rho\sigma}$ is independent of choices in this case. In general one needs to do some work to see how the numbers change with the choice of representive, and to vertify the chain-homorphism property (compare [AuBr2]).

2) *The four-dimensional approach.* We can generalise the problem to the case of a 4-manifold X which is, say, a cobordism from one 3-manifold Y_1 to another Y_2, and with a class $\gamma \in H_1(X)$. The 0-dimensional instanton moduli spaces over X give a chain map $\psi_X : C^i(Y_1) \to C^{i+n}(Y_2)$, say, with some normalisation of the grading. Then we use the 3-dimensional moduli spaces over X to define a chain map $\nu_X : C^i(Y_1) \to C^{i+n+3}(Y_2)$, with matrix entries defined by evaluating a representative for $\mu(\gamma)$ on the moduli spaces $\mathcal{M}(X)(\rho,\sigma)$, much as in the «ordinary' 4-dimensional theory [DK]. Taking X to be the tube $Y \times [0,1]$ gives the case at hand. The advantage of this 4-dimensional point of view is that the proofs of the properties we have asserted become straightforward combinations of those in the ordinary 4-dimensional theory, and in the ordinary Floer theory (*i.e.* showing that ψ_X is a chain map).
3) *The determinant line bundle approach.* This is the approach which is closest to the definition of the Fukaya groups. We fix a generic section of the determinant line bundle $\mathcal{L}$ of the surface $\gamma \times \mathbb{R}$, just as in Section 3, and then count the zeros on the 2-dimensional moduli space $\mathcal{M}_0(\rho,\sigma)$ to define a number $\hat{\nu}_{\rho\sigma}$. If $\mathcal{M}_0$ is compact this is just the homology pairing $< c_1(\mathcal{L}), \mathcal{M}_0(\rho,\sigma) >$.

It takes a little thought to see that these three aproaches are indeed equivalent. If the moduli space $\mathcal{M}_0(\rho,\sigma)$ is compact, so that the numbers obtained in all three cases are independent of choices, then the equivalence is a straightforward exercise in algebraic topology. Beyond this, one should also check that, with suitable choices in each case, the numbers can be made to agree even in the non-compact situation. The equivalence of (1) and (2) is quite easy, since one can choose representatives for the 3-dimensional cohomology which depend only upon the connection in a neighbourhood of a loop. In fact there is a canonical representative, given by the set of connections whose holonomy around the loop is the identity (compare [Th]). If we assume that this subset is transverse to our moduli spaces, then one sees that, in both (1) and (2), $\hat{\nu}_{\rho\sigma}$ is the count, with signs, of the instantons in $\mathcal{M}(\rho,\sigma)$ having holonomy 1 around $\gamma \times \{0\} \subset Y \times \mathbb{R}$. The same idea applies in (3). Suppose we have any codimension-3 representative U for the class ν, not containing any flat connections. Then in the space of connections over $Y \times \mathbb{R}$ with given flat limits ρ, σ, the subset V consisting of connections $\mathbf{A}$ over $Y \times \mathbb{R}$ such that $\mathbf{A}|_{Y\times\{t\}} \in U$ for some t, is of codimension 2 and represents the class $\mu(\gamma \times \mathbb{R})$. Thus the intersection $V \cap \mathcal{M}_0(\rho,\sigma)$ defines the pairing used in approach (3), but this intersection is canonically identified with $\mathcal{M}(\rho,\sigma) \cap U$. (In particular, one choice of V is to take the connections which have holonomy 1 around some $\gamma \times \{t\}$.)

As we mentioned above, the same ideas apply to define a product operation $\hat{\zeta} : HF^*(Y) \to HF^*(Y)$ for any ordinary cohomology class ζ which is an ordinary cup-product of elements $\mu(\alpha)$ for $\alpha \in H_*(Y)$. An advantage of the 4-dimensional approach above is that it is makes it easy to see that this construction is associative,*i.e.* that $\widehat{\mu(\alpha_1) \cup \mu(\alpha_2)}$ is the same as the composite $\widehat{\mu(\alpha_1)} \circ \widehat{\mu(\alpha_2)}$. Take representative for $\mu(\alpha_i)$ which depend only on the restriction to neighbourhoods of cycles α_i in $Y \times \mathbb{R}$. Different choices of cycles give chain-homotopic maps, at the chain level, and these induce the same map on Floer cohomology. Then move one representative α_1 towards $-\infty$ in the tube, and the other towards $+\infty$. The usual compactness and gluing arguments show that when the cycles are moved far apart in this way the associative property holds at the chain level.

4.3 Massey products and fibre bundles

We will now review some ideas from «ordinary» algebraic topology, beginning with the notion of a *Massey product* in cohomology. Let B be a space and $\alpha \in H^p(B), \beta \in H^q(B), \gamma \in H^r(B)$ be three cohomology classes. Suppose that the cup products $\alpha\beta$, $\beta\gamma$ are both zero. We choose cochains a, b, c representing the three classes, so by hypothesis we can write at the cochain level:

$$ab = dx, bc = dy,$$

where x, y are unique up to $H^{p+q-1}(B), H^{q+r-1}(B)$ respectively. We have then

$$d(xc) = abc = (-1)^p d(ay) \ ,$$

so the class $Z = xc + (-1)^{p-1} ay$ is a cocycle. The Massey product $\{a, b, c\}$ is defined to be the class of Z in

$$H^{p+q+r-1}(B)/aH^{q+r-1}(B) + cH^{p+q-1}(B).$$

This first Massey product is a secondary operation, defined in situations where the primary cup products vanish. If the first Massey products of suitable collections of classes vanish then one gets a tertiary operation, with more indeterminacy, and so on.

We now turn to consider bundles with fibre an Eilenberg-Maclane space. We fix attention on the case which will be relevant to us of the Eilenberg-Maclane space $K(\mathbb{Z}, 2) = \mathbf{CP}^\infty$, so the bundles are classified by the 3-dimensional cohomology of the base space. If ν is a class in $H^3(B)$ we get a corresponding fibration

$$\mathbf{CP}^\infty \to E_\nu \to B.$$

The relation between ν and E_ν can be seen in the E_3 term of the Serre spectral sequence for the fibration. We have a differential $d_3 : H^2(\mathbf{CP}^\infty) \to H^3(B)$ and $d_3 h = \nu$ where h is the generator of $H^2(\mathbf{CP}^\infty)$. More generally, $E_3^{pq} = H^p(B) \otimes H^q(\mathbf{CP}^\infty)$ and d_3 is defined by the cup product with ν on $H^*(B)$:

$$d_3(a \otimes h^q) = q\, a\nu \otimes h^{q-1}.$$

The cohomology ring of E_ν is an invariant attached to the space B and class ν. Massey products in the cohomology of B enter into the picture in two different ways: first, in the higher differentials in the Serre spectral sequence. Suppose

$a\nu = 0$ in $H^*(B)$ so that $d_3(a \otimes h^q) = 0$ and $a \otimes h^q$ lives to the E_5-term in the spectral sequence. Then we claim that

$$d_5(a \otimes h^q) = \{a, \nu, \nu\} \otimes h^{q-2} \in E_5^{p+5,2q-4}.$$

(The product $\nu\nu$ is 0 since ν has odd degree.) This is easy to see if one works with real coefficients and with differential forms (supposing B is a manifold). The forms on E are filtered by the «number of fibre variables', and this gives rise to the spectral sequence. We fix a closed representative 3-form on the base, which we also call ν. Then one can construct a 2-form $\overline{h}$ on the total space such that $d\overline{h} = \nu$ (really the lift of ν from B to E). If α is a representative for a and $\alpha \wedge \nu = dx$ then

$$d(\alpha \wedge \overline{h}^q + x\overline{h}^{q-1}) = (q-1)x \wedge \nu \wedge \overline{h}^{q-2}.$$

On the one hand the class of this gives, in the spectral sequence machinery, $d_5(a \otimes h^q)$, on the other hand $x \wedge \nu$ is the Massey product $\{a, \nu, \nu\}$, since the form $\nu \wedge \nu$ vanishes identically and we can take $y = 0$. We can go on in the same way to describe the higher differentials in the spectral sequence in terms of higher Massey products. To say that $a \otimes h^q$ lives to the E_{2r+1}-term in the spectral sequence is the same as to say that we can find forms x_i $i = 1, \dots, r-1$, with $x_1 = x$ above, on B such that

$$d(\alpha\overline{h}^q + x_1\overline{h}^{q-1} + x_2\overline{h}^{q-2} + \dots + x_{r-1}\overline{h}^{q-r+1}$$

contains only terms of order $q - r$ in $\overline{h}$. That is:

$$dx_i = \nu x_{i-1},\ i \leq r-1$$

The form νx_{r-1} is closed on B and the the differential $d_{2r+1}(a \otimes h^q)$ is the obstruction to extending the expansion one step further, *i.e.* to finding x_r such that $dx_r = \nu x_{r-1}$. Thus $d_{2r+1}(a \otimes h^q) = [\nu x_{r-1}] \otimes h^{q-r}$.

The second way in which Massey products on B enter is in determining the *cohomology ring* of E. To consider a simple case, suppose that we have $a \in H^p(B), b \in H^s(B)$ such that $a\nu$ and $b\nu$ vanish, so that $a\otimes h$ and $b\otimes h$ are annihilated by d_3 and live to define classes in $H^{p+2}(E), H^{s+2}(E)$. More precisely we have classes in the E_∞ filtrations of these groups, and we can choose extensions $\overline{ah}, \overline{bh}$ to genuine cohomology classes, with ambiguity in $H^{p+2}(B), H^{s+2}(B)$ respectively. Then we have a cup product of these classes in $H^{p+s+4}(E)$. It is easy to see that this product lies in the part of this cohomology, defined by the bottom three steps in the filtration, $E_\infty^{p+s\ 4}, E_\infty^{p+s+2\ 2}, E_\infty^{p+s+4\ 0}$. The first term of these three is invariantly defined — it is the cup product $(\overline{ah}\ \overline{bh})_0 = ab$. The other «terms» carry higher order information. The situation is rather hard to pin down because of the filtration. But suppose that the product ab is also zero, so that we get a new invariant part $(\overline{ah}\ \overline{bh})_1$ of $\overline{ah}\ \overline{bh}$ in the next term $E_\infty^{p+s+2\ 2} \equiv \ker \nu : H^{p+s+2}(B) \to H^{p+s+5}(B)$. A short calculation shows that this term is the combination of Massey products:

$$(\overline{ah}\ \overline{bh})_1 = \{\nu, a, b\} + \{\nu, b, a\}.$$

The ambiguity in these Massey products corresponds to the ambiguity in the choosing the extensions $\overline{ah}, \overline{bh}$ of $a \otimes h, b \otimes h$.

To sum up then, we can say that there is a dictionary for translating between the topology of the total space E_ν and the algebra of Massey products involving ν in the cochain complex of B, a familiar fact in the theory of minimal models and rational homotopy theory [Su]

4.4 Synthesis

We will now bring together the various ideas outlined in the previous three sections. We want to explain that the Fukaya-Floer groups $HFF_*(Y)$ can be viewed, formally, as «homology» groups of the total space of a $\mathbf{CP}^\infty$ bundle over the space $\mathcal{B}_Y$ of connections, where this should be interpreted in a similar spirit to the interpretation of the ordinary Floer groups as homology groups of $\mathcal{B}_Y$. This interpretation will make the Fukaya construction, and the gluing formulae, more transparent.

We begin with the Morse-theory point of view. We know that there is a homology spectral sequence $HF_p(Y) \otimes R_q \Rightarrow HFF_*(Y)$, which is obviously analogous to the Serre spectral sequence $H_*(B) \otimes H_*(\mathbf{CP}^\infty) \Rightarrow H_*(E_\nu)$ for ordinary homology, where the base is, say, a compact finite-dimensional manifold. (Of course $R_* = H_*(\mathbf{CP}^\infty)$.) To develop this analogy we consider how to define the Serre spectral sequence of $E_\nu \to B$ given a Morse function f on B. This follows ideas which have been developed by Furuta [Fur1], Fukaya [Fuk] and Austin and Braam [AuBr1]. We fix a connection on the bundle $E_\nu \to B$ and compatible metrics, and then lift f to a *Morse-Bott* function on the total space (the fact that the fibres are infinite dimensional is not a serious problem here). We need to know how to compute homology explicitly from a Morse-Bott function, and this is explained in the references above. For each critical point $x \in B$ of f one fixes a system of finite-codimensional submanifolds $V_x(l), l = 0, 1, 2, \dots$ in the fibre $\pi^{-1}(x) = \mathbf{CP}^\infty_x$ representing generators for the cohomology, and a complementary system of finite-dimensional manifolds $W_x(l)$ representing the homology. We could take these to be the ascending and descending manifolds of a perfect Morse function ϵ_x on the fibre. For each pair of critical points $x, y \in B$, we have a manifold $M_E(x, y)$ of gradient lines of the lifted function on $E = E_\nu$, which run from the fibre over x to the fibre over y, with end point maps

$$e_+ : M_E(x, y) \to \mathbf{CP}^\infty_x, \; e_- : M_E(x, y) \to \mathbf{CP}^\infty_y.$$

We can «cut down» the moduli spaces by imposing constraints that the end points lie in the chosen submanifolds, that is we look at

$$M(x, y; l_1, l_2) = e_+^{-1}\left(W_x(l_1) \cap e_-^{-1}(V_y(l_2))\right) \subset M_E(x, y).$$

If x, y have Morse index $p(x), p(y)$ respectively on B then the virtual dimension of the quotient of $M(x, y, l_1, l_2)$ by the obvious translation action is $p(x) - p(y) +$

$2(l_1 - l_2) + 1$, and we assume natural transversality conditions hold, so that when this number is 0 the space is a finite set of points which we can count with signs. This gives a number $n(x, y : l_1, l_2)$. The recipe for computing the homology of E_ν is to form a chain complex with an infinite set of generators $< x(l) > \; l = 0, 1 \ldots$ for each critical point x, and a differential with these numbers as matrix entries:

$$\partial < x(l_1) >= \sum n(x, y, l_1, l_2) < y(l_2) > .$$

The complex is graded by assigning dimension $p(x) + 2l$ to $< x(l) >$. One way of understanding that this gives the homology of E_ν is to see that if we perturb the Morse-Bott function by adding on small changes near the critical manifolds, extending the functions ϵ_x over tubular neighbourhoods, then this is the ordinary Morse complex of the resulting Morse function on E.

So far we have not used the fact that E was a fibre bundle — the discussion is valid for any Morse-Bott function with critical sets of the given form. In the case of a fibre bundle, with compatible metrics, the gradient lines in E are just horizontal lifts of those in B. This tells us that the moduli spaces $M(x, y)$ are empty if $p(x) \leq p(y)$ so we have a filtered complex, and the resulting spectral sequence gives what we are after, a Morse theory description of the Serre spectral sequence of the fibration. The data required to define the complex is just the collection of gradient line spaces $M_B(x, y)$ in B for critical points x, y together with the parallel transport of the fibration. That is, for each gradient curve $Z \in M_B(x, y)$ we have a holonomy map

$$\phi_Z : \mathbf{CP}^\infty_x \to \mathbf{CP}^\infty_y ,$$

and the number $n(x, y, l_1, l_2)$ is just the number of gradient curves Z such that $\phi_Z(W_x(l_1))$ meets $V_y(l_2)$; counted modulo translations and with signs, of course.

We can now compare the complex constucted above with the Fukaya complex. There certainly is a genuine $\mathbf{CP}^\infty$ bundle $\hat{\mathcal{E}}_\nu$ over $\mathcal{B}_Y$ corresponding to our class $\nu = \mu(\gamma)$ in $H^3(\mathcal{B}_Y)$ but it is more natural geometrically to use a slightly different space $\mathcal{E}_\nu$, which is the same from the point of view of rational homology. To define this we go back to our 3-manifold Y and form a new space $Y \cup_\gamma D^2$ by adjoining a thickened disc $D^2 \times D^2_\delta$ to an δ-neighbourhood of the loop γ. We let $\mathcal{E}$ be the space of gauge equivalence classes of connections over $Y \cup_\gamma D^2$ — there is no real problem in defining these, even though $Y \cup_\gamma D^2$ is not a manifold. Then there is a map π from $\mathcal{E}$ to $\mathcal{B}$ induced by the restriction of connections to $Y \subset Y \cup_\gamma D^2$. If we remove a subset $\mathcal{X}$ of infinite codimension, consisting of connections which are reducible on the attaching region, π becomes a fibration. The fibre is homotopy equivalent to the loop space of $SU(2)$, which has the rational cohomology of $\mathbf{CP}^\infty$. For simplicity we will ignore the difference between $\hat{\mathcal{E}}_\nu$ and $\mathcal{E}$. Similarly we will ignore the fact that the disc in $Y \cup_\gamma D^2$ has been thickened, and just regard the fibre, $\mathcal{B}_{D^2}(A)$ say, of π over a connection $A \in \mathcal{B}_Y$ as the space of connections over the disc with given boundary value, and we will ignore the infinite-codimensional subset $\mathcal{X}$: none of these will affect the main points we are trying to bring out.

We can now transport the recipe above from the finite dimensional case, taking the Chern-Simons function on $\mathcal{B}_Y$ in place of the Morse function f. Thus we choose submanifolds $V_\rho(l), W_\rho(l)$ (of finite codimension and dimension respectively) in the fibre $\mathcal{B}_{D^2}(\rho)$ over each critical point (flat connection) ρ. We choose the submanifolds $V_\rho(l)$ in the same way as in the Fukaya construction of Section 3. Thus we choose a collection of co-dimension 2 submanifolds $V_{\rho,i}$ which are compatible with submanifolds $V_{\rho,\sigma,i}$ on the space of connections with boundary values ρ, σ, under the gluing of connections. Then we let $V_\rho(l)$ be the intersection of the first l of the $V_{\rho,i}$. The remaining ingredient we need is a «connection» on the bundle $\mathcal{E}$ which will define holonomy maps between the different fibres. There is a tautological connection we can use here, which will simplify the construction considerably. Suppose $A(t)$ is a path in $\mathcal{B}_Y$ between ρ and σ. Using the Coulomb gauge we can make this into a connection $\mathbf{A}$ over $Y \times [0,1]$, in the way that is familiar in Floer's theory. Now suppose that a_ρ is a connection over the disc, whose boundary value matches ρ, *i.e.* a_ρ and ρ together define a point A_ρ in the fibre $\mathcal{B}_{D^2}(\rho)$. We choose a standard diffeomorphism which identifies $D^2 \cup_{S^1} (S^1 \times [0,1])$ with the disc D^2. We restrict $\mathbf{A}$ to $S^1 \times [0,1] \subset Y \times [0,1]$ and join this to the connection a_ρ to get a new connection a_σ over the disc, whose boundary value matches σ. The pair a_σ, σ define a point A_σ in the fibre $\mathcal{B}_{D^2}(\sigma)$. We define our tautological parallel transport map by saying that the holonomy $\phi_{A(t)}$ of the path $A(t)$ maps A_ρ to A_σ.

We can now see what our recipe gives in this situation. Consider first the case when ρ and σ have index-difference 3, so $\mathcal{M}_0(\rho,\sigma)$ is 2-dimensional. In our recipe we have chosen a particular connection $W_\rho(0) = \{a_\rho\}$ over the disc. If we follow through our recipe the number $n(\rho,\sigma;1,0)$ is just the number of instantons $\mathbf{A}$ in $\mathcal{M}_0(\rho,\sigma)$ such that the sum of $\mathbf{A}_\gamma = \mathbf{A}|_{S^1\times[0,1]}$ and a_ρ is a connection in the codimension-2 «jumping set' $V_\sigma(1)$. But the compatability of our choice of the V's with the gluing construction means that this is just the same as Fukaya's definition as the number of connections $\mathbf{A}$ such that $\mathbf{A}_\gamma$ lies in the codimension 2 subset $V_1(\rho,\sigma)$. Similarly, the number $n(\rho,\sigma;2,1)$ is just the number of instantons $\mathbf{A} \in \mathcal{M}_0(\rho,\sigma)$ such that the sum of some $\mathbf{A}_\gamma$ and a connection $a_\rho(w)$, where w runs over the 2-dimensional manifold $W_\rho(1)$, should lie on two codimension-2 jumping sets $V_{\sigma,1}$, $V_{\sigma,2}$ The compatability with gluing means that this is just the constraint that $\mathbf{A}_\gamma \in V_{\rho,\sigma,1}$ and $a_\rho(w)$ lies in the unique intersection point $W_\rho(1) \cap V_\rho(2)$, or that $\mathbf{A}_\gamma \in V_{\rho,\sigma,2}$ and $a_\rho(w)$ is fixed to be the unique intersection point of $W_\rho(1) \cap V_\rho(1)$. So $n(\rho,\sigma;2,1)$ is just the intersection number of $V_{\rho,\sigma,1}, V_{\rho,\sigma,2}$ in $\mathcal{M}_0(\rho,\sigma)$, which is precisely the same as Fukaya's definition. The same holds for all the other terms and we find that our recipe, mimicking the Morse-Bott theory in the finite dimensional case, produces the Fukaya complex of Y. This is the justification, from the Morse-theory point of view, for regarding $HFF_*(Y)$ as a kind of «homology» of $\mathcal{E}$.

We now turn to the other point of view and we shall explain why, for rather direct heuristic reasons, one should expect the homology of $\mathcal{E}$ to be the relevant space for our gluing problem. Suppose Y bounds a 4-manifold X_1 as before, and assume for simplicity that our relative class Σ_1 is represented by a disc, so we get a corresponding map from $Y \cup_\gamma D^2$ to X_1. We can pull back the infinite dimensional space of instantons over X_1 to get a subset $\tilde{\Lambda}_g \subset \mathcal{E}$ which is a lift of $\Lambda_g \subset \mathcal{B}$ over the fibration π. We want to see that the «homology» class of $\tilde{\Lambda}_g$ in $\mathcal{E}$ is just the right data to record the geometry in X_1 in order to be able to solve any instanton gluing problem. Consider our second 4-manifold X_2 with boundary $\overline{Y}$, and suppose we have a similar class Σ_2 reprsented by a disc in X_2. We write $\mathcal{E}_1, \mathcal{E}_2$ for the two copies of $\mathcal{E}$ associated to these two discs, so we get subsets $\tilde{\Lambda}_i \in \mathcal{E}_i$. We observe that the product over $\mathcal{B}_Y$, $\mathcal{E}_1 \times_{\mathcal{B}_Y} \mathcal{E}_2$ maps to the space $\mathcal{B}_\Sigma$ of connections over $\Sigma = \Sigma_1 \cup_\gamma \Sigma_2$ in an obvious way; so we get a 2-dimensional cohomology class $\mu(\Sigma)$ over $\mathcal{E}_1 \times_{\mathcal{B}_Y} \mathcal{E}_2$. On the other hand the instanton moduli space $\mathcal{M}(X)$ lifts to the product over $\mathcal{B}_Y$: $\mathcal{M}(X) \subset \mathcal{E}_1 \times_{\mathcal{B}} \mathcal{E}_2$, and the number we want to compute is the pairing of the homology class of this with $\mu(\Sigma)^m$. The instanton moduli space can be obtained as the «product over $\mathcal{B}$»

$$\mathcal{M}(X) = \Lambda_1 \times_{\mathcal{B}} \Lambda_2 \equiv \{(x \times y) \in \mathcal{E}_1 \times_{\mathcal{B}} \mathcal{E}_2 | \pi_1(x) = \pi_2(y)\}.$$

Thus we see that what we want is a homology pairing from the homology of $\mathcal{E}_1$ times the homology of $\mathcal{E}_2$ to the homology of $\mathcal{E}_1 \times_{\mathcal{B}} \mathcal{E}_2$, which is represented geometrically by the product of cycles over $\mathcal{B}$.

We can mimic all of this is in a finite dimensional situation. If $\pi_1 : E_1 \to B, \pi_2 : E_2 \to B$ are a pair of fibre bundles over a manifold B of dimension N there is a multiplication map

$$\times_B : H_i(E_1) \otimes H_j(E_2) \to H_{i+j-N}(E_1 \times_B E_2),$$

which is represented at the level of singular chains by the product over B, $(\lambda_1, \lambda_2) \mapsto \lambda_1 \times_B \lambda_2$. Suppose that E_1, E_2 are now $\mathbf{CP}^\infty$ bundles corresponding to classes $\nu, -\nu \in H^3(B)$. There is a 2-dimensional cohomology class h over the total space of $E_1 \times_B E_2$ (restricting to the «diagonal» class on each fibre $\mathbf{CP}^\infty \times \mathbf{CP}^\infty$) and so we get pairings

$$\sigma_m : H_i(E_1) \otimes H_{N+2m-j}(E_2) \to \mathbb{Z},\ \sigma_m(\lambda_1, \lambda_2) = < h^m, \lambda_1 \times_B \lambda_2 > .$$

Finally, to tie the two points of view together, one can check that if we represent the homology of E_1 by means of a Morse function f, and that of E_2 by $-f$, then the pairing σ_m we have defined above is represented by the obvious analogue of our definition in the Fukaya complex.

We will now examine the Fukaya constuction to see how the analogue of Massey products enter. First consider the spectral sequence

$$HF^*(Y) \otimes H^*(\mathbf{CP}^\infty) \Rightarrow HFF^*(Y).$$

It is clear from the discussion of 4.2 that the first differential d_3 is given by the cup product with the class ν, just as in the finite dimensional case. When we go to the next differential d_5 the story becomes more complicated. In the finite dimensional case this was given by the Massey product $\{a, \nu, \nu\}$ and the problem is that we cannot represent the cup-product ν^2 itself within the Floer complex, only its multiplication operation. We can define a chain map $\hat{\nu}$ on the Floer cochains as in 4.2, and the composite $(\hat{\nu})^2$ need not vanish, whereas we can suppose that $(\widehat{\nu^2})$ is zero. Thus we know that $(\hat{\nu})^2$ is chain-homotopic to zero, so we fix such a chain homotopy K:

$$(\hat{\nu})^2 = Kd + dK.$$

Now suppose that a is a Floer cochain in C^p and the product $\hat{\nu}a$ vanishes in cohomology. So we can write $\hat{\nu}a = dx$, for a Floer cochain x. Then

$$d\hat{\nu}x = \hat{\nu}dx = (\hat{\nu})^2 a = dKa,$$

and the class $Ka - \hat{\nu}x$ is a cochain, defining a class in $HF^{p+5}(Y)$. We denote this class by $\{a, \nu, \nu\}_{HF}$, and regard it as the analogue of a Massey product. To justify this we want to see that, if we use a suitable chain homotopy K, we get the differential in the spectral sequence:

$$d_5 a = \{a, \nu, \nu\}_{HF}.$$

Recall that in 4.2 we constructed the chain homotopy from $\widehat{\mu(\alpha)\mu(\beta)}$ to $\widehat{\mu(\alpha)} \circ \widehat{\mu(\beta)}$ by moving representative cycles to different ends of the tube. We do this now with cycles $\gamma \times \{t_1\}, \gamma \times \{t_2\}$ where $t_1 \to \infty, t_2 \to -\infty$. Representatives U_1, U_2 for ν over $\mathcal{B}_Y$ yield, by translation, families of codimension-3 submanifolds $U_1(t_1), U_2(t_2)$ in the space of connections over $Y \times \mathbb{R}$. The usual procedure for the construction of chain homotopies in Floer homology gives a chain homotopy K having matrix entries $K_{\rho\sigma}$ defined in terms of a 5-dimensional moduli space $\mathcal{M}(\rho, \sigma)$ by counting the number of connections $\mathbf{A} \in \mathcal{M}(\rho, \sigma)$ for which there are $t_1, t_2 \in \mathbb{R}$ such that $\mathbf{A}$ lies in $U_1(t_1)$ and $U_2(t_2)$. If we use the codimension-2 submanifolds V_i in the 4-dimensional space $\mathcal{M}_0(\rho, \sigma)$ associated to W_i in the manner of 4.2, then we see that $K_{\rho,\sigma}$ is precisely the intersection number of V_1 and V_2 in $\mathcal{M}_0$, as used in defining Fukaya's boundary operator. In sum, one sees that the chain homotopy

$$K : C^p(Y) \to C^{p+5}(Y)$$

is precisely the same as the component, D_5 say, of the Fukaya differential mapping $C^p \otimes R^4 = C^p$ to C^{p+5}. On the other hand, if one follows through the spectral sequence of the filtered Fukaya complex, one finds that the differential d_5 is precisely given by

$$d_5(a) = D_5(a) - \hat{nu}(x),$$

and since $D_5 = K$, this is $\{a, \nu, \nu\}_{HF}$ as we claimed.

We will now turn to the the pairing σ_m, which can be viewed as a variant of the cup product in the total space $\mathscr{E}$. For simplicity we just consider the case $m = 2$. The pairing σ_2 on $HFF_*(Y) \otimes HFF_*(\overline{Y})$ factors through $HFF_*^{(2)}(Y) \otimes HFF_*^{(2)}(Y)$. We fix attention on the groups in complementary dimensions i, j on which the pairing σ_2 is defined: the spectral sequence tells us that $HFF_i(Y)$ fits into an exact sequence

$$0 \to K_i \to HFF_i(Y) \to Q_{i-2} \to 0,$$

where K_i is the subgroup ker $\nu \subset HF_i(Y)$ and Q_{i-2} is the quotient

$$HF_{i-2}(Y)/\text{Im } \nu : HF_{i+1}(Y) \to HF_{i-2}(Y).$$

(The multiplication operation lowers degrees here, since we are using homology.) Taking account of the ordinary duality in Floer homology, we can describe the other group $HFF_j(\overline{Y})$ by an exact sequence:

$$0 \to Q_{i-2}^* \to HFF_j(\overline{Y}) \to K_i^* \to 0,$$

where

$$Q_{i-2}^* = \text{ker } \nu \subset HF^{i-2}(Y),$$

is dual to Q_{i-2} and

$$K_i^* = HF^i(Y)/\text{Im}\nu,$$

is dual to K_i. If we choose splittings of these sequences (working with rational coefficients), then the pairing σ_2 has four components

$$\sigma_{Q^*,K}, \quad \sigma_{Q,Q^*}, \quad \sigma_{K,K^*}, \quad \sigma_{Q,K^*}$$

say. It is easy to see that that the first three of these are intrinsic, independent of the splitting. The pairing $\sigma_{Q^*,K}$ on the subgroups is zero, and the next two are just given the canonical dual pairings on $Q \otimes Q^*, K \otimes K^*$. The remaining part σ_{Q,K^*} is given by a kind of Massey product, mixing the cup product on cohomology with the contraction on homology. Suppose we have a pair $a \in Q, b \in Q^*$, and let $\alpha \in C^{i-2}, \beta \in C_i$ be representatives. We can lift a, b to $HFF_i(Y), HFF_j(\overline{Y})$ by choosing classes $x \in C^i, y \in C_{i-2}$ such that $dx = \nu\alpha \in C^{i+1}$ and $\partial y = \nu\beta \in C_{i-3}$. The definition of σ_2 shows that the pairing of the lifted classes is just

$$\sigma_{Q,K^*}(a, b) = x \cdot \beta + y \cdot \alpha,$$

where the dot denotes the canonical duality between chains and cochains. Of course the ambiguity in this Massey product reflects the fact that σ_{Q,K^*} is not invariantly defined, it depends upon the choice of splittings.

While we have not done more here than spell out some of the formal algebro-topological aspects of the Fukaya construction, one might hope that the ideas would lead to deeper results in specific circumstances. For example, a famous result of Deligne *et al* [DGMS] asserts that if B is a compact Kähler manifold then all Massey products are zero: in particular the spectral sequence of our $\mathbf{CP}^\infty$ bundle degenerates after the E_3 term. Now suppose that Y is a product $Y = Z \times S^1$ where Z is a Riemann surface, and the $U(2)$ bundle P is lifted from Z. Then the work of Dostoglou and Salamon [DS] shows that the Floer homology $HF^*(Y)$ in the Yang-Mills situation is the same as that obtained in the symplectic situation, for the symplectic manifold N_Z of projectively flat connections over Z. This Floer cohomology can be regarded formally as the cohomology of the space LN_Z of loops in N_Z and this space is an infinite-dimensional Kähler manifold. This suggests that perhaps the spectral sequence defining Fukaya homology degenerates in the case of $Y = Z \times S^1$. In fact Dostoglou and Salamon show that the Floer cohomology of $Z \times S^1$ is the same as the ordinary cohomology of N_Z, but with the grading made $\mathbb{Z}/8$ periodic. The space N_Z is a compact Kähler manifold but the cup product which is relevant here is not the ordinary cup product on $H^*(N_Z)$ but rather a deformed product which has been discussed by Witten [W2] § 3. So the question is whether the Massey products defined by this deformation of the cup product vanish, in the same way as they do for the ordinary cup product on $H^*(N_Z)$.

5 Relations among polynomial invariants

When considering polynomial invariants for bundles $P \to X$, $Y \subset X$ such that $P_{|Y}$ is free of flat reductions we are dealing with a topological quantum field theory (Witten [W1], Atiyah [At2]). In the spirit of theoretical physics we will now pay special attention to the cylinder $Y \times \mathbb{R}$. Denote by $\mathcal{S}^*(H_*(Y;\mathbb{Q}))$ the super-algebra generated by $H_*(Y;\mathbb{Q})$. The μ-map follwed by cup products gives an algebra homomorphism:

$$\mathcal{S}^*(H_*(Y;\mathbb{Q})) \to H^*(\mathcal{B}_Y^*;\mathbb{Q})$$

and as we saw in the previous section, such elements give endomorphisms on the Floer homology which we denote by

$$\mathbf{c} : \mathcal{S}^*(H_*(Y;\mathbb{Q})) \to \mathrm{End}(HF_*(Y;\mathbb{Q})).$$

Our key observation here is that the gluing law states that $\mathbf{c}$ is an *algebra homomorphism*. The algebras $\mathcal{S}^*(H_*(Y;\mathbb{Q}))$ is infinite dimensional while $\mathrm{End}(HF_*(Y))$ is a finite dimensional matrix algebra, so $\mathbf{c}$ must have a big kernel. We see that $\mathbf{c}$ forces the following relations among invariants:

Proposition 1.

1) *Let* $\partial X_1 = Y$ *and* $P_1 \to X_1$ *such that* $P_{1|Y}$ *is free of flat reductions. If* $\sigma \in \ker\{\mathbf{c} : H_*(Y;\mathbb{Z}) \to H_*(X_1;\mathbb{Z})\}$ *then:*

$$\mathbf{c}(\sigma)\psi_{X_1} = 0$$

where ψ_{X_1} *is the relative polynomial invariant.*

2) *Let* $X = X_1 \cup_Y X_2$ *and* $P \to X$ *such that* $P_{|Y}$ *is free of flat reductions. For* $\sigma_i \in S^*(H_2(X_i;\mathbb{Q}))$ *and* $\sigma \in \ker \mathbf{c}$ *we have*

$$\psi_X(\sigma_1, \sigma, \sigma_2) = 0$$

The same considerations can be applied to the Fukaya-Floer groups, on which we can also define a cup-product action. In this case the second relation also will work when classes «go through the middle'. Summarizing, we see that a set of relations is associated to a splitting $X_1 \cup_Y X_2$, which depends only on Y. An example of a relation which is probably of this type was explained to us by Kronheimer (private communication). It would seem an interesting project to compute these relations in special cases.

References

[At1] M.F. Atiyah, *New Invariants for 3 and 4-dimensional manifolds*, Proc. Sympos. Pure Math. 48 (1988).

[At2] M.F. Atiyah, *Topological Quantum Field Theories*, Publ. Math. IHES, 68 (1988).

[APS] M. F. Atiyah, V. Patodi & I. M. Singer, *Spectral asymetry and Riemannian geometry II & III* , Math. Proc. Camb. Phil. Soc., 78 & 79 (1975), pp. 71–99, 405–432.

[AuBr1] D. Austin & P. Braam, *Equivariant Floer thoery and Donaldson Polynomials*, in preparation.

[AuBr2] D. Austin & P. Braam, *Morse-Bott theory and equivariant cohomology*, in this volume.

[DGMS] P. Deligne, P. Griffiths, J. Morgan & D. Sullivan, *Homotopy Theory of Compact Kähler Manifolds*, Inv. Math. 29 (1973), pp. 245–274.

[Do1] S.K. Donaldson, *Polynomial Invariants for smooth four-manifolds*, Topology 29 (1990), pp. 257–315.

[Do2] S.K. Donaldson, *Gluing techniques in the cohomology of moduli spaces*, to appear in the volume in honour of Milnor's sixtieth birthday, Publish and Perish.

[Do3] S.K. Donaldson, *The orientation of Yang-Mills moduli spaces and four-manifold topology*, J. Diff. Geom., 24 (1987), pp. 397–428.

[Do4] S.K. Donaldson, *Irrationality and the h-cobordism conjecture*, J. Diff. Geom. 26 (1987), pp. 141–168.

[DFK] S.K. Donaldson, M. Furuta & D. Kotschick, *Floer Homology Groups in Yang-Mills Theory*, in preparation.

[DK] S.K. Donaldson & P.B.K. Kronheimer, *The Geometry of 4-Manifolds*, Oxford University Press, Oxford, 1990.

[DS] S. Dostoglou & D. Salamon, *Instanton Homology and Symplectic Fixed Points*, preprint.

[Fl] A. Floer, *An instanton invariant for 3-manifolds*, Comm. Math. Phys. 118 (1988), 215–240.

[Fur1] M. Furuta, unpublished manuscript.

[Fur2] M. Furuta, *Morse theory and Thom-Gysin exact sequence*, preprint.

[Fuk] K. Fukaya, *Instanton Homology for Oriented 3-Manifolds*, to appear in Adv. Studies in Pure Mathematics, ed Y. Matsumoto & S. Morita.

[Sa] D. Salamon, *Morse theory, the Conley index and Floer homology*, Bull. London Math. Soc. 22 (1990), pp. 113–140.

[Su] D. Sullivan, *Infinitesimal computations in Topology*, Publ. IHES Vol 47 (1978), pp. 261–331.

[Tau] C.H. Taubes, *Gauge Theory on Asymptotically Periodic 4-Manifolds*, J. Diff. Geom. 25 (1987), pp. 363–430.

[Tau] C.H. Taubes, *A simplicial model for Donaldson-Floer theory*, in this volume.

[Th] M. Thaddeus, *Conformal Field Theory and the Cohomology of the Moduli Space of Stable Bundles*, J. Diff. Geom. 35, 1 (1992), pp. 131–150.

[W1] E. Witten, *Topological Quantum Field Theory*, Comm. Math. Phys., 117 (1988), pp. 353–386.

[W2] E. Witten, *Two dimensional gravity and intersection theory on moduli space*, Surveys in Diff. Geom. 1 (1991), pp. 243 ff.

P.J. Braam and S.K. Donaldson
The Mathematical Institute
23–29 St. Giles
Oxford OX1 3LB
UK

On generating families

M. Chaperon

In 1983, soon after Charles Conley and Eduard Zehnder proved ([13], slightly extended in [7]) the first of Vladimir Arnold's conjectures ([1], [2]) in global symplectic geometry, we discussed a very general statement about the number of intersections between two «exact» compact lagrangian submanifolds L_0 and L_1 of a symplectic manifold, isotopic through a hamiltonian isotopy. As far as I remember, Conley and I suggested looking for critical points of the action functional on the space of H^1-paths from L_0 to L_1, while Conley and Zehnder expressed the belief that, even though the L^2-gradient of this functional seemed to have few unbounded orbits, the Morse inequalities could be obtained by considering only the set of all bounded orbits. The three of us felt that very hard work and new ideas would be needed to carry out this program.

Andreas Floer was able to provide both [15] incredibly soon. He observed that the integral curves of our gradient, viewed as parametrised surfaces (they are paths in a space of paths), are pseudo-holomorphic curves; this enabled him to exploit Mikhael Gromov's brand new ideas on the subject [16], but he had to consider much harder cases than did Gromov and this was a personal masterpiece. It established its author among the very best mathematicians of our time, which his further work confirmed.

Before these crucial contributions by Gromov and Floer, I had tried something quite different ([8], [9]) to make the solution of such problems accessible to fourth year students. This «broken geodesics» construction was used and considered from various viewpoints by François Laudenbach, Jean-Claude Sikorav, Claude Viterbo, Yuri V. Chekanov, Alexander Givental and Patrice Le Calvez, among others.

The aim of this article is to generalise this construction to contact geometry. It bears the strong influence of my successors, to whom I shall try to do justice.

1 Introduction

Unless otherwise specified, all the objects we consider are smooth.

1.1 Jets

Let M be a manifold. For each $q \in M$ and each real function φ on M, the triple $(q, \varphi(q), d\varphi(q))$ is the 1-*jet* $j^1\varphi(q) = j^1_q\varphi$ of φ at q, and the set J^1M of all such jets is a manifold, identified to $T^*M \times \mathbf{R}$ by the map $(q,y,p) \mapsto ((q,p),y)$. This manifold can be viewed as a vector bundle over $J^0M := M \times \mathbf{R}$ with projection $\pi_M : (q,y,p) \mapsto (q,y)$, as a vector bundle over M with projection $\alpha_M : (q,y,p) \mapsto q$ and, of course, as a trivial vector bundle over T^*M. We call J^1M the space of 1-jets of real functions on M.

The 1-jet $j^1\varphi$ of a smooth $\varphi : M \to \mathbf{R}$ is a section of the «source» projection α_M. Conversely, if we denote by c_M the *contact form* $c_M(q, y, p) = dy - p\,dq$ on J^1M, a local section σ of α_M is *holonomic* (i.e. of the form $j^1\varphi$) if and only if $\sigma^* c_M = 0$; identifying a section of a fibre bundle with its image, this means that the tangent space to σ at each point x is contained in the *contact hyperplane* $\mathcal{H}_M(x) := \mathrm{Ker}(c_M(x))$. In other words, σ is an integral manifold of the *contact structure* $\mathcal{H}_M : x \mapsto \mathcal{H}_M(x)$.

1.2 Quasi-functions

Quasi-functions suggest a class of «generalised functions» on M, the *Legendrian submanifolds* of J^1M, which are the integral manifolds of $\mathcal{H}_M$ of dimension $\dim(M)$; this is the maximal possible dimension because the projection $J^1M \to T^*M$, restricted to an integral manifold L of $\mathcal{H}_M$, is an immersion j such that $j^*(dp \wedge dq) = 0$ (remember that $dy = p\,dq$ on L). The *critical points* of such a generalised function L are its intersections with $\{p = 0\}$, and a critical point is *non-degenerate* when the intersection is transversal.

Recall that the number of critical points of a real function on a closed (i.e. compact without boundary) manifold M is greater than the cuplength $c\ell(M)$ of M, and not less than the sum $b_*(M)$ of its Betti numbers — these estimates hold for cohomology with coefficients in any field. In order to get the same bounds for our generalised functions, we shall consider only those L's which are obtained from the zero section $j^1_M 0 = \{y = 0 \text{ and } p = 0\}$ of α_M by a smooth deformation through Legendre submanifolds of J^1M; Arnold [3] calls them *quasi-functions* on M — the 1-jet of a true function φ is a quasi-function, since it can be joined to the zero section by the smooth family $j^1(t\varphi)$ — and attributes the following theorem to Chekanov:

Theorem 1. *A quasi-function on a closed manifold M has more than $c\ell(M)$ critical points, and at least $b_*(M)$ if all of them are non-degenerate.*

1.3 Generating families

A *phase function* on M is a real function S on $M \times E$ for some finite dimensional real vector space E; it is called *quadratic* (a rather misleading way to say that it is non-degenerate quadratic at infinity) when there exists a non-degenerate quadratic form K on E such that $S(Q, v) = K(v)$ off a compact subset. We shall deduce Theorem 1 from the following well-known result:

Theorem 0. *A quadratic phase function on a closed manifold M has at least $c\ell(M) + 1$ critical points, and at least $b_*(M)$ if all of them are non-degenerate.*

Here is the link [18] with Theorem 1: a phase function $S : M \times E \to \mathbf{R}$ on M is a *generating family* (or generating phase function) of an immersion $i : L \to J^1M$ when the following two properties hold:

(i) the function $d_V S : (Q; v) \mapsto \partial_v S(Q; v)$ admits $0 \in E^*$ as a regular value; hence $\Sigma_S := d_V S^{-1}(0)$ is a submanifold and, setting $S_v(Q) := S(Q; v)$, the mapping $\Sigma_S \ni (Q; v) \mapsto j^1_Q S_v \in J^1M$ is an immersion i_S;

(ii) there exists a diffeomorphism h of L onto Σ_S such that $i = i_S \circ h$. Therefore, since i_S obviously is *legendrian* (i.e. $i_S^* c_M = 0$), so must i be.

When i is an embedding, condition (ii) just means that i_S is an embedding with the same image as i, and we call S a *generating family of the Legendrian submanifold* $i(L)$.

In this case i_S, restricted to the critical points of S, is a bijection onto the critical points of $i(L)$; hence Theorem 1 follows from Theorem 0 and

Theorem 2. *A quasi-function on a closed manifold M has a quadratic generating family.*

For example, the 1-jet of a function φ has a generating «family», namely φ itself, viewed as a function on $M \times \{0\}$; when φ has compact support, this generating family is quadratic!

I suppose Theorem 2 should be credited to Chekanov too, even though he did not (and apparently does not intend to) write anything on the subject. It was originally proved by Sikorav [21] for quasi-functions obtained by deforming the zero section through quasi-functions whose projection into T^*M is injective — or, in his words, for exact lagrangian submanifolds of T^*M obtained by deforming the zero section through exact lagrangian submanifolds; in this case, Theorem 1 is due to Helmut Hofer [17].

1.4 Contact transformations and isotopies

A *contact transformation* of J^1M is a (local) diffeomorphism g of J^1M which sends Legendrian submanifolds onto Legendrian submanifolds or, equivalently, which preserves the contact structure $\mathcal{K}_M$; this is expressed by the identity $c_M \wedge g^* c_M = 0$. Hence the flow of a vector field X on J^1M consists of contact transformations if and only if the Lie derivative $\mathcal{L}_X c_M$ satisfies $c_M \wedge \mathcal{L}_X c_M = 0$, in which case X is called a *Lie field.* As noticed by Sophus Lie, the mapping $X \mapsto \langle c_M , X \rangle$ is an isomorphism between Lie fields and real functions on J^1M; the function $-\langle c_M , X \rangle$ is the *hamiltonian* of the Lie field X with respect to c_M. In local coordinates, the Lie field $\vec{H}$ with hamiltonian $H(q, y, p)$ is

$$(1) \qquad \vec{H} = (\langle \partial_p H , p \rangle - H)\, \partial_y + \sum_{i=1}^{n} (\partial_{p_i} H\, \partial_{q_i} - (\partial_{q_i} H + p_i \partial_y H)\, \partial_{p_i}) \,.$$

Relationship with symplectic transformations of T^*M. A symplectic transformation g of T^*M can be lifted to a contact transformation $\tilde{g}$ of J^1M (unique up to translation in y-space) if and only if it is «globally canonical», i.e. $g^*(p\,dq) - p\,dq$ is exact. The corresponding $\tilde{g}$'s are precisely those which preserve not only $\mathcal{K}_M$, but the contact form c_M itself. The infinitesimal version of a globally canonical transformation is a hamiltonian vector field; the hamiltonians of the Lie fields lifting a given hamiltonian vector field are the lifts $H(q, p)$ of its hamiltonians.

An *isotopy* $(g_t)_{0\le t\le 1}$ of J^1M is a smooth family of diffeomorphisms $g_t : J^1M \to J^1M$ with $g_0 = \mathrm{Id}$; its (*infinitesimal*) *generator* is the smooth family $(\dot g_t)_{0\le t\le 1}$ of vector fields defined by

$$\frac{d}{dt} g_t(x) = \dot g_t(g_t(x)) \ ,$$

which shows that (g_t) is determined by $(\dot g_t)$ and by the initial condition $g_0(x) \equiv x$. In particular, we get a bijection between *compactly supported* isotopies (g_t) of J^1M (such that $g_t(x) = x$ off a compact subset) and *compactly supported* time-dependent vector fields $(X_t)_{0\le t\le 1}$ (such that $X_t(x) = 0$ off a compact subset).

Clearly, (g_t) is a *contact isotopy*, i.e. consists of contact transformations, if and only if each $(\dot g_t)$ is a Lie field. Hence, $(g_t) \mapsto (\langle c_M, \dot g_t\rangle)$ is a bijection between compactly supported contact isotopies (g_t) of J^1M and compactly supported smooth families $(H_t)_{0\le t\le 1}$ of real functions on J^1M (meaning that $(t,x) \mapsto H_t(x)$ is smooth and has compact support); the smooth family $(-\langle c_M, \dot g_t\rangle)$ is the *hamiltonian* of (g_t) with respect to c_M.

A *Legendrian isotopy* in J^1M is a smooth family $(j_t)_{0\le t\le 1}$ of Legendrian *embeddings* $L \hookrightarrow J^1M$; it has *compact support* if $j_t(x) \equiv j_0(x)$ off a fixed compact subset. The first step in the proof of Theorem 2 is the

Isotopy extension lemma. *For every compactly supported Legendrian isotopy (j_t) in J^1M, there exists a compactly supported contact isotopy (g_t) of J^1M extending (j_t), i.e. such that $j_t = g_t \circ j_0$ for $0 \le t \le 1$.*

In particular, a quasi-function on a closed manifold M is of the form $g_1(j^1_M 0)$ for some compactly supported contact isotopy (g_t) of J^1M (recall that $j^1_M 0$ denotes the zero section of α_M). As the zero function on $M \times \{0\}$ is a quadratic generating family of $j^1_M 0$, Theorem 2 follows from

Theorem 3. *Given a manifold M without boundary, let i be a Legendrian immersion into J^1M and let (g_t) denote a compactly supported contact isotopy of J^1M; if i has a quadratic generating family, then there exists a quadratic phase function S on $[0,1] \times M$ such that $S_t : (Q; v) \mapsto S(t, Q; v)$ is a generating family of $g_t \circ i$ for $0 \le t \le 1$.*

I do not know if this result can be attributed to Chekanov, but I am sure that it is due to Sikorav [21] in the symplectic case and that David Théret was able to prove it in general [22].

1.5 First order evolution of partial differential equations

They provide a good reason to study quasi-functions. Given a manifold M (which we assume closed for simplicity) and a smooth family $(H_t)_{0\le t\le 1}$ of real functions on J^1M, let us try to find a solution $(t,q) \mapsto u_t(q)$ of

$$(2) \qquad \partial_t u_t(q) + H_t(j^1_q u_t) = 0$$

which equals a given (smooth) $u_0 : M \to \mathbf{R}$ for $t = 0$. By (1), equation (2) means (in local coordinates) that the Lie derivative of the vector-valued function $F(t, q, y, p) := (y - u_t(q), p - du_t(q))$ with respect to $\partial_t + \vec{H}_t$ vanishes for $F = 0$ or, in other words, that $\partial_t + \vec{H}_t$ is tangent to $\bigcup\{t\} \times j^1 u_t$. Hence, if we denote by (g_t) the (local) contact isotopy with hamiltonian (H_t), our solution, if it is smooth, must be given by $j^1 u_t = g_t(j^1 u_0)$.

Since M is closed, this formula provides a genuine solution of our Cauchy problem for small enough t. When t increases, two unpleasant possibilities may occur:

(a) the contact transformation g_t is not defined at every point of $j^1 u_0$;

(b) the quasi-function $g_t(j^1 u_0)$, though well-defined, is not a section of α_M.

Assume that (a) does not happen for $0 \leq t \leq 1$ but that (b) takes place — which can be the case even in such simple examples as $M = \mathbf{T}^1$ and $H_t(q, y, p) = p^2$; if π_M denotes the natural projection of $J^1 M$ onto $J^0 M := M \times \mathbf{R}$, the *Cerf diagram* (*wavefront*, according to Arnold)

$$\bigcup_{0 \leq t \leq 1} \Big(\{t\} \times \pi_M(g_t(j^1 u_0)) \Big)$$

will not be the graph of a function $(t, q) \mapsto u_t(q)$, but a singular hypersurface with self-intersections, submanifolds of cuspidal points, etc.

In some examples (caustics), this hypersurface is the «physical» solution of the problem at hand; in other cases, the only acceptable solutions are functions or distributions on M. Then, we can construct such a solution as follows: by the isotopy extension lemma and Theorem 3, there exists a quadratic phase function S on $[0, 1] \times M$ such that $S_t : (Q; v) \mapsto S(t, Q; v)$ is a generating family of $g_t(j^1 u_0)$ for $0 \leq t \leq 1$; let $K : E \mapsto \mathbf{R}$ be the quadratic form such that $S(t, Q; v) = K(v)$ off a compact subset, and let $C > 0$ be large enough for every $S_{t,Q} : v \mapsto S_t(Q; v)$ to coincide with K on $E_C := S_{t,Q}^{-1}(-\infty, -C] = K^{-1}(-\infty, -C]$. For each (t, Q), consider the «minimax» critical value of $S_{t,Q}$ defined by

$$u(t, Q) = u_t(Q) := \inf_{\sigma} \max S_{t,Q} \circ \sigma\,, \tag{3}$$

where σ varies among all relative singular homology cycles whose class is a fixed generator of $H_j(E, E_C) \cong \mathbf{Z}$ and j is the index of K. It is quite easy to check that u *is a Lipschitz-continuous solution u of our problem* (this was suggested to me [11] by Sikorav); indeed, its definition implies that it is a Lipschitz-continuous section of the wavefront — hence u_t is smooth and satisfies $j^1 u_t = g_t(j^1 u_0)$ whenever the latter is a section of α_M, for example when $t = 0$; moreover, the 1-jet of u in the sense of distributions is its almost everywhere defined 1-jet in the usual sense, which is readily seen to satisfy (2).

Thus, we have constructed a generalised solution in the sense of analysis as a section of the generalised solution in the sense of geometry. The reader may wonder whether u depends on the choice of a particular quadratic generating family S; the answer is that *it does not*. This follows from the following important result, first stated by Viterbo ([23], Proposition 1.5) in the symplectic case:

Theorem (Théret [22]) *If* $S_1 : M \times E_1 \to \mathbf{R}$ *and* $S_2 : M \times E_2 \to \mathbf{R}$ *are two quadratic generating families of the same quasi-function on* M*, they are stably (right-) equivalent: there exist non-degenerate quadratic forms* K_j*, defined on finite dimensional real vector spaces* F_j $(j = 1, 2)$*, such that* $S_1(Q; v_1) + K_1(w_1) = S_2(Q; v_2) + K_2(w_2)$ *for some diffeomorphism* $(Q; v_1, w_1) \mapsto (Q; v_2(Q; v_1, w_1), w_2(Q; v_1, w_1))$.

Even though this proves that the minimax solution is somewhat unique, Tatiana Joukovskaia [19] has shown that the wavefront may have other Lipschitz-continuous sections solving our Cauchy problem. We do not know how to characterise the minimax solution, except in one case [19]: when every H_t is strictly convex with respect to p, $u_t(Q)$ is just the y-coordinate of the lowest point of the wavefront above (t, Q); indeed, $g_t(j^1 u_0)$ has a generating family which is *positive definite* at infinity, hence the minimax is a minimum. This implies at once that u satisfies the «entropy condition» ([20], [14]), which makes everyone believe in it...

The main tools for a theory of generic minimax functions are constructed in [19] (which contains a complete classification of their singularities and time evolution for $\dim(M) \leq 2$), yielding for example the following general bound for the complexity of the «shocks» exhibited by the derivative of our weak solution:

Theorem (Joukovskaia [19]) *A generic minimax function* u *on a manifold* B *is smooth except possibly along a closed stratified hypersurface; viewed in appropriate smooth local charts of* B*, this hypersurface and the graph of* u *are semi-algebraic.*

1.6 Guide for the reader

In the sequel, we construct generating families for compactly supported contact isotopies of $J^1\mathbf{R}^n$ (Theorem 4 below) and use them to prove Theorem 3. In a forthcoming paper [12], we shall show that the construction can be applied to isotopies whose support is not compact, yielding a very simple proof of two notoriously related theorems: Viterbo's celebrated statement that the Weinstein conjecture is true for hypersurfaces of $\mathbf{R}^{2n}$, and the fact that the Hofer-Zehnder capacity *is* a capacity.

2 Construction of generating families

2.1 Generating functions of contact transformations of $J^1\mathbf{R}^n$

We identify $J^1\mathbf{R}^n$ with the set of all $(q, y, p) \in \mathbf{R}^n \times \mathbf{R} \times (\mathbf{R}^n)^*$ and consider a global contact transformation $h : J^1\mathbf{R}^n \to J^1\mathbf{R}^n$.

The 1-jets of the affine maps $a_{z,p} : q \mapsto z + pq$, $(z, p) \in \mathbf{R} \times (\mathbf{R}^n)^*$ form a foliation of $J^1\mathbf{R}^n$; if h is C^1-close enough to the identity, it has the following properties:

(I) Each $h(j^1 a_{z,p})$ is a holonomic section $j^1\varphi_{z,p}$ of $\alpha_{\mathbf{R}^n}$. In other words, the graph of h is a set of $((q,y,p),(Q,Y,P))$ satisfying

$$(4)\qquad \begin{cases} y = z + pq \\ Y = \varphi(Q,z,p) := \varphi_{z,p}(Q) \\ P = \partial_Q\varphi(Q,z,p)\,. \end{cases}$$

(II) (Q,z,p) is a system of coordinates on the graph of h — note that this implies (I); hence, this graph is defined by (4) and by $q = q(Q,z,p)$.

As h preserves the contact structure, we have $dY = P\,dQ$ if and only if $dy = p\,dq$, i.e. $dz + q\,dp = 0$; since

$$dY - P\,dQ = (dz + q\,dp)\,\partial_z\varphi + \left(\partial_p\varphi - q\,\partial_z\varphi\right)dp\,,$$

we get

Proposition 2.1. *Under hypothesis* (II), *we have* $\partial_z\varphi(Q,z,p) \neq 0$ *everywhere — if h is close to the identity,* $\partial_z\varphi > 0$ *— and the mapping* $i_\varphi : (Q,z,p) \mapsto ((q,y,p),(Q,Y,P))$ *defined by* (4) *and*

$$(5)\qquad q = \frac{\partial_p\varphi(Q,z,p)}{\partial_z\varphi(Q,z,p)}$$

is an embedding whose image is the graph of h.

Definition. We call φ the «greek» *generating function* of h. If h is the identity, $\varphi(Q,z,p) = z + pQ$.

The «latin» generating function $f(Q,z,p) := \varphi(Q,z,p) - (z + pQ)$ of h has compact support if and only if the closure of $\{x : h(x) \neq x\}$ is compact; it is independent of z if and only if h is a lift of a symplectic transformation g of $T^*\mathbf{R}^n$ (in which case $f(Q,p)$ is a generating function of g in the sense of symplectic geometry). Note that h is $\mathbf{Z}^n$-periodic with respect to q if and only if f satisfies the identity $f(Q+m, z-pm, p) \equiv f(Q,z,p)$, $m \in \mathbf{Z}^n$.

2.2 Relationship with hamiltonians

Let (H_t) be the hamiltonian of a contact isotopy (h_t) of $J^1\mathbf{R}^n$. Assume that some h_t satisfies the hypotheses of 2.1, hence has a generating function φ_t; if we set $h_\tau(q, z+pq, p) = (Q_\tau, Y_\tau, P_\tau)$, (1) and (4) clearly imply that

$$(6)\qquad \varphi_t(Q_t,z,p) = Y_s + \int_s^t \dot{Y}_\tau\,d\tau = Y_s + \int_s^t \left(P_\tau\dot{Q}_\tau - H_\tau(Q_\tau,Y_\tau,P_\tau)\right)d\tau$$

for $0 \leq s \leq 1$. Assume that every h_s with s close to t has a generating function φ_s; differentiating (6) with respect to t and using (1) and (4), we get

$$(7)\qquad \left(\frac{d\varphi_t}{dt}\right)(Q_t,z,p) = -H_t(Q_t,Y_t,P_t)\,,$$

which implies

Proposition 2.2. *Let (h_t) be a contact isotopy of $J^1\mathbf{R}^n$, with hamiltonian (H_t). If h_s satisfies the hypotheses of 2.1 and H_s is nowhere negative for $0 \le s \le t$, then the latin generating function of h_t is nowhere positive.*

2.3 Two results from third year calculus

Global inverse mapping theorem. *For every Lipschitz-continuous mapping u of a Banach space E into itself with Lipschitz constant* $\operatorname{Lip} u < 1$, *the mapping* $\mathrm{Id} - u$ *is a bijection and* $\operatorname{Lip}\left((\mathrm{Id} - u)^{-1}\right) \le (1 - \operatorname{Lip} u)^{-1}$.

Call $h : E \to E$ a *diffeomorphism* when it is bijective and both h and h^{-1} are locally Lipschitz-continuous; our result implies that $\mathrm{Id} - u$ is a diffeomorphism. *If a diffeomorphism h is C^k (resp. analytic, holomorphic) in some open subset U, so is h^{-1} in $h(U)$.*

Global Cauchy-Lipschitz theorem. *Let $(X_t)_{0\le t\le 1}$ be a continuous family of continuous vector fields on a Banach space E such that all the* $\operatorname{Lip} X_t$ *are bounded by the same finite constant. Then, the maximal solutions of $\dot{x}(t) = X_t(x(t))$ are defined in the whole of $[0, 1]$. Denoting by $h_s^t(x)$ the value at time t of the solution equal to x at time s, we have the following:*

(i) *the mapping $(s, x, t) \mapsto h_s^t(x)$ is locally Lipschitz-continuous;*

(ii) $h_t^u \circ h_s^t = h_s^u$, $s, t, u \in [0, 1]$, *hence each h_s^t is a diffeomorphism whose inverse is h_t^s;*

(iii) *every h_s^t is globally Lipschitz-continuous, and* $\operatorname{Lip}(h_s^t - \mathrm{Id})$ *tends to* 0 *when* $|t - s| \to 0$, *uniformly with respect to s, t.*

Of course, this implies that (X_t) generates an isotopy (g_t), namely $g_t = h_0^t$ (hence $h_s^t = g_t \circ g_s^{-1}$); when (X_t) is smooth, so is (g_t).

2.4 Generating families of compactly supported contact isotopies

Theorem 4. *For each compactly supported isotopy (g_t) of $J^1\mathbf{R}^n$, there exists a quadratic phase function F on $[0, 1] \times J^1\mathbf{R}^n$ such that the function $\Phi_t : (Q, z, p; v) \mapsto z + pQ + F(t, Q, z, p; v)$ is a (greek) generating family of g_t, $0 \le t \le 1$, which means the following:*

(i) *0 is a regular value of $d_V\Phi_t : (Q, z, p; v) \mapsto \partial_v\Phi(t, Q, z, p; v)$, hence $\Sigma_{\Phi_t} := (d_V\Phi_t)^{-1}(0)$ is a submanifold;*

(ii) *the mapping $i_{\Phi_t} : \Sigma_{\Phi_t} \to (J^1\mathbf{R}^n)^2$ defined by $i_{\Phi_t}(Q, z, p; v) = ((q, y, p)\,(Q, Y, P))$, where*

$$(8)\qquad \begin{cases} q = \dfrac{\partial_p\Phi_t(Q, z, p; v)}{\partial_z\Phi_t(Q, z, p; v)} \\ y = z + pq \\ Y = \Phi_t(Q, z, p; v) \\ P = \partial_Q\Phi_t(Q, z, p; v) \end{cases}$$

is an embedding whose image is the graph of g_t (as before, we call F_t the «latin» generating family of g_t associated to Φ_t).

The proof takes three lemmas.

Lemma 1. *There exists* $\delta > 0$ *such that every* $h_s^t := g_t \circ g_s^{-1}$ *with* $|s - t| < \delta$ *satisfies* 2.1 (II) *and therefore has a latin generating function* f_s^t, *whose support in* $J^1\mathbf{R}^n$ *is compact.*

Proof. Setting $h_s^t(q,y,p) := (Q_s^t(q,y,p), Y_s^t(q,y,p), P_s^t(q,y,p))$, we wish to find δ such that, for $|s-t| < \delta$, the map

$$(q,y,p) \mapsto (Q_s^t(q,y,p), y - pq, p)$$

or, equivalently, the map

$$G_s^t : (q,z,p) \mapsto (Q_s^t(q,z+pq,p),z,p)$$

is a diffeomorphism of $J^1\mathbf{R}^n$ onto itself. By assertion (iii) in the global Cauchy-Lipschitz theorem, $\mathrm{Lip}(Q_s^t - q)$ tends to 0 when $s - t \to 0$, uniformly with respect to s, t. Hence, as $Q_s^t - q$ has compact support, there exists δ such that every G_s^t with $|s-t| < \delta$ satisfies $\mathrm{Lip}\,(G_s^t - \mathrm{Id}) < 1$ and therefore is a diffeomorphism by the global inverse mapping theorem. As the support of $R_s^t - \mathrm{Id}$ is compact, so is the support of f_s^t. □

By Lemma 1, if we pick $0 = a_0 < \ldots < a_{N+1} = 1$ ($N \in \mathbf{N}$) with $a_{j+1} - a_j < \delta$ then, for $0 \le t \le 1$, every $h_{ta_j}^{ta_{j+1}}$ has a compactly supported latin generating function which depends smoothly on t. Therefore, as the zero function on $J^1\mathbf{R}^n \times \{0\}$ is a (*quadratic*) latin generating family of Id, the following crucial lemma can be applied to $g = g_{ta_j}$ and $h = h_{ta_j}^{ta_{j+1}}$ for $0 \le j \le N$ and $0 \le t \le 1$:

Lemma 2. *Let* g, h *be compactly supported contact transformations of* $J^1\mathbf{R}^n$; *if* h *(resp.* g*) has a greek generating function* φ *(resp. a greek generating family* $\Phi : J^1\mathbf{R}^n \times E \to \mathbf{R}$*), then* $\Phi_1 : J^1\mathbf{R}^n \times E_1 \to \mathbf{R}$, *defined by*

$$\text{(9)} \qquad \begin{cases} E_1 := E \times \mathbf{R}^n \times (\mathbf{R}^n)^* \\ \Phi_1(Q,z,p;v,Q_0,p_1) := \varphi(Q, \Phi(Q_0,z,p;v) - p_1Q_0, p_1), \end{cases}$$

is a greek generating family of $h \circ g$.

Proof. Setting

$$(10)\qquad \begin{cases} z_1 = z_1(Q_0,z,p;v,p_1) := \Phi(Q_0,z,p;v) - p_1 Q_0 \\ P_0 = P_0(Q_0,z,p;v) := \partial_{Q_0}\Phi(Q_0,z,p;v) \\ q_1 = q_1(Q,z,p;v,Q_0,p_1) := \dfrac{\partial_{p_1}\varphi(Q,z_1,p_1)}{\partial_{z_1}\varphi(Q,z_1,p_1)} \\ q = q(Q,z,p;v) := \dfrac{\partial_p\Phi}{\partial_z\Phi} = \dfrac{\partial_{z_1}\varphi(Q,z_1,p_1)\,\partial_p\Phi}{\partial_{z_1}\varphi(Q,z_1,p_1)\partial_z\Phi} = \dfrac{\partial_p\Phi_1}{\partial_z\Phi_1}, \end{cases}$$

we have

$$(11)\qquad \begin{cases} \partial_v\Phi_1(Q,z,p;v,Q_0,p_1) = \partial_{z_1}\varphi(Q,z_1,p_1)\,\partial_v\Phi(Q_0,z,p;v) \\ \partial_{Q_0}\Phi_1(Q,z,p;v,Q_0,p_1) = \partial_{z_1}\varphi(Q,z_1,p_1)\,(P_0-p_1) \\ \partial_{p_1}\Phi_1(Q,z,p;v,Q_0,p_1) = \partial_{z_1}\varphi(Q,z_1,p_1)\,(q_1-Q_0)\,; \end{cases}$$

therefore, 0 is a regular value of $d_V\Phi_1$ if and only if it is a regular value of $(Q,z,p;v,Q_0,p_1) \mapsto (\partial_v\Phi(Q,z,p;v), P_0-p_1, q_1-Q_0)$, which is obviously the case because p_1 and q_1 vary freely and independently for fixed $(Q_0,z,p;v)$ — recall that $(Q_1,z_1,p_1) \mapsto (q_1,z_1,p_1)$ is a diffeomorphism.

Moreover, (11) tells us that $d_V\Phi_1 = 0$ if and only if $d_V\Phi = 0$, $p_1 = P_0$ and $q_1 = Q_0$ or, equivalently,

$$(q_1\,,\,z_1+p_1q_1\,,\,p_1) = g(q,z+pq,p)\,;$$

as φ is the generating function of h, this means that

$$(Q\,,\,\varphi(Q,z_1,p_1)\,,\,\partial_Q\varphi(Q,z_1,p_1)) = h(g(q,z+pq,p))\,,$$

which sets $(Q,\Phi_1,\partial_Q\Phi_1) = h\circ g(q,z+pq,p)$; hence, by the last equality (10), the image of i_{Φ_1} is the graph of $h\circ g$.

Finally, i_{Φ_1} is an embedding because it is obtained by composing i_Φ on the left with $(q,y,p,Q_0,Y_0,P_0) \mapsto (q,y,p,h(Q_0,Y_0,P_0))$ and, on the right, with $\Sigma_{\Phi_1} \ni (Q,z,p;v,Q_0,p_1) \mapsto (Q_0,z,p;v) \in \Sigma_\Phi$, whose inverse is given by $p_1 = \partial_{Q_0}\Phi$ and $Q =$ first component of $h(Q_0,\Phi,\partial_{Q_0}\Phi)$. □

As mentioned above, Lemma 2 provides inductively a latin generating family for each g_t, which depends smoothly on t. Hence Theorem 4 will be proved if we make this generating family quadratic.

Lemma 3. *Under the hypotheses of Lemma 2, denote by f, F and F_1 the latin versions of φ, Φ and Φ_1 respectively, and by K the non-degenerate quadratic form on E such that $F(Q_0,p,z;v) = K(v)$ off a compact subset (thus, F is assumed quadratic). Then, setting*

$$\begin{cases} G(Q,z,p;v,X,Y) = F_1(Q,z,p;v,Q+X,p+Y) \\ K_1(v,X,Y) = -YX + K(v)\,, \end{cases}$$

$(v,X,Y) \in E_1$, there exists a compactly supported $\theta : E_1 \to [0,1]$, equal to 1 near 0, such that $F_2 : (Q,z,p;w) \mapsto K_1(w) + \theta(w)\,(G(Q,z,p;w) - K_1(w))$ is a quadratic generating family of $h\circ f$.

Proof. Setting $(Q_0, p_1) := (Q+X, p+Y)$, (9) yields

$$(12)\qquad G = -YX + F(Q+X, z, p; v) + f(Q,\, \Phi(Q_0, z, p; v) - p_1 Q_0,\, p_1)\,;$$

hence $G(Q,z,p;w) - K_1(w)$ is uniformly bounded. Similarly, (10) can be written

$$\frac{d_V G}{\partial_{z_1}\varphi(Q, z_1, p_1)} = d_V F(Q+X, z, p; v)dv + (P_0 - p - Y)dX + (q_1 - Q - X)dY\,;$$

as g and h have compact support, $P_0 - p$ and $q_1 - Q$ are uniformly bounded, hence so is $d_V G - \partial_{z_1}\varphi(Q, z_1, p_1)\, dK_1$ because $\partial_{z_1}\varphi(Q, z_1, p_1)$ is bounded.

Now, $d_V F_2$ is the sum of

$$A := \big(1 - \theta(w) + \partial_{z_1}\varphi(Q, z_1, p_1)\theta(w)\big)\, dK_1(w)$$

and

$$B := \theta(w)\,\big(d_V G - \partial_{z_1}\varphi(Q, z_1, p_1)\, dK_1(w)\big) + \big(G(x; w) - K_1(w)\big)\, d\theta(w)\,;$$

choosing θ's with a given bound on $d\theta$, we get a uniform bound for the norm of B, valid for any such θ. As the minimum μ of $\partial_{z_1}\varphi$ lies in $(0, 1]$, we have

$$\begin{aligned} 1 - \theta + \partial_{z_1}\varphi\,\theta &= 1 + (\partial_{z_1}\varphi - 1)\theta \\ &\geq 1 + (\mu - 1)\theta \geq 1 + (\mu - 1)\max(\theta) \\ &\geq \mu > 0\,; \end{aligned}$$

thus, if we choose θ equal to 1 in a large enough subset, A will have a bigger norm than B for $\theta < 1$, hence, with the notation of Theorem 4, $i_{\Phi_2} = i_\Gamma$ (denoting by Φ_2 and Γ the greek versions of F_2 and G). As G clearly is a generating family of $h \circ f$ and F_2 is quadratic, this proves Lemma 3. □

If f and F depend smoothly on $t \in [0, 1]$, so does F_2 if we take the same θ for every t, hence Theorem 4 follows. □

Remarks. In the symplectic case, Lemma 3 is easier because $\partial_{z_1}\varphi(Q, z_1, p_1) \equiv 1$; moreover, the formula for the generating family of g_t, obtained by iterating (9), can be written down quite easily ([9], [10]) because the generating functions f_s^t do not depend on z. Sikorav's idea of isolating the induction step [21] is really useful in the contact case!

2.5 Proof of Theorem 3

The hypotheses and notation are those of Theorem 3, and *we embed M as a submanifold of* $\mathbf{R}^n$; let $J^1\mathbf{R}^n|_M$ be the set $M \times \mathbf{R} \times (\mathbf{R}^n)^*$ of all $j_q^1\varphi$ with $\varphi : \mathbf{R}^n \to \mathbf{R}$ and $q \in M$, and let $\rho_M : J^1\mathbf{R}^n|_M \to J^1 M$ be the projection $j_q^1\varphi \mapsto j_q^1(\varphi|_M)$, alias $(q, y, p) \mapsto (q, y, p|_{T_q M})$.

Lemma 4. *The generating family S of i can be extended to a quadratic phase function $\tilde{S}(Q;v)$ on $\mathbf{R}^n$, generating an immersion $i_{\tilde{S}}$ transversal to $J^1\mathbf{R}^n|_M$. The restriction of $i_{\tilde{S}}$ to $i_{\tilde{S}}^{-1}(J^1\mathbf{R}^n|_M) = \{(Q;v) \in \Sigma_{\tilde{S}} : Q \in M\}$ satisfies $\rho_M \circ (i_{\tilde{S}}|_{\{Q\in M\}}) = i_S$.*

Proof. First extend $S : M \times E \to \mathbf{R}$ to a quadratic $T : J^1\mathbf{R}^n \times E \to \mathbf{R}$. If $d_V T$ is transversal to 0, we can take $\tilde{S} := T$, which will automatically satisfy all our requirements. If $d_V T$ is not transversal to 0, the default of transversality will occur neither near $M \times E$ nor at infinity; hence we can perturb T a little so that the resulting $\tilde{S}$ still is quadratic and extends S but $d_V\tilde{S}$ is transversal to 0. □

Lemma 5. *There exists a compactly supported isotopy $(\tilde{g}_t)$ of $\mathbf{R}^n$ such that, for $0 \le t \le 1$, the restriction of $\tilde{g}_t \circ i_{\tilde{S}}$ to $i_{\tilde{S}}^{-1}(J^1\mathbf{R}^n|_M)$ takes its values in $J^1\mathbf{R}^n|_M$ and satisfies $\rho_M \circ (\tilde{g}_t \circ i_{\tilde{S}}|_{\{Q\in M\}}) = g_t \circ i_S$.*

Proof. Let (H_t) be the hamiltonian of (g_t) and let $\tilde{H}_t$ denote the hamiltonian of a contact isotopy $(\tilde{g}_t)$ of $J^1\mathbf{R}^n$; using local charts of $\mathbf{R}^n$ in which M is linear, it is easy to check that the following two properties are equivalent:

(i) for every $(t,x) \in [0,1] \times J^1\mathbf{R}^n|_M$, we have $\tilde{g}_t(x) \in J^1\mathbf{R}^n|_M$ and $\rho_M(\tilde{g}_t(x)) = g_t(\rho_M(x))$;

(ii) for every $t \in [0,1]$, the restriction of $\tilde{H}_t$ to $J^1\mathbf{R}^n|_M$ equals $H_t \circ \rho_M$.

Of course, by (ii), the support of such an isotopy $(\tilde{g}_t)$ cannot be compact unless $g_t \equiv \mathrm{Id}$. However, we can construct a *compactly supported* $(\tilde{H}_t)$ which satisfies (ii) not in the whole of $J^1\mathbf{R}^n|_M$, but in a very large subset, so as to get (i) for every (t,x) of the form $(t, i_{\tilde{S}}(Q;v))$. □

Proof of Theorem 3. Given $\tilde{S}$ and $(\tilde{g}_t)$ as in our two lemmas, let $\Phi_t(Q,z,p;v)$ be the greek version of the quadratic generating family of $\tilde{g}_t$ provided by Theorem 4. Then, the function $S_t(Q,v_1)$ defined by

$$\begin{cases} v_1 = (v_0, v, Q_0, p_1) \\ S_t(Q;v_1) = \Phi_t(Q,\, \tilde{S}(Q_0;v_0) - p_1Q_0\,,\, p_1\,;\, v)\,, \quad Q \in M \end{cases}$$

is a generating family of $g_t \circ i$, which can be made quadratic by the same device as in the proof of Lemma 3. □

Note. This proof — and, as far as I know, the theorem itself — is due to Théret... but the very clever use of Whitney's embedding theorem should be attributed to Chekanov (Sikorav told me about it around 1986) at least in the symplectic case [5] — Brunella uses Theorem 2 without proof in [6]. I do not know if the above would benefit from Benenti's general ideas [4] on symplectic reduction.

Added in proof. Yu. V. Chekanov has just written down his proof, which is slightly different from the above and should appear soon (in Russian) in *Functional analysis and its application.*

References

[1] V.I. Arnold, *Sur une propriété topologique des applications globalement canoniques de la mécanique classique*, C. R. Acad. Sci. Paris **261** (1965), 3719–3722.

[2] V.I. Arnold, *Mathematical methods of classical mechanics*, Springer-Verlag, 1980.

[3] V.I. Arnold, *First steps of symplectic topology*, Russian Math. Surveys, **41:6** (1986), 1–21.

[4] S. Benenti, *Symplectic Relations in Analytical Mechanics*, Atti Acad. Sci. Torino Cl. Sc. Fis. Mat. Natur. **117** (1983), Suppl. 1, 39–91.

[5] M. Brunella, *On a theorem of Sikorav*, L'Enseignement Mathématique **37** (1991), 83–87.

[6] M. Brunella, *Sur les familles génératrices des sous-variétés legendriennes*, C. R. Acad. Sc. Paris **314** (1992), 1049–1052.

[7] M. Chaperon, *Quelques questions de géométrie symplectique [d'après, entre autres, Poincaré, Arnold, Conley et Zehnder]*, Séminaire Bourbaki, Astérisque **105–106** (1983), 231–249.

[8] M. Chaperon, *Une idée du type «géodésiques brisées» pour les systèmes hamiltoniens*, C. R. Acad. Sc. Paris **298** (1984), 293–296.

[9] M. Chaperon, *An elementary proof of the Conley-Zehnder theorem in symplectic geometry*, in Braaksma, Broer et Takens, *Dynamical systems and bifurcations*, Springer Lecture Notes in Mathematics **1125** (1985), 1–8.

[10] M. Chaperon, *An old-fashioned method in the calculus of variations*, in P. H. Rabinowitz, A. Ambrosetti, I. Ekeland et E.J. Zehnder, *Periodic solutions of hamiltonian systems and related topics*, Reidel Publishing Company (1987), 93–98.

[11] M. Chaperon, *Lois de conservation et géométrie symplectique*, C. R. Acad. Sc. Paris **312** (1991), 345–348.

[12] M. Chaperon, *Some remarks on generating families*, to appear in the Proceedings of the 1993 Moscow Geometry Symposium.

[13] C.C. Conley, E. Zehnder, *The Birkhoff-Lewis fixed point theorem and a conjecture of V. I. Arnol'd*, Invent. Math. **73** (1983), 33–49.

[14] T. Debeneix, *Condition d'entropie pour certains systèmes hyperboliques quasilinéaires de n équations à n variables d'espace*, C. R. Acad. Sc. Paris **287** (1978), 763–766.

[15] A. Floer, *Morse theory for lagrangian intersections*, J. Diff. Geom. **28** (1988), 513–547.

The unregularized gradient flow of the symplectic action, Comm. in Pure and Appl. Math. **41** (1988), 775–813.

A relative Morse index for the symplectic action, Comm. in Pure and Appl. Math. **41** (1988), 393–407.

Witten's complex and infinite dimensional Morse theory, J. Diff. Geom. **30** (1989), 207–221.

Cuplength estimates for Lagrangian intersections, Comm. in Pure and Appl. Math. **42** (1989), 335–356.

[16] M. Gromov, *Pseudo-holomorphic curves in symplectic manifolds*, Invent. Math. **82** (1985), 307–347.

[17] H. Hofer, *Lagrangian embeddings and critical point theory*, Ann. Inst. H. Poincaré, Analyse non linéaire **2** n° 6 (1985), 407–462.

[18] L. Hörmander, *Fourier Integral Operators I*, Acta Math. **127** (1971), 79–183.

[19] T. Joukovskaïa, *Singularités de minimax et solutions faibles d'équations aux dérivées partielles*, Thèse, Université Denis Diderot (Paris 7), January 1994.

[20] P.D. Lax, *Hyperbolic systems of conservation laws II*, Comm. in Pure and Appl. Math. **10** (1957), 537–566

[21] J.C. Sikorav, *Problèmes d'intersections et de points fixes en géométrie hamiltonienne*, Comm. Math. Helvet. **62** N° 1 (1987), 62–73.

[22] D. Théret, *Thèse*, Université Denis Diderot (Paris 7), to appear.

[23] C. Viterbo, *Symplectic topology as the geometry of generating functions*, Math. Annalen **292** (1992), 685–710.

Marc Chaperon
Université Paris VII
UFR de mathématiques, Case 7012
2, place Jussieu
75251 Paris Cedex 05

Floer's infinite dimensional Morse theory and homotopy theory

R.L. Cohen, J.D.S. Jones, and G.B. Segal

1 Introduction

This paper is a progress report on our efforts to understand the homotopy theory underlying Floer homology. Its objectives are as follows:

(A) to describe some of our ideas concerning what exactly the Floer homology groups compute;

(B) to explain what kind of an object we think the «Floer homotopy type» of an infinite dimensional manifold should be;

(C) to work out, in detail, the Floer homotopy type in some examples.

We have not solved the problems posed by the underlying questions, but we do have a programme which we hope will lead to solutions. Thus it seems worthwhile to describe our ideas now, especially in a volume of papers dedicated to the memory of Andreas Floer. We plan to write a complete account of this approach to Floer homotopy theory in a future paper.

Floer homology arises in two different contexts, the study of curves and surfaces in symplectic manifolds, and gauge theory on three- and four-dimensional manifolds. In each of these contexts there are two different perspectives, which one can think of as «Hamiltonian» and «Lagrangian».

The theory began with Floer's proof of the Arnold conjecture. On a compact symplectic manifold M a Hamiltonian flow is generated by a Hamiltonian function $h : M \to \mathbb{R}$, and the stationary points of the flow are the critical points of h. Classical Morse theory tells us that there are at least as many such points as the dimension of the homology $H_*(M; \mathbb{R})$. Arnold conjectured that the same is true of the number of fixed points of a diffeomorphism $\varphi_1 : M \to M$ which arises from a *time dependent* Hamiltonian flow $\{\varphi_t\}_{0 \leq t \leq 1}$. The trajectories of such a flow are critical points of the «action functional» S_h on the space of paths $\gamma : [0,1] \to M$, where

$$S_h(\gamma) = \int_\gamma (pdq - hdt),$$

and $h : M \times \mathbb{R} \to \mathbb{R}$ is the varying Hamiltonian. The fixed points of φ_1 are therefore the critical points of S_h on the space $\mathcal{L}M$ of loops of length 1 in M. Thus Arnold's conjecture would follow from a version of Morse theory applicable to the function $S_h : \mathcal{L}M \to \mathbb{R}$, but relating its critical points to the homology of M rather than $\mathcal{L}M$. This was the theory that Floer developed.

In conventional Morse theory it is geometrically clear why the homotopy type of a manifold X is reflected in the disposition of the critical points of a function $f : X \to \mathbb{R}$; for example, if there are no critical points apart from the minimum, the gradient flow of f provides a contraction of X. At present there is no comparable homotopy-theoretic underpinning of Floer theory. Ordinary homology can be defined in a great variety of quite different ways, but one does not know how to define Floer's groups without using a Floer-Morse function. One of our purposes in this paper is to speculate about what exactly the Floer groups are describing, or what additional structure an infinite dimensional manifold such as $\mathcal{L}M$ must have for the groups to be defined. We should say at the outset, however, that we have not solved this problem.

One ingredient in the answer is clear. An important feature of a Floer-Morse function, such as S_h, is that at critical points its Hessian has infinitely many negative as well as positive eigenvalues. In fact at every point γ of $\mathcal{L}M$, the Hessian of S_h decomposes the tangent space $T_\gamma(\mathcal{L}M)$ into two parts $T_\gamma^\pm$ corresponding to the positive and negative eigenvalues, with a finite dimensional ambiguity coming from the zero eigenspace. Such an approximate splitting of the tangent bundle of an infinite dimensional manifold X, we shall call a *polarization* of X. A formal definition will be given in §2. The significance of the polarization becomes clearer if we turn to the second — or Lagrangian — perspective on Floer theory.

If we choose a Riemannian metric on M making it an almost complex manifold then we can consider *pseudo-holomorphic* maps $\varphi : \Sigma \to M$, where Σ is a Riemann surface. In Arnold's problem the gradient flow lines of the function $S_h : \mathcal{L}M \to \mathbb{R}$ when $h = 0$ are precisely the pseudo-holomorphic maps $S^1 \times \mathbb{R} \to M$. If a closed surface $\Sigma = \Sigma_1 \cup \Sigma_2$ is the union of two pieces intersecting in a common boundary circle then, because a pseudo-holomorphic map $\Sigma_i \to M$ is determined by its boundary values, the finite dimensional space $Z_\Sigma = \mathrm{Hol}(\Sigma; M)$ can be regarded as the intersection of two infinite dimensional submanifolds $Z_{\Sigma_i} = \mathrm{Hol}(\Sigma_i; M)$ of the loop space $\mathcal{L}M$. Here the notation Hol means pseudo-holomorphic maps. The tangent spaces to Z_{Σ_1} and Z_{Σ_2} are close — in a sense explained in §2 — to the positive and negative parts $T_\gamma^\pm$ of the polarization of $T_\gamma(\mathcal{L}M)$. Furthermore Z_{Σ_1} and Z_{Σ_2} define a cycle and a cocycle respectively in the Floer theory of $\mathcal{L}M$, and the pairing between them is, in good cases, the number of isolated pseudo-holomorphic maps $\Sigma \to M$.

This suggests (compare [4]) that Floer theory is the homology theory of *semi-infinite dimensional cycles* in a polarized manifold. There is a natural concept of semi-infinite dimensional differential forms on such a manifold, and one might hope to use them to give a de Rham definition of Floer homology; see [11] for interesting work along these lines. To define the Floer groups for an infinite dimensional manifold, it seems clear that more structure is needed than just the polarization of X. A crucial point seems to be that the critical manifold of a Floer-Morse function is *compact*, and it seems conceivable that some preferred class of compact subspaces of X should be an ingredient in the structure.

Though we cannot answer the question «what does the Floer homotopy type of a manifold depend on», we can do better with another question, «what sort of object is the Floer homotopy type of a manifold». Unfortunately one cannot hope that the Floer groups of X are the ordinary homology groups of a space associated to X, or even of a «stable space» or a «spectrum». We shall show that under reasonable hypotheses one can associate to the flow category, see [9], of a Floer-Morse function an object called a *pro-spectrum*. This is a technical homotopy-theoretic concept which has proved to be central in one of the deepest recent results of homotopy theory. As the pro-spectra involved arise in Floer theory on the projective spaces of polarized vector spaces and also the loop space $\mathcal{L}\mathbb{CP}^n$, it seems worthwhile to explain this result briefly.

For any positive integers $n > m$, let P_m^n be the space obtained from real projective space $\mathbb{RP}^n$ by collapsing the standard linear subspace $\mathbb{RP}^{m-1}$ to a point. Now fix two large positive integers p and q and consider the homotopy groups $\pi_{i+N}(P_{N-q}^{N+p})$ as $N \to \infty$. Notice that P_{N-q}^{N+p} is a space made of cells whose dimensions range from $N + p$ to $N - q$, so that for small $|i|$ we are looking at a kind of «middle dimensional» homotopy group. The following deep theorem of Lin [20], conjectured by Mahowald [21] and Adams [2], was a crucial step in determining the stable homotopy type of the classifying spaces of finite groups and proving the Segal conjecture; see [3] and [8] for surveys and further references.

Theorem *If N is a multiple of 2^{p+q}, then there is a map $S^{N-1} \to P_{N-q}^{N+p}$ which induces isomorphisms*

$$\pi_{i+N}(P_{N-q}^{N+p}) \cong \begin{cases} \mathbb{Z}/2^{a(N)} & \text{if } i = -1 \\ \pi_{i+N}(S^{N-1})_{(2)} & \text{if } i \neq -1 \text{ but } -p \ll i \ll q \end{cases}$$

where $\pi_{i+N}(S^{N-1})_{(2)}$ is the 2-primary component of the homotopy group $\pi_{i+N}(S^{N-1})$, and $a(N) \to \infty$ as $N \to \infty$.

In particular note that, when i is small compared to N, $\pi_{i+N}(P_{N-q}^{N+p})$ is independent of N, except that when $i = -1$ it tends to the 2-adic completion of $\pi_{N-1}(S^{N-1}) = \mathbb{Z}$.

There are natural inclusions $P_{N-q}^{N+p} \to P_{N-q}^{N+p+1}$ and collapsing maps $P_{N-q}^{N+p} \to P_{N-q+1}^{N+p}$. In addition, if N and M are both multiples of 2^{p+q} with $N \geq M$, there is a homotopy equivalence between P_{N-q}^{N+p} and the $(N-M)$-fold suspension $S^{N-M}P_{M-p}^{M+q}$. It follows that the system of spaces P_{N-q}^{N+p} form a pro-spectrum, and this is the prime example of a pro-spectrum.

Let us now outline the contents of this paper. In §2 we describe some of the homotopy-theoretic properties of polarized manifolds. In §3 we describe the flow categories of Morse functions. In §4 we analyse the flow category of $\mathcal{L}\mathbb{CP}^n$, and explain how to compactify this category. In §5 we describe a method of recovering the stable homotopy type of a finite dimensional manifold from the flow category of a Morse-Bott function. This method is not the same as that used in [9]; in spirit, it is related to the work of Franks [14]. We go on to show how a pro-spectrum can be associated to the idealized flow category of a Floer function. In §6 we describe how the ideas of §5 can be applied to the projective space of a polarized vector space, and to the area function on $\mathcal{L}\mathbb{CP}^n$, and we identify the corresponding pro-spectra. The most surprising point is that the pro-spectrum associated to the compactified flow category of the area function on $\mathcal{L}\mathbb{CP}^n$ is the complex analogue of the one occurring in Lin's theorem.

In an Appendix we give a very brief account, for non-experts, of some of the ideas which lead to the introduction of the stable category of spaces, the category of spectra, and the notion of a pro-spectrum.

2 Polarized manifolds

A polarization of a real topological vector space E is a *class* of decompositons of $E = E_+ \oplus E_-$ which do not differ too much among themselves. The main example arises when one has an unbounded self adjoint Fredholm operator $D : E \to E$. This splits E according to the positive and negative parts of the spectrum of D: we want to allow the ambiguity of assigning the 0-eigenspace arbitrarily to E_+ or E_-. The most convenient definition is as follows.

Definition 2.1 *A polarisation of E is a class $\mathcal{J}$ of linear operators $J : E \to E$, all congruent modulo the ideal of compact operators, and such that $J^2 = 1$ modulo compact operators. Further, $\mathcal{J}$ must not contain $+1$ or -1.*

If E is polarized we can define the restricted general linear group $GL_{\mathrm{res}}(E)$ which consists of all $g \in GL(E)$ which preserve the polarization. We can also define the restricted Grassmannian $Gr_{\mathrm{res}}(E)$, consisting of all the (-1)-eigenspaces of all $J \in \mathcal{J}$ such that $J^2 = 1$.

If E is a Banach space then $GL_{\mathrm{res}}(E)$ can be regarded as a closed subgroup of $GL(E)$ with the norm topology. But, in general, it is better to give it the topology for which $\{g_\alpha\}$ converges if both $\{g_\alpha\}$ and $\{g_\alpha^{-1}\}$ converge in the compact-open topology, and $[J, g_\alpha]$ converges in the uniform topology for some (and hence all) $J \in \mathcal{J}$.

A polarized manifold X is one whose tangent spaces $T_x X$ are polarized. More precisely, if X is modelled on E, the structural group of the tangent bundle of X is reduced to $GL_{\mathrm{res}}(E)$. In all of the examples we know the polarizations are *integrable*, that is X has an atlas $\{\varphi_\alpha : \mathcal{U}_\alpha \to X\}$ such that $D(\varphi_\beta \varphi_\alpha^{-1})(y) \in GL_{\mathrm{res}}(E)$ for all $y \in \varphi_\alpha(\mathcal{U}_\alpha)$; but we shall not need this. The two basic examples which arise in Floer theory are the following.

(i) An almost complex structure on a Riemannian manifold M defines a polarization of the loop space $\mathcal{L}M$. The tangent space T_γ at $\gamma \in \mathcal{L}M$ is the space of tangent vector fields to M along γ, and we have the self-adjoint operator

$jD/D\theta : T_\gamma \to T_\gamma$, where j is the almost-complex structure of M and $D/D\theta$ is covariant differentiation. The spectral decomposition of $jD/D\theta$ polarizes $\mathcal{L}M$.

(ii) The space $X = \mathcal{A}^*/\mathcal{G}$, where $\mathcal{A}^*$ is the space of irreducible connections on a complex vector bundle E with compact structural group G on a 3-manifold M and $\mathcal{G} = \mathrm{Aut}(E)$ is the gauge group of E, also carries a natural polarization. The tangent space to $\mathcal{A}^*$ at any point is $\Omega^1(M; \mathrm{End}(E))$, and that of $\mathcal{A}/\mathcal{G}$ at a connection A is the cokernel of

$$d_A : \Omega^0(M; \mathrm{End}(E)) \to \Omega^1(M; \mathrm{End}(E)).$$

If A is a flat connection then $d_A^2 = 0$, and the operator $*d_A$ induces a self-adjoint Fredholm operator, and hence a polarization, on the tangent space $T_A = \Omega^1/d_A\Omega^0$. With more work, one can define the polarization at all points of X.

In both these cases the polarization is the same as the polarization induced by a Floer-Morse function. In the first case, it is the area functional — the action functional S_h, described in the introduction, with $h = 0$. In the second case, it is the Chern-Simons functional.

For the usual topological vector spaces of analysis the group $GL(E)$ is contractible, and so the tangent bundle of a manifold X modelled on E carries no homotopy-theoretic information. The position is different when E is polarized. For the usual choices of E, the group $GL_{\mathrm{res}}(E)$ has the homotopy type of $\mathbb{Z} \times BO$, the classifying space for stable finite dimensional vector bundles; see [22]. The tangent bundle of a polarized manifold X is therefore described by a map $X \to BGL_{\mathrm{res}}(E)$, determined up to homotopy, which we call the *structural map* of X. By Bott periodicity the homotopy type of $BGL_{\mathrm{res}}(E)$ is U/O, where $U = \bigcup U(n)$ and $O = \bigcup O(n)$ are the infinite unitary group and orthogonal group respectively. The space U/O represents the functor KO^1. Its fundamental group is $\mathbb{Z}$ and its rational cohomology is an exterior algebra on generators of dimensions $4k+1$. Therefore, a polarized manifold X has characteristic classes in $H^{4k+1}(X; \mathbb{Q})$.

The most obvious information provided by the structural map $X \to U/O$ concerns the *grading* of Floer homology. At each point $x \in X$ we have the Grassmannian $Gr_x = Gr_{\mathrm{res}}(T_xX)$, whose connected components correspond to the integers $\mathbb{Z}$, though with no preferred choice of zero: two points of Gr_x have a well-defined *relative dimension*, but no absolute dimension [22]. The sets $\tilde{X}_x = \pi_0(Gr_x)$, as x varies, form a covering space $\tilde{X}$ of X, and when one goes around a path λ in X starting at x, the holonomy $\tilde{X}_x \to \tilde{X}_x$ shifts $\tilde{X}_x$ by the image of λ in $\pi_1(U/O) = \mathbb{Z}$. This means that for a particular polarized manifold X the «dimension» of a semi-infinite subspace of T_xX, and hence of a semi-infinite cycle, or Floer homology class, can be taken to be well-defined modulo the image of $\pi_1(X)$ in $\pi_1(U/O) = \mathbb{Z}$. On the covering space $\tilde{X}$ the dimension — or «virtual dimension» — is a well-defined element of $\mathbb{Z}$.

If M is an almost-complex manifold of real dimension $2m$, then the tangent bundle of M is classified by a map $\theta : M \to BU(m)$. The structural map of $X = \mathcal{L}M$ is easily seen to be the composite

$$\mathcal{L}M \xrightarrow{\mathcal{L}\theta} \mathcal{L}BU(m) \xrightarrow{\beta} U \longrightarrow U/O,$$

where β is the transgression. It is important that the structural map factorizes through U, i.e. the structural group of $T\mathcal{L}M$ is reduced to the *complex* restricted general linear group. This means, in particular, that the grading of Floer homology is always well defined modulo 2, for the map $\pi_1(U) \to \pi_1(U/O)$ is multiplication by 2.

The same is true in the gauge theory case, when X has the homotopy type of $\mathrm{Map}(M; BG)$. Then the structural map is the composite

$$\mathrm{Map}(M; BG) \to \mathrm{Map}(M; BU(k)) \to U \to U/O,$$

where the first map is induced by the representation $G \to U(k)$ which defines the bundle E, and the second map is the direct-image map in complex K-theory. (This map represents the element of $K^{-3}(\mathrm{Map}(M; BU(k))$ obtained by pulling back the tautological element of $K(BU(k))$ to $K(M \times \mathrm{Map}(M; BU(k)))$ by the evaluation map, and then «integrating» over the 3-dimensional manifold M, i.e. evaluating on the K-theory fundamental class in $K_3(M)$.)

From these descriptions of the structural map it is easy to compute its effect on π_1. If M is a simply connected almost complex manifold, we have $\pi_1(\mathcal{L}M) = \pi_2(M)$. The homomorphism

$$\pi_2(M) = \pi_1(\mathcal{L}M) \to \pi_1(U/O) = \mathbb{Z}$$

is the homomorphism defined by $2c_1(M)$, and the grading of Floer homology is well defined modulo its image. In the gauge theory case with structure group $G = SU(2)$, we have $\pi_1(\mathcal{A}^*/\mathcal{G}) = \mathbb{Z}$. The corresponding homomorphism is multiplication by 8, and the grading of Floer homology is well-defined modulo 8.

3 The flow category

Let us begin by considering a Morse-Bott function $f : X \to \mathbb{R}$ on a finite dimensional compact Riemannian manifold X. Morse-Bott means that the critical set F of f is a smooth manifold, and that the Hessian $D^2 f$ is non-degenerate on the fibres of the normal bundle to F in X.

In this situation we can define a category $\mathcal{C}_f$ whose objects are the critical points of f and whose morphisms from x to y are the piecewise gradient trajectories (or flow lines) γ of f from x to y. This means one permits γ to stop at intermediate critical points en route. More precisely, γ is a sequence $\gamma = (\gamma_0, \gamma_1, \ldots, \gamma_p)$ where $\gamma_i : \mathbb{R} \to X$ is a descending trajectory of grad(f) such that

$$\gamma_i(t) \to x_i^{\pm}, \qquad \text{as } t \to \pm\infty$$
$$x = x_0^{-}, \qquad x_i^{+} = x_{i+1}^{-}, \qquad x_p^{+} = y.$$

We identify two such sequences if they differ only by translating the parameters of the γ_i. The category $\mathscr{C}_f$ is a *topological* category [23], in that the sets $\mathrm{Ob}(\mathscr{C}_f)$ and $\mathrm{Mor}(\mathscr{C}_f)$, of objects and morphisms, have natural topologies, and the structure maps of $\mathscr{C}_f$ are continuous.

In the case where the gradient flow of the function f also satisfes an appropriate version of the Smale transversality condition (see [26] and [6]) these spaces have a great deal of extra structure which we now describe in detail. Let the critical values of f be $t_n > t_{n-1} \ldots > t_0$, and let the critical manifold with critical value t_i be F_i; then

$$\mathrm{Ob}(\mathscr{C}_f) = \coprod F_i.$$

If $j > i$, let the space of morphisms from points on F_j to points on F_i be $F(j,i)$. It is known (see [26] and [6]) that $F(j,i)$ is a compact manifold with corners. By a manifold with corners we mean a manifold modelled on the space $\mathbb{R}_+^d$, where $\mathbb{R}_+$ is the set of real numbers x with $x \geq 0$, and by the boundary of a manifold with corners we mean the set of points which in coordinate charts do not lie in the interior of $\mathbb{R}_+^d$.

If $I = (i_{k+1}, i_k, \ldots, i_0)$ is a sequence with

$$j = i_{k+1} > i_k > \ldots > i_0 = i,$$

let $F(I)$ be the part of $F(j,i)$ consisting of piecewise trajectories which stop at all of the F_{i_r} for $1 \leq r \leq k$. Then $F(j,i)$ is stratified by the $F(I)$, and $F(I)$ is a compact submanifold of codimension k in $F(j,i)$. Furthermore, $\partial F(I)$ is the union of the $F(J)$ with $J \supset I$. In a neighbourhood of a point of $F(I)$, the space $F(j,i)$ is modelled on $\mathbb{R}_+^k \times \mathbb{R}^{m-k}$ where $m = \dim F(j,i)$. Composition in the category maps $F(j,r) \times F(r,i)$ diffeomorphically to $F(j,r,i) \subset \partial F(j,i)$. Finally, the beginning and end point maps

$$F_j \xleftarrow{\pi_j} F(j,i) \xrightarrow{\pi_i} F_i$$

are transverse, i.e. they are either fibrations, or embeddings of submanifolds. In the following account we shall assume for simplicity that these maps are always fibrations, though it is not much harder to treat the general case.

We shall call a category of the type just described a *compact smooth* category. In the finite dimensional case it has one further basic property: it is *framed*, in the following sense.

Let E_i be the downward part of the tangent bundle to X along F_i. That is, E_i is the sub-vector-bundle of the normal bundle of F_i spanned by the eigenvectors of the Hessian $D^2 f$ corresponding to negative eigenvalues. The geometry of the flows gives us canonical isomorphisms of vector bundles on $F(j,i)$

$$\pi_j^* E_j \cong T_{ji} \oplus \mathbb{R} \oplus \pi_i^* E_i, \tag{3.1}$$

where T_{ji} is the tangent bundle along the fibres of the projection $\pi_j : F(j,i) \to F_j$. These isomorphisms are compatible with the compositions in the category $\mathscr{C}_f$. In the case when the function f has isolated critical points, the isomorphism (3.1) is a stable framing of the flow manifold in the usual sense.

To see that (3.1) holds, observe that $F(j,i)$ embeds in the sphere bundle $S(E_j)$ of E_j. (Actually the natural map $F(j,i) \to S(E_j)$ is not injective on $\partial F(j,i)$, but that is irrelevant to the present argument, and in any case the map can be made injective by a canonical small deformation.) The normal bundle to $F(j,i)$ in $S(E_j)$ can be identified canonically with $\pi_i^* E_i$, as its fibre at $\gamma \in F(j,i)$ consists of the piecewise trajectories emanating from $\pi_j(\gamma)$ which just miss F_i.

In the infinite dimensional situation which Floer considered, the function f always has a compact critical manifold F, and there is a flow category $\mathscr{C}_f$ in which each connected component of the space of morphisms is finite dimensional. Three new features, however, need to be considered.

(i) The function f is not usually single-valued. Usually it takes values in $\mathbb{R}/\mathbb{Z}$, but in principle it might be the indefinite integral of any closed 1-form representing a class α in $H^1(X;\mathbb{R})$. Floer theory seems to work well only in the *monotone* case where α is a multiple of the basic element of $H^1(X;\mathbb{Z})$ defined by the structural map $X \to U/O$ of the polarized manifold X. We shall confine ourselves to this case. Then f can be lifted to a map $\tilde{f} : \tilde{X} \to \mathbb{R}$, where $\tilde{X}$ is the infinite cyclic cover of X defined by $X \to U/O$. The critical set of $\tilde{f}$ is then an infinite disjoint union $\coprod_{i\in\mathbb{Z}} F_i$, where each F_i is compact, and is periodic in i with some finite period. Henceforth, when we speak of the flow category of a Floer function we shall mean the flow category of $\tilde{f}$.

(ii) The flow category is no longer framed. We still have the isomorphisms (3.1), but now the bundles E_i are infinite dimensional, and so give no information about the tangent bundle T_{ji}, except to give it a complex structure when the structural map $X \to U/O$ of the polarized manifold X lifts to U. This feature was pointed out long ago by Floer himself. We shall see that whether the flow category is framed is essentially the same question as whether the structural map $\tilde{X} \to U/O$ is homotopic to a constant map.

(iii) Because of the phenomenon of «bubbling», the flow spaces $F(j,i)$ are no longer compact. This is the most important difference from the finite dimensional case, and the hardest to handle. In the cases that we have studied in detail, there is a natural way to compactify the $F(j,i)$ so that one has a compact smooth category, but the precise relation between the categories before and after compactification is still not well-understood.

A topological category $\mathscr{C}$ has a realization $|\mathscr{C}|$ as a topological space; see [23]. In the case of a Morse-Bott-Smale function, i.e. a Morse-Bott function whose gradient flow satisfies an appropriate transversality condition, $|\mathscr{C}_f|$ is homeomorphic to X. These results are proved in [9].

It is striking that for the flow categories of the usual Floer functions, it still seems to be true that $|\mathscr{C}_f|$ is homotopy equivalent to X, if one does *not* compactify the category. Thus if $X = \mathscr{L}M$ is the loop space of a Kähler manifold M with $\pi_2(M) = \mathbb{Z}$ for which one knows that the inclusion

$$\underset{k}{\mathrm{Hol}}(S^2; M) \to \underset{k}{\mathrm{Map}}(S^2; M) \tag{3.2}$$

of holomorphic maps of degree k into smooth maps of degree k tends to a homotopy equivalences as $k \to \infty$, then, as we shall show in a future paper, $|\mathscr{C}_f| \simeq X$. The hypothesis is known to hold when X is $\mathbb{CP}^n$ [24], or more generally a Grassmannian [17]. Furthermore the appropriate version of (3.2) (taking account of the fact that π_2 is free abelian on more than one generator) is also true for a flag manifold [16].

A version of (3.2) is also true for the flow category arising in the context of gauge theory for a compact group G, in virtue of the corresponding homotopy approximation property for the inclusion

$$\underset{k}{\mathrm{Hol}}(S^2; \Omega G) \to \underset{k}{\mathrm{Map}}(S^2; \Omega G).$$

In fact to show that $|\mathscr{C}_f| \simeq X$ one needs only the weak version of this result — called the Atiyah-Jones conjecture [5] — which was proved by Taubes [27], and Gravesen [15], rather than the stronger version proved in [7] and [18].

We should also point out that for any compact smooth category $\mathscr{C}$ with

$$\mathrm{Ob}(\mathscr{C}) = \coprod F_i, \qquad \mathrm{Mor}(\mathscr{C}) = \coprod F(j,i),$$

the tangent bundles along the fibres of $F(j,i) \to F_j$ really define a *functor* from the topological category $\mathscr{C}$ to a topological category $\mathscr{V}$. This category $\mathscr{V}$ has one object; its morphisms $\mathrm{Mor}(\mathscr{V})$ are finite dimensional vector spaces; and the composition law is direct sum. The functor assigns to a morphism $\gamma \in F(j,i)$ the vector space $T_{ji,\gamma} \oplus \mathbb{R}$.

To be more precise, $\mathscr{V}$ is the topological semi-group

$$\mathscr{V} = \coprod_{p \geq 0} BGL_p(\mathbb{R}),$$

and the functor is a coherent collection of maps $F(j,i) \to \mathscr{V}$ which are classifying maps for the bundles $T_{ji} \oplus \mathbb{R}$. The realization $|\mathscr{V}| = B\mathscr{V}$ is the space U/O of §2, and so the functor $\mathscr{C} \to \mathscr{V}$ induces a map $|\mathscr{C}| \to U/O$ which, when $|\mathscr{C}| \simeq X$, is the structural map of the polarization of X.

4 The area function on $\mathcal{L}\mathbb{CP}^n$

We now analyse the flow category of the area functional on $\mathcal{L}\mathbb{CP}^n$. Our main objective is to show that even though this flow category is not compact it does have a natural compactification, which turns out to be the flow category of a function on an infinite dimensional complex projective space. For simplicity we describe the details for $\mathbb{CP}^1 = S^2 = \mathbb{C} \cup \infty$.

As we saw in §2, we must really consider the area functional on the universal cover $\tilde{\mathcal{L}}S^2$ of $\mathcal{L}S^2$. This is the space of smooth maps $S^1 \to S^2$, together with an extension to a smooth map $D^2 \to S^2$ which is well-defined up to homotopy relative to the boundary. If $f : D^2 \to S^2$ is a smooth map its area is given by

$$\int_{D^2} f^*\omega,$$

where ω is the standard symplectic 2-form on S^2. This gives smooth functions

$$\tilde{\mathcal{A}} : \tilde{\mathcal{L}}S^2 \to \mathbb{R}, \qquad \mathcal{A} : \mathcal{L}S^2 \to \mathbb{R}/4\pi\mathbb{Z}.$$

The critical points of $\mathcal{A}$ are the constant loops, and those of $\tilde{\mathcal{A}}$ are pairs (γ, n), where γ is a constant loop and n is the degree of the extension. Thus the critical manifold of $\mathcal{A}$ is S^2, and that of $\tilde{\mathcal{A}}$ is $S^2 \times \mathbb{Z}$.

It is easy to see, compare [12], that the gradient vector field of $\mathcal{A}$ at a loop γ is the vector field along γ given by $j\dot{\gamma}$, where j is the complex structure on TS^2. This means that flow lines of $\mathcal{A}$ are given by holomorphic maps $h : S^2 \to S^2$ in the following way. Consider the path in $\mathcal{L}S^2$ given by $t \mapsto h_t$ where

$$h_t(s) = h(e^{-t+is}).$$

(We have parametrized loops by the closed interval $[0, 2\pi]$.) Then h_t converges to the constant loop at $h(\infty) = a$ as $t \to -\infty$, and as $t \to \infty$ it converges to the constant loop at $h(0) = b$. This path h_t is a flow line of $\mathcal{A}$, and every flow line from a to b is of this form.

The holomorphic map h gives a natural exension of the loop h_t to the lower hemisphere of S^2 and this defines a path $\tilde{h}_t$ in $\tilde{\mathcal{L}}S^2$. This path $\tilde{h}_t$ is a flow line of $\tilde{\mathcal{A}}$ from (a, k) to $(b, 0)$, where k is the degree of h, and every such flow line arises in this way. The flow lines from $(a, n+k)$ to (b, n) are given by applying the appropriate covering translation to flow lines from (a, k) to $(b, 0)$.

Let $W(n, m) \subset \tilde{\mathcal{L}}S^2$ be the space of points which lie on flow lines from a critical point of the form (a, n) to one of the form (b, m). Thus a point is in $W(n, m)$ if and only if it is on a flow line of $\mathcal{A}$ which starts at level n and ends at level m. The above identification of the flow lines of $\mathcal{A}$ shows that

$$W(n, m) = \mathrm{Rat}_{n-m}$$

where Rat_{n-m} is the space of holomorphic maps, or rational functions, $h : S^2 \to S^2$ of degree $n - m$.

The space Rat_k is not compact and it is very important to understand this non-compactness. A rational function $h : S^2 \to S^2$ of degree k is given by $h = p/q$ where p and q are polynomials of degree $\leq k$ with no roots in common. Throughout we allow roots at infinity: thus if p has degree r with $r < k$, then we say p has $k - r$ roots at infinity. This is a convenient device which, for example, allows us to say that the zeroes of the rational function p/q are the roots of p, and its poles are the roots of q. Suppose we now take a sequence of rational functions $h_n = p_n/q_n$ where a root α_n of p_n converges to a root β_n of q_n. Then this sequence does not converge in Rat_k. This is the bubbling phenomenon for rational functions and we shall say that a bubble occurs at the point a which is the common limit of α_n and β_n.

The reason for this terminology is as follows. Suppose that $|\alpha_n - \beta_n|$ is extremely small, and that $\varepsilon > 0$ is also very small but much greater than $|\alpha_n - \beta_n|$. Let D be the disc of radius ε around β_n. Then p_n/q_n is almost constant on ∂D. Outside D the function p_n/q_n is almost equal to a rational function of degree $k - 1$, namely $(x - \beta_n)p_n/(x - \alpha_n)q_n$. The interior of D, however, is mapped by a map which is almost surjective with degree 1.

If we have a sequence of rational functions h_n in which a bubble occurs at either ∞ or 0, then the corresponding sequence of paths in $\tilde{\mathcal{L}}S^2$ converges to a piecewise flow line. However, if the bubble occurs at any other point the corresponding sequence of paths, no matter how it is parametrized, does not even converge to a path in $\tilde{\mathcal{L}}S^2$. Thus the flow category $\mathcal{C}_{\mathcal{A}}$ is not compact.

We now construct a compactification of $\mathcal{C}_{\mathcal{A}}$. Let $W = \mathbb{C}[z, z^{-1}]$ be the vector space of Laurent polynomials topologized as a space of maps $\mathbb{C}^\times \to \mathbb{C}$, where $\mathbb{C}^\times = \mathbb{C} \setminus 0$. Now the linear flow $z^n \mapsto e^{nt}z^n$ defines a flow Φ on $\mathbb{P}(\mathbb{C}^2 \otimes W)$. It is straightforward to check that the space of stationary points of Φ is $S^2 \times \mathbb{Z}$, where $S^2 \times n$ is the subspace $\mathbb{P}(\mathbb{C}^2 \otimes z^n) \subset \mathbb{P}(\mathbb{C}^2 \otimes W)$. Let W_m^n be the subspace of W spanned by z^i with $m \leq i \leq n$. The space of points which lie on piecewise flow lines of Φ, which go from level n to level m, is $\mathbb{P}(\mathbb{C}^2 \otimes W_m^n) = \mathbb{CP}^{2(n-m)+1}$, and since this space is compact the flow category $\mathcal{C}_\Phi$ is compact.

A pair (f_0, f_1) of elements of W with no roots in common except, possibly, at 0 and ∞ defines a map

$$f : \mathbb{C}^\times \to \mathbb{CP}^1 = S^2.$$

Since this map is algebraic it extends to a holomorphic map $f : \mathbb{C} \cup \infty = S^2 \to S^2$. Let $\mathcal{U}$ be the open subset of the projective space $\mathbb{P}(\mathbb{C}^2 \otimes W)$, defined by the pairs (f_0, f_1) with no roots in common in $\mathbb{C}^\times$. Then there is a map

$$i : \mathcal{U} \to \tilde{\mathcal{L}}S^2,$$

defined by restricting the holomorphic map $f : S^2 \to S^2$ given by the pair (f_0, f_1) to the unit circle and using the extension of this loop to the lower hemisphere to get an element of $\tilde{\mathscr{L}}S^2$. It is clear that $\mathscr{U} \subset \mathbb{P}(\mathbb{C}^2 \otimes W)$ is invariant under the flow Φ and that $i : \mathscr{U} \to \tilde{\mathscr{L}}S^2$ is equivariant with respect to the flow Φ on $\mathscr{U}$ and the gradient flow of $\mathscr{A}$ on $\tilde{\mathscr{L}}S^2$. Furthermore it is straightforward to check that $i : \mathscr{U} \to \tilde{\mathscr{L}}S^2$ defines an isomorphism of flow categories. In fact, this map i is a homotopy equivalence, as we will show in a future paper, but we do not need this for our present purpose.

The diagram

$$\tilde{\mathscr{L}}S^2 \xleftarrow{\ i\ } \mathscr{U} \xrightarrow{\ j\ } \mathbb{P}(\mathbb{C}^2 \otimes W)$$

(where j is the inclusion), together with the fact that i induces an isomorphism of flow categories, gives us an embedding of flow categories

$$\mathscr{C}_{\mathscr{A}} \to \mathscr{C}_{\Phi}.$$

The flow category $\mathscr{C}_{\Phi}$ is compact and $\mathscr{C}_{\mathscr{A}}$ is embedded as an open dense subcategory; therefore $\mathscr{C}_{\Phi}$ is a compactification of $\mathscr{C}_{\mathscr{A}}$. Moreover, it is natural to view the flow Φ on $\mathbb{P}(\mathbb{C}^2 \otimes W)$ as a «compactification» of the gradient flow of $\tilde{\mathscr{A}}$ on $\tilde{\mathscr{L}}S^2$.

The above compactification of $\mathscr{C}_{\mathscr{A}}$ gives a compactification of the space of rational functions $\mathrm{Rat}_k = W_{\mathscr{A}}(n+k, n)$ as $\mathbb{CP}^{2k+1}$. This is the precise analogue for rational functions of the Donaldson-Uhlenbeck compactification of the moduli space of instantons on a 4-manifold, compare [10], §4.4. To see the analogy, regard $\mathbb{CP}^{2k+1}$ as the projective space of the vector space of pairs of polynomials (p, q) where $\deg p, \deg q \leq k$. Then

$$\mathbb{CP}^{2k+1} = \bigcup_l \mathrm{Rat}_{k-l} \times \mathrm{SP}^l(S^2)$$

where $\mathrm{SP}^l(S^2)$ is the l-th symmetric product of S^2, i.e., the space of unordered sets of l not necessarily distinct points in S^2. To a pair of polynomials (p, q) we associate the rational function $f = p/q$ which has degree $k - l$, where p and q have l roots in common, and the point of $\mathrm{SP}^l(S^2)$ given by the l common roots allowing, as above, roots at infinity.

This construction of a compactification works equally well for the area functional on $\mathscr{L}\mathbb{CP}^n$, and it gives the flow category of the flow $\Phi^{(n)}$ on $\mathbb{P}(\mathbb{C}^{n+1} \otimes W)$ defined by the linear flow $v \otimes z^n \mapsto v \otimes e^{nt} z^n$ on $\mathbb{C}^{n+1} \otimes W$. It is striking that the compactification of the flow category of the area functional on the loop space of $\mathbb{CP}^n = \mathbb{P}(\mathbb{C}^{n+1})$ is given by the flow category of a function on $\mathbb{P}(\mathbb{C}^{n+1} \otimes \mathbb{C}[z, z^{-1}])$.

5 Morse theory and homotopy theory

The most important result of finite dimensional Morse theory is the relation between the topology of a compact manifold X and that of the manifold F of critical points of a smooth function $f : X \to \mathbb{R}$. It asserts that after changing the grading of the chain groups $C_*(F)$ appropriately, there is a differential $\tilde{d}$ on $C_*(F)$ whose homology is $H_*(X)$.

In fact more is true. Let F_i be the part of the critical set F at the ith critical level. We get a filtration of X

$$X_0 \subset X_1 \subset \ldots \subset X_n = X$$

by closed subspaces, where X_i consists of the points on downward piecewise trajectories emanating from F_i. The successive quotient spaces $Y_k = X_k/X_{k-1}$ are the Thom spaces $Y_k = F_k^{E_k}$ of the downward bundles E_k on F_k. Recall that the Thom space X^E of a bundle E over a compact space X is the one-point compactification E^+ of the total space of E.

This leads to the homological assertion above because of the following general principle. Although the homotopy type of a filtered space X is not determined by the quotients $Y_k = X_k/X_{k-1}$, nevertheless the *stable* homotopy type — more precisely, the homotopy type of the n-fold suspension S^nX — *is* determined by the Y_k together with certain maps between them.

Thus if $n = 1$, the Puppe construction for the inclusion $X_0 \to X_1$ tells us that the suspension SX_1 is obtained by attaching a cone $C(Y_1)$ on Y_1 to SX_0 by a map $\partial_1 : Y_1 = X_1/X_0 \to SX_0$:

$$SX_1 \simeq SX_0 \cup C(Y_1).$$

When $n = 2$, one finds that

$$S^2X_2 \simeq S^2X_0 \cup C(SY_1) \cup C^2(Y_2).$$

To reconstruct S^2X_0 in this way we need the maps $\partial_2 : Y_2 \to SY_1$ and $\partial_1 : Y_1 \to SX_0$, obtained from the Puppe construction, together with a null-homotopy of the composite $S\partial_1 \circ \partial_2$. Explicitly, this null-homotopy provides a map $SY_1 \cup_{\partial_2} C(Y_2) \to S^2(X_0)$ whose restriction to SY_1 is equal to $S\partial_1$, and the mapping cone of this map is homotopy equivalent to S^2X_2. In general, one finds that

$$S^nX_n \simeq S^nX_0 \cup C(S^{n-1}Y_1) \cup \cdots \cup C^n(Y_n).$$

To describe the maps and homotopies needed to reconstruct S^nX_n in this way requires some technology.

Let $\mathcal{J}$ be the topological category whose objects are the integers $\mathbb{Z}$, and whose non-identity morphisms $j \to i$, when $j > i$, form a space $J(j,i)$ which is the one-point compactification of the space of sequences of real numbers $\{\lambda_k\}_{k\in\mathbb{Z}}$ such that

$$\begin{aligned} \lambda_k &\geq 0, && \text{for all } k, \text{ and} \\ \lambda_k &= 0, && \text{unless } i < k < j. \end{aligned}$$

There are no non-identity morphisms unless $j > i$. Composition of morphisms is the map $J(k,j) \times J(j,i) \to J(k,i)$ induced by addition of sequences. Thus $J(j,i)$ is a compact space of dimension $j-i-1$, with ∞ as a distinguished base-point. If $j-i=1$, then $J(j,i)$ has just two points 0 and ∞; if $j-i=2$, then $J(j,i)$ is a closed interval $[0,\infty]$. Indeed, if $j-i \geq 2$, then $J(j,i)$ is homeomorphic to a disc of dimension $j-i-1$. We shall also consider the full subcategory $\mathcal{J}_a^b$ of $\mathcal{J}$ spanned by the objects $a, a+1, \ldots, b$.

There is a close relation between sequences of compact spaces

$$\underline{X} = \{X_a \to X_{a+1} \to \ldots \to X_b\}$$

and base-point-preserving covariant functors $Z : \mathcal{J}_a^b \to \mathcal{T}_*$. Here $\mathcal{T}_*$ is the category of compact spaces with base-point, and a base-point-preserving functor is one that maps ∞ in $J(j,i)$ to the zero map (i.e. the constant map with value the base-point) $Z(j) \to Z(i)$.

Let us assume for simplicity that the maps in $\underline{X}$ are inclusions. Then, the sequence $\underline{X}$ gives rise to a functor $Z : \mathcal{J}_a^b \to \mathcal{T}_*$ with

$$Z(i) = S^{b-i}(X_i/X_{i-1}),$$

for $i > a$, and

$$Z(a) = S^{b-a}(X_a^+) = (\mathbb{R}^{b-a} \times X_a)^+.$$

Here, if X is compact the notation X^+ means X with a disjoint base-point, denoted by ∞, adjoined, and if X is not compact it means the one-point compactification of X.

We will give the construction of the functor Z later, but for the moment, let us note two of its properties.

(i) The map $Z(i+1) \to Z(i)$ induced by the nontrivial morphism $i+1 \to i$ in $\mathcal{J}$ is the $(b-i-1)$-fold suspension of the map

$$X_{i+1}/X_i \to S(X_i/X_{i-1})$$

obtained by applying the Puppe construction to the inclusion $X_i/X_{i-1} \to X_{i+1}/X_{i-1}$.

(ii) The functor Z gives a map

$$J(i+2,i) \times S^{b-i-2}(X_{i+2}/X_{i+1}) \to S^{b-i}(X_i/X_{i-1}).$$

The space of morphisms $J(i+2,i)$ is the closed interval $[0,\infty]$, and this map is given by the $(b-i-2)$-fold suspension of a null-homotopy of the composite

$$X_{i+2}/X_{i+1} \to SX_{i+1}/X_i \to S^2X_i/X_{i-1}.$$

In the other direction, a functor $Z : \mathcal{J}_a^b \to \mathcal{T}_*$ has a realization $|Z|$ as a compact space. This is constructed from the disjoint union

$$\coprod_{a \leq i \leq b} Z(i) \wedge J(i, a-1)$$

by identifying the image of $Z(j) \times J(j,i) \times J(i,a-1)$ in $Z(j) \wedge J(j,a-1)$ with its image in $Z(i) \wedge J(i,a-1)$ whenever $a \leq i \leq j \leq b$. Notice that

$$Z(i) \wedge J(i,a-1) = C^{i-a}(Z(i))$$
$$|Z| = Z(a) \cup C(Z(a+1)) \cup \cdots \cup C^{b-a}(Z(b)).$$

If Z is the functor defined by a sequence of spaces $X_a \to \cdots \to X_b$, then the comparison between this decomposition of $|Z|$ and the decomposition of $S^{b-a}X_b$ described above leads, very naturally, to the following result.

Proposition 5.1 (i) *If Z is the functor associated to a sequence of compact spaces $\underline{X}$, then there is a canonical homotopy equivalence*

$$|Z| \simeq S^{b-a}(X_b^+).$$

(ii) *For any functor $Z : \mathcal{J}_a^b \to \mathcal{T}_*$, the homology $H_*(|Z|)$ can be calculated from the double complex*

$$C_*(Z) = \bigoplus_{a \leq i \leq b} C_*(Z(i)).$$

The proof of Proposition 5.1 is straightforward, given the construction of the functor $Z : \mathcal{J}_a^b \to \mathcal{T}_*$ from a sequence of compact spaces

$$\underline{X} = \{X_a \to X_{a+1} \to \ldots \to X_b\}.$$

To construct Z we shall, as above, assume that the maps in $\underline{X}$ are inclusions. If $a \leq i \leq b$, let Z_i be the open subspace of $X_b \times \mathbb{R}_+^{b-a}$ consisting of all points $(x; \lambda_a, \ldots, \lambda_{b-1})$ such that:

(i) $\lambda_r > 0$ if $r \geq i$, and
(ii) if $\lambda_r > 0$ then $x \in X_r$.

We now show that the one-point compactification Z_i^+ is homotopy equivalent to $S^{b-i}(X_i/X_{i-1})$. First note that Z_i is the product of a subset of $X_i \times \mathbb{R}_+^{i-a}$ with the extra factor $(0,\infty)^{b-i}$. The factor $(0,\infty)^{b-i}$ accounts for the suspension, and it is enough to consider Z_b^+. This space is obtained from X_a by attaching a cone $C(X_{a+1})$ on X_{a+1} and then a cone on $C(X_{a+2}) \subset C(X_{a+1})$, and so on. Contracting these cones in the standard way shows that Z_b^+ is homotopy equivalent to X_b/X_{b-1}.

If $j > i$ then Z_i is an open subset of Z_j, so there is a natural map $Z_j^+ \to Z_i^+$. Let W_{ij} be the open subspace of Z_i consisting of all points with $\lambda_j > 0$. Then $Z_i \subset W_{ij} \subset Z_j$, and so the map $Z_j^+ \to Z_i^+$ factors through W_{ij}^+. Furthermore there is a proper map $\mathbb{R}_+^{j-i-1} \times W_{ij} \to W_{ij}$ which simply adds the coordinates labelled $i+1, i+2, \ldots, j-1$. This induces $J(j,i) \wedge W_{ij} \to W_{ij}$, and hence $J(j,i) \wedge Z_j^+ \to Z_i^+$ giving us the desired functor.

Our next task is to see that the compact smooth framed category which arises in §3 from a Morse-Bott-Smale function gives rise to a functor $Z : \mathcal{J}_0^n \to \mathcal{T}_*$ such that

$$Z(i) = S^{n-i}(F_i^{E_i}).$$

This is just a version of the Pontryagin-Thom construction. We have seen that the manifold with corners $F(j,i)$ embeds in the sphere bundle $S(E_j)$, with normal bundle $\pi_i^* E_i$. Let us map it further into E_j with normal bundle $\pi_i^* E_i \oplus \mathbb{R}$. We can choose a map $F(j,i) \to \mathbb{R}_+^k$ inducing the stratification of $F(j,i)$, where $k = j-i-1$. This gives us an embedding of $F(j,i)$ in $E_j \times \mathbb{R}_+^k$, compatible with the boundary structure, with normal bundle

$$\nu_{ji} = \pi_i^* E_i \oplus \mathbb{R}^{j-i}.$$

In other words, we have maps

$$\mathbb{R}_+^k \times E_j \leftarrow \nu_{ji} \to E_i \times \mathbb{R}^{j-i},$$

where the first is an open inclusion, and the second is proper. Passing to the one-point compactifications, this gives

$$J(j,i) \wedge F_j^{E_j} \to \nu_{ji}^+ \to S^{j-i}(F_i^{E_i}),$$

and, after applying S^{n-j}, this is exactly what we need to define a base-point-preserving functor $Z : \mathcal{J}_0^n \to \mathcal{T}_*$ with $Z(i) = S^{n-i}(F_i^{E_i})$.

The method described above of reconstructing the manifold X from the data provided by the function f, is quite different from that used in [9]. It uses the framings of the spaces $F(j,i)$, and their compatibility under the composition law in $\mathcal{C}_f$, to recover the stable homotopy type of X. For example, in the case where f is a Morse-Smale function (that is one with isolated non-degenerate critical points whose gradient flow satisfies Smale's transversality condition) the method amounts to the following construction. Take a cell of dimension j for each critical point of index j; now one constructs a CW complex inductively, using the framings of the spaces of flow lines to give the maps needed to attach appropriate suspensions of these cells to the lower skeleta. In this way we recover the stable homotopy type of M. Thus, the construction is similar in spirit to the work of Franks [14]. Note that the Morse-Smale chain complex simply uses the framings of the zero dimensional spaces $F(i+1,i)$ to define the boundary map.

Now let us consider what happens when we apply this method to the infinite dimensional situations studied by Floer. Here we confine ourselves to summarising the basic points; we will give a complete account in a future paper.

Clearly, the first step is to consider functors Z defined on the whole category $\mathcal{J}$. If we have a functor $Z : \mathcal{J} \to \mathcal{T}_*$, then we get a functor $Z_a^b : \mathcal{J}_a^b \to \mathcal{T}_*$, and a compact space $|Z|_a^b = |Z_a^b|$ for each $a < b$. It is important to observe that, from the construction of the realizations, there are maps

$$S^{b'-b}|Z|_a^b \to |Z|_a^{b'}, \qquad |Z|_{a'}^b \to S^{a-a'}|Z|_a^b$$

when $a' \leq a < b \leq b'$. Such a system of spaces and maps defines a *pro-spectrum*; see the Appendix for a brief discussion of pro-spectra, and further references. Thus, in the case of a functor $Z : \mathcal{J} \to \mathcal{T}_*$, the output is a pro-spectrum, rather than a stable homotopy type.

If we have a compact smooth framed category with objects $\{F_i\}_{i\in\mathbb{Z}}$ we do not quite get a functor $\mathcal{J} \to \mathcal{T}_*$. The framing only provides us with «stable» or «virtual» bundles E_i instead of genuine finite dimensional vector bundles. The space $F_i^{E_i}$ is then an object in the «stable category» $\mathcal{S}$, which is described in the Appendix. Thus we get a functor $Z : \mathcal{J} \to \mathcal{S}$, with $Z(i) = S^{-i}(F_i^{E_i})$ and such a functor still defines a pro-spectrum.

What happens when we have a compact smooth category which is not framed? The essential point is to understand how to extract some kind of *stable map* $S^{2m}A \to B$ from a diagram of compact manifolds

$$A \xleftarrow{\pi_1} C \xrightarrow{\pi_2} B, \tag{5.2}$$

where π_1 is a fibration whose fibres are closed almost complex manifolds of (real) dimension $2m$. Evidently, we can lift π_1 to an embedding $C \to A \times \mathbb{C}^{m+p}$ for some p. Let the normal bundle be ν. Then we have

$$S^{2m+2p}(A^+) \to C^\nu, \qquad C \to B.$$

To proceed we must pass to a category in which C and C^ν are equivalent. If ν is trivialized, then $C^\nu = S^{2p}(C^+)$, and the usual stable category will serve. In general, we must do something more brutal, which we will digress to explain.

The stable category of compact spaces is described in the Appendix. In this category, two compact spaces X and Y become homotopy equivalent if their suspensions S^pX and S^pY are homotopy equivalent for large p. The notion of a *spectrum*, a sequence of spaces $\mathbf{K} = \{K_p\}$ with maps $S^qK_p \to K_{p+q}$, is also described in the Appendix. We are concerned here with *ring spectra*, where there are associative pairings $K_p \wedge K_q \to K_{p+q}$. The spheres themselves form a natural example. For a ring spectrum $\mathbf{K}$ we can define the $\mathbf{K}$-homotopy category: its objects are compact spaces and its morphisms from X to Y are

$$\mathrm{Mor}_{\mathbf{K}}(X,Y) = \lim_{p\to\infty}[S^pX, K_p \wedge Y].$$

Thus if $K_p = S^p$ this is the stable homotopy category.

There is an optimal spectrum $\mathbf{M}$ with the property that, for any p-dimensional complex vector bundle on a compact space C, the Thom space C^ν is *canonically* $\mathbf{M}$-homotopy equivalent to the suspension $S^{2p}(C^+)$. For this choice of $\mathbf{M}$, the diagram (5.2) induces a map $S^{2m}(A^+) \to B^+$ in the $\mathbf{M}$-homotopy category.

We can now carry out the Pontryagin-Thom construction for an arbitrary compact smooth category, and we shall obtain an object $|Z|_a^b$ of the $\mathbf{M}$-homotopy category for each pair of integers $a < b$. If $a' \le a < b \le b'$ there will be natural $\mathbf{M}$-maps

$$S^{b'-b}|Z|_a^b \to |Z|_a^{b'}, \qquad |Z|_{a'}^b \to S^{a-a'}|Z|_a^b.$$

This system of spaces and maps again defines a pro-spectrum, and it is our desired output.

For fixed a, the spaces $|Z|_a^b$ and $\mathbf{M}$-maps $S^{b'-b}|Z|_a^b \to |Z|_a^{b'}$ define an object $|Z|_a$ of the $\mathbf{M}$-homotopy category of spectra. Furthermore, for the flow categories of Floer functions $S^d|Z|_a^b$ and $|Z|_{a+d}^{b+d}$ are $\mathbf{M}$-homotopy equivalent, where d is the periodicity of the virtual dimension, compare §2. This gives a periodicity map $|Z|_a \to S^d|Z|_a$. Therefore the pro-spectrum which is the output of the construction is of a particularly simple kind; it is given by the inverse system of spectra

$$Z \leftarrow S^{-d}Z \leftarrow S^{-2d}Z \leftarrow \cdots$$

where $Z = |Z|_0$ and the map $S^{-d}Z \to Z$ is the periodicity map.

The spectrum $\mathbf{M}$ is traditionally called the MU-spectrum. The space M_{2p} is the Thom space of the universal $\mathbb{C}^p$ bundle on $BGL_p(\mathbb{C})$, and $M_{2p+1} = SM_{2p}$. If ν is a $\mathbb{C}^p$-bundle on C the classifying map $C \to BGL_p(\mathbb{C})$ extends to $C^\nu \to M_{2p}$. Putting this together with the projection $\nu \to C$ gives

$$C^\nu \to M_{2p} \wedge (C^+)$$

which is an $\mathbf{M}$-equivalence $C^\nu \to S^{2p}(C^+)$. This $\mathbf{M}$-equivalence does not depend on the choice of a classifying map for ν, though it would take us too far afield to explain that here: the point is the functoriality of the transformation $\mathscr{C} \to \mathscr{V}$ mentioned at the end of §3.

Spectra are the same things as generalized cohomology theories. If we define $h^p(X;\mathbf{K}) = \mathrm{Mor}_{\mathbf{K}}(X; S^p)$ then $h^*(-;\mathbf{K})$ is a cohomology theory, and the correspondence $\mathbf{K} \leftrightarrow h^*(-;\mathbf{K})$ is one-to-one. The theory $h^*(-;\mathbf{M})$, called complex cobordism, is universal among so-called complex oriented theories [25], which include ordinary cohomology and K-theory, and it determines them algebraically. So our construction gives us definitions of Floer cobordism and Floer K-theory, as well as Floer cohomology.

6 Floer theory for the projective space of a polarized vector space and $\mathscr{L}\mathbb{C}\mathbb{P}^n$

We now explain what the method described in the previous section gives in some infinite dimensional examples. Motivated by the fact that the compactification of the flow category of the area functional on $\mathscr{L}\mathbb{C}\mathbb{P}^n$, constructed in §4, is the flow category of a function on an infinite dimensional projective space, we begin by considering projective spaces.

Example 6.1 — Real projective space
Let V be the real vector space of sequences $x = \{x_n\}_{n\in\mathbb{Z}}$ with only a finite number of non-zero terms, topologized as the direct limit of its finite dimensional subspaces. We use the usual Hilbert norm $\|-\|$ on V; of course V is not complete in this norm. Let $S(V)$ be the sphere in V, and consider the function $f : S(V) \to \mathbb{R}$ defined by

$$f(x) = \sum_{n=-\infty}^{\infty} n x_n^2.$$

This descends to a function

$$f : \mathbb{P}(V) \to \mathbb{R}$$

with critical points $c_i = [\delta_i]$, $i \in \mathbb{Z}$, where δ_i is the i-th element in the standard basis for V. The gradient flow of f, with respect to the Hilbert norm, is the flow on $\mathbb{P}(V)$ defined by the linear flow ψ on V where $\psi_t(\delta_n) = e^{nt}\delta_n$. We could replace V by a space of sequences of suitably rapid decay, but this does not make any real difference.

The unstable manifold $W^u(c_i)$ and the stable manifold $W^s(c_i)$ are given by

$$\begin{aligned} W^u(c_i) &= \{[x] \in \mathbb{P}(V) : x_i \neq 0, x_j = 0 \quad \text{if } j < i\} \\ W^s(c_i) &= \{[x] \in \mathbb{P}(V) : x_i \neq 0, x_j = 0 \quad \text{if } j > i\}. \end{aligned}$$

Neither $W^u(c_i)$ nor $W^s(c_i)$ is finite dimensional, but the intersection $W^u(c_i) \cap W^s(c_j)$ is transverse, finite dimensional, and

$$\dim W^s(c_i) \cap W^u(c_j) = j - i \qquad \text{if } j \geq i.$$

From this it is not difficult to identify the flow category $\mathscr{C} = \mathscr{C}_f$ explicitly.

Now consider the question of whether this category is framed in the sense of §3. Fix a pair of integers $a < b$ and let $\mathscr{C}_a^b$ be the full subcategory generated by the critical points c_i with $a \leq i \leq b$. This category is the flow category of the function

$$f(x) = \sum_{n=a}^{b} n x_n^2$$

on $\mathbb{R}\mathbb{P}^{b-a} = \mathbb{P}(V_a^b)$, where V_a^b is the finite dimensional subspace of V with basis δ_i, $a \leq i \leq b$, and we are using the natural homogeneous coordinates

$x = [x_a, \dots, x_b]$ with $\sum x_n^2 = 1$ on $\mathbb{P}(V_a^b)$. Therefore, $\mathcal{C}_a^b$ is a framed category and each of the spaces of morphisms $F(j,i)$ with $a \le i < j \le b$ inherits a framing φ_a^b from the flow category $\mathcal{C}_a^b$. The framing φ_a^b comes from embedding $F(j,i)$ in the unstable sphere of the critical point c_j in $\mathbb{RP}^{b-a}$, which, since c_j has index $j-a$ in $\mathbb{RP}^{b-a}$, is a sphere of dimension $j-a-1$. In particular, this shows that the framings φ_a^b and $\varphi_a^{b'}$ are identical. However, the framings φ_a^b and $\varphi_{a'}^b$ are not the same, as we will show.

Let us work in the stable category $\mathcal{S}$, described in the Appendix, and define

$$|(\mathcal{C}_a^b, \varphi)| = S^{a-b}|Z|$$

where $Z : \mathcal{J}_a^b \to \mathcal{T}_*$ is the functor defined by the category $\mathcal{C}_a^b$ equipped with the framing φ. This has the effect of removing the suspensions which occur in the statement of Proposition (5.1) and it simplifies notation; it is a straightforward matter to keep track of the suspensions, if necessary.

The manifold $F(i+1,i)$ has dimension zero and, using φ_a^b, is framed in S^{i-a}, the unstable sphere of the critical point c_{i+1} in $\mathbb{RP}^{b-a}$. Thus it gives a map of spheres $S^{i-a} \to S^{i-a}$. This map is the relative attaching map between the $(i-a+1)$-cell and the $(i-a)$-cell in $|(\mathcal{C}_a^b, \varphi_a^b)| = \mathbb{RP}^{b-a}$. Therefore it has degree $1-(-1)^{i-a}$. Using the framing φ_{a-1}^b of $F(i+1,i)$ in the unstable S^{i-a+1} of the critical point c_{i+1} in $\mathbb{RP}^{b-a+1}$ we get a map $S^{i-a+1} \to S^{i-a+1}$ which is the relative attaching map between the $(i-a+2)$-cell and the $i+1-a$-cell of $|(\mathcal{C}_{a-1}^b, \varphi_{a-1}^b)| = \mathbb{RP}^{b-a+1}$; therefore it has degree $1-(-1)^{i-a+1}$. So the framings φ_a^b and φ_{a-1}^b produce different maps.

For $a' < a$ the framings $\varphi_{a'}^b$ and φ_a^b differ because the normal bundle to $\mathbb{RP}^{b-a}$ in $\mathbb{RP}^{b-a'}$, that is $(a-a')\eta$ where η is the real Hopf line bundle, is non-trivial. Furthermore, it is straightforward to check that

$$|(\mathcal{C}_a^b, \varphi_{a'}^b)| = (\mathbb{RP}^{b-a})^{(a-a')\eta}.$$

Since the framings $\varphi_{a'}^{b'}$ and $\varphi_{a'}^b$ agree it follows that

$$|(\mathcal{C}_a^b, \varphi_{a'}^{b'})| = (\mathbb{RP}^{b-a})^{(a-a')\eta}$$

whenever $a' \le a < b \le b'$.

From the construction of the realization there are maps

$$|(\mathcal{C}_a^b, \varphi_a^b)| \to |(\mathcal{C}_a^{b'}, \varphi_a^{b'})|, \qquad |(\mathcal{C}_{a'}^b, \varphi_{a'}^b)| \to |(\mathcal{C}_a^b, \varphi_{a'}^b)|$$

for $a' \le a < b \le b'$ which we now identify. The first is the inclusion

$$\mathbb{RP}^{b-a} \to \mathbb{RP}^{b'-a}$$

and the second is the map

$$\mathbb{RP}^{b-a'} \to (\mathbb{RP}^{b-a})^{(a-a')\eta}$$

obtained from the Pontryagin-Thom construction applied to the embedding $\mathbb{RP}^{b-a} \to \mathbb{RP}^{b-a'}$; we describe this construction briefly.

Suppose we have an embedding $P \to M$ of compact manifolds. Let ν be the normal bundle of the embedding and let N_P be an open tubular neighbourhood of P in M. Then the inclusion $N_P \to M$ is an open embedding and so it gives a map $N_P^+ \leftarrow M$ where N_P^+ is the one-point compactification of N_P. Now N_P^+ is just the Thom complex of ν and so the embedding $N \to M$ gives a map $M \to P^\nu$. Applied to the embedding $\mathbb{RP}^{b-a} \to \mathbb{RP}^{b-a'}$, with normal bundle $(a-a')\eta$, this gives the required map.

We now explain how to assemble the spaces $|(\mathscr{C}_a^b, \varphi_{a'}^{b'})|$ into a single object, a pro-spectrum, which correctly reflects the relation between the different spaces. To do this, we use the maps we have just described. The bundle $(a-a')\eta$ extends over $\mathbb{RP}^{b-a}$ and so, using the theory of Thom spaces of virtual bundles described in the Appendix, we can convert the map $\mathbb{RP}^{b-a'} \to (\mathbb{RP}^{b-a})^{(a-a')\eta}$ into a stable map

$$(\mathbb{RP}^{b-a'})^{-(a-a')\eta} \to \mathbb{RP}^{b-a}.$$

Therefore we can construct the sequence

$$\mathbb{RP}^{b-a} \leftarrow (\mathbb{RP}^{b-a+1})^{-\eta} \leftarrow (\mathbb{RP}^{b-a+2})^{-2\eta} \leftarrow \dots$$

in the stable category of compact spaces. As explained in the Appendix, this sequence defines a pro-spectrum.

Using the inclusions $\mathbb{RP}^{b-a} \to \mathbb{RP}^{b'-a}$, we can take the limit over b and then, using Thom spaces of virtual vector bundles over CW complexes of finite type (see the Appendix), we get the pro-spectrum defined by the sequence of Thom spectra

$$\mathbb{RP}^\infty \leftarrow (\mathbb{RP}^\infty)^{-\eta} \leftarrow (\mathbb{RP}^\infty)^{-2\eta} \leftarrow \cdots.$$

This pro-spectrum is the final output of the construction; it is the *Floer homotopy type* associated to the function $f : \mathbb{P}(V) \to \mathbb{R}$. In fact, this pro-spectrum, which is usually denoted by $\mathbb{RP}^\infty_{-\infty}$, is exactly the pro-spectrum which occurs in the theorem of Lin mentioned in the introduction.

Floer's method of associating a chain complex to the function $f : \mathbb{P}(V) \to \mathbb{R}$ gives the chain complex C_* with $C_p = \mathbb{Z}$ for all $p \in \mathbb{Z}$, and the boundary operator $\partial_p : C_p \to C_{p-1}$ is multiplication by $1 + (-1)^p$. It is easy to check that the homology of this chain complex is the same as the homology of the pro-spectrum $\mathbb{RP}^\infty_{-\infty}$ for any coefficient group. So the Floer groups do compute the homology of $\mathbb{RP}^\infty_{-\infty}$.

Example 6.2 — Complex projective space

Now consider the complex analogue of the previous example. Let W be the complex vector space of sequences $z = \{z_n\}_{n\in\mathbb{Z}}$ with only a finite number of non-zero terms, equipped with the direct limit topology, and Hilbert norm $\|-\|$. This time the function $S(W) \to \mathbb{R}$ defined by

$$z \mapsto \sum_{n=-\infty}^{\infty} n|z_n|^2$$

descends to a function $f : \mathbb{P}(W) \to \mathbb{R}$, and the flow of this function is exactly the flow $\Phi^{(0)}$ on the projective space $\mathbb{P}(W)$ which arose in §4. The construction used in the case of the real projective space $\mathbb{P}(V)$ shows that the Floer homotopy type associated to this function is the pro-spectrum $\mathbb{CP}^{\infty}_{-\infty}$ defined by the sequence of Thom spectra

$$\mathbb{CP}^{\infty} \leftarrow (\mathbb{CP}^{\infty})^{-\zeta} \leftarrow (\mathbb{CP}^{\infty})^{-2\zeta} \leftarrow \cdots$$

where ζ is the complex Hopf line bundle. Once more we find by direct computation that the Floer chain complex of f does indeed compute the cohomology of this pro-spectrum.

Example 6.3 — The area function on $\mathscr{L}\mathbb{CP}^n$

To associate a Floer homotopy type to the area functional $\mathscr{A}$ on $\mathscr{L}\mathbb{CP}^n$, we first compactify the flow category $\mathscr{C}_{\mathscr{A}}$ to give the flow category of the flow $\Phi^{(n)}$ on the projective space $\mathbb{P}(\mathbb{C}^{n+1} \otimes \mathbb{C}[z, z^{-1}])$, as in §4. Now the method of (6.1) and (6.2) gives the pro-spectrum defined by the sequence of Thom spectra

$$\mathbb{CP}^{\infty} \leftarrow (\mathbb{CP}^{\infty})^{-(n+1)\zeta} \leftarrow (\mathbb{CP}^{\infty})^{-2(n+1)\zeta} \leftarrow \cdots$$

where ζ is the complex Hopf line bundle. This pro-spectrum is $\mathbb{CP}^{\infty}_{-\infty}$.

The reason why $(n+1)\zeta$ appears in this construction is that the category $\mathscr{C}_a^b$ which occurs in this example is the flow category of a Morse-Bott-Smale function on $\mathbb{CP}^{(n+1)(b-a)}$, and the normal bundle to the embedding of $\mathbb{CP}^{(n+1)(b-a)}$ in $\mathbb{CP}^{(n+1)(b-(a-1))}$ is $(n+1)\zeta$.

As explained in the Appendix, the cohomology of $\mathbb{CP}^{\infty}_{-\infty}$ with integral coefficients is $\mathbb{Z}[u, u^{-1}]$, the ring of Laurent polynomials in u, where u has degree 2. Thus, as a group, this is one copy of $\mathbb{Z}$ in every even dimension. Computing the first Chern class of $\mathbb{CP}^n$ shows that the Floer homology $\mathrm{HF}_*(\mathscr{L}\mathbb{CP}^n)$ is $\mathbb{Z}/(2n+2)$ graded, and Floer [12] shows that these groups are $\mathbb{Z}$ in even degrees and 0 in odd degrees.

The relation between these groups is as follows. Let $e((n+1)\zeta)$ be the Euler class of the bundle $(n+1)\zeta$, which is the bundle which naturally occurs in the above sequence of Thom spectra; of course $e((n+1)\zeta) = u^{n+1}$. If we now set $e((n+1)\zeta)$ to be 1, we get Floer's groups with their $\mathbb{Z}/(2n+2)$ grading:

$$\mathrm{HF}_*(\mathscr{L}\mathbb{CP}^n) = \frac{\mathbb{Z}[u, u^{-1}]}{(u^{n+1} - 1)} = \frac{H^*(\mathbb{CP}^{\infty}_{-\infty})}{(e((n+1)\zeta) - 1)}.$$

The Floer cohomology of $\mathcal{L}\mathbb{CP}^n$ has a ring structure using the «pair of pants» product and what is more the above isomorphisms are isomorphisms of rings.

We close this section with some final comments.

(i) It seems very likely that the same method will work for the area functional on $\mathcal{L}Gr_k(\mathbb{C}^n)$ where $Gr_k(\mathbb{C}^n)$ is the Grassmannian of complex k planes in $\mathbb{C}^n$. In this case the ring strucuture of Floer cohomology ring has been computed, by Witten; it is the «deformed cohomology ring» of the Grassmannian. Once more, we expect the Floer homotopy type to come from an inverse system of Thom spectra, and to find the Floer cohomology ring is given by a formula similar to the one which arises in the case of $\mathbb{CP}^n$.

(ii) In the projective space examples, the pro-spectra $\mathbb{RP}^{\infty}_{-\infty}$ and $\mathbb{CP}^{\infty}_{-\infty}$ are indeed the natural candidates for the semi-infinite homotopy type of the polarized manifolds $\mathbb{P}(V)$ and $\mathbb{P}(W)$. The polarization of these projective spaces is defined by the natural polarization of V and W. For example the construction of $\mathbb{RP}^{\infty}_{-\infty}$ can be phrased so that it only depends on the polarization $V = V^- \oplus V^0 \oplus V^+$ where V^- has basis $\{\delta_i\}_{i<0}$, V^+ has basis $\{\delta_i\}_{i>0}$ and V^0, the «finite dimensional ambiguity», is the one dimensional space spanned by δ_0.

Let W be a finite dimensional subspace of V such that $W = W^- \oplus W^0 \oplus W^+$ where $W^\pm = W \cap V^\pm$ and $W^0 = W \cap V^0$. We refer to $W^0 \oplus W^+$ as the positive part of W and W^- as the negative part. To W we associate the Thom space $\mathbb{P}(W)^{-\xi}$ of the virtual bundle $-\xi$ where ξ is the vector bundle over $\mathbb{P}(W)$ defined by W^-. If we choose bases then we get an isomorphism $\xi \cong (\dim W^-)\eta$ where η is the Hopf line bundle. In the stable category $\mathbb{P}(W)^{-\xi}$ has one i-cell for each i with $-\dim W_- \leq i \leq \dim W - \dim W^-$.

Now suppose we have an inclusion $W_1 \to W_2$ which preserves the decompositions. If it is an isomorphism on the negative part, i.e. increases the positive part, then we get an obvious map

$$\mathbb{P}(W_1)^{-\xi_1} \to \mathbb{P}(W_2)^{-\xi_2}.$$

On the other hand, if it is an isomorphism on the positive part, i.e. increases the negative part, then we get a map

$$\mathbb{P}(W_1)^{-\xi_1} \leftarrow \mathbb{P}(W_2)^{-\xi_2}.$$

This map is constructed as follows. The bundle ξ_1 is a sub-bundle of ξ_2 restricted to $\mathbb{P}(W_1)$ and therefore the virtual bundle $-\xi_2$ restricted to $\mathbb{P}(W_1)$ is a sub-virtual-bundle of $-\xi_1$. This gives a map (in the stable category) of Thom spaces $\mathbb{P}(W_1)^{-\xi_1} \leftarrow \mathbb{P}(W_3)^{-\xi_2}$.

Thus we get a system of Thom spectra indexed by subspaces W of V. An inclusion which is an isomorphism on the negative part (i.e increases the positive part) gives a map in the same direction whereas an inclusion which is an isomorphism on the positive part (i.e. increases the negative part) gives a map in the other direction. Using the basis for V, we can use the subspaces V_a^b to reduce this system to one indexed by pairs of integers, and this gives the pro-spectrum $\mathbb{RP}^{\infty}_{-\infty}$. This construction fits in rather well with the coordinate free theory of spectra described in [19], where the spaces defining the spectrum are indexed by finite dimensional subspaces of an infinite dimensional real vector space, rather than the integers. Here we have spaces indexed by the finite dimensional subspaces of a polarized vector space.

Therefore, we are able to associate a «semi-infinite homotopy type to $\mathbb{P}(V)$ which depends only on the polarization of V, and the Floer function $f : \mathbb{P}(V) \to \mathbb{R}$ does indeed compute the «semi-infinite» cohomology of $\mathbb{P}(V)$. However the construction depends very heavily on special features of $\mathbb{P}(V)$.

Appendix — Spectra and pro-spectra

In stable homotopy theory it is convenient, and it greatly simplifies many arguments, to be able to work in a suitable stable category of spaces. The *stable category* $\mathcal{S}$ of finite CW complexes is defined, essentially, by inverting the suspension functor on the category of finite CW complexes. The objects of $\mathcal{S}$ are defined to be $S^n X$ where X is a finite CW complex, and $n \in \mathbb{Z}$. If n is positive then $S^n X$ is just the n-th suspension of X, but by allowing n to be negative we have introduced formal desuspensions of X. More precisely, the objects of $\mathcal{S}$ are pairs (X, n), where X is a finite CW complex and $n \in \mathbb{Z}$, modulo the equivalence relation generated by identifying (X, n) with $(S^n X, 0)$ if n is positive, and $S^n X$ is the equivalence class of (X, n). A map from $S^n X \to S^m Y$ is defined by giving a map $S^{n+k} X \to S^{m+k} Y$ where k is chosen large enough so that both $S^{n+k} X$ and $S^{m+k} Y$ are genuine CW complexes. Two maps are identified if they are equal after a suitably large number of suspensions. Thus,

$$\operatorname*{Map}_{\mathcal{S}}(S^n X, S^m Y) = \lim_{k \to \infty} \operatorname{Map}(S^{n+k} X, S^{m+k} Y)$$

where the maps in the direct system are given by suspension.

The main difficulty with this stable category is that it is not closed under direct and inverse limits. The category of *spectra* $\mathcal{SP}$ is defined, essentially, to be the category one obtains by formally adjoining direct limits of sequences in the stable category. Thus a spectrum $\mathbf{K}$ is given by a sequence of CW complexes $\{K_p\}$ and maps $SK_p \to K_{p+1}$. One thinks of the spectrum $\mathbf{K}$ as the direct limit of the sequence

$$K_0 \to S^{-1}K_1 \to \dots \to S^{-p}K_p \to S^{-p+1}K_{p+1} \to \cdots$$

in $\mathcal{S}$. Note that it is only necessary to define the spaces K_{p_n} and the maps $S^{p_{n+1}-p_n} K_{p_n} \to K_{p_{n+1}}$ for a strictly increasing sequence of integers p_n; for if

$p_n < m < p_{n+1}$ we can define $K_m = S^{m-p_n}K_{p_n}$ and use the identity map $SK_m \to K_{m+1}$ if $m+1 \neq p_{n+1}$.

It takes some work to define maps between spectra and set up a good category of spectra. The essential difficulty is the usual one when dealing with maps from a direct limit **K** to a direct limit **L**; it is certainly the case that a map of direct systems defines such a map but there are many different direct systems with limits **K** and **L**. However the technicalities involved in the definition of a suitable category of spectra are well-understood; see [1] and [19]. In particular [19] gives the definition from the «coordinate-free» point of view which leads to a category of spectra with all the good properties one could expect.

Spectra define generalised cohomology theories: if X is a finite CW complex

$$h^p(X;\mathbf{K}) = \lim_{k\to\infty} [S^{k-p}X; K_k].$$

The maps in the direct system are defined by

$$[S^{k-p}X, K_k] \to [S^{k-p+1}X, SK_k] \to [S^{k-p+1}X, K_{k+1}],$$

where the first map is given by suspension, and the second by the structure maps of the spectrum **K**. In the literature this group $h^p(X;\mathbf{K})$ is often denoted by $\mathbf{K}^p(X)$. Furthermore every generalised cohomology theory arises in this way. This is one of the main justifications for introducing spectra.

For some purposes the category of spectra is still not big enough. In the study of the Segal conjecture it becomes clear that one also needs pro-spectra, inverse systems of spectra. The most convincing argument for the necessity of pro-spectra is given in [2], pages 5–6. By definition, a *pro-spectrum* is a doubly indexed family of finite CW complexes $\{X_{p,q}\}$ equipped with maps

$$SX_{p,q} \to X_{p,q+1}, \qquad X_{p,q} \to X_{p,q-1}.$$

Thus if we fix p the sequence of spaces $X_{p,q}$ with structure maps $SX_{p,q} \to X_{p,q+1}$ form a spectrum $\mathbf{X}_p$ and the maps $X_{p,q} \to X_{p-1,q}$ give a map of spectra $\mathbf{X}_p \to \mathbf{X}_{p-1}$; so we have an inverse system of spectra

$$\cdots \leftarrow \mathbf{X}_{p-1} \leftarrow \mathbf{X}_p \leftarrow \cdots.$$

It is only necessary to define these spaces and maps for a strictly increasing sequences of integers p_n and q_n. Furthermore, to define a pro-spectrum it is sufficient to give the structure maps in the stable category $\mathcal{S}$.

The first example of a pro-spectrum which arises in the main text occurs in §5, where for any pair of integers a and b with $a < b$, we have spaces $|Z|_a^b$, and for $a' \leq a < b \leq b'$, we have maps

$$S^{b'-b}|Z|_a^b \to |Z|_a^{b'}, \qquad |Z|_{a'}^b \to S^{a-a'}|Z|_a^b.$$

The corresponding pro-spectrum is defined by $X_{a,b} = S^a|Z|_a^b$ and the natural structure maps. Note that since a can be negative, these spaces and maps really do lie in the stable category; as noted above, they nonetheless define a pro-spectrum.

A good example which illustrates these ideas and is of considerable importance in our approach to Floer homotopy type arises from the theory of Thom spaces. Recall that the Thom space X^E of a vector bundle E over a compact space X is defined to be the one-point compactification E^+ of the total space of E. Now consider the problem of defining the Thom space of a virtual vector bundle ξ over X. A virtual vector bundle is an element $\xi \in KO(X)$ and its dimension, which is an integer, is defined by the homomorphism $KO(X) \to KO(\mathrm{pt}) = \mathbb{Z}$ given by the inclusion of a point in X. We are assuming, of course, that X is connected. By standard properties of K-theory we can find a genuine vector bundle E over X such that

$$\xi = E - k \in KO(X),$$

where k is a trivial bundle of dimension k. Now X^ξ is defined to be the object $S^{-k}X^E$ in the stable category $\mathcal{S}$. It is not difficult to check that in $\mathcal{S}$ the homotopy type of X^ξ does not depend on the choice of E and k.

Now suppose that X is CW complex of finite type (this means that the n-skeleton $X^{(n)}$ of X is a finite CW complex for each n) and ξ is a virtual vector bundle on X. In this case we cannot necessarily choose a vector bundle E such that $\xi = E - k$, but we can choose vector bundles $E^{(n)}$ over $X^{(n)}$ such that

$$\xi^{(n)} = E^{(n)} - k_n \in KO(X^{(n)}),$$

where $\xi^{(n)}$ is the restriction of ξ to $X^{(n)}$. Furthermore these bundles can be chosen so that there is a bundle map $E^{(n)} + k_m - k_n \to E^{(m)}$, covering the inclusion of $X^{(n)} \to X^{(m)}$, which is an isomorphism on fibres. Thus we get maps

$$S^{k_{n+1}-k_n}(X^{(n)})^{E^{(n)}} \to (X^{(n+1)})^{E^{(n+1)}}.$$

and the spaces $\{(X^{(n)})^{E^{(n)}}\}$ with these maps define a spectrum. This is the Thom spectrum of ξ, denoted by X^ξ. Once more the homotopy type of X^ξ does not depend on the choices made in its definition. For example, the MU-spectrum used in §5 is the Thom spectrum of the universal bundle over BU.

On a compact space X, if E is a sub-vector-bundle of F, we get a map of Thom complexes $X^E \to X^F$. Similarily if X has finite type and ξ is a sub-virtual-vector-bundle of η, which means that there is a genuine vector bundle E such that $\xi + E = \eta$, then we get a map of Thom spectra

$$X^\xi \to X^\eta.$$

In particular, suppose that X has finite type, and E is a vector bundle over X. Then $-E$ is a virtual vector bundle over X, and if $k \geq 0$ then $-kE$ is a sub-virtual-vector-bundle of $(-k+1)E$. So we get maps $X^{-kE} \to X^{(-k+1)E}$ and we can form the inverse system of Thom spectra

$$X \leftarrow X^{-E} \leftarrow X^{-2E} \leftarrow \cdots$$

with k-th term X^{-kE} where $k \geq 0$. This inverse system of Thom spectra is a pro-spectrum which we will denote by $X^{-\infty E}$. The examples which occur in the main text are the cases where $X = \mathbb{RP}^\infty$ and E is the real Hopf line bundle; $X = \mathbb{CP}^\infty$ and E is the complex Hopf line bundle. These pro-spectra are denoted by $\mathbb{RP}^\infty_{-\infty}$ and $\mathbb{CP}^\infty_{-\infty}$ respectively.

The cohomology of a pro-spectrum $\mathbf{X}$ is defined as follows. The pro-spectrum is an inverse system of spectra

$$\cdots \leftarrow \mathbf{X}_{p-1} \leftarrow \mathbf{X}_p \leftarrow \cdots$$

and then

$$H^*(\mathbf{X}) = \lim_{p \to \infty} H^*(\mathbf{X}_p).$$

In good cases, that is where there is no $\lim^1$-term, we get

$$H^*(\mathbf{X}) = \lim_{p \to \infty} \lim_{\infty \leftarrow q} H^*(X_{p,q})$$

where the spaces $\{X_{p,q}\}$, with the appropriate structure maps, define the pro-spectrum $\mathbf{X}$.

In the case of the pro-spectrum $X^{-\infty E}$ defined by a vector bundle over a CW complex of finite type, it is straightforward to compute cohomology. If E is orientable it has an Euler class $e(E) \in H^{\dim(E)}(X;\mathbb{Z})$, and

$$H^*(X^{-\infty E};\mathbb{Z}) = H^*(X;\mathbb{Z})[e(E)^{-1}].$$

If E is not orientable then it has an Euler class in mod 2 cohomology, and the mod 2 cohomology of $X^{-\infty E}$ is given by the same formula. For $\mathbb{RP}^\infty_{-\infty}$ and $\mathbb{CP}^\infty_{-\infty}$ this shows that

$$H^*(\mathbb{RP}^\infty_{-\infty};\mathbb{Z}/2) = \mathbb{Z}/2[x,x^{-1}]$$
$$H^*(\mathbb{CP}^\infty_{-\infty};\mathbb{Z}) = \mathbb{Z}[u,u^{-1}].$$

Here x has degree 1 and corresponds to the Euler class of the real Hopf line bundle in $H^1(\mathbb{RP}^\infty;\mathbb{Z}/2)$; u has degree 2 and corresponds to the Euler class of the complex Hopf line bundle in $H^2(\mathbb{CP}^\infty;\mathbb{Z})$.

References

[1] J.F. Adams, *Stable homotopy and generalised homology*, Chicago Lecture Notes in Mathematics, The University of Chicago Press, Chicago and London, 1974.

[2] J.F. Adams, *Operations of the nth kind in K-theory, and what we don't know about* $\mathbb{RP}^\infty$, New developments in topology, LMS Lecture Note Series **11**, Cambridge University Press, Cambridge, 1974.

[3] J.F. Adams, *Graeme Segal's Burnside ring conjecture*, Bulletin of the AMS (New Series) **6**, 201–210.

[4] M.F. Atiyah, *New invariants of 3- and 4-dimensional manifolds*, The Mathematical Heritage of Hermann Weyl, Proceedings of Symposia in Pure Mathematics **48**, AMS, Providence, Rhode Island, 1988, 285–299.

[5] M.F. Atiyah and J.D.S. Jones, *Topological aspects of Yang-Mills theory*, Communications in Mathematical Physics **61**, 1978, 97–118.

[6] D.M. Austin and P.J. Braam, *Morse-Bott theory and equivariant cohomology*, this volume.

[7] C.P. Boyer, J.C. Hurtubise, B.M. Mann, and R.J. Milgram, *The topology of instanton moduli spaces I: The Atiyah-Jones conjecture*, preprint, 1992.

[8] G. Carlsson, *Segal's Burnside ring conjecture and the homotopy limit problem*, Homotopy Theory (Durham 1985), LMS Lecture Note Series **117**, 1987, 6–34.

[9] R.L. Cohen, J.D.S. Jones, and G.B. Segal, *Morse theory and classifying spaces*, preprint, 1992.

[10] S.K. Donaldson and P.B. Kronheimer, *The Geometry of Four-Manifolds*, Oxford Mathematical Monographs, Oxford University Press, Clarendon Press, Oxford, 1990.

[11] B. Feigen and E. Frenkel, *Affine Kac-Moody algebras and semi-infinite flag manifolds*, Communications in Mathematical Physics **128**, 1990, 161–189.

[12] A. Floer, *Morse theory for Lagrangian intersections*, Journal of Differential Geometry **28**, 1988, 513–547.

[13] A. Floer, *An instanton invariant for* 3-*manifolds*, Communications in Mathematical Physics **118**, 1988, 215–240.

[14] J.M. Franks, *Morse-Smale flows and homotopy theory*, Topology **18**, 1979, 119–215.

[15] J. Gravesen, *On the topology of spaces of holomorphic maps*, Acta Mathematica **162**, 1989, 249–286.

[16] M.A. Guest, *Topology of the space of absolute minima of the energy functional*, American Journal of Mathematics **106**, 1984, 21–42.

[17] F. Kirwan, *On spaces of maps from Riemann surfaces to Grassmannians and the cohomology of moduli spaces of vector bundles*, Arkiv Math. **122**, 1985, 221–275.

[18] F. Kirwan, *Geometric invariant theory and the Atiyah-Jones conjecture*, preprint, Oxford 1992 (to appear).

[19] L.G. Lewis, J.P. May, and M. Steinberger (with contributions by J.E. McClure), *Equivariant stable homotopy theory*, Springer Lecture Notes in Mathematics **1213**, Springer-Verlag, Berlin, Heidelberg, New York, 1986.

[20] W.H. Lin, D.M. Davis, M.E. Mahowald, J.F. Adams, *Calculation of Lin's Ext groups*, Math. Proc. Camb. Phil. Soc. **87**, 1980, 459–469.

[21] M.E. Mahowald, *The metastable homotopy of* S^n Memoirs Amer. Math. Soc. **72**, AMS, Providence, Rhode Island, 1967.

[22] A.N. Pressley and G.B. Segal, *Loop groups*, Oxford Mathematical Monographs, Oxford University Press, Clarendon Press, Oxford, 1986.

[23] G.B. Segal *Classifying spaces and spectral sequences*, Publications I.H.E.S **34**, 1968, 105–112.

[24] G.B. Segal, *The topology of spaces of rational functions*, Acta Mathematica **143**, 1979, 39–72.

[25] G.B. Segal, *Elliptic Cohomology (after Landweber-Stong, Ochanine, Witten and others)*, Séminaire Bourbaki 1987–1988, Astérisque **161–162**, 1988, 191–202.

[26] S. Smale, *On gradient dynamical systems*, Annals of Mathematics **74**, 1961, 199–206.

[27] C.H. Taubes, *The stable topology of self-dual moduli spaces*, Journal of Differential Geometry **29**, 1989, 163–230.

R.L. Cohen
Mathematics Department,
Stanford University,
Stanford,
California CA 94305,
USA

G.B. Segal
DPMMS,
Cambridge University,
16 Mill Lane,
Cambridge CB2 1SB,
England

J.D.S. Jones
Mathematics Institute,
Warwick University,
Coventry CV4 7AL,
England

Periodic solutions of elliptic type for strongly nonlinear Hamiltonian systems

G. Dell'Antonio
B. D'Onofrio
I. Ekeland

Summary. It is shown that, for autonomous Hamiltonian systems, every convex energy level which is symmetric with respect to the origin carries a periodic solution of elliptic type. An extension is given to nonautonomous systems.

I The main result

In a celebrated experiment, way back in 1950, Fermi, Pasta and Ulam studied numerically a system of 64 nonlinear springs, each one coupled to its two neighbours (see [8] for the original paper, and [2] for a discussion). The global Hamiltonian was of the following type:

$$H(p,q) = \sum_i \frac{1}{2} p_i^2 + \sum_i \left[\frac{1}{2}(q_i - q_{i-1})^2 + \frac{\varepsilon}{4}(q_i - q_{i-1})^4\right]$$

with appropriate boundary conditions.

This was one of the first numerical experiments ever conducted, and it made immediately clear how important the newly invented electronic computer was going to be in the study of dynamical systems. In fact, Fermi, Pasta and Ulam were looking for chaos: they were expecting quasi-random behaviour, in accordance with Birkhoff's ergodic hypothesis. What they observed instead was quasi-periodic behaviour, which today is attributed to KAM theory, since the nonlinearity is small.

The result we shall presently describe will imply that quasi-periodic behaviour is present in a large class of strongly nonlinear oscillators. It is for instance the case for Hamiltonians of the following type:

$$H(p,q) = \sum_i \frac{1}{2} p_i^2 + \sum_{i,j} V(q_i - q_j),$$

provided the interaction potential V is convex, even, increases to infinity at infinity, and satisfies some generic conditions.

More precisely, let us endow $\mathbb{R}^{2n} = \mathbb{R} \times \mathbb{R}^n$ with its standard symplectic structure, and consider a Hamiltonian $K \in \mathscr{C}(\mathbb{R}^{2n}, \mathbb{R})$. In the standard basis, Hamilton's equations are written:

$$\dot{x} = JH'(x)$$

where

$$H' = \left(\frac{\partial H}{\partial x_i}\right)_{i=1,\dots,2n}$$

$$J = \begin{pmatrix} 0 & I \\ -I & 0 \end{pmatrix}.$$

Let $x(t)$ be a periodic solution, so that $x(0) = x(T)$ for some $T > 0$. The linearized equation around x is written

$$\dot{y} = JH''(x(t))y$$

with obvious notations. The fundamental solution (or matrizant) $R(t)$ is given by

$$\dot{R} = JH''(x(t))R$$
$$R(0) = I.$$

The T-periodic solution $x(t)$ is called *elliptic* if all the eigenvalues of $R(T)$ are on the unit circle (of course 1 is always an eigenvalue, with even multiplicity). The (Morse) index of $x(t)$ is defined in the book 7, to which we refer for the relevant theory.

Let S be a C^2 hypersurface in $\mathbb{R}^{2n}$, bounding a compact convex set with non-empty interior. We can find many different Hamiltonians H and energy levels h such that

$$S = \{x \mid H(x) = h\}$$

but they are all equivalent, in the sense that they lead to the same Hamiltonian trajectories on S (up to time reparametrization): see 7, Prop. I.6.12. For this reason, we will choose H to be positively homogeneous of degree α, and denote it by H_α to remind us of this fact:

$$S = \{x \mid H_\alpha(x) = 1\}$$
$$H_\alpha(\lambda x) = \lambda^\alpha H(x) \quad \forall \lambda > 0.$$

In practice, we will use H_2 and H_α for $1 < \alpha < 2$. Note that the index of a closed trajectory x does not depend on the choice of α in the range $]1,2]$ (see 7, Prop. I.7.5). The eigenvalues of $R(T)$ do not depend on the choice of the Hamiltonian H, although R and T do (see 7, Prop. I.6.12).

Our main result is as follows:

Theorem 1. *Assume S is symmetric with respect to the origin and bounds a compact convex domain with non-empty interior. Assume moreover that $H_2''(x)$ is positive definite for all $x \in S$. Then S carries a closed Hamiltonian trajectoty $\bar{x}(t)$ which is elliptic and has index zero.*□

The requirement that $H_2''(x)$ be positive definite is a nondegeneracy condition on the curvature of S. By compactness it follows that

$$\| y \| R^{-2} \leq \frac{1}{2}(H''(x)y, y) \quad \forall x \in S$$

for some $R > 0$. The value 0 for the index refers to the Hamiltonians H_α, $1 < \alpha < 2$, and implies an upper bound for the length of $\bar{x}(t)$, namely (see 7, Prop. V.1.8):

$$\int_0^T \left(J\bar{x}, \frac{d\bar{x}}{dt}\right) dt = \oint (J\bar{x}, d\bar{x}) \leq \pi R^2.$$

Since $\bar{x}(t)$ is elliptic, we will be able to apply KAM theory as soon as some additional nondegeneracy conditions are satisfied. This will imply the existence of invariant tori near $\bar{x}(t)$, filling up a set of positive measure on S so that the Hamiltonian flow on S will not be ergodic. Since these nondegeneracy conditions are known to be generic, we have indeed found a large class of strongly nonlinear oscillators where quasi-periodic motion appears.

In a earlier result (6, 7 Theorem V.1.10) Ekeland has found an elliptic trajectory of index zero without assuming S to be symmetric, but requiring instead a pinching assumption on the curvature. Writing by compactness

$$\frac{1}{2}(H''(x)y, y) \leq \| y \|^2 r^{-2}$$

he assumed $Rr^{-1} < \sqrt{2}$. In unrelated work, Dell'Antonio, D'Onofrio and Ekeland (3, 4, 5) have found elliptic or non-hyperbolic solutions for various kinds of Hamiltonian systems.

II. Proof of the main result

We fix $\alpha \in]1, 2]$ and H_α such that

$$S = \{x \mid H_\alpha(x) = 1\}$$

$$H_\alpha(\lambda x) = \lambda^\alpha H(x).$$

Since S is symmetric with respect to the origin, H_α is even:

$$H_\alpha(-x) = H_\alpha(x).$$

We will look for a solution $\bar{x}(t)$ which is antiperiodic; that is, there should exist some $\frac{T}{2} > 0$ such that

$$(1) \qquad \begin{cases} \dfrac{d\bar{x}}{dt} = JH'_\alpha(\bar{x}(t)) \\ H_\alpha(\bar{x}(t)) = 1 \quad \forall t \\ \bar{x}(0) + \bar{x}\left(\dfrac{T}{2}\right) = 0. \end{cases}$$

To do this, by standard procedure, we seek a $\frac{1}{2}$-periodic solution of $\dot{x} = JH'_\alpha(x)$, and then rescale it back to the energy level $H = 1$ (see 7, Lemma I.7.2 and §V.1). In other words, we replace the fixed-energy problem (1) with the fixed-period problem (2):

$$\text{(2)} \qquad \begin{cases} \dfrac{dx}{dt} = JH'_\alpha(x(t)) \\ x(0) + x\left(\dfrac{1}{2}\right) = 0. \end{cases}$$

The latter is equivalent to the variational problem (3)

$$\text{(3)} \qquad \begin{cases} F'(x) = 0 \\ x \in E \end{cases}$$

with

$$F(x) = \int_0^{\frac{1}{2}} \left[\frac{1}{2}(J\dot{x}, x) + H_\alpha(x)\right] dt$$

$$E = \left\{x \in W^{1,\beta}\left(0, \frac{1}{2}; \mathbb{R}^{2n}\right) \mid x(0) + x\left(\frac{1}{2}\right) = 0\right\}.$$

By the Clarke duality formula (7, Prop. II.4.6) problem (3) turns out to be equivalent to the following:

$$\text{(4)} \qquad \begin{cases} \Phi'(x) = 0 \\ x \in E \end{cases}$$

with

$$\Phi(x) = \int_0^{\frac{1}{2}} \left[\frac{1}{2}(J\dot{x}, x) + H^*_\alpha(-J\dot{x})\right] dt$$

where H^*_α is the Legendre transform of H_α (or Fenchel conjugate; see 7 Prop. II.1.7). Note that it is positively homogeneous of degree β, with $\frac{1}{\alpha} + \frac{1}{\beta} = 1$, so that $\beta > 2$.

Setting $u = \dot{x} \in L^\beta$, we find

$$0 = x(0) + x\left(\frac{1}{2}\right) = 2x(0) + \int_0^{\frac{1}{2}} u\, dt$$

so that $x(0) = -\frac{1}{2}\int_0^{\frac{1}{2}} u\,dt$ and problem (4) can be rewritten as follows:

$$\text{(5)} \qquad \begin{cases} \varphi'(u) = 0 \\ u \in L^\beta \end{cases}$$

with

$$\varphi(u) = \int_0^{\frac{1}{2}} \left[\frac{1}{2} \left(Ju, \int_0^t u\,ds - \frac{1}{2} \int_0^{\frac{1}{2}} u\,ds \right) + H_\alpha^*(-Ju) \right] dt.$$

The functional φ is the sum of two terms. The first one is compact and quadratic, the second is convex and has degree $\beta > 2$. By standard arguments (see 7 Theorem III.2.1) it attains its minimum, and this minimum is not 0 (see / Prop. III.3.1); one can also refer to the original paper of Clarke-Ekeland [1].

We have thus found a non-constant $x \in W^{1,\beta}$ such that $\Phi'(x) = 0$, and since it is a minimizer, the quadratic form

$$Q(y, y) = \frac{1}{2} \int_0^{\frac{1}{2}} [(J\dot{y}, y) + ([H_\alpha^*]''(-J\dot{x})J\dot{y}, J\dot{y})]dt$$

must be non-negative definite on E.

So x solves problem (2). We extend it by symmetry to $[0, 1]$:

$$x\left(t + \frac{1}{2}\right) = -x(t) \quad \text{for} \quad 0 \leq t \leq \frac{1}{2}.$$

Since H_α is even, x is a solution of $\dot{x} = JH_\alpha'(x)$ on $[0, 1]$, and it is 1-periodic. In addition,

$$A(t) = [H_\alpha^*]''(-J\dot{x}(t))$$

is $\frac{1}{2}$-periodic. We may therefore apply the index theory and the iteration formula of 7, §I.4 and 5, to the linear Hamiltonian system $\dot{y} = JA(t)y$, and introduce the Bott map $j_{\frac{1}{2}}$ from the unit circle $\mathbb{U}$ to $\mathbb{N}$ (see 7 Def. I.5.3). Since Q is non-negative definite on E, we must have

$$\text{(6)} \qquad j_{\frac{1}{2}}(-1) = 0.$$

On the other hand, looking at the situation near 1 on the unit circle $\mathbb{U}$, we have

$$\text{(7)} \qquad j_{\frac{1}{2}}(e^{\pm i\varepsilon}) \geq n \quad \text{for} \quad |\varepsilon| \quad \text{small}.$$

This is proved in 7, Prop. I.5.10 and 11. It can also be seen directly, by testing the Hermitian from Q on the complex $2n$-dimensional space L_ε of $W^{1,\beta}(0; \frac{1}{2}; \mathbb{C}^{2n})$ spanned by the functions $z(t) = \xi e^{i\varepsilon t}$, for $\xi \in \mathbb{C}^{2n}$. We get:

$$Q(z,z) = \frac{\varepsilon}{4}(iJ\xi,\xi)\frac{\varepsilon^2}{2}\left(\int_0^{\frac{1}{2}} A(t)dt, J\xi, J\xi\right)$$

and for $\varepsilon \to 0$ we find that the restriction of Q to L_ε has the same index as the Hermitian form iJ on $\mathbb{C}^{2n}$, namely n. Since $z\left(\frac{1}{2}\right) = e^{i\varepsilon/2}z(0)$, this shows that $j_{\frac{1}{2}}(e^{i\varepsilon/2})$ is at least n.

Comparing (6) and (7) we see that $j_{\frac{1}{2}}(\omega)$ must drop from at least n to 0 as one travels from $+1$ to -1 along the upper unit circle. But the only jumps of $j_{\frac{1}{2}}(\omega)$ are at the eigenvalues of $R\left(\frac{1}{2}\right)$: they are downwards at a Krein-positive eigenvalue, upwards at a Krein-negative one (see 7, Prop. I.5.9). Since there are exactly $2n$ eigenvalues of $R\left(\frac{1}{2}\right)$, the only possibility is that they are all on the unit circle, the Krein-positive ones on the upper half-circle and the Krein-negative ones downstairs. Note, however, that -1 must be an eigenvalue of $R\left(\frac{1}{2}\right)$ (see 7 Prop. I.5.11 for $j_{\frac{1}{2}}(-1) = 0$ whereas $j_{\frac{1}{2}}(-e^{i\varepsilon}) > 0$ for $|\varepsilon|$ small).

So all the eigenvalues of $R\left(\frac{1}{2}\right)$ have modulus 1. Since $R(1) = R\left(\frac{1}{2}\right)^2$ (remember that $A(t)$ is $\frac{1}{2}$-periodic) the same is true for $R(1)$. So $x(t)$ is a 1-periodic solution of elliptic type.

To compute its index, we note that $j_{\frac{1}{2}}(e^{i\varepsilon})$ cannot be greater than n, otherwise $j_{\frac{1}{2}}(1)$ would be greater than 0, since $j_{\frac{1}{2}}(\omega)$ can fall at most by n, the number of eigenvalues upstairs. So $j_{\frac{1}{2}}(1) = 0$ by 7 Prop. I.5.11 and we can use the iteration formula of 7 Cor. I.5.4 to compute the index of x:

$$j_1(1) = j_{\frac{1}{2}}(-1) + j_{\frac{1}{2}}(1) = 0.$$

Everything is proved. All that remains to do is to rescale x to the energy level $H = 1$ by setting

$$\bar{x}(t) = h^{-1/\alpha}x\left(h^{\frac{2-\alpha}{\alpha}}t\right)$$

where $h = H(x(t))$. Then $\bar{x}$ is a T-periodic solution, with $T = h^{-\frac{2-\alpha}{\alpha}}$, lies on S is elliptic and has index 0.

III. The nonautonomous case

Using the same method, it is not difficult to prove the following:

Theorem 2. *Let $H \in \mathcal{C}(\mathbb{R} \times \mathbb{R}^{2n}, \mathbb{R})$ and $T > 0$ be such that:*

(a) *for every x fixed, H is $\frac{T}{2}$-periodic in t*

$$H\left(t + \frac{T}{2}, x\right) = H(t,x)$$

(b) *for every t fixed, H is convex and even in x:*

$$H''(t,x) \;\; \textit{positive definite}$$

$$H(t,-x) = H(t,x)$$

(c) *H is subquadratic at infinity*

$$H(t,x) \parallel x \parallel^{-2} \quad \textit{as} \quad \parallel x \parallel \to \infty.$$

If $H''(t,0) > \frac{2\pi}{T} I$, then the system $\dot{x} = JH'(t,x)$ has a non-constant T-periodic solution of elliptic type.□

Bibliography

1. F. Clarke et I. Ekeland, «Hamiltonian trajectories having prescribed minimal period», *Comm. Pure App. Math.* **33** (1980), p. 103–116.
2. N. Cooper, ed., «From Cardinal to Chaos», Cambridge University Press 1989.
3. G. Dell'Antonio, B. D'Onofrio, I. Ekeland, «Stability form index estimates for periodic solutions of Lagrangian systems», submitted to *Journal of Differential Equations.*
4. B. D'Onofrio, I. Ekeland, «Hamiltonian systems with elliptic periodic orbits», *Nonlinear Analysis TMA.*
5. B. D'Onofrio, I. Ekeland, «Morse theory and the existence of periodic solutions of elliptic type», in *Variational methods*, Berestycki, Ekeland and Coron ed., Birkhäuser, Progress in Nonlinear Differential Equations and their Applications, 1990, p. 447–454.
6. I. Ekeland, «An index theory for periodic solutions of convex Hamiltonian systems», Proceedings for Symposia in Pure Mathematics, **45** (1986), p. 395–423.
7. I. Ekeland, «Convexity methods in Hamiltonian Mechanics», Springer Verlag 1990.
8. E. Fermi, J. Pasta, S. Ulam, «Studies of nonlinear problems», in Enrico Fermi: Collected papers» vol II, p. 978–998, University of Chicago Press, 1965.

Gianfausto Dell'Antonio
Dept. of Mathematics
University of Rome
Piazzale Aldo Moro
I-00185 Roma
Italy

Biancamaria D'Onofrio
Dept. of Mathematics
University of Rome
Piazzale Aldo Moro
I-00185 Roma
Italy

Ivar Ekeland
Université Paris-Dauphine
Place du Marechal de
Lattre de Tassigny
F-75775 Paris Cedex 16
France

Topology of 2-knots in $\mathbb{R}^4$ and symplectic geometry

Y. Eliashberg*

1 Introduction

We show in this paper that there exists a deep relationship between the differential topology of S^2-knots in $\mathbb{R}^4$ and their symplectic geometry. In particular, we use symplectic tools to define a real-valued topological invariant of a knotted S^2 in $\mathbb{R}^4$ (see Section 3.4 below). Here are the main results which motivate this definition.

1.1 The unknottedness of positive and Lagrangian surfaces

The 2-dimensional complex space $\mathbb{C}^2$ with coordinates $z_1 = x_1 + iy_1$, $z_2 = x_2 + iy_2$ we will identify with the symplectic space $(\mathbb{R}^4, \omega = dx_1 \wedge dy_1 + dx_2 \wedge dy_2)$. Given a domain $\Omega \subset \mathbb{R}^4$ with the boundary $\partial\Omega$, we say that a 2-disc $(\Delta, \partial\Delta) \subset (\Omega, \partial\Omega)$ is *unknotted* if it is isotopic relative to $\partial\Delta$ to a disc which spans $\partial\Delta$ in $\partial\Omega$.

Theorem 1.1.A *Suppose that a 2-disc* $(\Delta, \partial\Delta) \subset (\mathbb{R}^4_+ = \{y_2 \geq 0\}, \partial\mathbb{R}^4_+ = \mathbb{R}^3)$ *intersects* $\mathbb{R}^3$ *transversely along the circle* $\partial\Delta = \{|z_1| = 1, z_2 = 0\}$. *If* $\omega|_\Delta \geq 0$ *then* Δ *is unknotted.*

Theorem 1.1.A′ *Let* $(\Delta, \partial\Delta) \subset (B, \partial B)$ *be a 2-disc in a round 4-ball* $B \subset \mathbb{C}^2$. *Suppose that* $\omega|_\Delta \geq 0$, $\partial\Delta$ *is unknotted in* ∂B *and transversal to the distribution of complex tangencies to* ∂B *(see 2.1 below). Then* Δ *is unknotted.*

Remarks. a) As it will be shown (see Theorem 2.4.A), Theorems 1.1.A and 1.1.A′ hold in a stronger form: if $\omega|_\Delta > 0$ then Δ is *symplectically unknotted.*

b) It is unclear if the statement of Theorem 1.1.A′ remains true without an assumption that $\partial\Delta$ is unknotted in ∂B. In other words, *are there symplectically sliced knots*?

Let us denote by U the domain $\{0 \leq y_2 \leq 1\} \subset \mathbb{R}^4$, by S_0 and S_1 the circles $\{|z_1| = 1, z_2 = 0\}$ and $\{|z_1| = 1, z_2 = i\}$ respectively. We denote by Λ the Lagrangian cylinder $\{|z_1| = 1, x_2 = 0, 0 \leq y_2 \leq 1\}$ and by j the inclusion $\Lambda \hookrightarrow U$. Denote by L the ray $\{z_1 = 0, x_2 = 0, y_2 > 1\}$ and by D_1 the disc $\{|z_1| \leq 1, z_2 = i\}$ bounded by S_1.

Theorem 1.1.B *Any Lagrangian embedding* $\varphi : \Lambda \to U$ *which coincides with* j *near* $\partial\Lambda$ *is isotopic to* j *relative* $\partial\Lambda$. *Moreover,* φ *and* j *are Lagrangian isotopic relative* $\partial\Lambda$ *as embeddings into* $\mathbb{R}^4_+ \setminus (L \cup D_1)$.

I do not know if φ and j are Lagrangian isotopic *inside* the domain U.

---

*) Partially supported by NSF grant DMS–9006179

1.2 An invariant for S^2-knots in $\mathbb{R}^4$

The next theorem strengthens Theorem 1.1.A by showing that nonnegative 2-discs cannot be knotted not only in the half-space but even in the complement of a sufficiently large polydisc. Let us denote by $\mathfrak{D}(a,b)$ the polydisc $\{|z_1| \leq a, |z_2| \leq b\} \subset \mathbb{C}^2$.

Theorem 1.2.A *Let $a > 4$, $b > 2$. Suppose that a disc $(\Delta, \partial\Delta) \subset (\mathbb{C}^2 \setminus \mathrm{Int}\,\mathfrak{D}(a,b), \partial\mathfrak{D}(a,b))$ intersects $\partial\mathfrak{D}(a,b)$ transversely along the circle $\partial\Delta = \{|z_1| = 1, z_2 = b\}$. If $\omega|_\Delta \geq 0$ then the disc $(\Delta, \partial\Delta) \subset (\mathbb{C}^2 \setminus \mathrm{Int}\mathfrak{D}(a,b), \partial\mathfrak{D}(a,b))$ is unknotted.*

Let α be an isotopy class of embeddings $S^2 \to \mathbb{R}^4$. We say that the class α admits a (a,b)-realization for $a > 1, b > 0$ if α can be represented by an embedded sphere $S = \Delta \cup D \subset \mathbb{R}^4$ where $D = \{|z_1| \leq 1, z_2 = b\}$ and Δ is a 2-disc as in 1.2.A, i.e., $(\Delta, \partial\Delta) \subset (\mathbb{C}^2 \setminus \mathrm{Int}D(a,b), \partial\mathfrak{D}(a,b))$ and $\omega|_\Delta \geq 0$. Thus Theorem 1.2.A can be reformulated by saying that *no one non-trivial class α admits a (a,b)-realization if $a > 4, b > 2$*. It is interesting to confront this result with the following

Theorem 1.2.B *For any isotopy class α of embeddings $S^2 \to \mathbb{R}^4$ there exists $a > 1, b > 0$ such that α admits a (a,b)-realization.*

In Section 3 we use extremal, for a given knot, values of a and b to define a real valued function Ψ_α which depends only on the isotopy class α of a knot. We call Ψ_α the *symplectic shape* of the knot α. Very little is known about the function Ψ_α beyond its existence and non-triviality guaranteed by Theorems 1.2.A and 1.2.B (see Theorem 3.4.A).

1.3 Plan of the paper

In the next section we discuss the method of filling with holomorphic discs and use it to prove theorems 1.1.A, 1.1.A$'$, 1.1.B, and 1.2.A. In Section 3 we prove Theorem 1.2.B and define and discuss the invariant Ψ_α.

Results of the present paper grew out of the symplectic study of the so-called Gluck's surgery (see [Gl] and 3.5 below) which we attempted a couple of years ago together with K. Luttinger. Our joint result at this direction is formulated in Section 3.5 below. I want to mention that K. Luttinger recently independently developed these ideas in a different direction and found an obstruction for Lagrangian realizations of isotopy classes of embeddings of the 2-torus into $\mathbb{R}^4$ (see [Lu]).

Added in proof: Since this paper was submitted a number of new results about topology of Lagrangian 2-knots were proven. We refer the reader to the paper {Y. Eliashberg and L. Polterovich, The problem of Lagrangian knots in four-manifolds, to appear in Proceed. of Georgia Int. Top. Conf.} for a survey of recent results.

2 Holomorphic filling. Proof of the main results

2.1 Symplectic and holomorphic preliminaries

Let (X, ω, J) be a symplectic almost complex manifold. The almost complex structure J is said to be *tamed* (see [Gr]) by the symplectic form ω if $\omega(T, JT) > 0$ for any nonzero tangent vector $T \in T(X)$. We say that J is *calibrated* by ω (see [HL]) if, in addition, J leaves ω invariant, i.e. $\omega(JT_1, JT_2) = \omega(T_1, T_2)$ for any $T_1, T_2 \in T(X)$. Note that the standard complex structure on $\mathbb{C}^2 = \mathbb{R}^4$ is calibrated by the standard symplectic structure ω. Let Σ be a cooriented real hypersurface in X. Each tangent space $T_x(X), x \in X$, contains a maximal complex ($J-$invariant) subspace $T_x^J(X)$ which has real codimension 1 in $T_x(X)$. If ν is a normal vector field to Σ in X which defines the given coorientation of Σ then the field $J\nu$ defines a coorientation of $T^J(X)$ in $T(X)$. There exists a 1-form α on X such that $\alpha|_{T^J(X)} = 0$ and $\alpha(J\nu) > 0$. The form α is defined uniquely up to multiplication by a positive function. Then the 2-form $d\alpha|_{T^J}$ is defined up to multiplication by the same function. The hypersurface Σ is called (strictly) J-convex (or strictly plurisubharmonic if the structure J is integrable) if the quadratic form $T \mapsto d\alpha(T, JT)$, $T \in T^J(X)$, is positive definite. The distribution $T^J(X) \subset T(X)$ is, in this case, a contact structure, i.e. a completely non-integrable tangent hyperplane distribution.

If the form $d\alpha(T, JT)$ vanishes identically on Σ then Σ is called J-*flat* (or Levi-flat). In this case the distribution $T^J(\Sigma) \subset T(\Sigma)$ integrates to a codimension 1 foliation with J-holomorphic leaves.

Suppose now that $\dim X = 4$. A generic 2-surface $F \subset X$ have isolated points where the tangent plane is a complex line. These complex points generically are either elliptic or hyperbolic. If F is oriented then they can also be divided by positive and negative. We denote by $e_\pm(F), h_\pm(F)$ numbers elliptic and hyperbolic, positive and negative. Set $d_\pm(F) = e_\pm(F) - h_\pm(F)$.

Let Σ be the J-convex boundary of a domain $\Omega \subset X$ and let $F \subset \Omega$ be a surface with boundary $\partial F \subset \partial\Omega$. Suppose that ∂F is transversal to the contact structure $T^J(\Sigma)$. We say that F has a *complex-like* orientation if its oriented boundary defines the same coorientation of $T^J(\Sigma)$ in $T(\Sigma)$ as boundaries of holomorphic curves in $\Omega, \partial\Omega$. For a surface F with the complex-like orientation the number $d_-(F)$ is an invariant which depends only on the boundary ∂F, the Euler characteristic of $\chi(F)$ and the homology class of F in $H_2(\Omega, \partial F)$ (see [HE]). If the surface F is required to be tangent to Σ along ∂F (for instance, if F is contained in Σ) then the number $d_+(F)$ is also invariant and we have $d_+(F) + d_-(F) = \chi(F)$ (see [HE]).

If F is contained in a J-convex hypersurface Σ (which does not necessarily bounds a domain) then the intersection $T^J(\Sigma)$ with $T(F)$ integrates outside complex points of F to a 1-dimensional foliation which is called *characteristic*. Note that any holomorphic disc Δ with $\partial\Delta \subset F$ which approaches to Σ from inside is necessarily transversal to Σ along $\partial\Delta$, $\partial\Delta$ is contained in the totally real part of F and is transversal to the characteristic foliation of F.

According to a theorem of Bishop (see [Bi]) a punctured neighborhood of an elliptic complex point $p \in F$ in an integrable complex manifold has a canonical foliation by circles which span disjoint embedded holomorphic discs. A family $\{\Delta_t\}, t \in]0, a]$, of holomorphic discs with $\partial\Delta_t \subset F$ is called a *Bishop family* if discs Δ_t converge to the elliptic point $p \in F$ when $t \to 0$. For a sufficiently small $\varepsilon > 0$ discs $\{\Delta_t\}$, $0 < t \leq \varepsilon$, from a Bishop family fill a J−flat 3-ball B bounded by Δ_ε and a neighborhood of the elliptic point in the surface F.

2.2 Technique of filling with holomorphic discs

Filling with holomorphic discs was studied and applied by several authors (see [BG], [BK], [Gr], [El3]). Gromov was the first who apllied this technique for almost complex manifolds where instead of integrability he used the taming condition. This technique exists as a method: I do not know the general theorem which would embrace all cases where the filling technique works. In this section we describe a filling result suitable for the purposes of the present paper.

Let (X, J) be an almost complex 4-manifold where the almost complex structure J is tamed by a symplectic form ω. Let F be a 2-dimensional surface with the boundary ∂F. Suppose that J is integrable near F, and F has exactly one complex point $p \in F$ which is elliptic. Suppose further that ∂F bounds a J-holomorphic disc Δ so that the union $S = F \cup \Delta$ is an embedded piecewise smooth 2-sphere. We say that the pair (F, Δ) *can be filled with J-holomorphic discs* if there exists an embedded J-flat 3-ball $B \subset X$ which spans S such that Int$B$ is foliated by J-holomorphic discs. In other words, the Bishop family starting from p consists of disjoint embedded discs and includes Δ.

Suppose from now on that $X = \mathbb{R}^4$, ω is the standard symplectic structure and J is an almost complex structure tamed by ω and standard at ∞.

Proposition 2.2.A *Suppose that (F, Δ) are as above. Let $\Sigma \subset \mathbb{R}^4$ be a strictly J-convex hypersurface (possibly, with boundary). Suppose that $F \subset$ IntΣ, Δ is transversal to Σ along $\partial\Delta$ and approaches Σ from inside, and discs from any Bishop family starting at the elliptic point $p \in F$ do not intersect Σ at their interior points. Then (F, Δ) admits a filling by J-holomorphic discs.*

Proof. The proof follows standard filling arguments (comp [BG], [Gr], [El3]). We start the Bishop family $\{\Delta_t\}$ from the elliptic point $p \in F$. Local index arguments show that if Δ_{t_0} is an embedded disc then the family can be extended for $t \in]0, t_0 + \epsilon[$. The J-convexity of Σ and the fact that discs Δ_t do not intersect Σ at their interior points guarantee that all discs Δ_t are transversal to Σ (from inside) along $\partial\Delta_t$, and that $\partial\Delta_t \subset F$ is embedded and transversal to the characteristic foliation on F. Then McDuff's result from [McD] together with the standard topological arguments shows that all discs Δ_t are embedded and disjoined. The transversality of $\partial\Delta_t$ to the characteristic foliation excludes also possible cusp-degenerations of J-holomorphic discs Δ_t at their boundaries, and therefore, Gromov's compactness theorem (see [Gr]) guarantees that the Bishop's family $\{\Delta_t\}$ can be developed until one of the discs, say Δ_{t_0}, hits the disc Δ. Again, McDuff's theorem shows that Δ_{t_0} and Δ cannot have common interior points. Let us show that Δ_{t_0} and Δ cannot be tangent at a boundary point either. Indeed, suppose that Δ and Δ_{t_0} are tangent at an isolated point $q \in \partial\Delta \cap \partial\Delta_{t_0}$. Without loss of generality we can assume that the surface F is real analytic. Let us recall that the complex structure J is standard near F. As F is totally real, the holomorphic discs Δ and Δ_t for $t \leq t_0$ extends to bigger discs $\widetilde{\Delta}$ and $\widetilde{\Delta}_t$ which are invariant with respect to a local antiholomorphic involution near F which leaves F fixed. Discs $\widetilde{\Delta}_t$ for t close to t_0 should also intersect $\widetilde{\Delta}$ because the local intersection index of $\widetilde{\Delta}_{t_0}$ and $\widetilde{\Delta}$ at q is positive. There are no intersections of $\widetilde{\Delta}_t, t < t_0$, and $\widetilde{\Delta}$ on F. Therefore, $\widetilde{\Delta}_t$ intersects $\widetilde{\Delta}$ at 2 points $Z_t, Z'_t \in \widetilde{\Delta}$ which are symmetric with respect to the antiholomorphic involution. Now remember that all discs $\Delta_t, t \leq t_0$, and Δ approach Σ from the same side. Hence, one of the intersection points, say Z_t, belongs to $\Delta_t \cap \Delta$ and the other to $(\widetilde{\Delta}_t \backslash \Delta_t) \cap (\widetilde{\Delta} \backslash \Delta)$. But Δ_t and Δ should be disjoint and, therefore, Δ_{t_0} and Δ cannot be tangent. Thus the disc Δ_{t_0} coincides with Δ. □

Notice that if the hypersurface Σ bounds a domain $\Omega \subset \mathbb{R}^4$, then for any J-holomorphic disc $\Delta \subset \Omega$ with $\partial\Delta \subset \Sigma$, we have $\mathrm{Int}\Delta \cap \Sigma = \emptyset$. Therefore, Proposition 2.2.A implies

Corollary 2.2.B *Let (F, Δ) be as above and Ω be a domain in $\mathbb{R}^4$ with strictly J-convex boundary Σ. Suppose that $F \subset \Sigma$ and $\Delta \subset \Omega$. Then (F, Δ) can be filled with holomorphic discs.*

2.3 Positive vs. non-negative

The goal of this section is the following

Lemma 2.3.A *Let F be a connected 2-surface with boundary in a 4-dimensional symplectic manifold (X, ω). Suppose that $\omega|_F$ is positive near ∂F and non-negative elsewhere. Then F can be C^∞-approximated by a surface $\widetilde{F}$ which coincides with F near $\partial F = \partial\widetilde{F}$ and such that $\omega|_{\widetilde{F}} > 0$.*

Let us choose an almost complex structure on X which is calibrated by ω, i.e. J is tamed by ω and $\omega(JU, JV) = \omega(U, V)$ for any $U, V \in T(X)$.

The proof of 2.3.A is divided below by few steps.

Lemma 2.3.B *Let $h : F \to X$ be a totally real embedding and $\mathcal{F}$ a 1-dimensional foliation of $\widehat{F} = h(F)$. Then h extends to an embedding $H : V = F \times [-\varepsilon, \varepsilon] \to X$ such that $\widehat{V} = H(V)$ admits a Lagrangian foliation $\mathcal{L}$ which intersects $\widehat{F}$ transversely along $\mathcal{F}$.*

Proof. Let us extend $\mathcal{F}$ to a codimension 1 foliation $\mathcal{S}$ on a neighborhood G of $\widehat{F}$ in X such that leaves of $\mathcal{S}$ are perpendicular to $\mathcal{F}$. Let $\mathcal{K}$ be the 1-dimensional foliation on G which is formed by characteristics of ω restricted to leaves of $\mathcal{S}$. Thus $\mathcal{K}$ foliates leaves of the foliation $\mathcal{S}$. Then the characteristics intersecting leaves of $\mathcal{F}$ fill a 3-dimensional collar $\widehat{V} \simeq \widehat{F} \times [-\varepsilon, \varepsilon]$ foliated by Lagrangian leaves intersecting $\widehat{F}$ transversely along leaves of $\mathcal{F}$. $\square$

The proof of the next lemma 2.3.C is straightforward.

Lemma 2.3.C *Let C be a compact set, $C_1, C_2 \subset C$ be two disjoint closed subsets of C, U be an open subset of $\mathbb{R}^2 \times C$ and $\varphi: U \to \mathbb{R}$ be a smooth function such that $\frac{\partial \varphi}{\partial x_1}(x_1, x_2, t) \geq 0$ for all $(x_1, x_2, t) \in U$. Then for any neighborhood $G \subset U$ of the set $K = \left\{\frac{\partial \varphi}{\partial x_1} = 0\right\}$ there exists a function $\tilde{\varphi}: U \to \mathbb{R}$ which coincides with φ on $(U \setminus G) \cup (U \cap (\mathbb{R}^2 \times C_2))$ and such that $\frac{\partial \tilde{\varphi}}{\partial x_1} > 0$ in $U \cap (\mathbb{R}^2 \times C_1)$ and $\frac{\partial \tilde{\varphi}}{\partial x_1} \geq 0$ everywhere in U.*

An oriented curve F is said to be *positively semitransversal* to a cooriented codimension 1 foliation $\mathcal{F}$ if at each its point Γ either tangent to $\mathcal{F}$ or intersects $\mathcal{F}$ in a positive direction with respect to the coorientation of $\mathcal{F}$.

Lemma 2.3.D *Let U be a neighborhood of the interval $I = \{0 \leq x_1, \leq 1, x_2 = 0\}$ in $\mathbb{R}^2$ and $\mathcal{F}_t, t \in [0,1]$, be a family of cooriented 1-dimensional foliations on U. Suppose that I is positively semitransveral to $\mathcal{F}_t$ for all $t \in [0,1]$ (the orientation of I is given by the coordinate x_1). For $\varepsilon > 0$ let K_ε be the set $\{(x,t) \mid x \in I, t \in [\varepsilon, 1-\varepsilon], \mathcal{F}_t$ is tangent to I at the point $x\}$. Then for any neighborhood G of K in $I \times [0,1]$ there exists a family of embeddings $j_t: I \to \mathbb{R}^2$, $t \in [0,1]$, such that j_t is positively semitransversal to $\mathcal{F}_t$ and C^∞-close to the inclusion $j: I \subset \mathbb{R}^2$ for all $t \in [0,1]$, $j_t(x) = j(x)$ for $(x,t) \notin G$ and j_t is transversal to $\mathcal{F}_t$ for $t \in [\varepsilon, 1-\varepsilon]$.*

Proof. Choose open coverings $\bigcup_1^N U_i$ of U and $\bigcup_{j=1}^M V_j$ of $[0,1]$ such that for each $i = 1, \ldots, N$ and $j = 1, \ldots, M$ we have that either $U_i \cap I$ is transversal to $\mathcal{F}_t$ for $t \in V_j$ or there exists a family (α_t, β_t), $t \in V$ of coordinate systems on U_i such that the foliation $\mathcal{F}_t|_{U_i}$ is given by the equation $d\alpha_t = 0$, and $d\beta_t|_I$ does not vanish. Thus for a pair U_i, V_j of the second type there exists a family of smooth functions f_t, $t \in V_j$, such that the equation $\alpha_t = f_t(\beta_t)$ defines $I \cap U_i$ for each $t \in V_j$. The semitransversality of $I \cap U_i$ to $\mathcal{F}_t$, $t \in V_j$, means that $\frac{\partial f_t}{\partial \beta_t} \geq 0$, $t \in V_j$. Thus to finish the proof we can apply 2.3.B as an extension lemma consequently for each $j = 1, \ldots, M$ and $i = 1, \ldots, N$. $\square$

Let F be a compact surface with boundary ∂F. A 1-dimensional foliation $\mathcal{S}$ on F is called *short* (see [GE]) if all its leaves are compact and it does not have closed leaves. Thus all leaves of $\mathcal{S}$ are arcs with end points on ∂F.

Lemma 2.3.E *Let $\mathcal{S}$ be a short foliation on a surface F and $(\mathcal{F}, \mathcal{G})$ be a $(2,1)$-flag foliation on $V = F \times [-\varepsilon, \varepsilon]$ (i.e., leaves of the 1-dimensional foliation $\mathcal{G}$ foliate leaves of the 2-dimensional foliation $\mathcal{F}$). Suppose that: $\mathcal{F}$ is transversal to F and $\mathcal{F} \cap F = \mathcal{S}$; for each leaf Φ of $\mathcal{F}$ the leaf $S = \Phi \cap F$ of $\mathcal{S}$ is positively* *semitransversal to $\mathcal{G}|_\Phi$; F is transversal to $\mathcal{G}$ near ∂F. Then there exists a surface $\widetilde{F} \subset V$ which is C^∞-close to F, coincides with F near $\partial F = \partial\tilde{F}$ and transversal to $\mathcal{G}$ everywhere.*

Proof. The space of connected leaves of the short foliation $\mathcal{S}$ is a finite graph Γ. For each vertex $p \in \Gamma$ take the leaf $\mathcal{S}_p$ of $\mathcal{S}$ which corresponds to the point p and use 2.3.D to deform the surface F near $\mathcal{S}_p$ in such a way that $\mathcal{S}_p$ become transversal to the foliation G restricted to the leaf $\mathcal{F}_p$ of $\mathcal{F}$ through the point p. We can do that in such a way that the perturbed surface $\widetilde{F}$ still satisfies the conditions of 2.3.E. To finish the proof we apply again 2.3.D to each edge of Γ as a parameter space. $\square$

Proof of 2.3.A. Choose any short foliation $\mathcal{S}$ of the surface F. Let us denote by K the set $\{x \in F | \omega|_F(x) = 0\}$. Notice that F is totally real near K. Thus making F smaller, if necessary, we can think that it is totally real everywhere. According to 2.3.B the inclusion $i: F \hookrightarrow X$ extends to an embedding $H: F \times [-\varepsilon, \varepsilon] \to X$ such that $\widehat{V} = H(V)$ carries a Lagrangian foliation $\mathcal{F}$ which is transversal to F and intersects F along $\mathcal{S}$. The characteristic foliation $\mathcal{G}$ of $\widehat{V}$ (i.e., foliation obtained by integrating the line field $\mathrm{Ker}\,\omega|_{\widehat{V}}$) is tangent to $\mathcal{F}$. The condition $\omega|_F \geq 0$ means that $\mathcal{G}$ is positively semitransversal to F. Therefore we can apply 2.3.E to the $(2,1)$-flag $(\mathcal{F}, \mathcal{G})$ and perturb F in V to a surface $\widetilde{F}$ which coincides with F near ∂F and transversal to $\mathcal{G}$ which is equivalent to the inequality $\omega|_{\widetilde{F}} > 0$. $\square$

2.4 Proof of theorems 1.1.A and 1.1.A′

Proof of Theorem 1.1.A′. According to 2.3.A we can arrange that $\omega|_\Delta > 0$. There exists an almost complex structure J on B which is tamed by ω, standard near ∂B and such that the disc $\Delta \subset B$ is J-holomorphic. By the assumption the circle $\partial\Delta$ bounds a disc $\mathcal{D} \subset \partial B$. Holomorphicity of Δ implies that $\partial\mathcal{D}$ is transversal to the standard contact structure $T^J(\partial B) \subset T(\partial B)$. Note that $d_-(\Delta) = 0$. Then according to 2.1 we have $d_-(\mathcal{D}) = 0$ and $d_+(\mathcal{D}) = 1$. It is possible to arrange via a C^0-small isotopy of $\mathcal{D}$ in ∂B (see [El4]) that the new disc, still denoted by $\mathcal{D}$, has exactly one complex point $e \in \mathcal{D}$ which is positive and elliptic. Thus we can apply 2.2 and fill the pair $(\mathcal{D}, \Delta)$ by holomorphic discs. In other words, there exists a J-flat embedded 3-ball $V \subset B$ with $\partial V = \mathcal{D} \cup \Delta$ which provides us with the required isotopy between Δ and $\mathcal{D}$. $\square$

Proof of Theorem 1.1.A. Take an $\varepsilon > 0$ and consider the symplectic contraction $c_\varepsilon: \mathbb{C}^2 \to \mathbb{C}^2$ defined by the formula $c_\varepsilon(z_1, z_2) = \big((1-\varepsilon)z_1, (1-\varepsilon)(z-i)+i\big)$. The image $\Delta_\varepsilon = c_\varepsilon(\Delta)$ is a nonnegative disc bounded by the circle $S_\varepsilon = \{|z_1| = 1-\varepsilon,\ z_2 = i\varepsilon\}$. Let C_ε be the annulus $\{1 \geq |z_1| \geq 1-\varepsilon,\ x_2 = 0,\ y_2 = \varphi(|z_1|)\}$ where the function $\varphi: [1-\varepsilon, 1] \to [0, \varepsilon]$ has the following properties:

$\varphi(1-\varepsilon)=\varepsilon$; $\varphi(1)=0$; $\varphi(t)=\frac{\varepsilon}{2}$ for $t\in\left[1-\frac{2\varepsilon}{3},1-\frac{\varepsilon}{3}\right]$;

φ is strictly decreasing on $\left[1-\varepsilon,1-\frac{2\varepsilon}{3}\right]\cup\left[1-\frac{\varepsilon}{3},1\right]$.

Thus C_ε is an ω-positive annulus bounded by circles $S_0=\{|z_1|=1,\, z_2=0\}$ and $S_\varepsilon=\{|z_1|=1-\varepsilon,\, z_2=i\varepsilon\}$. Then the union $\Delta'_\varepsilon=C_\varepsilon\cup\Delta_\varepsilon$ is an embedded piecewise smooth disc with a corner along the circle S_ε. One can smooth Δ'_ε into a nonnegative disc Δ''_ε via a perturbation in an arbitrary small neighborhood of the circle S_ε. Clearly Δ''_ε is isotopic to Δ. Moreover, if $\omega|_\Delta>0$ then $\omega|_{\Delta''_\varepsilon}>0$ and discs Δ''_ε and Δ are isotopic among ω-positive discs. Consider now the ball $B_R=\{|z_1|^2+|z_2-iR|^2=R^2\}$ where the radius $R=(\varepsilon^2-2\varepsilon+2)/2\varepsilon$ is chosen in such a way that the circle $S_{\varepsilon/2}=\{|z_1|=1-\varepsilon/2,\, z_2=i\varepsilon/2\}$ is contained in the sphere ∂B_R. If ε is chosen sufficiently small than the disc Δ''_ε is contained in Int B_ε and therefore ∂B_ε intersects the disc Δ''_ε along the circle $S_{\varepsilon/2}$. The disc $\widetilde{\Delta}_\varepsilon=\Delta''_\varepsilon\cap B_R$ is holomorphic near $\partial\widetilde{\Delta}_\varepsilon$ and, therefore, to finish the proof we can apply Theorem 1.1.A$'$. □

Note that the isotopy provided by the construction in proofs of 1.1.A and 1.1.A$'$ has the property that it preserves the positivity of the form $\omega|_\Delta$ if this form was positive from the beginning. Thus we have

Theorem 2.4.A *Let $\Delta,\Delta'\subset\mathbb{R}^4_+$ be 2-discs bounded by the circle $S_0=\{|z_1|=1,\, z_2=0\}\subset\partial\mathbb{R}^4_+$. Suppose that* Int$\Delta$, Int$\Delta'\subset$ Int$\mathbb{R}^4_+$ *and that discs Δ and Δ' are transversal to $\partial\mathbb{R}^4_+$. If $\omega|_\Delta,\ \omega|_{\Delta'}>0$ then there exists a compactly supported and fixed on $\partial\mathbb{R}^4_+$ Hamiltonian isotopy $h_t\colon\mathbb{R}^4_+\to\mathbb{R}^4_+$, $t\in[0,1]$, with $h_0=id$ and $h_1(\Delta)=\Delta'$.*

Theorem 1.1.A$'$ has a similar symplectic ramification.

2.5 Proof of Theorem 1.1.B.

We use here notations of Section 1 and introduce some new notations. For a sufficiently small $\varepsilon>0$ let us fix a smooth function $\gamma\colon[0,1+\varepsilon^2]\to[\varepsilon,1]$ such that

$$\gamma(t)=\begin{cases}\varepsilon & \text{for } t\le 1-3\varepsilon\\ 1 & \text{for } t\ge 1-\varepsilon^2\end{cases}$$

and $\int\limits_0^{1-\varepsilon^2}\gamma(t)dt=\varepsilon(1+\varepsilon^2)$.

We denote by V_ε the domain $\{1-\varepsilon\le|z_1|\le 1+\varepsilon^2,\, 0\le y_2\le 1+\varepsilon^2,\, |x_2|\le\gamma(y_2)\}$. Note that there exists an area preserving map ψ of the domain $\{0\le y_2\le 1+\varepsilon^2,\ |x_2|\le\gamma(y_2)\}$ onto the domain $\{0\le y_2\le 1+\varepsilon^2,\ |x_2|\le\varepsilon\}$. This map can be chosen to be the identity for $y_2\le 1-3\varepsilon$. Therefore the map Ψ defined by the formula

$$\Psi(x_1,y_1,x_2,y_2)=(x_1,y_1,\psi(x_2,y_2))$$

is a symplectomorphism of V_ε onto $W_\varepsilon = \{1-\varepsilon \le |z_1| \le 1+\varepsilon^2,\ 0 \le y_2 \le 1+\varepsilon^2,\ |x_2| \le \varepsilon\}$. For $t \in [-1,1], \delta \in [-\varepsilon^2/2, \varepsilon]$ let us denote by $D_{t,\delta}$ the disc $\{|z_1| \le 1-\delta 2\varepsilon,\ x_2 = t,\ y_2 = 1-\delta\}$, by $D'_{t,\delta}$ the disc $\{|z_1| \le 1-\varepsilon,\ x_2 = t,\ y_2 = 1-\delta\} \subset D_{t,\delta}$ and by $A_{t,\delta}$ the annulus $D_{t,\delta} \setminus \mathrm{Int} D'_{t,\delta}$. Consider a family of embeddings

$$\beta_\delta \colon [0, 1+\varepsilon^2] \to [0, 1+\varepsilon^2],\ \delta \in [-\varepsilon^2, \varepsilon]\ ,$$

with the following properties:

$$\beta_\delta(0) = 0 \text{ for all } \delta \in [-\varepsilon^2, \varepsilon]\ ;$$

$$\beta_\delta(u) = u \text{ for } u \in [1-\varepsilon, 1+\varepsilon^2], \delta \in [3\varepsilon/4, \varepsilon]\ ;$$

$$\beta_\delta(1-u) = \varepsilon(1-u)^2 \text{ for } u \in [-\varepsilon^2, \varepsilon],\ ,0 \le \delta - u \le \varepsilon^2/2\ ;$$

$$\beta_\delta(1+\varepsilon^2) = 1+\varepsilon^2 \text{ for } \delta \in [0, \varepsilon]\ .$$

Now we define the embedding $B \colon W_\varepsilon \to W_\varepsilon$ by the formula $B(z_1, x_2, y_2) = (z_1, x_2, \beta_{1-|z_1|}(y_2))$. The map B is not a symplectic embedding but it is symplectic on each slice $W_\varepsilon \cap \{x_2 = t\}$, $|t| \le \varepsilon$. Using the symplectomorphism $\Psi : V_\varepsilon \to W_\varepsilon$ we can transfer the map B to V_ε. Thus we get a map $\widetilde{B} = \Psi^{-1} \circ B \circ \Psi \colon V_\varepsilon \to V_\varepsilon$. We denote by $\widetilde{A}_{t,\delta}$ the image $\widetilde{B}(A_{t,\delta}) \subset W_\varepsilon$, $t \in [-1,1]$, $\delta \in [-\varepsilon^2/2, \varepsilon]$. Note that all annuli $\widetilde{A}_{t,\delta}$ are ω-positive and holomorphic near $\partial \widetilde{A}_{t,\delta}$.

We start with the following obvious

Lemma 2.5.A *Any Lagrangian embedding $\varphi : \Lambda \to U$ which is fixed near $\partial\Lambda$ extends for some $\varepsilon > 0$ to a symplectic embedding $\Phi : W_\varepsilon \to U$ fixed on $W_\varepsilon \cap (\{0 \le y_2 \le \varepsilon\} \cup \{1-\varepsilon \le y_2 \le 1+\varepsilon^2\})$.*

For each $t \in [-1,1]$, $\delta \in [-\varepsilon^2, \varepsilon]$ let $\Delta_{t,\delta}$ be a smooth ω-positive disc $\Delta_{t,\delta} = D'_{t,\delta} \cup \Phi(\tilde{A}_{t,\delta}) \subset U$. All these discs are disjoint and holomorphic near their boundaries $\partial\Delta_{t,\delta}$, $t \in [-1,1], \delta \in [-\varepsilon^2, \varepsilon]$. Let Ω be the domain $\{y_2 = \varepsilon|z_1|^2\} \subset \mathbb{R}^4$. Ω has a strictly pseudoconvex boundary $\partial\Omega$ which contains boundaries $\partial\Delta_{t,\delta}$ of discs $\Delta_{t,\delta}, t \in [-1,1], \delta \in [-\varepsilon^2, \varepsilon]$. Moreover, if $\varepsilon > 0$ is chosen sufficiently small then all discs $\Delta_{t,\delta}$ are contained in Ω. There exists a function $\alpha \colon [-1,1] \to [-\varepsilon, \varepsilon]$ such that for each $t \in [-1,1]$ the circle $\partial\Delta_{t,0}$ lies in the hyperplane $\{x_2 = \alpha(t)\}$. We denote by S_t the 2-disc spanned by $\partial\Delta_{t,0}$ in $\partial\Omega \cap \{x_2 = \alpha(t)\}$. Notice that each S_t, $t \in [-1,1]$, has exactly one (elliptic) complex point $p_t = (0, 0, \alpha(t), 0)$.

Let us choose an almost complex structure J on Ω such that:

- J is tamed by ω;
- J is standard near $\partial\Omega$ and on the domain $G = \{y_2 \ge 1, |x_2| \le 1, |z_1| \le 1 - \frac{\varepsilon}{2}\}$;

– all disks $\Delta_{t,\delta}$, $t \in [-1,1]$, $\delta \in [-\varepsilon^2, \varepsilon]$, are J-holomorphic.

According to Proposition 2.2.A each pair $(S_t, \Delta_{t,0})$ can be filled with J-holomorphic discs, i.e., there exists a family of disjoint embedded J-flat 3-balls $H_t \subset \Omega$, $t \in [-1,1]$, such that $\partial H_t = S_t \cup \Delta_{t,0}$. Notice that for each $t \in [-1,1]$ the ball H_t contains all J-holomorphic discs $\Delta_{t,\delta}$ for $\delta \in [0,\varepsilon]$ and does not intersect with discs $\Delta_{t,\delta}$ for $\delta \in [-\varepsilon^2, 0]$. Denote by L the ray $\{y_2 > 1, x_2 = 0, z_1 = 0\} \subset G$.

In the proof of Proposition 2.5.C and Theorem 1.2.A below, we will need the following Lemma 2.5.B which is proved by Federer (in a more general setting of minimal surfaces instead of holomorphic curves) in [Fe] and was used for similar purposes by Gromov in [Gr].

Lemma 2.5.B *Let C be a properly embedded holomorphic curve in the open ball B of radius r in $\mathbb{C}^n$. Suppose that C contains the center of B. Then Area $C \geq \pi r^2$.*

Remark. It would be possible get better estimates for parameters a and b in the Theorem 1.2.A if one were able to prove some generalizaton of this lemma. For instance, suppose that the curve C contains not the center but a point $p \in B$ at the distance λr, ,$0 < \lambda < 1$, from the center of the ball B. Is it true that Area $C \geq \pi(1-\lambda^2)r^2$?

Proposition 2.5.C

$$H_0 \cap L = \emptyset\,.$$

Proof. Suppose that a point $q \in L$ belongs to the intersection $L \cap H_0$. Then q belongs to a holomorphic disc $\Delta \subset H_0$, $\partial\Delta \subset S_0$. The disc Δ is clearly different from discs $\Delta_{0,\delta} \subset H_0$, $-\varepsilon^2 \leq \delta \leq \varepsilon$. In particular, $\int\limits_{\Delta} \omega < \int\limits_{\Delta_{0,\varepsilon}} \omega = \pi\left(1 - \frac{\varepsilon}{2}\right)^2$.

Let us consider the intersection $\widehat{\Delta} \subset \Delta \cap G$. Then $\int\limits_{\Delta} \omega > \int\limits_{\widehat{\Delta}} \omega$. But $\int\limits_{\widehat{\Delta}} \omega =$ Area $\widehat{\Delta}$ because $\widehat{\Delta}$ is holomorphic with respect to the standard complex structure on G. On the other hand Δ does not intersect the bottom $C = G \cap \{y_2 = 1\}$ of the domain G. Indeed C is contained in $\bigcup\limits_{t\in[-1,1]} \Delta_{t,0}$ and, therefore, is disjoint from Δ. Hence $\widehat{\Delta}$ can be considered as a closed analytic subset of the bigger domain $G' = \{|z_1| \leq 1 - \varepsilon/2, |x_2| \leq 1\}$ in the standard complex space $\mathbb{C}^2$. Let K be the ball of radius $1 - \varepsilon/2$ centered at the point $q \in L \cap \Delta$. Then $\Delta \cap K$ is a properly embedded holomorphic curve passing through the center of the ball K. Then according to Lemma 2.5.B we have Area $\widehat{\Delta} \geq$ Area $(\Delta \cap K) \geq \pi(1-\varepsilon/2)^2$. Hence, we get a contradiction:

$$\pi(1-\varepsilon/2)^2 \leq \text{ Area } \widehat{\Delta} = \int\limits_{\widehat{\Delta}} \omega < \int\limits_{\Delta} \omega < \pi(1-\varepsilon/2)^2\,.$$

□

Let us denote by W the hypersurface $\{|z_1| \leq 1, x_2 = 0, 0 \leq y_2 \leq 1\} \subset U$.

Proposition 2.5.D *The Lagrangian embedding $\varphi\colon \Lambda \to U$ from 1.1.B extends to a symplectic embedding $\tilde{\varphi}\colon W \to \mathbb{R}^4_+ \setminus L$ which coincides with the inclusion near $\partial U \cap W$.*

Proof. Let E be the part of the hypersurface $\tilde{\varphi}\big(W_\varepsilon \cap \{x_2 = 0\} \cup \{y_2 \leq 1\}\big)$ enclosed between $\varphi(\Lambda)$ and $\Delta_{0,0}$. Then $E \cap \Omega \subset \bigcup_{\delta \in [-\epsilon^2, 0]} \Delta_{0,\delta}$. Therefore, $E \cap H_0 = \Delta_{0,0}$ and the union $E \cup H_0$ is a smooth hypersurface $\tilde{W} \subset \mathbb{R}^4_+ \setminus L$ which is diffeomorphic to W and which admits a foliation by ω-positive discs with boundaries in $\varphi(\Lambda)$. This implies that there exists a symplectomorphism $\varphi'\colon W \to \tilde{W}$ such that φ is the identity near $W \cap \{0 \leq y_2 \leq \varepsilon\}$ and it differs from the identity on $W \cap \{y_2 = 1\} = D_{0,0}$ by an area preserving diffeomorphism $f\colon D_{0,0} \to D_{0,0}$.

Let $H\colon D_{0,0} \times [0,1] \to \mathbb{R}$ be a time-dependent Hamiltonian generating f and equal to O on $\partial D_{0,0} \times [0,1]$. We may choose H to be identically 0 near $t = 0$ and $t = 1$. Let us define a hypersurface $\tilde{\tilde{W}}$ as follows:

$$\tilde{\tilde{W}} \cap \{y_2 \geq \varepsilon\} = \tilde{W} \cap \{y_2 \geq \varepsilon\}$$

$$\tilde{\tilde{W}} \cap \{0 \leq y_2 \leq \varepsilon\} = \{x_2 = H(z_1, y_2/\varepsilon)\}\,.$$

Then W is symplectomorphic to $\tilde{\tilde{W}}$ via a symplectomorphism $\tilde{\varphi}$ fixed near $W \cap \partial U$. □

End of the proof of 1.1.B. The first part of Theorem 1.1.B concerning the existence of a smooth isotopy between φ and j follows immediately from 2.5.D. Remember that the embedding $\tilde{\varphi}$ constructed in 2.5.D concides with the inclusion on $\{0 \leq x_2 \leq \varepsilon\} \cap W$. For any function $\gamma\colon [0,\varepsilon] \to \mathbb{R}$ which is equal to 0 near the ends of the interval there exists a symplectic (but not Hamiltonian!) isotopy of $W \cap \{0 \leq y_2 \leq \varepsilon\}$ into the hypersurface $\{x_2 = \gamma(y_2), 0 \leq y_2 \leq \varepsilon\}$. Using, if necessary, this deformaton for a suitable function γ we can arrange that

$$\int_\Gamma (x_1 dy_1 + x_2 dy_2) = 0, \text{ where } \Gamma = \tilde{\varphi}(\{z_1 = 0, x_2 = 0, 0 \leq y_2 \leq 1\}).$$

After this adjustment we can arrange via a compactly supported in U Hamiltonian isotopy that the embedding $\tilde{\varphi}$ from 2.5.D coincides with the inclusion on $W \cap \{|z_1| \leq \varepsilon\}$ for a sufficiently small $\varepsilon > 0$. This follows from the fact that all embeddings of a 1-dimensional manifold into an exact symplectic manifold of dimension > 2 are Hamiltonian isotopic (via a compactly supported isotopy) provided that they are homotopic and have the same indegrals of the primitive of the symplectic form. For each $\lambda < 1$ we denote by c_λ the symplectic contraction $U \to U$ given by the formula

$$(z_1, x_2, y_2) \mapsto (\lambda z_1, \lambda^2 x_2, y_2).$$

Notice that c_λ maps W into itself. Therefore, we can define a Lagrangian isotopy φ_t, $t \in [\varepsilon, 1]$, between $\varphi_\varepsilon = j$ and $\varphi_1 = \varphi$ in $\mathbb{R}^4_+ \setminus L$ by the formula

$$\varphi_t = c_t^{-1} \circ \tilde{\varphi} \circ c_t.$$

□

2.6 Proof of Theorem 1.2.A.

Let λ_ε be the symplectic contraction $(z_1, x_2, y_2) \mapsto \big((1-\varepsilon)z_1, x_2, (1-\varepsilon)^2 y_2\big)$. Suppose that $\varepsilon > 0$ is chosen sufficiently small, so that the open disc $\operatorname{Int}\big(\lambda_{2\varepsilon}(\Delta)\big)$ still does not intersect $\mathfrak{D}(a,b)$. Choose a function $\alpha\colon [1-2\varepsilon, 1-\varepsilon] \to [\varepsilon, b]$ such that $\alpha(1-2\varepsilon) = b$ and $\lambda(t) = \varepsilon$ for $t \in [\varepsilon, 3\varepsilon/2]$. Set $C = \big\{y_2 = 0,\ x_2 = \alpha\big(|z_1|\big),\ 1-2\varepsilon \le |z_1| \le 1-\varepsilon\big\}$. Then a piecewise smooth disc $\widetilde{\Delta} = \lambda_{2\varepsilon}(\Delta) \cup C$ is holomorphic near $\partial\widetilde{\Delta}$ and we have $\omega|_{\widetilde{\Delta}} \ge 0$. According to Lemma 2.3.A we can perturb $\widetilde{\Delta}$ into a new disc, still denoted by $\widetilde{\Delta}$, in such a way that: $\widetilde{\Delta}$ coincides with $\lambda_{2\varepsilon}(\Delta)$ outside $\mathfrak{D}(a,b)$; $\omega|_{\widetilde{\Delta}} > 0$; $\widetilde{\Delta}$ is holomorphic near its boundary $\partial\widetilde{\Delta}$; $\widetilde{\Delta} \cap \mathfrak{D}(a,b) \subset \{|y_2| < \varepsilon\}$.

Set the following notations:

$$\begin{aligned}
\Omega &= \{x_2 \ge \varepsilon|z_1|^2/(1-\varepsilon)^2\},\\
\Sigma &= \partial\Omega = \{x_2 = \varepsilon|z_1|^2/(1-\varepsilon)^2\} \cap \mathfrak{D}(a,b),\\
A_{c,d} &= \{|z_1| \le c,\ |y_2| \le d\};\\
\Sigma_{c,d} &= A_{c,d} \cap \Sigma;\\
G &= \mathfrak{D}(a,b) \setminus (A_{1,\varepsilon} \cap \Omega)\\
S &= \{y_2 = 0,\ |z_1| \le 1-\varepsilon\} \cap \Sigma.
\end{aligned}$$

Let us choose an almost complex structure J on $\mathbb{R}^4$ such that:

- J is tamed by ω;
- J is standard on G, near Σ and at infinity;
- disc $\widetilde{\Delta}$ is J-holomorphic.

Now Theorem 1.2.A is an immediate corollary of the following

Lemma 2.6.A *The pair $(S, \tilde{\Delta})$ can be filled with J-holomorphic discs.*

Proof. Let $q \in S$ be the elliptic complex point of S. According to Proposition 2.2.A it is sufficient to prove that for a Bishop family of J-holomorphic discs $(\Delta_t, \partial\Delta_t) \subset (\mathbb{R}^4, S)$, $t \in]0, \alpha]$, developing from the point q we have $\operatorname{Int} \Delta_t \cap \Sigma_{1,\varepsilon} = \emptyset$. First observe that for any $t \in]0, \alpha]$we have $\int\limits_{\Delta_t} \omega < \int\limits_S \omega = \pi(1-\varepsilon)^2$. Suppose that there exists a disc Δ_β, $\beta \in]0, \alpha]$, which intersects $\Sigma_{1,\varepsilon}$. Discs Δ_t for $t \in]0, \beta]$ fill the J-flat 3-ball $V \subset \mathbb{R}^4$ with $\partial V = S \cup \Delta_\beta$. Consider two cases:

a) $V \cap (\Sigma_{3,2} \setminus \Sigma_{2,1}) \neq \emptyset$.
In other words, there exists a disc $\Delta' = \Delta_{t_0}$, $t_0 \in]0, \beta]$, and a point $p \in \Delta' \cap [\Sigma_{3,2} \setminus \Sigma_{2,1})$. In this case the ball B of radius $1-\varepsilon$ with the center at the point p is contained inside the domain G where the almost complex structure J is standard. Hence we can apply 2.5.B and conclude that $\operatorname{Area}(\Delta_{t_0} \cap B) > \pi(1-\varepsilon)^2$. Thus we get a contradiction:

$$\pi(1-\varepsilon)^2 < \operatorname{Area}(\Delta_{t_0} \cap B) < \int\limits_{\Delta_{t_0}} \leq \pi(1-\varepsilon)^2.$$

b) $V \cap (\Sigma_{3,2} \setminus \Sigma_{2,1}) = \emptyset$.
Let X be a component of the intersection $V \cap \Sigma$ which has a non-empty intersection with $\Sigma_{1,\varepsilon}$. Set $\gamma = \inf\{t \in]0, \beta] \mid \Delta_t \cap X \neq \emptyset\}$. Denote by Ω_- the domain $(\mathfrak{D}(a,b) \setminus \Omega) \cap A_{3,2}$. The J-holomorphic curve Δ_γ is tangent to $\Sigma_{2,1}$ and approaches Σ form the concave side near tangency points. Thus the J-holomorphic curve $C_\gamma = \Delta_\gamma \cap A_{3,2}$ is contained in $\overline{\Omega}_-$. Notice that the almost complex structure J is standard near $\overline{\Omega}_-$. Therefore, C_γ can be considered as a closed analytic submanifold of the domain $\widehat{\Omega} = \mathfrak{D}(a,b) \cap A_{3,2}$ with the *standard* complex structure. Take a point $p \in C_\gamma \cap X \subset \Sigma_{2,1}$. Then the ball B of radius $1-\varepsilon$ is contained in $\widehat{\Omega}$ and hence we can apply Lemma 2.5.B which gives the inequality $\operatorname{Area}(C_\gamma \cap B) > \pi(1-\varepsilon)^2$ which again contradicts to the inequality $\operatorname{Area}(C_\gamma \cap B) = \int\limits_{C_\gamma \cap B} \omega < \int\limits_{\Delta_\gamma} \omega < \pi(1-\varepsilon)^2$.

□

3 S^2-knot invariants

3.1 Normalization of S^2-knots

An embedding $\varphi : S^2 \to (\mathbb{R}^4, \omega)$ will be called ω-normalized (or just *normalized*) if $\varphi^*\omega = \lambda\omega_0$ where $\lambda > 0$ is a constant and $\omega_0 = i^*\omega$ is a pull-back of ω by the inclusion $i: S^2 \hookrightarrow \mathbb{R}^4$ as the unit sphere in $\mathbb{R}^3 \subset \mathbb{R}^4$. If $\lambda = 1$ then we call the embedding φ *special normalized.*

The following statement is a version of J. Moser's theorem (see [Mo] and also [GrEl]).

Lemma 3.1.A *Let ω be a 2-form on S^2. Suppose that $\int_{S^2} \omega = 0$ and $\omega = h\omega_0$ for a function $h > 0$. Then there exists a diffeomorphism $\varphi: S^2 \to S^2$ which is fixed on the equator $S^1 \subset S^2$ and such that $\varphi^*\omega = \lambda\omega_0$ for a constant $\lambda > 0$.*

Remark. The diffeomorphism φ in 3.1.A can be chosen canonically so that the statement of 3.1.A remains true for any family of forms.

The following Proposition 3.1.B and its multi-parametric generalization 3.1.B (see Section 3.3 below for the proof) are key points in the proof of Theorem 1.2.B.

Proposition 3.1.B *For any embedding $f: S^2 \to \mathbb{R}^4$ there exists a C^0-small isotopy $f_t: S^2 \to \mathbb{R}^4$ and a diffeotopy $\varphi_t: S^2 \to S^2$, $t \in [0,1]$, such that $\varphi_0 = id$, $f_0 = f$ and the embedding $f \circ \varphi_1: S^2 \to \mathbb{R}^4$ is normalized.*

Proposition 3.1.B′ *Let K be a compact set and $L, L \subset K$, its closed subset. Let $f^s: S^2 \to \mathbb{R}^4$, $s \in K$, be a family of embeddings. Suppose that f^s is normalized when $s \in L$. Then there exist families $f_t^s: S^2 \to \mathbb{R}^4$ of embeddings and $\varphi_t^s: S^2 \to S^2$ of diffeomorphisms, $(s,t) \in K \times [0,1]$, such that $\varphi_t^s = id$ and $f_t^s = f^s$ for $(s,t) \in L \times [0,1] \cup K \times 0$, $f_1^s \circ \varphi_1^s$ are normalized for $s \in K$, f_t^s are C^0-close to f^s for all $(s,t) \in K \times [0,1]$.*

Let us denote by $\mathcal{E}$ the space of all embeddings $S^2 \to \mathbb{R}^4$, by $\mathcal{N}$ the subspace of $\mathcal{E}$ which consists of normalized embeddings, by $\mathcal{D}$ the space of diffeomorphisms $S^2 \to S^2$ and by $\mathcal{D}_0, \mathcal{D}_0 \subset \mathcal{D}$, its subgroup which fixes the form ω_0. Let $c : \mathcal{N} \times \mathcal{D}/\mathcal{D}_0 \to \mathcal{E}$ be the map defined by the formula

$$(f, \varphi) \mapsto f \circ \varphi,$$

$f \in \mathcal{N}, \varphi \in \mathcal{D} \pmod{\mathcal{D}_0}$.

Proposition 3.1.B′ implies

Proposition 3.1.C *The map c is a homotopy equivalence.*

Special normalized embeddings differs from normalized embeddings only by rescaling. Therefore, the Proposition 3.1.C above holds also for special normalized embeddings.

Notice that the isotopy in the class of special normalized embeddings can be realized via an ambient Hamiltonian isotopy. In other words, if $f_t : S^2 \to \mathbb{R}^4$, $t \in [0,1]$, is a family of special normalized embeddings then there exists a compactly supported (time-dependent) Hamiltonian function $H : \mathbb{R}^4 \times [0,1] \to \mathbb{R}$ such that $f_t = i_t \circ f_0$ where $i_t : \mathbb{R}^4 \to \mathbb{R}^4$, $t \in [0,1]$, is an isotopy defined by the Hamiltonian H.

Thus 3.1.C implies

Corollary 3.1.D *Let $[\mathscr{E}]$ be the set of isotopy classes of embeddings $S^2 \to \mathbb{R}^4$. Then each class α of $[\mathscr{E}]$ can be represented by a special normalized embedding. Any two such representatives are isotopic via special normalized isotopy. In other words, there is a compactly supported Hamiltonian diffeomorphism $\Psi : \mathbb{R}^4 \to \mathbb{R}^4$ such that $g = \Psi \circ f$.*

In particular, *isotopic special normalized knots have symplectomorphic complements.* Thus the knot complement with its symplectic structure is the invariant of the (differential) isotopical class of the knot.

3.2 Standardization of singularities

Let V be a 3-manifold with a 1-dimensional oriented foliation $\mathscr{F}$ and let $S \subset V$ be an oriented and cooriented surface in V. Let $\Sigma^1(S, \mathscr{F})$ be the subset of S where $\mathscr{F}$ is tangent to S. Then for a generic S we have the following picture of tangency of $\mathscr{F}$ to S (see [El1]). The set $\Sigma^1(S, \mathscr{F})$ is a smooth (not necessarily connected) curve which divides S into two parts: $S = S_+ \cup S_-$, $\partial S_+ = \partial S_- = \Sigma^1(S, \mathscr{F})$. At points of Int S_+ (resp. Int S_-) the orientation of $\mathscr{F}$ coincides with (resp. is opposite to) the coorientation of S. There is a finite subset $\Sigma^{11}(S, \mathscr{F}) \subset \Sigma^1(S, \mathscr{F})$ where $\mathscr{F}$ is tangent to $\Sigma^1(S, \mathscr{F})$. The set $\Sigma^{11}(S, \mathscr{F})$ can be divided into two parts $\Sigma^{11} = \Sigma^{11}_+ \cup \Sigma^{11}_-$. The set Σ^{11}_+ (resp. Σ^{11}_-) consists of points where the orientation of $\mathscr{F}$ coincides with the orientation of $\Sigma^1(S, \mathscr{F})$ as the boundary of the oriented surface S_+ (resp. S_-).

Let $\nu(\mathscr{F})|_S$ be the normal bundle to $\mathscr{F}$ in $T(V)$ restricted to S. We denote by $e(S, \mathscr{F})$ the Euler number of this bundle. In other words, $e(S, \mathscr{F})$ is the obstruction to the construction on S of a vector field tangent to V and transversal to $\mathscr{F}$. We denote by $c_\pm(S, \mathscr{F})$ the number of points in $\Sigma^{11}_\pm(S, \mathscr{F})$ and by $\chi(F)$ the Euler characteristic of a surface F. The following proposition is proved in [El1].

Proposition 3.2.A *a) If $S, V, \mathscr{F}$ are as above then the following equality holds*

$$|\chi(S_+) - \chi(S_-) - c_+(S, \mathscr{F}) + c_-(S, \mathscr{F})| = |e(S, \mathscr{F})|.$$

b) Let $\Gamma \subset S$ be a non-empty curve which divides S into S' and S'', Γ^1_+, Γ^1_- finite subsets of Γ. Let $c_\pm$ be numbers of points in $\Gamma^1_\pm$. Suppose that

$$\left|\chi(S') - \chi(S'') - c_+ + c_-\right| = \left|e(S, \mathscr{F})\right|.$$

Then there exists a C^0-small diffeomorphism $f : V \to V$ and a diffeomorphism $\varphi : S \to S$ such that

$$\Sigma^1(f(S), \mathscr{F}) = \varphi(\Gamma),\ \Sigma^{11}_+(f(S), \mathscr{F}) = \varphi(\Gamma^1_\pm).$$

Corollary 3.2.B *Suppose that $e(S, \mathscr{F}) = 0$. Then S can be C^0-perturbed to a surface $\widetilde{S}$ such that $\Sigma^1(\widetilde{S}, \mathscr{F})$ is an equator and both sets $\Sigma^{11}_\pm(S, \mathscr{F})$ consist of single points.*

If the surface $S \subset V$ and the foliation $\mathcal{F}$ satisfy the conclusion of Corollary 3.2.B we say that $(S, \mathcal{F})$ is an *adapted pair*. The following multiparametric version of 3.2.B also follows from the results in [El1] and [El2].

Proposition 3.2.B′ *Let K be a compact set and $L, L \subset K$, be its closed subset. Let $\mathcal{F}^s$, $s \in K$, be a family of 1-dimensional foliations on a 3-manifold V and $S, S \subset V$, be a 2-dimensional embedded sphere. Suppose that $e(S, \mathcal{F}^s) = 0$ for all $s \in K$ and that for $s \in L$ the pair $(S, \mathcal{F}^s)$ is adapted. Then there exists a family of C^0-small diffeomorphisms $h_s : V \to V$, $s \in K$, such that $h_s = id$ for $s \in L$ and pairs $\big(h_s(S), \mathcal{F}^s\big)$ are adapted for all $s \in K$.*

3.3 Proof of 3.1.B and 1.2.B.

Proof of Theorem 3.1.B. The embedding $f : S^2 \to \mathbb{R}^4$ can be extended to an embedding $F : S^2 \times [-\varepsilon, \varepsilon] \to \mathbb{R}^4$ such that $F|_{S^2 \times 0} = f$. The image $V = F\big(S^2 \times [-\varepsilon, \varepsilon]\big)$ carries a 1-dimensional characteristic foliation $\mathcal{F}$ (defined by integrating the line field $\mathrm{Ker}(\omega|_V) \subset T(V)\big)$. Fix orientations of $S = F(S)$ and $\mathcal{F}$. Suppose that S is generic in the sense of Section 3.2. Choose a coorientation of S in such a way that $\omega|_{\mathrm{Int}\, S_+} > 0$. Notice that the bundle $\nu(\mathcal{F})|_S = \big(T(V)/\mathrm{Ker}(\omega)\big)|_S$ is trivial. This follows from two facts: all embeddings $S^2 \to \mathbb{R}^4$ are regularly homotopic and all normal framings of an embedded S^2 in $\mathbb{R}^4$ are homotopic. Therefore, we can apply Corollary 3.2.B and perturb S inside V in such a way that for the new surface $\widetilde{S}$ the pair $(\widetilde{S}, \mathcal{F})$ is normalized, i.e., we get that: $\Sigma^1(\widetilde{S}, \mathcal{F})$ is an equator which divides $\widetilde{S}$ into two hemispheres $\widetilde{S}_+$ and $\widetilde{S}_-$; sets $\Sigma_1^{11}(\widetilde{S}, \mathcal{F})$ consist of single points. But then $\omega|_{\mathrm{Int}\widetilde{S}_+} > 0$, $\omega|_{\mathrm{Int}\widetilde{S}_-} < 0$ and the form ω has simple zeroes along $\partial\widetilde{S}_+ = \partial\widetilde{S}_-$. This means according to Moser's lemma 3.1.A that the embedding $\widetilde{S} \hookrightarrow V \hookrightarrow \mathbb{R}^4$ is equivalent to a normalized one. □

The Proposition 3.1.B′ can be proved similarly using 3.2.B′ instead of 3.2.B.

Proof of Theorem 1.2.B. According to Proposition 3.1.A any isotopy class φ can be represented by a normalized embedding. Let S be the unit sphere in $\mathbb{R}^4$. As it follows from 3.1 there exists a neighborhood $V \subset \mathbb{R}^4$ of S and a symplectic embedding $F : V \to \mathbb{R}^4$ such that $F|_S$ represents the class α. Let us consider a piecewise smooth sphere $\Sigma = D \cup \Delta \subset \mathbb{R}^4 = \mathbb{C}^2$ where $D = \big\{|z_1| \le 1,\, z_2 = 0\big\}$, $\Delta = \big\{|z_1|^2 \le 1,\, y_2 = 0,\, |z_1|^2 + x_2^2 = 1,\, y_2 \ge 0\big\}$. There exists a symplectic embedding Φ of a neighborhood $U \subset \mathbb{R}^4$ of Σ into the neighborhood $V \subset \mathbb{R}^4$ of S such that $\Phi(\Sigma)$ is C^0-close to S. Thus the isotopy class α can be represented by the embedding $f = F \circ \Phi|_\Sigma : \Sigma \to \mathbb{R}^4$. By the construction this embedding extends to the neighborhood $U \supset \Sigma$. In particular, there exists $\varepsilon > 0$ such that f extends to a symplectic embedding of $\Delta \cup (P_\varepsilon = \big\{|z_1| \le 1 + \varepsilon,\, |z_2 - \varepsilon| \le \varepsilon\big\})$. But all symplectic embeddings of a polydisc into $\mathbb{R}^4$ are Hamiltonian isotopic. Therefore, we can think that $F \circ \Phi|_{P_\varepsilon} = id$. That means, by the definition, that α admits a $(1 + \varepsilon, \varepsilon)$-representation. □

3.4 Symplectic shape of a knot

Let α be a (smooth) isotopy class of embeddings $S^2 \to \mathbb{R}^4$. Define the function $\Psi_\alpha : [1, \infty[\to \mathbb{R}_+$ by the formula

$$\Psi_\alpha(t) = \inf \left\{ 1/u^2 \mid \alpha \text{ admits a } (\sqrt{t}, u)\text{-realization} \right\}.$$

We call the function Ψ_α the *symplectic shape* of the knot α. The properties of Ψ_α are listed in the following

Theorem 3.4.A

a) *Ψ_α is an invariant of the smooth isotopy class of the knot α;*

b) *Ψ_α is a monotonically increasing bounded function: there exists a constant N_α such that $\Psi_\alpha(t) \le N_\alpha$ for all $t \ge 1$;*

c) *If α is a trivial knot then $\Psi_\alpha \equiv 0$;*

d) *If a knot α is non-trivial then $\Psi_\alpha(16) \ge \frac{1}{4}$.*

Proof. a) and c) are immediate from the definition, d) is a reformulation of Theorem 1.2.A. Thus we have to prove only b). According to Theorem 1.2.B the knot α admit a (a, b)-realization for some $a > 1$, $b > 0$. Thus α is represented by a sphere $S = D \cup \Delta$ where $D = \{|z_1| \le 1,\ z_2 = b\}$ and Δ is an ω-positive disc which lies in the complement of the polydisc $\mathfrak{D}(a, b)$. Suppose that the union $\mathfrak{D}(a, b) \cup \Delta$ is contained in the ball of radius R. Let us denote by P_r the polydisk $\{|z_1 - R - r| \le r,\ |z_2| \le b\}$, by D_r the disc $\{|z_1 - R - r| \le r,\ z_2 = b\} \subset \partial P_r$, and by D' the disc $\{|z_1| \le r,\ z_2 = b\} \subset \partial\mathfrak{D}(a, b)$. Let Γ be an embedded path which connects points $A \in \partial D'$ and $B \in \partial D_r$ with complex coordinates (a, b) and $(R + r - a, b)$ respectively. We can choose Γ in such a way that Γ do not intersect $\Delta \cup \operatorname{Int}(\mathfrak{D}(a, b) \cup P_r)$. Let C be a 2-dimensional collar of Γ, such that: $\omega|_C > 0$; the disc $\widetilde{D}_r = D' \cup C \cup D_r$ is a smooth embedded disc which lies outside $\Delta \cup \operatorname{Int}(\mathfrak{D}(a, b) \cup P_r)$; there exists a neighborhood $U \supset \mathfrak{D}(a, b) \cup P_r$ such that $C \cap (U \setminus (\mathfrak{D}(a, b) \cup P_r)) = \{z_2 = b,\ |y_1| \le \varepsilon\} \cap (U \setminus (\mathfrak{D}(a, b) \cup P_r))$. The collar C has a neighborhood G which is diffeomorphic to the product $C \times D(\varepsilon)$ where $D(\varepsilon)$ is a disc of radius ε in complex line $\{z_2 = 0\}$ and the symplectic structure on the product $C \times D(\varepsilon)$ is the product of forms $\omega|_C$ and $\omega|_{D(\varepsilon)}$. G can be chosen in such a way that $G \cap (U \setminus (\mathfrak{D}(a, b) \cup P_r)) = \{|z_2 - b + \delta| \le \delta,\ |y_2| \le \varepsilon\} \cap (U \setminus (\mathfrak{D}(a, b) \cup P_r))$ for some $\delta > 0$. Then the domain

$$G'_r = \{|z_1| \le a,\ |z_2 - b + \varepsilon| \le \varepsilon\} \cup G \cup \{|z_1 - R - r| \le r,\ |z_2 - b + \varepsilon| \le \varepsilon\}$$

is symplectomorphic to the polydisc $\mathfrak{D}(r', \varepsilon)$ for some $r' > r$ and does not intersect Δ. This implies that the knot α has a (r', ε)-representation where r' can be chosen arbitrarily big and $\varepsilon > 0$ is independent of r'. Hence $\Psi_\alpha(t) \le \frac{1}{\varepsilon^2}$ for all $t \ge 1$. □

Almost nothing is known about the function Ψ_α beyond Theorem 3.4.A. It is clearly possible to improve the lower bound given by the inequality 3.4.A d) and in examples it is easy to produce upper bounds for Ψ_α but it is, probably, difficult to compute Ψ_α exactly. Some examples, actually, suggest that the function Ψ_α is a constant (which, of course, depends on the knot α). In this case the only information which is carried by this function would be the number $\Psi_\alpha(1)$.

3.5 Symplectic shape and Gluck's construction

Herman Gluck's construction (see [Gl]) is a Dehn-type surgery along a two-knot in $\mathbb{R}^4$ (or S^4). Let $f: S^2 \to \mathbb{R}^4$ be an embedding. Then the tubular neighborhood U of the image $S = f(S^2)$ is trivial: $U \simeq S^2 \times D^2$. The boundary $\partial U = S^2 \times S^1$ admits, essentially, only one non-trivial diffeomorphism $h : \partial U \to \partial U$ which can be described as follows. Let $\gamma : S^1 \to SO_3$ be a loop representing the non-trivial element of $\pi_1(SO_3)$. We define the diffeomorphism $h : \partial U \to \partial U$ by the formula $h(x,t) = \big(\gamma(t)(x), t\big)$, $(x,t) \in S^2 \times S^1 = \partial U$ (we consider elements of SO_3 represented by rotations of S^2). Notice that the diffeomorphism $h \circ h$ is isotopic to the identity. Gluck's surgery along the knot $f : S^2 \to \mathbb{R}^4$ is defined as the manifold $X_f = (\mathbb{R}^4 \setminus \operatorname{Int} U) \bigcup_h U$. The diffeomorphism h induces an isomorphism on the fundamental group and homology groups. It follows that there exists a homotopy equivalence $X_f \to \mathbb{R}^4$ which is equal to the identity at infinity. Thus Gluck's construction may be a potential source of counterexamples to the smooth 4-dimensional Poincaré conjecture. We say that the Gluck's surgery along a knot $f : S^2 \to \mathbb{R}^4$ is trivial if there exists a fixed at infinity diffeomorphism $X_f \to \mathbb{R}^4$.

Theorem 3.5.A *(joint with K. Luttinger). Suppose that an embedding $f : S^2 \to \mathbb{R}^4$ represents an isotopy class α such that $\Psi_\alpha(1) < 1$. Then the Gluck's surgery along the knot f is trivial.*

The idea of the proof is the following. First we normalize the embedding f using Proposition 3.1.B. The inequality $\Psi_\alpha(1) < 1$ guarantees the existence of symplectically big enough neighborhood $U \supset f(S^2)$. This allows us to extend the symplectic structure $f^*\omega$ from the boundary ∂U to a symplectic form on U. Therefore, X_f admits a symplectic structure which coincides with the standard symplectic $\mathbb{R}^4$ at infinity. According to a theorem of Gromov (see[Gr]) this implies that X_f is diffeomorphic to $\mathbb{R}^4$.

Unfortunately, the value of this result is questionnable at this point: I do not know if there are any non-trivial S^2-knots α with $\Psi_\alpha(1) < 1$.

References

[BG] E. Bedford, B. Gaveaux, Envelopes of holomorphy of certain 2-spheres in $\mathbb{C}^2$, *Amer. J. of Math.*, **105** (1983), 975–1009.

[BK] E. Bedford, W. Klingenberg, On the envelope of holomorphy of a 2-sphere in $\mathbb{C}^2$, *J. of Amer. Math. Soc.* **4** (1991).

[Be] D. Bennequin, Entrelacements et équations de Pfaff, *Astérisque* **107–108** (1983), 83–61.

[Bi] E. Bishop, Differentiable manifolds in complex euclidean space, *Duke Math. J.*, **32** (1965), 1–22.

[El1] Y. Eliashberg, On singularities of folding type, *Izvestya AN SSSR*, ser. math. **34** (1970), 1110–1126.

[El2] Y. Eliashberg, Surgery of singularities of smooth mappings, *Izvestya AN SSSR*, ser. mat., **36** (1972), 1321–1347.

[El3] Y. Eliashberg, Filling by holomorphic discs and its applications, *London Math. Soc. Lect. Notes Series*, **151** (1990), part II, 45–68.

[El4] Y. Eliashberg, Contact 3-manifolds, twenty years since J. Martinet's work, *Ann. Inst. Fourier*, **92** (1992), 1–2 (165–192).

[Fe] H. Federer, Geometric measure theory, Springer, 1969.

[Gl] H. Gluck, The embedding of two-spheres into the four-spheres, *Trans AMS*, **104** (1962), 308–333.

[Gr] M. Gromov, Pseudoholomorphic curves in symplectic manifolds, *Inv. Math.*, **82** (1985), 307–347.

[GrEl] M. Gromov and Y. Eliashberg, Construction of a smooth mapping with the prescribed Jacobian, *Funkt. Anal e Pril.*, **7** (1973), 33–40.

[Lu] K. Luttinger, Lagrangian tori in $\mathbb{R}^4$, preprint, 1992, to appear in Journal of Diff. Geometry.

[HE] V. Harlamov and Y. Eliashberg, On the number of complex points of a real surface in a complex one, *Proc. LITC–82*, 143–148.

[HL] R. Harvey and H.B. Lawson, Calibrated geometries, Acta Mathematica, **148** (1982), 47–157.

[McD] D. McDuff, Singularities of J-holomorphic curves in almost complex 4-manifolds, *The Journal of Geometric Analysis*, **3** (1992), 249–266.

[Mo] J. Moser, On the volume elements of a manifold, *Trans. Amer. Math. Soc.* **120** (1965), 286–294.

Yakov Eliashberg
Stanford University
Stanford
CA 94305, USA

The ends of the monopole moduli space over $\mathbb{R}^3\#$ (homology sphere): Part I

K.D. Ernst

Abstract. Let K be a homology sphere with respect to integer coefficients. Denote by K' the connected sum of K with $\mathbb{R}^3$. In this paper (Part I) and a sequel (Part II) we study the three different ends of the moduli space $\mathfrak{M}(K')$ of charge one monopoles on K' in the Prasad-Sommerfield limit. These ends consist of monopoles of large mass, small mass and monopoles whose main part of the energy is far away from the neck.

Part I gives a complete description of the large mass end, the end consisting of monopoles which are located far away from the neck, as well as a partial discussion of the small mass end. Part II gives a complete description of the small mass end.

1 Introduction

Monopoles in the Prasad-Sommerfield limit are pairs consisting of a connection and a Higgs field. They fulfill a first-order differential equation called the Bogomol'nyi equation, as well as an asymptotic condition on the Higgs field. The YMH-functional gives the energy for the monopoles. See [JT] for an introduction to monopoles.

The YMH-functional in three dimensions is not conformally invariant. A change in scale results in a change of the mass and the energy. Monopoles with small mass also have small energy, which is spread out over a large region of $\mathbb{R}^3$. Large mass monopoles have large energy whose main part is concentrated near a point. Here we restrict ourselves to monopoles of charge one. On $\mathbb{R}^3$, such monopoles are determined by their center up to gauge equivalence. Thus, the moduli space $\mathfrak{M}_m\left(\mathbb{R}^3\right)$ of mass m monopoles equals $\mathbb{R}^3$. Varying the mass gives a 4-dimensional moduli space $\mathfrak{M}\left(\mathbb{R}^3\right)$ which equals $\mathbb{R}^3 \times \mathbb{R}^+$.

Now replace $\mathbb{R}^3$ by K'. K' is the connected sum of $\mathbb{R}^3$ with K. K is a homology sphere with respect to integer coefficients. Through the whole paper we keep the radius of the neck fixed in the connected sum.

In [F1] Floer announced several results concerning the moduli space of monopoles on asymptotically Euclidean manifolds. Some of these results describe the ends of this moduli space. Altogether there are three such ends consisting of monopoles with large mass, monopoles whose centers drift out on K', and monoples with small mass.

The purpose of this paper is to give a detailed discussion of the large mass end of $\mathfrak{M}_m(K')$ as well as the end of $\mathfrak{M}_m(K')$ which consists of monopoles located far away from the neck. Furthermore, we give here a partial discussion of the small mass end. In a sequel to this paper [E2], also referred to as Part II, we give a complete description of the small mass moduli space $\mathfrak{M}_m(K')$.

In [E1] we apply the results of this paper to manifolds with asymptotically Euclidean ends. Assuming a certain bound on the negative part of the Ricci curvature, we show that these manifolds are diffeomorphic to $\mathbb{R}^3$.

In Sections 2–3 we are concerned with constructing monopoles which lie in several ends of the moduli space $\mathfrak{M}(K')$. The techniques used are based on the ones developed by Taubes and described in [JT]. In this situation the deformation operators in the self-dual directions do not have cokernels. This will be different in Part II. There we encounter a 3-dimensional obstruction for the self-dual perturbation. This obstruction is similar to the one occurring for self-dual connections on four-manifolds with indefinite intersection matrix. See [T5].

In Sections 5–10 we study the moduli space problem. We use the methods developed in [D1] to investigate the large mass moduli space. For the ends of the moduli space consisting of small mass monopoles or monopoles located far away from the neck, we follow the techniques developed by Donaldson in [D2] and [DK] as closely as possible. The fact that we are dealing with a noncompact manifold brings some additional difficulties.

The author would like to thank C. Taubes for suggesting this problem and for helpful remarks.

1.1 Statement of results

First consider the end of the monopole moduli space which consists of large mass monopoles. The main part of the energy of a large mass monopole is concentrated at a point on K'. Monopoles, with most of their energy concentrated in balls smaller than the injectivity radius of K', are in one-to-one correspondence with points on K'. This situation is similar to the concentration of instantons on four-manifolds [D1].

Theorem 1.1 *There is an m_1 such that for $m \geq m_1$, $\mathfrak{M}_m(K')$ is smooth and a diffeomorphism Ψ_1*

$$\Psi_1 : \mathfrak{M}_m(K') \longrightarrow K' .$$

Monopoles whose main part of the energy is located far away from the neck feel only a small influence of the connected sum. They are similar to monopoles on $\mathbb{R}^3$.

Denote by D_R the complement in $\mathbb{R}^3$ of a ball of radius R around the origin. Consider D_R also as a subset of K'. Assume that Q is a monopole on K' with most of the energy located on D_R. We define in Section 9 a center for such monopoles. Denote by $\mathfrak{W}_{m,R}$ the set of all monopoles with center on D_R.

Theorem 1.2 *Fix $0 < m_0 < m_1 < \infty$. There is a R_0 such that for $m \in [m_0, m_1]$ and $R \geq R_0$, $\mathfrak{W}_{m,R}$ is smooth and there is a diffeomorphism Ψ_2*

$$\Psi_2 : \mathfrak{W}_{m,R} \longrightarrow D_R .$$

The energy or, in other words, the value of the YMH-functional of a small mass monopole $C = (\Phi, A)$ is small. Thus the L^2-norm of F_A is small and A is close to a flat connection.

Denote by $\mathfrak{R}(K')$ the space of flat connections over K' up to gauge equivalence. $\mathfrak{R}(K')$ decomposes as

$$\mathfrak{R}(K') = \theta \coprod V,$$

where θ is the trivial flat connection and V is a compact space parametrizing irreducible flat connections. Irreducible flat connections are well separated from the trivial flat connection. This indicates that, for small m, the moduli space splits up into a disjoint union

$$\mathfrak{M}_m(K') = \mathfrak{M}_{m,\theta}(K') \coprod \mathfrak{M}_{m,V}(K')$$

where the connection part of monopoles in $\mathfrak{M}_{m,\theta}(K')$, respectively in $\mathfrak{M}_{m,V}(K')$ is close to θ, respectively to an element in V.

Small mass monopoles have their energy spread out over a large region of K'. They feel only a small influence of the connected sum and behave like monopoles on $\mathbb{R}^3$. One can expect that all elements of $\mathfrak{M}_{m,\theta}(K')$ arise by gluing a charge one monopole $C_x \in \mathfrak{M}_m(\mathbb{R}^3)$ to a configuration (Φ, θ) with energy zero on K'. Such a gluing requires a bundle identification. For $SU(2)$ bundles these identifications are parametrized by $SO(3)$. However if we glue $C_x \in \mathfrak{M}_m(\mathbb{R}^3)$ to (Φ, θ) the resulting configuration is independent up to gauge equivalence of the chosen bundle identification, suggesting that $\mathfrak{M}_{m,\theta}(K')$ is parametrized by $\mathbb{R}^3$.

Theorem 1.3 *There is an m_0 such that, for $m \leq m_0$, the small mass moduli space splits up into two well separated pieces*

$$\mathfrak{M}_m(K') = \mathfrak{M}_{m,\theta}(K') \coprod \mathfrak{M}_{m,V}(K').$$

Furthermore $\mathfrak{M}_{m,\theta}(K')$ is diffeomorphic to $\mathbb{R}^3$ and $\mathfrak{M}_{m,V}(K')$ is a compact set.

In Part II we will give a complete description of the compact set $\mathfrak{M}_{m,V}(K')$.

1.2 Outline of the proofs

Let us first explain the strategy of the proof of Theorem 1.2 and 1.3. In Chapter 7 of [DK] (see also [D2]), open sets of moduli spaces of self-dual connections over connected sums of four-manifolds are described in terms of the moduli spaces over the individual manifolds. In the proof of Theorem 1.2 and 1.3 we follow these methods. But unlike in the mentioned reference, we keep here the radius of the neck fixed.

We call configurations which can be perturbed into self-dual ones, almost self-dual. For an almost self-dual configuration C, a pertubation Ψ is constructed in Section 2 such that $C + \Psi$ is self-dual. Ψ has the form $P\xi$ where P is the right inverse to the derivative in the self-dual directions at C. ξ is the unique fixed point of a certain map.

Our next task is to construct almost self-dual configurations, which approximate monopoles in the small mass end of $\mathfrak{M}_m(K')$ and in the end consisting of monopoles located far away from the neck. This is done in Section 3. These configurations are parametrized by the spaces D_R and $\mathbb{R}^3$. Again D_R denotes the complement of a ball of radius R centered at the origin in $\mathbb{R}^3$.

Denote by X one of these parameter spaces. The constructed, almost self-dual configurations are denoted by $\mathcal{A}_m(X)$. The construction gives the map

$$\begin{aligned} J : X &\longrightarrow \mathcal{A}_m(X) \subset \mathcal{C}_m \\ \varphi &\longmapsto C_\varphi \end{aligned} \tag{1.1}$$

where $\mathcal{C}_m$ denotes the space of mass m configurations. Composing the map J with the self-dual perturbation gives monopoles on K'. We can express that by the map

$$\begin{aligned} I : X &\longrightarrow \mathfrak{M}_m(K') \\ \varphi &\longmapsto C_\varphi + P_\varphi \xi \,. \end{aligned} \tag{1.2}$$

Denote by $\mathcal{U}_m(\nu, X)$ a certain tubular neighborhood of $\mathcal{A}_m(X)$ in $\mathcal{C}_m$, the space of mass m configurations. ν labels the «thickness» of this tubular neighborhood. Making ν sufficiently large ensures that all constructed monopoles are contained in that neighborhood

$$I(X) \subset \mathcal{U}_m(\nu, X)\,. \tag{1.3}$$

Let $Q \in \mathcal{U}_m(\nu, X)$. In Section 6 and 7 we show that after a suitable gauge transformation u, we can write Q in the form

$$Q = u(C_\varphi + P_\varphi \xi) \tag{1.4}$$

for a certain $\varphi \in X$ and a certain $\xi \in \Omega^1$. Remark that ξ is here not necessarily a fixed point of the map which produces self-dual perturbations. However in case Q is self-dual, $P\xi$ is a self-dual pertubation. As such pertubations are unique, (1.4) shows that Q is contained in the image of the map I. In other words

$$I(X) = \mathcal{U}_m(\nu, X) \cap \mathfrak{M}_m(K')\,.$$

In Section 8 we show that the map I is injective. Thus I gives a diffeomorphism from the parameter space X to the monopoles in the tubular neighborhood $\mathcal{U}_m(\nu, X)$ of $J(X)$ and

$$X \simeq \mathcal{U}_m(\nu, X) \cap \mathfrak{M}_m(K')\,.$$

It remains to show that all monopoles in the ends are a priori contained in the tubular neighborhood. More specifically, we show in Section 9 that

$$\mathfrak{W}_{m,R} \subset \mathcal{U}_m(\nu, D_R) \quad \text{for} \quad R \geq R_0(\nu)\,.$$

For small monopoles, we show in Section 10 that

$$\mathfrak{M}_m\,(K') = \mathfrak{M}_{m,\theta}(K') \coprod \mathfrak{M}_{m,V}(K')\,.$$

The proof of Theorem 1.3 is concluded by showing that $\mathfrak{M}_{m,V}(K')$ is compact and that

$$\mathfrak{M}_{m,\theta}(K') \subset \mathcal{U}_m(\nu, \mathbb{R}^3)\,.$$

This gauge fixing strategy (1.4) becomes more difficult for large mass monopoles. For this reason we choose for this end of the moduli space another approach to prove Theorem 1.1. This approach is based on the one explained in [D1]. We show that every large mass monopole Q has a center $p(Q)$. Furthermore the map

$$p : \mathfrak{M}_m\,(K') \longrightarrow K'\,,$$

giving for each monopole its center is a covering map. From Theorem 1.2 it follows that $\mathfrak{M}_m\,(K')$ is a one-fold covering of K'. Thus $\mathfrak{M}_m\,(K')$ is diffeomorphic to K' (Sections 4 and 5).

1.3 YMH-configurations

Let A be a connection on a trivial $su(2)$-bundle over $\mathbb{R}^3$ or K' and Φ a $su(2)$-valued function. The YMH-functional for the configuration $C = (A, \Phi)$ is

$$\mathrm{YMH}(C) = \frac{1}{2}\left(\|F_A\|_2^2 + \|d_A\Phi\|_2^2\right)\,.$$

The formal variational equations for this functional are the YMH equations. The configuration space $\mathcal{C}_m$ consists of C's with $C \in L^2_{1,loc}$, $\mathrm{YMH}(C) < \infty$ and $\lim_{|x|\to\infty} |\Phi|(x) = m$. All the configurations we use in this paper satisfy some additional conditions. They are listed in Definition 2.1.

$\mathcal{C}_m$ splits up into components corresponding to the charge of the configurations. In [G] the charge l is shown to be an integer and equal to

$$l = \frac{1}{4\pi m}\,(*F_A, d_A\Phi)_2\,.$$

In this paper only charge one and charge zero configurations appear. Let $\mathcal{G}$ be the space of $L^2_{2,loc}$ gauge transformations. $\mathcal{G}$ acts on $\mathcal{C}_m$ yielding a quotient space $\mathcal{C}_m/\mathcal{G}$. In [F2] and [T4], the space $\mathcal{C}_m/\mathcal{G}$ is shown to be a smooth Banach manifold.

Every configuration defines a norm for pairs $\xi = (\xi_0, \xi_1)$ where ξ_0 is a su(2)-valued zero form and ξ_1 is a $su(2)$-valued one form

$$\|\xi\|_c^2 := \|\nabla_A \xi\|_2^2 + \| [\Phi, \xi] \|_2^2 . \tag{1.5}$$

Kato's and a Sobolev inequality show that this is indeed a norm. We denote with L_c the completion of the space of smooth compactly supported forms in the $\|.\|_c$-norm.

We will also make use of the decomposition of ξ_j in transversal and longitudinal components with respect to the Higgs field. For $j \in (0, 1)$.

$$\begin{aligned} \xi_j &= \xi_{j,L} + \xi_{j,T} \\ \xi_{j,L} &= \Phi(\Phi, \xi_j)|\Phi|^{-2} . \end{aligned} \tag{1.6}$$

Let $B := \{x \in K', |\Phi| \leq \frac{1}{2}m\}$. Assume that $Vol(B) \leq (100m)^{-3}$. Then using the decomposition (1.6) it is shown in Lemma 6.6 of [T1] that

$$\| [\xi, \eta] \|_2 \leq D_0 m^{-1/2} \|\xi\|_c \|\eta\|_c \tag{1.7}$$

and

$$\| [\xi, \xi] \|_2 \leq D_0 m^{-1/2} \|\xi\|_c \|\xi\|_6 . \tag{1.8}$$

1.4 Monopoles

The first-order nonlinear equation

$$d_A \Phi - *F_A = 0 \tag{1.9}$$

is the Bogomol'nyi equation. Its solutions are called monopoles and they are automatically solutions of the YMH-equations.

By the Bogomol'nyi equation and partial integration, monopoles must have zero energy on a compact manifold.

Monopoles are also called self-dual configurations. In comparison with [JT] we will call charge one monopoles what is labeled in that reference as $N = -1$ monopoles. $G_C := d_A \Phi - *F_A$ is the anti-self-dual part of C, $\widehat{G_C} := d_A \Phi + *F_A$ is the self-dual part. For a charge one monopole

$$\text{YMH}(C) = 4\pi m + \frac{1}{2}\|G_C\|_2^2 . \tag{1.10}$$

The value of the YMH-functional is also called the energy of the configuration.

The equation $G_C = 0$ also has abelian solutions, called Dirac monopoles. Use on $\mathbb{R}^3$ the polar coordinates (r, χ, θ). σ^3 is the third Pauli spin matrix. In polar gauge the Dirac monopole centered at the origin has the form

$$\widehat{\mathfrak{D}}_p = \left(m - \frac{1}{r} , (1 - \cos\theta) d\chi\right) \frac{i}{2}\sigma^3 . \tag{1.11}$$

However $\widehat{G}_{\widehat{\mathfrak{D}}_p}$ is not square integrable. Thus the Dirac monopoles have infinite energy. The finite energy monopoles approximate asymptotically a Dirac monopole. However the nonabelian parts cancel the singularity of $\widehat{\mathfrak{D}}_p$ at the origin.

1.5 Deformation complex

The complex for monopoles corresponding to the deformation complex of anti-self-dual connections is

$$\Omega^0 \xrightarrow{d_c^1} \Omega^0 \oplus \Omega^1 \xrightarrow{d_c^2} \Omega^1 . \tag{1.12}$$

There

$$d_c^1(\chi) := ([\Phi,\chi], d_A\chi) \tag{1.13}$$

is the linear part of the deformation with respect to the gauge transformation $\exp(\chi)$, and

$$d_c^2(\varphi,a) := d_A\varphi - [\Phi,a] - *d_A a \tag{1.14}$$

is the linear, self-dual part of the deformation (φ,a). Also important are the adjoints of these operators

$$(d_c^1)^*(\varphi,a) = -[\Phi,\varphi] + d_A^* a \tag{1.15}$$

$$(d_c^2)^*(a) = (d_A^* a, [\Phi,a] - *d_A a) \ . \tag{1.16}$$

The relation between d_c^1, d_c^2 and ∇_A is given by following Weitzenböck-like formulas: For $\chi \in \Omega^0$

$$(d_c^1)^* d_c^1 \chi = \nabla_A^* \nabla_A \chi - [\Phi,[\Phi,\chi]] ; \tag{1.17}$$

for $a \in \Omega^1$

$$d_c^2 (d_c^2)^* a = \nabla_A^* \nabla_A a - [\Phi,[\Phi,a]] - \{G_c, a\} + \{Ric, a\} . \tag{1.18}$$

And for $u = (\varphi,a) \in \Omega^0 \oplus \Omega^1$

$$d_c^1 (d_c^1)^* u + (d_c^2)^* d_c^2 u = \Big(\nabla_A^* \nabla_A \varphi - [\Phi,[\Phi,\varphi]] + *[\widehat{G_c}, *a] , \tag{1.19}$$

$$\nabla_A^* \nabla_A a - [\Phi,[\Phi,a]] - \{\widehat{G_c}, u\}_1 + \{Ric, a\} \Big)$$

where $\{\widehat{G_c}, u\}_1 = -[\widehat{G_c},\varphi] - *[\widehat{G_c}, a]$. *Ric* is the Ricci curvature and the $\{,\}$ denotes a certain bilinear form. It is convenient to take $(d_c^1)^*$ and d_c^2, respectively d_c^1 and $(d_c^2)^*$, together with one operator. Let L_C be the space explained in (1.5). Set

$$\delta_c := (d_c^1)^* \oplus d_c^2 : L_c(\Omega^0 \oplus \Omega^1) \longrightarrow L^2(\Omega^0 \oplus \Omega^1) \tag{1.20}$$

$$\delta_c^* := d_c^1 \oplus (d_c^2)^* : L_c(\Omega^0 \oplus \Omega^1) \longrightarrow L^2(\Omega^0 \oplus \Omega^1) . \tag{1.21}$$

In [T2] it is shown that δ_c is a bounded Fredholm operator with

$$\text{index } \delta_c = 4l \tag{1.22}$$

where l is the charge of the configuration C.

Finally we recall here a well-known Sobolev theorem.

Lemma 1.4 *For $\Psi \in C_0^\infty(K')$ there is a constant such that*

$$\|\Psi\|_6 \leq \text{ const } \|\nabla\Psi\|_2 .$$

We denote the smallest constant which satisfies this equation by S_3 and call it the Sobolev constant. Remark that S_3 depends on K'.

Some notations:
Below several cutoff functions will be used. Let α respective β be the cutoffs with $0 \leq \alpha(x), \beta(x) \leq 1$,

$$\alpha(x) = 0 \text{ for } |x| \leq 1 \text{ and } \alpha = 1 \text{ for } |x| \geq \frac{3}{2} \tag{1.23}$$

$$\beta(x) = 1 \text{ for } |x| \leq 1 \text{ and } \beta = 0 \text{ for } |x| \geq \frac{3}{2} . \tag{1.24}$$

For $r > 0$ define $\alpha_{x,r} = \alpha(x/r)$ respectively $\beta_{x,r} = \beta(x/r)$.

2 Perturbations of almost self-dual configurations

The purpose of this section is to perturb almost self-dual configurations to self-dual ones. To do that we construct first a right inverse P for the operator δ_c defined in (1.20). Then we bound the operator norm of P and use a fixed point theorem.

2.1 The operators δ_c and δ_c^*

Before we start with the detailed discussion of the perturbations let us state some facts about the operators δ_c and δ_c^*. Define the set of configurations

Definition 2.1 *Let C be a configuration on K' which fulfills the following properties:*

1. $C \in L^2_{1,loc}$.
2. $YHM(C) < \infty$.
3. $\lim_{|x| \to \infty} |\Phi|(x) = m$.
4. *G_c is compactly supported and smooth.*
5. *Let $B := \{x \in K', |\Phi|(x) \leq \frac{1}{2}m\}$. We assume $Vol(B) \leq (100m)^{-3}$.*
   Define $\mathscr{C}_m(K')$ to be the set of configurations which fulfill these properties.

The configurations which will be constructed in Sections 3 fulfill all these properties. We do not assume C to be smooth as we sometimes have to work with gauges where C is singular along a string. Condition 5) ensures that (1.7) and (1.8) are fulfilled.

Lemma 2.2 *Take* $C \in \mathscr{C}_m(K')$.

1. *Denote by* C_0^∞ *compactly supported smooth sections on* K'. *The operators*

$$\delta_c\,,\ \delta_c^*\ :\ C_0^\infty(\Omega^0 \oplus \Omega^1) \longrightarrow C_0^\infty(\Omega^0 \oplus \Omega^1)$$

extend to bounded, Fredholm operators from L_c *to* L^2.

2. *Assume* δ_c^* *has no kernel on* $L_c(\Omega^0 \oplus \Omega^1)$. *Then there is a constant* $\mu > 0$ *which depends on* C *such that for all* $\Psi \in L_c(\Omega^0 \oplus \Omega^1)$

$$\|(\delta_c)^*\Psi\|_2 \geq \mu \|\Psi\|_c \tag{2.1}$$

where $\|.\|_c$ *is the norm defined in (1.5).*

Using Lemma 2.2 we define

Definition 2.3 *Let* C *fulfill the assumptions of Lemma 2.2. Define* μ_c *to be the largest constant* $\mu > 0$ *which satisfies (2.1).*

We will see in Subsection 4.1 that μ_c^2 is the lowest eigenvalue of the bounded form

$$(\Psi, \eta) \longmapsto (\delta_c^*\Psi, \delta_c^*\eta)$$

for Ψ, $\eta \in L_c$.

For smooth configurations on $\mathbb{R}^3$, Lemma 2.2 is proved in Section 7 of [T2]. This proof generalizes to configurations in $\mathscr{C}_m(K')$ on K'. To make this discussion more self contained, we include here the proof of the second statement of Lemma 2.2.

Proof of the second Statement of Lemma 2.2: For $\Psi = (\Psi_0, 0)$, (2.1) follows from (1.17). Thus we only have to show that there is a $\mu > 0$ such that

$$\|(d_c^2)^*\xi\|_2 \geq \mu \|\xi\|_c \tag{2.2}$$

for $\xi \in L_c(\Omega^1)$. To prove that we proceed as in Lemma 7.6 of [T2]. Suppose on the contrary that no such $\mu > 0$ exists. Then there exists a sequence $\{\xi_j\} \in L_c$ with

$$\|(d_c^2)^*\xi_j\|_2 \longrightarrow 0 \tag{2.3}$$

and

$$\|\xi_j\|_c = 1\,. \tag{2.4}$$

Due to (2.4) there exists an L_c-weak limit ξ of $\{\xi_j\}$; also $\xi_j \to \xi$ strongly in L^2 on bounded domains (Rellich lemma). Estimate

$$\begin{aligned} \|(d_c^2)^*\xi\|_2^2 &= \left(d_c^2(d_c^2)^*(\xi - \xi_j), \xi\right) + \left((d_c^2)^*\xi_j, (d_c^2)^*\xi\right) \\ &\leq (\nabla_c(\xi - \xi_j), \nabla_c\xi) + (\{G_c, \xi - \xi_j\}, \xi) \\ &\quad + (\{Ric, \xi - \xi_j\}, \xi) + \|(d_c^2)^*\xi_j\|_2 \|(d_c^2)^*\xi\|_2\,. \end{aligned} \tag{2.5}$$

The first three terms on the RHS of (2.5) converge to zero for $j \to \infty$ by the L_c-weak convergence and L^2-strong convergence on compact domains. The last term goes to zero by (2.3). Thus $(d_c^2)^*\xi = 0$. By assumption $(d_c^2)^*$ has no kernel; thus $\xi = 0$. By (1.18)

$$\|\xi_j\|_c^2 \leq \text{ const } \left(\|(d_c^2)^*\xi_j\|_2^2 + |(\{G_c, \xi_j\}, \xi_j)| + |(\{Ric, \xi_j\}, \xi_j)|\right) . \quad (2.6)$$

By (2.3) and strong L^2 convergence on compact domains, the RHS of (2.6) converges to zero which contradicts (2.4). Thus there exists a $\mu > 0$ such that (2.2) is true.

2.2 The linearized equation

Perturbing configurations into monopoles amounts to solving a certain non-linear equation. Consider first a linear equation which is in some sense the leading part of that nonlinear equation.

Define the second-order differential operator:

$$H_c = \delta_c \delta_c^* = (H_{0,c}, H_{2,c}) : C_0^\infty(\Omega^0 \oplus \Omega^1) \longrightarrow C_0^\infty(\Omega^0 \oplus \Omega^1) . \quad (2.7)$$

Proposition 2.4 *Assume $C \in \mathscr{C}_m(K')$ with $(\delta_c)^*$ injective on L_c. For $\Psi \in L^2 \cap L^{6/5} \cap C^\infty(\Omega^0 \oplus \Omega^1)$, the equation*

$$H_c u = \Psi \quad (2.8)$$

has a solution $u \in L_c(\Omega^0 \oplus \Omega^1)$ with

$$\|u\|_c \leq S_3 \, \mu_c^{-2} \|\Psi\|_{6/5} \quad (2.9)$$

where S_3 is the Sobolev constant on K' described in Lemma 1.4.

Proof of Proposition 2.4: The existence of the solution u is proved using variational techniques similar to those in Chapter 4 of [JT]. Consider the functional

$$S[u] := \frac{1}{2}\|\delta_c^* u\|_2^2 - (u, \, \Psi)_2$$

on L_c. Solutions to (2.8) are formally critical points for $S[u]$. $S[u]$ satisfies the coercive bounds

$$S[u] \geq \frac{1}{2}\mu_c \|u\|_c^2 - \|u\|_6 \|\Psi\|_{6/5} \geq (\frac{1}{2}\mu_c - \epsilon)\|u\|_c^2 - \epsilon^{-1} \text{ const } \|\Psi\|_{6/5}^2$$

and

$$S[u] \leq \text{ const } \|u\|_c^2 + \text{ const } \|\Psi\|_{6/5}^2 .$$

Furthermore $S[u]$ is strictly convex and differentiable. The calculus of variations implies that (2.8) has a weak solution u. Using elliptic regularity we can show that $u \in C^\infty$. Thus u is also a strong solution.

By assumption δ_c^* has no kernel. So we can use Lemma 2.2 to estimate

$$\|H_c^{-1}\Psi\|_c^2 \leq \mu_c^{-2}\|\delta_c^* H_c^{-1}\Psi\|_2^2 = \mu_c^{-2}(H_c^{-1}\Psi, \Psi)$$
$$\leq S_3 \mu_c^{-2} \|H_c^{-1}\Psi\|_c \|\Psi\|_{6/5}$$

This implies (2.9).

2.3 Almost self-dual configurations

Configurations which can be perturbed to self-dual ones are called almost self-dual. To be precise, let us state the following definition:

Definition 2.5 *Let $C \in \mathscr{C}_m(K')$. C is called almost self-dual if there is an N_0 such that following conditions are satisfied:*

1. *δ_c^* has no kernel on $L_c(\Omega^0 \oplus \Omega^1)$. Thus by Lemma 2.2 there is a constant $\mu_c > 0$.*
2. *$\mu_c^{-5} N_0^{-1/2} \leq D(K')$ where $D(K')$ is a constant depending only on the manifold K'.*
3. *The anti-self-dual energy of C is relatively small*

$$\|G_c\|_p \leq N_0^{-1/2} m^{2-\frac{3}{p}} \text{ for } 6/5 \leq p \leq 2 .$$

4. $\|G_c\|_c \leq N_0^{-3/2} m^{3/2}$.

Denote here with C an almost self-dual configuration. As $\delta_c : L_c \to L^2$ is a surjective operator it has a right inverse P

$$\delta_c P\xi = \xi \text{ for all } \xi \in C^\infty(\Omega^0 \oplus \Omega^1) . \tag{2.10}$$

Fix P by setting

$$P(\Psi_0, \Psi_1) := \delta_c^* H_c^{-1}(\Psi_0, \Psi_1) . \tag{2.11}$$

Lemma 2.6 *For an almost self-dual configuration C*

$$P : L^{6/5}(\Omega^0 \oplus \Omega^1) \longrightarrow L^2(\Omega^0 \oplus \Omega^1)$$

and

$$P : L^2 \cap L^{6/5}(\Omega^0 \oplus \Omega^1) \longrightarrow L_c(\Omega^0 \oplus \Omega^1)$$

are bounded operators. Denote by S_3 the Sobolev constant on K' described in Lemma 1.4. For $\Psi \in L^{6/5}(\Omega^0 \oplus \Omega^1)$

$$\|P\Psi\|_2 \leq S_3\, \mu_c^{-1} \|\Psi\|_{6/5} \tag{2.12}$$

and for $\Psi \in L^2 \cap L^{6/5}(\Omega^0 \oplus \Omega^1)$

$$\|P\Psi\|_c \leq S_3 \mu_c^{-1}\Big(\|\Psi\|_2 + m\mu_c^{-1}\|\Psi\|_{6/5}\Big) \leq S_3\, \mu_c^{-2}\Big(\|\Psi\|_2 + m\|\Psi\|_{6/5}\Big) . \tag{2.13}$$

Proof of Lemma 2.6: (2.12) follows from Proposition 2.4. Remark that

$$\delta_c^* \Psi = \delta_c \Psi + 2[\Phi, \Psi] .$$

Use this to derive (2.13):

$$\begin{aligned} \|P\Psi\|_c &\le \mu_c^{-1} \|(\delta_c)^* P\Psi\|_2 \le \mu_c^{-1} \Big(\|\delta_c P\Psi\|_2 + \| [\Phi, P\Psi] \|_2 \Big) \\ &\le \mu_c^{-1} \Big(\|\Psi\|_2 + m\|P\Psi\|_2 \Big) \le \text{ const } \mu_c^{-1} \Big(\|\Psi\|_2 + m\mu_c^{-1} \|\Psi\|_{6/5} \Big) \end{aligned}$$

2.4 The self-dual equation

Perturbing the configuration C to a monopole amounts to solving the equation

$$G_{C+\Psi} = d_c^2(\phi, a) + [a, \phi] - *(a \wedge a) + G_C = 0 \qquad (2.14)$$
$$\text{for a perturbation } \Psi = (\phi, a) .$$

A fixed point method on Ω^1 will give us a solution of the form $\Psi = P(0, \xi)$ with $\xi \in \Omega^1$.

For $\xi \in L^{6/5} \cap L^2 \cap L_c(\Omega^1)$ define the norm

$$\|\xi\|_B := m\|\xi\|_{6/5} + \|\xi\|_2 + m^{-1}\|\xi\|_c . \qquad (2.15)$$

We make the following convention: For $\xi \in \Omega^1(K')$ set $P\xi := P(0, \xi)$

Proposition 2.7 *Let C be an almost self-dual configuration. Then (2.14) has a solution of the form $\Psi = P\xi$ with*

$$\|\Psi\|_c \le \text{ const } \big(\|\xi\|_2 + m\|\xi\|_{6/5} \big) \le \text{ const } \|G_c\|_B \qquad (2.16)$$

$$\|\Psi\|_2 \le \text{ const } \|\xi\|_{6/5} \le \text{ const } m^{-1} \|G_c\|_B \qquad (2.17)$$

$$\|\xi\|_c \le \text{ const } m\|G_c\|_B . \qquad (2.18)$$

Furthermore Ψ is the only such solution which satisfies (2.16–18) and can be written in the form $\Psi = P\xi$. Also for $C = (\Phi, A)$; $\Psi = (\Psi_0, \Psi_1)$

$$\lim_{|x| \longrightarrow \infty} |\Phi + \Psi_0|(x) = m . \qquad (2.19)$$

(2.19) ensures that the perturbation does not change the mass of the configuration.

The following proposition gives a C^0 bound on the perturbation. We will make use of this proposition in Sections 9 and 10.

Proposition 2.8

1. *Let C be an almost self-dual configuration on $\mathbb{R}^3$. Then*

$$\|\Psi\|_\infty = \|P\xi\|_\infty \leq \text{ const } m^{1/2}\|G_c\|_B \,. \tag{2.20}$$

2. *Let C be an almost self-dual configuration on K'. Then there is an m_1 such that (2.20) holds for $m \geq m_1$.*
3. *For an almost self-dual configuration C on K'*

$$\|\Psi\|_\infty = \|P\xi\|_\infty \leq \text{ const } \left((m^2+1)\|G_c\|_{6/5} + m\|G_c\|_2 + \|G_c\|_c\right).$$

Defer the proof of this proposition for the moment.

Proof of Proposition 2.7: Set $\Psi = P\xi = P(0,\xi)$. P is by definition the right inverse to δ_c on $\Omega^0 \oplus \Omega^1$. Thus

$$\delta_c P(0,\xi) = (0,\xi) = d_c^2 P(0,\xi)\,.$$

Plugging this into (2.14) gives

$$\xi + \{P\xi, P\xi\} + G_C = 0\,. \tag{2.21}$$

The brackets denote here a certain quadratic form. Define the function

$$f(\xi) := -\{P\xi, P\xi\} - G_C\,. \tag{2.22}$$

The fixed points of f are solutions of (2.21). Let B_c be the following space

$$B_c := \left\{\xi \in L^{6/5} \cap L^2 \cap L_c(\Omega^1) \mid \|\xi\|_B \leq 2\|G_c\|_B\right\}\,. \tag{2.23}$$

Lemma 2.9 *For an almost self-dual configuration C, f maps B_c into B_c.*

Proof of Lemma 2.9: For $\xi \in B_c$

$$\|f(\xi)\|_2 \leq \text{ const } \mu_c^{-4} N_0^{-1/2}\left(\|\xi\|_2 + m\|\xi\|_{6/5}\right) + \|G_c\|_2 \tag{2.24}$$

also

$$\begin{aligned} m\|f(\xi)\|_{6/5} &\leq m\|\{P\xi, P\xi\}\|_{6/5} + \|G_c\|_B \\ &\leq m\|P\xi\|_3\|P\xi\|_2 + \|G_c\|_B \leq m\|P\xi\|_6^{1/2}\|P\xi\|_2^{3/2} + \|G_c\|_B \\ &\leq \text{ const } \mu_c^{-3} m^{-1/2}\|\xi\|_B^2 + \|G_c\|_B \leq \text{ const } \mu_c^{-3} N_0^{-1/2}\|\xi\|_B + \|G_c\|_B \end{aligned} \tag{2.25}$$

for the L_C-norm part estimate,

$$
\begin{aligned}
m^{-1}\|f(\xi)\|_c &\leq \mu_c^{-1}m^{-1}\|\delta_c^*\{P\xi,P\xi\}\|_2 + \|G_c\|_c \\
&\leq 2\mu_c^{-1}m^{-1}\Big(\|\{\xi,P\xi\}\|_2 + \|\{[\Phi,P\xi],P\xi\}\|_2\Big)+\|G_c\|_c \\
&\leq \text{ const } \mu_c^{-1}m^{-3/2}\Big(\|\xi\|_c\|P\xi\|_c + m\|P\xi\|_c^2\Big) + \|G_c\|_c \qquad (2.26)\\
&\leq \text{ const } \mu_c^{-3}N_0^{-1/2}\|\xi\|_c \\
&+ \text{ const } \mu_c^{-5}N_0^{-1/2}m\Big(\|\xi\|_2 + m\|\xi\|_{6/5}\Big) + \|G_c\|_c\,.
\end{aligned}
$$

(2.24–26) and Property 3) of Definition 2.5 implies

$$\|f(\xi)\|_B \leq \|\xi\|_B\,.$$

Lemma 2.9 follows now.

Lemma 2.10 *For* $\xi_1,\,\xi_2\in B_c$

$$\|f(\xi_1)-f(\xi_2)\|_B \leq \frac{1}{2}\|\xi_1-\xi_2\|_B\,. \qquad (2.27)$$

Lemma 2.10 is proved in a similar way as Lemma 2.9 using Property 3) of Definition 2.5. Lemmas 2.9–10 say that f is a contracting map on the complete set B_c. By the contracting mapping principle f has a fixed point $\xi\in B_c$. This implies (2.16–18).

Theorem 10.3 in Chapter 4 of [JT] implies that the absolute value of Φ converges for $|x|\to\infty$. As $\Psi\in L^6$ the absolute value of $|\Phi|(x)$ has to converge to m. This completes the proof of Proposition 2.7.

We will make use of the following lemma in the proof of Proposition 2.8.

Lemma 2.11 *Denote by* L_{cc} *the completion of the space* $C_0^\infty(\Omega\oplus\Omega^1)$ *in the norm* $\|\Psi\|_c+\|\nabla_c^2\Psi\|_2$.

1. *For* $\Psi=(\Psi_0,\Psi_1)\in L_{cc}$ *a zero-one form on* $\mathbb{R}^3$ *following the Sobolev inequality holds*

$$\|\Psi\|_\infty \leq \text{ const } \Big(m^{1/2}\|\Psi\|_c + m^{-1/2}\|\nabla_c^2\Psi\|_2\Big) \qquad (2.28)$$

with const *independent of* m.

2. *Furthermore there is an* m_1 *such that for* $m\geq m_1$ *and* $\Psi\in L_{cc}$ *a zero-one form on* K', *(2.28) is also true.*

Proof of Lemma 2.11: The first statement follows straight from a Sobolev inequality. For the second statement use the cutoff $\beta_{x,m^{-1}}$. This cutoff interpolates between 1 and 0 over an annulus of radius m^{-1} around x: see (1.24). Assume that m^{-1} is smaller than the injectivity radius of K'. Applying (2.28) to $\beta_{x,m^{-1}}\Psi$ gives the second statement.

Proof of Proposition 2.8: Let C be a configuration as described in Statement 1) or 2) of Proposition 2.8. From Lemma 2.11 follows

$$\|P\xi\|_\infty \leq \text{ const } \left\{m^{1/2}\|\xi\|_2 + m^{3/2}\|\xi\|_{6/5} + m^{-1/2}\|\nabla_c^2 P\xi\|_2\right\}. \tag{2.29}$$

Use (1.13) to rewrite the third term on the RHS of (2.29)

$$\begin{aligned}\|\nabla_c^2 P\xi\|_2 &\leq \|\xi\|_c + \|\{\widehat{G}_c, P\xi\}\|_2 \\ &\leq \|\xi\|_2 + \text{ const } m^{-1/2}\|\widehat{G}_c\|_c\|P\xi\|_c \\ &\leq \text{ const } \left(\|\xi\|_c + m\|\xi\|_2 + m^2\|\xi\|_{6/5}\right).\end{aligned} \tag{2.30}$$

(2.29) and (2.30) yield Statement 1) and 2). Using the Sobolev inequality (2.28) without the mass factors gives Statement 3).

3 Construction of almost self-dual configurations

In this section we construct almost self-dual configurations which are located far away from the neck or have a small mass.

Denote here by X one of the spaces D_R or $\mathbb{R}^3$. We are able to construct almost self-dual configurations in neighborhoods of the ends of the moduli space. They will be parametrized by X. Denote them by $\mathcal{A}_m(X)$. We require $\mathcal{A}_m(X)$ to have the following properties:

1. $\mathcal{A}_m(X) \subset \mathcal{C}_m(K')$ where $\mathcal{C}_m(K')$ is described in Definition 2.1.
2. $\mathcal{A}_m(X)$ is parametrized by X. This means there is a diffeomorphism

$$\begin{aligned} J : X &\longrightarrow \mathcal{A}_m(X) \\ \varphi &\longmapsto C_\varphi . \end{aligned}$$

3. For $N \gg 1$ there is an $R(N)$ respectively an $m(N)$, such that for $R \geq R(N)$, respectively $m \leq m(N)$, and $C \in \mathcal{A}_m(X)$

$$\|G_c\|_p \leq \text{ const } N^{-1/2} m^{2-\frac{3}{p}} \text{ for } 6/5 \leq p \leq 2$$

and

$$\|G_c\|_c \leq \text{ const } N^{-3/2} m^{3/2}.$$

4. For $C \in \mathcal{A}_m(X)$, C is almost self-dual in the sense of Definition 2.5. Furthermore the constant μ_C^{-1} described in Definition 2.3 can be uniformly bounded independently of m, R and $\varphi \in X$.
5. Composing the map J with the self-dual perturbation now gives for each parameter $\varphi \in X$ a monopole. Express this with the map I:

$$I : X \longrightarrow \mathfrak{M}_m$$
$$\varphi \longmapsto C_\varphi + P_\varphi \xi_\varphi$$

I is shown to be injective in Section 8.

The purpose of this section is to construct for each of the parameter spaces D_R and $\mathbb{R}^3$ such a space $\mathcal{A}_m(X)$.

Starting points in the construction are the charge one monopoles on $\mathbb{R}^3$ and the charge zero monopoles $C_{\theta,m} := (\frac{i}{2} m\sigma^3, \theta)$ on K'.

First let us recall the expression of the charge one monopole on $\mathbb{R}^3$. Denote by $\widehat{C}_{m,p}$ the charge one monopole in polar gauge centered at the origin. $\widehat{C}_{m,p} = (\widehat{\Phi}_{m,p}, \widehat{A}_{m,p})$ is explicitly given by

$$\widehat{\Phi}_{m,p}(x) = mf(mr)\, i/2\, \sigma^3 \tag{3.1}$$

$$\widehat{A}_{m,p}(x) = (1 - \cos\theta)\, d\chi\, i/2\, \sigma^3 + g(mr) i/2 \{b\, d\theta + c \sin\theta\, d\chi\} \tag{3.2}$$

where $f(r) = \coth r - 1/r$; $g(r) = r(\sinh r)^{-1}$; $b = \sin\chi\,\sigma^1 - \cos\chi\,\sigma^2$ and $c = \cos\chi\,\sigma^1 + \sin\chi\,\sigma^2$.

In this gauge the transversal part decays exponentially and the longitudinal part approximates a Dirac monopole for $|x| \to \infty$. However this gauge has also a drawback. As $|d\chi| = \frac{1}{r \sin\theta}$, the first term of the gauge field $(1 - \cos\theta) d\chi$ is singular along the string $\cos\theta = -1$. This is also called the Dirac string.

Alternatively we can express the charge one monopole in the cartesian gauge. Denote by $\widehat{C}_{m,c} = (\widehat{\Phi}_{m,c}, \widehat{A}_{m,c})$ the monopole centered at the origin expressed in this gauge. Remark that $\widehat{C}_{m,c}$ is smooth and spherically symmetric.

$$\widehat{\Phi}_{m,c}(x) = m\, f(mr) \vec{n} \cdot \vec{e} \tag{3.3}$$

$$\widehat{A}_{m,c}(x) = m\, h(mr)\, (\vec{n} \times \vec{e})\, d\vec{x} \tag{3.4}$$

where $\vec{e} = i/2\, \vec{\sigma}$, $\vec{\sigma}$ denote the Pauli spin matrices, $\vec{n} = \frac{\vec{x}}{|\vec{x}|}$ and $h(r) = \sinh^{-1} r - 1/r$.

For C a charge one monopole on $\mathbb{R}^3$, (1.17–18) show that the constant μ_C described in Definition 2.3 is uniformly bounded from below.

Denote by $C_{\theta,m} := (\frac{1}{2} m\sigma^3, \theta)$ the mass m charge zero monopole on K'. The following lemma ensures that for these configurations μ_C is uniformly bounded from below.

Lemma 3.1 *There is a constant $\mu > 0$ independent of m such that for the mass m charge zero monopole $C_{\theta,m} := (\frac{i}{2}m\sigma^3, \theta)$*

$$\|\delta^*_{C_{\theta,m}}\Psi\|_2 = \|\delta_{C_{\theta,m}}\Psi\|_2 \geq \mu\|\nabla_{C_{\theta,m}}\Psi\|_2 \tag{3.5}$$

for all $\Psi \in L_c(\Omega^0 \oplus \Omega^1)$

Proof of Lemma 3.1: Write $\Psi = (\Psi_0, \Psi_1)$ in components consisting of zero and one forms and remark the following identity

$$\delta_c\delta_c^*\Psi = \delta_c^*\delta_c\Psi = \left(d_\theta^* d_\theta \Psi_0 + m^2\Psi_{0,T},\ (d_\theta + d_\theta^*)^2\Psi_1 + m^2\Psi_{1,T}\right). \tag{3.6}$$

Taking the inner product on both sides of (3.6) with Ψ gives

$$\|\delta_c^*\Psi\|_2^2 = \|\delta_c\Psi\|_2^2 = \|\nabla_\theta\Psi_0\|_2^2 + \|(d_\theta + d_\theta^*)\Psi_1\|_2^2 + m^2\|\Psi_T\|_2^2. \tag{3.7}$$

Ellipticity of the operator $d_\theta + d_\theta^* : \Omega^1 \to \Omega^0 \oplus \Omega^1$ and the fact that K is a homology sphere imply

$$\|\nabla_\theta\Psi_1\|_2 \leq \text{ const } \|(d_\theta + d_\theta^*)\Psi_1\|_2. \tag{3.8}$$

From (3.7–8) follows

$$\|\nabla_{C_{\theta,m}}\Psi\|_2 \leq \text{ const } \|\delta^*_{C_{\theta,m}}\Psi\|_2 = \text{ const } \|\delta_{C_{\theta,m}}\Psi\|_2 \tag{3.9}$$

where const is independent of m.

Definition 3.2 *Define μ_0 to be the largest constant $\mu > 0$ which satisfies (3.5). The proof of Lemma 3.1 shows that μ_0 is independent of m.*

Before we start with the constructions of almost self-dual configurations, let us remark two special cutoffs. These cutoffs are the 3-dimensional analogues of the one described in Lemma 7.2.10 of [DK].

Lemma 3.3 *Let $r > 0$ and $1 \gg \epsilon > 0$. Assume Ω is an annulus centered at x of radii $[\epsilon r, \epsilon^{-1}r]$ which fulfills one of the conditions:*

1. *Ω is contained in the $\mathbb{R}^3$ part of K'. This means especially that r is large and that ϵr is greater than the radius of the neck of K'.*
2. *$r\epsilon^{-1}$ is smaller than the injectivity radius of K'. In this case Ω can be centered anywhere on K' and r is very small.*

   There are cutoffs $\tilde\alpha$, $\tilde\beta$ such that

$$\tilde\alpha(y) = 0 \ \ \textit{and} \ \ \tilde\beta = 1 \ \ \textit{for} \ \ |x - y| \leq \epsilon r$$

$$\tilde\alpha(y) = 1 \ \ \textit{and} \ \ \tilde\beta = 0 \ \ \textit{for} \ \ |x - y| \geq \epsilon^{-1} r$$

with

$$\|\nabla\tilde\alpha\|_3 \leq \text{ const } (\log\epsilon^{-1})^{-2/3}$$

$$\|\nabla\tilde\beta\|_3 \leq \text{ const } (\log\epsilon^{-1})^{-2/3}.$$

The reason this works is that the L^3 norm fails to control $|\nabla\tilde{\alpha}|$. Also the L^3 norm of one forms is conformally invariant. A usual cutoff on the cylindrical blowup gives the desired result. The details of the proof are the same as in Lemma 7.2.10 of [DK].

3.1 Monopoles far away from the neck

Take the charge one mass m monopole $\widehat{C}_{x,m,p}$ in polar gauge on $\mathbb{R}^3$, centered at x. We perform cutoffs to get a monopole $C_{1,x}$ on $\mathbb{R}^3$ which fulfills the properties:

1. The center x of the monopole $C_{1,x}$ has at least distance R to the origin of $\mathbb{R}^3$.
2. Let $k \gg 1$ be a large constant. Over the ball centered at the origin of radius k, denote it by $B_{0,k}$, $C_{1,x}$ should equal the flat configuration $C_{\theta,m} = (\frac{i}{2}m\sigma^3, \theta)$. Denote by C_x the almost self-dual configuration on K' which we will define in this subsection. In Lemma 3.4 we establish a uniform lower bound on μ_{C_x}. To proof this lemma it is convenient to use a cutoff of the form described in Lemma 3.3. Furthermore on the domain where α equals zero, $C_{1,x}$ should equal $C_{\theta,m}$.

   A large k puts us in situation 1 of Lemma 3.3. The minimal size of k is independent of R and m, it will be determined in Lemma 3.4. Apart from this lemma we will always treat k as a constant.

   Arrange the situation in such a way that the Dirac string does not intersect the ball $B_{0,2k}$.

For $|x| \geq R$ define the configuration $C_{1,x} = (\Phi_{1,x}, A_{1,x})$.

$$\Phi_{1,x}(y)=m\left(1+\alpha_{0,k}(y)\frac{1}{m|y-x|}\right.$$
$$\left.+\alpha_{0,k}(y)\left(\coth(m|y-x|)-1\right)\right)\frac{i}{2}\sigma^3 \tag{3.10}$$

$$A_{1,x}(y)=\alpha_{0,k}(y)(1-\cos\theta')d\chi\frac{i}{2}\sigma^3$$
$$+\alpha_{0,k}(y)g(m|y-x|)\frac{i}{2}(b'd\theta+c'\sin\theta'd\chi) \tag{3.11}$$

The primes on the RHS denote the transformed angles.

Now glue this configuration to the charge zero configuration $C_{\theta,m}$ on K'. Denote the resulting configuration by C_x. Estimate the anti-self-dual part of C_x. As $|G_{C_x}|_\infty \leq$ const R^{-1} we have the estimates

$$\|G_c\|_p \leq \text{ const } R^{-1} \tag{3.12}$$

$$\|G_c\|_c \leq \text{ const } R^{-1}. \tag{3.13}$$

Let $m \in [m_0, m_1]$. Remark that in this subsection we leave the mass boundaries m_0 and m_1 fixed. For $N \gg 1$ there is an $R(N)$ such that C_x satisfies properties 4 and 5 of Definition 2.5.

The next task is to get a uniform lower bound for μ_{C_x}. μ_{C_x} is the constant described in Definition 2.3. Uniform here means independent of m, R and x. For $R \to \infty$, the configuration C_x drifts out on K' and thus converges weakly. That means in C^∞ over compact subsets of K' to $C_{\theta,m}$.

On four manifolds weak convergence of instantons implies convergence of the spectra of the operators associated to these instantons. The same is true for monopole configurations on K'. Weak convergence of C_x to $C_{\theta,m}$ implies that μ_{C_x} converges to μ_0. μ_0 is the constant from Definition 3.2. For our purposes it is enough to show

Lemma 3.4 *There is an R_1 such that for $R \geq R_1$*

$$\|\delta^*_{C_x}\Psi\|_2 \geq \frac{\mu_0}{10}\|\nabla_{C_x}\Psi\|_2\,,$$

for all $\Psi \in L_{C_x}(\Omega^0 \oplus \Omega^1)$.

Proof of Lemma 3.4: Let $\Psi \in L_c(\Omega^0 \oplus \Omega^1)$. Decompose $\Psi = \Psi_1 + \Psi_2$ where $\Psi_1 = \tilde{\alpha}\Psi$ with $\tilde{\alpha}$ a cutoff of the sort described in Lemma 3.3. For $\epsilon > 0$ choose $l(\epsilon)$ such that $\epsilon l(\epsilon)$ is larger than the radius of the neck of K'. ϵ will be fixed at the end of this proof. The k from (3.10) will then be fixed such that $\epsilon^{-1} \leq k$. Assume that $\tilde{\alpha}$ interpolates between 0 and 1 over the annulus $[\epsilon l, \epsilon^{-1} l]$ on the $\mathbb{R}^3$ part of K' centered at the neck. By Lemma 3.3

$$\|\nabla\tilde{\alpha}\|_3 \leq \text{ const } \left(\log \epsilon^{-1}\right)^{-2/3}. \tag{3.14}$$

Now estimate

$$\begin{aligned}
\|\delta^*_c\Psi\|_2^2 &= \|\delta^*_c\Psi_1\|_2^2 + \|\delta^*_c\Psi_2\|_2^2 + 2\,(\delta^*_c\Psi_1, \delta^*_c\Psi_2) \\
&\geq \|\delta^*_c\Psi_1\|_2^2 + \|\delta^*_c\Psi_2\|_2^2 - 2\Big(\|\nabla\tilde{\alpha}\|_3\|\Psi\|_6 + \|\delta^*_c\Psi\|_2\Big)\|\delta^*_c\Psi_2\|_2 \\
&\geq \|\delta^*_c\Psi_1\|_2^2 + \|\delta^*_c\Psi_2\|_2^2 - \text{ const } \|\nabla\tilde{\alpha}\|_3^2\|\Psi\|_c^2 - 2\|\delta^*_c\Psi\|_2^2 - \frac{1}{2}\|\delta^*_c\Psi_2\|_2^2\,.
\end{aligned}$$

We can assume $\mu_0 \leq 1$. Thus

$$\begin{aligned}
\|\delta^*_c\Psi\|_2^2 &\geq 1/6\|\delta^*_c\Psi_1\|_2^2 + 1/12\|\delta^*_c\Psi_2\|_2^2 - \text{ const } \|\nabla\tilde{\alpha}\|_3^2\|\Psi\|_c^2 \\
&\geq \frac{1}{12}\mu_0^2\Big(\|\nabla_c\Psi\|_2^2 - 2(\nabla_c^2\Psi_1, \Psi_2)\Big) - \text{ const } \|\nabla\tilde{\alpha}\|_3^2\|\Psi\|_c^2\,.
\end{aligned} \tag{3.15}$$

The supports of Ric and Ψ_1 as well as the supports of G_c, Ψ_1 and Ψ_2 are disjoint. From (1.17–18)

$$\left(\nabla_c^2\Psi_1, \Psi_2\right) = \left(\delta_c\delta^*_c\Psi_1, \Psi_2\right)$$

Plugging this into (3.15) gives

$$\|\delta_c^* \Psi\|_2^2 \geq \frac{\mu_0^2}{50}\|\Psi\|_c^2 - \text{ const } \left(\log \epsilon^{-1}\right)^{-4/3} \|\Psi\|_c^2 . \tag{3.16}$$

Now fix ϵ by the condition const $\left(\log \epsilon^{-1}\right)^{-4/3} \leq \frac{\mu_0^2}{100}$. Thus

$$\|\delta_{C_x}^* \Psi\|_2 \geq \frac{\mu_0}{10}\|\Psi\|_c \tag{3.17}$$

and the proof of Lemma 3.4 is complete.

Lemma 3.4 and (3.12–13) imply

Proposition 3.5 *There is an R_1 such that for $R \geq R_1$ the configurations C_x are almost self-dual.*

3.2 Small mass monopoles parametrized by $\mathbb{R}^3$

Let $\widehat{C}_{x,c} = (\widehat{\Phi}_{x,c}, \widehat{A}_{x,c})$ be a mass m monopole on $\mathbb{R}^3$ centered at x in cartesian gauge. Let $k \gg 1$. Over the annulus centered at the origin in $\mathbb{R}^3$ of radius k, we would like to interpolate a flat configuration $\widehat{C}_{x,c}$. For the same reason as in the previous subsection, k has to be large. But we can choose it independent of m and x. The exact size of k is fixed in Lemma 3.6. Apart from this lemma we will treat k as a constant.

For this interpolation it is favorable to have a small connection matrix $\widehat{A}$ over the region $B_{0,k}$. In case the curvature of the connection is small enough, the exponential gauge or the radial gauge as it is sometimes called (c.f. [U]) gives such a connection matrix. If $\widehat{A}$ is in the exponential gauge near the origin, then

$$|\widehat{A}(x)| \leq \frac{1}{2}|x| \max_{|y|\leq|x|} |F_{\widehat{A}}(y)| \tag{3.18}$$

and

$$|\nabla_{\widehat{A}}\widehat{A}|(x) \leq \frac{1}{2} \max_{|y|\leq|x|} \left(|F_{\widehat{A}}|(y) + |x|\,|\nabla_{\widehat{A}}F_{\widehat{A}}|(y)\right). \tag{3.19}$$

Denote with g_1 the gauge transformation which interpolates between the exponential gauge and the identity over the annulus $\Omega_{x,[3k,4k]}$. Define

$$\widehat{C}_{x,e,c} := g_1(\widehat{C}_{x,c}). \tag{3.20}$$

Let α_k be the cutoff interpolating between 0 and 1 over the annulus $[k, 2k]$. Define the configuration $C_{1,x}$:

$$\begin{aligned} C_{1,x}(y) &:= (\Phi_{1,x}(y), A_{1,x}(y)) \\ &:= \Big(\alpha_k(y)\widehat{\Phi}_{x,e,c}(y) + \big(1-\alpha_k(y)\big)\widehat{\Phi}_{x,e,c}(0)\,,\ \alpha_k(y)\widehat{A}_{x,e,c}(y) \\ &\qquad + \big(1-\alpha_k(y)\big)\theta\Big) \end{aligned} \tag{3.21}$$

where $\widehat{\Phi}_{x,e,c}(0)$ is the value of $\widehat{\Phi}_{x,e,c}$ at the origin of $\mathbb{R}^3$.

$|\widehat{\Phi}_{x,e,c}(y) - \widehat{\Phi}_{x,e,c}(0)|$ is gauge independent. It is convenient to estimate this expression in polar gauge. This leads to

$$|\left(\widehat{\Phi}_{x,e,c}(y) - \widehat{\Phi}_{x,e,c}(0)\right)| \le \text{ const } m^2 . \tag{3.22}$$

Similarly we get

$$|\nabla_{\widehat{C}_{x,e,c}}\left(\widehat{\Phi}_{x,e,c}(y) - \widehat{\Phi}_{x,e,c}(0)\right)| \le \text{ const } m^2 . \tag{3.23}$$

From (3.18) respectively (3.19) follows

$$|\left(\widehat{A}_{x,e,c}(y) - \theta\right)| \le \text{ const } m^2 \text{ for } y \in B_{0,4k} \tag{3.24}$$

and

$$|\nabla_{\widehat{C}_{x,e,c}}(\widehat{A}_{x,e,c}(y) - \theta)| \le \text{ const } m^2 \text{ for } y \in B_{0,4k} \,. \tag{3.25}$$

Let $C_0 := \left(\widehat{\Phi}_{x,e,c}(0), \theta\right)$. Remark that C_0 is a flat configuration whose mass is between 0 and m. Put C_0 on K'. Now glue $C_{1,x}$ to C_0. Denote the resulting configuration by C_x. Using (3.22–25) one shows

$$\|G_{C_x}\|_p \le \text{ const } m^2 \tag{3.26}$$

$$\|G_{C_x}\|_c \le \text{ const } m^2 . \tag{3.27}$$

For $N \gg 1$ there is an $m(N)$ such that C_x satisfies Properties 4 and 5 of Definition 2.5. In the same way as in Lemma 3.4 we show

Lemma 3.6 *There is an m_0 such that for $m \le m_0$*

$$\mu_{C_x} \ge \frac{\mu_0}{10} .$$

This implies

Proposition 3.7 *There is an m_0 such that for $m \le m_0$ the configurations C_x are almost self-dual.*

This gives the map

$$J : \mathbb{R}^3 \longrightarrow \mathcal{A}_m(\mathbb{R}^3)$$
$$x \longmapsto C_x$$

where $\mathcal{A}_m(\mathbb{R}^3)$ is described at the beginning of this section.

4 Eigenvalues and large mass monopoles

In this section we prove a technical result which we will use in the next section to investigate the large mass moduli space. Namely we show, that for a large mass monopole Q, the operator $\delta_Q\delta_Q^*$ has no kernel and that its lowest eigenvalue is uniformly bounded from below.

Definition 4.1 *Let $Q \in \mathfrak{M}_m(K')$. The smallest eigenvalue of the operator $\delta_Q\delta_Q^*$ is the largest constant μ_1 which satisfies*

$$\|\delta_Q^*\Psi\|_2 \geq \mu_1\|\nabla_Q\Psi\|_2 \tag{4.1}$$

for all $\Psi \in L_Q(\Omega^0 \oplus \Omega^1)$.

In Definition 3.2 we defined μ_0. μ_0 as the smallest eigenvalue of the operator $\delta_{C_{\theta,m}}\delta_{C_{\theta,m}}^*$ where $C_{\theta,m}$ is the flat configuration $C_{\theta,m} = (\frac{i}{2}m\sigma^3, \theta)$.

We remarked in the last section that weak convergence of monopoles Q implies convergence of the spectra of the operators $\delta_Q\delta_Q^*$. In the limit $m \to \infty$ a large mass monopole «bubbles off». But it leaves a Dirac monopole behind whose Higgs field has infinite mass. Thus in the limit $m \to \infty$, Q does not converge weakly to the flat configuration $C_{\theta,m}$. Nevertheless we conjecture that μ_1, the lowest eigenvalue of the operator $\delta_Q\delta_Q^*$ converges to μ_0, the lowest eigenvalue of the operator $\delta_{C_{\theta,m}}\delta_{C_{\theta,m}}^*$.

For our purposes it is enough to show the following weaker result.

Theorem 4.2 *There is an m_1 such that for $m \geq m_1$ and $Q \in \mathfrak{M}_m(K')$*

$$\|\delta_Q^*\Psi\|_2 \geq \frac{\mu_0}{20}\|\nabla_Q\Psi\|_2 .$$

Assume the Higgs field of Q has a zero at x_0. Let $r > 0$ be smaller than the injectivity radius of K'. Denote by B the ball of radius r around x_0.

To prove Theorem 4.1, we would like to apply the same strategy as we used in the proof of Lemma 3.4. Namely, we decompose Ψ into two parts, one with support on B and the other one with support in the complement of B. Using the cutoffs of Lemma 3.3, we can control the mixed terms.

Proposition 4.3 *Let $\epsilon > 0$ with $\epsilon^{-1}r$ smaller than the injectivity radius of K'. ϵ will be fixed below. Denote by α the cutoff described in Lemma 3.3 which interpolates between 0 and 1 over the annulus $[\epsilon r, \epsilon^{-1}r]$. Remark that*

$$\|\nabla\alpha\|_3 \leq \text{ const } (\log\epsilon^{-1})^{-3/2} . \tag{4.2}$$

Then there is an m_1 such that for $m \geq m_1$ and $Q \in \mathfrak{M}_m(K')$

$$\|\delta_Q^*(\alpha\Psi)\|_2 \geq \frac{\mu_0}{10}\|\nabla_Q(\alpha\Psi)\|_2 \tag{4.3}$$

for all $\Psi \in L_Q(\Omega^0 \oplus \Omega^1)$.

We defer the proof of Proposition 4.3 to Subsection 4.2.

Proof of Theorem 4.2 assuming Proposition 4.3: Let $\beta = (1-\alpha)$. Decompose Ψ into $\alpha\Psi$ and $\beta\Psi$

$$\|\delta_Q^*\Psi\|_2^2 = \|\delta_Q^*(\alpha\Psi)\|_2^2 + \|\delta_Q^*(\beta\Psi)\|_2^2 + 2\Big(\delta_Q^*(\alpha\Psi), \delta_Q^*(\beta\Psi)\Big)\,. \tag{4.4}$$

Using Proposition 4.3 and (1.18) to estimate (4.4) by

$$\begin{aligned}&\geq \frac{\mu_0^2}{100}\|\nabla_Q(\alpha\Psi)\|_2^2 + \|\nabla_Q(\beta\Psi)\|_2^2\\ &\quad - |(\{Ric, \beta\Psi\}, \beta\Psi)| - 2\Big(\|\nabla\alpha\|_3\|\Psi\|_6 + \|\delta_Q^*\Psi\|_2\Big)\|\delta_Q^*(\beta\Psi)\|_2\,,\end{aligned}$$

this leads to

$$\begin{aligned}3\|\delta_Q^*\Psi\|_2^2 \geq \frac{\mu_0^2}{100}&\|\nabla_Q(\alpha\Psi)\|_2^2 + \frac{1}{2}\|\nabla_Q(\beta\Psi)\|_2^2\\ &- \|Ric\|_3\epsilon^{-1}r\|\beta\Psi\|_6^2 - \text{ const } \left(\log\epsilon^{-1}\right)^{-4/3}\|\Psi\|\,.\end{aligned} \tag{4.5}$$

From (4.5) follows

$$\|\delta_Q^*\Psi\|_2^2 \geq \left(\frac{\mu_0^2}{300} - \text{ const } \epsilon^{-1}r\|Ric\|_3 - \text{ const } \left(\log\epsilon^{-1}\right)^{-4/3}\right)\|\nabla_Q\Psi\|_2^2\,. \tag{4.6}$$

Choose $\epsilon^{-1}r$ small relative to $\|Ric\|_3$ and ϵ absolutely small. Remark that we require $m^{-1/2} < \epsilon r$ for Proposition 4.3. Imposing these conditions in (4.6) proves Theorem 4.2.

4.1 A priori estimates

Denote here again by B the ball of radius r around x_0, where x_0 is a zero of the Higgs field of Q, and r will be chosen below. Using the self-dual equation we establish in this subsection some properties of Q on the region $K' \setminus B$.

Lemma 4.4 *Make the same assumptions on ϵ, r and α as in Proposition 4.3. Choose $l \gg 1$ with $lm^{-1} \leq \epsilon r$. Then*

1. $\|\alpha * F_A\|_2 \leq$ const $l^{-1/2}m^{1/2}$
2. $\|\nabla_Q(\alpha * F_A)\|_2 \leq$ const $l^{-1/2}m^{1/2}$.

Lemma 4.5 *Make again the same assumptions as in Proposition 4.3. Then*

$$|\Phi|(x) \geq \frac{m}{2} \text{ for } x \in K' \setminus B_{\epsilon r, x_0}\,.$$

Proof of Lemma 4.4: Let

$$h: B_{m^{-1}\epsilon^{-1}r,0} \subset \mathbb{R}^3 \longrightarrow B_{\epsilon^{-1}r,x_0} \subset K' \tag{4.7}$$

be the inverse coordinate map scaled by the factor m^{-1}. Denote by Q_1 the pullback of Q by the map h. Set on $B_{m^{-1}\epsilon^{-1}r,0} \subset \mathbb{R}^3$ the induced metric from K'. In the limit $m \to \infty$, Q_1 converges in C^∞ on compact subsets of $\mathbb{R}^3$ to the unit mass monopole. l is defined in the assumptions to this lemma. On the ball $B_{l,0} \subset \mathbb{R}^3$, Q_1 has $4\pi - O(l^{-1})$ energy. Thus

$$\|\alpha * F_A\|_2 \le \|F_A\|_{2,B_{lm^{-1},x_0}} \le \text{ const } l^{-1/2} m^{1/2}$$

This proves Statement 1).

Remark that $\delta_Q^* \delta_Q(*F_A) = 0$. Thus from (1.19) we get

$$\nabla_Q^2(*F_A) = \{\widehat{G}_Q, *F_A\} + \{Ric, *F_A\}\,. \tag{4.8}$$

Taking on both sides of (4.8) the inner product with $\alpha^2(*F_A)$ gives the estimate

$$\begin{aligned} \|\nabla_Q(*\alpha F_A)\|_2^2 \le \text{ const } \|\widehat{G}_Q\|_{3,K'\setminus B_{lm^{-1}}} \|\alpha * F_A\|_3^2 + \|(\nabla\alpha)F_A\|_2^2 \\ + \|\nabla_Q(\alpha * F_A)\|_2 \Big(\|\alpha F_A\|_2 + \|(\nabla\alpha)F_A\|_2\Big) + \|Ric\|_\infty \|\alpha * F_A\|_2^2\,. \end{aligned} \tag{4.9}$$

Statement 1) now implies

$$\|\nabla_Q(*\alpha F_A)\|_2 \le \text{ const } l^{-1/2} m^{1/2}\,.$$

Proof of Lemma 4.5: Set

$$w(x) := \frac{1}{2}\left(1 - \frac{|\Phi|^2(x)}{m^2}\right)\,. \tag{4.10}$$

From the self-dual equation follows

$$\Delta w(x) = -m^{-2}|d_A\Phi|^2(x) = -m^{-2}|F_A|^2(x)\,. \tag{4.11}$$

Let $B_1 := B_{\frac{1}{5}\epsilon r,x_1}$ be a ball of radius $\frac{1}{5}\epsilon r$ centered at x_1. Assume B_1 is contained in $K' \setminus B_{lm^{-1},x_0}$. We would like to estimate $|w|(x)$ on B_1. Denote with β a cutoff centered at x_1 interpolating between 1 and 0 over the annulus $[\frac{1}{10}\epsilon r, \frac{1}{5}\epsilon r]$.

Take on both sides of (4.11) the inner product with $\beta^2 w$. Now convolute both sides with the Green function of the 3-dimensional Laplacian on K'. In a Gaussian coordinate system the Green function of the Laplacian on K' approximates the Green function of the Laplacian on $\mathbb{R}^3$ over B_1. This leads to the estimate

$$\begin{aligned} |\beta w|^2(x) \le m^{-2} \int \frac{(\beta|F_A|^2, \beta w)}{|x-y|} d^3y + 2m^{-2} \int \frac{(\nabla\beta\nabla w, \beta w)}{|x-y|} d^3y \\ + m^{-2} \int \frac{(\Delta\beta w, \beta)}{|x-y|} d^3y\,. \end{aligned}$$

Thus

$$|\beta w|(x) \le \text{ const } l^{-1}\,.$$

This implies Lemma 4.5.

4.2 A uniform lower bound on μ_Q

In this subsection we prove Proposition 4.3. Consider first the transversal part of $\alpha\Psi$:

Lemma 4.6 *Make the same assumptions on α as in Proposition 4.3. Then there is a constant D_1 such that*

$$\|\delta_Q^*(\alpha\Psi_T)\|_2 \geq \left(1 - D_1 m^{-1/2}\right)\|\nabla_Q(\alpha\Psi_T)\|_2 \,.$$

Proof of Lemma 4.6: By (1.18)

$$\begin{aligned}\|\delta_Q^*(\alpha\Psi_T)\|_2^2 &\geq \|\nabla_Q(\alpha\Psi_T)\|_2^2 - \left|\left(\{Ric, \alpha\Psi_T\}, \alpha\Psi_T\right)\right| \\ &\geq \|\nabla_Q(\alpha\Psi_T)\|_2^2 - \|Ric\|_3\|\alpha\Psi_T\|_6\|\alpha\Psi_T\|_2 \,.\end{aligned} \tag{4.12}$$

Estimate the last factor in the last term on the RHS of (4.12) with Lemma 4.5:

$$\|\alpha\Psi_T\|_2 \leq \text{ const } \|\alpha\frac{[\Phi,\Psi_T]}{|\Phi|}\|_2 \leq \text{ const } m^{-1}\|\nabla_c(\alpha\Psi_T)\|_2 \,;$$

thus

$$(4.12) \geq \|\nabla_\alpha(\alpha\Psi_T)\|_2^2\left(1 - \text{ const } \|Ric\|_3 m^{-1}\right).$$

Assume Q is a monopole whose Higgs field points to a constant direction in $su(2)$. Then we say that Q is in polar gauge. The following lemma shows that it is always possible to arrange this situation.

Lemma 4.7 *For $Q \in \mathfrak{M}_m(K')$ there is a gauge transformation u such that the Higgs field of $u(Q)$, Φ_1 has the form*

$$\Phi_1 = mf(x)\sigma^3$$

Proof of Lemma 4.7: In general Φ has the form

$$\Phi(x) = mf(x)\vec{g}(x)\vec{\sigma} \tag{4.13}$$

where $\vec{g}$ takes values in the unit 2-sphere S^2 and $\vec{\sigma}$ is the vector of Pauli spin matrices $\vec{\sigma} = (\sigma^1, \sigma^2, \sigma^3)$. Assume $\vec{g}(x)$ is described in polar coordinates by $\left(\theta(x), \chi(x)\right)$. The following gauge transformation is given in Chapter 4 of [JT]:

$$u(x) = \cos\theta/2 + i\sin\theta/2\left(-\sin\chi\,\sigma^1 + \cos\chi\,\sigma^2\right) \tag{4.14}$$

A computation shows $\Phi_1 := u\Phi u^{-1} = mf(x)\sigma^3$.

Lemma 4.8 *Assume Q is in polar gauge. Then there is an m_1 such that for $m \geq m_1$*

$$\|\delta_Q^*(\alpha\Psi_L)\|_2 \geq \frac{\mu_0}{2}\|\nabla_Q(\alpha\Psi_L)\|_2$$

for $\Psi \in L_Q(\Omega^0 \oplus \Omega^1)$. μ_0 is the constant described in Definition 3.2.

Proof of Lemma 4.8: Let $Q = (\Phi, A)$. We compare Q and $C_{\theta,m} = (\frac{i}{2}m\sigma^3, \theta)$ on the support of α. This gives the estimate

$$\begin{aligned}\|\delta_Q^*(\alpha\Psi_L)\|_2 &\geq \|\delta_{C_{\theta,m}}^*(\alpha\Psi_L)\|_2 - \|\,[A, \alpha\Psi_L]\,\|_2 \\ &\geq \mu_0\|\nabla_{C_{\theta,m}}(\alpha\Psi_L)\|_2 - \|\alpha_{1/2}A_T\|_3\|\alpha\Psi_L\|_6\,,\end{aligned} \tag{4.15}$$

where $\alpha_{1/2}$ interpolates between 0 and 1 over the annulus $[1/2\epsilon r, \epsilon r]$.

Let B be the ball of radius r centered at a zero of the Higgs field Q. For $x \in K' \setminus B$ we have the pointwise estimates

$$|\,[\Phi, A]\,|(x) \leq 2m^{-1}|\,[\Phi, [\Phi, A]]\,|(x) \tag{4.16}$$

and

$$|A_T|(x) \leq 2m^{-1}|\,[\Phi, A]\,|(x) \leq 4m^{-1}|(d_A\Phi)_T|(x)\,. \tag{4.17}$$

The a priori estimates on the L^2 and L_Q norm of $\alpha * F_A = \alpha d_A\Phi$ in Lemma 4.4 imply

$$\begin{aligned}\|\alpha_{1/2}A_T\|_3 &\leq 4m^{-1}\|\alpha_{1/2}(d_A\Phi)_T\|_3 = 4m^{-1}\|\alpha_{1/2} * F_A\|_2 \\ &\leq 4m^{-1}\|\alpha_{1/2} * F_A\|_2^{1/2}\|\alpha_{1/2} * F_A\|_6^{1/2} \leq \text{ const } l^{-1}\,.\end{aligned} \tag{4.18}$$

Now use (4.18) and (4.15)

$$\begin{aligned}\|\delta_Q^*(\alpha\Psi_L)\|_2 &\geq \left(\mu_0 - \text{ const } l^{-1}\right)\|\nabla_{C_{\theta,m}}(\alpha\Psi_L)\|_2 \\ &\geq \left(\mu_0 - \text{ const } l^{-1}\right)\|\nabla_Q(\alpha\Psi_L)\|_2\end{aligned}$$

which implies Lemma 4.8.

Proof of Proposition 4.3: As μ_Q is independent of the choice of gauge in Q we can assume Q to be in polar gauge.

$$\begin{aligned}\|\delta_Q^*(\alpha\Psi)\|_2^2 &\geq \|\delta_Q^*(\alpha\Psi_T)\|_2^2 + \|\delta_Q^*(\alpha\Psi_L)\|_2^2 - 2\|\delta_Q^*(\alpha\Psi_T)\|_2\|\delta_Q^*(\alpha\Psi_L)\|_2 \\ &\geq \frac{\mu_0^2}{4}\left(\|\nabla_Q(\alpha\Psi_T)\|_2^2 + \|\nabla_Q(\alpha\Psi_L)\|_2^2\right) - 2\|\delta_Q^*(\alpha\Psi_T)\|_2\|\delta_Q^*(\alpha\Psi_L)\|_2\,.\end{aligned} \tag{4.19}$$

We used Lemma 4.6, respectively Lemma 4.8, to estimate the first two terms on the RHS from below. For the last term estimate

$$\|\delta_Q^*(\alpha\Psi_T)\|_2 = \|\delta_Q^*\Big(\frac{[\Phi,\alpha\Psi]}{|\Phi|}\Big)\|_2$$

$$\leq \|\frac{[d_A\Phi,\alpha\Psi]}{|\Phi|}\|_2 + \|\frac{|d_A\Phi|}{|\Phi|^2}[\Phi,\alpha\Psi]\|_2 + \|\frac{[\Phi,\delta_Q^*(\alpha\Psi)]}{|\Phi|}\|_2$$

by (1.7) and Lemma 4.4

$$\leq \text{ const } l^{-1/2}m^{-1}\|\nabla_Q(\alpha\Psi)\|_2 + 2\|\delta_Q^*(\alpha\Psi)\|_2 \,. \tag{4.20}$$

(4.19) and (4.20) give

$$\|\delta_Q^*(\alpha\Psi)\|_2^2 \geq \Big(\frac{\mu_0^2}{50} - \text{ const } l^{-1/2}m^{-1} - \text{ const } \big(\log\epsilon^{-1}\big)^{-4/3}\Big)\|\nabla_Q(\alpha\Psi)\|_2 \,. \tag{4.21}$$

Proposition 4.3 follows from (4.21).

5 The large mass moduli space covers K'

So far we have been concerned with constructing monopoles over K' which lie in the ends of the moduli space. In the next 6 sections we will investigate the moduli space itself.

We are able to use the same strategy to investigate the small mass end of the moduli space and the end consisting of monopoles which drift out on $\mathbb{R}^3$. This is the gauge fixing method developed in [D2]. See also [DK].

The advantage of this method is that it is very general and that it can be used to understand quite different aspects of the moduli space. Its drawback is that it requires the construction map $I : X \longrightarrow \mathfrak{M}_m(K')$. Furthermore we have to distinguish different configurations in the L_c as well as in the L^2-norm. This does not cause any difficulties for small mass monopoles and monopoles located far away from the neck.

The construction of the map $I : K' \rightarrow \mathfrak{M}_m(K')$ for large mass monopoles poses some difficulties. In order to distinguish large mass monopoles in the L^2 norm we would have to deal with a «moduli space» of Dirac monopoles.

We can avoid all these problems by choosing an entirely different approach to investigate the large mass moduli space. This approach is based on the one developed for anti-self-dual connections on four manifolds in [D1]. In fact the techniques used in [D1] carry over to our situation with only minor modifications.

In Section 3 we constructed large mass monopoles which are located far away from the neck. Thus the large mass moduli space $\mathfrak{M}_m(K')$ is non empty. The main part of the energy of a large mass monopole is concentrated in a small ball around a point. In Subsection 5.1 we define the center of a monopole as the center of its energy. This gives us the center map p

$$p : \mathfrak{M}_m(K') \longrightarrow K' .$$

Subsection 5.2 contains the main result of this section. It says that the map p is a covering map; thus $\mathfrak{M}_m(K')$ is a covering of K'.

To conclude the proof of Theorem 1.1, it remains to show that p is a one-fold covering. But this follows directly from Theorem 1.2. Theorem 1.2 shows that this is the case for the part of the large mass moduli space which is located far away from the neck. This theorem is proved in Section 9.

5.1 The center map

We define here the center of a large mass monopole as the center of its energy. Introduce here the energy density of a configuration Q:

$$f_Q(y) := |d_A\Phi|^2(y) + |F_A|^2(y) . \tag{5.1}$$

For s smaller than the injectivity radius of K', the following function is well defined:

$$R_Q(x,s) := \int_{K'} \beta_{x,s}(y) f(y) d\mu . \tag{5.2}$$

$R_Q(x,s)$ tells how much energy the configuration Q has in a ball of radius s around x.

Lemma 5.1 *Let Q be an almost self-dual configuration. There is an m_1 such that for $m \geq m_1$, an $x \in K'$ and an $s(x)$ with*

$$R_Q(x, s(x)) = 2\pi m . \tag{5.3}$$

Proof of Theorem 5.1: The Higgs field of Q has at least one zero, let's say at x_o. Let $r > 0$ be the injectivity radius of K'. Let

$$h : B_{m^{-1}r,0} \subset \mathbb{R}^3 \longrightarrow B_{x_0,r} \subset K'$$

be the inverse coordinate map scaled by the factor m^{-1}. Denote by Q_1 the pullback of Q by the map h. Use on $B_{m^{-1}r,0}$ the induced metric from K'. In the limit $m \to \infty$, Q_1 converges in C^∞ on compact subsets of $\mathbb{R}^3$ to a unit mass L^2-almost self-dual configuration. For sure there is an m_1 such that 3π of energy of Q_1 is concentrated on $B_{m_1^{-1}r}$. Lemma 5.1 now follows.

Definition 5.2 *Make the same assumptions as in Lemma 5.1. Define the center $p(Q)$ of Q as the $x \in K'$ for which $s(x)$ is minimized in (5.3). Remark that the center is unique and that p is a smooth map.*

Lemma 5.3 *Let Q_t be a smooth path of almost self-dual configrations. Then*

$$|\frac{d}{dt}p(Q_t)| \leq \text{ const } m^{-1}\|\frac{d}{dt}\widehat{G}_{Q_t}\|_2 \,.$$

Lemma 5.3 is proved similar to Corollary 17 in [D1].

5.2 The center map is a covering map

Proposition 5.4 *There is an m_1 such that for $m \geq m_1$*

$$dp : T\mathfrak{M}_m(K')|_Q \longrightarrow TK'|_{p(Q)} \tag{5.4}$$

is an isomorphism for $Q \in \mathfrak{M}_m(K')$.

Corollary 5.5 *Make the same assumptions as in Proposition 5.4. Then p is surjective.*

Proof of Corollary 5.5: $\mathfrak{M}_m(K')$ is complete. As p is a local diffeomorphism, $p(\mathfrak{M}_m(K')) \subset K'$ is also complete. From Section 3 we know that there are large mass monopoles located far away from the neck. Thus $\mathfrak{M}_m(K')$ is not empty. From this it follows that $p(\mathfrak{M}_m(K'))$ is a closed nonempty subset of K'. Assume that p is not surjective. Then $p(\mathfrak{M}_m(K'))$ would have a boundary in K'. But this is impossible as $\mathfrak{M}_m(K')$ and K' are locally diffeomorphic. Thus p is surjective.

The weak compactness theorem implies that p is a proper map. Proposition 5.4 and Corollary 5.5 imply the main result of this section.

Theorem 5.6 *Under the same assumptions as Propostion 5.4,*

$$p : \mathfrak{M}_m(K') \longrightarrow K'$$

is a covering map.

It remains to prove Proposition 5.4. Both spaces in (5.4) have the same dimension. Thus we only have to show that dp is injective. As in [D1] we construct an approximate right inverse α to dp and show

$$|(dp\,\alpha)\zeta - \zeta| < |\zeta| \text{ for all } \zeta \in T_Q\mathfrak{M}_m(K') \,. \tag{5.5}$$

An infinitesimal translation which is supported near $p(Q)$ is expected to give us this approximate right inverse. Define on K' the vector field

$$V_\zeta := \beta_{m^{-1/2},p(Q)}\zeta^j\frac{\partial}{\partial x^j} \,, \tag{5.6}$$

where x^j are local geodesic coordinates. Denote by h_t the flow of V_ζ. Lift this flow to the space of configurations. Define the one parameter family of configurations

$$Q_t := h_t(Q) \,. \tag{5.7}$$

Similar as in [D1], we show

Lemma 5.7

$$\frac{d}{dt}\|G_{Q_t}\|_2 \leq \text{ const } m^{1/2}|\zeta|$$

$$\frac{d}{dt}\|G_{Q_t}\|_{6/5} \leq \text{ const } m^{-1/2}|\zeta|\,.$$

Lemma 5.8 *There is an* $\epsilon > 0$ *such that for* $t \leq \epsilon$ *the configurations* Q_t *are almost self-dual.*

Proof of Lemma 5.8: Lemma 5.7 ensures that the anti-self dual part of the energy is sufficiently small. From Subsection 5.2 we know that the smallest eigenvalue μ_1 of $\delta_Q\delta_Q^*$ is uniformly bounded from below. A small t ensures that the smallest eigenvalue of $\delta_{Q_t}\delta_{Q_t}^*$ is in a neighborhood of μ_1, thus also uniformly bounded from below.

Denote with

$$\widehat{Q}_t := Q_t + \Psi_t = Q_t + P_t\xi_t \tag{5.8}$$

the self-dual perturbation of Q_t. Define the tangent vector

$$\alpha(\zeta) := \frac{d}{dt}\widehat{Q}_t|_{t=0} \in T_{[Q]}\mathfrak{M}_m\,. \tag{5.9}$$

It remains to show that α satisfies (5.5). From Lemma 5.7 follows

$$\|\frac{d}{dt}(\widehat{G}_{\widehat{Q}_t} - \widehat{G}_{Q_t})\|_2 \leq \text{ const } m^{-1/2}|\zeta|$$

and Lemma 5.3 implies now

$$|\,(dp\,\alpha)\zeta - \frac{d}{dt}p(Q_t)\,| \leq \text{ const } m^{-1/2}|\zeta|\,. \tag{5.10}$$

Furthermore a computation shows that

$$|\frac{d}{dt}p(Q_t) - \zeta| \leq \text{ const } m^{-1/2}|\zeta| \tag{5.11}$$

(5.5) follows now from (5.10) and (5.11). This proves Proposition 5.4.

6 A Fredholm operator

In this section we will discuss a technical result, which will be used in Section 7 for the construction of some special gauges.

Let $\mathcal{A}_m(X)$ be a set of almost self-dual configurations, constructed in Section 3. For $C \in \mathcal{A}_m(X)$, μ_c^{-1} is bounded uniformly, independent of C. To shorten the notation, from now on we treat μ_c^{-1} as a constant.

For $C \in \mathcal{A}_m(X)$ define the operator

$$\begin{aligned} T : \Omega^0 \times \mathbb{R} \times \mathbb{R}^3 \times \Omega^1 &\longrightarrow \Omega^0 \oplus \Omega^1 \\ (\chi, f, \zeta, \xi) &\longmapsto d_c^1 \chi + f d_c^1 \Phi + l(\zeta) + P\xi \end{aligned} \tag{6.1}$$

where

$$l(\zeta) := \frac{\partial}{\partial s} C_{(\varphi, s\zeta)}|_{s=0} \tag{6.2}$$

is the directional derivative of C_φ at $\varphi \in X$ in the direction ζ; $\zeta \in T_\varphi X \simeq \mathbb{R}^3$. Remark that $|l(\zeta)|(r)$ decays like $O(r^{-2})$, thus it is in L^2.

The purpose of T is to generate all deformations of C in the space of connections. The first term of T generates all gauge transformations which decay at infinity. This will become apparent when we choose the norm for the argument space. The second term generates the single gauge transformation with respect to Φ. This gauge transformation is asymptotically constant.

The third term of T gives the deformation in the three self-dual directions. The last term represents deformations produced by the operator P, the right inverse of δ_c. P was defined in (2.11). We use here the convention $P\xi = P(0, \xi)$.

The main result of this section is that these are all the possible deformations of the operator T. In other words, the operator T is an isomorphism. We show that T is an isomorphism by showing that T is Fredholm, injective and that its index is zero. Also important is that the norm of $\|T^{-1}\|$ is bounded by a constant independent of m and C. In Section 7 we will consider an operator DM which differs from T by a small amount. The above result implies that DM is also an isomorphism.

Although the general strategy in this section is the same for all three parameter spaces the technical details vary. First we discuss the D_R parameter space. The $\mathbb{R}^3$ parameter space is discussed in Subsection 6.4.

6.1 The Fredholm operator for the D_R parameter space

Restrict in this subsection to the case where X is the parameter space D_R. The mass of the monopoles considered in this subsection lies between the bounds m_0 and m_1. As these bounds are fixed, we treat in this subsection the mass as a constant.

Let B_1 be the closure of

$$C_0^\infty(\Omega^0) \times \mathbb{R} \times \mathbb{R}^3 \times C_0^\infty(\Omega^1)$$

in the norm

$$\begin{aligned}\|(\chi,f,\zeta,\xi)\|_{B_1} := \|\chi\|_c + \|f d_c^1\Phi\|_2 + \|l(\zeta)\|_2 + \|P\xi\|_2 \\ + \|d_c^1\chi\|_c + \|f d_c^1\Phi\|_c + \|l(\zeta)\|_c + \|\xi\|_2 \,.\end{aligned} \tag{6.3}$$

Also, let E_1 be the closure of $C_0^\infty(\Omega^0) \oplus C_0^\infty(\Omega^1)$ in the norm

$$\|\eta\|_{E_1} := \|\eta\|_2 + \|\eta\|_c \tag{6.4}$$

The main result of this subsection is

Theorem 6.1 *There is an R_0 such that for $R \geq R_0$ and $C \in \mathcal{A}_m(D_R)$*

$$T : B_1 \longrightarrow E_1$$

is a bounded, injective Fredholm operator with index $T = 0$. Furthermore the norm of its inverse can be uniformly bounded by a constant independent of R.

Before turning to the proof of this theorem, we list some estimates on $l(\zeta) := \partial_\zeta C_\zeta$ which will be used repeatedly below. These estimates follow from explicit computations. Remark that they would not hold if we would not treat m as a constant.

Lemma 6.2 *The norms $|\zeta|$, $\|l(\zeta)\|_c$, $\|\nabla l(\zeta)\|_c$ and $\|l(\zeta)\|_\infty$ are all equivalent.*

In order to proof Theorem 6.1 we decompose the operator T. Consider first the part of T which involves infinite dimensional operators. Define

$$T_1(\chi,\xi) := d_c^1\chi + P\xi \,. \tag{6.5}$$

To simplify the notation, denote again with B_1 the obvious restriction of the above Banach space B_1, defined in (6.3), to the domain of T_1. The statement for the operator T_1 corresponding to Theorem 6.1 is

Lemma 6.3 *There is a constant R_0 such that for $R \geq R_0$ and $C \in \mathcal{A}_m(D_R)$*

$$T_1 : B_1 \longrightarrow E_1$$

is a bounded injective Fredholm operator with index $T_1 = -4$. In particular, there is a constant D independent of R such that

$$\|(\chi,\xi)\|_{B_1} \leq D\,\|T_1(\chi,\xi)\|_{E_1} \,. \tag{6.6}$$

Proof of Lemma 6.3: Obviously the operator T_1 is bounded. Take the equation

$$T_1 = d_c^1\chi + P\xi \tag{6.7}$$

and apply the operator d_c^2 on both sides. This gives the estimate

$$\|\xi\|_2 \leq \text{ const } (\|T_1\|_c + \|\,[G_c,\chi]\,\|_2) \tag{6.8}$$

by (1.5)

$$\|\xi\|_2 \leq \text{ const } (\|T_1\|_c + \|G_c\|_c\|\chi\|_c) \tag{6.9}$$

and

$$\|d_c^1\chi\|_c + \|\xi\|_2 \leq \text{ const } \left(\|T_1\|_c + \|P\xi\|_2 + R^{-1}\|\chi\|_c\right). \tag{6.10}$$

In the next step we show that there is a constant D independent of the monopole C and thus especially also independent of R, such that for all $\xi \in L^2 \cap L^{6/5}(\Omega^1)$

$$\|\chi\|_c \leq D\|T_1(\chi,\xi)\|_2\,. \tag{6.11}$$

Applying the operator Pd_c^2 on both sides of (6.5) gives

$$P\xi = Pd_c^2T_1 - P[G_c,\chi]\,. \tag{6.12}$$

Use (6.12) to estimate

$$\begin{aligned}\|\chi\|_c &\leq \|T_1\|_2 + \|P\xi\|_2 \\ &\leq \|T_1\|_2 + \|Pd_c^2T_1\|_2 + \|P[G_c,\chi]\,\|_2\,.\end{aligned} \tag{6.13}$$

Estimate the second term on the RHS of (6.13) by

$$\|Pd_c^2T_1\|_2^2 = (Pd_c^2T_1, T_1)\,.$$

Thus

$$\|Pd_c^2T_1\|_2 \leq \text{ const } \|T_1\|_2\,. \tag{6.14}$$

For the third term of (6.13) remark that

$$\|P[G_c,\chi]\|_2 \leq \text{ const } \|[G_c,\chi]\|_{6/5} \leq \text{ const } \|G_c\|_{3/2}\|\chi\|_c \leq \text{ const } R^{-1}\|\chi\|_c\,. \tag{6.15}$$

From (6.13–15)

$$\|\chi\|_c \leq \text{ const } \|T_1\|_2$$

and

$$\|\chi\|_c + \|P\xi\|_2 \leq \text{ const } \|T_1\|_2\,. \tag{6.16}$$

(6.10) and (6.16) imply (6.6).

6.2 An index calculation

To proof Lemma 6.3 it remains to calculate the index of the operator T_1. Introduce here some notation. For $\Psi = (\Psi_0, \Psi_1)$ denote

$$\nabla_c \Psi = ([\Phi, \Psi], \nabla_A \Psi) \tag{6.17}$$

and

$$\nabla_c^2 \Psi = -[\Phi, [\Phi, \Psi]] - \nabla_A^2 \Psi . \tag{6.18}$$

Define the norm

$$\|\Psi\|_{cc} := \|\nabla_c^2 \Psi\|_2 + \|\nabla_c \Psi\|_2 \tag{6.19}$$

and denote the corresponding space by L_{cc}. The fact that C is almost self-dual, Lemma 2.2 and (1.18) imply

Lemma 6.4 *The norms $\|\Psi\|_{cc}$ and $\|\delta_c \delta_c^* \Psi\|_2 + \|\delta_c^* \Psi\|_2$ are equivalent.*

Remember $B_1(\Omega^1)$, the space of one forms on K' in the norm

$$\|\xi\|_{B_1} := m^{1/2}\|P\xi\|_2 + m^{-1/2}\|\xi\|_2 .$$

Define

$$\begin{aligned} F_1 : B_1(\Omega^1) &\longrightarrow L_{cc}(\Omega^1) \\ \xi &\longmapsto H_{2,c}^{-1}\xi := (d_c^2 d_c^{2*})^{-1}\xi . \end{aligned}$$

A calculation shows that F_1 is bounded, injective and continuous in both directions. Furthermore F_1 is surjective. Let

$$\begin{aligned} F_2 : L_{cc}(\Omega^0 \oplus \Omega^1) &\longrightarrow L_c(\Omega^0 \oplus \Omega^1) \\ \Psi &\longmapsto \delta_c^* \Psi . \end{aligned}$$

Remark that $T_1(\chi, \xi) = F_2(\chi, F_1(\xi))$ and index T_1 =index F_2.

For C a smooth finite energy configuration on $\mathbb{R}^3$, the operator

$$\begin{aligned} \widehat{F}_2 : L_c &\longrightarrow L^2 \\ \Psi &\longmapsto \delta_c^* \Psi \end{aligned}$$

is discussed in [T2]. This is a Fredholm operator with index $(\delta_c^*) = -4$. Using the excision principle for indices, one can show that this is also the case for $C \in \mathscr{C}_m(K')$, where the set of configurations $\mathscr{C}_m(K')$ is described in Definition 2.1. As F_2 and $\widehat{F}_2$ have no kernel, the following lemma concludes the proof of Lemma 6.3.

Lemma 6.5 $\dim \operatorname{coker} F_2 = \dim \operatorname{coker} \widehat{F}_2$

1) $\operatorname{coker}\widehat{F}_2 \subset \operatorname{coker} F_2$

Remark that $\dim \operatorname{coker}\widehat{F}_2 = \dim\ker(\widehat{F}_2)^*$ where $(\widehat{F}_2)^* : L_c \longrightarrow L^2$. Let $\Psi = (\Psi_0, \Psi_1) \in L_c(\Omega^0 \oplus \Omega^1)$. Assume $\delta_c\Psi = 0$. By (1.19)

$$\|\nabla_c^2\Psi\|_2 \leq \|(\nabla_c^2\Psi_0, \nabla_c^2\Psi_1 + \nabla_A\nabla_A^*\Psi_1)\|_2 \leq \text{ const } \left(\|\widehat{G}_c\|_3 + \|Ric\|_3\right) \|\Psi\|_c\,; \tag{6.20}$$

thus $\Psi \in L_{cc}$ and $\Psi \in \ker(F_2)^*$.

2) $\operatorname{coker} F_2 \subset \operatorname{coker}\widehat{F}_2$

Assume $\Psi \in \operatorname{coker} F_2$. Thus $\Psi \in L_c(\Omega^0 \oplus \Omega^1)$ and for η a compactly supported section of $C^\infty(\Omega^0 \oplus \Omega^1)$

$$0 = (\nabla_c\Psi, \nabla_c\delta_c^*\eta) = \left(\Psi, \delta_c^*\nabla_c^2\eta\right) + (\Psi, [\nabla_c, \delta_c^*]\nabla_c\eta) + (\nabla_c\Psi, [\nabla_c, \delta_c^*]\eta)\ .$$

Write this as

$$0 = \left(\delta_c\Psi, \nabla_c^2\eta\right) + (\Psi, K_1\nabla_c\eta) + (\nabla_c\Psi, K_2\eta)\ ,$$

where K_1 and K_2 are degree zero operators arising from commutators. Write $\nabla_c^2\eta = \Phi$ with $\Phi \in L^2(\Omega^0 \oplus \Omega^1)$. Thus

$$(\delta_c\Psi, \Phi) - \left(\Phi, (\nabla_c^2)^{-1}\left(\nabla_c(K_1\Psi)\right)\right) + \left(\Phi, \nabla_c^{-2}(K_2\nabla_c\Psi)\right) = 0\,.$$

Write this as

$$\delta_c\Psi + f(\Psi) = 0$$

where

$$f : L_c(\Omega^0 \oplus \Omega^1) \longrightarrow L_{cc}(\Omega^0 \oplus \Omega^1)\,.$$

Thus $\delta_c\Psi \in L_{cc}$ or $\nabla_c^2\Psi \in L_c$. Partial integration gives

$$0 = (\nabla_c^2\Psi, \delta_c^*\eta) = (\delta_c\nabla_c^2\Psi, \eta)$$

$$\nabla_c^2\Psi \in \ker(\widehat{F}_2)^*\,.$$

Lemma (6.3) follows now and the proof of Lemma 6.3 is complete.

6.3 The finite dimensional operator

In this subsection we finish the proof of Theorem 6.1. So far we understand the infinite dimensional part T_1 of the operator T. Now the task is to investigate what happens when we add the finite dimensional part T_2.

$$T(\chi, f, \zeta, \xi) = T_1(\chi, \xi) + T_2(f, \zeta) \tag{6.21}$$

where

$$T_2(f, \zeta) = f d_c^1\Phi + l(\zeta)\,.$$

Decompose T_2 as

$$T_2(f,\zeta) = Q_1(f,\zeta) + Q_2(f,\zeta)$$

where

$$Q_1(f,\zeta) = (1 - P\delta_c)T_2$$

and

$$Q_2(f,\zeta) = P\delta_c T_2\,.$$

Remark that $(1 - P\delta_c)$ is a projection on the kernel of δ_c. Thus $\mathrm{Im}(R_1) \subset \operatorname{coker}\delta_c^*$ and the images of T_1 and Q_1 are L^2-orthogonal.

Lemma 6.6

$$\|Q_2(f,\zeta)\|_{E_1} \leq \text{ const } R^{-1}\|(f,\zeta)\|_{B_1}$$

Lemma 6.7 *There is a constant D independent of C such that*

$$\|(f,\zeta)\|_{B_1} \leq D\|T_2\|_{E_1}\,.$$

Defer the proofs of these lemmas for the moment. Lemma 6.3 and 6.7 imply

$$\|(\chi,f,\zeta,\xi)\|_{B_1} \leq \text{ const } (\|T_1\|_{E_1} + \|T_2\|_{E_1})\,. \tag{6.22}$$

By the L^2-orthogonality of the images of T_1 and Q_1 and Lemma 6.6

$$(\|T_1\|_2 + \|T_2\|_2) \leq \text{ const } \left(\|T\|_2 + \|(f,\zeta)\|_{B_1}\right)\,. \tag{6.23}$$

Use (1.19) to estimate

$$\begin{aligned} |(T_1,Q_1)_c| &\leq |(T_1,\nabla_c^2 Q_1)_2| \\ &\leq |(T_1,\{\widehat{G}_c,Q_1\})| + \|T_1\|_2\|d_A d_A^* Q_1\|_2 + |(T_1,\{Ric,Q_1\})| \end{aligned} \tag{6.24}$$

Remark that the supports of T_2 and Ric are disjoint. Thus

$$\begin{aligned} |(T_1,\{Ric,Q_1\})| &\leq \left|\Big(T_1,\{Ric,(1-P\delta_c)T_2\}\Big)\right| = |(T_1,\{Ric,Q_2\})| \\ &\leq \|Ric\|_{3/2}\|T_1\|_c\|Q_2\|_c \end{aligned} \tag{6.25}$$

Estimate $\|Q_2\|_c$ by Lemma 6.6. This gives

$$(6.25) \leq \text{ const } \|Ric\|_{3/2}\|T_1\|_c^2 + R^{-2}\|(f,\zeta)\|_{B_1}^2\,. \tag{6.26}$$

Thus by (6.26)

$$\begin{aligned} (6.24) \leq{} & \text{ const } \epsilon^{-1}\|T_1\|_2^2 + \epsilon\|Q_1\|_c^2 + \|Q_1\|_2^2 \\ & + \text{ const } \epsilon\|T_1\|_c\|Q_1\|_c + R^{-2}\|(f,\zeta)\|_{B_1}^2\,. \end{aligned}$$

From this follows

$$\begin{aligned} \|T_1\|_c^2 + \|T_2\|_c^2 \leq{} & \|T\|_c^2 + \text{ const } (\|T\|_2^2 + \|Q_2\|_2^2 \\ & + \text{ const } \|Q_2\|_c^2) + R^{-2}\|(f,\zeta)\|_{B_1}^2 \end{aligned} \tag{6.27}$$

(6.22–27) and Lemma 6.6 imply

$$\|(\chi,f,\zeta,\xi)\|_{B_1} \leq \text{ const } \|T\|_{E_1}$$

and Theorem 6.1 follows for the D_R parameter space.

Proof of Lemma 6.6: Estimate

$$\|P\delta_c T_2\|_{E_1} \leq \|\delta_c T_2\|_2 + \|\delta_c T_2\|_{6/5}\,.$$

Assume C is centered at x. $\delta_C T_2$ is supported away from the neck. Pulling back to $\mathbb{R}^3$ gives

$$\|\delta_C T_2\|_{E_1} \leq \text{ const } \|\delta_{C_1} T_{2,C_1}\|_{E_1}\,,$$

where C_1 is given in (3.10–11). Write $\widehat{C}$ for $\widehat{C}_{x,p}$-the monopole on $\mathbb{R}^3$ in polar gauge centered at x defined in (3.1–2). Use also the notation $\widehat{l}(\zeta) := \frac{\partial}{\partial s}\widehat{C}(x, s\zeta)|_{s=0}$ and $T_{2,\widehat{C}} = f d^1_{\widehat{C}}\widehat{\Phi} + \widehat{l}(\zeta)$. Then

$$\delta_{C_1} T_{2,C_1} = \delta_{\widehat{C}} T_{2,\widehat{C}} + \delta_{\widehat{C}}(T_{2,C_1} - T_{2,\widehat{C}}) + [C - \widehat{C}, T_{2,C_1}]\,. \tag{6.28}$$

By the self-dual equations

$$\delta_{\widehat{C}} d^1_{\widehat{C}} \widehat{\Phi} = 0\,. \tag{6.29}$$

As translations on $\mathbb{R}^3$ are deformations in self-dual directions for which the gauge does not change, we have

$$\delta_{\widehat{C}} \widehat{l}(\zeta) = 0\,. \tag{6.30}$$

Thus the first term on the RHS of (6.28) is zero. Now estimate

$$\begin{aligned} &\|\delta_{C_1} T_{2,C_1}\|_2 + \|\delta_{C_1} T_{2,C_1}\|_{6/5} \\ &\leq \text{ const } \left(\|\delta_{\widehat{C}}(T_{2,C_1} - T_{2,\widehat{C}})\|_2 + \|\,[C_1 - \widehat{C}, T_{2,C_1}]\,\|_2\right) \\ &+ \text{ const } \left(\|\delta_{\widehat{C}} T_{2,C_1} - T_{2,\widehat{C}}\|_{6/5} + \|\,[C - \widehat{C}, T_{2,C_1}]\,\|_{6/5}\right)\,. \end{aligned} \tag{6.31}$$

$\widehat{C}$ is the self-dual perturbation of C_1. The difference between $C_1 - \widehat{C}$ is just the perturbation term which is estimated in Sections 2 and 3.1. Thus

$$\|C_1 - \widehat{C}\|_{E_1} \leq \text{ const } R^{-1}\,. \tag{6.32}$$

Using this to estimate (6.31) gives

$$(6.31) \leq \text{ const } R^{-1}\|(f, \zeta)\|_{B_1} \tag{6.33}$$

and Lemma 6.6 follows.

Proof of Lemma 6.7: For a zero form χ write here $P\chi := P(\chi, 0)$. Set

$$T_2(f, \zeta) = f d^1_c \Phi + \left(1 - P(d^1_c)^*\right) l(\zeta) + P(d^1_c)^* l(\zeta)\,. \tag{6.34}$$

Remark that

$$\begin{aligned} &\left(f d^1_c \Phi, \left(1 - P(d^1_c)^*\right) l(\zeta)\right) \\ &= \left(f(\Phi - \frac{i}{2} m\sigma^3), (d^1_c)^* \left(1 - P(d^1_c)^*\right) l(\zeta)\right) = 0\,. \end{aligned}$$

Thus, the images of the first two operators on the RHS in (6.34) are L^2-orthogonal. Now proceed similarly as above and the result follows.

6.4 The Fredholm operator for the $\mathbb{R}^3$ parameter space

Remark that in the case of small mass monopoles all estimates have to be uniform with respect to m. From this point of view it would be tempting also to use mass weighted norms in this subsection in the sense that we weight the

$$\|\chi\|_c + \|fd_c^1\Phi\|_2 + \|l(\zeta)\|_2 + \|P\xi\|_2$$

term by a factor $m^{1/2}$ and the

$$\|d_c^1\chi\|_c + \|fd_c^1\Phi\|_c + \|l(\zeta)\|_c + \|\xi\|_2$$

term by a factor $m^{-1/2}$. However this strategy fails. In Section 7 we have to estimate $\|\chi\|_\infty$ for $\chi \in \Omega^0(K')$. Lemma 2.11 gives, for large masses m, the estimate

$$\|\chi\|_\infty \leq \text{ const } \left(m^{1/2}\|\chi\|_c + m^{-1/2}\|\nabla_c^2\chi\|_2\right) \tag{6.35}$$

This estimate garantees uniformity with respect to m when mass weighted norms are used. For small masses (6.35) does not hold. Instead we have to use

$$\|\chi\|_\infty \leq \text{ const } \left(\|\chi\|_c + \|\nabla_c^2\chi\|_2\right). \tag{6.36}$$

Using the estimate (6.36) has the drawback that the analysis is not uniform anymore with respect to m. On the other hand using not weighted norms together with the fact that m is small garantees uniformity with respect to m. Thus we use the B_1 and the E_1 norm defined in Subsection 6.1 for the D_R parameter space.

The lemma corresponding to Lemma 6.2 is

Lemma 6.8

1. $\|l(\zeta)\|_c \leq$ const $m^{1/2}|\zeta| \leq$ const $\|l(\zeta)\|_c$
2. $\|\nabla_c l(\zeta)\|_c \leq$ const $m^{3/2}|\zeta| \leq$ const $\|\nabla_c l(\zeta)\|_c$
3. $\|l(\zeta)\|_\infty \leq$ const $m|\zeta|$.

The analogue of Theorem 6.1 is now

Theorem 6.9 *There is an m_0 such that, for $m \leq m_0$ and $C \in \mathscr{C}_m(\mathbb{R}^3)$,*

$$T : B_1 \longrightarrow E_1$$

is a bounded, injective Fredholm operator with index $T = 0$. Furthermore the norm of its inverse can be uniformly bounded by a constant independent of C and m.

The proof of Theorem 6.9 follows the lines of the proof of Theorem 6.1, with the modification that we keep now track of the mass dependence.

7 Some special gauges

Introduce here neighborhoods of the almost self-dual configurations $\mathcal{A}_m(X)$ constructed in Section 3. These neighborhoods consist of equivalence classes of configurations whose gauge orbit has a short distance to $C_\varphi := J(\varphi)$ for a parameter φ.

$$J : X \longrightarrow \mathcal{A}_m(X)$$

is the map describing the construction of monopoles discussed in Section 3. For the same reason as in Section 6 we have to proceed here differently for the various parameter spaces.

7.1 Gauge fixing for the D_R parameter space

For the same reason as in Subsection 6.1, we can in this subsection treat m as a constant.

Definition 7.1 *Denote by $\mathcal{U}_m(\nu, D_R)$ the set of all configurations in $\mathcal{A}(D_R)$ which fulfill the property: For $Q \in \mathcal{U}_m(\nu, D_R)$ there exists a gauge transformation $u \in \mathcal{G}$ and a parameter $\varphi \in D_R$ such that*

$$\|u(Q) - C_\varphi\|_2 + \|\nabla_{u(Q)}(u(Q) - C_\varphi)\|_2 < \nu \tag{7.1}$$

ν labels the «thickness» of these tubular neighborhoods $\mathcal{U}_m(\nu, D_R)$. I is the composition of the map

$$J : D_R \longrightarrow \mathcal{A}_m(D_R)$$

and the self-dual perturbation P:

$$I : D_R \longrightarrow \mathfrak{M}_m \tag{7.2}$$

$$\varphi \longmapsto C_\varphi + P_\varphi \xi$$

Making the tubular neighborhoods $\mathcal{U}_m(\nu, D_R)$ thick enough ensures that the constructed monopoles $I(D_R)$ are contained in them.

Lemma 7.2 *Let $\nu > 0$. There is an $R_1(\nu)$ such that for $R \geq R_1(\nu)$*

$$I(D_R) \subset \mathcal{U}_m(\nu, D_R)\,. \tag{7.3}$$

Lemma 7.2 follows from (3.12–13).

Theorem 7.3 *There is a $\nu_0 > 0$ such that, for $\nu \leq \nu_0$, there is an $R_1(\nu)$ such that for $R \geq R_1(\nu)$ and $Q \in \mathcal{U}_m(\nu, D_R)$, we have that: there exists a gauge transformation $u \in \mathcal{G}$, $\xi \in L^2 \cap L^{6/5}(\Omega^1)$ and a parameter $\varphi \in D_R$ such that*

$$u(Q) = C_\varphi + P_\varphi \xi \tag{7.4}$$

and

$$\|\xi\|_{6/5} + \|\xi\|_2 \leq \text{ const } \nu\,. \tag{7.5}$$

In case ξ is a fixed point of f, Q has to be self-dual. As ξ is uniquely determined, this implies that up to gauge transformations all self-dual configurations in $\mathcal{U}_m(\nu, D_R)$ can be constructed with I. Express this in the following corollary:

Corollary 7.4 *Make the same assumptions as in Theorem 7.3. Then*

$$I(D_R) = \mathcal{U}_m(\nu, D_R) \cap \mathfrak{M}_m .$$

Proof of Theorem 7.3: Take $[Q] \in \mathcal{U}_m(\nu, D_R)$ and $Q \in [Q]$ with

$$\|Q - C_\varphi\|_2 + \|\nabla_Q(Q - C_\varphi)\|_2 \leq \nu .$$

Set $q := Q - C_\varphi$ and $Q_t := C_\varphi + tq$. Remark that $Q_t \in \mathcal{U}_m(\nu, D_R)$ for $t \in [0, 1]$. In Chapter 7 of [DK] the corresponding gauge fixing problem for anti-self-dual connections is approached with the continuity method. We follow here along these lines:

Definition 7.5 *Let $\delta > 0$ be a constant. The exact size of δ will be fixed at the end of this section. Define $S \subset [0, 1]$ to be the set of all times t, for which there exists a gauge transformation u_t, a parameter $\varphi \in X$ and a 1-form ξ such that*

$$u_t(Q_t) = C_\varphi + P_\varphi \xi \tag{7.6}$$

with

$$\|\xi\|_{6/5} + \|\xi\|_2 < \delta . \tag{7.7}$$

Obviously $0 \in S$. We will show that S is open and closed in the interval $[0, 1]$. Thus $S = [0, 1]$ and this implies Theorem 7.3.

Lemma 7.6 *S is closed in [0,1].*

Proof of Lemma 7.6: First we show that (7.7) is also a closed condition. Assume $t \in S$ and that $u_t(Q_t) = C_\varphi + P_\varphi \xi$ with ξ fulfilling (7.7). From

$$u_t G_{Q_t} u_t^{-1} = G_{C_\varphi} + d^2_{C_\varphi}(P_\varphi \xi) + \{P_\varphi \xi, P_\varphi \xi\}$$

it follows that

$$\xi = u_t G_{Q_t} u_t^{-1} - G_{C_\varphi} - \{P_\varphi \xi, P_\varphi \xi\} . \tag{7.8}$$

We use (7.8) to estimate the L^2 and $L^{6/5}$ norms of ξ.

$$\|\xi\|_2 \leq \|G_{Q_t}\|_2 + \|G_{C_\varphi}\|_2 + \text{ const } \|P_\varphi \xi\|_c^2 \leq \text{ const } (\nu + R^{-1} + \delta^2) .$$

Fix δ by the condition

$$\delta \geq 8 \text{ const } (\nu + R^{-1}) .$$

This implies

$$\|\xi\|_2 \leq \frac{1}{4}\delta . \tag{7.9}$$

Similarly we show

$$\|\xi\|_{6/5} \leq \frac{1}{4}\delta . \tag{7.10}$$

(7.9) and (7.10) show that (7.7) is also a closed condition. Now proceed as in Section 7.2.4 of [DK] to show that S is closed.

As S is compact we can define

$$D(u) := \sup_{t \in S} \|d_c^1 u_t\|_3 \tag{7.11}$$

Lemma 7.7 *S is open in* $[0, 1]$.

Assume for $t_0 \in [0, 1]$,

$$u_{t_0}(Q_{t_0}) = C_{\varphi_0} + P_{\varphi_0}\xi_0$$

with $\|\xi_0\|_2 < \delta$ and $\|\xi_0\|_{6/5} < \delta$. We want to show that for t close to t_0, we also have such a gauge. Define the map

$$M : \Omega^0_{K'} \times \mathbb{R} \times \mathbb{R}^3 \times \Omega^1_{K'} \longrightarrow \Omega^0_{K'} \oplus \Omega^1_{K'} \tag{7.12}$$

$$M(\chi, f, \zeta, \eta) := \exp(\chi + f\Phi)\left(C_{(\varphi_0,\zeta)} + P_{[\zeta]}(\xi_o + \eta)\right) - u_{t_0}(Q_{t_0}) .$$

Remark that $M(0, 0, 0, 0) = 0$ and that M is a smooth map from C_0^∞ sections of the first space to C_0^∞ sections of the second and extends to a smooth map from B_1 to E_1. These spaces were introduced in (6.3) and (6.4).

Lemma 7.8 *$DM|_{(0,0,0,0)} : B_1 \longrightarrow E_1$ is an isomorphism continuous in both directions.*

Remark that the leading term of DM is the operator T of Section 6. Defer the proof of Lemma 7.8 for the moment.

Proof of Lemma 7.7 assuming Lemma 7.8: Remark that

$$\|uqu^{-1}\|_{E_1} \leq \text{ const } \nu\,(1 + D(u)) \tag{7.13}$$

where $D(u)$ is the constant defined in (7.11). Assume $|t - t_0| < \epsilon$. By (7.13)

$$\|u_{t_0}(Q_t - Q_{t_0})u_{t_0}^{-1}\|_{E_1} \leq \text{ const } \epsilon\nu\,(1 + D(u)) . \tag{7.14}$$

By Lemma 7.8 and the implicit function theorem there is an $\epsilon > 0$ such that $u_{t_0}(Q_t - Q_{t_0})u_{t_0}^{-1}$ is in the image of M. Thus there exists (χ, f, ζ, η) with

$$u_{t_0}(Q_t - Q_{t_0})u_{t_0}^{-1} = \exp(\chi + f\Phi)\left(C_{(\varphi_0,\zeta)} + P_{[\zeta]}(\xi_o + \eta)\right) - u_{t_0}(Q_{t_0})$$

or

$$Q_t = \left(u_{t_0}^{-1} \exp(\chi + f\Phi)\right)\left(C_{(\varphi_0,\zeta)} + P_{[\zeta]}(\xi_o + \eta)\right) . \tag{7.15}$$

Furthermore a suitable choice of ϵ gives

$$\|\xi_0 + \eta\|_{6/5} + \|\xi_0 + \eta\|_2 < \delta$$

Thus $t \in S$ and Lemma 7.7 follows.

Proof of Lemma 7.8:

$$DM|_{(0,0,0,0)}(\chi,f,\zeta,\eta) = T(\chi,f,\zeta,\eta) + [P\xi,\chi+f\Phi] + \partial_\zeta(P_{[\zeta]}\xi)\,. \tag{7.16}$$

From Section 6 we know that T is an isomorphism which is continuous in both directions. So it remains to show that the last two terms do not destroy this property, or in other words, that they have a small operator norm compared to the one of T.

Estimate the E_1 norm of the second term on the RHS of (7.16). For the L^2 part of this norm we get

$$\begin{aligned} \|\,[P\xi,\chi+f\Phi]\,\|_2 \le\ & \text{const } \|P\xi\|_2^{1/2}\|P\xi\|_6^{1/2}(\|\chi\|_c+|f|) \\ \le\ & \text{const } (\|\xi\|_2+\|\xi\|_{6/5})\|(\chi,f,\zeta,\xi)\|_{B_1}\,. \end{aligned} \tag{7.17}$$

And for the L_c part of this norm estimate

$$\begin{aligned} \|\,[P\xi,\chi+f\Phi]\,\|_c \le\ & \|P\xi\|_c(\|\chi\|_\infty+|f|) \\ & + \text{ const } \|P\xi\|_2^{1/2}\|P\xi\|_6^{1/2}(\|d_c^1\chi\|_c+f\|d_c^1\Phi\|_c)\,. \end{aligned} \tag{7.18}$$

Use (6.36) to estimate $\|\chi\|_\infty$ on the RHS of (7.18). This gives

$$(7.18) \le \text{ const } \left(\|\xi\|_2+\|\xi\|_{6/5}\right)\|(\chi,f,\zeta,\eta)\|_{B_1}\,. \tag{7.19}$$

(7.17) and (7.19) imply

$$\|\,[P\xi,\chi+f\Phi]\,\|_{E_1} \le \text{ const } \delta\|(\chi,f,\zeta,\xi)\|_{B_1}\,. \tag{7.20}$$

We can make the RHS of (7.20) arbitrary small by choosing δ small. Thus the operator norm of the second term on the RHS of (7.16) is small.

For the third term on the RHS of (7.16) write

$$\partial_\zeta P_\zeta\xi = \left(\partial_\zeta(d^2_{C_\zeta})^*\right)H_{C_\zeta}^{-1}\xi + P_\zeta(\partial_\zeta H_{C_\zeta})H_{C_\zeta}^{-1}\xi \tag{7.21}$$

Remember $l(\zeta) := \partial_\zeta C_\zeta$. Thus

$$(7.21) = -*[l(\zeta),*H_{C_\zeta}^{-1}\xi] + P_\zeta\,[l(\zeta),P_\zeta\xi] - P_\zeta d_c^2 * [l(\zeta),*H_{C_\zeta}^{-1}\xi] \tag{7.22}$$

Use this decomposition to estimate the E_1 norm of $\partial_\zeta P_\zeta\xi$. For the L^2-norm we get

$$\begin{aligned} \|\partial_\zeta P_\zeta\xi\|_2 \le\ & \text{const } \left(\|\,[l(\zeta),*H_{C_\zeta}^{-1}\xi]\|_2 + \|P_\zeta\,[l(\zeta),P_\zeta\xi]\,\|_2\right) \\ \le\ & \text{const } (\|l(\zeta)\|_2+\|l(\zeta)\|_6)\|H_{C_\zeta}^{-1}\xi\|_c + \text{ const } \|\,[l(\zeta),P_\zeta\xi]\,\|_{6/5} \end{aligned}$$

By Lemma 6.2

$$\leq \text{ const } \delta \, \|(\chi, f, \zeta, \eta)\|_{B_1} \tag{7.23}$$

And for the c-norm of the thrid term of (7.16) estimate

$$\begin{aligned}\|\partial_\zeta P_\zeta \xi\|_c \leq & \text{ const } \Big(\|\nabla_c^2 l(\zeta)\|_2 \|\xi\|_{6/5} + (\|l(\zeta)\|_2 + \|l(\zeta)\|_6)(\|\xi\|_2 + m\|\xi\|_{6/5}) \Big) \\ & + \text{ const } \|l(\zeta)\|_\infty \|\xi\|_{6/5}\end{aligned} \tag{7.24}$$

Again use (6.4) and Lemma 6.2 to estimate (7.24):

$$(7.24) \leq \text{ const } \delta \|\xi\|_{B_1} \, . \tag{7.25}$$

(7.23) and (7.25) imply

$$\|\partial_\zeta P_\zeta \xi\|_{E_1} \leq \text{ const } \delta \|(\chi, f, \zeta, \eta)\|_{B_1} \, . \tag{7.26}$$

(7.26) shows that the operator norm of the last term on the RHS of (7.16) is small.

By Theorem 6.1 the operator norm of T^{-1} is bounded by a universal constant. Thus for sufficiently small δ, DM is an isomorphism continuous in both directions. This completes the proof of Theorem 7.3.

7.2 Gauge fixing for the $\mathbb{R}^3$ parameter space

In Subsection 3.2 we constructed small mass almost self-dual configurations which are parametrized by the space $\mathbb{R}^3$. For these $C \in \mathscr{C}_m(\mathbb{R}^3)$ we showed in Section 3 that

$$\|G_c\|_2 \leq \text{ const } m^2$$

$$\|G_c\|_{6/5} \leq \text{ const } m^2 \, .$$

Define in this case the neighborhood $U_m(\nu, \mathbb{R}^3)$ as

Definition 7.9 *Denote by $U_m(\nu, \mathbb{R}^3)$ the set of all configurations in $\mathscr{A}_m(\mathbb{R}^3)$ which fulfill the property: For $Q \in U_m(\nu, \mathbb{R}^3)$ there exists a gauge transformation $u \in \mathscr{G}$ and a parameter $\varphi \in \mathbb{R}^3$ such that*

$$\|u(Q) - C_\varphi\|_2 + \|\nabla_{u(Q)}(u(Q) - C_\varphi)\|_2 < \nu \, . \tag{7.27}$$

Again by choosing m small with respect to ν we can ensure that

$$I(\mathbb{R}^3) \subset U_m(\nu, \mathbb{R}^3) \, . \tag{7.28}$$

The theorem corresponding to Theorem 7.3 is

Theorem 7.10 *There is a ν_0 such that for $\nu \leq \nu_0$ there is an $m_0(\nu)$ such that for $m \leq m_0(\nu)$ and $Q \in \mathcal{U}_m(\nu, \mathbb{R}^3)$ we have that: there is a gauge transformation $u \in \mathcal{G}$, a $\xi \in L^2 \cap L^{6/5}(\Omega^1)$ and a parameter $\varphi \in \mathbb{R}^3$ with*

$$Q = u(C_\varphi + P_\varphi \xi) \tag{7.29}$$

and

$$\|\xi\|_{6/5} + \|\xi\|_2 \leq \text{ const } \nu \,. \tag{7.30}$$

Corollary 7.11

$$I(\mathbb{R}^3) = \mathcal{U}_m(\nu, \mathbb{R}^3) \cap \mathfrak{M}_m \,.$$

Theorem 7.10 and Corollary 7.11 are proved in a similar way as Theorem 7.3 and Corollary 7.4.

8 Different parameters yield different monopoles

The map $J : X \to \mathcal{A}_m(X)$ is injective. The purpose of this section is to show that composition with the self-dual perturbation does not destroy this property. In other words, the map $I : X \to \mathfrak{M}_m$ is injective. From Section 7 we know that $X = \mathcal{U}_m(\nu, X) \cap \mathfrak{M}_m$. Injectivity of I implies that

$$I : X \longrightarrow \mathcal{U}_m(\nu, X) \cap \mathfrak{M}_m$$

is a diffeomorphism. A similar problem for anti-self-dual connections is discussed in Section 7.2.6 of [DK] or also in [D2]. We follow along these lines.

Proposition 8.1 *The map $I : X \to \mathfrak{M}_m$ is injective.*

We show first that the map I is a local diffeomorphism. Use in this section the notation $a = P\xi$, $a_\zeta = P_\zeta \xi$ and $b_\zeta = C_\zeta - C$ where ζ is in the tangent space of the parameter space. Assume we have a gauge transformation u with

$$C + a = u(C + b_\zeta + a_\zeta) \,. \tag{8.1}$$

Write this also in the form

$$d_C^1 u = -[a, u] + u(a_\zeta - a + b_\zeta) \,. \tag{8.2}$$

For the same reasons as in the last two sections we have to proceed slightly differently in the proof of Proposition 8.1 for the various parameter spaces.

8.1 Injectivity for monopoles parametrized by D_R

The first goal is to show that I is a local diffeomorphism. Thus assume that $|\zeta|$ is small. The maximal size of $|\zeta|$ will be determined below. In this subsection we do not require uniformity with respect to m. To simplify the notation we treat here m as a constant.

Lemma 8.2

$$\|u-1\|_\infty \leq \text{ const } \left(\|a_\zeta - a\|_{E_1} + |\zeta|\right)$$

Proof of Lemma 8.2: Similar to (6.36) we have

$$\|u-1\|_\infty \leq \text{ const } \left(\|d_c^1 u\|_2 + \|d_c^1 u\|_c\right) . \tag{8.3}$$

Similar to Lemma 6.2 one shows

$$\|b_\zeta\|_{E_1} \leq \text{ const } |\zeta| \leq \text{ const } \|b_\zeta\|_{E_1}$$

where $\|.\|_{E_1}$ is the norm defined in (6.4). By (8.2)

$$\begin{aligned} \|d_c^1 u\|_2 &\leq \|\,[a,u]\,\|_2 + \text{ const } \|a_\zeta - a + b_\zeta\|_2 \\ &\leq \text{ const } \left(\|a\|_c \|d_c^1 u\|_2 + \|a_\zeta - a\|_2 + |\zeta|\right) . \end{aligned} \tag{8.4}$$

Use also (8.2) to estimate the c-norm of $d_c^1 u$

$$\begin{aligned} \|d_c^1 u\|_c &\leq \|\,[a,u]\,\|_c + \|u(a_\zeta - a + b_\zeta)\|_c \\ &\leq \|\,[\nabla_c a, u]\,\|_2 + \|\,[a, d_c^1 u]\,\|_2 + \|d_c^1 u(a_\zeta - a + b_\zeta)\|_2 + \|a_\zeta - a + b_\zeta\|_c \\ &\leq \|u-1\|_\infty R^{-1} + \text{ const } R^{-1}\|d_c^1 u\|_c + \|\,[a,u](a_\zeta - a + b_\zeta)\|_2 \\ &\quad + \frac{1}{2}\|\,[a_\zeta - a + b_\zeta, a_\zeta - a + b_\zeta]\,\|_2 + \|a_\zeta - a + b_\zeta\|_c . \end{aligned}$$

Remark that

$$\|a_\zeta - a + b_\zeta\|_c \leq \text{ const } .$$

Thus

$$\|d_c^1 u\|_c \leq \text{ const } \left(\|u-1\|_\infty R^{-1} + \|a_\zeta - a\|_c + |\zeta|\right) . \tag{8.5}$$

(8.4) and (8.5) imply Lemma 8.2.

Lemma 8.3

$$\|a_\zeta - a\|_{E_1} \leq \text{ const } R^{-1}\|l(\zeta)\|_{E_1}|\zeta| .$$

Proof of Lemma 8.3: Recall $a_\zeta = P_\zeta \xi_\zeta$ where ξ_ζ solves the equation (2.21)

$$\xi_\zeta = -\{P_\zeta \xi_\zeta, P_\zeta \xi_\zeta\} - G_{C_\zeta}\,. \tag{8.6}$$

Taking the ∂_ζ derivative of (8.6) gives

$$\partial_\zeta \xi_\zeta = -2\Big\{(\partial_\zeta P_\zeta)\xi_\zeta + P_\zeta(\partial_\zeta \xi_\zeta), P_\zeta \xi_\zeta\Big\} - \partial_\zeta G_{C_\zeta}\,. \tag{8.7}$$

We will need this to estimate

$$\|\partial_\zeta a_\zeta\|_{E_1} \leq \|(\partial_\zeta P_\zeta)\xi_\zeta\|_{E_1} + \|P_\zeta(\partial_\zeta \xi_\zeta)\|_{E_1}\,. \tag{8.8}$$

By Section 7 the first term on the RHS of (8.8) is estimated by

$$\leq \text{ const } R^{-1}\|l(\zeta)\|_{E_1}\,.$$

For the L_C-norm part of the second term on the RHS of (8.8), use the equation (8.7)

$$\begin{aligned} \|P_\zeta(\partial_\zeta \xi_\zeta)\|_c &\leq \Big(\|\partial_\zeta \xi_\zeta\|_2 + \|\partial_\zeta \xi_\zeta\|_{6/5}\Big) \\ &\leq \text{ const } \Big(\|(\partial_\zeta P_\zeta)\xi_\zeta\|_c + \|\partial_\zeta \xi_\zeta\|_2\Big)\|\xi_\zeta\|_2 \\ &+ \text{ const } \|P_\zeta \xi_\zeta\|_2^{3/2}\|(\partial_\zeta P_\zeta)\xi_\zeta\|_c^{1/2} + \|\partial_\zeta G_{C_\zeta}\|_2 \\ &\leq \text{ const } R^{-1}\|l(\zeta)\|_c \end{aligned} \tag{8.9}$$

Estimate the L^2-norm similarly. This gives

$$\|\partial_\zeta a_\zeta\|_{E_1} \leq \text{ const } R^{-1}\|l(\zeta)\|_{E_1}\,.$$

Integration over a path from a to a_ζ yields Lemma 8.3.

Lemma 8.2 and 8.3 imply

$$\|u-1\|_\infty \leq \text{ const } |\zeta|\,. \tag{8.10}$$

For $|\zeta|$ enough small we can write $u = \exp(\chi)$ with

$$\|\chi\|_\infty \leq \text{ const } \Big(\|\chi\|_c + \|\nabla_c^2 \chi\|_2\Big)\,. \tag{8.11}$$

We know from Section 6

$$|\zeta| + \|\chi\|_\infty \leq \text{ const } \|l(\zeta) + d_c^1 \chi\|_{E_1}\,.$$

Generalizing this inequality by including nonlinear terms gives

$$|\zeta| + \|\chi\|_\infty \leq \text{ const } \|b(\zeta) + u^{-1} d_c^1 u\|_{E_1} + O(|\zeta|^2) + O(\|\chi\|_\infty \|\chi\|_{B_1})\,;$$

by (8.11)

$$\leq \text{ const } \|b(\zeta) + u^{-1} d_c^1 u\|_{E_1} + O(|\zeta|^2)$$

and from (8.2)

$$\leq \text{ const } \Big(\|u^{-1}[a,u]\|_{E_1} + \|a_\zeta - a\|_{E_1}\Big) + O(|\zeta|^2)\,.$$

Thus, for ζ small enough, say $\zeta < \delta$

$$|\zeta| + \|\chi\|_\infty \leq \text{ const } \|a_\zeta - a\|_{E_1} \leq \text{ const } R^{-1}|\zeta|^2\,.$$

Thus $\zeta = 0$ and $u = 1$. This tells us that I is a local diffeomorphism. As mentioned above the map J is injective.

Lemma 8.4 *There is a $\epsilon > 0$ such that for $\varphi_1, \varphi_2 \in X$, $u \in \mathcal{G}$ with $dist(\varphi_1, \varphi_2) > \delta$, we have*

$$\|J(\varphi_1) - u(J(\varphi_2))\|_6 > \epsilon m^{1/2}.$$

The lemma follows from considering the energy distribution $|E|^2(x) := |F_A|^2(x) + |d_A\Phi|(x)$.

Now assume $I(\varphi_1) = u(I(\varphi_2))$. Thus

$$\|J(\varphi_1) - u(J(\varphi_2))\|_6 \leq \text{ const } R^{-1}.$$

Lemma 8.4 implies that for R large enough, $\text{dist}(\varphi_1, \varphi_2) \leq \delta$. As I is locally injective it follows that $\varphi_1 = \varphi_2$. Thus I is injective and

$$I : X \longrightarrow U_m(\nu, X) \cap \mathfrak{M}_m$$

is a diffeomorphism. This completes the proof of Proposition 8.1 for monopoles parametrized by D_R.

8.2 Injectivity for small mass monopoles parametrized by $\mathbb{R}^3$

In contrast to the last subsection, all estimates have to be uniform with respect to m in this subsection. Assume here for the moment that

$$\|b_\zeta\|_{E_1} = \|b_\zeta\|_2 + \|b_\zeta\|_c < \epsilon \tag{8.12}$$

where ϵ will be determined below. The analogues of Lemma 8.2 and Lemma 8.3 are respectively

Lemma 8.5

$$\|u - 1\|_\infty \leq \text{ const } \left(\|a_\zeta - a\|_{E_2} + \|b_\zeta\|_{E_2}\right).$$

Lemma 8.6

$$\|a_\zeta - \|_{E_2} \leq \text{ const } m^{1/2}\|l(\zeta)\|_{E_2}|\zeta|.$$

These two lemmas are proved similar as in the previous subsection. Lemma 8.5 combined with Lemma 8.6 imply

$$\|u - 1\|_\infty \leq \text{ const } \|b_\zeta\|_{E_1}. \tag{8.13}$$

Thus for ϵ small enough in (8.12) we can write u in terms of an element of the Lie algebra $u = \exp(\chi)$. From (8.13) and (6.4) follows

$$\|\chi\|_\infty \leq \text{ const } \|\chi\|_{B_1} \leq \text{ const } \|b_\zeta\|_{E_1}. \tag{8.14}$$

As

$$\|b_\zeta\|_{E_1} \leq \text{ const } \|l(\zeta)\|_{E_1} \leq \text{ const } \|b_\zeta\|_{E_1},$$

and by Section 6 we get the estimate

$$\|b_\zeta\|_{E_1} + \|\chi\|_\infty \leq \text{ const } \|(\zeta, \chi)\|_{B_1} \leq \text{ const } \|l(\zeta) + d_c^1\chi\|_{E_1}.$$

Similar to the previous subsection, we conclude

$$\|b_\zeta\|_{E_1} + \|\chi\|_\infty \leq \text{ const } m^{1/2}\|b_\zeta\|_{E_1}|\zeta|.$$

Thus $\zeta = 0$. Similar as above, we prove that I is a global diffeomorphism.

9 Moduli space far away from the neck

We describe in this section the part of the moduli space which consists of monopoles located far away from the neck. Of course we have to make precise what we mean by the location of a monopole. For that we use the following procedure.

Take a monopole Q on K' for which most of the energy is located far away from the neck. Cut out Q to an almost self-dual configuration Q_1 on $\mathbb{R}^3$. Now perturb Q_1 to a monopole $\widehat{Q}_1 = Q_1 + P\xi$ on $\mathbb{R}^3$. As a monopole on $\mathbb{R}^3$, $\widehat{Q}_1$ has a unique center. Define the center of Q to be the one of $\widehat{Q}_1$. To show that Q is almost self-dual we need some a priori estimates on Q. Remember that D_R denotes the subset of K' whose points have at least the distance R to the neck. Define

$$\mathfrak{M}_{m,R} := \left\{[Q] \in \mathfrak{M}_m \mid \text{YMH}(Q)|_{D_R} \geq 2\pi m\right\} .$$

Denote in this section with $\Omega_{[r_1,r_2]}$ the annulus on the $\mathbb{R}^3$ part of K' centered at the neck with radii $[r_1, r_2]$. k denotes the constant from Section 3.1.

Lemma 9.1 *For $\nu > 0$ there is an $R_2(\nu)$ such that for $R \geq R_2(\nu)$ and $[Q] \in \mathfrak{M}_{m,R}$ there is a $Q = (\Phi, A) \in [Q]$ with*

1. $|Q - C_{0,m}|(x) < \nu$ for $x \in \Omega_{[k,4k]}$
2. $|\nabla_{C_{0,m}}(Q - C_{0,m})|(x) < \nu$ for $x \in \Omega_{[k,4k]}$ *where* $C_{0,m} = (\frac{i}{2}m\sigma^3, \theta)$.

Proof of Lemma 9.1: Let $Q_j \in \mathfrak{M}_{R_j}$ with $R_j \to \infty$. Thus Q_j is a mass m monopole on K' whose energy is drifting out. By the weak compactness theorem Q_j converges over compact domains on K' to a flat mass m configuration. All flat mass m configurations are gauge equivalent to $C_{0,m} := (\frac{i}{2}m\sigma^3, \theta)$. In particular, this means that there is a $Q_j \in [Q]$ such that Q_j is C^∞ close to $C_{0,m}$ over B_k for j large enough.

Assume from now on that Q is in the gauge described in Lemma 9.1. Let α be the cutoff which interpolates between 0 and 1 over the annulus $\Omega_{[2k,4k]}$. Define following configuration on $\mathbb{R}^3$:

$$\Phi_1 := \alpha_{2k}\Phi + (1 - \alpha_{2k})\frac{i}{2}m\sigma^3 \tag{9.1}$$

$$A_1 := \alpha_{2k}A + (1 - \alpha_{2k})\theta\, . \tag{9.2}$$

For $R \geq R_2(\nu)$ Lemma 9.1 implies

$$\|G_{Q_1}\|_p \leq \text{ const } \nu \tag{9.3}$$

$$\|G_{Q_1}\|_c \leq \text{ const } \nu\, . \tag{9.4}$$

As Q_1 is a configuration on $\mathbb{R}^3$, from (9.3–4) it follows that Q_1 is almost self-dual. Denote by

$$\widehat{Q}_1 := Q_1 + P\xi$$

its self-dual perturbation. $\widehat{Q}_1$ is a charge one, mass m configuration on $\mathbb{R}^3$. It has a unique center. Denote it by $\phi(Q)$. This gives the map

$$\begin{aligned}\phi : \mathfrak{M}_{m,R} &\longrightarrow \mathbb{R}^3 \\ Q &\longmapsto \phi(Q)\,.\end{aligned} \tag{9.5}$$

Now define

$$\mathfrak{W}_{m,R} := \left\{ Q \in \mathfrak{M}_{m,R/2} \mid |\phi(Q)| \geq R \right\}\,. \tag{9.6}$$

This is the set of monopoles whose center is on D_R.

Proposition 9.2 *For $\nu > 0$ there is an $R_3(\nu)$ such that for $R \geq R_3(\nu)$ and $Q \in \mathfrak{W}_{m,R}$ there is a gauge transformation g with*

$$\|g(Q) - C_x\|_2 + \|\nabla_{C_x}\Big(g(Q) - C_x\Big)\|_2 \leq \text{ const } \nu$$

where $x = \phi(Q)$ and C_x denotes the almost self-dual configuration on K' constructed in Section 3.1.

This proposition has the consequence

Corollary 9.3 *Make the same assumptions as in Proposition 9.2. Then*

$$\mathfrak{W}_{m,R} \subset U_m(\nu, D_R)\,.$$

From Sections 3 and 6–8 we know

$$D_R \approx I(D_R) \approx U_m(\nu, D_R) \cap \mathfrak{W}_{m,R}$$

where $\approx$ denotes diffeomorphisms. Corollary 9.3 implies now

$$D_R \approx \mathfrak{W}_{m,R}$$

which is the content of Theorem 1.2. Furthermore Theorem 5.6 and Theorem 1.2 imply Theorem 1.1.

It remains to prove Proposition 9.2.

Proof of Proposition 9.2: Let Q be a monopole in $\mathfrak{W}_{m,R}$. Assume Q is in the gauge of Lemma 9.1. Let x be the center of Q. Denote with $\widehat{C}_{x,p}$ the mass m monopole on $\mathbb{R}^3$ in polar gauge centered at x. $\widehat{C}_{x,p}$ is explicitly given in (3.1–2). Then there is a gauge transformation g_1 on $\mathbb{R}^3$ such that

$$g_1(\widehat{Q}_1) = \widehat{C}_{x,p}\,. \tag{9.7}$$

We arranged in Lemma 9.1 that Q is approximately in polar gauge over Ω. Thus we expect g_1 to be close to the identity over Ω, or dg_1 to be close to zero over said region. Indeed for points in K' whose distance to the neck is bigger than k, we have the estimate

$$|dg_1|(x) \leq |A_{Q_1} - A_Q|(x) + |P_{Q_1}\xi|(x) + |A_Q - A_\theta|(x) + |A_Q - A_{\widehat{C}_{x,p}}|(x)\,. \quad (9.8)$$

We estimate the absolute value of P_{Q_1} by Proposition 2.8 and inequalities (9.3–4)

$$\|P_{Q_1}\xi\|_\infty \leq \text{ const } \nu\,. \quad (9.9)$$

Thus

$$|dg_1|(x) \leq \text{ const } \nu \text{ for } x \in \Omega\,. \quad (9.10)$$

Through a constant gauge transformation which differs from the identity in the C^0-norm by less than *const* ν we can arrange that at a point in $\Omega_{[k,2k]}$, $g_1 = 1$. Integration gives

$$|1 - g_1|(x) \leq \text{ const } \nu \text{ for } x \in \Omega\,. \quad (9.11)$$

Thus, in this region we can write $g_1 = \exp\chi$. Now define

$$g_2(x) := \exp\left(\alpha_k(x)\chi(x)\right)\,. \quad (9.12)$$

In the obvious way extend g_2 to a gauge transformation on K'. Set $x := \phi(Q)$ and denote by C_x the configuration constructed in Subsection 3.1.

Lemma 9.4 *Make the same assumptions as in Proposition 9.2. Then*

$$\|g_1(Q) - C_x\|_{2,D_{2k}} \leq \text{ const } \nu \quad (9.13)$$

$$\|\nabla_{C_x}(g_1(Q) - C_x)\|_{2,D_{2k}} \leq \text{ const } \nu \quad (9.14)$$

where D_{2k} denotes the points on the $\mathbb{R}^3$ part on K' whose distance to the neck is larger than $4k$.

Proof of Lemma 9.4: On the region D_{2k} we have the identity

$$g_1(Q) - C_x = g_1(Q - Q_1 + P\xi)g_1^{-1} + \widehat{C}_{x,p} - C_x\,. \quad (9.15)$$

Remark that $Q - Q_1$ is compactly supported. Use (9.13), Lemma 9.1 and Proposition 2.8 to estimate

$$\|g_1(Q) - C_x\|_{2,D_{2k}} \leq \text{ const } \nu$$

which proves (9.13). We also use (9.15) to estimate (9.14). In contrast to the previouse case we have to deal with derivatives of the gauge transformation g. Estimate

$$\begin{aligned}
&\|g_1(Q-Q_1-P\xi)g_1^{-1}\|_{C_x,D_{2k}} \leq \text{ const } \|g_1(Q-Q_1-P\xi)g_1^{-1}\|_{\widehat{C}_x,D_{2k}} \\
&\leq \text{ const } \left(\| [g_1^{-1}d^1_{\widehat{C}_x}g_1, Q-Q_1-P\xi] \|_{2,D_{2k}} + \|Q-Q_1-P\xi\|_{\widehat{C}_x,D_{2k}}\right) \\
&\leq \text{ const } m^{-1/2}\left(\|d^1_{\widehat{C}_x}g_1\|_{6,D_{2k}} + \|d^1_{\widehat{C}_x}g_1\|^{1/2}_{6,D_{2k}} \| [\Phi, g_1^{-1}d^1_{\widehat{C}_x}g_1] \|^{1/2}_{2,D_{2k}}\right) \\
&\quad \times (\|Q-Q_1\|_c + \|\xi\|_2) + \text{ const } \|Q-Q_1-P\xi\|_{\widehat{C}_x,D_{2k}} \, .
\end{aligned} \tag{9.16}$$

Remark that

$$\|d^1_{\widehat{C}_x}g_1\|_{6,D_{2k}} + \| [\Phi, g_1^{-1}d^1_{\widehat{C}_x}g_1] \|_{2,D_{2k}} \leq \text{ const } R^{-1} \, . \tag{9.17}$$

Thus

$$(9.16) \leq \text{ const } \nu$$

and

$$\|\nabla_{C_x}(g_1(Q)-C_x)\|_{2,D_{2k}} \leq \text{ const } \nu \, .$$

Lemma 9.5 *Make the same assumptions as in Proposition 9.2. Then*

$$\|g_2(Q)-C_x\|_{2,\Omega_{[k,2k]}} \leq \text{ const } \nu \tag{9.18}$$

$$\|\nabla_{C_x}(g_2(Q)-C_x)\|_{2,\Omega_{[k,2k]}} \leq \text{ const } \nu \, . \tag{9.19}$$

Proof of Lemma 9.5: To shorten the notation denote in this proof $\Omega_{[k,2k]}$ by Ω. By (9.10)

$$\|d\chi\|_{\infty,\Omega} \leq \text{ const } \nu$$

and

$$\|\chi\|_{\infty,\Omega} \leq \text{ const } \nu$$

Using this and Lemma 9.1

$$\|g_2(Q)-C_x\|_{2,\Omega} \leq \| [g_2, Q] \|_{2,\Omega} + \|dg_2g_2^{-1}\|_{2,\Omega} + \|Q-C_x\|_{2,\Omega} \leq \text{ const } \nu \, . \tag{9.20}$$

Remark that

$$\|dg_2g_2^{-1}\|_{c,\Omega} \leq \text{ const } \nu$$

and

$$\|d^1_cg_2\|_{c,\Omega} \leq \text{ const } \nu$$

and estimate

$$\begin{aligned}
\|g_2(Q)-C_x\|_{C_x,\Omega} &\leq \| [d^1_cg_2, Q] \|_{C_x,\Omega} + \| [g_2, \nabla_cQ] \|_{C_x,\Omega} \\
&\quad + \| [g_2, Q]g_2^{-1}d^1_cg_2g_2^{-1}\|_{C_x,\Omega} \leq \text{ const } \nu \, .
\end{aligned} \tag{9.21}$$

(9.20) and (9.21) imply Lemma 9.5.

10 Small mass moduli space

Let $Q = (\Phi, A)$ be a small mass monopole. As F_A is small, A has to be close to a flat connection. Let $\mathfrak{R}(K')$ denote the space of flat connections on K' up to gauge equivalence. $\mathfrak{R}(K')$ splits into two pieces

$$\mathfrak{R}(K') = \theta \coprod V$$

where θ is the trivial flat connection and V is a compact set parametrizing irreducible flat connections.

For $b \in V$ denote by C_b the flat configuration $(0, b)$. Let $C_\theta = (0, \theta)$.

Lemma 10.1 *For $\nu > 0$ there is an $m_0(\nu)$ such that for $m \leq m_0(\nu)$ and $Q \in \mathfrak{M}_m(K')$ there is an $a \in \mathfrak{R}(K')$ with*

$$|Q - C_a|(x) + |\nabla_{C_a}(Q - C_a)|(x) \leq \nu \text{ for } x \in B'_{4k}$$

where k is the constant from Subsection 3.2. $B'_{4k} \subset K'$ is the connected sum of the ball of radius $4k$ centered at the origin in $\mathbb{R}^3$ with K.

Lemma 10.1 follows from the weak compactness theorem. Lemma 10.1 implies that the space of small mass monopoles splits up into several sectors. This leads to the definition:

Definition 10.2 *Make the same assumptions as in Lemma 10.1. Define $\mathfrak{M}_{m,\theta}(K')$ and $\mathfrak{M}_{m,V}(K')$ to be the set of monopoles which are respectively close to C_θ, and to C_b for a $b \in V$, in the sense of Lemma 10.1. We can write $\mathfrak{M}_m(K')$ as a disjoint union*

$$\mathfrak{M}_m(K') = \mathfrak{M}_{m,\theta}(K') \coprod \mathfrak{M}_{m,V}(K').$$

First we investigate the sector $\mathfrak{M}_{m,\theta}(K')$. This is the part of the small mass moduli space whose connection part is close to θ. Most of the energy of small mass monopoles is located on the $\mathbb{R}^3$ part of K'. For $Q \in \mathfrak{M}_m(K')$ we would like to define its center. Using cutoffs we get for Q an almost self-dual, charge one configuration Q_1 on $\mathbb{R}^3$. We define the center of Q to be the center of the self-dual perturbation of Q_1.

To define Q_1 we have to interpolate between the connection part of Q and the flat connection θ over Ω_k. For the same reason as in Subsection 3.2 we transform Q into the exponential gauge over the annulus Ω_k. Choose an $x_0 \in \Omega_k$ and define

$$\begin{aligned} Q_1(y) &:= (\Phi_1(y), A_1(y)) \\ &:= \Big(\alpha_k(y)\Phi(y) + (1 - \alpha_k(y))\Phi(x_0), \alpha_k(y)A(y) + (1 - \alpha_k(y))\theta\Big) \end{aligned} \tag{10.1}$$

Lemma 10.3

$$|G_{Q_1}|(y) + |\nabla_{Q_1} G_{Q_1}|(y) \leq \text{ const } m^2 .$$

Lemma 10.3 is proved in a similar way as (3.22–25).

As Q_1 is a configuration on $\mathbb{R}^3$, Lemma 10.3 implies that Q_1 is almost self-dual.

Let $\widehat{Q}_1 = Q_1 + P\xi$ be its self-dual perturbation on $\mathbb{R}^3$. $\widehat{Q}_1$ has a unique center; denote it by x_Q. This gives a center for all small mass monopoles. Express this by the map x

$$x : \mathfrak{M}_m(K') \longrightarrow \mathbb{R}^3 \qquad (10.2)$$

$$Q \longmapsto x_Q .$$

Now we can proceed in a similar way as in Section 9 to investigate the neighborhood of $\mathfrak{M}_{m,\theta}(K')$. This leads to

Proposition 10.4 *For $\nu > 0$ there is an $m_0(\nu)$ such that for $m \leq m_0$ and $Q \in \mathfrak{M}_{m,\theta}(K')$ there is a gauge transformation g with*

$$\|g(Q) - C_x\|_2 + \|\nabla_{C_x}\Big(g(Q) - C_x\Big)\|_2 \leq \text{ const } \nu$$

where x is the center of Q and C_x is the configuration constructed in Section 3.2.

Remark that Proposition 10.3 implies $\mathfrak{M}_{m,\theta}(K') \subset U_m(\nu, \mathbb{R}^3)$. Together with the results from Sections 7 and 8 this shows

Corollary 10.5 *There is an m_0 such that for $m \leq m_0$, $\mathfrak{M}_{m,\theta}(K')$ is diffeomorphic to $\mathbb{R}^3$.*

The following proposition concludes the proof of Theorem 1.3.

Proposition 10.6 *The set $\mathfrak{M}_{m,V}(K')$ is compact.*

Proof of Proposition 10.6: Let Q_j be a sequence of monopoles in $\mathfrak{M}_m(K')$ whose centers drift out on K'. By the weak compactness theorem Q_j converges weakly to a flat mass m monopole. Up to gauge equivalence $C_{\theta,m} = (\frac{i}{2}m\sigma^3, \theta)$ is the only flat mass m monopole on K'. Thus Q_j converges in C^∞ to $C_{\theta,m}$. From this we conclude that $Q_j \in \mathfrak{M}_{m,\theta}(K')$ and $\mathfrak{M}_{m,V}(K')$ is a compact set.

References

[D1] Donaldson, S.K., *An Application of Gauge Theory to four dimensional Topology*. Jour. Diff. Geom. 18 (1983), 279–315

[D2] Donaldson, S.K. (1986), Connections, Cohomology and the Intersection Forms of 4-Manifolds. Jour. Diff. Geom. 24, 275–341

[DK] Donaldson, S.K., Kronheimer P.B., *The Geometry of Four-Manifolds*. Oxford University Press, 1990.

[E1] Ernst, K. (1992).: Three Manifolds with small negative Ricci Curvature and asymptotically Euclidean Ends; preprint

[E2] Ernst, K. (1992).: The Ends of the Monopole Moduli Space over $\mathbb{R}^3$#(Homology Sphere): Part II; in this volume.

[F1] Floer, A. (1987).: Monopoles on asymptotically Euclidean 3-Manifolds. Bulletin AMS Volume 16, 125.

[F2] Floer, A. (1987).: The configuration space of Yang-Mills-Higgs theory on asymptotically flat manifolds; in this volume.

[F3] Floer, A. (1987).: Construction of Monopoles on asymptotically flat Manifolds; in this volume.

[G] Groisser, D. (1984) Integrality of the monopole number in $SU(2)$ Yang-Mills-Higgs theories on $\mathbb{R}^3$. Commun. Math. Phys. 93, 367

[JT] Jaffe, A., Taubes, C., (1980). *Vortices and Monopoles*. Boston: Birkhäuser.

[T1] Taubes, C. (1982). The existence of a non-minimal solution to the SU(2) Yang-Mills-Higgs equations on $\mathbb{R}^3$. Part 1. Commun. Math. Phys. 86, 257; Part 2. Commun. Math. Phys. 86, 299.

[T2] Taubes, C. (1983). Stability in Yang-Mills Theories. Commun. Math. Phys. 91, 235–263.

[T3] Taubes, C. (1982). Self-dual connections on non-self-dual 4-manifolds. Jour. Diff. Geom. 17, 139–170.

[T4] Taubes, C. (1989). A Note and Erratum Concerning «Min-Max Theory for the Yang-Mills-Higgs Equations». Commun. Math. Phys. 122, 609–613.

[T5] Taubes, C. (1984). Self-dual Connections on 4-Manifolds with indefinite Intersection Matrix. Jour. Diff. Geom. 19, 517–560.

[U] Uhlenbeck, K. (1982). Removable Singularities in Yang-Mills fields. Commun. Math. Phys. 83, 11–29.

The ends of the monopole moduli space over $\mathbb{R}^3\#$ (homology sphere): Part II

K.D. Ernst

Abstract. Let K be a homology three-sphere with respect to integer coefficients. With K' we denote the connected sum of K and $\mathbb{R}^3$. Let $\mathfrak{M}(K')$ be the moduli space of charge one monopoles on K' in the Prasad-Sommerfield limit. We continue here our study of the ends of $\mathfrak{M}(K')$. Especially we give a complete description of the moduli space of small mass monopoles.

1 Introduction

Let K' denote the connected sum of the homology three-sphere K with $\mathbb{R}^3$. In this paper we continue the study of the ends of the moduli space of monopoles on K'.

$\mathfrak{M}(K')$ has three types of ends. These ends consist of monopoles with large mass, monopoles whose energy is located far away from the neck and with small mass. The first two types of ends are completely described in [E] (to which we will also refer as Part I). We give here a complete description of the small mass moduli space.

For an introduction to monopoles see [JT]. The beginning of Part I contains some brief remarks about monopoles. Through this paper we will use the same notation as in Part I.

The author would like to thank C. Taubes for suggesting this problem and for helpful remarks.

1.1 Statement of results

The value of the YMH-functional of small mass monopoles is small. Thus the L^2-norm of F_A is small and A has to be close to a flat connection. Denote the space of flat connections up to gauge equivalence by $\mathfrak{R}(K')$. A priori, $\mathfrak{R}(K')$ has the form

$$\mathfrak{R}(K') = \theta \coprod V. \tag{1.1}$$

Here θ is the trivial flat connection and V is a compact set parametrizing irreducible flat connections. We showed in Part I:

Theorem 1.1 *There is an m_0 such that for $m \leq m_0$ the moduli space $\mathfrak{M}_m(K')$ splits up into two well separated pieces:*

$$\mathfrak{M}_m(K') = \mathfrak{M}_{m,\theta}(K') \coprod \mathfrak{M}_{m,V}(K') \tag{1.2}$$

where $\mathfrak{M}_{m,\theta}(K')$ is diffeomorphic to $\mathbb{R}^3$ and $\mathfrak{M}_{m,V}(K')$ is a compact set.

$\mathfrak{M}_{m,\theta}(K')$ and $\mathfrak{M}_{m,V}(K')$ consist of small mass monopoles whose connection part is close to the trivial flat connection or to an irreducible flat connection, respectively.

The purpose of this paper is to investigate $\mathfrak{M}_{m,V}(K')$. We impose a further assumption on K':

Definition 1.2 *We say $\mathfrak{R}(K')$ is generic if $\mathfrak{R}(K')$ consists of a finite number of points, i.e.*

$$\mathfrak{R}(K') = \{\theta, b_1, \ldots, b_{k_0}\} \tag{1.3}$$

where θ is the trivial flat connection and b_j, for $1 \leq j \leq k_0$, are the irreducible flat connections.

We assume from now on that $\mathfrak{R}(K')$ is generic. Our purpose is to prove:

Theorem 1.3 *There is an m_0 such that for $m \leq m_0$, $\mathfrak{M}_{m,V}(K')$ splits up into well separated pieces*

$$\mathfrak{M}_{m,V}(K') = \mathfrak{M}_{m,1}(K') \coprod \mathfrak{M}_{m,2}(K') \coprod \cdots \coprod \mathfrak{M}_{m,k_0}(K') \tag{1.4}$$

where the connection part of the elements in $\mathfrak{M}_{m,j}(K')$ is close to $b_j$up to a gauge transformation. Furthermore for $1 \leq j \leq k_0$, $\mathfrak{M}_{m,b_j}$ is diffeomorphic to $SO(3)$.

The proof of Theorem 1.3 is completed in Section 9.

1.2 Outline of the proofs

Let $C = (\Phi, A)$ be a monopole configuration. As in Part I, we denote by δ_C the operator

$$\delta_C : L_C(\Omega^0 \oplus \Omega^1) \longrightarrow L^2(\Omega^0 \oplus \Omega^1) \tag{1.5}$$

$$\delta_C(\Psi_0, \Psi_1) = \left(-[\Phi, \Psi_0] + d_A^* \Psi_1, d_A \Psi_0 - [\Phi, \Psi_1] - *d_A \Psi_1\right).$$

By C_j we denote the flat mass zero configuration $(0, b_j)$ on K'. In Section 2 we show that the operator $\delta_{C_j} \delta_{C_j}^*$ has a 3-dimensional kernel.

Let C_x be the charge one, mass m monopole on $\mathbb{R}^3$ centered at x. In order to glue C_x to C_j over K', we need to identify two $su(2)$-bundles. The space $SO(3)$ parametrizes such bundle identifications. Denote the resulting configuration on K' by $C^1_{j,x,\rho}$ where ρ is the identification parameter in $SO(3)$.

$C^1_{j,x,\rho}$ is a mass m charge one configuration on K' whose energy is centered at x. $G_C := d_A \Phi - *F_A$ is the anti-self-dual part of a configuration. The L^2 and the $L^{6/5}$ norms of the anti-self-dual part of $C^1_{j,x,\rho}$ are small. More precisely $\|G_{C^1_{j,x,\rho}}\|_2$ and $\|G_{C^1_{j,x,\rho}}\|_{6/5}$ are of order $O(m^2)$.

We show in Section 3 that the operator $\delta_{C^1_{j,x,\rho}} \delta^*_{C^1_{j,x,\rho}}$ has three small eigenvalues. Let H_C be the 3-dimensional space spanned by the corresponding eigenvectors. This is in sharp contrast to Part I where the lowest eigenvalue of the operators $\delta_C \delta_C^*$ arising from monopole configurations C was always uniformly bounded away from zero.

To perturb $C^1_{j,x,\rho}$ to a self-dual configuration, we need to invert the operator $\delta_{C^1_{j,x,\rho}} \delta^*_{C^1_{j,x,\rho}}$. Thus H_C occurs as an obstruction to the self-dual perturbation. Section 4 discusses the perturbation of $C^1_{j,x,\rho}$ to a configuration $C+P\xi$ whose anti-self-dual part $G_{C+P\xi}$ vanishes on the orthogonal complement of H_C.

To obtain a self-dual configuration, the anti-self-dual part $G_{C+P\xi}$ has to vanish on H_C. We can formulate this in terms of the vanishing of a section of a bundle over $\mathbb{R}^3 \times SO(3)$. This is done in Section 5.

In Section 6 we establish the existence of a subset N of the parameter space $\mathbb{R}^3 \times SO(3)$ on which the section arising from the obstruction vanishes. Furthermore N is diffeomorphic to $SO(3)$. This constructs a family of monopoles parametrized by $SO(3)$. These monopoles are centered close to the neck of K'.

The rest of this paper adresses the moduli space problem. As in Part I we use here the gauge-fixing method developed in [D] and [DK].

Let $\mathcal{U}^1_{m,j}(\nu)$ be a tubular neighborhood of the almost self-dual configurations constructed in Section 6. ν labels the «thickness» of this neighborhood. In Section 7 we prove that the configurations in $\mathcal{U}^1_{m,j}(\nu)$ can be expressed in terms of a special gauge.

We use this gauge fixing to show that $\mathcal{U}^1_{m,j}(\nu) \cap \mathfrak{M}_m(K')$ is diffeomorphic to $SO(3)$. It remains to show that the corresponding sector of the moduli space is contained in the neighborhood

$$\mathfrak{M}_{m,j}(K') \subset \mathcal{U}^1_{m,j}(\nu)\,. \tag{1.6}$$

To prove (1.6) we have to show in particular that monopoles in $\mathfrak{M}_{m,j}(K')$ are «centered» close to the neck of K'.

Small mass monopoles also have a small energy. Thus $F_A = d_A\Phi \approx 0$. This implies that A is close to a flat connection, say $A \approx b_j$, and that $d_{b_j}\Phi \approx 0$. As b_j is irreducible, Φ has to be close to zero over a neighborhood of the neck.

A charge one configuration which is centered far away from the neck and whose Higgs field is close to zero over a neighborhood of the neck requires more energy than a charge one monopole which is centered at the neck. In the space of charge one configurations, monopoles have the smallest energy. From that piont of view we could say that an attractive force acts on elements of $\mathfrak{M}_{m,j}(K')$. This attractive force has the consequence that all monopoles in $\mathfrak{M}_{m,j}(K')$ are centered at the neck.

Rather than computing this attractive force directly, we work in Section 8 and 9 with the obstruction to self-duality to show that $Q \in \mathfrak{M}_{m,j}(K')$ is centered at the neck.

As b_j is an irreducible connection there is no flat mass m monopole with $m > 0$ whose connection part equals b_j. We exploit this fact in Section 8 where we show that any charge zero configuration whose energy is small and whose connection part is close to b_j cannot be close to a self-dual configuration. In particular such a configuration has an obstruction to self-duality which has at least the size μm where $\mu > 0$ is a universal constant.

In Section 9 we use the results from Section 8 to show that the Higgs field of $Q \in \mathfrak{M}_{m,j}(K')$ has to be close to zero in a neighborhood of the neck. This implies that Q is «centered» close to the neck. Furthermore this implies that Q is contained in the neighborhood $\mathcal{U}^1_{m,j}(\nu)$. Thus $\mathfrak{M}_{m,j}(K')$ is indeed diffeomorphic to $SO(3)$.

2 Spectra and charge zero configurations

In Definition 1.2 we assumed that $\mathfrak{R}(K')$, the space of flat connections on K', consists up to gauge equivalence of a finite set of points

$$\mathfrak{R}(K') = \{\theta, b_1, \ldots, b_{k_0}\}. \tag{2.1}$$

We choose here the b_j's in a gauge such that they restrict to the trivial flat connection on the $\mathbb{R}^3$ part of K'.

Define the flat, charge zero, mass zero monopoles

$$C_j = (0, b_j). \tag{2.2}$$

Proposition 2.1 *The operator*

$$\delta^*_{C_j} : L_{C_j}(\Omega^1) \longrightarrow L^2(\Omega^0 \oplus \Omega^1)$$

has a 3-dimensional kernel.

In the proof of Proposition 2.1 we use following lemma:

Lemma 2.2 *Given a $\sigma \in su(2)$ there exists a function $u : K' \to su(2)$ with*

$$\nabla^2_{b_j} u = 0 \tag{2.3}$$

and

$$\lim_{|x| \longrightarrow \infty} u(x) = \sigma. \tag{2.4}$$

Proof of Lemma 2.2: Denote by L_{b_j} the completion of $C_0^\infty(\Omega^0)$, the smooth compactly supported functions on K', in the norm

$$\|v\|_{b_j} := \|\nabla_{b_j} v\|_2 .$$

Denote with r_0 the radius of the neck of K'. Let α_0 be a cutoff which interpolates between 0 and 1 over the annulus $[2r_0, 4r_0]$. Take a $\sigma \in su(2)$, $v \in L_{b_j}$ and set

$$u = v + \alpha_0\,\sigma . \tag{2.5}$$

Consider following functional on L_{b_j}:

$$S[v] := \int_{K'} |\nabla_{b_j} v|^2 + 2(\nabla_{b_j} v, d\alpha_0\,\sigma) . \tag{2.6}$$

At least formally, critical points of the functional S satisfy the equation (2.3). By the calculus of variations (c.f. Chapter 6 of [JT]) and the elliptic regularity arguments, there exists a $v \in L_{b_j}$ with $\nabla_{b_j}^2 v + d^* d\alpha_0\,\sigma = 0$. This implies Lemma 2.2.

For $1 \leq k \leq 3$, let $\sigma_k \in su(2)$ be the Pauli spin matrices. For $\sigma_k \in su(2)$, denote the corresponding function u from Lemma 2.2 by u_k. Set

$$\eta_{0,k} := \nabla_{b_j} u_k = d_{b_j} v + d\alpha_0 \sigma_k . \tag{2.7}$$

Note that

$$\delta_{C_j}^* \eta_{0,k} = 0$$

The next lemma implies that the kernel of $\delta_{C_j}^*$ is nonempty.

Lemma 2.3 *The one form $\eta_{0,k}$ is contained in L_{C_j}. Furthermore $\eta_{0,k}$ is unequal to zero.*

Proof of Lemma 2.3: The first statement follows from (2.5) and the fact that $v \in L_{b_j}$. To prove the second statement we assume the contrary, namely that $\eta_{0,k} = 0$. Thus on the $\mathbb{R}^3$ part of K', u_k fulfills the equation $du_k = 0$. Thus u_k equals σ_k on this region. Now we glue K to S^3 instead of $\mathbb{R}^3$ and we obtain a compact manifold K''. On K'' the function u_k is unequal to zero and fulfills the equation $d_{b_j} u_k = 0$. As b_j is an irreducible connection this is impossible and gives the desired contradiction.

We will use the following proposition to show that the kernel of the operator $\delta_{C_j}^*$ is exactly 3-dimensional. Furthermore this proposition will be used in Section 6.

Proposition 2.4 *For $1 \leq k \leq 3$ there is a vector $\vec{a}_k \in \mathbb{R}^3$, $\vec{a}_k \neq 0$ such that for* $|\vec{x}| \gg \max(1, r_0)$

$$\eta_{0,k}(\vec{x}) = d_{b_j} v_k(\vec{x}) = \frac{\vec{a}_k \cdot \vec{\sigma}}{|\vec{x}|^2} + O(|\vec{x}|^{-3}). \tag{2.8}$$

Furthermore the vectors $\vec{a}_k \in \mathbb{R}^3$ are a basis for $\mathbb{R}^3$.

Proof of Proposition 2.4: Let α_1 be a cutoff which interpolates between 0 and 1 over the annulus $[4r_0, 8r_0]$. Remark that $d\alpha_0$ and α_1 have disjoint supports.

$$\delta_{C_j}\delta^*_{C_j}(\alpha_1\eta_{0,k}) = \Delta\alpha_1\eta_{0,k} \tag{2.9}$$

Apply the Greens function for the 3-dimensional Laplacian to both sides of (2.9).

$$\alpha_1\eta_{0,k}(x) = \int \frac{\Delta\alpha_1(y)\eta_{0,k}(y)}{|x-y|}. \tag{2.10}$$

Integration by parts yields

$$\alpha_1\eta_{0,k}(x) = \int \frac{*\Big(\nabla\alpha_1(y) \wedge *\eta_{0,k}(y)\Big)}{|x-y|^2} dy. \tag{2.11}$$

Define the radially symmetric part of η_0 as

$$\widehat{\eta}_{0,k}(\vec{x}) := \eta_{0,k}(|\vec{x}|\vec{n}_1) \tag{2.12}$$

where $\vec{n}_1$ is the unit normal in the x_1-direction. By (2.11)

$$\begin{aligned} &\alpha_1\Big(\eta_{0,k}(\vec{x}) - \widehat{\eta}_{0,k}(\vec{x})\Big) \\ &= \int \nabla\alpha(\vec{y})\eta_{0,k}(\vec{y})\left(\frac{1}{|\vec{x}-\vec{y}|^2} - \frac{1}{|\,|\vec{x}|\vec{n}_1 - \vec{y}|^2}\right)dy \leq \text{ const } |\vec{x}|^{-3} \end{aligned} \tag{2.13}$$

Furthermore

$$\begin{aligned} \widehat{\eta}_{0,k}(\vec{x}) &= \int \nabla\alpha(\vec{y})\widehat{\eta}_{0,k}(\vec{y})\frac{1}{|\,|\vec{x}|\vec{n}_1 - \vec{y}|^2}dy \\ &\leq \frac{1}{|\vec{x}|^2}\int \nabla\alpha(\vec{y})\eta_{0,k}(\vec{y})dy + O(|x|^{-3}). \end{aligned} \tag{2.14}$$

From (2.13) and (2.14) follows

$$\eta_{0,k}(\vec{x}) = \frac{1}{|\vec{x}|^2}\int \nabla\alpha(\vec{y})\eta_{0,k}(\vec{y})dy + 0(|\vec{x}|^{-3}). \tag{2.15}$$

For $|\vec{x}| \gg \max(1, r_0)$, define

$$\vec{a}_k := \int \nabla\alpha_1(y)\eta_{0,k}(y)dy\,. \tag{2.16}$$

In the next step we show that the vectors $\vec{a}_k$, $1 \leq j \leq 3$ are a basis for $\mathbb{R}^3$. Assume that

$$\sum_k \lambda_k \vec{a}_k = 0\,.$$

By (2.15)

$$\sum_k \lambda_k \eta_{0,k}(x) = O(|\vec{x}|^{-3})\,. \tag{2.17}$$

Integration by parts gives

$$\begin{aligned} \|\sum_k \lambda_k \eta_{0,k}\|_2^2 &= \Big(\sum_k \lambda_k d_{b_j}^* d_{b_j} u_k, \sum_l \lambda_l u_l\Big) \\ &+ \lim_{r\to\infty} \int_{S_r} \Big(\sum_k \lambda_k \eta_{0,k}, \sum_l u_l\Big) r^2\, d\Omega \end{aligned} \tag{2.18}$$

The first term on the RHS of (2.18) is zero as $d_{b_j}^* d_{b_j} u_k = 0$. From (2.17) it follows that the second term is also zero. Thus $\sum_k \lambda_k \eta_{0,k} = 0$. This implies $\lambda_k = 0$ for $1 \leq k \leq 3$ and the vectors $\vec{a}_k$ are a basis for $\mathbb{R}^3$. This concludes the proof of Proposition 2.4.

The next lemma completes the proof of Proposition 2.1.

Lemma 2.5 *The kernel of $\delta_{C_j}^*$ is at most 3-dimensional.*

Proof of Lemma 2.5: Let $\Psi \in L_{C_j}(\Omega^0 \oplus \Omega^1)$ with $\delta_{C_j}^* \Psi = 0$. By Definition 1.2 the set of configurations on K' consists up to gauge equivalence of a finite set of points. This implies that the first cohomology group of K' with coefficients in b_j is zero. In other words the first cohomology of the complex

$$\Omega^0 \xrightarrow{d_{b_j}} \Omega^1 \xrightarrow{d_{b_j}} \Omega^2$$

is zero. This is also denoted by $H_{b_j}^1(K') = 0$. This has the consequence that $\Psi = d_{b_j} u$ for a certain u with $u \in L_{1,loc}^2$ and $du \in L^2$. A function u with these properties can be written in the form

$$u = u_0 + \sum_k l_k \sigma_k \tag{2.19}$$

where $u_0 \in L_{b_j}$ and l_k are real numbers. The space L_{b_j} is defined in the proof of Lemma 2.2. Assume for the moment that the l_k's are zero. Thus $u \in L_{b_j}$. Integration by parts yields

$$0 = (d^*_{b_j} d_{b_j} u, u) = \|d_{b_j} u\|_2^2 + \lim_{r\to\infty} \int_{|x|=r} (u, \Psi)\, r^2\, d\Omega\,. \tag{2.20}$$

The assumption $u = u_0$ and Proposition 2.4 imply that the last term on the RHS of (2.20) is zero. Thus $d_{b_j} u = 0$. This shows that the kernel of $\delta^*_{C_j}$ is spanned by the one forms constructed in Lemma 2.2.

3 Spectra and charge one configurations

First we construct charge one configurations whose connection part is close to a nontrivial flat connection. These small mass configurations are parametrized by $\mathbb{R}^3 \times SO(3)$ where the $\mathbb{R}^3$ parameter determines the center of energy and the $SO(3)$ parameter identifies two $su(2)$-bundles.

For our purposes it is enough to study configurations which are parametrized by $B_r \times SO(3)$ where $B_r \subset \mathbb{R}^3$ is the ball of radius r centered at the origin. $r \ll m^{-1}$, the exact size of r will be determined in Section 7.

Let $x \in \mathbb{R}^3$ with $|x| \leq r$. We start with the charge one monopole on $\mathbb{R}^3$ centered at x in cartesian gauge. Denote it by $\widehat{C}_{x,c} = (\widehat{\Phi}_{x,c}, \widehat{A}_{x,c})$. The explicit formula for $\widehat{C}_{x,c}$ is given in (3.3–4) of Part I.

Choose $R \gg 1$ with $Rm \ll 1$. The exact size of R will be determined in Section 6. Proposition 3.2 will also give a restriction on the size of R. As in Section 2 let α_R be a cutoff which interpolates between 0 and 1 over the annulus of radii $[R, 2R]$ centered at the origin. Define

$$C_{1,x}(y) := \left(\alpha_R(y)\widehat{\Phi}_{x,c}(y), \alpha_R(y)\widehat{A}_{x,c}(y) - (1-\alpha_R(y))\theta\right). \tag{3.1}$$

Note that

$$\|G_{C_{1,x}}\|_p \leq \text{ const } m^2 R^{3/p-1} \tag{3.2}$$

and

$$\|\nabla_{C_{1,x}} G_{C_{1,x}}\|_2 \leq \text{ const } m^2 R^{-1/2}. \tag{3.3}$$

Now pull the configuration $C_{1,x}$ back to K' and glue it over K' to C_j. This gluing requires the identification of two $su(2)$-bundles. Such identifications are described by constant gauge transformations or alternatively by elements of $SO(3)$. For $\rho \in SO(3)$ denote the corresponding charge one configuration on K' by $C^1_{j,x,\rho}$. Denote by $C_{j,\rho}$ the charge zero configuration we obtain after applying the corresponding constant gauge transformation to C_j.

For $1 \leq k \leq 3$ let $\eta_{0,k,\rho}$ span the 3-dimensional kernel of $\delta_{C_{j,\rho}}\delta^*_{C_{j,\rho}}$. Note that

$$\eta_{0,k,\rho}(\vec{x}) = \frac{\vec{a_k} \cdot \rho(\vec{\sigma})}{|\vec{x}|^2} + O(|\vec{x}|^{-3}). \tag{3.4}$$

We can also obtain $C^1_{j,x,\rho}$ by gluing $C_{1,x}$ to $C_{j,\rho}$ using the trivial bundle identification.

This gives the map

$$J : B_r \times SO(3) \longrightarrow \mathscr{C}_m(K') \tag{3.5}$$
$$(x, \rho) \longmapsto C^1_{j,x,\rho}.$$

Note that $C^1_{j,x,\rho}$ fulfills (3.2–3).

In order to perturb $C^1_{j,x,\rho}$ to a self-dual configuration we have to identify the small eigenvalues of the operator $\delta_{C^1_{j,x,\rho}}\delta^*_{C^1_{j,x,\rho}}$. We will make use of the following definition:

Definition 3.1 *We say $C \in \mathscr{C}_m(K')$ fulfills the Spectral Property I, if there is an L_C-orthogonal projection Π_C onto a 3-dimensional subspace of $L_C(\Omega^0 \oplus \Omega^1)$ with the following properties:*

1. *For $\gamma > 0$ there is an $m(\gamma)$ such that for $m \leq m(\gamma)$ and for $\Psi \in \Pi_C L_C(\Omega^0 \oplus \Omega^1)$*
$$\|\delta^*_C \Psi\|_2 < \gamma \|\nabla_C \Psi\|_2 \tag{3.6}$$
2. *There is a constant $\mu_1 > 0$ and an m_1 such that for $m \leq m_1$ and $\Psi \in \Pi^\perp_C L_C(\Omega^0 \oplus \Omega^1)$*
$$\|\delta^*_C \Psi\|_2 \geq \mu_1 \|\nabla_C \Psi\|_2 . \tag{3.7}$$

Proposition 3.2 *The configurations $C^1_{j,x,\rho}$ fulfill the Spectral Property I described in Definition 3.1.*

Proof of Proposition 3.2: Choose $\epsilon > 0$ and $L \gg 1$ with $\epsilon L \geq 2r_0$, where r_0 is the radius of the neck of K'. Furthermore we impose the condition $\epsilon^{-1}L \leq R$, where R is the radius introduced at the beginning of this section. We will choose L and ϵ in this proof. These choices can be made independently of m and C. Let α_L be a cutoff which interpolates between 0 and 1 over the annulus $[\epsilon L, \epsilon^{-1}L]$. Let $\beta_L := 1 - \alpha_L$. As explained in Lemma 3.3 of Part I, we can assume that the cutoff α_L has the property

$$\|\nabla \alpha_L\|_3 \leq \text{ const } \left(\log \epsilon^{-1}\right)^{-2/3}. \tag{3.8}$$

To shorten the notation, we introduce the abbreviation

$$l(\epsilon) := \text{ const } \left(\log \epsilon^{-1}\right)^{-2/3}. \tag{3.9}$$

Lemma 3.3 *For* $1 \leq k \leq 3$, *let* $\eta_{0,k,\rho}$ *be a basis of the kernel of* $\delta_{C_{j,\rho}}\delta^*_{C_{j,\rho}}$. *Then*

$$\|\nabla_{C^1_{j,x,\rho}}(\beta_L \eta_{0,k,\rho})\| \geq \Big(1 - l(\epsilon)\Big)\|\nabla_{C^1_{j,x,\rho}}\eta_{0,k,\rho}\|_2 .$$

Proof of Lemma 3.3: By the Weitzenböck formula

$$\begin{aligned}\|\nabla_{C^1_{j,x,\rho}}(\alpha_L \eta_{0,k,\rho})\|_2^2 &= \|\delta^*_{C^1_{j,x,\rho}}(\alpha_L \eta_{0,k,\rho})\|_2^2 + \Big(\{G_{C^1_{j,x,\rho}}, \alpha_L \eta_{0,k,\rho}\}, \alpha_L \eta_{0,k,\rho}\Big)\\ &\leq \|\nabla \alpha_L\|_3^2 \|\eta_{0,k,\rho}\|_6^2 + \|G_{C^1_{j,x,\rho}}\|_{3/2}\|\alpha_L \eta_{0,k,\rho}\|_6^2.\end{aligned}$$

It follows that

$$\|\nabla_{C^1_{j,x,\rho}}(\alpha_L \eta_{0,k,\rho})\|_2 \leq l(\epsilon)\|\nabla_{C^1_{j,x,\rho}}\eta_{0,k,\rho}\|_2$$

and

$$\begin{aligned}\|\nabla_{C^1_{j,x,\rho}}(\beta_L \eta_{0,k,\rho})\|_2 &\geq \|\nabla_{C^1_{j,x,\rho}}\eta_{0,k,\rho}\|_2 - \|\nabla_{C^1_{j,x,\rho}}(\alpha_L \eta_{0,k,\rho})\|_2\\ &\geq \Big(1 - l(\epsilon)\Big)\|\nabla_{C^1_{j,x,\rho}}\eta_{0,k,\rho}\|_2\end{aligned}$$

which proves Lemma 3.3.

Consider now

$$\begin{aligned}\|\delta^*_{C^1_{j,x,\rho}}(\beta_L \eta_{0,k,\rho})\|_2 &\leq \|\nabla \beta_L\|_3 \|\eta_{0,k,\rho}\|_6 \leq l(\epsilon)\|\nabla_{C^1_{j,x,\rho}}\eta_{0,k,\rho}\|_2\\ &\leq \frac{l(\epsilon)}{1 - l(\epsilon)}\|\nabla_{C^1_{j,x,\rho}}(\beta_L \eta_{0,k,\rho})\|_2.\end{aligned}$$

For any $\gamma > 0$ we can choose an $m(\gamma)$ and an R such that $\epsilon^{-1}L \leq R$ and $Rm^{-1} \ll 1$. This proves that $C^1_{j,x,\rho}$ fulfills (3.5).

We prove that $C^1_{j,x,\rho}$ fulfills (3.6) by assuming the opposite and deriving a contradiction. Assume that there is a vector Ψ which is $\nabla_{C^1_{j,x,\rho}}$-orthogonal to the vectors $\eta_{0,k,\rho}$ and for which (3.6) is not satisfied. In other words, for $\gamma > 0$ there is an $m(\gamma)$ such that

$$\|\delta^*_{C^1_{j,x,\rho}}\Psi\| \leq \gamma\|\nabla_{C^1_{j,x,\rho}}\Psi\|_2. \tag{3.10}$$

As in Lemma 3.3, we show

Lemma 3.4

$$\|\nabla_{C^1_{j,x,\rho}}(\beta_L \Psi)\|_2 \geq \Big(1 - l(\epsilon)\Big)\|\nabla_{C^1_{j,x,\rho}}\Psi\|_2 .$$

Now estimate

$$\begin{aligned}\|\delta^*_{C_j}(\beta_L \Psi)\|_2 = \|\delta^*_{C^1_{j,x,\rho}}(\beta_L \Psi)\|_2 &\leq \|\nabla \beta_L\|_3 \|\Psi\|_6 + \|\beta_L \delta^*_{C^1_{j,x,\rho}}\Psi\|_2\\ &\leq \|\nabla \beta_L\|_3 \|\Psi\|_6 + \gamma\|\nabla_{C^1_{j,x,\rho}}\Psi\|_2\\ &\leq \frac{l(\epsilon) + \gamma}{1 - l(\epsilon)}\|\nabla_{C^1_{j,x,\rho}}(\beta_L \Psi)\| = \frac{l(\epsilon) + \gamma}{1 - l(\epsilon)}\|\nabla_{C_j}(\beta_L \Psi)\|\end{aligned}$$

which contradicts the results of Section 2. Proposition 3.2 now follows.

4 Perturbations

Let $C^1_{j,x,\rho}$ be one of the configurations constructed in the previous section. Proposition 3.2 says that the operator $\delta_{C^1_{j,x,\rho}}\delta^*_{C^1_{j,x,\rho}}$ has three small eigenvalues. Hence we can not invert this operator on the total space $\Omega^1(K')$.

Let $\Pi_{C^1_{j,x,\rho}}$ be the $\nabla_{C^1_{j,x,\rho}}$-orthogonal projection onto the eigenspace spanned by the eigenvectors corresponding to the three small eigenvalues. On

$$\Pi^{\perp}_{C^1_{j,x,\rho}}\Omega^1(K')$$

we can apply the same self-dual perturbation as in Part I. This gives a configuration $C+\Psi$, whose anti-self-dual part $G_{C+\Psi}$ vanishes on $\Pi^{\perp}_C\Omega^1(K')$. For $C+\Psi$ to be self-dual, $G_{C+\Psi}$ has to vanish on $\Pi_C\Omega^1(K')$. This vanishing condition is also called the obstruction, a strategy copied from the one used for self-dual connections in [T1] and Section 7.2.2 of [DK].

First let us introduce a more general concept of almost self-dual configurations:

Definition 4.1 *We call a configuration almost self-dual if the following two conditions are satisfied:*

1. *C fulfills the Spectral Property I described in Definition 3.1.*
2. *For* $6/5 \leq p \leq 6$
$$\|G_C\|_p \leq \text{ const } m^2$$
and
$$\|\nabla_C G_C\|_2 \leq \text{ const } m^2 .$$

Remark that the configurations $C^1_{j,x,\rho}$ from Section 3 are almost self-dual.

Let C denote here an almost self-dual configuration. For $\xi \in L_C(\Omega^1)$ set

$$\xi_1 := \xi - \Pi_1\xi . \tag{4.1}$$

The operator δ_C maps onto $\Pi^{\perp}_C\Omega^1(K')$. Thus in the same way as in Section 2 of Part I, we can define a right inverse P for the operator δ_C on the space $\Pi^{\perp}_C\Omega^1(K')$. Thus

$$\delta_C P\xi_1 = \xi_1$$

for all $\xi \in \Omega^1(K')$.

Set $\Psi := P\xi_1$. We get following equation for ξ_1:

$$G_{C+\Psi} = \xi_1 + \{P\xi_1, P\xi_1\} + G_C = 0 . \tag{4.2}$$

As in Part I, we solve (4.2) by a fixed point method. Let

$$f : L^2 \cap L_C \cap L^{6/5}\Pi^{\perp}_1(\Omega^1) \longrightarrow L^2 \cap L_C \cap L^{6/5}\Pi^{\perp}_1(\Omega^1) \tag{4.3}$$
$$f(\xi_1) := \Pi^{\perp}_1\Big(\{P\xi_1, P\xi_1\} + G_C\Big) .$$

Lemma 4.2 *f has a unique fixed point. Denote it by ξ_0. ξ_0 fulfills the bounds*

$$\|\xi_0\|_2 + \|\nabla\xi_0\|_2 + \|\xi_0\|_{6/5} \leq \text{ const } m^2 .$$

Lemma 4.2 is proved in the same way as in Section 2 of Part I.

This leads us to the map

$$\begin{aligned} h : \mathscr{C}_m &\longrightarrow \Pi_C\Omega^1 \\ C &\longmapsto \Pi_C\Big(\{P\xi_0, P\xi_0\} + G_C\Big) . \end{aligned} \tag{4.4}$$

We can summrize this in

Proposition 4.3 *The configuration $C + P\xi_0$ is self-dual if and only if*

$$h(C) = 0.$$

In the next two sections we will investigate the map h for the small mass configurations constructed in Section 3.

5 The obstruction bundle

Proposition 3.2 gives us a bundle

$$su(2) \longrightarrow H \longrightarrow B_r \times SO(3) .$$

The fibers are spanned by the three eigenvectors corresponding to the three small eigenvectors of the operator $\delta_{C^1_{j,x,\rho}} \delta^*_{C^1_{j,x,\rho}}$ with small eigenvalues. Proposition 4.3 motivates the study of the section

$$\begin{aligned} h : B_r \times SO(3) &\longrightarrow \Pi_{C^1_{j,x,\rho}} \Omega^1 \\ (x,\rho) &\longmapsto \Pi_{C^1_{j,x,\rho}}\Big(\{P\xi_1, P\xi_1\} + G_{C^1_{j,x,\rho}}\Big) . \end{aligned} \tag{5.1}$$

Theorem 5.1 *There is a three manifold $N \subset B_r \times SO(3)$ which is diffeomorphic to $SO(3)$ on which the section h vanishes. Furthermore for $(x,\rho) \in N$*

$$\|C^1_{j,O,\rho} - C^1_{j,x,\rho}\|_2 + \|\nabla_C\Big(C^1_{j,O,\rho} - C^1_{j,x,\rho}\Big)\|_2 \leq \text{ const } m^2 \tag{5.2}$$

where O denotes the origin of $\mathbb{R}^3$.

Using Proposition 4.3 we can formulate this as

Corollary 5.2 *There is a map*

$$I : SO(3) \longrightarrow \mathfrak{M}_m(K')$$

which constructs a 3-parameter family of small mass monopoles whose connection part is close to a nontrivial flat connection.

We will show in Section 7 that I is injective.

To prove Theorem 5.1 we use a similar strategy as in [T1]. We approximate h by a section which we know explicitly and whose zero set is diffeomorphic to the one of h.

Let $h_1(x,\rho) = \Pi_{C_{j,\rho}} h(x,\rho)$. $\Pi_{C_{j,\rho}}$ is the $\nabla_{C_{j,\rho}}$-orthogonal projection onto the kernel of the operator $\delta^*_{C_{j,\rho}}$.

Proposition 5.3

$$|h(x,\rho) - h_1(x,\rho)| \leq l(\epsilon)\,.$$

Before we turn to the proof of Proposition 5.3, let us compare the eigenvectors corresponding to the small eigenvalues of the operator $\delta_{C^1_{j,x,\rho}}\delta^*_{C^1_{j,x,\rho}}$ with the kernel of the operator $\delta_{C_{j,\rho}}\delta^*_{C_{j,\rho}}$.

Lemma 5.4 *Let* $\eta_{0,k,\rho}$ *be an* $\nabla_{C_{j,\rho}}$*-orthonormal basis of* $\Pi_{C_{j,\rho}}\Omega^1(K')$*. There is a* $\nabla_{C_{j,\rho}}$*-orthonormal basis* $\eta_{1,k,\rho}$ *of* $\Pi_{C^1_{j,x,\rho}}\Omega^1(K')$ *such that*

$$\|\nabla_{C_{j,\rho}}(\eta_{0,k,\rho} - \eta_{1,k,\rho})\|_2 \leq \text{ const } l(\epsilon)\,. \tag{5.3}$$

Proof of Lemma 5.4: Choose the cutoff β_L as in the proof of Proposition 3.2. Remember the convention

$$l(\epsilon) := \text{ const } (\log \epsilon^{-1})^{-2/3}\,.$$

Note that

$$\|\delta^*_{C^1_{j,x,\rho}}(\beta_L \eta_{0,k,\rho})\|_2 = \|\delta^*_{C_{j,\rho}}(\beta_L \eta_{0,k,\rho})\|_2 \leq \text{ const } \|\nabla \beta_L\|_3 \leq \text{ const } l(\epsilon)\,. \tag{5.4}$$

We show, as in Lemma 3.2, that

$$\|\nabla_{C^1_{j,x,\rho}}(\beta_L \eta_{0,k,\rho})\|_2 = \|\nabla_{C_{j,\rho}}(\beta_L \eta_{0,k,\rho})\|_2 \geq \Big(1 - l(\epsilon)\Big)\,. \tag{5.5}$$

Furthermore

$$|\Big(\nabla_{C^1_{j,x,\rho}}(\beta_L \eta_{0,k}), \nabla_{C^1_{j,x,\rho}}(\beta_L \eta_{0,l})\Big)| \leq \delta_{kl} + l(\epsilon)\,. \tag{5.6}$$

Apply now the Gram-Schmidt orthonormalization method to the vectors $\beta_L \eta_{0,k,\rho}$. Denote the resulting vectors by $\eta_{1,k,\rho}$. It follows from (5.5) and (5.6) that

$$\|\nabla_{C^1_{j,x,\rho}}(\eta_{0,k,\rho} - \eta_{1,k,\rho})\|_2 \leq l(\epsilon). \tag{5.7}$$

By (5.5)

$$\|\delta^*_{C^1_{j,x,\rho}} \eta_{1,k}\|_2 \leq l(\epsilon).$$

Thus the $\eta_{1,k,\rho}$ are functions in $\Pi_{C^1_{j,x,\rho}} \Omega^1(K')$ which satisfy (5.3).

Proof of Proposition 5.3:

$$|h(x,\rho) - h_1(x,\rho)| = |(1 - \Pi_{C_{j,\rho}})\Pi_{C^1_{j,x,\rho}} \Psi|$$

where

$$\Psi = \{P\xi_1, P\xi_1\} + G_{C^1_{j,x,\rho}}.$$

Write the projection $\Pi_{C^1_{j,x,\rho}}$ as

$$\Pi_{C^1_{j,x,\rho}} \xi = \sum_k (\eta_{0,k,\rho}, \xi)\eta_{0,k,\rho} \tag{5.8}$$

and the projection $\Pi_{C^1_{j,x,\rho}}$ as

$$\Pi_{C^1_{j,x,\rho}} \xi = \sum_k (\eta_{1,k,\rho}, \xi)\eta_{1,k,\rho}. \tag{5.9}$$

It follows from Lemma 5.4 that

$$|(1 - \Pi_{C_{j,\rho}})\Pi_{C^1_{j,x,\rho}} \Psi| \leq l(\epsilon).$$

This implies Proposition 5.3.

Now we describe the section which we know explicitly:

$$h_0 : B_r \times SO(3) \longrightarrow su(2) \tag{5.10}$$

$$h_0(\vec{x}, \rho) := \frac{4\pi m^2}{6} D_1\, \rho(\vec{x}) \cdot \vec{\sigma},$$

where

$$D_1 := \int d\alpha_R\, dr.$$

Remark that $\frac{1}{100} \leq D_1 \leq 100$. $\vec{\sigma}$ denotes the vector of the Pauli spin matrices $\vec{\sigma} = (\sigma_1, \sigma_2, \sigma_3)$. Obviously we have $h_0^{-1}(0) = SO(3)$. Remark that dh_0 is surjective.

Proposition 5.5

$$|h_1(\vec{x},\rho) - h_0(\vec{x},\rho)| \leq \text{ const } \left(m^{5/2} + m^3 R + m^2 R^{-1/2}\right).$$

The proof of Proposition 5.5 is the content of the next section. Proposition 5.5 and Proposition 5.3 imply

Corollary 5.6 *For $\delta > 0$ there is an $m(\delta)$ such that for $m \leq m(\delta)$*

$$|h(x,\rho) - h_0(x,\rho)| < \delta . \tag{5.11}$$

The fact that dh_0 is surjective, (5.11) and the Implicit Function Theorem imply that the zero set of h is diffeomorphic to $SO(3)$.

6 Approximation of the obstruction

Recall the sections where

$$h_0(\vec{x},\rho) = \frac{4\pi m^2}{6} D_1\, \rho(\vec{x}) \cdot \vec{\sigma} \tag{6.1}$$

$$h_1(\vec{x},\rho) = \Pi_{C_{j,\rho}}\left(\{P\xi, P\xi\} + G_C\right). \tag{6.2}$$

$\Pi_{C_{j,\rho}}$ is the $\nabla_{C_{j,\rho}}$-orthogonal projection onto the kernel of $\delta_{C_{j,\rho}}\delta^*_{C_{j,\rho}}$. The content of this section is to show that h_0 and h_1 differ by $O(m^{5/2} + m^3 R + m^2 R^{-1/2})$. This will imply Proposition 5.5.

Let

$$h_2(\vec{x},\rho) = \Pi_{C_{j,\rho}}\left(d\alpha_R \Phi - *(d\alpha_R \wedge A)\right) \tag{6.3}$$

Lemma 6.1

$$|h_1(\vec{x},\rho) - h_2(\vec{x},\rho)| \leq O(m^4 R^3)$$

Proof of Lemma 6.1:

$$\begin{aligned}
&|h_1(\vec{x},\rho) - h_2(\vec{x},\rho)| \\
&\leq |\Pi_{C_{j,\rho}}\left(\{P\xi, P\xi\}\right)| + |\Pi_{C_{j,\rho}}\left((\alpha_R^2 - \alpha_R)[\Phi, A] - (\alpha_R^2 - \alpha_R) * A \wedge A\right)| \\
&\leq \text{ const } m^{-1/2} \|P\xi\|_c^2 + O(m^4 R^{3/2}) \leq O(m^4 R^3).
\end{aligned}$$

As in (5.8) we write the projection $\Pi_{C_{j,\rho}}$ in terms of the eigenvectors $\eta_{0,k,\rho}$ of the kernel

$$h_2(\vec{x},\rho) = \sum_k \left(\alpha'_R \Phi - *(\alpha'_R \wedge A), \eta_{0,k,\rho}\right)\eta_{0,k,\rho}. \tag{6.4}$$

From now on we treat the contributions arising from the Higgs field Φ and the connection A separately. Define

$$h_3(\vec{x},\rho) = \sum_k \Big(d\alpha_R \Phi, \eta_{0,k,\rho}\Big)\eta_{0,k,\rho} \tag{6.5}$$

and

$$h_4(\vec{x},\rho) = -\sum_k \Big(*(d\alpha_R \wedge A), \eta_{0,k,\rho}\Big)\eta_{0,k,\rho}\,. \tag{6.6}$$

Lemma 6.2

$$|h_3(\vec{x},\rho) - h_0(\vec{x},\rho)| \leq O(m^3 R + m^2 R^{-1/2})\,.$$

Proof of Lemma 6.2: The Higgs field in cartesian gauge has the form

$$\Phi_{\vec{x}}(\vec{y}) = mf(m|\vec{x}-\vec{y}|)\frac{\vec{y}-\vec{x}}{|\vec{y}-\vec{x}|}\cdot\vec{\sigma} \tag{6.7}$$

where $f(r) = \coth r - \frac{1}{r}$. Remark that for $R \ll m^{-1}$

$$|f(m|\vec{x}-\vec{y}|) - \frac{m|\vec{x}-\vec{y}|}{6}| \leq O(m^2|\vec{x}-\vec{y}|)\,. \tag{6.8}$$

This follows from

$$h_3(\vec{x},1) = \frac{m^2}{6}\sum_k \int \Big(d\alpha_R(\vec{y})(\vec{y}-\vec{x})\cdot\vec{\sigma}, \eta_{0,k,\rho}\Big)dy\,\eta_{0,k,\rho} + O(m^3 R)\,. \tag{6.9}$$

It follows from Proposition 2.4 that for $|\vec{y}| \gg \max(1, r_0)$

$$\eta_{0,k,\rho}(\vec{y}) = \frac{\vec{a}_k\cdot\rho(\vec{\sigma})}{|\vec{y}|^2} + O(|\vec{y}|^{-3}) \tag{6.10}$$

Plugging this into (6.9) leads to

$$h_3(\vec{x},1) = \frac{m^2}{6}\sum_k \int \Big(d\alpha_R(\vec{y})(\vec{y}-\vec{x})\cdot\vec{\sigma}, \frac{\vec{a}_k\cdot\rho(\vec{\sigma})}{|\vec{y}|^2}\Big)dy\,\eta_{0,k,\rho} \\ + O(m^3 R + m^2 R^{-1/2}) \tag{6.11}$$

We integrate in (6.11) over a symmetric domain. Thus the term in the integral involving $\vec{y}$ drops out

$$h_3(\vec{x},1) = -\frac{4\pi m^2}{6}\sum_k \rho(\vec{x})\cdot\vec{a}_k \int d\alpha_R\, dr + O(m^3 R + m^2 R^{-1/2})\,.$$

Set $D_1 := \int d\alpha_R\, dr$. Remark that $\frac{1}{100} \leq D_1 \leq 100$. Lemma 6.2 follows from (6.12).

Lemma 6.3

$$|h_4(\vec{x},\rho)| \le O(m^3 R + m^2 R^{-1/2}).$$

Proof of Lemma 6.5: In cartesian gauge $A_{\vec{x}}$ takes the form

$$A_{\vec{x}}(\vec{y}) = \frac{m^2}{6}\Big((\vec{y}-\vec{x})\times\vec{\sigma}\Big)\cdot d\vec{y} + O(m^3R). \tag{6.12}$$

Thus

$$h_4(\vec{x},1) = -\frac{m^2}{6}\sum_k \int d\alpha_R\left(*dr\wedge\Big((\vec{y}-\vec{x})\times\vec{\sigma}\Big)\cdot d\vec{y},\eta_{0,k,\rho}\right)dy\,\eta_{0,k,\rho} + O(m^3R) \tag{6.13}$$

Using again (6.10) we obtain

$$h_4(\vec{x},1) = \frac{m^2}{6}\sum_k \int d\alpha_R\Big(*dr\wedge(\vec{x}\times\vec{\sigma})\cdot d\vec{y}, \frac{\vec{a}_k\cdot\rho(\vec{\sigma})}{|\vec{y}|^2}dr\Big)dy\eta_{0,k,\rho} \tag{6.14}$$

$$= -\frac{m^2}{6}\sum_l\sum_k (\rho(\vec{x})\times\vec{a}_l)_k\Big(*(dr\wedge dy_k),dr\Big) + O(m^3R + m^2R^{-1/2}) \tag{6.15}$$

Remark that $\Big(*(dr\wedge dy_k), dr\Big) = 0$. Lemma 6.3 now follows from (6.16).

Lemmas 6.1–6.3 imply

$$|h_0(\vec{x},\rho) - h_1(\vec{x},\rho)| \le O(m^{5/2} + m^3R + m^2R^{-1/2}). \tag{6.16}$$

(6.16) implies Proposition 5.5.

7 Some special gauges

Consider the map

$$J_j : N \longrightarrow \mathscr{C}_m(K') \tag{7.1}$$
$$(x,\rho) \longmapsto C^1_{j,x,\rho}$$

where N is the set of Theorem 5.1 and $\mathscr{C}_m(K')$ denotes the configuration space. $C^1_{j,x,\rho}$ is close to the flat connection b_j over a neighborhood of the neck. J_j is described in Section 3 and N in Theorem 5.1. Define the neighborhood $\mathcal{U}^1_{m,j}(\nu)$ of $J_j(N)$ in the space of charge one configurations.

Definition 7.1 *Denote by $\mathcal{U}^1_{m,j}(\nu)$ the set of charge one, mass m configurations Q which fulfill the property: There exists a gauge transformation $u \in \mathcal{G}$ and a $\rho \in SO(3)$ such that*

$$\|u(Q) - C^1_{j,O,\rho}\|_2 + \|\nabla_{C^1_{j,O,\rho}}\left(u(Q) - C^1_{j,O,\rho}\right)\|_2 < \nu \tag{7.2}$$

where O denotes the origin of $\mathbb{R}^3$.

Theorem 7.2 *There is a $\nu_0 > 0$ such that, for $\nu \leq \nu_0$, there is an $m_0(\nu)$ and for $m \leq m_0(\nu)$ we have: For any configuration $Q \in \mathcal{U}^1_{m,j}(\nu)$ there is a gauge transformation $u \in \mathcal{G}$,a $\xi \in \Pi^{\perp}_Q L_Q(\Omega^1)$ and parameters $(x, \rho) \in B_r \times SO(3)$ with*

$$Q = u\left(C^1_{j,x,\rho} + P_{j,x,\rho}\xi\right) \tag{7.3}$$

and

$$\|\xi\|_{6/5} + \|\xi\|_2 \leq \text{ const } \nu\,.$$

As in Section 8 of Part I we show that I is injective.

Proposition 7.3 *There is an m_0 such that for $m \leq m_0$ the map*

$$I : SO(3) \longrightarrow \mathfrak{M}_m(K')$$

is injective.

Proposition 5.1 says that N is diffeomorphic to $SO(3)$. Theorem 7.2 and Proposition 7.3 imply

Corollary 7.4 *Under the assumptions of Theorem 7.2 and Proposition 7.3*

$$I_j(N) = \mathcal{U}^1_{m,j}(\nu) \cap \mathfrak{M}_m(K')\,.$$

Furthermore $\mathcal{U}^1_{m,j}(\nu) \cap \mathfrak{M}_m(K')$ is diffeomorphic to $SO(3)$.

The proof of Theorem 7.2 is very similar to the proof of Theorem 7.3 in Part II. For this reason we give here only a brief outline of the proof.

Again we use the method of continuity. Pick a $Q \in \mathcal{U}^1_{m,j}(\nu)$. Thus there is a $C^1_{j,O,\rho}$ such that after applying a suitable gauge transformation on Q

$$\|Q - C^1_{j,O,\rho}\|_2 + \|\nabla_{C^1_{j,O,\rho}}(Q - C^1_{j,O,\rho})\|_2 < \nu\,.$$

For $t \in [0, 1]$ let Q_t be the interpolation between Q and C_ρ. Let $S \subset [0, 1]$ be the set of times for which Q_t can be written in a gauge of Theorem 7.1. S is nonempty as $0 \in S$. As in Part I we show that S is closed.

It remains to show that S is open. Assume that $t \in S$. Then there is an $x_0 \in \mathbb{R}^3$, $\rho_0 \in SO(3)$, $\xi_0 \in \Pi^{\perp} L_{Q_{t_0}} \Omega^1$ and a $u_t \in \mathcal{G}$ with

$$C^1_{j,x_0,\rho_0} + P_{j,x_0,\rho_0}\xi_0 = u_{t_0}(Q_{t_0})\,. \tag{7.4}$$

Consider the following map:

$$M : \Omega^0_{K'} \times \mathbb{R} \times \mathbb{R}^3 \times \mathbb{R}^3 \times \Pi^{\perp}_{C}\Omega^1_{K'} \longrightarrow \Omega^0_{K'} \oplus \Omega^1_{K'} \tag{7.5}$$

$$M(\chi, f, \zeta_1, \zeta_2, \xi) = \exp(\chi + f\Phi)\Big(C^1_{j,x_0+\zeta_1,\rho_0+\zeta_2} + P_{j,x_0+\zeta_1,\rho_0+\zeta_2}(\xi_0 + \xi)\Big) - u_{t_0}(Q_{t_0})$$

The derivative of M takes the form

$$\begin{aligned} DM|_{(0,0,0,0,0)}(\chi, f, \zeta_1, \zeta_2, \xi) &= T(\chi, f, \zeta_1, \zeta_2, \xi) \\ &+ [P\xi, \chi + f\Phi] + \partial_{\zeta_1}(P_{x,\rho}\xi) + \partial_{\zeta_2}(P_{x,\rho}\xi) \end{aligned} \tag{7.6}$$

where

$$T(\chi, f, \zeta_1, \zeta_2, \xi) = d^1_c\chi + f d^1_c\Phi + l_1(\zeta_1) + l_2(\zeta_2) + P\xi \tag{7.7}$$

and

$$l_1(\zeta_1) = \frac{\partial}{\partial s} C^1_{j,x+s\zeta_1,\rho} \tag{7.8}$$

is the deformation of $C^1_{j,x,\rho}$ in the x direction and

$$l_2(\zeta_2) = \frac{\partial}{\partial s} C^1_{j,x,\rho+s\zeta_2} \tag{7.9}$$

is the deformation of $C^1_{j,x,\rho}$ in the ρ direction.

Similar to Lemma 7.2.49 of [DK], we show

Lemma 7.5

$$\|\chi\|_\infty + |\zeta_2| \le \text{ const } \Big(\|d^1_c\chi + l_2(\zeta_2)\|_2 + \|d^1_c\chi + l_2(\zeta_2)\|_c\Big)\,.$$

This motivates the definition of the norms

$$\begin{aligned} \|(\chi, f, \zeta_1, \zeta_2, \xi)\|_B &= \|d^1_c\chi + l_2(\zeta_2)\|_2 + \|l_1(\zeta_1)\|_2 + \|f d^1_c\Phi\|_2 + \|P\xi\|_2 \\ &+ \|d^1_c\chi + l_2(\zeta_2)\|_c + \|l_1(\zeta_1)\|_c + \|f d^1_c\Phi\|_c + \|\xi\|_2 \end{aligned} \tag{7.10}$$

and

$$\|\xi\|_E := \|\xi\|_2 + \|\xi\|_c\,. \tag{7.11}$$

Let B denote the closure of $\mathcal{C}^\infty_0(\Omega^0) \times \mathbb{R} \times \mathbb{R}^3 \times \mathbb{R}^3 \times \mathcal{C}^\infty_0 \Pi^{\perp}_c(\Omega^1)$ in the B-norm and let E denote the closure of $\mathcal{C}^\infty_0(\Omega^0) \oplus \mathcal{C}^\infty_0(\Omega^1)$ in the E-norm.

Define the operator

$$\begin{aligned} T_1 : B &\longrightarrow E \\ T_1(\chi, f, \zeta_2) &:= d^1_c\chi + l_2(\zeta_2) + P\xi \end{aligned} \tag{7.12}$$

Lemma 7.6 *There is an m_0 such that for $m \leq m_0$*

$$T_1 : B \longrightarrow E$$

is a bounded, injective Fredholm operator with index $T_1 = -4$. In particular, there is a constant D independent of m such that

$$\|(\chi, \zeta_1, \xi)\|_B \leq D \|T_1(\chi, \zeta_1, \xi)\|_E \,.$$

Lemma 7.6 is proved similar as Lemma 7.2.52 in [DK] and Lemma 6.3 in Part I.

The supports of $l_1(\zeta_1)$ and $l_2(\zeta_2)$ are disjoint. In particular, l_1 and l_2 are orthogonal. Adding the term $l_1(\zeta_1) + fd_c^! \Phi$ to T_1 reduces the dimension of the cokernel of T_1 by 4. Thus $T = T_1 + l_1(\zeta_1) + fd_c^! \Phi$ has no cokernel. Similar to Section 6 of Part I, we conclude

Proposition 7.7 *There is an m_0 such that for $m \leq m_0$*

$$T : B \longrightarrow E$$

is a bounded injective Fredholm operator with index $T = 0$. In particular, there is a constant D independent of m such that

$$\|(\chi, f, \zeta_1, \zeta_2, \xi)\|_B \leq D \|T(\chi, f, \zeta_1, \zeta_2, \xi)\|_E \,.$$

Similar to Section 7 of Part I, we show that T is the leading part of $DM|_{(0,0,0,0,0)}$. In particular

$$\|DM|_{(0,0,0,0,0)} - T\| \leq \text{ const } m \,. \tag{7.13}$$

Proposition 7.8 shows that for m small enough, the operator $DM|_{(0,0,0,0,0)}$ is an isomorphism continuous in both directions. In particular, the operator $DM|_{(0,0,0,0,0)}$ maps onto a neighborhood of zero in $\Omega^0_{K'} \oplus \Omega^1_{K'}$. This implies Proposition 7.5.

8 A lower bound on the obstruction for charge zero configurations

In this section we prove an a priori lower bound on the obstruction h. This will be useful in the next section.

Definition 8.1 *We say $C \in \mathscr{C}_m(K')$ fulfills the Spectral Property II, if there is an L_C-orthogonal projection Π_C onto a 3-dimensional subspace of $L_C(\Omega^0 \oplus \Omega^1)$ with the following properties:*

1. *For $\gamma > 0$ there is an $m(\gamma)$ such that for $m \leq m(\gamma)$ and for $\Psi \in \Pi_C L_C(\Omega^0 \oplus \Omega^1)$*

$$\|\delta_C^* \Psi\|_2 < \gamma \|\nabla_C \Psi\|_2 \tag{8.1}$$
$$\|\delta_C \Psi\|_2 < \gamma \|\nabla_C \Psi\|_2 . \tag{8.2}$$

2. *There is a constant $\mu_1 > 0$ and an m_1 such that for $m \leq m_1$ and $\Psi \in \Pi_C^{\perp} L_C(\Omega^0 \oplus \Omega^1)$*

$$\|\delta_C^* \Psi\|_2 \geq \mu_1 \|\nabla_C \Psi\|_2 \tag{8.3}$$
$$\|\delta_C \Psi\|_2 \geq \mu_1 \|\nabla_C \Psi\|_2 . \tag{8.4}$$

For $R \gg 1$, introduce following mass m, charge zero configurations

$$C_{m,\theta} = \left(\alpha_R \frac{i}{2} m \sigma^3, \theta \right) . \tag{8.5}$$

Now glue $C_{m,\theta}$ to the configuration $(0, b_j)$ on K'. As in Section 3 this gluing requires the identification of two $su(2)$ bundles. These bundle indentifications are parametrized by $SO(3)$. Choose a $\rho \in SO(3)$ and denote the resulting configuration by $C^1_{j,x,\rho}$. We remark that

$$\|G_{C^0_{m,j,\rho}}\|_{6/5} + \|G_{C^0_{m,j,\rho}}\|_2 \leq \text{ const } m . \tag{8.6}$$

Proposition 8.2 *The configuration $C^0_{m,j,\rho}$ fulfills the Spectral Property II described in Definition 8.1.*

Proposition 8.2 is proved similarly to Proposition 3.2.

In the next step we define a neighborhood of the configurations $C^0_{m,j,\rho}$ in the space of charge zero configurations:

Definition 8.3 *Denote by $\mathcal{U}^0_{m,j}(\nu)$ the set of charge zero, mass m configurations Q which fulfill the following properties: There is a gauge transformation u and a parameter $\rho \in SO(3)$ such that*

$$\|u(Q) - C^0_{m,j,\rho}\|_2 + \|\nabla_{C^0_{m,j,\rho}} \left(u(Q) - C^0_{m,j,\rho} \right) \|_2 \leq \nu . \tag{8.7}$$

Furthermore assume that

$$\|G_{u(Q)}\|_2 + \|\nabla_{u(Q)} G_{u(Q)}\|_2 < \text{ const } m . \tag{8.8}$$

The main result of this section is:

Theorem 8.4 *There is a constant* $\mu > 0$ *and an* m_0 *such that for* $m \leq m_0$ *and* $Q \in \mathcal{U}^0_{m,j}(\nu)$

$$|h(Q)| \geq \mu m \tag{8.9}$$

where h *is the obstruction described in (4.4).*

A calculation shows that Theorem 8.4 is fulfilled for the configurations $C^0_{m,j,\rho}$. To prove this theorem for all the configurations in the neighborhood $\mathcal{U}^0_{m,j}(\nu)$, we use a gauge fixing method similar to the one in Section 7 but in this case for charge zero configurations.

Proposition 8.5 *There is a* $\nu_0 > 0$ *such that for* $\nu \leq \nu_0$ *there is an* $m_0(\nu)$ *such that for* $m \leq m_0(\nu)$ *we have: For any configuration* $Q \in \mathcal{U}^0_{m,j}(\nu)$ *there is a gauge transformation* $u \in \mathcal{G}$, *a* $\xi \in \Pi^{\perp}_{C}\Omega^1$ *and a parameter* $\rho \in SO(3)$ *with*

$$Q = u\left(C^0_{m,j,\rho} + P_{C^0_{m,j,\rho}}\xi\right). \tag{8.10}$$

Furthermore

$$\|\xi\|_{6/5} + \|\xi\|_2 \leq \text{ const } \nu. \tag{8.11}$$

Proposition 8.5 is proved in a similar way as Theorem 7.2. Again we use the method of continuity. Let us just point out the main difference in the two proofs. In (7.7) we defined the operator T. To prove Proposition 8.5, we use the operator

$$\begin{aligned} &T_1 : \Omega^0_{K'} \times \Pi^{\perp}_{C}\Omega^1_{K'} \times \mathbb{R} \longrightarrow \Omega^0_{K'} \oplus \Omega^0_{K'} \\ &T_1(\chi, \xi, f) = d^1_{C^1_{j,x,\rho}}\chi + P_{m,j,\rho}\xi + f d^1_{C^1_{j,x,\rho}}\Phi \end{aligned} \tag{8.12}$$

T_1 is injective and has a 3-dimensional cokernel. The term

$$l_2(\zeta_2) := \frac{\partial}{\partial s} C^0_{j,x,\rho+s\zeta_2} \tag{8.13}$$

produces the deformations of $C^1_{j,x,\rho}$ in the three ρ directions. The operator

$$\begin{aligned} &T : \Omega^0_{K'} \times \Pi^{\perp}_{C}\Omega^1_{K'} \times \mathbb{R} \times \mathbb{R}^3 \longrightarrow \Omega^0_{K'} \oplus \Omega_{K'} \\ &T := T_1 + l_2 \end{aligned} \tag{8.14}$$

is an isomorphism. Again T is the leading part of the deformation of the map M corresponding to the one in (7.5).

Proposition 8.6 *With the same assumption and notation as in Proposition 8.5 we can improve (8.11) to*

$$\|\xi\|_{6/5} + \|\xi\|_2 \leq \text{ const } m. \tag{8.15}$$

Proof of Proposition 8.6: From (8.10) we get

$$\xi = G_{u^{-1}(Q)} - G_{C^1_{j,x,\rho}} - \{P\xi, P\xi\} \tag{8.16}$$

From (8.11) and (8.16) it follows that

$$\|\xi\|_{6/5} + \|\xi\|_2 \leq \text{ const } (m + \nu^2)\,. \tag{8.17}$$

Now iterate (8.17). This leads us to the estimate (8.15).

Proof of Theorem 8.4: Let $Q \in U^0_{m,j}(\nu)$. By Proposition 8.5 we can write Q in the form

$$Q = u\Big(C^0_{m,j,\rho} + P_{C^0_{m,j,\rho}}\xi\Big) \tag{8.18}$$

and

$$G_{u^{-1}(Q)} = G_{C^0_{m,j,\rho}} + \xi + \Big\{P_{C^0_{m,j,\rho}}\xi, P_{C^0_{m,j,\rho}}\xi\Big\}. \tag{8.19}$$

Denote by $\Pi_{u^{-1}(Q)}$ respective $\Pi_{C^0_{m,j,\rho}}$ the projections onto the 3-dimensional spaces introduced in Definition 8.1. Remark that

$$\|\Pi_{u^{-1}(Q)} - \Pi_{C^0_{m,j,\rho}}\| \leq \text{ const } m\,. \tag{8.20}$$

Use (8.19) and (8.20) to estimate the obstruction by

$$\begin{aligned} |h_{u^{-1}(Q)}| &= |\Pi_{u^{-1}(Q)} G_{u^{-1}(Q)}| \geq |\Pi_{C^0_{m,j,\rho}} G_{u^{-1}(Q)}| - O(m^2) \\ &\geq |\Pi_{C^0_{m,j,\rho}} G_{C^0_{m,j,\rho}}| - \text{ const } m^{-1/2} \|P_{C^0_{m,j,\rho}}\xi\|_c^2 - O(m^2) \end{aligned}$$

By (8.6)

$$\geq |\Pi_{C^0_{m,j,\rho}} G_{C^0_{m,j,\rho}}| - O(m^{3/2})\,.$$

Let Ψ_k, $1 \leq k \leq 3$ be an orthonormal basis of $\Pi_{C^0_{m,j,\rho}}$. Note that

$$\begin{aligned} \Pi_{C^0_{m,j,\rho}} G_{C^0_{m,j,\rho}} &= \sum_k \Big(G_{C^0_{m,j,\rho}}, \Psi_k\Big)\Psi_k \\ &= \frac{1}{2} m \int_R^{2R} d\alpha_0\, r^2\, dr \sum_k (i\sigma_3, \Psi_k)\Psi_k = m\mu\,. \end{aligned}$$

Since the Ψ_k are a basis

$$|\sum_k (i\sigma_3, \Psi_k)| > 0$$

and

$$|\Pi_{C^0_{m,j,\rho}} G_{C^0_{m,j,\rho}}| = \mu m$$

with $\mu > 0$ and μ independent of m.

9 The small mass moduli space

Theorem 1.1 said that the small mass moduli space $\mathfrak{M}_m(K')$ splits up into two well separated pieces $\mathfrak{M}_{m,\theta}(K')$ and $\mathfrak{M}_{m,V}(K')$. We assume in this paper that $\mathfrak{N}(K')$ is generic. This expression is explained in Definition 1.2. From the weak compactness theorem, it follows that $\mathfrak{M}_{m,v}(K')$ splits up into k_0 pieces:

Lemma 9.1 *There is an m_0 such that for $m \leq m_0$, $\mathfrak{M}_{m,V}(K')$ splits up into well separated pieces:*

$$\mathfrak{M}_{m,V}(K') = \mathfrak{M}_{m,1}(K') \coprod \mathfrak{M}_{m,2}(K') \coprod \cdots \coprod \mathfrak{M}_{m,k_0}(K') .$$

Denote by Ω the annulus of radii $[R_0, 2R_0]$ around the origin of $\mathbb{R}^3$. R is the radius used in Section 3. The absolute value of Φ is small over Ω.

Theorem 9.2 *There is a constant D and an m_0 such that for $m \leq m_0$, $Q = (\Phi, A) \in \mathfrak{M}_{m,j}(K')$, $1 \leq j \leq k_0$ and $x \in \Omega$*

$$|\Phi|(x) \leq Dm^2 . \tag{9.1}$$

Theorem 9.3 *For $\nu > 0$ there is a $m_0(\nu)$ such that for $m \leq m_0(\nu)$*

$$\mathfrak{M}_{m,j}(K') \subset \mathcal{U}^1_{m,j}(\nu) .$$

Note that Theorem 9.3 together with Corollary 7.4 implies Theorem 1.3.

The contents of the rest of this section are the proof of Theorem 9.2 and Theorem 9.3. As a first step to prove Theorem 9.2, we use scaling arguments to establish narrow bounds for the Higgs field over Ω.

Lemma 9.4 *Let $Q = (\Phi, A) \in \mathfrak{M}_{m,j}(K')$. There is a $\gamma \in [0, 1]$ such that for $x \in \Omega$*

$$\gamma m - O(m^2) \leq |\Phi|(x) \leq \gamma m + O(m^2) .$$

Our next task is to gain from the monopole Q a charge zero configuration Q_0 which lies in the neighborhood $\mathcal{U}^0_{m,j}(\nu)$ of the configuration $C^0_{m,j,\rho}$.

For that assume that Q is in the exponential gauge over the region Ω. Pick a $x \in \Omega$. Assume $|\Phi(x_1)| = \gamma m$ and define

$$Q_0(y) = \Big(\alpha_R(y)\Phi(x_1) + (1 - \alpha_R(y))\Phi(y), \alpha_R(y)\theta + (1 - \alpha_R(y))A(y)\Big) \tag{9.2}$$

Similar as in Section 3.2 of Part I we show:

Lemma 9.5

$$\|G_{Q_0}\|_2 \leq \text{ const } m^2 \tag{9.3}$$

and

$$|h(Q_0)| \leq \text{ const } m^2 . \tag{9.4}$$

Using similar methods as in Section 9 and 10 of Part I, we show:

Proposition 9.6 *For $\nu > 0$ there is an $m_0(\nu)$ such that for $m \leq m_0(\nu)$ there is a gauge transformation $u \in \mathcal{G}$ and $\rho \in SO(3)$ with*

$$\|u(Q_0) - C^0_{\gamma m,j,\rho}\|_2 + \|\nabla_{C^0_{\gamma m,j,\rho}} \Big(u(Q_0) - C_{\gamma m,j,\rho}\Big) \|_2 \leq \nu , \tag{9.5}$$

and furthermore

$$Q_0 \in \mathcal{U}^0_{\gamma m,j}(\nu) .$$

Proof of Theorem 9.2: Since $Q_0 \in \mathcal{U}^0_{\gamma m,j}(\nu)$ we can use Theorem 8.4 to get the estimate

$$|h(Q_0)| \geq \mu\gamma m .$$

It follows from (9.4) that

$$\gamma \leq \frac{\text{const } m}{\mu} .$$

This implies Theorem 9.2.

Proof of Theorem 9.3: We proceed here similarly to Section 9 and 10 of Part I. Again denote by Ω the annulus of radii $[R, 2R]$ around the neck. Let $Q \in \mathcal{U}^1_{m,j}(\nu)$ and choose Q to be in the exponential gauge over Ω. Pick a $x_1 \in \Omega$ and define the configuration

$$\begin{aligned} Q_1(y) &:= \Big(\Phi_1(y), A(y)\Big) \\ &:= \Big(\alpha_R(y)\Phi(y) + (1 - \alpha_R(y))\Phi(x_1), \alpha_R(y)A(y) + (1 - \alpha_R(y))\theta\Big) . \end{aligned} \tag{9.6}$$

Consider Q_1 as a configuration on $\mathbb{R}^3$. Note that

$$\|G_{Q_1}\|_{6/5} + \|G_{Q_1}\|_2 \leq \text{ const } m^2 .$$

Let $Q_1 + P\xi$ be the self-dual perturbation of Q_1 on $\mathbb{R}^3$. Denote by C_x the charge one, mass m monopole on $\mathbb{R}^3$ in cartesian gauge centered at x. There is a gauge transformation $u \in \mathcal{G}$ with

$$u(Q_1 + P\xi) = C_x . \tag{9.7}$$

Similarly to Section 9 of Part I, this leads to the estimate

$$\|u(Q_1) - C_x\|_\infty = \|P\xi\|_\infty \leq \text{ const } m^2 . \tag{9.8}$$

Denote by Φ_1 the Higgs field of Q_1 and by Φ_x the Higgs field of C_x. From Theorem 9.2 it follows that

$$\|\Phi_1\|_{\infty,\Omega} \leq \text{ const } m^2 . \tag{9.9}$$

(9.8) and (9.9) imply

$$\|\Phi_x\|_{\infty,\Omega} \leq \text{ const } m^2 . \tag{9.10}$$

From (9.10) follows that $|x| \leq D$ where D is a constant independent of m. This implies

$$\|u(Q_1) - C_O\|_2 + \|\nabla_{C_O}\Big(u(Q_1) - C_O\Big)\|_2 \leq \text{ const } m^2 \tag{9.11}$$

where O denotes the origin of $\mathbb{R}^3$. Using the methods from Section 9 of Part I Theorem 9.3 now follows.

References

[D] Donaldson, S.K. (1986). Connections, Cohomology and the Intersection Forms of 4-Manifolds. Jour. Diff. Geom. **24**, 275–341

[DK] Donaldson, S.K., Kronheimer P.B., (1990). *The Geometry of Four-Manifolds*. Oxford University Press.

[E] Ernst, K. (1992). The Ends of the Monopole Moduli Space over $\mathbb{R}^3$#(Homology Sphere): Part I; in this volume.

[JT] Jaffe, A., Taubes, C. (1980). *Vortices and Monopoles*. Boston: Birkhäuser.

[T1] Taubes, C. (1984). Self-dual Connections on 4-Manifolds with indefinite Intersection Matrix. Jour. Diff. Geom. **19**, 517–560.

K.D. Ernst
Goldmann-Sachs International
Peterborough Court
133 Fleet Street
London EC4A2B
England

Using Floer's exact triangle to compute Donaldson invariants

R. Fintushel[1] and R.J. Stern[2]

1 Introduction

In Floer's 1990 ICM talk ([F3]), he stated that «with luck, one may be able to analyse the change of the Donaldson polynomial under handle addition. The exact triangle for surgery on M may be considered as one step in this direction.» It is the purpose of this note to elaborate on this comment and to illustrate this circle of ideas by computing the 0-degree Donaldson invariant for the $K3$ surface. This invariant was first calculated by Donaldson [D3] using stable bundles. It was recently calculated by Kronheimer [K] who reduced the calculation to a count of representations of a related orbifold fundamental group. Our purpose here is to illustrate how Floer's exact triangle can be implemented to compute Donaldson invariants without resorting to any underlying complex structure. Such cut and paste techniques were also utilized in [FS4] where we constructed irreducible 4-manifolds not homotopy equivalent to any complex surface. The methods in this note complement those of [FS3] where it is shown that, for any 4-manifold X homotopy equivalent to the $K3$ surface and containing the Brieskorn homology 3-sphere $\Sigma(2,3,7)$, certain values of the degree 10 Donaldson polynomial invariant are odd.

2 A Mayer-Vietoris principle and Floer's exact triangle

Suppose we are given an oriented simply connected 4-manifold $X = W \cup_\Sigma V$ with $\partial W = \Sigma = -\partial V$, an integral homology 3-sphere. For simplicity we assume that the character variety $\chi(\Sigma) = \mathrm{Hom}(\pi_1(\Sigma), SO(3))/\text{conjugacy}$ consists of isolated points. (This assumption will hold for all the situations which we shall consider.) The general idea for calculating the Donaldson invariant of X is that if one stretches out a collar $\Sigma \times [-r, r]$ to have infinite length, the invariant of X can be computed as a sum of products $q_M[\lambda] \cdot q_B[\lambda]$ of relative invariants where $[\lambda]$ runs over the character variety $\chi(\Sigma)$. However this is not precisely correct and requires an understanding of the Floer homology of the homology 3-sphere Σ which we now review. (See [F1], [F2], [FS2], and [DFK] for more details.)

The Floer chain groups are free abelian groups, graded mod 8 and generated by $\chi(\Sigma) \setminus [\vartheta]$ (where ϑ denotes the trivial $SO(3)$ representation). A particular

1) Partially supported by NSF Grant DMS9102522

2) Partially supported by NSF Grant DMS9002517

character $[\lambda]$ gives rise to a generator $\langle\lambda\rangle$ of the chain group C_n if n is the mod 8 index of the anti-self-duality operator on $\Sigma \times \mathbf{R}$ with asymptotic values $[\lambda]$ at $-\infty$ and $[\vartheta]$ at ∞. The boundary operator of the complex is defined by $\partial\langle\lambda\rangle = \sum n([\lambda],[\beta])\langle\beta\rangle$ where $[\beta]$ runs over the characters generating C_{n-1}, and $n([\lambda],[\beta])$ is a signed count of the connected components of the 1-dimensional moduli space $\mathcal{M}^1_{\Sigma\times\mathbf{R}}([\lambda],[\beta])$. Floer homology $HF_*(\Sigma)$ is the (mod 8 graded) homology of this complex. Orientation reversal from Σ to $-\Sigma$ causes a change in grading from i to $-3-i$ but, of course, no change in the character variety; so there is a canonical identification of $C_i(\Sigma)$ with $C_{-3-i}(-\Sigma)$. Here, for a manifold N with boundary, $\mathcal{M}^n_N[\lambda]$ denotes the n-dimensional component of the oriented moduli space of finite-action anti-self-dual $SO(3)$ connections on a bundle E (which will be clear from the context) over $N_+ = N \cup (\partial N \times \mathbf{R}^+)$ with a metric which is cylindrical on the end and limits asymptotically to a flat connection over ∂N corresponding to $[\lambda] \in \chi(\partial N)$.

If W is a simply connected 4-manifold with $\partial W = \Sigma$ and E is an $SO(3)$ bundle over W, there is a (degree 0) Donaldson invariant q_W with values in $HF_n(\Sigma)$ where $n = -2w_2(E)^2 - 3(1+b^+_W)$. It is defined by

$$q_W = \sum q_W([\lambda])\langle\lambda\rangle$$

where $[\lambda]$ runs over the characters generating C_n and $q_W([\lambda])$ is the signed count of points in $\mathcal{M}^0_W[\lambda]$, and where this moduli space is assumed to be compact. It is an illuminating exercise to see that q_W is a cycle, and so defines a class in $HF_n(\Sigma)$.

2.1 Theorem *(Donaldson; cf. [A], [DFK]). Let $X = W \cup_\Sigma V$ be a simply connected 4-manifold with $\partial W = \Sigma$ and $\partial V = -\Sigma$. Suppose that W and V are also simply connected. Let q_X be a degree 0 Donaldson invariant corresponding to a bundle P over X. If*

(i) $b^+_W > 0$ *or* $w_2(P_W) \neq 0$, *and*

(ii) $b^+_V > 0$ *or* $w_2(P_V) \neq 0$

then

$$q_X = \langle q_W, q_V\rangle$$

where the pairing is the «Kronecker» pairing of $HF_*(\Sigma)$ with $HF_{-3-*}(-\Sigma)$.

The trivial representation ϑ fails to show up in the formula because the assumptions (i) and (ii) allow one to invoke the arguments of Donaldson's connected sum theorem [D3].

Floer homology can also be defined for 3-manifolds M which are homology equivalent to $S^2 \times S^1$ [F2]. This is done by taking, for C_*, the free chain complex generated by the characters $[\lambda] \in \chi(M)$ whose corresponding flat $SO(3)$ bundle V_λ is the nontrivial bundle over M, i.e. the bundle with $w_2(V_\lambda) \neq 0$. All such representations are irreducible, and so one cannot «compare» λ to ϑ to obtain an absolute grading. Instead there are only relative gradings. In either case, the ambiguity of the gradings arises from the formal dimensions of moduli spaces $\mathcal{M}_{M\times\mathbf{R}}([\lambda],[\lambda])$ which can be computed from the bundle over $M \times S^1$ obtained by identifying the ends. Since for a homology $S^2 \times S^1$ we have $H^1(M;\mathbf{Z}_2) = \mathbf{Z}_2$, there are two such bundles. For each, $p_1 \equiv (w_2)^2 \bmod 4$. But $(w_2)^2 \equiv 0 \bmod 2$ since $M\times S^1$ has an even intersection form, and so $\dim \mathcal{M}_{M\times\mathbf{R}}([\lambda],[\lambda]) \equiv 0 \bmod 4$. Thus the chain complex C_* is (relatively) graded mod 4 as are the resulting Floer homology groups $HF_*(M)$.

Let C be a simply connected cobordism with boundary $\partial C = Y_1 - Y_0$ where each Y_i is a homology sphere or homology $S^2 \times S^1$, and suppose that E is an $SO(3)$ bundle over C which is nontrivial over each boundary component which is a homology $S^2 \times S^1$. It is shown in [F1] that the Donaldson invariant of C which comes from the 0 dimensional components of the moduli space of anti-self-dual connections on E induces a homomorphism in Floer homology

$$C : HF_*(Y_0) \to HF_{*'}(Y_1)$$

where $*' = -2w_2(E)^2 - 3b_C^+ + *$.

Given a homology 3-sphere Σ and a knot K in Σ, let $S_r(\Sigma, K)$ denote the 3-manifold obtained from an r-framed surgery on the knot K. Then there is a natural cobordism B_K (the trace of the surgery) with negative definite intersection form (-1) from Σ to the homology 3-sphere $S_{-1}(\Sigma, K)$. Likewise, there is a cobordism C_K with intersection form (0) from $S_{-1}(\Sigma, K)$ to $S_0(\Sigma, K)$, a homology $S^2 \times S^1$, and there is a cobordism D_K with intersection form (0) from $S_0(\Sigma, K)$ back to Σ. Using the trivial $SO(3)$ bundle over B_K and the nontrivial $SO(3)$ bundles over C_K and D_K, there are induced homomorphisms on Floer homology. In [F2] (cf. [BD], [F3]) it is shown that the following triangle is exact:

$$\begin{array}{ccccc} & & HF_*(S_0(\Sigma,K)) & & \\ & {}^{D_K}\swarrow & & \nwarrow^{C_K} & \\ HF_*(\Sigma) & & \xrightarrow{B_K} & & HF_*(S_{-1}(\Sigma,K)) \end{array}$$

This is known as Floer's exact triangle. Note that the gradings of both $HF_*(\Sigma)$ and $HF_*(S_{-1}(\Sigma, K))$ are well-defined, while the relative grading for $HF_*(S_0(\Sigma, K))$ is fixed in this diagram by the requirement that the homomorphism C_K preserve the grading. Because $H^1(S_0(\Sigma, K); \mathbf{Z}_2) = \mathbf{Z}_2$, there are two ways to glue together the bundles over C_K and D_K to form one over $C_K \cup D_K$. Thus there are two distinct $SO(3)$ bundles over the union $B_K \cup C_K \cup D_K$ which restrict to the correct bundles over each of the cobordisms B_K, C_K, and D_K. These bundles induce endomorphisms of $HF_*(\Sigma)$ of degrees -1 and -5. It is common practice to use the bundle which induces the degree -1 endomorphism. Then it follows that D_K carries $HF_*(\Sigma_{K_0})$ to $HF_{*-1}(S_{-1}(\Sigma, K))$. Thus, the Floer exact triangle induces a long exact sequence of Floer homology groups where the connecting homomorphism has degree -1.

3 Floer's exact triangle and 0-degree Donaldson invariants

Let X be an oriented simply connected 4-manifold and consider an $SO(3)$ bundle E over X with $p_1(E) = -\frac{3}{2}(1+b^+)$. Then the formal dimension of the moduli space $\mathcal{M}_E$ of anti-self-dual connections is 0. For a generic metric on X, the Donaldson invariant $q_E \in \mathbf{Z}$ counts (with sign) the number of points in $\mathcal{M}_E$. Recall that the Pontryagin number p_1 does not completely determine E; the missing data is the second Stiefel-Whitney class $w_2(E) \in H^2(X;\mathbf{Z}_2)$ with ${w_2}^2 \equiv p_1 \bmod 4$. The $SO(3)$ bundles over X with $p_1(E) = -\frac{3}{2}(1+b^+)$ define degree 0 Donaldson invariants, which give a function

$$q_X : \mathcal{C}_X \longrightarrow \mathbf{Z}$$

where $\mathcal{C}_X = \{\eta \in H^2(X;\mathbf{Z}_2) \mid \eta \neq 0,\ \eta^2 \equiv -\frac{3}{2}(1+b^+) \bmod 4\}$.

Now suppose that the intersection form Q_X for X^4 decomposes as $Q_X = D \oplus r\mathbf{H} \oplus E$ where $\mathbf{H}$ denotes a hyperbolic pair and D and E are negative definite forms not diagonalizable over $\mathbf{Z}$. This algebraic decomposition can be realized topologically as follows. According to [FT] there are simply connected smooth 4-manifolds W_1, W_2, and W_3 with $M = W_1 \cup_{\Sigma_1} W_2 \cup_{\Sigma_{r+1}} W_3$ where the Σ_i are homology 3-spheres with $\partial W_1 = \Sigma_1$, $\partial W_2 = -\Sigma_1 \coprod \Sigma_{r+1}$, and $\partial W_3 = \Sigma_{r+1}$. Furthermore, the intersection forms Q_{W_i} for the bounded manifolds W_i satisfy $Q_{W_1} = D$, $Q_{W_2} = r\mathbf{H}$, and $Q_{W_3} = E$. An ambitious goal would be to compute the relative Donaldson invariants for the W_i, and then paste them together as modeled in [A] and [DFK] (see (2.1)) to recover the Donaldson invariants for M.

As we shall see, the techniques of [FS1] can be implemented to show that certain 0-degree relative Donaldson polynomials for W_1 and W_3 are ± 1. To see how the Floer exact triangle can be used to compute the relative Donaldson invariants for W_2, further decompose W_2 as $W_2 = H_1 \cup_{\Sigma_2} H_2 \cup \cdots \cup_{\Sigma_r} H_r$ where the Σ_j's are homology spheres with $\partial H_j = -\Sigma_j \coprod \Sigma_{j+1}$ and with $Q_{H_j} = \mathbf{H}$. Further, suppose that each H_j is composed of two 2-handles $h_{j,1}$ and $h_{j,2}$ attached to $\Sigma_j \times I$ with the new boundary of $H_{j,1} = \Sigma_j \times I \cup h_{j,1}$ a homology $S^2 \times S^1$ and $H_j = H_{j,1} \cup H_{j,2}$. Then each $H_{j,1}$ induces a homomorphism C_K in some Floer exact sequence for which this data completely determines Σ, D_K and B_K. (See §4 below.) Similarly, each $H_{j,2}$ induces a homomorphism D_K in another Floer exact sequence for which $S_{-1}(\Sigma, K)$, B_K and C_K are determined. Thus, an understanding of these Floer exact sequences and their homomorphisms would yield a computation of $q_X : \mathcal{C}_X \longrightarrow \mathbf{Z}$ for certain elements of $\mathcal{C}_X$.

We illustrate these ideas by computing the 0-degree Donaldson invariant for the $K3$ surface. Here, the self-diffeomorphism group of $K3$ acts transitively on the elements of $\mathcal{C}_{K3}$ [M] so that the Donaldson invariant is constant on $\mathcal{C}_{K3}$. Our goal is to show that $q_{K3} \equiv \pm 1$.

4 The decomposition of $K3$

The intersection form Q_{K3} for the $K3$ surface decomposes as $Q_{K3} = E_8 \oplus 3H \oplus E_8$. We realize this decomposition topologically as follows. First, let B be the Milnor fiber for the $(2,3,5)$ Brieskorn singularity. It is a plumbing manifold whose intersection form is E_8 (negative definite) and whose boundary $\partial B = P$ is diffeomorphic to the Poincaré homology 3-sphere. We realize the above algebraic decomposition Q_{K3} by $K3 = B \cup C \cup B$ where C is obtained by attaching six 2-handles h_i, $1 \leq i \leq 6$, to P. To more explicitly describe C view P as the boundary of the handlebody of Figure 1.

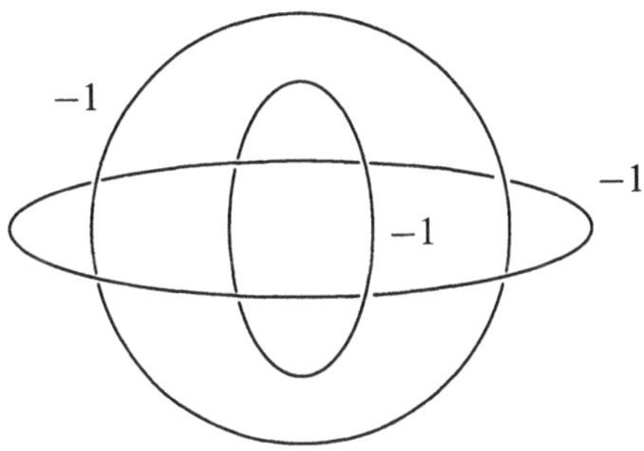

Fig. 1

(See [Hr],[HKK], and [G] for a more detailed discussion of such pictures.) Let $M\{p,q,r\}$ denote the result of p, q, r surgery on the 3 components of the Borromean rings so that $P = M\{-1,-1,-1\}$. Since the Borromean rings have an obvious 3-fold symmetry, $M\{p,q,r\}$ does not depend on the ordering of the surgery coefficients.

To obtain C we attach six 2-handles to P as in Figure 2.

It is shown in [FS4] that $K3 = B \cup C \cup B$. Note that if $P_0 = P$, $M_k = \partial_+ C_k$, and $P_k = \partial_+ D_k$ where $C_k = P_{k-1} \cup h_{2k-1}$ and $D_k = M_k \cup h_{2k}$, $1 \leq k \leq 3$, then each

$$Q_{C_k \cup D_k} = \begin{pmatrix} 0 & 1 \\ 1 & -2 \end{pmatrix}$$

which is equivalent to $\mathbf{H}$. Furthermore the manifolds $M_1 = M\{0,-1,-1\}$, $M_2 = M\{1,0,-1\}$, and $M_3 = M\{1,1,0\}$ have the integral homology of $S^2 \times S^1$, while $P_1 = M\{1,-1,-1\}$, $P_2 = M\{1,1,-1\}$, and $P_3 = M\{1,1,1\} = -P$ are integral homology 3-spheres. Let $\mathcal{L}$ denote the left-handed trefoil knot, $\mathcal{E}$ the figure-eight knot, and $\mathcal{T}$ the right-handed trefoil knot. By blowing down appropriate ± 1 curves in the surgery descriptions of the $M\{p,q,r\}$, we get the following simple surgery descriptions of these 3-manifolds:

$P_0 = P$ is the result of -1-surgery on $\mathcal{L}$,

P_1 is the result of both a $+1$-surgery on $\mathcal{L}$ and a -1 surgery on $\mathcal{E}$,

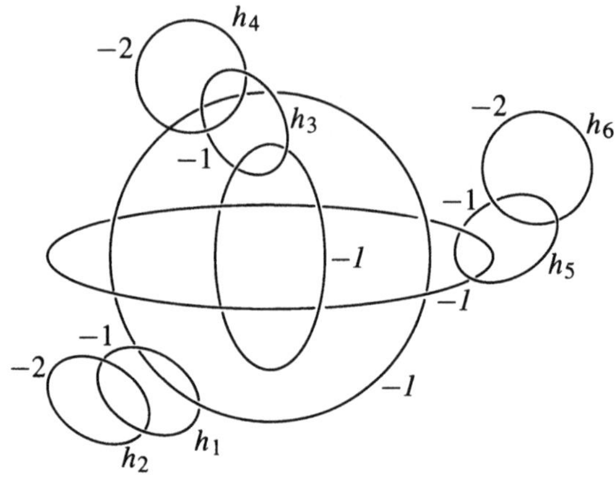

Fig. 2

P_2 is the result of both a $+1$-surgery on $\mathscr{E}$ and a -1-surgery on $\mathscr{T}$,
M_1 is the result of 0-framed surgery on $\mathscr{L}$,
M_2 is the result of 0-framed surgery on $\mathscr{E}$, and
M_3 is the result of 0-framed surgery on $\mathscr{T}$.

Note that if Σ is a homology 3-sphere then

$$\Sigma = S_{-1}(S_1(\Sigma, K), K')$$

where K' is the belt circle of the 2-handle attached to Σ. Furthermore, it is easily checked that

$$S_0(\Sigma, K) = S_0(S_1(\Sigma, K), K')$$

Thus we may rewrite our surgery descriptions of the 3-manifolds P_i and M_j as follows:

$$P_0 = S_{-1}(S^3, \mathscr{L})$$

$$P_1 = S_{-1}(S^3, \mathscr{E}) \text{ and } S^3 = S_{-1}(P_1, \mathscr{L}')$$

$$P_2 = S_{-1}(S^3, \mathscr{T}) \text{ and } S^3 = S_{-1}(P_2, \mathscr{E}')$$

$$M_1 = S_0(S^3 \mathscr{L}) = S_0(P_1, \mathscr{L}')$$

$$M_2 = S_0(S^3, \mathscr{E}) = S_0(P_2, \mathscr{E}')$$

$$M_3 = S_0(S^3, \mathscr{T}) = S_0(P_3, \mathscr{T}')$$

5 The relative Donaldson invariants for the definite manifold B

The relative Donaldson invariants of B take their values in the Floer homology groups of its boundary, P. It is shown in [FS2] that (with our orientation conventions)

$$HF_i(P) = \begin{cases} \mathbf{Z} & i = 1,5 \\ 0 & i \neq 1,5 \end{cases}$$

Let $\langle\alpha\rangle \in HF_1(P)$ be a generator. Our goal in this section is to show that $q_B = \pm\langle\alpha\rangle$. But this follows almost immediately from the proof of the main theorem of [FS1]. For, let $e \in H^2(B;\mathbf{Z})$ be a cohomology class with square $Q_B(e,e) = -2$ and let $E = L \oplus \mathbf{R}$ be the reducible $SO(3)$-bundle over B such that the Euler class of L is e. The moduli space $\mathcal{M}_B(\vartheta)$ of anti-self-dual connections in E with trivial asymptotic condition is a 1-manifold whose connected components are circles, open arcs J_i, half-open arcs K_j, and closed arcs L_k. The endpoints of both the K_j and L_k correspond to reducible connections. Since there is a unique equivalence class of reductions of E (cf. [FS1]), there are no L_k and exactly one K_j. Each of the open ends of the J_i and K_1 corresponds to popping off a one dimensional moduli space $\mathcal{M}^1_{P\times\mathbf{R}}([\alpha],[\vartheta])$ and leaving a point moduli space $\mathcal{M}^0_B([\alpha])$. The number of these ends (counted with orientation) is thus

$$\pm 1 = \#\mathcal{M}^0_B([\alpha]) \cdot \#\mathcal{M}^1_{P\times\mathbf{R}}([\alpha],[\vartheta]),$$

where «#» denotes a count with signs.

$$q_B = \#\mathcal{M}^0_B([\alpha])\,\langle\alpha\rangle = \pm\langle\alpha\rangle \tag{5.1}$$

6 The relative Donaldson invariants for the indefinite manifold C

In §4 we decomposed C as $C = C_1 \cup D_1 \cup C_2 \cup D_2 \cup C_3 \cup D_3$. Each of the cobordisms C_i and D_i induce homomorphisms

$$\begin{aligned} C_i &: HF_*(P_{i-1}) \to HF_*(M_i) \\ D_i &: HF_*(M_i) \to HF_{*-1}(P_i) \end{aligned}$$

which (recalling the surgery descriptions given at the end of §4) fit into the following six Floer triangles.

$$\begin{array}{ccc} & HF_*(M_1 = S_0(S^3,\mathcal{L})) & \\ \swarrow & & \nwarrow^{C_1} \\ 0 = HF_*(S^3) & \longrightarrow & HF_*(P = S_{-1}(S^3,\mathcal{L})) \end{array} \tag{6.1}$$

$$\begin{array}{ccc} & HF_*(M_1 = S_0(P_1,\mathcal{L}')) & \\ {}^{D_1}\swarrow & & \nwarrow \\ HF_*(P_1) & \longrightarrow & HF_*(S^3 = S_{-1}(P_1,\mathcal{L}')) = 0 \end{array} \tag{6.2}$$

$$\begin{array}{ccc} & HF_*(M_2 = S_0(S^3, \mathscr{E})) & \\ \swarrow & & \nwarrow^{C_2} \\ 0 = HF_*(S^3) & \longrightarrow & HF_*(P_1 = S_{-1}(S^3, \mathscr{E})) \end{array} \tag{6.3}$$

$$\begin{array}{ccc} & HF_*(M_2 = S_0(P_2, \mathscr{E}')) & \\ D_2 \swarrow & & \nwarrow \\ HF_*(P_2) & \longrightarrow & HF_*(S^3 = S_{-1}(P_2, \mathscr{E}')) = 0 \end{array} \tag{6.4}$$

$$\begin{array}{ccc} & HF_*(M_3 = S_0(S^3, \mathscr{T})) & \\ \swarrow & & \nwarrow^{C_3} \\ 0 = HF_*(S^3) & \longrightarrow & HF_*(P_2 = S_{-1}(S^3, \mathscr{T})) \end{array} \tag{6.5}$$

$$\begin{array}{ccc} & HF_*(M_3 = S_0(P_3, \mathscr{T}')) & \\ D_3 \swarrow & & \nwarrow \\ HF_*(P_3 = -P) & \longrightarrow & HF_*(S^3) = 0 \end{array} \tag{6.6}$$

By the exactness of these triangles, all the homomorphisms C_1, D_1, C_2, D_2, C_3 and D_3 are isomorphisms. Keep in mind that the gradings of the $HF_*(M_i)$ are relative and only fixed by differing conventions determined by each exact triangle. Pasting these cobordisms together (using the analysis already present in [F1]) we have that

$$C : HF_*(P) \to HF_{*'}(-P)$$

is an isomomorphism. It remains to sort out what bundles we have and to determine the relationship between the gradings $*$ and $*'$.

Consider the cobordism $C_1 \cup D_1$ between P and P_1. As we pointed out in §3, there are two ways to glue together the bundles over C_1 and D_1 arising in the Floer exact triangles. Denote them by E_1 and E_2. Since $w_2(E_i) \neq 0$ on $H_2(M_1; \mathbf{Z}_2)$, which is generated by the core of the 2-handle h_1, these bundles are characterized by the conditions

$$\langle w_2(E_1), [h_2] \rangle = 0$$
$$\langle w_2(E_2), [h_2] \rangle \neq 0$$

Thus the Poincaré dual of $w_2(E_1)$ is $[h_2]$ so that $w_2(E_1)^2 \equiv 0 \bmod 4$, and the Poincaré dual of $w_2(E_2)$ is $[h_1] + [h_2]$ so that $w_2(E_2)^2 \equiv 2 \bmod 4$. The same is true for the other cobordisms. Thus, there are several w_2's that can be utilized for our computations. Fortunately, we only need one, so choose a bundle E' over C with $w_2(E'|_{M_i}) \neq 0$ and with $w_2(E')^2 \equiv 2 \bmod 4$. Then C induces the isomorphism

$$C : HF_*(P) \to HF_{*-5}(-P) \tag{6.7}$$

Pasting E' over C with two copies of the bundle over B from section §5 we get an $SO(3)$-bundle E over $K3$ with $w_2(E)^2 \equiv 2 \bmod 4$. Applying (2.1) together with the calculations (5.1) and (6.7), we have that

$$q_{K3}(w_2(E)) = \langle C(q_B), q_B \rangle = \pm \langle \langle \alpha \rangle, \langle \alpha \rangle \rangle = \pm 1.$$

References

[A] M. Atiyah, *New invariants of 3 and 4 dimensional manifolds*, in 'The Mathematical Heritage of Hermann Weyl', Proceedings Symposia Pure Math **48**(1988), pp. 285–299.

[BD] P. Braam and S. Donaldson, *Floer's exact triangle*, in this volume.

[D1] S. Donaldson, *Anti-self-dual Yang-Mills connections on complex algebraic surfaces and stable vector bundles*, Proc. Lon. Math. Soc. **3** (1985), 1–26.

[D2] S. Donaldson, *Connections, cohomology, and the intersection forms of 4-manifolds*, J. Diff. Geom. **24** (1986), 275–341

[D3] S. Donaldson, *Polynomial invariants for smooth 4-manifolds*, Topology **29** (1990), 257–315.

[DFK] S. Donaldson, M. Furuta, and D. Kotschick, *Floer homology groups in Yang-Mills theory*, in preparation.

[FS1] R. Fintushel and R. Stern, $SO(3)$ *connections and the topology of* 4-*manifolds*, Jour. Diff. Geom. **20** (1984), 523–539.

[FS2] R. Fintushel and R. Stern, *Instanton homology of Seifert fibered homology three spheres*, Proc. London Math. Soc. **61** (1990), 109–137.

[FS3] R. Fintushel and R. Stern, *Homotopy* $K3$ *surfaces containing* $\Sigma(2,3,7)$, Jour. Diff. Geom. **34** (1991), 255–265.

[FS4] R. Fintushel and R. Stern, *Surgery in cusp neighborhoods and the geography of irreducible* 4-*manifolds*, to appear in Inventiones Math.

[F1] A. Floer, *An instanton invariant for 3-manifolds*, Commun. Math. Phys. **118** (1988), 215–240.

[F2] A. Floer, *Instanton homology, surgery, and knots*, in «Geometry of Low-Dimensional Manifolds; I», London Math. Soc. Lecture Notes Ser. **150**, 97–114.

[F3] A. Floer, *Elliptic methods in variational problems*, Proceedings of the International Congress of Mathematicians, Kyoto, Japan, 1990, The Mathematical Society of Japan, Springer Verlag (1991) p. 87–97.

[FT] M. Freedman and L. Taylor, Λ-*splitting* 4-*manifolds*, Topology **16** (1977), 181–184.

[G] R. Gompf, *Nuclei of elliptic surfaces*, Topology **30** (1991), 479–511.

[Hr] J. Harer, *On handlebody structures for hypersurfaces in* $\mathbf{C}^3$ *and* $\mathbf{CP}^3$, Math. Ann. **238** (1978), 51–58.

[HKK] J. Harer, A. Kas, and R. Kirby, *Handlebody decompositions of complex surfaces*, Mem. Amer. Math. Soc. **62** (1986).

[K] P.B. Kronheimer, *Instanton invariants and flat connections on the Kummer surface*, Duke Math. J. **64** (1991), 229–241.

[M] T. Matumoto, *On the diffeomorphisms of a K3 surface*, in 'Algebraic and Topological Theories' Kinosaki (1986), 616–621.

Department of Mathematics,
Michigan State University
East Lansing,
Michigan 48824, USA

Department of Mathematics,
University of California
Irvine,
California 92717, USA

A symplectic fixed point theorem for toric manifolds

A.B. Givental

In this paper, by a toric manifold we mean a non-singular symplectic quotient $M = \mathbb{C}^n//T^k$ of the standard symplectic space by a linear torus action. Such a toric manifold is in fact a complex Kähler manifold of dimension $n-k$. We denote $p(M)$ and $c(M)$ the cohomology class of the Kähler symplectic form and the first Chern class of M respectively. They both are effective, that is, Poincaré-dual to some holomorphic hypersurfaces. We call a *homology* class in $H_2(M, \mathbb{Z})$ *effective* if it has non-negative intersection indices with fundamental cycles of all compact holomorphic hypersurfaces in M, and denote $\mathscr{E}$ the set of all non-zero effective homology classes. Our main result is the following

Theorem *Let M be a compact toric manifold with an integer class $p(M)$ of the symplectic form. Then for any hamiltonian diffeomorphism $h : M \to M$*

(i) *the number of its fixed points is not less than*

$$\max_{\gamma \in \mathscr{E}} \langle c(M), \gamma \rangle / \langle p(M), \gamma \rangle \ ,$$

(ii) *the total multiplicity of its fixed points is not less than* $\dim H^*(M, \mathbb{C})$.

This formulation deserves some discussion.

1. By definition a hamiltonian diffeomorphism is the time-1 map of a non-autonomous hamiltonian system on M. In his original formulation of the symplectic fixed point conjecture, V. Arnold [Ar1],[Ar2] called such h «homologous to the identity». The conjecture itself says that «on a compact symplectic manifold M the number of fixed points of such h is not less than the number of critical points of some function on M». One may consider the theorem as cup-length and Betti-sum estimates for the numbers of fixed points and thus as a confirmation of Arnold's conjecture for toric manifolds. In fact all compact toric manifolds are simply-connected and therefore hamiltonian transformations form just the identity component in the symplectomorphism group of M.

2. The second statement is included in the theorem only for the sake of completeness: It follows from Lefschetz's theorem and holds for any diffeomorphism from the identity component (not necessarily hamiltonian and without the assumption that p is integer of course): any toric manifold has holomorphic cell decompositions and thus its Betti sum equals its Euler characteristic.

---

*) The present research is partially supported by NSF grant DMS 90-01089.

3. The first statement of the theorem is absolutely non-trivial. But it is not very general either. It happens very often that $\langle c(M),\gamma\rangle \leq \langle p(M),\gamma\rangle$ for all $\gamma \in \mathscr{E}$, and the conclusion that *h has a fixed point* still follows from Lefschetz's theorem.

Example Our theorem implies that *a hamiltonian transformation of* $S^2 \times S^2$ *has at least two fixed points provided that the symplectic areas of the factors have integer ratio* and yields nothing new against Lefschetz's theorem if the ratio is fractional.

4. The most general results so far on symplectic fixed points cover a broad variety of symplectic manifolds satisfying however the following monotonicity restriction: a symplectic manifold M is called *monotone* if its symplectic class and first Chern class are positively proportional on spherical 2-cycles in M:

$$\exists \mu \in [0,\infty]\ \forall \gamma : S^2 \longrightarrow M \qquad \langle p(M),\gamma\rangle = \mu\langle c(M),\gamma\rangle.$$

Example $S^2 \times S^2$ is monotone iff the symplectic areas of the factors are equal.

Arnold's conjecture was confirmed for monotone compact symplectic manifolds by

A. Floer [Fl1] for $\mu = 0$ (1987),
A. Floer [Fl2] for $0 < \mu < \infty$ (1989), and
H. Hofer and D. Salamon [HoS] for $\mu = \infty$ (that is, $c(M) = 0$) (1991).

As far as I know our main theorem (along with some results in [HoS]) gives the first examples of non-trivial symplectic fixed point results for *non-monotone* symplectic manifolds.

5. Although all the symplectic fixed point theorems give rise to the Morse-type estimate (for the total multiplicity # of fixed points) that was expected according to Arnold's conjecture (# $\geq$ Betti-sum(M)), the cup-length-type estimates for the number of geometrically distinct fixed points sometimes lead to weaker results than expected. The first discrepancy of this sort arose in Y.-G. Oh's paper [Oh] on fixed points on $T^{2k} \times \mathbb{C}P^m$ where a lower bound for the fixed point number was found to be $\max(m+1, 2k+1)$ instead of the cup-length bound $m+2k+1$ for the critical point number on this manifold. Floer's theorem [Fl2] on strictly monotone symplectic manifolds ($0 < \mu < \infty$) gives another example of this kind. In his theorem the lower estimate for the fixed point number is the greatest common divisor of all values of the first Chern class of the symplectic manifold. Applied to monotone toric manifolds ($c(M) = \mu p(M)$, p is primitive, $\mu \in \mathbb{N}$) this gives $\#(h) \geq \mu$ (which is in fact worse than the cup-length estimate $\dim(M) + 1$ for critical points of functions unless $M \simeq \mathbb{C}P^m$ in which case $\mu = 1$).

We see that our main theorem gives a straightforward generalization of Floer's theorem to non-monotone toric manifolds — and inherits the aforementioned discrepancy as well.

Now let me say a few general words about the proof of our theorem. It is not a secret anymore that symplectic fixed point theorems are Morse-theoretic results for action functionals on spaces of loops in symplectic manifolds. The main difficulty is not that the loop spaces are infinite-dimensional, but rather comes from the fact that both Morse index and coindex of action functionals at critical points are infinite. This means that the Morse complex one should construct from the critical points has nothing to do with usual homotopy invariants of the loop spaces. In order to handle the problem one has to construct a sort of *semi-infinite homology theory* (the term comes from graded Lie algebra theory [F], see also [V]).

A general approach to such a homology theory for symplectic loop spaces leads to Floer homology. Floer's construction depends on Gromov's compactness theorem for moduli spaces of holomorphic curves in almost-complex manifolds [Gro] and seems to fail beyond the monotonicity assumption. It is rather a matter of faith whether this failure is only technical.

The idea of the present paper was to try a more elementary approach in a particular case (beyond the monotonicity restriction) where the approach would do. The method we mean is the *finite-dimensional approximation* of action functionals, and it usually works if the symplectic manifold in question can be obtained somehow from a symplectic vector space. For instance, the pioneer Conley-Zehnder theorem [CZ] on symplectic tori (as a twist of fate, they are not toric manifolds) exploits truncations of Fourier series of loops lifted to the universal covering of the torus. Another idea, by B. Fortune and A. Weinstein [FW], was to represent $\mathbb{C}P^m$ as a symplectic reduction of $\mathbb{C}^{m+1}$ by a circle action and to look for critical loops of a suitable invariant action functional in $\mathbb{C}^{m+1}$. We borrow this idea replacing the circle by a torus and thus come to the category of toric manifolds.

As a finite-dimensional approximation we use discrete loops (rather than Fourier truncation) — a method suggested by M. Chaperon [Cha] and Yu. Chekanov [Che] (see also [LS]) — but in a modified form successfully exploited in my preceding papers [Gi1], [Gi2]. However it turns out that in the non-monotone case action functionals do not have a *representative* (in some sense) finite-dimensional approximation, and we face the necessity of constructing the *semi-infinite cohomology as a direct limit over an exhausting sequence of finite-dimensional approximations*.

Another complicating circumstance is that our action functionals are circle-valued maps rather than usual functions, and therefore we deal with *Morse-Novikov theory of multi-valued functionals*: our semi-infinite cohomology bears some important algebraic super-structure — an action of a lattice of *covering transformations*.

Furthermore, since our hamiltonian transformations «live» in $\mathbb{C}^n$ instead of $\mathbb{C}^n//T^k$ their fixed points form whole T^k-orbits, and thus we deal with *equivariant Morse theory* and T^k-equivariant cohomology respectively.

Finally, in order to obtain a lower estimate for the number of critical points one has not only to construct a suitable cohomology but also to prove that it is not trivial or, in other words, to calculate the cohomology (for some special cases at least). It turns out that the toric manifolds are constructive enough in order to make such a computation possible, but the answer and the computation itself depend on all the geometrical combinatorics of Newton polyhedra that usually accompanies the theory of toric manifolds.

One could pronounce the final sentence to our elementary approach in either of the following opposite ways:

- It is fascinatingly profound: it fits together symplectic semi-infinite equivariant Morse-Novikov critical point theory and algebraic geometry on spectra of some cohomological algebras in terms of combinatorial geometry on Newton polyhedra;
- It is discouragingly complicated: it mixes together symplectic, semi-infinite, equivariant, Morse-Novikov critical point theory with algebraic geometry on spectra...

Leaving this choice to the reader's taste, I would like to express my gratitude to all participants of the symplectic topology section at the AMS meetings in Baltimore and of the Berkeley-Davis-Santa Cruz-Stanford symplectic seminar where preliminary versions of this work were presented, and especially to Ya. Eliashberg, D. Fuchs, V. L. Ginzburg, D. McDuff, Y.-G. Oh, and A. Weinstein for numerous stimulating discussions.

1 Toric manifolds

We recall here a definition and some facts on symplectic toric manifolds. The basic reference for us will be the last chapter of the remarkable book [Au] by M. Audin, *The Topology of Torus Actions on Symplectic Manifolds*. Our approach here to toric manifolds is only a little different from that in the book (basically we dualize notations) and readers will easily recognize corresponding formulations in the book.

Let $\mathbb{C}^n$ denote the standard coordinate $2n$-dimensional real vector space provided with the standard symplectic form

$$\text{imaginary part}\left(-\frac{1}{2}\sum dz_i \wedge d\bar{z}_i\right) .$$

The maximal real torus T^n acts on $\mathbb{C}^n$ by hamiltonian linear transformations

$$z_i \longmapsto \exp(2\pi\sqrt{-1}t_i)z_i \ , \qquad i = 1, \dots, n.$$

This action is generated by n quadratic hamiltonians

$$\pi|z_1|^2, \dots, \pi|z_n|^2 .$$

By definition,[1]) *a toric symplectic manifold is a symplectic quotient of* $\mathbb{C}^n$ *by a subtorus* $T^k \subset T^n$.

In more detail, one fixes a subtorus $T^k \subset T^n$ and considers the momentum map of its action on $\mathbb{C}^n$,

$$P : \mathbb{C}^n \longrightarrow \mathbb{R}^{k*} = (\text{Lie } T^k)^* \ .$$

Then one picks a regular value p of the momentum map and defines the corresponding toric orbifold

$$M_p = P^{-1}(p)/T^k \ ,$$

which automatically inherits a symplectic form from $\mathbb{C}^n$.

In even more detail, the momentum map splits as:

$$\begin{array}{ccc} \mathbb{C}^n & \longrightarrow & \mathbb{R}^{n*} = (\text{Lie} T^n)^* \\ & & \downarrow \pi \\ & & \mathbb{R}^{k*} = (\text{Lie} T^k)^* \end{array}$$

where π is the projection dual to the embedding Lie $T^k \subset$ Lie T^n, and the horizontal arrow is the momentum map $(\pi|z_1|^2, \dots, \pi|z_n|^2)$ of the T^n-action. Its image is the first orthant Π in $\mathbb{R}^{n*}$,

$$\Pi = \{t \in \mathbb{R}^{n*} | t_i \geq 0, i = 1, \dots, n\} \ .$$

Unless the opposite is specified, $\mathbb{R}^n$ and $\mathbb{R}^k$ will always mean Lie T^n and Lie T^k respectively, $\mathbb{Z}^n \subset \mathbb{R}^n$, $\mathbb{Z}^k \subset \mathbb{R}^k$ denote the kernels of the exponential maps $t \mapsto \exp(2\pi t)$: Lie $T^m \to T^m$. Thus $\mathbb{R}^n$ and $\mathbb{Z}^n$ really have the canonical positive basis $e_1, \dots, e_n$ (while the notations $\mathbb{R}^k$ and $\mathbb{Z}^k$ are a bit frivolous).

Most properties of toric manifolds can be formulated in combinatorial terms of the projection $\pi : \Pi \to \pi(\Pi)$.

1. Regular momentum values

[2])The irregular locus of P consists of projections of all $(k-1)$-dimensional faces of Π.

2. Dimension

For a regular p, $\dim_{\mathbb{R}} P^{-1}(p)/T^k = 2(n-k)$.

---

1) As I learned from V.L. Ginzburg (with reference to [KT]) the usual algebraic construction of toric manifolds that begins with a fan in $\mathbb{R}^{n-k}$ is wider than our definition which works only for symplectic toric manifolds.

2) Be careful: the corresponding formulation on p.163 in [Au] contains a misprint which can make the following two pages in the book confusing!

3. Compactness

Toric varieties $P^{-1}(p)/T^k$ are compact (they are or are not for all $p \in \pi(\Pi)$ simultaneously) iff

$$(\ker \pi) \cap \Pi = \{0\} .$$

4. Smoothness

For a regular $p \in \pi(\Pi)$ the toric variety $P^{-1}(p)/T^k$ is non-singular iff projections to $\mathbb{R}^{k*}$ of all the k-dimensional faces of Π that cover p are isomorphisms over $\mathbb{Z}$ (that is the corresponding integer $k \times k$-determinants equal ± 1).

5. Betti sum

The Betti sum of a toric manifold M_p equals the number of the k-dimensional faces of Π that cover p when projected to $\mathbb{R}^{k*}$. In fact it is the number of vertices of the convex polyhedron

$$\pi^{-1}(p) \cap \Pi \subset \mathbb{R}^{n*} \tag{1.1}$$

This polyhedron is identified with the image of the momentum map of the quotient hamiltonian action of T^n/T^k on the toric manifold M_p. Its vertices are in one-to-one correspondence with critical points of some *perfect* Morse function on M_p — the hamiltonian of any dense 1-parametric subgroup in the quotient torus [At],[G].

6. Cohomology

Let $\mathbb{Z}[u]$ denote the *graded* algebra of polynomials on $\mathbb{R}^n$ (with integer coefficients), $u = (u_1, \dots, u_n)$ be its n generators of *degree* 2. We introduce two ideals in $\mathbb{Z}[u]$:

I = the ideal of $\mathbb{R}^k \subset \mathbb{R}^n$,

J = the ideal of the following union Σ of coordinate $(n-k)$-subspaces in $\mathbb{R}^n$: *a coordinate subspace $\mathbb{R}^{n-k}$ is in Σ if its annihilator $(\mathbb{R}^{n-k})^{\perp} \subset \mathbb{R}^{n*}$ covers P when projected to $\mathbb{R}^{k*}$.*

Another description:

I is generated by $n-k$ linear equations of $\mathbb{R}^k$ in $\mathbb{R}^n$,

J is spanned by the monomials $u^m = u_1^{m_1} \cdot \ldots \cdot u_n^{m_n}$ whose exponents $m \in \mathbb{Z}_+^n \subset \mathbb{Z}^n$ being considered as linear functionals on $\mathbb{R}^{n*}$ assume strictly positive values on the (vertices of the) convex polyhedron (1.1).

In fact, J depends only on the component of p in the regular value locus of the momentum map P.

The cohomology ring of the toric manifold M_p is canonically isomorphic to the graded quotient algebra

$$H^*(M_p, \mathbb{Z}) \cong \mathbb{Z}[u]/(I+J) \tag{1.2}$$

The fact that the (complexified) quotient algebra $\mathbb{C}[u]/(I+J)$ has a finite $\mathbb{C}$-dimension reflects the geometric transversality of $\mathbb{R}^k \subset \mathbb{R}^n$ to all the coordinate

$(n-k)$-subspaces that constitute Σ (and whose total number equals the Betti sum of M_p).

The isomorphism (1.2) is induced by the composition

$$H^*(BT^n, \mathbb{Z}) \longrightarrow H^*(BT^k, \mathbb{Z}) \longrightarrow H^*(M_p, \mathbb{Z})$$

where BT^n, BT^k are classifying spaces of the tori, the left arrow is induced by the embedding $T^k \subset T^n$, the right arrow is induced by the classifying map $M_p \to BT^k$ of the principal T^k-bundle $P^{-1}(p) \to P^{-1}(p)/T^k$, and the characteristic class algebras of T^n and T^k are identified with $\mathbb{Z}[u]$ and $\mathbb{Z}[u]/I$ respectively.

7. Symplectic periods

In particular the above identifications determine epimorphisms

$$\begin{array}{ccc} \mathbb{R}^{k*} & \longrightarrow & H^2(M_p,\ \mathbb{R}) \\ \cup & & \cup \\ \mathbb{Z}^{k*} & \longrightarrow & H^2(M_p,\ \mathbb{Z}) \end{array} \tag{1.3}$$

Using (1.3) we indicate the cohomology class of the symplectic form in the cohomology of the toric symplectic manifold (M_p, ω):

$$[\omega] = \text{ image of } p \in \mathbb{R}^{k*} \text{ under } (1.3).$$

8. Complex structure

A more traditional viewpoint on toric varieties is that they are complex-algebraic compactifications of the complex torus $(\mathbb{C}^\times)^{n-k}$. This torus is the quotient $(\mathbb{C}^\times)^n/(\mathbb{C}^\times)^k$ of the complement to the coordinate cross in $\mathbb{C}^n$ by the complexified torus T^k. In order to construct the compactification one defines M_p as a quotient $(\mathbb{C}^n \backslash \mathcal{U}_p)/(\mathbb{C}^\times)^k$ of a bigger complex subspace where

$$\mathcal{U}_p = \text{ union of those coordinate subspaces in } \mathbb{C}^n$$

whose momentum image in $\mathbb{R}^{k*}$ does not contain p. This description provides M_p with a complex structure and stratification which do not depend on p unless p crosses the singular value locus of the momentum map P (when it does the manifold M_p changes itself).

9. First Chern class

Denote $(e_1^*, \ldots, e_n^*)$ the basis in $\mathbb{R}^{n*}$ dual to the standard basis $(e_1, \ldots, e_n)$ in $\mathbb{R}^n$. Then

$$\mathbb{R}^{k*} \ni \Sigma\pi(e_i^*) \longmapsto c(M_p) \in H^2(M_p, \mathbb{Z})$$

where $c(M_p)$ is the first Chern class of the tangent bundle of M_p, and the arrow is the above epimorphism (1.3).

Proof. Generators of the $(\mathbb{C}^\times)^{n-k}$-action on M_p become linear dependent holomorphic vector fields on strata of codimension 1. These strata correspond to hyperplane walls of Π and form a divisor in M_p Poincaré-dual to $c(M)$.

Notice that a codimension-1 stratum is empty iff the corresponding wall $\langle e_i\rangle^\perp$ does not cover p in which case however $\pi(e_i^*)$ is in the kernel of the epimorphism (1.3) too.

The fact that a wall $\langle e_i\rangle^\perp$ does not cover p actually means that the representation of M_p as a toric manifold is not *minimal* in the following sense: one can obtain the same symplectic manifold as a symplectic quotient of $\mathbb{C}^{n-1}$ by T^{k-1}. Thus without loss of generality *we will assume further that our toric manifold M_p is minimal, that is, p is the image of all hyperplane walls of Π, and therefore (1.3) is an isomorphism*

$$\mathbb{R}^{k*} \simeq H^2(M_p, \mathbb{R}) .$$

10. Effective classes

The set $\mathscr{E} \subset \mathbb{Z}^k = H_2(M_p, \mathbb{Z})$ of effective homology classes is the intersection of the lattice with the 1st orthant in $\mathbb{R}^n$:

$$\gamma \in \mathscr{E} \Longleftrightarrow \langle e_i^*, \gamma\rangle \geq 0 \quad \text{for } i = 1, \ldots, n. \tag{1.4}$$

Indeed, any holomorphic hypersurface in M_p contracts to a positive linear combination of the codimension-1 strata by the action of the imaginary part of the complex quotient torus $T_{\mathbb{C}}^n/T_{\mathbb{C}}^k$, and these strata are Poincaré dual to $\pi(e_i^*)$.

Example $S^2 \times S^2$. Let an action of T^2 on $\mathbb{C}^4$ be generated by $\pi(|z_1|^2 + |z_2|^2)$ and $\pi(|z_3|^2 + |z_4|^2)$. The corresponding projection $\pi : \mathbb{R}^{4*} \to \mathbb{R}^{2*}$ is given by the matrix

$$\begin{bmatrix} 1 & 1 & 0 & 0 \\ 0 & 0 & 1 & 1 \end{bmatrix}$$

The image of the first orthant and the corresponding first Chern class are shown in Figure 1. The toric manifold $P^{-1}(p)/T^2$ is $S^2 \times S^2$ provided with a product symplectic form $\omega = \omega_1 \oplus \omega_2$ with periods

$$p = \left(\int_{S^2} \omega_1, \int_{S^2} \omega_2\right) .$$

It is monotone only if the areas of factors are equal. Effective homology classes form the first quadrant on the dual lattice $\mathbb{Z}^2$.

It is convenient sometimes to choose a basis in $\mathbb{Z}^k$ and determine the projection $\pi : \mathbb{R}^{n*} \to \mathbb{R}^{k*}$ by an integer $k \times n$-matrix $\pi = (\pi_{ij})$ as in the example. In terms of this matrix, compactness of the toric manifold means that, upon a suitable choice of the basis in $\mathbb{Z}^k$, the matrix has non-negative entries with positive sum in each column. The 1st Chern class $c(M_p)$ is represented in $\mathbb{R}^{k*}$ by the total sum of all columns. Smoothness of M_p (for regular p) means that each k-minor of the matrix such that p is the convex hull of its columns is unimodular (det $= \pm 1$).

Now we are ready to give a combinatorial reformulation of our main theorem.

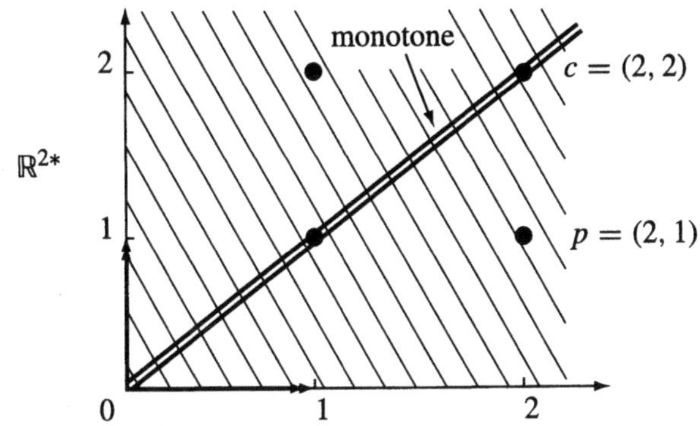

Fig. 1

Theorem 1.1 *Let* $m = (m_1, \ldots, m_n) \in \mathbb{Z}^n_+ \cap \mathbb{Z}^k$ *be a non-negative point in the lattice* $\mathbb{Z}^k \subset \mathbb{Z}^n$, $m_i \geq 0$, $m \neq 0$, $p \in \mathbb{Z}^{k*}$ *be a primitive (that is,* $GCD(\langle p, \ell\rangle | \ell \in \mathbb{Z}^k) = 1$*) regular value of* $\pi : \Pi \to \mathbb{R}^{k*}$. *Then for any hamiltonian diffeomorphism* $M_p \to M_p$ *the number of its fixed points is not less than* $(m_1 + \ldots + m_n)/\langle p, m\rangle$.

Example $S^2 \times S^2$. For $p = (p_1, p_2)$ with $p_1 \geq p_2 > 0$, $m = (m_1, m_1, m_2, m_2)$ Theorem 1 gives rise to

$$2(m_1 + m_2)/(p_1 m_1 + p_2 m_2) = 2/\left[p_2 + \frac{(p_1 - p_2)m_1}{m_1 + m_2}\right] \leq 2$$

where the equality holds only if $p_2 = 1$ and $m_1 = 0$ (or $p_1 = p_2$).

Our proof of Theorem 1.1 in §6 is based on the machinery of semi-infinite equivariant Morse theory developed in §2–6 and on properties of some cohomological algebras described below.

Let $\mathbb{C}[u, u^{-1}] = \mathbb{C}[u_1^{\pm 1}, \ldots, u_n^{\pm 1}]$ be the algebra of polynomial functions on the complex torus $\mathbb{C}^n \backslash$(coordinate cross) $= (\mathbb{C}\backslash 0)^n$. It contains the polynomial algebra $\mathbb{C}[u]$ and we will treat $\mathbb{C}[u, u^{-1}]$ as a $\mathbb{C}[u]$-module. Let us introduce the following submodules in $\mathbb{C}[u, u^{-1}]$:

$I =$ the ideal of $\mathbb{C}^k \subset \mathbb{C}^n$ in $\mathbb{C}[u]$ (here $\mathbb{C}^k \subset \mathbb{C}^n$ is the complexification of $\mathbb{R}^k \subset \mathbb{R}^n$);

$J_r = (u^m, m \in \mathbb{Z}^k | \langle p, m\rangle \geq r)_{\mathbb{C}[u]}$ (here p is a regular integer primitive value of $\pi : \Pi \to \mathbb{R}^{k*}$).

Proposition 1.2 *The quotient algebra* $\mathbb{C}[u]/(I + \mathbb{C}[u] \cap J_r)$ *is finite dimensional.*

Proof. It is a funny combination of Hilbert's zeroes theorem with geometry on Newton lattices. Let us think of integer points $m \in \mathbb{Z}^n$ in $\mathbb{C}^n$ as of exponents of basic monomials $u^m \in \mathbb{C}[u, u^{-1}]$. This «Newton lattice» $\mathbb{Z}^n$ contains a sublattice $\mathbb{Z}^k$ (integer points in $\mathbb{C}^k \subset \mathbb{C}^n$) and a half-lattice

$$\mathbb{Z}_r^k = \{m \in \mathbb{Z}^k | \langle p, m\rangle \geq r\}$$

of monomials generating J_r. We assert at first that *if* $\mathbb{C}^k$ *has a non-trivial intersection with a coordinate subspace* $\mathbb{C}^d \subset \mathbb{C}^n$ *then the halflattice* $\mathbb{Z}_r^k$ *does too.* Indeed, if $\mathbb{Z}_r^k \cap \mathbb{C}^d = \varnothing$ then $p|\mathbb{C}^k \cap \mathbb{C}^d = 0$, and if $\dim \mathbb{C}^k \cap \mathbb{C}^d = \ell > 0$ then the π-image in $\mathbb{R}^{k*}$ of the orthogonal coordinate subspace $(\mathbb{R}^d)^\perp \subset \mathbb{R}^{n*}$ has positive codimension ℓ and contains p, in which case p would be an *irregular* value of $\pi : \Pi \to \mathbb{R}^{k*}$.

Now let us consider a coordinate subspace $\mathbb{C}^d$ that does meet $\mathbb{Z}_r^k$ at some point m_0 (see Figure 2). The first orthant in $\mathbb{Z}^d \subset \mathbb{C}^d$ intersects with the first orthant shifted to m_0. This means that $(u^{m_0}\mathbb{C}[u]) \cap \mathbb{C}[u]$ *contains a monomial that* (as a function on $\mathbb{C}^d$) *does not vanish in* $\mathbb{C}^d \backslash$(coordinate cross).

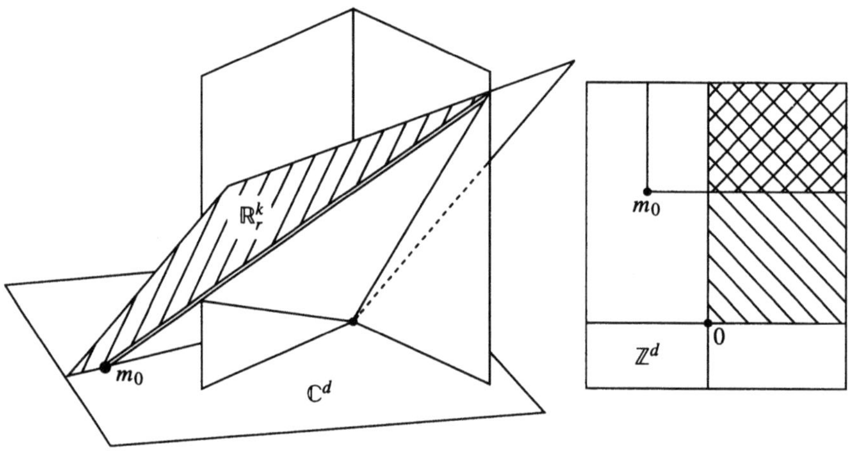

Fig. 2

Now let us consider in $\mathbb{C}^n$ the zero set of the ideal $\mathbb{C}[u] \cap J_r$. What has been said implies that this set is contained in the union of those coordinate subspaces $\mathbb{C}^d$ in $\mathbb{C}^n$ which have trivial intersection with $\mathbb{C}^k$. In other words: *the zero set of* $I + \mathbb{C}[u] \cap J_r$ *contains at most one point — the origin.* Now Hilbert's theorem completes the proof.

Later we will apply this information in the following form. Let $\mathcal{J}_r$ mean the image of J_r in the quotient algebra $\mathcal{R} = \mathbb{C}[u, u^{-1}]/\mathbb{C}[u, u^{-1}]I$, $\mathcal{J}$ a $\mathbb{C}[u]$-submodule such that $\mathcal{J}_{r_+} \subset \mathcal{J} \subset \mathcal{J}_{r_-}$ for some $r_+ > r_-$.

Corollary 1.3 *There exists* $q \in \mathcal{R}$ *such that* $q \notin \mathcal{J}$ *but* $u_1 q, \ldots, u_n q \in \mathcal{J}$.

Indeed, obviously $1 \notin J_1$ and thus there is a monomial u^{-m} which is not in $J_{r_-} + I\mathbb{C}[u, u^{-1}]$. Since

$$u^{-m}\mathbb{C}[u]/(J_{r_+} \cap u^{-m}\mathbb{C}[u] + u^{-m}I)$$

is finite-dimensional one can choose q as a monomials of maximal degree in $u^{-m}\mathbb{C}[u]$ among those whose image in $\mathcal{R}$ is still not in $\mathcal{J}$.

Remark 1.4 The total Betti sum estimate of the total multiplicity of fixed points could also be obtained by Morse-theoretic tools. But instead of the cohomology algebra $\mathbb{C}[u]/(I+J)$ of the toric manifold (See (1.2)) one would then face another interesting «semi-infinite cohomology algebra», namely $\mathcal{J}_0/\mathcal{J}_1$. It is a module (probably free) over

$$R_p = \mathrm{Span}(u^m, m \in \mathbb{Z}^k \subset \mathbb{Z}^n | \langle p, m \rangle = 0)$$

and its rank equals the total Betti sum of the toric manifold M_p.

2 Least action principle

The principle says that fixed points of hamiltonian transformations correspond to critical points of action functionals on loop spaces. Following [FW] we intend to consider loops in $\mathbb{C}^n$ lifting a hamiltonian isotopy of a toric manifold $M_p = \mathbb{C}^n//T^k$ up to a homogeneous equivariant hamiltonian isotopy of $\mathbb{C}^n$. In this section we formulate the infinite dimensional Morse-theoretic problem that arises in this way. By *homogeneity* we mean $\mathbb{R}_+^\times$-equivariance with respect to the dilatation group action in $\mathbb{C}^n$, and we will talk of homogeneous functions, vector fields and diffeomorphisms in $\mathbb{C}^n$, keeping in mind that they are smooth probably only in $\mathbb{C}^n \backslash 0$.

Proposition 2.1 *A hamiltonian isotopy h^t of a compact toric manifold $M_p = \mathbb{C}^n//T^k$ can be lifted up to a T^k-equivariant homogeneous hamiltonian isotopy in $\mathbb{C}^n$.*

Proof. Consider the Poisson quotient map $\mathbb{C}^n \to \mathbb{C}^n/T^k$ to the Poisson variety $\mathbb{C}^n/T^k$. Its Cazimir functions are components of the momentum map $P : \mathbb{C}^n \to \mathbb{C}^n/T^k \to \mathbb{R}^{k*}$ and its symplectic leaves are our toric varieties M_s, $s \in \mathbb{R}^{k*}$. We start with a hamiltonian function on a smooth symplectic leaf M_p and extend it somehow to a regular function on $\mathbb{C}^n/T^k$ homogeneous of degree 2 with respect to $\mathbb{R}_+^\times$-action. The extension is possible due to the fact that near M_p the symplectic leaves form a fibration. Then we pull back the extended hamiltonian function to $\mathbb{C}^n$. The lifted function is homogeneous of degree 2 in $\mathbb{C}^n$ and T^k-invariant. This means that its hamiltonian flow commutes with $\mathbb{R}_+^\times$- and T^k-actions, preserves $P^{-1}(p)$ and thus projects to the original flow on $M_p = P^{-1}(p)/T^k$.

Now let $\mathcal{H}^t$ denote a homogeneous, degree 2, T^k-invariant hamiltonian on $\mathbb{C}^n \times [0,1]$. We define *the action functional*

$$\mathcal{A} : \mathcal{L}\mathbb{C}^n \times \mathbb{R}^k \longrightarrow \mathbb{R} \tag{2.1}$$
$$\mathcal{A} = \oint pdq - \oint \mathcal{H}^t dt - \lambda_1 \oint \mathcal{P}_1 dt - \ldots - \lambda_k \oint \mathcal{P}_k dt$$

on the product of the loop space $\mathcal{L}\mathbb{C}^n$ and $\mathbb{R}^k = \text{Lie } T^k$, where pdq is a potential for the symplectic form in $\mathbb{C}^n$, $\mathcal{P} = (\mathcal{P}_1, \ldots, \mathcal{P}_k)$ are components of the momentum map of the T^k-action on $\mathbb{C}^n$, $\lambda = (\lambda_1, \ldots, \lambda_k) \in \mathbb{R}^k$ are Lagrange multipliers. The action functionals

$$\mathcal{A}_\lambda = \mathcal{A}|\mathcal{L}\mathbb{C}^n \times \{\lambda\} \tag{2.2}$$

are homogeneous degree 2 T^k-invariant functions on the vector space $\mathcal{L}\mathbb{C}^n$. We are going to indicate some relation between critical points of $\{\mathcal{A}_\lambda\}$ and fixed points of the time-1 hamiltonian transformation h induced by the hamiltonians $\mathcal{H}^t$ on the toric manifold M_p.

At first a fixed point corresponds to a time-1 closed trajectory of the hamiltonian isotopy h^t on M_p ($h = h^1$). Since h^t is lifted up to a T^k-invariant isotopy in $\mathbb{C}^n$, its trajectory in M_p is lifted to a trajectory in $P^{-1}(p)$. Such a lifting is unique up to a shift by the T^k-action (so it is not unique!) Even if the trajectory in M_p is closed, its liftings can be disclosed. But they become closed if we replace $\mathcal{H}^t$ by $\mathcal{H}^t + \langle \mathcal{P}, \lambda_0 \rangle$ with appropriate λ_0. This modification means that the $\mathcal{H}^t$-isotopy is accompanied by the action of the 1-parametric subgroup $\exp(t\lambda_0) \subset T^k$, generated by $\lambda_0 \in \mathbb{R}^k = \text{Lie } T^k$.

The choice of such a «closing» λ_0 is not unique: the resulting time-1 transformation $\exp(\lambda_0)$ will not change if we add to λ_0 any integer vector $m \in \mathbb{Z}^k \subset \mathbb{R}^k$.

Summarizing, we can say that a fixed point of h on M_p corresponds to a $\mathbb{Z}^k$-lattice of T^k-orbits of closed trajectories in $P^{-1}(p)$ of hamiltonians $\mathcal{H}^t + \langle \mathcal{P}, \lambda_0 + \mathbb{Z}^k \rangle$.

According to the Least Action Principle, closed trajectories of hamiltonians $\mathcal{H}^t + \langle \mathcal{P}, \lambda \rangle$ are critical points of action functionals $\mathcal{A}_\lambda$ on $\mathcal{L}\mathbb{C}^n$. Due to homogeneity of $\mathcal{A}_\lambda$, all the critical points have critical value 0 (Euler's formula for homogeneous functions). Now we have to extract those critical loops that are situated in $P^{-1}(p)$.

Let $\mathcal{S}$ denote the sphere of all rays ($\mathbb{R}_+^\times$-orbits) in the loop space $\mathcal{L}\mathbb{C}^n \backslash \{\text{zero loop}\}$, $A_\lambda \subset \mathcal{S}$ — (the rays on) the zero cone of the homogeneous function $\mathcal{A}_\lambda$, $A \subset \mathcal{S} \times \mathbb{R}^k$ is the union of $\{A_\lambda\}_{\lambda \in \mathbb{R}^k}$,

$$A = [\mathcal{A}^{-1}(0) \backslash (0 \times \mathbb{R}^k)] / \mathbb{R}_+^\times \ . \tag{2.3}$$

We pull back the linear function $p : \mathbb{R}^k \to \mathbb{R}$ to $\mathcal{S} \times \mathbb{R}^k$ and denote by $\hat{p} : A \to \mathbb{R}$ its restriction to A. Both A and $\hat{p}$ are T^k-invariant.

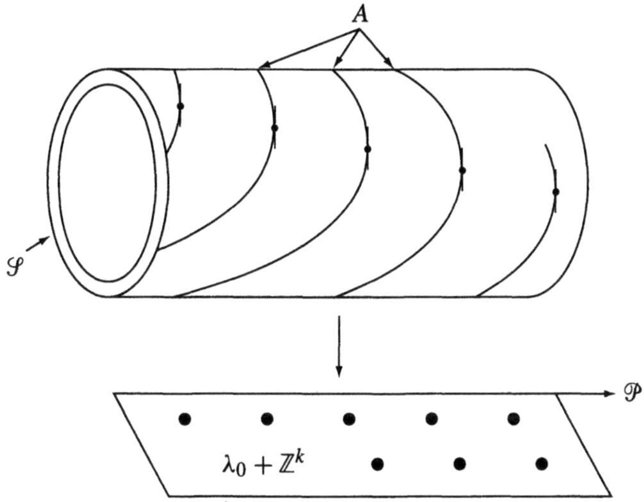

Fig. 3

Proposition 2.2 *Fixed points of $h : M_p \to M_p$ are in one-to-one correspondence with $\mathbb{Z}^k$-lattices of critical T^k-orbits of the function $\hat{p}$ (see Figure 3)*

Proof. Let us notice first that A is non-singular. Indeed, critical points of $\mathcal{A}$ are those closed trajectories of hamiltonians $\mathcal{H}^t + \langle \mathcal{P}, \lambda \rangle$ which satisfy the constraint $\oint \mathcal{P} = 0$. A critical trajectory is situated on a level of the momentum map (since $\mathcal{H}^t$ are T^k-invariant) and the constraint

$$\oint \mathcal{P} = \overrightarrow{\text{const}} \tag{2.4}$$

implies the point-wise one

$$\mathcal{P} = \overrightarrow{\text{const}}\,. \tag{2.5}$$

In particular, if $\overrightarrow{\text{const}} = 0$, the trajectory must be the zero loop since $P^{-1}(0) = 0$ (compactness of our toric manifolds!, see §1, point 3).

Now, a ray in $\mathcal{A}^{-1}(0) \times \mathbb{R}^k$ is critical for $\hat{p}$ if and only if it is critical for corresponding $\mathcal{A}_\lambda$ (recall that non-zero critical points of $\mathcal{A}_\lambda$ always form rays on the zero cone $\mathcal{A}_\lambda = 0$) and the derivative of $\mathcal{A}$ in λ is proportional to p, that is, along this ray

$$\oint \mathcal{P} \sim p\,.$$

Together with (2.5) this means that the ray contains a closed trajectory satisfying $\mathcal{P} \equiv p$, and vice versa.

3 Fronts and their generating families

We formulate here some technical statements allowing us to carry out homotopies in families of functions looking at their discriminants. This technique containing transversality and Morse-theoretic arguments is clear by itself and becomes especially obvious after the book *Stratified Morse Theory* by M. Goresky and R. MacPherson [GM] where we refer the reader for details.

Let us consider a family $f : X \times \Lambda \to \mathbb{R}$ of functions f_λ on a compact manifold X where $\Lambda \ni \lambda$ is a parametrizing manifold. In our applications Λ will be simply a Euclidean space, and $f \in C^1$ with Lipschitz derivatives (so that Morse-theoretic gradient flow deformations still apply).

We will consider restrictions of such a family to compact submanifolds $\Gamma \subset \Lambda$ with boundary $\partial\Gamma$ and study homotopy types of sets

$$F_\Gamma^+ = \{(x, \lambda) \in X \times \Lambda | \lambda \in \Gamma,\ f(x, \lambda) \geq 0\}$$
$$F_\Gamma^- = \{(x, \lambda) \in X \times \Lambda | \lambda \in \Gamma,\ f(x, \lambda) \leq 0\}.$$

Assuming that $f^{-1}(0)$ is non-singular let us define *the front* Φ of the family f:

$$\Phi = \{\lambda \in \Lambda | f_\lambda^{-1}(0) \text{ is singular}\}.$$

It is better to think of the front as of a singular hypersurface in Λ provided at every point with tangent hyperplane(s) (not unique in general). In fact, the front can be obtained by the following contact geometry construction [00], v.1. In the contact manifold $PT^*(X \times \Lambda)$ of all *contact elements* (= tangent hyperplanes) on $X \times \Lambda$ we consider a Legendrian submanifold L of such elements tangent to the hypersurface $f^{-1}(0)$. Intersection of L with the submanifold $\mathcal{P}$ of all *vertical* contact elements (those tangent to fibers of $X \times \Lambda \to \Lambda$) is, generally speaking, a subvariety $\hat{\Phi}$ in $\mathcal{P}$. This subvariety parametrizes the front Φ in the following two-step way. At first we project $\hat{\Phi}$ to the space $PT^*(\Lambda)$ of contact elements on Λ ($\mathcal{P} \to PT^*(\Lambda)$ projects a vertical contact element on $X \times \Lambda$ to a contact element on Λ). The image $\hat{L}$ is a Legendrian subvariety of the contact structure on $PT^*(\Lambda)$ in the sense that it is integral at its non-singular points. Then $\hat{L}$ projects to the front Φ in the base Λ. By a tangent hyperplane to Φ at Λ, we mean a contact element from $\hat{L} \cap PT_\lambda^*(\Lambda)$.

For *generic* f its front Φ actually is a hypersurface because $\hat{L}$ is an immersed Legendrian submanifold in this case. In general if f were smooth Φ would have zero measure due to Sard's lemma (Φ is the critical value locus for the projection $f^{-1}(0) \to \Lambda$). Our C^1-assumption is not sufficient for that, but in our applications, Φ will still have zero measure since all critical points of $f^{-1}(0) \to \Lambda$ will appear to be C^∞-points of $f^{-1}(0)$.

We call a submanifold in Λ *transversal* to the front Φ if, at its every intersection point with Φ, it is transversal to all the tangent hyperplanes to Φ.

From the definitions we obtain

Proposition 3.1 *A submanifold* $\Gamma \subset \Lambda$ *is transversal to the front* Φ *if and only if* $X \times \Gamma$ *is transversal to* $f^{-1}(0)$.

Corollary 3.2 *Suppose that* f *is* C^∞ *at all critical points with zero critical values of all functions* f_λ. *Let* $\Gamma_t = \rho^{-1}(t)$ *be non-singular levels of some smooth map* $\rho : \Gamma \to \mathbb{R}^m$. *Then almost all* Γ_t *are transversal to* Φ.

Applying standard Morse theoretic arguments we come to the following proposition.

Proposition 3.3 *Suppose that* f *and* Γ *deform in a way that* $f^{-1}(0)$ *remains non-singular and both* Γ *and* $\partial\Gamma$ *remain transversal to the front of* f. *Then homotopy types of* F_Γ^+, F_Γ^- *do not change during the deformation.*

Corollary 3.4 *If* f *varies in a way that its front* Φ *does not change, and* Γ *and* $\partial\Gamma$ *are transversal to* Φ, *then homotopy types of* $F_\Gamma^\pm$ *do not change.*

Remarks 3.5

(1) If f is invariant under a fiberwise action of a compact Lie group (say a torus) on X then all the mentioned homotopies can be chosen equivariant.

(2) Under assumptions made, not only homotopy types of $F_\Gamma^\pm$ remain unchanged but also those of the subdivision $X \times \Gamma = F_\Gamma^+ \cup F_\Gamma^-$ by $f^{-1}(0)$, so that we can replace $F_\Gamma^\pm$ in the proposition and its corollary by the pairs $(F_\Gamma^\pm, F_{\partial\Gamma}^\pm)$, or by $f_\Gamma^{-1}(0)$ (or by something else). We will refer to homotopy types of all such spaces as *the homotopy type of the function* $f|_{X\times\Gamma}$ itself, keeping in mind that it is the homotopy type with respect to its *zero level* only.

(3) All the formulated statements remain valid if we replace the manifold Γ with boundary by a more complicated stratified manifold (say the surface of a cube). In this case one should improve the transversality definition: Γ is said to be transversal to Φ if each of its strata is.

Example: Action functionals. We may consider the family

$$\mathcal{A} = \{\mathcal{A}_\lambda\} : \mathcal{S} \times \mathbb{R}^k \longrightarrow \mathbb{R}$$

of action functionals restricted to the unit sphere $\mathcal{S}$ in the loop space of $\mathcal{L}\mathbb{C}^n$ (at least formally — $\mathcal{S}$ is infinite-dimensional and non-compact). Its front $\Phi \subset \mathbb{R}^k$ consists of those λ for which the time-1 map of the hamiltonian $\mathcal{H}_t + \langle \mathcal{P}, \lambda \rangle$ has non-trivial fixed points. It will be essential later that the front is periodic in the sense that it is the lifting to $\mathbb{R}^k$ of some front in $\mathbb{R}^k/\mathbb{Z}^k$ (notice that the family $\mathcal{A}_\lambda$ is *not* periodic in λ). Roughly speaking we will use the fact that the front in $\mathbb{R}^k/\mathbb{Z}^k$ has «bounded geometry» since it is compact.

According to Propositions 2.2 and 3.1, fixed points of our original hamiltonian transformation on M_p correspond to tangency events of levels $p = \text{const}$ of the linear function $p : \mathbb{R}^k \to \mathbb{R}$ with the front Φ. For integer p the tangency points form whole $(k-1)$-dimensional lattices (due to periodicity of Φ) and our objective will be to «count» the number of such tangency lattices between two non-singular levels, say $p^{-1}(0)$ and $p^{-1}(1)$.

4 Generating functions of hamiltonian maps

In this section we construct finite-dimensional approximations to action functionals described in §2. Given a hamiltonian isotopy h^t of a symplectic manifold (M, ω) we (following M. Chaperon [Cha]) represent it as a composition $h_N \circ \ldots \circ h_1$ of «small» hamiltonian transformations and (following [Gi1]) define two symplectomorphisms $h^{(N)}, q^{(N)}$ of $M^{(N)} = (M \times \ldots \times M, \omega \oplus \ldots \oplus \omega)$ to itself:

$$h^{(N)} = (h_1, h_2, \ldots, h_N) \quad \text{the component-wise map}$$

$$q^{(N)} : (x_1, \ldots, x_N) \longmapsto (x_2, \ldots, x_N, x_1) \quad \text{the cyclic shift.}$$

Obviously fixed points of the time-1 map h^1 are in one-to-one correspondence with solutions of the equation $q^{(N)}(\vec{x}) = h^{(N)}(\vec{x})$ and with intersection points of graphs of $q^{(N)}$ and $h^{(N)}$. These graphs are Lagrangian submanifolds in $(M^{(N)} \times \bar{M}^{(N)}, \omega^{(N)} \ominus \omega^{(N)})$ and the second one is close to the diagonal $\Delta \subset M^{(N)} \times \bar{M}^{(N)}$. A neighborhood of Δ is symplectomorphic to $T^*\Delta$ and thus the graph H of $h^{(N)}$ represents in $T^*\Delta$ the differential of some function $\mathcal{H} : \Delta \to \mathbb{R}$. (This function is called the *generating function* of H or $h^{(N)}$.) Now assume that $(M, \omega) \simeq \mathbb{C}^n$. Then $M^{(N)} \times \bar{M}^{(N)}$ has a global structure of the cotangent bundle $T^*\Delta$ and the graph Q of the cyclic shift also has a generating function $\mathcal{Q}$. Thus the fixed points correspond to critical points of $\mathcal{F} = \mathcal{Q} - \mathcal{H}$. This is the idea of our construction.

In our actual situation described in §2, we begin with a *homogeneous* hamiltonian isotopy $h^t : \mathbb{C}^n \to \mathbb{C}^n$ and subdivide it into the *even* number $2N$ of «small» parts $(h_1, \ldots, h_{2N})$ so that $H = H_1 \times \ldots \times H_{2N}$ is a *conical* Lagrangian graph of the component-wise map $(\mathbb{C}^n)^{2N} \to (\mathbb{C}^n)^{2N}$. By a technical reason we define $q^{(N)}$ as the cyclic shift twisted by the central symmetry:

$$q(z_1, \ldots, z_{2N}) = (z_2, \ldots, z_{2N}, -z_1), \qquad z_1, \ldots z_{2N} \in \mathbb{C}^n ,$$

so that $Q = \text{graph}\, q^{(N)}$ is a Lagrangian subspace in $(\mathbb{C}^n \times \bar{\mathbb{C}}^n)^{2N}$.

Now we introduce in $(\mathbb{C}^n \times \bar{\mathbb{C}}^n)^{2N}$ a structure of the cotangent bundle over the diagonal $\Delta(\simeq \mathbb{C}^{2nN})$ choosing Δ itself for its zero section and $(\mathbb{C}^n \times \bar{\mathbb{C}}^n)^{2N} \to \Delta$ along the anti-diagonal

$$(z_1, w_1, \ldots, z_{2N}, w_{2N}) \longmapsto \left(\frac{z_1 + w_1}{2}, \ldots, \frac{z_{2N} + w_{2N}}{2} \right)$$

for the canonical projection $T^*\Delta \to \Delta$. What was said above implies that $\mathcal{F} = Q - \mathcal{H}$ is a well defined *homogeneous degree 2 function on Δ whose critical points are in one-to-one correspondence with fixed points of the hamiltonian transformation* $(-\mathrm{id}) \circ h^1$. We call it the *generating function* of h^1.

Remarks 4.1

(1) $\mathcal{Q}$ is a non-degenerate quadratic form on Δ: it follows from the fact that the Lagrangian subspace Q is transversal to the diagonal and anti-diagonal since id and $(-\mathrm{id})^{2N+1}$ do not have non-trivial fixed points (this justifies our choice of «even» subdivisions and the «twisting»).

(2) $\mathcal{H}$ is well defined provided that $h_1, \dots, h_{2N}$ are sufficiently close to identity and is a homogeneous degree 2 C^1-function with Lipschitz derivatives since $H_1, \dots, H_{2N}$ are conical Lagrangian subspaces in $\mathbb{C}^n \times \bar{\mathbb{C}}^n$ smooth outside the origin. This implies also that $\mathcal{H}$ is C^∞ outside the «coordinate cross» in $(\mathbb{C}^n)^{2n}$.

(3) Notice that a critical point of $\mathcal{F} = \mathcal{Q} - \mathcal{H}$, that is, a discrete loop

$$(z_1, w_1, z_2, w_2, \dots, z_{2N}, w_{2N}) =$$
$$= (z_1, h_1(z_1), h_1(z_1), h_2(h_1(z_1)), \dots, h_{2N}(\dots(h_1(z_1)\dots) = -z_1)$$

coincides with the zero loop if at least one of the coordinates $z_i - w_i$ vanishes (since h_i are small). This means that $\mathcal{F}$ is smooth (C^∞) at every non-trivial critical point.

Proposition 4.2 *Suppose that all the transformations $h_1, \dots, h_{2N}$ commute with the T^k-action on $\mathbb{C}^n$. Then the generating function $\mathcal{F} : \Delta \simeq (\mathbb{C}^n)^{2N} \to \mathbb{R}$ is T^k-invariant relative to the simultaneous action of the torus on all 2N factors $\mathbb{C}^n$.*

Proof. All the ingredients of the above construction — the symplectic structure in $\mathbb{C}^n$, graphs of h_i, $\pm$id: $\mathbb{C}^n \to \mathbb{C}^n$ and of the cyclic shift — are invariant with respect to the simultaneous action of the torus on all factors in $(\mathbb{C}^n \times \bar{\mathbb{C}}^n)^{2N}$.

In particular this proposition means that critical points of $\mathcal{F}$ actually occur as whole T^k*-orbits of critical rays* (due to the homogeneity and T^k-invariance) similar to those of action functionals $\mathcal{A}_\lambda$.

Notice also that $-\mathrm{id} : \mathbb{C}^n \to \mathbb{C}^n$ commutes with the T^k-action and therefore induces symplectic transformations on toric manifolds M_p. Such a transformation is *hamiltonian*. This means that any estimate for fixed points of all its compositions with hamiltonian transformations on M_p will give rise to the same estimate for all hamiltonian transformations themselves; our twisting does not affect the resulting supply of symplectomorphisms.

The generating function $\mathcal{F}$ that we have constructed serves as a finite-dimensional approximation for a single action functional $\mathcal{A}_0$. Now we will take into account the effect of the Lagrange multipliers.

For a given $\lambda \in \mathbb{R}^k = \mathrm{Lie}T^k$, we put $t = \exp(\lambda/2N_1)$ where N_1 is big enough to make the transformation t sufficiently close to identity and apply the above construction of the generating function to the decomposition

$$h_{2N_2} \circ h_{2N_2-1} \circ \ldots \circ h_2 \circ h_1 \circ \underbrace{t \circ \ldots \circ t}_{2N_1 \text{ times}}$$

of our time-1 hamiltonian transformation $h^1 : \mathbb{C}^n \to \mathbb{C}^n$ composed with linear transformation $\exp(\lambda)$ (and twisted by $-\mathrm{id}$ of course). We denote the resulting generating function as

$$\mathcal{F}_\lambda^{(N)} : (\mathbb{C}^n)^{2N} \longrightarrow \mathbb{R}, \qquad N = N_1 + N_2.$$

It will serve us as a finite-dimensional substitute for the action functional $\mathcal{A}_\lambda$.

What we can see from this construction is that it does not work *simultaneously* for all $\lambda \in \mathbb{R}^k$: N_1 must grow with λ. We will assume further that, for given h^1, its decomposition $h_{2N_2} \circ \ldots \circ h_1$ is fixed, and for given $N > N_2$, the family $\mathcal{F}_N = \{\mathcal{F}_\lambda^{(N)}\}$ is defined over a compact domain $\Lambda_N \subset \mathbb{R}^k$ of parameter values. For the sake of definiteness let us consider Λ_N to be a cube in $\mathbb{R}^k$ (centered at the origin) of the size growing linearly with N. Thus we come to *an exhausting sequence of generating families*

$$\mathcal{F}_N : \mathbb{C}^{2Nn} \times \Lambda_N \longrightarrow \mathbb{R}, \qquad \Lambda_N \subset \mathbb{R}^k, \quad \bigcup_N \Lambda_N = \mathbb{R}^k.$$

It is a family of homogeneous degree 2, T^k-invariant functions on $(\mathbb{C}^n)^{2N}$ provided with the component-wise action of the torus.

Now let $\mathcal{S}_N$ denote a unit sphere S^{4nN-1} in $\mathbb{C}^{2Nn}$ or, to be more invariant, the sphere of all real rays in $\mathbb{C}^{2Nn}\backslash 0$. Each homogeneous function $\mathcal{F}_\lambda^{(N)}$ determines a subdivision of $\mathcal{S}_N$ into positive and negative parts

$$F_N^{\pm} = \{(x, \lambda) \in \mathcal{S}_N \times \Lambda_N | \quad \mathcal{F}_\lambda^{(N)}|_x \geq 0 (\text{resp.} \leq 0)\}$$

and their intersection

$$F_N^0 = (\mathcal{F}_N^{-1}(0) \backslash \Lambda_N)/\mathbb{R}_+^\times = F_N^+ \cap F_N^- .$$

We denote by $\hat{p}_N$ the following composed function on F_N^0:

$$\hat{p}_N : F_N^0 \subset \mathcal{S}_n \times \Lambda_N \longrightarrow \Lambda_N \subset \mathbb{R}^k \xrightarrow{p} \mathbb{R}$$

where $p \in \mathbb{R}^{k*} = (\mathrm{Lie}T^k)^*$ is the linear function that determines our toric manifold M_p.

Proposition 4.3 *F_N^0 is a non-singular submanifold in $\mathcal{S}_n \times \Lambda_N$. Critical T^k-orbits of the function $\hat{p}_N$ correspond to fixed points of the hamiltonian transformation $-\mathrm{id} \circ h_{2N_2} \circ \ldots \circ h_1$ pushed forward to M_p.*

Remark The correspondence is not one-to-one. Actually the critical orbits of the functions $\hat{p}_N$ fit some $\mathbb{Z}^k$-lattices (and exhaust them as $N \to \infty$) in the following sense: a fixed point of our hamiltonian transformation on M_p corresponds to a finite set of critical T^k-orbits of $\hat{p}_N$ on F_N^0. This critical set is situated in $\mathcal{S}_N \times \Lambda_N$ over the part of some lattice $\lambda_0 + \mathbb{Z}^k$ in $\mathbb{R}^k$ that fits the cube $\Lambda_N \subset \mathbb{R}^k$. Proposition 4.3 reduces our symplectic fixed point problem to the lower estimate problem for the number of such lattices in $\mathbb{R}^k$. This is a sort of equivariant Lusternik-Schnirelman-Morse-Novikov problem.

Proof. We will look for non-zero critical points of the function

$$\mathcal{F}_N : \mathbb{C}^{2Nn} \times \Lambda_N \longrightarrow \mathbb{R}$$

and of its restrictions to levels of the linear function

$$\mathbb{C}^{2Nn} \times \Lambda_N \longrightarrow \Lambda_N \xrightarrow{p} \mathbb{R} .$$

Let us begin with the generating function $\mathcal{F}_{\lambda_0}^{(N)} : \mathbb{C}^{2Nn} \times \{\lambda_0\} \to \mathbb{R}$. By construction its critical point x_0 corresponds to a fixed point of the map

$$-\mathrm{id} \circ h_{2N_2} \circ \ldots \circ h_1 \circ t \circ \ldots \circ t : \mathbb{C}^n \longrightarrow \mathbb{C}^n .$$

Let z_0 be such a fixed point. It also corresponds to an intersection point of the conical Lagrangian submanifolds Q and H_{λ_0} in $(\mathbb{C}^n \times \bar{\mathbb{C}}^n)^{2N}$. Let us calculate now partial derivatives of $\mathcal{F}_N$ in $\lambda_1, \ldots, \lambda_k$ at the point z_0.

Lemma *$\frac{\partial \mathcal{F}_N}{\partial \lambda_i}(x_0, \lambda_0) = -2N_1 \mathcal{P}_i(z_0)$ where $\mathcal{P}_1, \ldots, \mathcal{P}_n : \mathbb{C}^n \to \mathbb{R}$ are quadratic hamiltonians generating the action on $\mathbb{C}^n$ of coordinate 1-parametric subgroups in T^k.*

Indeed:

(1) The generating function $\int pdq$ of a *conical* Lagrangian submanifold coincides with the restriction of the quadratic form $pq/2$ to this Lagrangian submanifold.

(2) Its derivative along a linear hamiltonian flow at the point $(p, q) = (0, q_0)$ equals $-\mathcal{P}(0, q_0)$ where $\mathcal{P}$ is the quadratic hamiltonian of the flow.

(3) Choose (p, q)-coordinates on $(\mathbb{C}^n \times \bar{\mathbb{C}}^n)^{2N} = T^*\Delta$ in a way that $p = 0$ is an equation of Q, $q =$ const are equations of fibers in $T^*\Delta$. Then $\mathcal{F}_\lambda^{(N)} = \mathcal{Q} - \mathcal{H}_\lambda = -\int pdq|H_\lambda$.

(4) H_λ is obtained from H_{λ_0} by the simultaneous action of T^k on the first $2N_1$ factors $\bar{\mathbb{C}}^n$ in $(\mathbb{C}^n \times \bar{\mathbb{C}}^n)^{2N}$. Hamiltonians of this action are $-\mathcal{P}_i(w_1) - \ldots - \mathcal{P}_i(w_{2N_1})$, $i = 1, \ldots, k$.

Combining (1)–(4) we find

$$\frac{\partial \mathcal{F}_N}{\partial \lambda_i}(x_0, \lambda_0) = -\mathcal{P}_i(tz_0) - \mathcal{P}_i(t^2 z_0) - \ldots = -2N_1 \mathcal{P}_i(z_0) \,. \tag{4.1}$$

Formula (4.1) should be compared with (2.4), (2.5). The lemma means that
(1) critical points of $\mathcal{F}_N$ correspond to the fixed points z_0 that satisfy the constraints

$$\mathcal{P}_1(z_0) = \ldots = \mathcal{P}_k(z_0) = 0;$$

(2) critical points of the restriction $\mathcal{F}_N|_{p=\text{const}}$ correspond to the fixed points z_0 that satisfy the constraints

$$(\mathcal{P}_1(z_0), \ldots, \mathcal{P}_k(z_0)) \sim p = (p_1, \ldots, p_k).$$

The first system of constraints holds only for $z_0 = 0$ (since some linear combination of $\mathcal{P}_i$ is positive definite due to compactness of M_p, see §1). The second one implies

$$\exists \mu > 0 : \mu z_0 \in P^{-1}(p) \,,$$

that is, selects fixed points of $(-\text{id}) \circ h^1$ on M_p. Similarly to Proposition 2.2, the homogeneity of $\mathcal{F}_\lambda^{(N)}$ completes the proof.

Our construction of generating functions depends on the decomposition

$$h^1 = (-\text{id}) \circ h_{2N} \circ \ldots \circ h_1 \,. \tag{4.2}$$

We describe below two moves of the decomposition which change the homotopy type of the corresponding subdivision $\mathcal{S}_N = F^+ \cup F^-$ in a controllable way, provided at least that $F^+ \cap F^-$ is non-singular (that is, if h^1 does not have non-trivial fixed points).

(1) If the decomposition (4.2) varies in a way that h^1 remains unchanged and all h_i remain small then the homotopy type of the subdivision $\mathcal{S}_N = F^+ \cup F^-$ does not change.

(2) Suppose that the hamiltonian isotopy h^t of $\mathbb{C}^n$ consist of two parts and the first one is a loop decomposed into $2K$ small pieces, $\text{id} = h_{2K} \circ \ldots \circ h_1$. One may compare the generating function $\mathcal{F}^{(N)}$ of the whole isotopy to the generating functions $\mathcal{F}^{(N-K)}$ and $\mathcal{G}^{(K)}$ of the parts

$$h^1 = (-\text{id}) \circ h_{2N} \circ \ldots \circ h_{2K+1}, (-\text{id}) \circ h_{2K} \circ \ldots \circ h_1 \,, \quad N > K.$$

We assert that $\mathcal{F}^{(N)}$ *can be deformed to* $\mathcal{F}^{(N-K)} \oplus \mathcal{G}^{(K)}$ *in a way such that the homotopy type of the subdivision* $\mathcal{S}_N = F^+ \cup F^-$ *does not change during the deformation.*

The first statement is obvious. The second one is based on the following deformation Q^ε of the Lagrangian subspace $Q = \{w_1 = z_2, w_2 = z_3, \ldots, w_{2N} = -z_1\}$:

$$\begin{aligned} z_1 + w_{2K} &= \varepsilon(z_{2K+1} - w_{2N}) \\ w_{2N} + z_{2K+1} &= \varepsilon(w_{2K} - z_1) \end{aligned} \tag{4.3}$$

For $\varepsilon = 1$, (4.3) is equivalent to two of the equations for Q: $z_1 = -w_{2K}$, $w_{2K} = z_{2K+1}$. For $\varepsilon = 0$, we obtain $Q_1 \times Q_2$ instead of Q where Q_1 and Q_2 correspond to cyclic shifts in $(\mathbb{C}^n)^{2K}$ and $(\mathbb{C}^n)^{2(N-K)}$ respectively. Since $H = H_1 \times H_2$ by the very definition, the generating function $\mathcal{F}^{(N)}$ converts into $\mathcal{F}^{(N-K)} \oplus \mathcal{G}^{(K)}$ at $\varepsilon = 0$. In between, for $0 < \varepsilon < 1$, at intersections of $Q^\varepsilon \cap H$ we have $w_{2K} = z_1$ (since the first part of the hamiltonian isotopy forms a loop) and thus (4.3) turns into

$$z_{2K+1} = \frac{1}{\varepsilon} w_{2K}, \qquad z_1 = -\varepsilon w_{2N} .$$

This means that the critical rays of $\mathcal{F}_\varepsilon^{(N)}$ are the same as those of $\mathcal{F}^{(N)}$ and implies that the deformation of $\mathcal{F}^{(N)}$ to $\mathcal{F}^{(N-K)} \oplus \mathcal{G}^{(K)}$ does not change the homotopy type of the subdivision $\mathcal{S}_N = F^+ \cup F^-$, at least if $\mathcal{F}^{(N)}$ does not have such critical rays at all.

Now let us consider the family (in the sense of §3)

$$\mathcal{F}_N | \mathcal{S}_N \times \Lambda_N \longrightarrow \mathbb{R} .$$

It was designed in a way that its front $\Phi_N \subset \Lambda_N$ would consist of those $\lambda \in \Lambda_N$ for which $(-\mathrm{id}) \circ h^1 \circ \exp(\lambda) : \mathbb{C}^n \longrightarrow \mathbb{C}^n$ does not have non-trivial fixed points. Therefore $\Phi_N = \Phi \cap \Lambda_N$ where Φ is the front of a suitable family $\{\mathcal{A}_\lambda\}$ of action functionals (that is, independent on the decomposition). Applying our two moves to decompositions

$$(-\mathrm{id}) \circ h_{2N_2} \circ \ldots \circ h_1 \circ t_{2N_1} \circ \ldots \circ t_1, \qquad t_i \in T^k$$

with $t_{2N_1} \circ \ldots \circ t_1 = \exp(\lambda)$, we conclude that

> $\mathcal{F}_{N+K}|\Lambda_N + m$ *with* $m \in \mathbb{Z}^k$ *deforms fiberwise into* $\mathcal{F}_N|_{\Lambda_N} \oplus \mathcal{G}_m^{(K)}$ *(where* $\mathcal{G}_m^{(K)}$ *is the generating function of the «discrete loop»* $t \circ \ldots \circ t$ *with* $t = \exp(m/2K)$*) in such a way that the front* $\Phi_N \subset \Lambda_N$ *of the family remains unchanged during the deformation* (see Figure 4).

Combining this statement with the results of §3 (see Corollary 3.2 and Remarks 3.4 and 4.1.3) and assuming that the cubes Λ_N are chosen transversal to the (periodic, zero measure) front $\Phi \subset \mathbb{R}^k$, we obtain the following

Proposition 4.4 *Let $m \in \mathbb{Z}^k$ and $\Lambda_N + m \subset \Lambda_{N+K}$. Let $\mathcal{G}_m^{(K)} : \mathbb{C}^{2Kn} \to \mathbb{R}$ be the Hermitian generating form of the «twisted discrete loop» $(-\mathrm{id}) \circ t \circ \ldots \circ t$ (2K times), $t = \exp(m/2K) \in T^k$. Let $\mathcal{F}_N$ and $\mathcal{F}_{N+K}$ be the generating families of the same exhausting sequence. Then the restricted family*

$$\mathcal{F}_{N+K}|(\Lambda_N + m)$$

is fiberwise homotopy equivalent to the fiberwise suspension

$$\mathcal{F}_N \oplus_{\Lambda_N} \mathcal{G}_m^{(K)}$$

in the sense that they have the same fronts and, for all $\Gamma \subset \Lambda_N$ transversal to the front $\Phi_N \subset \Lambda_N$, there is a simultaneous T^k-equivariant homotopy equivalence of the subdivisions $\mathcal{S}_N \times \Gamma = F^+ \cup F^-$ determined by

$$\mathcal{F}_{N+K}|_{\Gamma+m} \qquad \text{and} \qquad \mathcal{F}_N|_\Gamma \oplus_\Gamma \mathcal{G}_m^{(K)} .$$

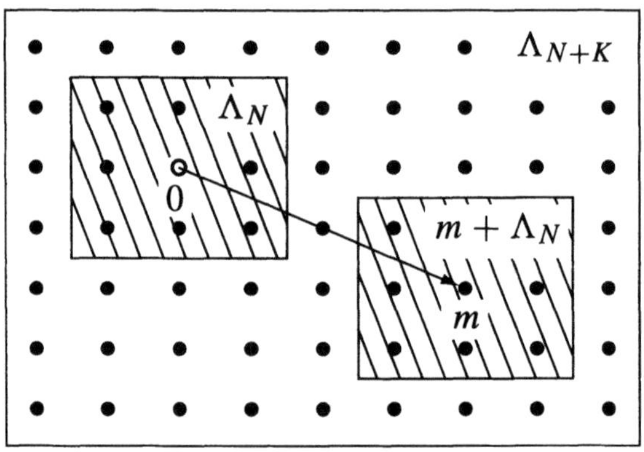

Fig. 4

This homotopy equivalence will serve us as a basis for constructing a «semi-infinite limit» of our finite-dimensional approximations and at the same time for introducing a $\mathbb{Z}^k$-action (Novikov's structure) in this limit.

4.5 Examples: Quadratic generating families. We consider here «generating families» with $N = N_1$ (which actually generate nothing since $N_2 = 0$).

(1) $k = n = 1$. The torus T^1 acts by $\exp(2\pi\lambda)$ on $\mathbb{C}^1$. A straightforward computation shows that $\mathcal{Q} : \mathbb{C}^{2N} \to \mathbb{R}$ is an Hermitian form with the matrix $i(1-U)/(1+U)$ which is a Cayley transform of the unitary matrix of the cyclic shift

$$U = \begin{bmatrix} 0 & 1 & & \\ & \ddots & \ddots & \\ & & & 1 \\ -1 & & & 0 \end{bmatrix}$$

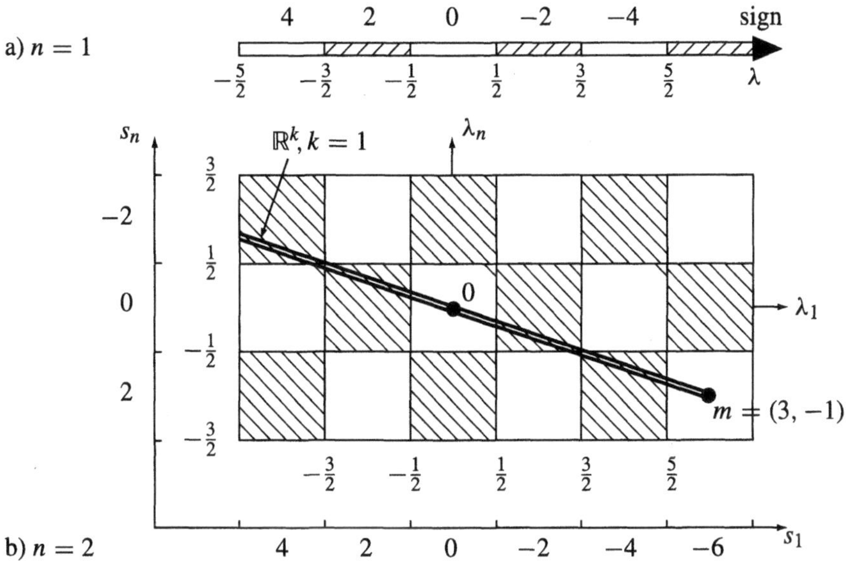

Fig. 5

The spectrum of $\mathcal{Q}$ is $\tan(\frac{\pi}{2N}\ell)$ where $-N < \ell < N$ is *odd integer*. The matrix of $\mathcal{H}$ is Cayley's transform of the *scalar* matrix $t = \exp(\pi\lambda/N)$, that is, $\tan(\frac{\pi\lambda}{2N})$. We find that the maximal cube Λ_N is the interval $-N < \lambda < N$, and the *signature* of the Hermitian generating family $\mathcal{F}_N = \mathcal{Q} - \mathcal{H}$ depends linearly on λ (as displayed on Figure 5a),

$$\text{sign}\mathcal{F}_\lambda^{(N)} = -2\ell \ \text{ for } \ \ell - \frac{1}{2} < \lambda < \ell + \frac{1}{2}, \quad \ell \in \mathbb{Z}. \tag{4.4}$$

(2) $k = n > 1$. It is the direct sum of n copies of example 1. The representation of the torus T^n in $\mathbb{C}^{2nN}$ decomposes by its n coordinate characters into n isotypical $2N$-dimensional components. The *equivariant signature* of an invariant Hermitian form (by definition it is the collection of signatures of its restrictions to isotypical components of the representation) is in this case simply a vector $\text{sign} = (s_1, \ldots, s_n)$ of n even integers. Such a vector-function $\lambda \mapsto \text{sign}\mathcal{F}_\lambda^{(N)}$ is the direct sum of n copies of (4.4) (see Figure 5b for $k = n = 2$).

(3) $k < n$. The generating family in this case is obtained by restriction of the generating family of example 2 to the subspace $\mathbb{R}^k \subset \mathbb{R}^n$ in the parameter space.

(4) The Hermitian form $\mathcal{G}_m^{(K)}$ from Proposition 4.4 is just the member of the family of example 2 with $\lambda = m \in \mathbb{Z}^k \subset \mathbb{R}^n$, and its T^n-equivariant signature can also be found on the diagram 5b.

5 Equivariant cohomology

We formulate here following [Hs] some facts about torus-equivariant cohomology and then describe the topological properties of generating families.

Let X be a topological space provided with an action of a compact Lie group G. Equivariant cohomology $H_G^*(X)$ is defined as the usual singular cohomology $H^*(X_G)$ of the homotopy quotient

$$X_G = (X \times EG)/G .$$

Here $EG \to BG$ is the universal principal G-bundle over the classifying space BG. The canonical projection

$$(X \times EG)/G \longrightarrow EG/G$$

provides $H_G^*(X)$ with a module structure over the characteristic class algebra $H^*(BG)$ playing the role of the coefficient ring $H_G^*(pt)$ in equivariant cohomology theory.

In this paper we need only equivariant cohomology of torus actions. In this case the coefficient ring $H^*(BT^n)$ is a polynomial algebra in n variables $u = (u_1, \dots, u_n)$ of degree 2 since the universal bundle of the circle T^1 is the Hopf bundle $S^\infty \to \mathbb{C}P^\infty$. We will assume all singular cohomology to be with complex (rather than integer) coefficients so that $H^*(Y) \stackrel{\text{def}}{=} H^*(Y, \mathbb{C})$ and thus

$$H_{T^n}^*(pt) = \mathbb{C}[u_1, \dots, u_n] .$$

We list below some properties of T^n-equivariant cohomology (see [Hs]).

1. Let X be a T^n-space, $T^k \subset T^n$ a subgroup. Then the homotopy quotients X_{T^n} and X_{T^k} form the bundle

$$X_{T^k} \xrightarrow{T^n/T^k} X_{T^n}$$

since ET^n can be considered as ET^k if provided with the subgroup action. The spectral sequence of this bundle has E_2-term

$$E_2^{p,q} = H_{T^n}^p(X) \otimes H^q(T^n/T^k) \tag{5.1}$$

and converges to $H_{T^k}^*(X)$. The algebra $H^*(T^n/T^k)$ is the exterior algebra with $n-k$ generators of degree 1. Their transgression images in $H^2(BT^n)$ generate in $\mathbb{C}[u_1, \dots, u_n]$ the ideal I of the subspace $\mathbb{R}^k = \text{Lie } T^k \subset \text{Lie } T^n = \mathbb{R}^n$. It will be important for us that the E_2-term (5.1) is nothing but *the Koszul complex* (see [GrH]) *of these $n-k$ degree 2 elements in the algebra $H_{T^n}^*(X)$.*

2. *Künneth spectral sequence* for equivariant cohomology is obtained when one considers the product $X \times Y$ of two G-spaces as a $G \times G$-space provided with the action of the diagonal subgroup $G \subset G \times G$. It has the E_2-term

$$E_2^{*,q} = \mathrm{Tor}^q_{H^*(BG\times BG)}(H^*(BG) \otimes H^*(X_G \times Y_G)) \ .$$

In the special case when $H^*_G(Y)$ *is a free* $H^*_G(pt)$-module, the spectral sequence reduces to one line $q = 0$ *provided at least that G is a torus*. Indeed, a resolution of $H^*(X_G \times Y_G)$ over $H^*(BG \times BG) = \mathbb{C}[u, v]$ is just the tensor product of a resolution of $H^*(X_G)$ over $\mathbb{C}[u]$ and a free $\mathbb{C}[v]$-module $H^*(Y_G)$ while $\mathbb{C}[u, v]$-module $H^*(BG)$ is $\mathbb{C}[u, v]/(u_i = v_i)$. This results in $\mathrm{Tor}^q = 0$ for $q > 0$. We conclude that

$$H^*_G(X \times Y) = H^*_G(X) \otimes H^*_G(Y)$$

if G is a torus and $H^*_G(Y)$ is free.

Our objective now is to study equivariant cohomology of some spaces related to invariant homogeneous functions and their families.

We begin with a geometric fact well known in singularity theory (cf. [AVG], v.2). Let $\mathcal{F} : \mathbb{C}^M \to \mathbb{R}$ be a homogeneous degree 2 function, $\hat{\mathcal{F}} : \mathbb{C}^{M+1} \to \mathbb{R}$ be its suspension

$$\hat{\mathcal{F}}(x, z) = \mathcal{F}(x) + |z|^2 \ .$$

We want to compare homotopy types of the corresponding ray spaces $\hat{F}_\pm$ and $F_\pm$.

Proposition 5.1 $\hat{F}_+ \approx F_+ * S^1$, $\hat{F}_- \approx F_-$ *where $*$ means the join of topological spaces and* $S^1 = \{z \in \mathbb{C} \,|\, |z| = 1\}$. *If $\mathcal{F}$ is invariant relative to a T^1-action on $\mathbb{C}^M$, then the homotopy equivalences $\approx$ can be done equivariant with respect to the diagonal action on $\mathbb{C}^{M+1}$: $t(x, z) = (tx, tz)$. If $\mathcal{F}$ depends on additional parameters then the homotopy depends continuously on them.*

Proof. Let us begin with the case when z is a *real* variable. In Figure 6 the north semi-sphere D of rays in $\mathbb{C}^M \oplus \mathbb{R}$ is shown. It splits into $\hat{F}_+$ and $\hat{F}_-$. Every meridional semicircle (through the north pole $P = (0, 1)$) meets $\hat{F}_+$ along an arc about the pole. The meridional contraction from the pole towards the equatorial sphere ∂D contracts $\hat{F}_-$ to F_-. On the other hand, the same contraction transforms the pair $(\hat{F}_+ \cap D, F_+)$ to (D, F_+). Obviously the pair (D, F_+) is homotopy equivalent to (CF_+, F_+) where CF_+ is the cone (from P) over its bottom $F_+ \subset \partial D$.

Now, if z is a complex variable, we have a polar circle S^1, and the ray sphere in $\mathbb{C}^{M+1}$ is the union of identical semi-spheres D, parametrized by S^1, with common equator ∂D. From what was said above we conclude that $\hat{F}_+$ is homotopy equivalent to

$$S^1 \times CF_+/((z, x) \sim (z', x) \text{ for } x \in F_+) \ ,$$

that is, to $S^1 * F_+$.

Since all the homotopies above are carried out in a canonical way, they respect group actions and parametric dependence. □

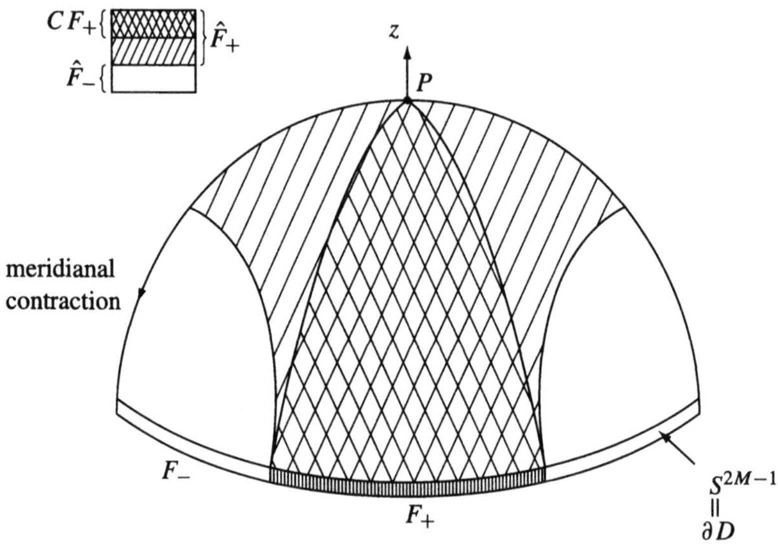

Fig. 6

It is time now to disclose the sort of equivariant cohomology that we intend to use in the proof of our main theorem. Let us consider the family

$$\mathcal{F}_N : \mathbb{C}^M \times \Lambda_N \longrightarrow \mathbb{R} , \qquad M = 2nN ,$$

of homogeneous T^k-invariant functions parametrized by the cube Λ_N. Let

$$\mathcal{S}_N \times \Lambda_N = F^+ \cup F^-$$

be the decomposition of the ray space into its positive and negative parts (with respect to $\mathcal{F}_N$), and

$$\partial F^\pm = F^\pm \cap (\mathcal{S}_N \times \partial\Lambda_N)$$

— the parts of $F^\pm$ situated over the boundary of the cube Λ_N. We intend to deal with the *relative T^k-equivariant cohomology* (with complex coefficients)

$$H^*_{T^k}(F^-, \partial F^-) .$$

In order to figure out the algebraic structure of this module it is convenient to introduce the pair (CF^-, F^-), where $CF^- \subset \mathbb{C}^M \times \Lambda_N$ means *the fiberwise cone* over F^-, that is, the cone of the projection map $F^- \to \Lambda_N$. It is natural to call the quotient space CF^-/F^- *the fiberwise suspension* and denote it by ΣF^- referring to the quotient point F^- as the distinguished one.

From the long exact sequence of the pair (CF^-, F^-)

$$\longrightarrow H^{*-1}_{T^k}(F^-, \partial F^-) \longrightarrow H^*_{T^k}(\Sigma F^-, \Sigma\partial F^-) \longrightarrow$$
$$\longrightarrow H^*_{T^k}(CF^-, C\partial F^-) \longrightarrow$$

one extracts the short exact segments

$$0 \longrightarrow \tilde{H}^{*-1}_{T^k}(F^-, \partial F^-) \longrightarrow H^*_{T^k}(\boldsymbol{\Sigma} F^-, \boldsymbol{\Sigma}\partial F^-) \longrightarrow J_F \longrightarrow 0 \qquad (5.2)$$

where J_F is the *kernel* of the homomorphism

$$H^*_{T^k}(\Lambda_N, \partial\Lambda_N) \longrightarrow H^*_{T^k}(F^-, \partial F^-) ,$$

and $\tilde{H}^*$ denotes its *cokernel*. We will call this cokernel the *reduced* equivariant cohomology. Since $\Lambda_N/\partial\Lambda_N$ is homeomorphic to a sphere with the trivial torus action we find $H^*_{T^k}(\Lambda_N, \partial\Lambda_N)$ to be *a free $H^*_{T^k}(pt)$-module of rank 1 generated by the fundamental cocycle of the sphere, and J_F is a submodule in this module.* The exact segments (5.2) reduce computation of $H^*_{T^k}(F^-, \partial F^-)$ to that of J_F and $\tilde{H}^*_{T^k}(F^-, \partial F^-)$. We state below how the suspension of Proposition 4.4 affects these data.

Proposition 5.2 *Let $\hat{\mathcal{F}}_{N+K} = \mathcal{F}_N \oplus_{\Lambda_N} \mathcal{G}^{(K)}_m$ be the fiberwise direct sum of the family $\mathcal{F}_N$ with the Hermitian form $\mathcal{G}^{(K)}_m : \mathbb{C}^{2Kn} \to \mathbb{R}$ of Proposition 4.4 and $\hat{F}^-, F^-$ denote the corresponding ray spaces. Then*

$$\tilde{H}^*_{T^k}(\hat{F}^-, \partial\hat{F}^-) = \tilde{H}^*_{T^k}(F^-, \partial F^-) \otimes I^{K+m} , \qquad (5.3)$$

$$J_{\hat{F}} = J_F \otimes I^{K+m} ,$$

where $\otimes$ means the tensor product of $H^*_{T^k}(pt)$-modules, and I^{K+m} is the principal ideal in $H^*_{T^k}(pt)$ generated by the monomial $u^{K+m} = u_1^{K+m_1} \cdot u_2^{K+m_2} \cdot \ldots \cdot u_n^{K+m_n}$ (recall that $H^*_{T^k}(pt)$ is a quotient of $H^*_{T^n}(pt) = \mathbb{C}[u_1, \ldots, u_n]$).

Proof. It consists of the following three kinds of arguments.

(1) An equivariant decomposition of the Hermitian form $\mathcal{G}^{(k)}_m$ into the direct sum of one-dimensional forms and multiple application of Proposition 5.1 shows that $\hat{F}^-$ is *the fiberwise join $F^- *_{\Lambda_N} G^-$* of F^- with the negative ray space G^- of the Hermitian form $\mathcal{G}^{(K)}_m$.

(2) Equivariant cohomology of joins. Let $\mathbb{Y}$ be a family of T^k-spaces parametrized by Λ, X be another T^k-space, and $X *_\Lambda \mathbb{Y}$ be their fiberwise join. The fiberwise suspension operation $\boldsymbol{\Sigma}$ transforms functorially fiberwise joins into *tensor products*:

$$\boldsymbol{\Sigma}(X *_\Lambda \mathbb{Y}) \simeq \Sigma X \otimes \boldsymbol{\Sigma}\mathbb{Y}$$

where $\otimes$ is the product operation in the category of punctured spaces,

$$(A, pt) \otimes (B, pt) = A \times B / (A \times pt) \cup (pt \times B) .$$

Now Proposition 5.2 reduces to the (relative) equivariant Künneth formula provided that $H^*_{T^k}(\Sigma G^-, pt)$ is a free $H^*_{T^k}(pt)$-module (see point 2 above).

(3) The following computation of $H^*_{T^k}(\Sigma G^-, pt)$ completes the proof.

Proposition 5.3 $J_G = I^{K+m}, \tilde{H}^*_{T^k}(G^-) = 0.$

We consider here $\mathcal{G}_m^{(K)}$ formally as a «family» parametrized by a point so that $J_G \subset H^*_{T^k}(pt)$.

Proof. inductive application of Proposition 5.1. Let us begin with the S^1-invariant form $a|z|^2$ on $\mathbb{C}^1$. Then

$$G^- = \begin{cases} S^1 & a \leq 0 \\ \varnothing & a > 0, \end{cases}$$

and a simple straightforward computation gives

$$H^*_{S^1}(\Sigma G^-, pt) = \begin{cases} u\mathbb{C}[u] & a \leq 0 \\ \mathbb{C}[u] & a > 0 \,. \end{cases}$$

The same answer is still valid for the «multiple» circle action $(e^{i\phi}, z) \mapsto e^{in\phi}z$, $n \in \mathbb{Z}$ (it is essential here that we use rational — or even complex — coefficients instead of integers).

Now suppose that T^k acts on $\mathbb{C}^1$ through its character $\chi : T^k \to S^1$ and denote by u an equation of $\ker\chi$ in Lie T^k (identifying $H^*_{T^k}(pt)$ with $\mathbb{C}[\text{Lie } T^k]$). Then

$$H^*_{T^k}(\Sigma G^-, pt) = \begin{cases} uH^*_{T^k}(pt) & a \leq 0 \\ H^*_{T^k}(pt) & a > 0 \,. \end{cases}$$

Let us now consider the generating Hermitian form $\mathcal{G}_m^{(K)}$ of the «discrete loop» $-id \circ t \circ \ldots \circ t$ ($2K$ times), where $t = \exp(\pi i m/K)$ is a hamiltonian transformation of $\mathbb{C}^1$. Then $\mathcal{G}_m^{(K)}$ has $K - m$ positive and $K + m$ negative squares (cf. Example 4.5). Applying Proposition 5.1 together with the equivariant Künneth formula, we find

$$H^*_{T^k}(\Sigma G^-, pt) = u^{K+m} H^*_{T^k}(pt) \,.$$

In the end we notice that $\mathcal{G}_m^{(k)}$ in Proposition 5.3 corresponds to the transformation

$$t = \text{diag}(e^{\pi i m_1/K}, \ldots, e^{\pi i m_n/K})$$

on $\mathbb{C}^n$ and is the direct sum of n $2K$-dimensional Hermitian forms considered above. Applying Proposition 5.1 and the Künneth formula once again we conclude that

$$H^*_{T^k}(\Sigma G^-, pt) = u_1^{K+m_1} \ldots u_n^{K+m_n} H^*_{T^k}(pt) \,,$$

where $u_1, \ldots, u_n$ are images in $H^*_{T^k}(pt)$ of generators in $H^*_{T^n}(pt)$. □

Our last step in this section is an asymptotic computation of the relative equivariant cohomology $H^*_{T^k}(F^-, \partial F^-)$ for the *quadratic* generating family $\mathcal{F}_N : \mathbb{C}^{2nN} \times \Lambda_N \to \mathbb{R}$ of Example 4.5 *in the limit* $N \to \infty$. To my taste it is the most elegant point in the proof of our main theorem.

We begin with the case $k = n$, that is, $\{\Lambda_N\}$ is a sequence of cubes exhausting the parameter space $\mathbb{R}^n = \text{Lie } T^n$. Our objective is to restrict the quadratic families $\mathcal{F}_N$ to *a convex compact r-dimensional polyhedron* $\Gamma \subset \mathbb{R}^n$ and calculate

$$\mathcal{H}^*_{T^n}(\Gamma) = \lim_{N\to\infty} u^{-N} \cdot H^{*+2N}_{T^n}(F_N^-|_\Gamma, F_N^-|_{\partial\Gamma}) \,.$$

(We will see in the next section how such a limit corresponds to some semi-infinite cohomology.)

The answer is that the reduced part of this cohomology is trivial ($\tilde{\mathcal{H}}^*_\Gamma = 0$) and the corresponding «kernel» module J_Γ can be easily described in terms of Newton diagrams in the same space $\mathbb{R}^n$. This means that we should interpret the lattice $\mathbb{Z}^n$ in the parameter (!) space $\mathbb{R}^n$ as the lattice of exponents of monomials $u^\ell = u_1^{\ell_1} \dots u_n^{\ell_n}$. To be more precise, *we introduce in* $\mathbb{R}^n$ *new coordinates* $\mu_1 = \lambda_1 - \frac{1}{2}, \dots, \mu_n = \lambda_n - \frac{1}{2}$, *associate an integer point* $\mu = \ell$ *with the monomial* $u^\ell \in \mathbb{C}[u, u^{-1}]$ *and attach some open Newton diagram* Δ_Γ *to the polyhedron* Γ:

$$\Delta_\Gamma = \{\mu \in \mathbb{R}^n | \exists \gamma \in \Gamma : \mu_i > \gamma_i, i = 1, \dots, n\} = \Gamma + \mathbb{R}^n_+$$

(see Figure 7a).

Proposition 5.4

$$\mathcal{H}^*_{T^n}(\Gamma) = H^r(\Gamma, \partial\Gamma) \otimes_{\mathbb{C}} \mathbb{C}[u, u^{-1}]/J_\Gamma \,,$$

where J_Γ *is the* $\mathbb{C}[u]$*-submodule in* $\mathbb{C}[u, u^{-1}]$ *generated by the monomials* u^ℓ *with* $\ell \in \Delta_\Gamma$.

Proof. The idea is to apply Leray spectral sequences [GrH]. The Leray spectral sequence of a map $X \xrightarrow{\pi} \Lambda$ converges to the cohomology of the «total space», $H^*(X)$, and has

$$E_2^{p,q} = H^p(\Lambda, \mathcal{J}^q)$$

where $\mathcal{J}^*$ is the local coefficient sheaf whose fiber at $\lambda \in \Lambda$ is $\mathcal{J}^*_\lambda = H^*(\pi^{-1}(\lambda))$. The same construction applies to equivariant cohomology provided that the group action on X is fiberwise.

In our case $\Lambda = \Gamma \subset \mathbb{R}^n$, $X = F^-$, and the equivariant cohomology of fibers has just been described in Proposition 5.3:

$$\mathcal{J}_\lambda = \mathbb{C}[u]/(u^{N+m}) \,, \qquad m = [\lambda + \tfrac{1}{2}] = [\mu + 1] \,,$$

(see Examples 4.5.1, 4.5.2). We will find below

$$E_2^{p,*} = H^p(\Gamma, \partial\Gamma; \mathcal{J}^*)$$

and observe that it is non-trivial only for $p = r$. This will mean in particular that $E_2^{*,*} = E_\infty^{*,*} = H^*_{T^n}(F^-|_\Gamma, F^-|_{\partial\Gamma})$.

We will carry out our computation of $E_2^{*,*}$ in the limit of large N. This makes convenient the following «re-grading» of the sheaf $\mathcal{J}^*$:

$$\hat{\mathcal{J}}^*_\lambda = \lim_{N\to\infty} u^{-N}\mathbb{C}[u]/(u^m) = \mathbb{C}[u,u^{-1}]/\mathbb{C}[u]\cdot u^m \ .$$

The idea of our computation of $H^*(\Gamma, \partial\Gamma; \hat{\mathcal{J}}^*_\lambda)$ is to decompose the constant sheaf $\mathbb{C}[u,u^{-1}]$ into one-dimensional monomial components $\mathbb{C}\cdot u^\ell$ and apply the following pretty obvious general lemma.

Lemma *Let* $\mathbb{C}$ *be a constant sheaf on a topological space* X, $\mathbb{I}_Y$ *be the subsheaf vanishing on the closed subspace* $Y \subset X$, $\mathbb{C}/\mathbb{I}_Y$ *be the quotient sheaf supported at* Y. *Then*

$$H^*(X, \mathbb{I}_Y) = H^*(X,Y)\ , \qquad H^*(X, \mathbb{C}/\mathbb{I}_Y) = H^*(Y)\ .$$

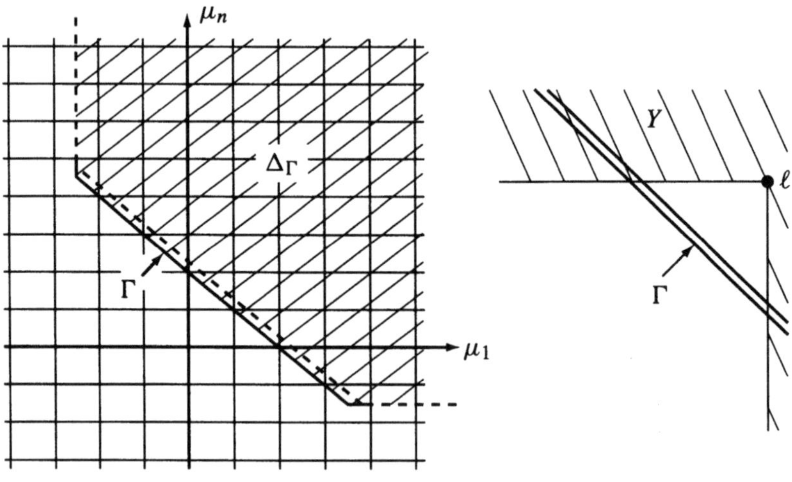

a) b)

Fig. 7

The monomial $u^\ell \in \mathbb{C}[u,u^{-1}]$ is contained in the $\mathbb{C}[u]$-submodule (u^m) only iff $\ell \geq m$. This event is supported in the open «last orthant» $\mu_j < \ell_j$, $j = 1, \ldots, n$. Let Y be its complement in $\mathbb{R}^n$. Applying the lemma we find (Figure 7b)

$$H^*(\Gamma, \partial\Gamma; \mathbb{C}/\mathbb{I}_Y|_\Gamma) = \begin{cases} \mathbb{C} & * = r \text{ and } \Gamma \subset Y \\ 0 & \text{otherwise} \end{cases}$$

since $\Gamma/\partial\Gamma$ is an r-dimensional sphere and $\Gamma \cap Y$ contracts to $\partial\Gamma$ if $\Gamma \not\subset Y$. Taking the direct sum over all the monomials u^ℓ, we get

$$H^*(\Gamma, \partial\Gamma; \hat{\mathcal{J}}^*) = H^r(\Gamma, \partial\Gamma) \otimes_{\mathbb{C}} \mathbb{C}[u, u^{-1}]/J_\Gamma$$

where the $\mathbb{C}[u]$-submodule J_Γ is spanned by those monomials u^ℓ whose «last orthant» $\{\mu|\ell + \mathbb{R}^n_-\}$ meets Γ. Or, in other words, (Figure 7a)

$$J_\Gamma = \langle u^\ell | \ell \in \mathbb{Z}^n : \ell \in \Gamma + \mathbb{R}^n_+ \rangle .$$

Now let us consider the case where Γ_N is a sequence of, say, $k-1$-dimensional cubes exhausting a $(k-1)$*-dimensional affine subspace* Γ *in* $\mathbb{R}^n$ *parallel to the subspace* $\mathbb{R}^k \cap p^{-1}(0) \subset \mathbb{R}^k \subset \mathbb{R}^n$. Then $J_{\Gamma_N} \subset J_{\Gamma_{N+1}} \subset \dots$ and the union is the submodule $J_\Gamma \subset \mathbb{C}[u, u^{-1}]$ spanned by the monomials u^ℓ supported at the Newton diagram $\Delta_\Gamma = \Gamma + \mathbb{R}^n_+$. The submodule J_Γ should be compared to

$$J_r = \langle u^\ell, \ell \in \mathbb{Z}^k \subset \mathbb{Z}^n | \langle p, \ell \rangle \geq r \rangle_{\mathbb{C}[u]}$$

considered in the end of §1, and one concludes that they are basically the same (up to a shift of the origin in the lattice $\mathbb{Z}^n$). To be more precise, J_r ($r \in \mathbb{R}$) and J_Γ (Γ is parallel to $\mathbb{R}^k \cap p^{-1}(0)$) majorate each other:

$$\begin{aligned} &\forall \Gamma\, \exists r_+ > r_- : J_{r_+} \subset J_\Gamma \subset J_{r_-}, \\ &\forall r\, \exists \Gamma_+, \Gamma_- : J_{\Gamma_+} \subset J_r \subset J_{\Gamma_-}. \end{aligned} \tag{5.5}$$

Let us denote $\mathcal{J}_\Gamma$ (resp. $\mathcal{J}_r$) projections of J_Γ (resp. J_r) to $\mathcal{R} = \mathbb{C}[u, u^{-1}]/I \cdot \mathbb{C}[u, u^{-1}]$, where I is the ideal of $\mathbb{C}^k \subset \mathbb{C}^n$ in $\mathbb{C}[u]$. The same inclusions (5.5) hold for $\mathcal{J}_\Gamma, \mathcal{J}_r$ of course.

Corollary 5.5 ($k < n$).

$$\lim_{N \to \infty} \mathcal{H}^*_{T^k}(\Gamma_N) = H^{k-1}(\Gamma, \infty) \underset{\mathbb{C}}{\otimes} \mathcal{R}/\mathcal{J}_\Gamma.$$

Proof. T^k-equivariant cohomology is related to T^n-equivariant cohomology by means of Serre's spectral sequence whose E_2-term is the Koszul complex (5.1). For a regular linear function $p : \mathbb{R}^k \to \mathbb{R}$ the $n-k$ equations of $\mathbb{R}^k$ in $\mathbb{R}^n$ form a regular sequence in the algebra

$$\mathbb{C}[u, u^{-1}]/\mathbb{C}[u, u^{-1}]J_\Gamma$$

since it is true for J_r instead of J_Γ (as it was shown in §1) and due to inclusions (5.5). Therefore the Koszul complex has only 0-dimensional (in our grading) homology

$$E_2^{*,0} = (\mathbb{C}[u, u^{-1}]/(\mathbb{C}[u, u^{-1}]I + J_\Gamma)) \underset{\mathbb{C}}{\otimes} H^{k-1}(\Gamma, \infty),$$

and the spectral sequence degenerates, that is $E_2 = E_\infty$.

6 Semi-infinite cohomology

We combine here Propositions 4.4 and 5.2 in order to define semi-infinite cohomology related with our generating families, describe its Morse-theoretic properties and prove Theorem 1.1.

Throughout this section $p : \mathbb{R}^k \to \mathbb{R}$ is a regular primitive integer value of the momentum map of the T^k-action in $\mathbb{C}^n$.

Let $\mathcal{F}_N : \mathcal{S}_N \times \Lambda_N \to \mathbb{R}$ be generating families (of the hamiltonian transformation $h^1 : M_p \to M_p$) constructed in §4 so that $\Lambda_N \subset \Lambda_{N'} \subset \ldots \subset \mathbb{R}^k$ form an exhausting sequence of cubes. Let

$$\Gamma_N(\nu) = \Lambda_N \cap p^{-1}(\nu)$$

be their intersection with hyperplane levels of the linear function p. Let $F_N^-(\nu)$, $\partial F_N^-(\nu)$ denote negative ray spaces of the families $\mathcal{F}_N|_{\Gamma_N(\nu)}$, $\mathcal{F}_N|_{\partial\Gamma_N(\nu)}$. For $N < N'$ we have natural homomorphisms

$$H^*_{T^k}(F^-_{N'}(\nu)|_{\Gamma_N(\nu)}, F^-_{N'}(\nu)|_{\partial\Gamma_N(\nu)}) \to H^*_{T^k}(F^-_{N'}(\nu), \partial F^-_{N'}(\nu)). \tag{6.1}$$

For generic ν the hyperplane $p^{-1}(\nu)$ is transversal to the front of the generating families $\mathcal{F}_N$ and, in accordance with Proposition 4.4, there is an equivariant homotopy equivalence of the families

$$\mathcal{F}_{N'}|\Gamma_N(\nu) \text{ and } (\mathcal{F}_N \oplus \mathcal{G}_0^K)|\Gamma_N(\nu), \quad K = N' - N.$$

Together with isomorphisms (5.3), (5.4) of Proposition 4.4 (applied to the families restricted to $\Gamma_N(\nu)$) this gives rise to the raising dimensions monomorphisms (6.2), (6.3):

$$\begin{array}{ccc} H^{*+2N}_{T^k}(F^-_N(\nu), \partial F^-_N(\nu)) & \to & H^{*+2N'}_{T^k}(F^-_{N'}(\nu)|_{\Gamma_N(\nu)}, F^-_{N'}(\nu)|_{\partial\Gamma_N(\nu)}) \\ \uparrow & & \uparrow \\ H^{*+2N}_{T^k}(\Gamma_N(\nu), \partial\Gamma_N(\nu)) & \xrightarrow{u^K \times} & H^{*+2N'}_{T^k}(\Gamma_N(\nu), \partial\Gamma_N(\nu)). \end{array} \quad \begin{array}{c}(6.2)\\ \\ (6.3)\end{array}$$

Combining (6.1) and (6.2) we obtain homomorphisms

$$H^{*+2N}_{T^k}(F^-_N(\nu), \partial F^-_N(\nu)) \to H^{*+2N'}_{T^k}(F^-_{N'}(\nu), \partial F^-_{N'}(\nu)). \tag{6.4}$$

Definition 6.1 We define *semi-infinite equivariant cohomology*

$$\mathcal{H}^*_{T^k}(F^-(\nu)) = \lim_{N\to\infty} H^{*+2N}_{T^k}(F^-_N(\nu), \partial F^-_N(\nu))$$

as the direct limit over the directed system (6.4).

Let us list some algebraic properties of $\mathcal{H}^*_{T^k}$.

1. Spaces $\mathcal{H}^*_{T^k}(F^-(\nu))$ descend the structure of graded $H^*_{T^k}(pt)$-modules of ordinary equivariant cohomology.

2. The direct limit $\mathcal{H}^*_{T^k}(\Gamma(\nu))$ of

$$H^{*+2N}_{T^k}(\Gamma_N(\nu), \partial\Gamma_N(\nu)) \xrightarrow{u^K \times} H^{*+2N'}_{T^k}(\Gamma_{N'}(\nu), \partial\Gamma_{N'}(\nu))$$

is isomorphic to $H^*(\mathbb{R}^{k-1}, \infty) \otimes_{\mathbb{C}} \mathcal{R}$, where

$$\mathcal{R} = \mathbb{C}[u, u^{-1}]/\mathbb{C}[u, u^{-1}]I,$$

(I is the ideal of $\mathbb{C}^k \subset \mathbb{C}^n$ in $\mathbb{C}[u]$) and $\mathbb{R}^{k-1} = \Gamma(\nu) = p^{-1}(\nu)$. The diagram (6.2–6.3) gives rise to the «augmentation homomorphisms» of $\mathbb{C}[u]$-modules

$$\mathcal{H}^*_{T^k}(\Gamma(\nu)) \to \mathcal{H}^*_{T^k}(F^-(\nu)). \tag{6.5}$$

We denote its kernel $\mathcal{J}^*(F^-(\nu))$ and cokernel $\tilde{\mathcal{H}}^*_{T^k}(F^-(\nu))$ and call the latter the *reduced* semi-infinite cohomology.

3. Our computation in the end of §5 shows that for the quadratic generating families of Example 4.5.3 (let us denote them here $\mathcal{G}_N$) $\tilde{\mathcal{H}}^*(\mathcal{G}^-(\nu)) = 0$ and

$$\mathcal{J}^*(G^-(\nu)) = \mathcal{J}_{\Gamma(\nu)} \quad \text{(in notations of §5)}. \tag{6.6}$$

In the very bottom of all our constructions there is a homogeneous degree 2 hamiltonian in $\mathbb{C}^n$. Such a hamiltonian $\mathcal{H}$ can be bounded by quadratic hamiltonians, $\mathcal{H}_- < \mathcal{H} < \mathcal{H}_+$. It is easy to see that this leads to an estimate of ray spaces

$$G^-_N(\nu_-) \subset F^-_N(\nu) \subset G^-_N(\nu_+)$$

for some $\nu_- < \nu < \nu_+$ and to homomorphisms

$$\mathcal{H}^*_{T^k}(G^-(\nu_+)) \to \mathcal{H}^*_{T^k}(F^-(\nu)) \to \mathcal{H}^*_{T^k}(G^-(\nu_-)).$$

Together with (6.5), (6.6) and (5.5) this gives rise to the estimate

$$\mathcal{J}_{r_+} \subset \mathcal{J}^*(F^-(\nu)) \subset \mathcal{J}_{r_-} \tag{6.7}$$

(where $\mathcal{J}_{r_\pm}$ are images of $J_{r_\pm}$ of (5.5) in $\mathcal{R}$, see §1). In particular $\mathcal{J}^*(F^-(\nu))$ is a non-trivial submodule in $\mathcal{R}$.

4. Applying Proposition 4.4 and 5.2 in a manner similar to our definition of semi-infinite cohomology but with non-zero $m \in \mathbb{Z}^k$, we obtain isomorphisms

$$\mathcal{U}_m : \mathcal{H}^*_{T^k}(F^-(\nu)) \xrightarrow{\simeq} \mathcal{H}^*_{T^k}(F^-(\nu + p(m))$$

which form the lattice's action: $\mathcal{U}_m \circ \mathcal{U}_{m'} = \mathcal{U}_{m+m'}$. This is nothing but Novikov's action of the fundamental group $H_2(M_p, \mathbb{Z})$ of the loop space $\mathcal{L}M_p$. Notice that the sublattice $\mathbb{Z}^{k-1} = p^{-1}(0) \cap \mathbb{Z}^k$ acts on the semi-infinite cohomology space $\mathcal{H}^*_{T^k}(F^-(\nu))$ itself providing it with structure of $\mathbb{C}[\mathbb{Z}^{k-1}]$-module.

The action of $\mathcal{U}_m$ on $\mathcal{R}$ is just the multiplication by $u^m = u_1^{m_1} \dots u_n^{m_n}$. Therefore *multiplication by u^m with $m \in \mathbb{Z}^k \subset \mathbb{Z}^n$ is an isomorphism*

$$\mathcal{U}^m : \mathcal{J}^*(F^-(\nu)) \xrightarrow{\simeq} \mathcal{J}^*(F^-(\nu + p(m)). \tag{6.8}$$

In particular $\mathcal{J}^*(F^-(\nu))$ is invariant with respect to such multiplication if $p(m) = 0$.

Now let us consider the action functional $\mathfrak{p} : A \to \mathbb{R}$ of §2. According to Proposition 2.2 its (lattices of) critical orbits correspond to fixed points of the hamiltonian transformation $h^1 : M_p \to M_p$.

Proposition 6.2 *Suppose that the function $\mathfrak{p}$ does not have critical values on the closed segment $[\nu_0, \nu_1] \subset \mathbb{R}$. Then $\mathcal{H}^*_{T^k}(F_-(\nu_0)) \simeq \mathcal{H}^*(F_-(\nu_1))$.*

Proposition 6.3 *Suppose that the function $\mathfrak{p}$ has only one critical value within the segment $[\nu_0, \nu_1]$ and that all the critical T^k-orbits on this level are isolated. Let $v \in H^*_{T^k}(pt)$ be a positive degree element and $q_0 \in \mathcal{H}^*_{T^k}(F^-(\nu_0))$, $q_1 \in \mathcal{H}^*_{T^k}(F^-(\nu_1))$ be images of the same $q \in \mathcal{R}$ under the homomorphism* (6.5). *Then $q_0 = 0$ implies $vq_1 = 0$.*

Proof. Proof of Proposition 4.1 is based of course on Proposition 4.3 and «gradient flow» deformations between $F_N^-(\nu_0)$ and $F_N^-(\nu_1)$. But we should prevent their «topology» from «flowing out» through their boundaries $\partial F_N^-(\nu)$.

Lemma *If ν_* is a regular value of $\mathfrak{p}$ then there exists a non-zero segment $[\nu_* - \varepsilon, \nu_* + \varepsilon]$ and an exhausting sequence of cubes $\Lambda_N \subset \Lambda_{N+1} \subset \dots \subset \mathbb{R}^k$ such that all $\Gamma_N(\nu), \partial\Gamma_N(\nu)$ remain transversal to the front $\Phi \subset \mathbb{R}^k$ for ν from the segment.*

Applying this lemma and Proposition 3.3, we obtain a sequence of equivariant homotopy equivalences

$$(F_N^-(\nu_* - \varepsilon), \partial F_N^-(\nu_* - \varepsilon)) \approx (F_N^-(\nu_* + \varepsilon), \partial F_N^-(\nu_* + \varepsilon))$$

and thus an isomorphism of semi-infinite cohomology $\mathcal{H}^*_{T^k}(F^-(\nu_* - \varepsilon)) \simeq \mathcal{H}^*_{T^k}(F^-(\nu_* + \varepsilon))$. Choosing a finite subcovering of $[\nu_0, \nu_1]$ by such segments completes the proof.

Our proof of the lemma is based on the fact that the front Φ of the family $\mathcal{A} = \{\mathcal{A}_\lambda\}$ of action functionals is $\mathbb{Z}^k$-periodic. The hyperplane $\Gamma(\nu_*) = p^{-1}(\nu_*)$ is transversal to Φ since ν_* is regular value of p (Proposition 3.1). Due to Corollary 3.2 we can choose $k-1$ coordinate hyperplanes in $\Gamma(\nu_*)$ such that they and all their intersections are transversal to $\Phi \cap \Gamma(\nu_*)$. Then all their integer translations are also transversal to Φ (periodicity!). Now we can choose an exhausting sequence of growing integer cubes framed by this coordinate net. Corresponding $\Gamma_N(\nu)$ and $\partial\Gamma_N(\nu)$ will be transversal to Φ for $\nu = \nu_*$ and thus for all ν in some neighborhood of ν_* (we use here openness of transversality provided that $\Phi/\mathbb{Z}^k$ is compact).

The same arguments apply even if ν_* is an isolated critical value of p, but in this case $\Gamma(\nu_*)$ itself is tangent to Φ. In the proof of Proposition 4.2 one may assume without loss of generality that

(1) $\nu_0 = \nu_* - \varepsilon$, $\nu_1 = \nu_* + \varepsilon$,

(2) there is only one (lattice of) critical orbit(s) on the level ν_*,

(3) $(F_N^-(\nu_0), \partial F_N^-(\nu_0))$is embedded into$(F_N^-(\nu_1), \partial F_N^-(\nu_1))$ as the complement to a neighborhood of the critical orbit(s) (use isotopy of Proposition 3.3 outside of this neighborhood),

(4) the semi-infinite cohomology class $q \in \mathcal{R} \simeq \mathcal{H}^*_{T^k}(\mathbb{R}^{k-1}, \infty)$ is represented by a non-zero cohomology class

$$q \in H^*_{T^k}(F_N^-(\nu_1), \partial F_N^-(\nu_1))$$

which vanishes when restricted to $(F_N^-(\nu_0), \partial F_N^-(\nu_0))$.

The latter implies that q_1 is the image of some $\alpha \in H^*_{T^k}(F_N^-(\nu_1), F_N^-(\nu_0))$ under the connecting homomorphism δ in the long exact sequence

$$\to H^{*-1}_{T^k}(F_N^-(\nu_1), F_N^-(\nu_0)) \xrightarrow{\delta} H^*_{T^k}(F_N^-(\nu_1), \partial F_N^-(\nu_1)) \longrightarrow$$

$$\to H^*_{T^k}(F_N^-(\nu_0), \partial F_N^-(\nu_0)) \to \dots$$

The crucial point now is that in a neighborhood of the critical orbit the T^k-action is *free* — due to non-singularity of our initial toric manifold in fact. This implies that the equivariant cohomology $H^*_{T^k}(F_N^-(\nu_1), F_N^-(\nu_0))$ is isomorphic to the usual cohomology $H^*(F_N^-(\nu_1)/T^k, F_N^-(\nu_0)/T^k)$ of the quotient neighborhood of our critical orbit. The coefficient algebra $H^*_{T^k}(pt)$ acts trivially in the usual cohomology! Thus

$$v\alpha = 0 \text{ and } vc_1 = \delta(v\alpha) = 0.$$

Remark 6.4 In a similar manner one can show that in the long exact sequence of semi-infinite cohomology

$$\to \mathcal{H}^*_{T^k}(F^-(\nu_1), F^-(\nu_0)) \to \mathcal{H}^*_{T^k}(F^-(\nu_1)) \to \mathcal{H}^*_{T^k}(F^-(\nu_0)) \longrightarrow$$

the left term is a free rank 1 $\mathbb{C}[\mathbb{Z}^{k-1}]$-module provided that the segment $[\nu_0, \nu_1]$ contains only one $\mathbb{Z}^{k-1}$-lattice of *non-degenerate* critical T^k-orbits.

Proof of Theorem 1.1. Our ship is fully loaded now.

Let $p : \mathbb{R}^k \to \mathbb{R}$ be a primitive integer value of the momentum map $\mathbb{C}^n \to \mathbb{R}^{k*}$. We are going to count critical levels of the function $\hat{p} : A \to \mathbb{R}$ between two non-singular levels ν and $\nu + 1$. On every critical level there is at least one $\mathbb{Z}^{k-1}$-lattice of critical T^k-orbits, all corresponding to the same fixed point of the hamiltonian transformation $-\mathrm{id} \circ h^1$ on our toric manifold M_p (Propositions 2.2, 4.3). To the same fixed point there correspond therefore ℓ critical levels of $\hat{p}$ between regular levels ν and $\nu + \ell$.

Let $m = (m_1, \dots, m_n) \in \mathbb{Z}^k \subset \mathbb{Z}^n$ be positive, that is $m_j \geq 0$. We will show that the number $\#(m)$ of critical values of the function $\hat{p}$ between regular values ν and $\hat{\nu} = \nu + \langle p, m \rangle$ is not less than $\langle c, m \rangle$ $(= m_1 + \ldots + m_n)$. This would imply that the total number of fixed points is not less than $\langle c, m \rangle / \langle p, m \rangle$.

Applying Corollary 1.3 to $\mathcal{J} = \mathcal{J}^*(F^-(\nu))$ (see (6.7)) we find some $q \in \mathfrak{R}$ such that $q \notin \mathcal{J}$ but $u_1 q, \dots, u_n q \in \mathcal{J}$. If $\#(m) < m_1 + \ldots + m_n$ then $u_1^{m_1} \ldots u_n^{m_n} q \in \mathcal{J}^*(F^-(\hat{\nu}))$ (Propositions 6.2, 6.3). On the other hand multiplication by u^m is the isomorphism (6.8): $\mathcal{J}^*(F^-(\nu)) \to \mathcal{J}^*(F^-(\hat{\nu}))$ and thus $u^m q \notin \mathcal{J}^*(F^-(\hat{\nu}))$.

This contradiction completes the proof.

Remark 6.5 One could prove the Morse-type estimate $\# \geq \dim H^*((M_p)$ for the number of non-degenerate fixed points of a hamiltonian transformation (without a reference to the Lefschetz theorem) combining Remark 6.4 with the fact that $\mathbb{C}[\mathbb{Z}^{k-1}]$-module

$$\mathcal{J}^*(F^-(\nu))/\mathcal{J}^*(F^-(\nu + 1))$$

has the same rank as $\mathcal{J}_0/\mathcal{J}_1$ (equal to $\dim H^*(M_p)$, see Remark 1.4).

References

[Ar1] V.I. Arnold, *Sur une propriété topologique des applications globalement canonique de la méchanique classique*, C.R. Acad. Sci. Paris 261 (1965), 3719–3722.

[Ar2] V.I. Arnold, *Mathematical methods of classical mechanics*, 2nd ed, Springer, N.Y. 1989.

[AVG] V.I. Arnold, A.N. Varchenko, S.M. Gusein-Zade, *Singularities of differentiable maps*, Birkhäuser, Boston, v.1 (1985), v.2 (1988).

[At] M. Atiyah, *Convexity and commuting hamiltonians*, Topology 23:1 (1984), 1–15.

[Au] M. Audin, *The topology of torus actions on symplectic manifolds*, Birkhäuser, Basel, 1991.

[Cha] M. Chaperon, *Une idée du type «géodésique brisées» pour les systèmes hamiltoniens*, C.R. Acad. Sci. Paris 298 (1984), 293–296.

[Che] Yu.V. Chekanov, *Legandrova teorija Morsa*, Uspekhi Mat. Nauk 42:4 (1987), 139–141.

[CZ] C.C. Conley, E. Zehnder, *The Birkhoff-Lewis fixed point theorem and a conjecture of V.I. Arnold*, Invent. Math. 73:1 (1983), 33–49.

[F] B.L. Feigin, *Semi-infinite cohomology of Kac-Moody and Virasoro Lie algebras*, Russian Math. Surveys, 39:2 (1984), 155–156.

[Fl1] A. Floer, *Morse theory for Lagrangian intersections*, J. Diff. Geom. 28 (1988), 513–547.

[Fl2] A. Floer, *Symplectic fixed points and holomorphic spheres*, Commun. Math. Phys. 120 (1989), 575–611.

[FW] B. Fortune, A. Weinstein, *A symplectic fixed point theorem for complex projective spaces*, Bull. AMS 12:1 (1985), 128–130.

[G] V.A. Ginsburg, *Equivariant cohomology and Kähler geometry*, Funct. Anal. Appl. 21:4 (1987), 271–283.

[Gi1] A.B. Givental, *Periodic mappings in symplectic topology*, Funct. Anal. Appl. 23:4 (1989), 287–300.

[Gi2] A.B. Givental, *Nonlinear generalization of the Maslov index*, Adv. in Sov. Math., v.1, AMS, Providence, RI, 1990, 71–103.

[GM] M. Goresky, R. MacPherson, *Stratified Morse theory*, Springer, N.Y., 1987.

[GrH] P. Griffiths, J. Harris, *Principles of algebraic geometry*, Wiley, N.Y., 1978.

[Gro] M. Gromov, *Pseudo-holomorphic curves in symplectic manifolds*, Invent. Math. 82:2 (1985), 307–347.

[HoS] H. Hofer, D.A. Salamon, *Floer homology and Novikov rings*, in this volume.

[Hs] W.Y. Hsiang, *Cohomology theory of topological transformation groups*, Springer, N.Y., 1975.

[KT] Y. Karshon, S. Tolman, *The moment map and line bundles over presymplectic toric manifolds*, Preprint, Harvard univ., 1992, 16 p.

[LS] F. Laudenbach, J.-C. Sikorav, *Présidence d'intersections avec la section nulle au course d'une isotopie hamiltonienne dans une fibré cotangent*, Invent. Math. 82:2 (1985), 349–358.

[Oh] Y.-G. Oh, *A symplectic fixed point theorem on* $T^{2n} \times \mathbb{C}P^k$, Math.-Z. 203:4 (1990), 535–552.

[V] A.A. Voronov, *Semi-infinite De Rham cohomology*, Preprint, MSRI, 1992, 29 p.

Alexander B. Givental
Department of Mathematics
University of California
Berkeley, California 94720, USA

Floer homology and Novikov rings

H. Hofer and D.A. Salamon

Abstract. We prove the Arnold conjecture for compact symplectic manifolds under the assumption that either the first Chern class of the tangent bundle vanishes over $\pi_2(M)$ or the minimal Chern number is at least half the dimension of the manifold. This includes the important class of Calabi-Yau manifolds. The key observation is that the Floer homology groups of the loop space form a module over Novikov's ring of generalized Laurent series. The main difficulties to overcome are the presence of holomorphic spheres and the fact that the action functional is only well defined on the universal cover of the loop space with a possibly dense set of critical levels.

1 Introduction

Let (M, ω) be a $2n$-dimensional compact symplectic manifold and consider the time-dependent Hamiltonian differential equation

$$\dot{x}(t) = X_H(t, x(t)). \tag{1}$$

Here the vector field $X_H : S^1 \times M \to TM$ is associated to the 1-periodic Hamiltonian function $H : S^1 \times M \to \mathbb{R}$ via $i_{X_H}\omega = dH$. Throughout we identify $S^1 = \mathbb{R}/\mathbb{Z}$. We denote by

$$\mathcal{P}(H)$$

the set of all contractible 1-periodic solutions $x(t) = x(t+1)$ of (1). The Arnold conjecture states that if these 1-periodic solutions are all nondegenerate, then their number can be estimated below by the sum of the Betti numbers of M [1], [2]. This conjecture was first proved by Conley and Zehnder for the $2n$-torus [5]. In [8] Floer proved the Arnold conjecture for **monotone** symplectic manifolds. These are manifolds for which the cohomology class of the symplectic form ω over $\pi_2(M)$ is a non-negative multiple of the first Chern class $c_1 \in H^2(M)$ of the tangent bundle (with a suitable almost complex structure)

$$\int_{S^2} v^*\omega = \lambda \int_{S^2} v^*c_1$$

for $v : S^2 \to M$. Floer constructed a chain complex from the periodic solutions of (1) and the solutions $u : \mathbb{R} \times S^1 \to M$ of the partial differential equation

$$\frac{\partial u}{\partial s} + J(u)\frac{\partial u}{\partial t} - \nabla H(t, u) = 0 \tag{2}$$

which satisfy the limit condition

$$\lim_{s \to \pm\infty} u(s, t) = x^{\pm}(t) \tag{3}$$

with $x^{\pm} \in \mathcal{P}(H)$. Here $J : TM \to TM$ is an almost complex structure on M which is compatible with ω. This means that

$$g_J(v,w) = \omega(v, Jw)$$

defines a Riemannian metric on M. We denote by $\mathcal{J}(M,\omega)$ the space of all smooth almost complex structures on M which are compatible with ω. Let

$$\mathcal{M}(x^-, x^+; H, J)$$

denote the space of all smooth solutions of (2) and (3). For a generic Hamiltonian this space is a finite dimensional manifold. It decomposes into different components and we denote the local dimension of $\mathcal{M}(x^-, x^+; H, J)$ near u by $\mu(u)$. The **minimal Chern number** of (M,ω) is the integer $N \geq 0$ defined by $N\mathbb{Z} = c_1(\pi_2(M))$. The Conley-Zehnder index defines a map $\mu : \mathcal{P}(H) \to \mathbb{Z}/2N\mathbb{Z}$ and in [21] it is shown that

$$\mu(u) = \mu(x^+) - \mu(x^-) \pmod{2N}$$

for $u \in \mathcal{M}(x^-, x^+; H, J)$. Floer's cochain complex is defined by

$$C^k = \bigoplus_{\mu(x)=k \pmod{2N}} \mathbb{Z}\langle x\rangle$$

and the (x,y)-entry of the coboundary operator $\delta : C^k \to C^{k+1}$ is given by the number of one dimensional components of $\mathcal{M}(x,y;H,J)$ whenever the index difference is 1 modulo $2N$. This number is finite whenever the manifold (M,ω) is monotone and the connecting orbits are counted with appropriate signs. In [8] Floer proved that $\delta^2 = 0$. The homology of this cochain complex is called the **Floer cohomology** of the pair (H,J) and will be denoted by

$$HF^*(M,H,J) = \frac{\ker\delta}{\operatorname{im}\delta}.$$

The Floer cohomology groups are graded modulo $2N$. In [8] Floer proved that these cohomology groups in fact agree with the integral cohomology of the underlying manifold M

$$HF^k(M,H,J) = \bigoplus_{j=k+n \pmod{2N}} H^j(M),$$

and this proves the Arnold conjecture.

We generalize these ideas to $2n$-dimensional compact symplectic manifolds such that for every $A \in \pi_2(M)$

$$3 - n \leq c_1(A) < 0 \qquad \Longrightarrow \qquad \omega(A) \leq 0.$$

Such symplectic manifolds are called **weakly monotone**. They have the property that for a generic almost complex structure there is no pseudo-holomorphic sphere with negative Chern number.

Lemma 1.1. *A comact symplectic manifold (M, ω) is weakly monotone if and only if one of the following conditions is satisfied.*

(a) $\omega(A) = \lambda c_1(A)$ *for every* $A \in \pi_2(M)$ *where* $\lambda \geq 0$ *(M is* **monotone***).*

(b) $c_1(A) = 0$ *for every* $A \in \pi_2(M)$.

(c) *The minimal Chern number* $N \geq 0$ *defined by* $c_1(\pi_2(M)) = N\mathbb{Z}$ *is greater than or equal to* $n - 2$.

Proof: Assume M is weakly monotone but does not satisfy either of the conditions (a), (b), (c). Since (b) and (c) do not hold there there exists a homotopy class $A \in \pi_2(M)$ with

$$3 - n \leq c_1(A) < 0.$$

Since M is weakly monotone this implies

$$\omega(A) \leq 0.$$

Denote $\lambda = \omega(A)/c_1(A) \geq 0$. Since (a) does not hold there exists a homotopy class $A' \in \pi_2(M)$ such that $\omega(A') \neq \lambda c_1(A')$ and hence $\omega(A')c_1(A) - \omega(A)c_1(A') \neq 0$. Change the sign of A', if necessary, to obtain

$$\omega(A)c_1(A') - \omega(A')c_1(A) > 0.$$

Then the class $B = c_1(A')A - c_1(A)A'$ satisfies

$$c_1(B) = 0, \qquad \omega(B) > 0.$$

Hence for $k > 0$ sufficiently large the class $A + kB$ satisfies

$$3 - n \leq c_1(A + kB) < 0, \qquad \omega(A + kB) > 0.$$

This contradicts the definition of weak monotonicity. □

The three cases in the previous lemma are not disjoint and Floer proved the Arnold conjecture in the case (a). In extending Floer homology to manifolds which satisfy (b) or (c) one encounters two difficulties. The first is the presence of non-constant J-holomorphic spheres with Chern number $c_1 \leq 0$. Such J-holomorphic spheres cannot exist in the monotone case. They obstruct the compactness of the one dimensional components of the space of connecting orbits with bounded energy. We overcome this difficulty by taking account of the fact that the space of points lying on J-holomorphic spheres of Chern number $c_1 = 0$ form roughly speaking a subset of codimension 4. Moreover, our assumptions guarantee that generically there are no J-holomorphic spheres with negative Chern number. The J-holomorphic spheres of Chern class $c_1 = 1$ also obstruct compactness if they intersect periodic solutions of (1). However such intersections do not exist generically since these spheres form a set of codimension 2. As a consequence we can prove in the weakly monotone case that the space of one dimensional connecting orbits with a bound on the energy is compact. The second difficulty arises from sequences of connecting orbits with index difference 1 whose energy converges to infinity. We take account of these sequences by constructing a suitable coefficient ring Λ_ω which algebraically incorporates the period map

$$\pi_2(M) \to \mathbb{R} : A \mapsto \int_A \omega.$$

Such a ring was used by Novikov in his generalization of Morse theory for closed 1-forms [17]. An exposition of Novikov homology with applications in symplectic geometry can be found in the thesis of Sikorav [22].

For a large class of weakly monotone symplectic manifolds we shall prove that the Floer cohomology groups agree with the cohomology of the manifold M with coefficients in Λ_ω. This class includes all manifolds which satisfy (a) or (b) but in the case (c) we must assume in addition that the minimal Chern number is $N \geq n$.[1)] We conjecture that the Floer cohomology groups always agree with the cohomology of M. A generalization to arbitrary symplectic manifolds seems to require a better understanding of J-holomorphic spheres with negative Chern number. We conjecture that the presence of such spheres will have no algebraic ramifications for our setup.

The present work was inspired in part by a recent lecture of M.F. Atiyah on duality for Calabi-Yau manifolds. We also point out that the results of this paper were in principle anticipated by A. Floer. In [8] he observed the relevance of Novikov's ring for the construction of Floer cohomology and indicated a compactness argument involving the codimension of the space of holomorphic spheres.

2 Holomorphic spheres

Let (M, ω) be a $2n$-dimensional compact symplectic manifold and let $J \in \mathcal{J}(M, \omega)$. A J-holomorphic sphere is a smooth map $v : S^2 \to M$ such that

$$dv \circ i = J \circ dv$$

where i is the standard complex structure on $S^2 = \mathbb{C} \cup \{\infty\}$.

Let $c_1 \in H^2(M; \mathbb{Z})$ denote the first Chern class of the tangent bundle of M considered as a complex vector bundle with a complex structure J which is compatible with ω. The complex isomorphism class of TM is independent of the choice of $J \in \mathcal{J}(M, \omega)$. Consider the homomorphisms $\phi_\omega : \pi_2(M) \to \mathbb{R}$ and

1) After this paper was written our proof of the Arnold conjecture has been extended by Kaoru Ono to the case $N \geq n - 2$ (cf. [18]). In particular this includes all compact symplectic manifolds of dimensions 4 and 6.

$\phi_{c_1} : \pi_2(M) \to \mathbb{Z}$ defined by integration of ω and c_1 over a sphere and define the abelian group

$$\Gamma = \frac{\pi_2(M)}{\ker \phi_{c_1} \cap \ker \phi_\omega}.$$

In the following we shall denote by $c_1(A)$ and $\omega(A)$ the integrals of c_1 and ω over the class A. Given $A \in \Gamma$, denote by $\mathcal{M}(A; J)$ the space of all J-holomorphic spheres which represent the class A. A smooth map $v : S^2 \to M$ is called **simple** if $v = w \circ \phi$ with $\phi : S^2 \to S^2$ implies $\deg \phi = 1$. Denote by

$$\mathcal{M}_s(A; J)$$

the subspace of simple J-holomorphic spheres in the class A. This space is a finite dimensional manifold for a generic almost complex structure $J \in \mathcal{J}(M, \omega)$.

More precisely, let $\mathcal{S} = W^{1,p}(S^2, M)$ with $p > 2$ and for $v \in \mathcal{S}$ let $\Lambda_v \to S^2$ denote the vector bundle whose fibre at $z \in S^2$ is the space of complex anti-linear maps $T_z S^2 \to T_{v(z)} M$. Consider the bundle

$$\mathcal{E} \to \mathcal{S}$$

whose fibre $\mathcal{E}_v = L^p(\Lambda_v)$ at $v \in \mathcal{S}$ consists of all L^p sections η of the vector bundle Λ_v. In local coordinates such a section is a 1-form of the form $\eta = \xi ds - J(v)\xi dt$. The $\bar{\partial}$ operator defines a Fredholm section of the bundle $\mathcal{E}$ given by

$$\bar{\partial}_J(v) = dv + J \circ dv \circ i = \left(\frac{\partial v}{\partial s} + J(v) \frac{\partial v}{\partial t} \right) ds + \left(\frac{\partial v}{\partial t} - J(v) \frac{\partial v}{\partial s} \right) dt.$$

By definition the J-holomorphic spheres are the zeros of this section. It follows from elliptic regularity that every J-holomorphic sphere is smooth. Thus every J-holomorphic sphere can be represented by a smooth function $v : \mathbb{C} \to M$ which satisfies the partial differential equation

$$\frac{\partial v}{\partial s} + J(v) \frac{\partial v}{\partial t} = 0.$$

Such a function extends to S^2 if and only if it has finite energy

$$E(v) = \frac{1}{2} \int_{\mathbb{C}} \left(\left| \frac{\partial v}{\partial s} \right|^2 + \left| \frac{\partial v}{\partial t} \right|^2 \right) dsdt = \int_{\mathbb{C}} v^* \omega < \infty.$$

The linearization of $\bar{\partial}_J$ at a zero $v \in \mathcal{M}(A; J)$ defines a Fredholm operator $D\bar{\partial}_J(v) : W^{1,p}(v^*TM) \to L^p(\Lambda_v)$ whose Fredholm index, by the Riemann-Roch theorem, is given by

$$\text{index } D\bar{\partial}_J(v) = 2n + 2 \int_{S^2} v^* c_1.$$

The space $\mathcal{M}(A; J)$ can only be expected to be a manifold if $\bar{\partial}$ is transversal to the zero section. This means that the operator $D\bar{\partial}_J(v)$ is onto for every $v \in \mathcal{M}(A; J)$. To achieve this we must perturb J and restrict ourselves to the class of simple curves. For the latter McDuff has proved the following result [14].

Lemma 2.1. *Assume that $v : S^2 \to M$ is a nonconstant J-holomorphic sphere. Then there are only finitely many points $z_1, \ldots, z_m \in S^2$ such that $dv(z_j) = 0$. If v is simple then there exists a point $z_0 \in S^2$ such that $dv(z_0) \neq 0$ and $v(z) \neq v(z_0)$ for every $z \neq z_0$.*

The tangent space $T_J\mathcal{J}$ to $\mathcal{J} = \mathcal{J}(M,\omega)$ is the vector space of smooth sections $X \in C^\infty(\mathrm{End}(TM,J,\omega))$ where $\mathrm{End}(TM,J,\omega) \subset \mathrm{End}(TM)$ is the bundle over M whose fibre at $x \in M$ is the space $\mathrm{End}(T_xM, J_x, \omega_x)$ of linear endomorphisms $Y : T_xM \to T_xM$ which satisfy

$$JY + YJ = 0, \qquad \omega_x(Y\xi,\eta) + \omega_x(\xi, Y\eta) = 0.$$

Following Floer [6] we choose a sufficiently rapidly decreasing sequence $\varepsilon_k > 0$ and denote by $C^\infty_\varepsilon(\mathrm{End}(TM,J,\omega))$ the subspace of those smooth sections $Y \in C^\infty(\mathrm{End}(TM,J,\omega))$ for which

$$\|Y\|_\varepsilon = \sum_{k=0}^{\infty} \varepsilon_k \|Y\|_{C^k(M)} \quad < \quad \infty.$$

This defines a separable Banach space of smooth sections which for a suitable choice of the sequence ε_k is dense in $L^2(\mathrm{End}(TM,J,\omega))$ [6]. For any sufficiently small section $Y \in C^\infty(\mathrm{End}(TM,J,\omega))$ we have $J\exp(-JY) \in \mathcal{J}(M,\omega)$. Given $J_0 \in \mathcal{J}(M,\omega)$ and $\delta > 0$ denote by $\mathcal{U}_\delta(J_0)$ the set of all $J = J_0 \exp(-J_0Y) \in \mathcal{J}(M,\omega)$ with $\|Y\|_\varepsilon \leq \delta$. We also denote by

$$\mathcal{J}_{\mathrm{reg}} = \mathcal{J}_{\mathrm{reg}}(M,\omega)$$

the set of all smooth almost complex structures $J \in \mathcal{J}(M,\omega)$ such that $D\bar{\partial}_J(v)$ is onto for every simple J-holomorphic sphere $v : S^2 \to M$. Using Lemma 2.1 one can prove the following result [14].

Theorem 2.2. *For every $J_0 \in \mathcal{J}(M,\omega)$ the set $\mathcal{J}_{\mathrm{reg}}(A) \cap \mathcal{U}_\delta(J_0)$ is generic in the sense of Baire (i.e. a countable intersection of open dense sets).*

As a consequence of the implicit function theorem the space $\mathcal{M}_s(A;J)$ is a finite dimensional manifold of dimension

$$\dim \mathcal{M}_s(A;J) = 2n + 2c_1(A)$$

whenever $J \in \mathcal{J}_{\mathrm{reg}}$. The differentiable structure on this manifold is independent of the choice of the analytic setup. Moreover, it follows from Lemma 2.1 that the group $\mathbb{G} = \mathrm{PSL}(2,\mathbb{C})$ of biholomorphic maps of the sphere acts freely on $\mathcal{M}_s(A;J)$. Hence the quotient $\mathcal{M}_s(A;J)/\mathbb{G}$ is a finite dimensional manifold of dimension

$$\dim \mathcal{M}_s(A;J)/\mathbb{G} = 2n + 2c_1(A) - 6.$$

In particular this space must be empty whenever $J \in \mathcal{J}_{\mathrm{reg}}$ and $c_1(A) < 3 - n$.

Proposition 2.3. *Assume that (M, ω) is a weakly monotone $2n$-dimensional compact symplectic manifold. Let $J \in \mathcal{J}_{\mathrm{reg}}(M, \omega)$ and $A \in \Gamma$.*

(i) *If $c_1(A) < 0$ then $\mathcal{M}(A; J) = \emptyset$.*

(ii) *If $n = 2$ and $A \neq 0$ with $c_1(A) = 0$ then $\mathcal{M}(A; J) = \emptyset$.*

(iii) *If $n = 2$ and $c_1(A) = 1$ then the moduli space $\mathcal{M}(A; J)/\mathbb{G}$ is a finite set.*

Proof: To prove statement (i) note first that $\mathcal{M}_s(B; J) = \emptyset$ for every $B \in \Gamma$ with $c_1(B) < 0$. If $c_1(B) < 3 - n$ then this follows from the above dimension formula and if $c_1(B) \geq 3 - n$ then this follows from the fact that M is weakly monotone and hence $\omega(B) \leq 0$. Now assume that $c_1(A) < 0$ and $v \in \mathcal{M}(A; J)$. Then there exists a simple J-holomorphic map $w : S^2 \to M$ and a rational map $\phi : S^2 \to S^2$ of degree k such that $v = w \circ \phi$. Let $B \in \Gamma$ denote the equivalence class of w. Then $kc_1(B) = c_1(A) < 0$. Since $A \neq 0$ we have

$$0 < E(v) = \int_{S^2} v^*\omega = k \int_{S^2} w^*\omega = kE(w),$$

hence $k > 0$, and hence $c_1(B) < 0$. This implies $\mathcal{M}_s(B; J) = \emptyset$ in contradiction to $w \in \mathcal{M}_s(B; J)$.

We prove statement (ii). In the case $n = 2$ it follows from the dimension formula that $\mathcal{M}_s(B; J) = \emptyset$ for every $B \in \Gamma$ with $c_1(B) = 0$. Now the same argument as above shows that if $\mathcal{M}(A; J) \neq \emptyset$ for some nonzero $A \in \Gamma$ with $c_1(A) = 0$ then there exists a $B \in \Gamma$ with $c_1(B) = 0$ such that $\mathcal{M}_s(B; J) \neq \emptyset$.

To prove statement (iii) note first that every sphere $v : S^2 \to M$ in the homotopy class A is simple. Hence the space $\mathcal{M}(A; J)$ is a 6-dimensional manifold for $J \in \mathcal{J}_{\mathrm{reg}}$. Now every J-holomorphic curve $v \in \mathcal{M}(A; J)$ has energy $E(v) = \omega(A)$. By Gromov's compactness theorem the space $\mathcal{M}(A; J)/\mathbb{G}$ can only fail to be compact if J-holomorphic curves of nonpositive Chern number bubble off. (See for example [11], [19], and the appendix of this paper.) By (i) and (ii) such spheres do not exist. □

In the following we denote by

$$M_k(c; J)$$

the set of points $x \in M$ such that there exists a non-constant J-holomorphic sphere $v : S^2 \to M$ such that $c_1(v) \leq k$, $E(v) \leq c$ and $x \in v(S^2)$.

Proposition 2.4. *Let (M, ω) be a weakly monotone compact symplectic manifold of dimension $2n \geq 6$ and assume that $J \in \mathcal{J}_{\mathrm{reg}}$. Then the set $M_k(c; J)$ is compact for every $c > 0$ and every integer k.*

Proof: Take a sequence $x_\nu = v_\nu(z_\nu)$ converging to $x \in M$ where $v_\nu : S^2 \to M$ is a sequence of J-holomorphic spheres with $c_1(v_\nu) \le k$ and $E(v_\nu) \le c$. We must prove that $x \in M_k(c; J)$. To see this assume without loss of generality that the v_ν all represent the same homotopy class A. By Theorem A.1 a subsequence of v_ν converges to a finite collection of J-holomorphic spheres $v^j : S^2 \to M$ for $j = 1, \ldots, \ell$ whose connected sum represents the homotopy class A. In particular

$$\sum_{j=1}^{\ell} E(v^j) = \omega(A), \qquad \sum_{j=1}^{\ell} c_1(v^j) = c_1(A)$$

and $x \in v^j(S^2)$ for some j. Since $E(v^j) > 0$ and $c_1(v^j) \ge 0$ for all j the statement follows. □

For every $A \in \Gamma$ the space $\mathcal{M}_s(A; J) \times S^2$ is a manifold of dimension $2n + 2c_1(A) + 2$ on which the group $\mathbb{G}$ acts freely by

$$\phi^*(v, z) = (v \circ \phi, \phi^{-1}(z))$$

for $(v, z) \in \mathcal{M}_s(A; J) \times S^2$ and $\phi \in \mathbb{G}$. Hence the quotient is a manifold of dimension

$$\dim \mathcal{M}_s(A; J) \times_{\mathbb{G}} S^2 = 2n + 2c_1(A) - 4.$$

Note that the evaluation map

$$e_A : \mathcal{M}_s(A; J) \times_{\mathbb{G}} S^2 \to M : (v, z) \mapsto v(z)$$

is well defined. In particular the set $M_0(\infty; J)$ is a countable union of images of smooth maps defined on manifolds of dimension $2n - 4$. Hence the set $M_0(\infty; J)$ is roughly speaking a subset of M of codimension 4. Similarly the set $M_1(\infty; J)$ is a subset of codimension 2.

3 Transversality and compactness

Throughout this section we assume that (M, ω) is a weakly monotone compact symplectic manifold of dimension $2n \ge 4$ and $J \in \mathcal{J}_{\mathrm{reg}}(M, \omega)$ is a regular almost complex structure compatible with ω. Let $C^\infty_\varepsilon(S^1 \times M)$ denote the Banach space of all smooth functions $h : S^1 \times M \to \mathbb{R}$ such that

$$\|h\|_\varepsilon = \sum_{k=0}^{\infty} \varepsilon_k \|h\|_{C^k(S^1 \times M)} \quad < \quad \infty.$$

Here $\varepsilon_k > 0$ is a sufficiently rapidly decreasing sequence such that C^∞_ε is dense in L^2. The space $C^\infty_\varepsilon(S^1 \times M)$ can be viewed as a closed subspace of a suitable separable Banach space and is therefore separable. Given any smooth Hamiltonian function $H_0 : S^1 \times M \to \mathbb{R}$ we denote by

$$\mathcal{U}_\delta(H_0)$$

the set of all Hamiltonians H with $\|H - H_0\|_\varepsilon < \delta$.

Theorem 3.1. *There is a generic set $\mathcal{H}_0 \subset \mathcal{U}_\delta(H_0)$ such that the following holds for $H \in \mathcal{H}_0$.*

(i) *Every periodic solution $x \in \mathcal{P}(H)$ is non-degenerate.*
(ii) $x(t) \notin M_1(\infty; J)$ *for every $x \in \mathcal{P}(H)$ and every $t \in \mathbb{R}$.*

Proof: Let $A \in \Gamma$ with $c_1(A) \in \{0,1\}$. Denote by $\mathcal{B}$ the Hilbert manifold of contractible $W^{1,2}$ loops $x : S^1 \to M$ and consider the bundle $\mathcal{E} \to \mathcal{B}$ whose fibre at $x \in \mathcal{B}$ is the Hilbert space of L^2-vector fields along x. Let the section $\mathcal{F} : \mathcal{B} \times \mathcal{U}_\delta(H_0) \to \mathcal{E}$ be defined by

$$\mathcal{F}(x, H) = \dot{x} - X_H(t, x).$$

The differential of this section at a zero (x, H) is the linear operator

$$D\mathcal{F}(x, H) : W^{1,2}(x^*TM) \times C^\infty_\varepsilon(S^1 \times M) \to L^2(x^*TM)$$

given by

$$D\mathcal{F}(x, H)(\xi, h) = A_x\xi - X_h(t, x)$$

where the operator $A_x : W^{1,2}(x^*TM) \to L^2(x^*TM)$ is given by

$$A_x\xi = \nabla\xi - \nabla_\xi X_H(t, x).$$

We prove that $d\mathcal{F}(x, H)$ is onto. Since A_x is a Fredholm operator (of Fredholm index zero) it suffices to show that $d\mathcal{F}(x, H)$ has a dense range. To see this note that every C^∞ vector field $\eta \in C^\infty(x^*TM)$ can be written in the form $\eta(t) = X_h(t, x(t))$ for some smooth function $h : S^1 \times M \to \mathbb{R}$. Hence every L^2-vector field $\eta \in L^2(x^*TM)$ can be approximated by a sequence of vector fields of the form $X_h(t, x(t))$ with $h \in C^\infty_\varepsilon(S^1 \times M)$.

Thus we have proved that $\mathcal{F}$ intersects the zero section of $\mathcal{E}$ transversally. Hence the set

$$\mathcal{P} = \{(x, H) \in \mathcal{B} \times \mathcal{U}_\delta(H_0) : \mathcal{F}(x, H) = 0\}$$

is a separable infinite dimensional Banach manifold. A Hamiltonian $H \in \mathcal{U}_\delta(H_0)$ is a regular value of the projection $\mathcal{P} \to \mathcal{U}_\delta(H_0)$ onto the second factor if and only if every periodic solution $x \in \mathcal{P}(H)$ is non-degenerate (or equivalently the operator A_x is onto). By the Sard-Smale theorem [23] the set $\mathcal{H}'_0 \subset \mathcal{U}_\delta(H_0)$ of regular values is generic in the sense of Baire.

We prove that the evaluation map

$$e_t : \mathcal{P} \to M, \qquad e_t(x, H) = x(t)$$

is a submersion for every $t \in S^1$. The tangent space $T_{(x,H)}\mathcal{P}$ is the Banach space of all pairs $(\xi, h) \in W^{1,2}(x^*TM) \times C_\varepsilon^\infty(S^1 \times M)$ such that

$$A_x\xi = X_h(t,x).$$

The differential of e_t at (x, H) is the linear functional

$$De_t : T_{(x,H)}\mathcal{P} \to T_{x(t)}M, \qquad De_t(\xi, h) = \xi(t)$$

and we must prove that this map is onto. Given $\xi_0 \in T_{x(t)}M$ choose any $\xi \in W^{1,2}(x^*TM)$ such that $\xi(t) = \xi_0$. Since A_x is a Fredholm operator the set of all vector fields $\eta \in \operatorname{range} A_x$ which are of the form form $X_h(t, x(t))$ with $h \in C_\varepsilon^\infty(S^1 \times M)$ is dense in the range of A_x. Hence there exists a sequence $(\xi_\nu, h_\nu) \in T_{(x,H)}\mathcal{P}$ such that $A_x\xi_\nu$ converges to $A_x\xi$. Since there is an estimate

$$\inf_{A_x\zeta=0} \|\xi - \zeta\|_{W^{1,2}} \le c\|A_x\xi\|_{L^2},$$

we can choose ξ_ν as to converge to ξ in the $W^{1,2}$ norm. Hence $\xi_\nu(t)$ converges to $\xi(t) = \xi_0$. This shows that De_t has a dense range. Since $T_{x(t)}M$ is finite dimensional De_t is onto.

Now consider the evaluation map

$$\mathcal{M}_s(A; J) \times_G S^2 \times S^1 \times \mathcal{P} \to M \times M : ([v, z], t, x, H) \mapsto (v(z), x(t)).$$

Since $e_t : \mathcal{P} \to M$ is a submersion this map is transversal to the diagonal Δ_M in $M \times M$. Hence the space

$$\mathcal{N} = \{([v, z], t, x, H) : v(z) = x(t), (x, H) \in \mathcal{P}\}$$

is an infinite dimensional Banach submanifold of $\mathcal{M}_s(A; J) \times_G S^2 \times S^1 \times \mathcal{P}$ of codimension $2n$.

The projection

$$\mathcal{N} \to \mathcal{U}_\delta(H_0) : ([v, z], t, x, H) \mapsto H$$

is a Fredholm map. The Fredholm index of this projection is $2c_1(A) - 3$. In the case $c_1(A) \le 1$ this number is negative and hence the regular values of the projection $\mathcal{N} \to \mathcal{U}_\delta(H_0)$ are those which are not in the image. Now $\mathcal{N}$ is separable and hence it follows from the Sard-Smale theorem [23] that the set $\mathcal{H}_0(A)$ of regular values is of the second category in the sense of Baire. We conclude that the 1-periodic solutions of (1) do not touch the J-holomorphic spheres in the class A for $H \in \mathcal{H}_0(A)$. The required set $\mathcal{H}_0$ is the is the intersection of $\mathcal{H}_0'$ with the sets $\mathcal{H}_0(A)$ where A ranges over the countable set of those $A \in \Gamma$ for which $c_1(A) \le 1$.
□

By the previous theorem there exists a Hamiltonian function $H_0 : S^1 \times M \to \mathbb{R}$ such that all contractible 1-periodic solutions of (1) are nondegenerate and do not intersect the set $M_1(\infty; J)$. We denote by

$$\mathcal{V}_\delta(H_0)$$

the set of all Hamiltonians H which satisfy $\|H - H_0\|_\varepsilon < \delta$ and agree with H_0 up to second order on the contractible 1-periodic solutions of (1). If $\delta > 0$ is sufficiently small then the contractible 1-periodic solutions of H_0 agree with those of H for every $H \in \mathcal{V}_\delta(H_0)$.

Theorem 3.2. *There is a generic set $\mathcal{H}_1 \subset \mathcal{V}_\delta(H_0)$ such that the following holds for $H \in \mathcal{H}_1$.*

(i) *The moduli space $\mathcal{M}(x^-, x^+; H, J)$ of connecting orbits is a finite dimensional manifold for all $x^\pm \in \mathcal{P}(H)$.*
(ii) *$u(s,t) \notin M_0(\infty; J)$ for every $u \in \mathcal{M}(x^-, x^+; H, J)$ with $\mu(u) \le 2$ and every $(s,t) \in \mathbb{R} \times S^1$.*

Proof: We fix a pair $x^\pm \in \mathcal{P}(H_0)$ with index difference 1 or 2 modulo $2N$. We also choose $A \in \Gamma$ with $c_1(A) = 0$. Following the same line of argument as in the proof of Theorem 3.1, denote by $\mathcal{B}$ the Banach manifold of $W^{1,p}$ maps $u : \mathbb{R} \times S^1 \to M$ which satisfy the limit condition (3) in the $W^{1,p}$ sense with $p > 2$ (see [6] and [8]). Consider the bundle $\mathcal{E} \to \mathcal{B}$ whose fibre at $u \in \mathcal{B}$ is the Banach space of L^p-vector fields along u. Let the section $\mathcal{F} : \mathcal{B} \times \mathcal{V}_\delta(H_0) \to \mathcal{E}$ be defined by

$$\mathcal{F}(u, H) = \frac{\partial u}{\partial s} + J(u)\frac{\partial u}{\partial t} - \nabla H(t, u).$$

The differential of this section at a zero (u, H) is the linear operator

$$D\mathcal{F}(u, H) : W^{1,p}(u^*TM) \times T_H\mathcal{V}_\delta(H_0) \to L^p(u^*TM)$$

given by

$$D\mathcal{F}(u, H)(\xi, h) = D_u\xi - \nabla h(t, u)$$

where the operator $D_u : W^{1,p}(u^*TM) \to L^p(u^*TM)$ is given by

$$D_u\xi = \nabla_s\xi + J(u)\nabla_t\xi + \nabla_\xi J(u)\frac{\partial u}{\partial t} - \nabla_\xi \nabla H(t, u).$$

This is a Fredholm operator of Fredholm index $\mu(u)$. In [21] using results in [10] it is proved that $\mathcal{F}$ intersects the zero section of $\mathcal{E}$ transversally. Hence the set

$$\mathcal{M}(x^-, x^+; J) = \{(u, H) \in \mathcal{B} \times \mathcal{V}_\delta(H_0) : \mathcal{F}(u, H) = 0\}$$

is a separable infinite dimensional Banach manifold.

A Hamiltonian $H \in \mathcal{U}_\delta(H_0)$ is a regular value of the projection

$$\mathcal{M}(x^-, x^+; J) \to \mathcal{V}_\delta(H_0) : (u, H) \mapsto H$$

if and only if the operator D_u is onto for every $u \in \mathcal{M}(x^-, x^+; H, J)$ and hence $\mathcal{M}(x^-, x^+; H, J)$ is a manifold whose dimension is the Fredholm index of D_u. By the Sard-Smale theorem [23] the set $\mathcal{H}_2(x^-, x^+) \subset \mathcal{V}_\delta(H_0)$ of regular values is generic in the sense of Baire.

Now as in the proof of Theorem 3.1 the evaluation map

$$e_t : \mathcal{M}(x^-, x^+; J) \to M, \qquad e_t(u, H) = u(0, t)$$

is a submersion for every $t \in S^1$. The evaluation map at the point (s, t) is also a submersion due to the obvious action of $\mathbb{R}$ on the solutions of (2). Since e_t is a submersion it follows that the evaluation map

$$\mathcal{M}_s(A; J) \times_G S^2 \times S^1 \times \mathcal{M}(x^-, x^+; J) \to M \times M$$

given by

$$([v, z], t, u, H) \mapsto (v(z), u(0, t))$$

is transversal to the diagonal Δ_M. Hence the space

$$\mathcal{N} = \{([v, z], t, u, H) : v(z) = u(0, t), (u, H) \in \mathcal{M}(x^-, x^+; J)\}$$

is a Banach submanifold of $\mathcal{M}_s(A; J) \times_G S^2 \times S^1 \times \mathcal{M}(x^-, x^+; J)$ of codimension $2n$.

The projection

$$\mathcal{N} \to \mathcal{V}_\delta(H_0) : ([v, z], u, t; H) \mapsto H$$

is a Fredholm map. The Fredholm index of this projection is

$$2c_1(A) + \mu(u) - 3.$$

In the case $c_1(A) = 0$ and $\mu(u) \leq 2$ this number is negative. Denote by $\mathcal{H}_3(x^-, x^+, A)$ the set of regular values of the projection $\mathcal{N} \to \mathcal{V}_\delta(H_0)$ and define

$$\mathcal{H}_1(x^-, x^+, A) = \mathcal{H}_2(x^-, x^+) \cap \mathcal{H}_3(x^-, x^+, A).$$

The required set $\mathcal{H}_1$ is the countable intersection of the sets $\mathcal{H}_1(x^-, x^+, A)$ where $x^\pm$ runs over all pairs in $\mathcal{P}(H_0)$ with index difference 1 or 2 and A runs over all spheres of Chern number 0. □

For any pair (H, J) denote by

$$\mathcal{M}_k(c, H, J)$$

the set of all point $x = u(s,t) \in M$ where $u : \mathbb{R} \times S^1 \to M$ is a contractible solution of (2) with energy

$$E(u) = \frac{1}{2}\int_{-\infty}^{\infty}\int_0^1 \left(\left|\frac{\partial u}{\partial s}\right|^2 + \left|\frac{\partial u}{\partial t} - X_H(t,u)\right|^2 \right) dtds \leq c$$

and $\mu(u) \leq k$. If the operator D_u is onto for every solution of u of (2) and (3) then $\mathcal{M}_k(c, H, J) = \emptyset$ for $k < 0$ and the set

$$\mathcal{M}_0(H) = \mathcal{M}_0(c, H, J)$$

consists of all points lying on a contractible 1-periodic solution of (1). Let

$$\mathcal{H}_{\mathrm{reg}}(J)$$

denote the set of all smooth Hamiltonian functions $H : S^1 \times M \to \mathbb{R}$ such that the contractible 1-periodic solutions of (1) are nondegenerate, the operator D_u is onto for every contractible solution $u : \mathbb{R} \times S^1 \to M$ of (2) and (3), and

$$\mathcal{M}_0(H) \cap \mathcal{M}_1(\infty; J) = \emptyset, \qquad \mathcal{M}_2(\infty; H, J) \cap \mathcal{M}_0(\infty; J) = \emptyset.$$

It follows from Theorem 3.1 and Theorem 3.2 that for every $J \in \mathcal{J}_{\mathrm{reg}}$ the set $\mathcal{H}_{\mathrm{reg}}(J)$ is dense in $C^\infty(S^1 \times M)$ with respect to the topology of uniform convergence with all derivatives.

Theorem 3.3. *Assume $J \in \mathcal{J}_{\mathrm{reg}}(M, \omega)$ and $H \in \mathcal{H}_{\mathrm{reg}}(J)$. Then the sets $\mathcal{M}_1(c; H, J)$ and $\mathcal{M}_2(c; H, J)$ are compact for every $c > 0$.*

Proof: We prove that there is positive number $\hbar > 0$ such that

$$E(v) \geq \hbar, \qquad E(u) \geq \hbar$$

for every nonconstant J-holomorphic sphere $v : S^2 \to M$ and every nontrivial s-dependent solution $u : \mathbb{R} \times S^1 \to M$ of (2). For nonconstant J-holomorphic spheres this follows from Gromov's compactness. Now assume that there is a sequence of solutions u_ν of (2) with $0 \neq E(u_\nu) \to 0$.

We prove that $\partial u_\nu / \partial s$ converges to zero uniformly on $\mathbb{R}^2$ as ν tends to ∞. Otherwise there would exist a sequence (s_ν, t_ν) such that $|\partial u_\nu / \partial s(s_\nu, t_\nu)| \geq \delta > 0$. Assume without loss of generality that $s_\nu = 0$. Since $E(u_\nu)$ converges to zero no bubbling can occur and hence a subsequence of $u_\nu(s,t)$ converges with its derivatives uniformly on compact sets to a solution $u : \mathbb{R} \times S^1 \to M$ of (2) with

$|\partial u_\nu/\partial s(0,t^*)| \geq \delta$ and $E(u) = 0$. But the latter implies that $u(s,t) \equiv x(t)$ in contradiction to the former.

Thus we have proved that $\partial u_\nu/\partial s$ converges to zero uniformly and, passing to a subsequence, that $u_\nu(s,t)$ converges with its derivatives uniformly on compact sets to a periodic solution $x(t)$ of (1). We prove that $u_\nu(s,t)$ converges to $x(t)$ uniformly on $\mathbb{R}^2$. To see this choose $\varepsilon > 0$ such that $d(x,y) = \sup_t d_M(x(t), y(t)) < \varepsilon$ for every $y \in \mathcal{P}(H)$, $y \neq x$. Then there exists a $\delta > 0$ such that for every C^1-function $z : S^1 \to M$ with $d(z,x) = \varepsilon$ we have $\sup_t |\dot{z}(t) - X_H(t,z)| \geq \delta$. (Otherwise, by the Arzela-Ascoli theorem, there would exist a periodic solution $y \in \mathcal{P}(H)$ with $d(x,y) = \varepsilon$.) Now choose ν sufficiently large such that $|\partial u_\nu/\partial s| < \delta$. Then it follows that $d_M(u_\nu(s,t), x(t)) < \varepsilon$ for all s and t.

Thus we have proved that $u_\nu(s,t)$ converges to $x(t)$ uniformly on $\mathbb{R}^2$. In particular this implies that u_ν satisfies the limit condition (3) with $x^- = x^+ = x$ and u_ν represents a trivial homology class. Now the energy of u_ν is given by

$$\begin{aligned} E(u_\nu) &= \int_{-\infty}^{\infty} \int_0^1 \left| \frac{\partial u_\nu}{\partial s} \right|^2 dtds \\ &= \int_{-\infty}^{\infty} \int_0^1 \left\langle \frac{\partial u_\nu}{\partial s}, \nabla H(t,u_\nu) - J(u_\nu) \frac{\partial u_\nu}{\partial t} \right\rangle dtds \\ &= \int_{-\infty}^{\infty} \int_0^1 \omega \left(\frac{\partial u_\nu}{\partial s}, \frac{\partial u_\nu}{\partial t} \right) dtds + \int_{-\infty}^{\infty} \frac{d}{ds} \int_0^1 H(t,u_\nu)\, dtds \\ &= 0. \end{aligned}$$

The last identity follows from Stokes' theorem and contradicts our assumption that $E(u_\nu) \neq 0$.

Now let u_ν be a sequence of solutions of (2) and (3) with fixed limits such that

$$\mu(u_\nu) = 2, \qquad E(u_\nu) \leq c.$$

Assume without loss of generality that $E(u_\nu)$ converges. Using a standard argument as in [20] one can show that there exist a subsequence (still denoted by u_ν), periodic solutions $x^- = x^0, x^1, \ldots, x^{\ell-1}, x^\ell = x^+$ (not necessarily distinct), and connecting orbits $u^j \in \mathcal{M}(x^j, x^{j-1}; H, J)$ for $j = 1, \ldots, \ell$ with total energy

$$\sum_{j=1}^{\ell} E(u^j) \leq c$$

such that the following holds. Given any sequence $s_\nu \in \mathbb{R}$ the sequence $v_\nu(s,t) = u_\nu(s+s_\nu, t)$ has a subsequence which converges modulo bubbling either to $u^j(s+s^j, t)$ for some s^j or to $x^j(t)$ for some j. Here convergence modulo bubbling means that there exist finitely many points in $\mathbb{R} \times S^1$ such that v_ν converges with its derivatives uniformly on every compact subset of the complement of these points. Moreover, every u^j is such a limit and no other connecting orbit can be approximated by u_ν in this way.

We prove that bubbling cannot occur. By Theorem A.1 there are only finitely many J-holomorphic spheres which can bubble off in our limit process. Denote these spheres by $v^1, \ldots, v^m$. It follows from Theorem A.1 that

$$\sum_{j=1}^{m} E(u^j) + \sum_{j=1}^{\ell} E(v^j) = \lim_{\nu\to\infty} E(u_\nu)$$

and

$$\sum_{j=1}^{m} \mu(u^j) + \sum_{j=1}^{\ell} 2c_1(v^j) = 2.$$

Since there is no J-holomorphic sphere with negative Chern number this implies that $\mu(u^j) \le 2$ for every j. The key point in our argument is that, again by Theorem A.1, the spheres v^j together with the connecting orbits u^j and the periodic solutions x^j form a connected family. So if bubbling occurs then one of the spheres v^j must intersect one of the connecting orbits u^j or one of the periodic solutions x^j. Since $\mathcal{M}_2(\infty; H, J) \cap \mathcal{M}_0(\infty; J) = \emptyset$ there must be a j with $c_1(v^j) > 0$. This implies

$$\sum_{j=1}^{m} \mu(u^j) \le 0.$$

But since $H \in \mathcal{H}_{\mathrm{reg}}(J)$ there is no nonconstant connecting orbit u with $\mu(u) \le 0$. Hence $x^- = x^+$, $m = 1$, and one of the spheres v^j must intersect the periodic solution $x^\pm$ contradicting the fact that $\mathcal{M}_0(H) \cap \mathcal{M}_1(\infty; J, H) = \emptyset$. The same argument works for $\mu(u_\nu) = 1$ and this proves the theorem. □

4 Generalized Laurent series

Let Γ be a group with a **weight homomorphism** $\phi : \Gamma \to \mathbb{R}$ and let $\mathbb{F}$ be an integral domain. Consider the $\mathbb{F}$-module

$$\Lambda = \Lambda(\Gamma, \phi; \mathbb{F})$$

of all functions $\Gamma \to \mathbb{F} : A \mapsto \lambda_A$ such that the set

$$\{A \in \Gamma : \lambda_A \ne 0,\ \phi(A) < c\}$$

is finite for every constant $c \in \mathbb{R}$. This space is a ring with the product given by the convolution

$$(\lambda * \theta)_A = \sum_{B\in\Gamma} \lambda_B \theta_{B^{-1}A}.$$

This is a finite sum and $\lambda * \theta \in \Lambda$. The unit element is the delta function $\delta : \Gamma \to \mathbb{F}$ defined by $\delta_{\mathbb{1}} = 1$ and $\delta_A = 0$ for $A \neq \mathbb{1}$. In the case $\mathbb{F} = \mathbb{Z}$ and $\phi = 0$ the module $\Lambda(\Gamma, 0; \mathbb{Z})$ is the group ring.

Now assume that $\phi : \Gamma \to \mathbb{R}$ is injective. Then $\Lambda(\Gamma, \phi; \mathbb{F})$ is an integral domain. Moreover, the group Γ is necessarily isomorphic to a free abelian group with finitely many generators. Hence we assume $\Gamma = \mathbb{Z}^m$ and

$$\phi(k_1, \ldots, k_m) = \sum_{j=1}^{m} \omega_j k_j = \omega \cdot k$$

where the ω_j are positive and rationally independent. In this case we can identify $\Lambda(\Gamma, \phi; \mathbb{F}) = \Lambda(\omega; \mathbb{F})$ with the space of formal power series

$$f(t) = \sum_k a_k t^k$$

where $t = (t_1, \ldots, t_m)$, $k = (k_1, \ldots, k_m)$, and $t^k = t_1^{k_1} \cdots t_m^{k_m}$. The coefficients are subject to the condition

$$\#\{k : a_k \neq 0,\ \omega \cdot k \leq c\} < \infty$$

for every constant $c \in \mathbb{R}$. We call such a power series f a **generalized Laurent series**. The map $d : \Lambda \setminus \{0\} \to \mathbb{R}$ defined by

$$d(f) = \inf\{\omega \cdot k : a_k \neq 0\}$$

for $f = \sum_k a_k t^k$ is a homomorphism

$$d(fg) = d(f) + d(g).$$

Here we have used the fact that $\mathbb{F}$ has no zero divisors. Moreover, it follows from our definitions that for $f \neq 0$ there exists a unique $k = k(f) \in \mathbb{Z}^m$ such that $\omega \cdot k = d(f)$. We call the term $a_{k(f)} t^{k(f)}$ the **leading term** of f and $a_{k(f)}$ the **leading coefficient**.

Theorem 4.1. *Assume that the numbers $\omega_j > 0$ are rationally independent. Then $f \in \Lambda(\omega; \mathbb{F})$ is invertible if and only if the leading coefficient is invertible in $\mathbb{F}$. In particular, $\Lambda(\omega; \mathbb{F})$ is a field if and only if $\mathbb{F}$ is a field.*

Proof: Assume without loss of generality that the leading term is 1. We shall construct a sequence $g_\nu = 1 + q_\nu \in \Lambda$ such that $d(q_\nu) > 0$ and

$$\lim_{\nu\to\infty} d(q_\nu) = \infty, \qquad \lim_{\nu\to\infty} d(g_\nu \cdots g_2 g_1 f - 1) = \infty.$$

For any such sequence the infinite product

$$g = \prod_{\nu=1}^{\infty} g_\nu \in \Lambda$$

is well defined and is the inverse of f.

The set $\{\omega \cdot k : a_k \neq 0\}$ can be written as an increasing sequence $d_0(f) < d_1(f) < d_2(f) < \cdots$ converging to infinity with $d_0(f) = d(f) = 0$. Since the ω_j are rationally independent there exists a unique sequence $k_\nu \in \mathbb{Z}^m$ such that $\omega \cdot k_\nu = d_\nu(f)$. Assume that $f \neq 1$ and define

$$g_1(t) = 1 - a_{k_1} t^{k_1}$$

to obtain

$$d_1(g_1) = d_1(f), \qquad d_1(g_1 f) > d_1(f).$$

The last inequality follows from the fact that either $d_1(g_1 f) = d_2(f)$ or $d_1(g_1 f) = 2d_1(f)$. More generally we find that $d_\nu(g_1 f) \in \Sigma(f)$ for every ν where $\Sigma(f)$ denotes the set of finite linear combinations of the $d_\nu(f)$ with nonnegative integer coefficients. Hence

$$\Sigma(g_1 f) \subset \Sigma(f).$$

Now proceed by induction and construct a (possibly finite) sequence $g_\nu \in \Lambda$ with leading term 1 such that

$$d_1(g_\nu) = d_1(g_{\nu-1} \cdots g_1 f), \qquad d_1(g_\nu \cdots g_1 f) > d_1(g_{\nu-1} \cdots g_1 f)$$

Moreover $\Sigma(g_\nu \cdots g_1 f) \subset \Sigma(g_{\nu-1} \cdots g_1 f)$ and hence $d_1(g_\nu)$ is a strictly increasing sequence in $\Sigma(f)$. Hence $d_1(g_\nu)$ converges to infinity and this proves the theorem. □

Theorem 4.2. *Assume that the numbers $\omega_j > 0$ are rationally independent. Then $\Lambda(\omega; \mathbb{F})$ is a principal ideal domain if and only if $\mathbb{F}$ is a principal ideal domain.*

Proof: First assume that $\Lambda = \Lambda(\omega, \mathbb{F})$ is a PID. If $I \subset \mathbb{F}$ is an ideal then $I\Lambda$ is an ideal in Λ and is therefore generated by a single element f. Assume without loss of generality that $d(f) = 0$ and let a_0 be the leading coefficient of f. Then every $a \in I$ can be viewed as a generalized Laurent series in $I\Lambda$. Hence $a = gf$ for some $g \in \Lambda$ and hence $a = xa_0$ for some $x \in \mathbb{F}$.

Conversely, suppose that $\mathbb{F}$ is a PID and let $I \subset \Lambda$ be an ideal. Then the set $I_0 \subset \mathbb{F}$ of leading coefficients of elements of I is an ideal in $\mathbb{F}$. Hence I_0 is generated by a single element a_0. Choose $f_0 \in I$ with $d(f_0) = 0$ and leading coefficient a_0. We prove that I is generated by f_0. Let $f \in I$ and assume without loss of generality that $d(f) = 0$. Since the leading coefficient of f is an element of I_0 it must be a multiple of a_0. Hence there exists a $b_0 \in \mathbb{F}$ such that

$$d(f - b_0 f_0) > d(f) = 0.$$

Note in fact that either $f - b_0 f_0 = 0$ or $d(f - b_0 f_0) \in \Sigma(f) \cup \Sigma(f_0) \subset \Sigma(f) + \Sigma(f_0)$. More generally

$$\Sigma(f - b_0 f_0) \subset \Sigma(f) + \Sigma(f_0).$$

Since $f - b_0 f_0 \in I$ its leading coefficient is again a multiple of a_0 and hence there exists a $b_1 \in \mathbb{F}$ and a $k_1 \in \mathbb{Z}^m$ such that

$$d(f - b_0 f_0 - b_1 t^{k_1} f_0) > d(f - b_0 f_0) = k_1 \cdot \omega.$$

Proceed by induction and construct a sequence

$$g_\nu = b_0 + b_1 t^{k_1} + \cdots + b_\nu t^{k_\nu} \in \Lambda$$

such that

$$d(f - g_\nu f_0) > d(f - g_{\nu-1} f_0) = k_\nu \cdot \omega.$$

Then $k_\nu \cdot \omega$ is a strictly increasing sequence in $\Sigma(f) + \Sigma(f_0)$. Hence $k_\nu \cdot \omega$ converges to infinity and hence g_ν converges to $g \in \Lambda$ with $f = g f_0$. □

5 Floer cohomology

Let (M, ω) be a $2n$-dimensional compact weakly monotone symplectic manifold with a regular almost complex structure $J \in \mathcal{J}_{\mathrm{reg}}(M, \omega)$ in the sense of Section 2 and let $H \in \mathcal{H}_{\mathrm{reg}}(J)$ be a regular Hamiltonian in the sense of Section 3. Let $\mathcal{L}$ denote the space of contractible loops $x : S^1 \to M$. The universal cover of $\mathcal{L}$ is the set of equivalence classes of pairs (x, u) where $x \in \mathcal{L}$ and $u : D \to M$ is a smooth disc such that $u(e^{2\pi i t}) = x(t)$. Two such pairs (x, u) and (y, v) are equivalent if $x = y$ and u is homotopic to v with fixed boundary. We shall, however, use a weaker equivalence relation

$$[x, u_0] \equiv [x, u_1] \quad \Longleftrightarrow \quad \int_D {u_0}^* c_1 = \int_D {u_1}^* c_1, \qquad \int_D {u_0}^* \omega = \int_D {u_1}^* \omega.$$

Here c_1 denotes a 2-form which represents the first Chern class of TM. The definition of the equivalence relation is independent of the choice of this form. For simplicity of notation we fix a point $x_0 \in M$ and assume that all discs satisfy $u(0) = x_0$. We denote by $\widetilde{\mathcal{L}}$ the set of all such equivalence classes. The group

$$\Gamma = \frac{\pi_2(M)}{\ker \phi_{c_1} \cap \ker \phi_\omega}$$

acts on the space $\widetilde{\mathcal{L}}$ via $[x,u] \mapsto [x, A\#u]$. Here $A\#u$ denotes the equivalence class of the connected sum $v\#u$ for $v \in A$. More explicitly, choose $v \in A$ with $v(0) = v(\infty) = x_0$ and define the connected sum by $v\#u(z) = v((1-2|z|)^{-1}z)$ for $|z| \leq 1/2$ and $v\#u(z) = u((2-|z|^{-1})z)$ for $|z| \geq 1/2$. Note in fact that

$$\mathcal{L} = \widetilde{\mathcal{L}}/\Gamma.$$

Let $\widetilde{\mathcal{P}}(H)$ denote the subset of those pairs $[x,u] \in \widetilde{\mathcal{L}}$ where $x \in \mathcal{P}(H)$ is a contractible 1-periodic solution of (1). These are the critical points of the **action functional** $a_H : \widetilde{\mathcal{L}} \to \mathbb{R}$ defined by

$$a_H([x,u]) = \int_D u^*\omega + \int_0^1 H(t,x(t))\,dt.$$

The solutions of (2) are the gradient flow lines of a_H. Given $[x^\pm, u^\pm] \in \widetilde{\mathcal{P}}(H)$ denote by

$$\mathcal{M}([x^-,u^-],[x^+,u^+];H,J)$$

the set of those connecting orbits $u \in \mathcal{M}(x^-,x^+;H,J)$ for which $[x^+, u^-\#u] \equiv [x^+,u^+]$. More explicitly we introduce the function $\beta(s) = (1+s^2)^{-1/2}s$. Then the connected sum $u^-\#u : D \to M$ is defined by $u^-\#u(z) = u^-(e^{4\pi}z)$ for $|z| \leq e^{-4\pi}$ and $u^-\#u(e^{2\pi(\beta(s)+it-1)}) = u(s,t)$ for $(s,t) \in \mathbb{R}^2$. It follows from the energy identity in the proof of Theorem 3.3 that

$$E(u) = a_H([x^+,u^+]) - a_H([x^-,u^-])$$

for $u \in \mathcal{M}([x^-,u^-],[x^+,u^+],H;J)$. Note that

$$a_H([x,A\#u]) - a_H([x,u]) = \omega(A).$$

for every $[x,u] \in \widetilde{\mathcal{L}}$ and every $A \in \Gamma$.

The dimension of the space $\mathcal{M}([x^-,u^-],[x^+,u^+],H;J)$ can be expressed in terms of the Conley-Zehnder index which is defined as follows. Given $[x,u] \in \widetilde{\mathcal{P}}(H)$ choose a symplectic trivialization $\Phi(t) : \mathbb{R}^{2n} \to T_{x(t)}M$ of x^*TM which extends over the disc u. Now linearize the Hamiltonian differential equation (1) along $x(t)$ to obtain a path of symplectic matrices

$$\Psi(t) = \Phi(t)^{-1} d\psi_t(x(0))\Phi(0) \in Sp(2n;\mathbb{R}).$$

Here the symplectomorphism $\psi_t : M \to M$ denotes the time-t-map of (1). Then $\Psi(0) = \mathbb{1}$ and $\Psi(1)$ is conjugate to $d\psi_1(x(0))$ so that $\det(\mathbb{1} - \Psi(1)) \neq 0$. The homotopy class of the path Ψ subject to these conditions is independent of the choice of the trivialization and is determined by the **Conley-Zehnder index** $\mu([x,u]) = \mu(\Psi)$ (see [5]). In [21] it is shown that the dimension of the space of connecting orbits is given by the formula

$$\dim \mathcal{M}([x^-,u^-],[x^+,u^+];H,J) = \mu([x^+,u^+]) - \mu([x^-,u^-]).$$

Moreover, the Conley-Zehnder index satisfies the identity

$$\mu([x,A\#u]) - \mu([x,u]) = 2c_1(A).$$

In particular the Conley-Zehnder index $\mu(x)$ of a periodic solution $x \in \mathcal{P}(H)$ of (1) is well defined modulo $2N$ where N is the minimal Chern number of (M,ω).

Denote by $\widetilde{\mathcal{P}}_k(H)$ the subset of all $[x,u] \in \widetilde{\mathcal{P}}(H)$ with $\mu([x,u]) = k$ and consider the cochain complex whose k-th cochain group

$$C^k = C^k(H)$$

consists of all functions $\widetilde{\mathcal{P}}_k(H) \to \mathbb{Z}_2 : [x,u] \mapsto \xi_{[x,u]}$ for which there are only finitely many nonzero entries in every region of finite symplectic action, i.e.

$$\#\left\{[x,u] \in \widetilde{\mathcal{P}}_k(H) : \xi_{[x,u]} \neq 0,\, a_H([x,u]) \geq c\right\} \quad < \quad \infty$$

for every constant $c > 0$. This is an infinite dimensional vector space over $\mathbb{Z}_2$ but a finite dimensional vector space over the field $\Lambda_\omega = \Lambda(\Gamma_0, \phi_\omega; \mathbb{Z}_2)$ where

$$\Gamma_0 = \frac{\ker \phi_{c_1}}{\ker \phi_{c_1} \cap \ker \phi_\omega}.$$

The scalar multiplication of $\xi \in C^k$ with $\lambda \in \Lambda_\omega$ is given by

$$(\lambda * \xi)_{[x,u]} = \sum_{A \in \Gamma_0} \lambda_A \xi_{[x,(-A)\#u]}.$$

The reader may check that this is a finite sum for every $[x,u] \in \widetilde{\mathcal{P}}(H)$ and that $\lambda * \xi \in C^k$. The dimension of C^k as a vector space over Λ_ω is precisely the number of contractible 1-periodic solutions $x \in \mathcal{P}(H)$ of (1) with Conley-Zehnder index $\mu(x) = k (\mathrm{mod}\, 2N)$. An explicit basis is a set of pairs $[x, u_x]$ with one representative for each periodic solution. Thus we may identify C^k with the vector space

$$C^k = \bigoplus_{\substack{x \in \mathcal{P}(H) \\ \mu(x) = k (\mathrm{mod}\, 2N)}} \Lambda_\omega x.$$

Now it follows from Theorem 3.3 that the space $\mathcal{M}([x^-, u^-], [x^+, u^+]; H, J)$ consists of finitely many orbits (modulo time shift) whenever

$$\mu([x^+, u^+]) - \mu([x^-, u^-]) = 1.$$

We denote by $n_2([x^-, u^-], [x^+, u^+])$ the number modulo 2 of these connecting orbits. The coboundary map $\delta^k : C^k \to C^{k+1}$ is defined by

$$\delta[x, u] = \sum_{\mu([y,v])=k+1} n_2([x, u], [y, v])\, [y, v]$$

for $[x, u] \in \widetilde{\mathcal{P}}(H)$ with $\mu([x, u]) = k$. By Theorem 3.3 there are only finitely many connecting orbits from x to y in every region of finite energy and hence $\delta[x, u] \in C^k$. The reader may check that δ is a Λ_ω-linear map.

Theorem 5.1. $\delta \circ \delta = 0$.

Proof: The argument is the same as in Floer's original work [8], [15] and we shall only sketch the main idea. We must prove that

$$\sum_{\mu([y,v])=k} n_2([x, u], [y, v]) \cdot n_2([y, v], [z, w]) = 0$$

whenever $[x, u], [z, w] \in \widetilde{\mathcal{P}}(H)$ with

$$\mu([x, u]) = k - 1, \qquad \mu([z, w]) = k + 1.$$

This follows by examining the ends of the 1-manifold $\mathcal{M}([x, u], [z, w]; H, J)/\mathbb{R}$. Here $\mathbb{R}$ acts by translation in the s-variable. Since all the connecting orbits in $\mathcal{M}([x, u], [z, w]; H, J)$ have index $\mu = 2$ and energy $E = a_H([z, w]) - a_H([x, u])$ it follows from the proof of Theorem 3.3 that no bubbling can occur. Hence the usual glueing argument shows that the ends of the 1-manifold $\mathcal{M}([x, u], [z, w]; H, J)/\mathbb{R}$ are in one-to-one correspondence with the pairs of connecting orbits from $[x, u]$ to $[y, v]$ and from $[y, v]$ to $[z, w]$ [10], [15]. Hence the number of such pairs is even and this proves the theorem. □

The cohomology groups

$$HF^k(M, \omega, H, J; \mathbb{Z}_2) = \frac{\ker \delta^k}{\operatorname{im} \delta^{k-1}}$$

are called the **Floer cohomology groups** of the quadruple (M, ω, H, J) with coefficients in $\mathbb{Z}_2$. The Floer cohomology groups are finite dimensional vector spaces over the field Λ_ω and they are graded modulo $2N$

$$HF^k = HF^{k+2N}.$$

It follows by the same arguments as in [8] that these groups are independent of the choice of the Hamiltonian H and the almost complex structure J used to define them (see also [15] and [21]). This is stated more precisely in the next theorem.

Theorem 5.2. *Given $J^\alpha, J^\beta \in \mathcal{J}_{\mathrm{reg}}(M,\omega)$ and $H^\alpha \in \mathcal{H}_{\mathrm{reg}}(J^\alpha)$, $H^\beta \in \mathcal{H}_{\mathrm{reg}}(J^\beta)$ there exists a natural vector space homomorphism*

$$HF^{\beta\alpha} : HF^*(M,\omega,H^\alpha,J^\alpha;\mathbb{Z}_2) \to HF^*(M,\omega,H^\beta,J^\beta;\mathbb{Z}_2)$$

which preserves the grading by the Conley-Zehnder index. If (H^γ, J^γ) is any other such pair then

$$HF^{\gamma\beta} \circ HF^{\beta\alpha} = HF^{\gamma\alpha}, \qquad HF^{\alpha\alpha} = \mathrm{id}\,.$$

In particular, $HF^{\beta\alpha}$ is an isomorphism.

Proof: Choose a regular homotopy $J_s \in \mathcal{J}(M,\omega)$ of almost complex structures connecting $J_0 = J_\alpha$ to $J_1 = J_\beta$ such that the parametrized versions of the results of Section 2 hold. We may assume that J_s extends to a smooth function on $s \in \mathbb{R}$ such that $J_s = J_\alpha$ for $s \leq 0$ and $J_s = J_\beta$ for $s \geq 1$. The analogue of Theorem 2.2 now states that the set

$$\mathcal{M}_s(A;\{J_s\}) = \{(s,v) : v \in \mathcal{M}_s(A;J_s)\}$$

is a manifold of dimension

$$\dim \mathcal{M}_s(A;\{J_s\}) = 2n + 2c_1(A) + 1$$

for a generic family $\{J_s\}$. This space will be empty in the case $c_1(A) < 3 - n$. In the case $c_1(A) = 0$ the set of pairs (s,p) such that p is a point on a J_s-holomorphic curve in the class A is roughly speaking a set of codimension 4 in $\mathbb{R} \times M$.

The results of Section 3 now apply to the solutions of the equation

$$\frac{\partial u}{\partial s} + J(s,u)\frac{\partial u}{\partial t} - \nabla H(s,t,u) = 0 \tag{4}$$

where $H(s,t,x) = H_s(t,x)$ is a generic homotopy of Hamiltonians connecting $H_0 = H_\alpha$ to $H_1 = H_\beta$. In particular, the solutions with index difference 0 or 1 do not intersect the set $M_0(\infty;\{J_s\})$ of J_s-holomorphic curves of Chern number 0. (Note that J_s-holomorphic spheres bubble off at parameter values (s,t).) The solutions of (4) determine a cochain homomorphism from the Morse complex of the pair (H^α, J^α) to the Floer chain complex of the pair (H^β, J^β).

It follows by the same arguments as in Floer's original work that this cochain homomorphism induces an isomorphism on cohomology. For details we refer to [8], [9], [15], and [21]. □

In the terminology of Conley the above theorem states that the Floer homology goups corresponding to regular pairs (H,J) form a connected simple system. They can be viewed as the *middle dimensional cohomology* of the Γ-cover $\widetilde{\mathcal{L}}$ of the space $\mathcal{L}$ of contractible loops in M with coefficients in the field Λ_ω viewed as a representation of Γ. In Section 7 we shall prove that these cohomology groups are naturally isomorphic to the homology groups of M with coefficients in Λ_ω.

6 Morse inequalities

Let $H : M \to \mathbb{R}$ be a Morse function and choose a regular almost complex structure $J \in \mathcal{J}_{\mathrm{reg}}(M, \omega)$ such that the gradient flow of H is of Morse-Smale type. If H is sufficiently small then the periodic solutions of (1) are precisely the critical points of H. For each such critical point we fix a disc $u_x(re^{i\theta}) = \gamma_x(r)$ where $\gamma_x(0) = x_0$ and $\gamma_x(1) = x$. The equivalence class $[x, u_x]$ is independent of the choice of the path γ_x. These equivalence classes form a natural basis of the cochain complex C^*. In [21] it was proved that if the second derivatives of H are sufficiently small then the Conley-Zehnder index of the pair (x, u_x) is given by

$$\mu(x) = \mathrm{ind}_H(x) - n$$

where $\mathrm{ind}_H(x)$ denotes the Morse index of x as a critical point of H. Note that every solution u of (2) which is independent of t is a gradient flow line of H

$$\frac{d}{ds}u(s) = \nabla H(u(s)). \tag{5}$$

Since (5) is a Morse-Smale flow there are only finitely many connecting orbits from x to y whenever $\mu_H(y) - \mu_H(x) = 1$. We denote the number of these connecting orbits modulo 2 by $n_2(x, y)$. Let C^k denote the $\mathbb{Z}_2$-vector space generated by the critical points x of H with Morse index $\mu_H(x) = k(\mathrm{mod}\, 2N)$ and let

$$\delta^k : C^k \to C^{k+1}$$

denote the linear map whose (x, y) entry is the number $n_2(x, y)$ whenever the index difference is 1. Then δ is a coboundary operator and its homology agrees with the cohomology of M

$$H^k(C, \delta) = \bigoplus_{j=k(\mathrm{mod}\, 2N)} H^j(M; \mathbb{Z}_2).$$

(See [24] and [20].) It follows from the universal coefficient theorem that

$$H^k(C \otimes \Lambda_\omega, \delta \otimes \mathbb{1}) = \bigoplus_{j=k(\mathrm{mod}\, 2N)} H^j(M; \mathbb{Z}_2) \otimes \Lambda_\omega.$$

Now the Morse function $H : M \to \mathbb{R}$ will in general not be regular in the sense of Section 3 since there may be solutions $u : \mathbb{R}^2 \to M$ of (2) and (3) which are not independent of t. We shall, however, prove that these nontrivial solutions cannot occur with index difference 1 provided that H is sufficiently small. This leads to the following theorem which in the monotone case was proved by Floer [8]. We postpone the proof to Section 7.

Theorem 6.1. *Let (M,ω) be a compact symplectic manifold of dimension $2n$. Assume either that M is monotone or $c_1(\pi_2(M)) = 0$ or the minimal Chern number is $N \geq n$. Then for every $J^\alpha \in \mathcal{J}_{\mathrm{reg}}(M,\omega)$ and every $H^\alpha \in \mathcal{H}_{\mathrm{reg}}(J^\alpha)$ there exists a natural isomorphism*

$$HF^\alpha : HF^k(M,\omega,H^\alpha,J^\alpha;\mathbb{Z}_2) \to \bigoplus_{j=k(\mathrm{mod}\,2N)} H^{j+n}(M;\Lambda_\omega)$$

If (H^β,J^β) is any other such pair then $HF^\beta \circ HF^{\beta\alpha} = HF^\alpha$.

Remark 6.2. If the manifold M admits a Morse function which has only critical points of even index then the conclusions of Theorem 6.1 are obviously satisfied. However, in this case the Arnold conjecture already follows from the Lefschetz fixed point theorem.

Given any Hamiltonian $H : S^1 \times M \to \mathbb{R}$ with nondegenerate 1-periodic solutions, we define the numbers

$$p_k = \#\{x \in \mathcal{P}(H) : \mu(x) = k\ (\mathrm{mod}\,2N)\}.$$

Note that $p_k = p_{k+2N}$. By Theorem 6.1 these numbers are related to the $2N$-periodic Betti numbers

$$b_k = \sum_{j=k\ (\mathrm{mod}\,2N)} \dim H^j(M;\mathbb{Z}_2).$$

via the Morse inequalities.

Theorem 6.3. *Assume that (M,ω) satisfies the requirements of Theorem 6.1 and let $H : S^1 \times M \to \mathbb{R}$ be a Hamiltonian with nondegenerate 1-periodic solutions. If $N = 0$ then we have*

$$p_\ell - p_{\ell-1} + \cdots + \pm p_{-k} \geq b_{n+\ell} - b_{n+\ell-1} + \pm b_{n+\ell-k}$$

for any two integers k and ℓ and equality holds for ℓ and $-k$ sufficiently large. In the case $N \neq 0$ the morse inequalities are satisfied when $\ell - k$ even and equality holds for $\ell - k = 2N - 1$. In both cases we have

$$p_k \geq b_{n+k}$$

for every integer k.

Remark 6.4.

(i) The Floer homology groups can be defined with coefficients in $\mathbb{Z}$ (rather than $\mathbb{Z}_2$) or in any other principal ideal domain. In that case the number $n([x^-,u^-],[x^+,u^+])$ must be defined by counting the connecting orbits with appropriate signs as in [9].

(ii) The Conley-Zehnder index of a nondegenerate periodic solution $x \in \mathcal{P}(H)$ of (1) satisfies the identity

$$\operatorname{sign}\det(\mathbb{1} - d\psi_1(x(0)) = (-1)^{\mu(x)+n}.$$

Hence we recover the Lefschetz fixed point formula

$$(-1)^n\chi(M) = \sum_{k=0}^{2N-1}(-1)^k \dim_{\Lambda_\omega} HF^k(M,\omega,H,J;\mathbb{Z}_2) = \sum_{x\in\mathcal{P}(H)}(-1)^{\mu(x)}$$

from our Morse inequalities.

(iii) Poincaré duality can be expressed in the form

$$HF^k(M,\omega,H,J;\mathbb{Z}_2) = HF^{-k}(M,\omega,-H,J;\mathbb{Z}_2).$$

To see this replace $x \in \mathcal{P}(H)$ by $y(t) = x(-t)$ and $u \in \mathcal{M}(x^-,x^+;H,J)$ by $v(s,t) = u(-s,-t)$. This duality will remain valid in cases where the Floer cohomology groups are not isomorphic to the cohomology of M.

(iv) An interesting class of symplectic manifolds is where the first Chern class vanishes. In this case the Conley-Zehnder index of a nondegenerate periodic solution $x \in \mathcal{P}(H)$ is a well defined integer and hence the Floer homology groups are graded by the integers. If in addition the complex structure J is integrable then (M,ω,J) is called a **Calabi-Yau manifold**. A Calabi-Yau metric is one where the first Chern form of the curvature vanishes. Manifolds of this type have received considerable interest in the recent physics literature [3], [4].

(v) Our construction can be used to recover the Novikov-homology groups associated to a closed 1-form α on a compact Riemannian manifold M with nondegenerate zeros [17]. In this case the associated ring is $\Lambda_\alpha(\mathbb{F}) = \Lambda(\Gamma,\phi_\alpha;\mathbb{F})$ where $\phi_\alpha : \pi_1(M) \to \mathbb{R}$ is the homomorphism induced by α and $\Gamma = \pi_1(M)/\ker\phi_\alpha$. The Novikov homology groups can be recovered as the homology groups of a chain complex generated by the zeros of α. The boundary operator is determined by the one dimensional connecting orbits in the covering space $\tilde{M} \to M$ with fiber Γ. As in the case of Calabi-Yau manifolds the index of a zero is well defined. This construction is a natural generalization of Witten's approach to Morse theory for functions [24]. The details will be carried out elsewhere.

7 Proof of Theorem 6.1

In order to compute the Floer cohomology groups we must study the partial differential equation (2) with a time independent Hamiltonian function $H : M \to \mathbb{R}$

$$\frac{\partial u}{\partial s} + J(u)\frac{\partial u}{\partial t} - \nabla H(u) = 0. \tag{6}$$

The gradient flow lines $\gamma(s)$ of H appear as solutions $u(s,t) = \gamma(s)$ of (6) which are independent of t. If the gradient flow of H is of Morse-Smale type then the gradient flow lines determine a chain complex whose homology agrees with the homology of M [24], [7], [20]. As in [8] we shall use this fact to prove that the Floer cohomology groups are naturally isomorphic to the cohomology of M. The main problem is to prove that every solution of (6) with relative Morse index 1 must be independent of t. We outline a proof here and refer the reader to [10] for complete details.

Fix an almost complex structure $J \in \mathcal{J}_{\text{reg}}(M,\omega)$ which is regular in the sense of Section 3. For every Hamiltonian function H which is sufficiently small the 1-periodic solutions $x \in \mathcal{P}(H)$ are constant (Yorke's estimate) and hence agree with the critical points of H. Moreover, for a generic Hamiltonian function H, the gradient flow with respect to the metric induced by J is of Morse-Smale type. Now every solution $u(s,t) = u(s,t+1)$ of (6) with limits

$$\lim_{s\to\pm\infty} u(s,t) = x^{\pm}, \qquad dH(x^{\pm}) = 0, \tag{7}$$

determines a homotopy class in $\pi_2(M)$. The energy and the index of u are given by

$$E(u) = H(x^+) - H(x^-) + \int u^*\omega \tag{8}$$

and

$$\mu(u) = \text{ind}_H(x^+) - \text{ind}_H(x^-) + 2\int u^*c_1. \tag{9}$$

Equation (8) follows from the energy identity in the proof of Theorem 3.3 and the fact that the limit orbits $x^{\pm}$ are critical points of H. Equation (9) follows from the relation between the Conley-Zehnder index and the Morse index when H is sufficiently small.

Lemma 7.1. *Fix $J \in \mathcal{J}_{\text{reg}}(M,\omega)$ and let $H : M \to \mathbb{R}$ be a smooth Morse function whose gradient flow is of Morse-Smale type with respect to the metric induced by J. Then there exists a constant $\tau_0 > 0$ such that every solution $u(s,t) = u(s,t+1)$ of*

$$\frac{\partial u}{\partial s} + J(u)\frac{\partial u}{\partial t} - \tau\nabla H(u) = 0 \tag{10}$$

and (7) with $\tau < \tau_0$, and

$$\int u^*\omega \le 0$$

is independent of t.

Proof: By Gromov's compactness there exists a constant $\hbar > 0$ such that

$$E(v) = \int v^*\omega \geq \hbar$$

for every nonconstant J_0-holomorphic sphere $v : S^2 \to M$. Now assume that the statement of the lemma were false. Then there would exist a sequence of solutions u_ν of (6) and (3) with $\tau = \tau_\nu$ converging to zero, $J = J_\nu$ converging to J_0, and with

$$\int {u_\nu}^*\omega \leq 0, \qquad \frac{\partial u_\nu}{\partial t} \not\equiv 0.$$

Passing to a subsequence we may assume without loss of generality that the limit points $x^\pm$ are independent of ν. Choose a sequence of integers k_ν converging to infinity such that

$$k_\nu \tau_\nu \to \tau > 0.$$

Choose τ so small that the index formula (9) holds for τH_0 and that

$$\tau \left(H_0(x^+) - H_0(x^-)\right) \leq \frac{\hbar}{2}.$$

Now define

$$v_\nu(s,t) = u_\nu(k_\nu s, k_\nu t) = v_\nu(s, t + 1/k_\nu).$$

Then it follows from equation (8) that

$$\begin{aligned} E(v_\nu) &= k_\nu \tau_\nu \left(H_0(x^+) - H_0(x^-)\right) + k_\nu \int {u_\nu}^*\omega \\ &\leq k_\nu \tau_\nu \left(H_0(x^+) - H_0(x^-)\right) \\ &\leq \hbar/2. \end{aligned}$$

Hence bubbling cannot occur and it follows, as in the proof of Theorem 3.3, that v_ν converges to a finite collection $v^1, \ldots, v^m$ of J_0-gradient flow lines of τH_0 connecting x^j to x^{j+1} where $x^0, \ldots, x^m$ are critical points of H_0 with $x^0 = x^-$ and $x^m = x^+$. Now in [21] it is shown that for τ sufficiently small the t-independent solutions of (10) are isolated in the space of all solutions. Hence $v_\nu(s,t) \equiv v_\nu(s)$ for ν sufficiently large in contradiction to our assumption. This proves Lemma 7.1. □

The following example shows that the statement of Lemma 7.1 becomes false without the assumption $\int u^*\omega \leq 0$.

Example 7.2. Consider the symplectic manifold $M = S^2 = \mathbb{C} \cup \{\infty\}$ with the standard symplectic form

$$\omega = \frac{dx \wedge dy}{\left(x^2 + y^2 + 1\right)^2}.$$

Here we denote by $z = x + iy$ a point in S^2 and by $\zeta = \xi + i\eta$ an associated tangent vector. The standard complex structure is multiplication by $i = \sqrt{-1}$ and the induced metric is

$$\langle \zeta, \zeta' \rangle = \frac{\xi\xi' + \eta\eta'}{\left(x^2 + y^2 + 1\right)^2}.$$

The gradient of the Hamiltonian function

$$H(x, y) = \frac{x^2 + y^2 - 1}{x^2 + y^2 + 1}$$

with respect to this metric is $\nabla H(z) = 4z$. Hence equation (6) reads

$$\frac{\partial u}{\partial s} + i\frac{\partial u}{\partial t} = 4\tau u.$$

The critical points of H are $z = 0$ and $z = \infty$ and we require u to satisfy the limit condition

$$\lim_{s \to -\infty} u(s, t) = 0, \qquad \lim_{s \to +\infty} u(s, t) = \infty.$$

Explicit solutions are given by

$$u_k(s, t) = e^{4\tau s} e^{2\pi k(s+it)}, \qquad \pi k + 2\tau > 0.$$

This example is rather degenerate since all the integral curves of the Hamiltonian vector field $X_H(z) = -4iz$ are periodic with the same period $T = \pi/2$. The critical values $\tau = -k\pi/2$ are those where the integral curves of τX_H are of period 1. In particular, the fixed points 0 and ∞ are nondegenerate as 1-periodic solutions of τX_H if and only if $2\tau/\pi \notin \mathbb{Z}$. Their Conley-Zehnder index (with the constant disc) is

$$\mu(0) = -1 - 2[2\tau/\pi], \qquad \mu(\infty) = 1 + 2[2\tau/\pi].$$

Moreover we have

$$\int u_k{}^* \omega = \pi k, \qquad \int u_k{}^* c_1 = 2k$$

and hence

$$E(u_k) = \pi k + 2\tau, \qquad \mu(u_k) = 4[k + 2\tau/\pi] + 2$$

whenever $\pi k + 2\tau > 0$ and $2\tau/\pi \notin \mathbb{Z}$. It is also of interest to consider the solutions u_{-k} with $\pi k - 2\tau > 0$. These are connecting orbits from ∞ to 0 and they satisfy

$$\int u_{-k}{}^*\omega = \pi k, \qquad \int u_{-k}{}^* c_1 = 2k$$

and

$$E(u_{-k}) = \pi k - 2\tau, \qquad \mu(u_{-k}) = 4[k - 2\tau/\pi] + 2.$$

In particular, there is no solution with $\int u^*\omega < 0$ whenever $0 < \tau < \pi/2$. But there are always solutions with $\int u^*\omega > 0$ however small we choose τ. These solutions are stable and cannot be destroyed by a perturbation. They all have positive index $\mu(u) \geq 2$.
In the usual coordinates $(x_1, x_2, x_3) \in S^2 \subset \mathbb{R}^3$, related to $z = (1-x_3)^{-1}(x_1+ix_2)$ via stereographic projection, the symplectic form is $\omega = (4x_3)^{-1} dx_2 \wedge dx_1$ and the Hamiltonian is the height function $H(x) = x_3$.

Our aim is to prove that the situation of the previous example is the general one, i.e. for a generic almost complex structure J every solution $u(s,t) = u(s,t+1)$ of (6) and (7) with $\mu(u) \leq 1$ must be independent of t. Since the Hamiltonian function is required to be time independent it is more difficult to prove that the solutions of (6) and (7) form finite dimensional manifolds and we can only do this for simple solutions. This difficulty is similar to the one that arises in the study of J-holomorphic curves and was discussed in Section 3. A function $u : \mathbb{R} \times S^1 \to M$ is called **simple** if for every integer $m > 1$ there exists a point $(s,t) \in \mathbb{R}^2$ such that $u(s, t + 1/m) \neq u(s,t)$. For any two critical points $x^{\pm}$ of H denote by

$$\mathcal{M}_s(x^-, x^+, H, J)$$

the space of all simple solutions of (6) and (7). We point out that the gradient flow lines of H are not simple solutions of (6).

In [10] it is proved, roughly speaking, that given a sufficiently small Morse function $H : M \to \mathbb{R}$ there exists a generic set

$$\mathcal{J}_0 = \mathcal{J}_0(H) \subset \mathcal{J}(M, \omega)$$

of almost complex structures on M such that for every $J \in \mathcal{J}_0$ the simple solutions u of (6) and (7) are regular in the sense that the linearized operator D_u is surjective. (See the proof of Theorem 3.2 for the definition of D_u.) This implies that the moduli space $\mathcal{M}_s(x^-, x^+, H, J)$ is a finite dimensional manifold whose local dimension near u is the number $\mu(u)$ given by (9). This result is stated more precisely in the following theorem which is proved in [10], Theorem 7.2.

Theorem 7.3. *There exist a (sufficiently small) Morse function $H_0 : M \to \mathbb{R}$, an open set $\mathcal{J} = \mathcal{J}(H) \subset \mathcal{J}(M,\omega)$, and a generic set $\mathcal{J}_0 \subset \mathcal{J}_0(H) \subset \mathcal{J}(H)$ (in the sense of Baire with respect to the C^∞-topology) such that the following holds for every $J \in \mathcal{J}_0$.*

(i) *Every nonconstant periodic solution*

$$x(t) = x(t+T)$$

of the Hamiltonian differential equation

$$\dot{x} = X_H(x)$$

has period

$$T > 1.$$

(ii) *The moduli space $\mathcal{M}_s(x^-, x^+, H/m, J)$ is a manifold of local dimension*

$$\dim_u \mathcal{M}_s(x^-, x^+, H/m, J) = \operatorname*{ind}_H(x^+) - \operatorname{ind}(x^-) + 2\int u^* c_1$$

near u for any two critical points $x^\pm$ of H and any integer $m \geq 1$.

The assertions of the previous theorem hold in fact for an open and dense set of sufficiently small Morse functions on M but in [10] it is not proved for all Morse functions. Also, it is not proved in [10] that the set $\mathcal{J}(H)$ can be chosen dense in $\mathcal{J}(M,\omega)$.

Now the almost complex structure $J \in \mathcal{J}_0(H)$ can be chosen such that in addition to (i) and (ii) the following conditions are satisfied.

(iii) $J \in \mathcal{J}_{\mathrm{reg}}(M,\omega)$ is regular in the sense of Section 2.
(iv) The gradient flow of H with respect to the metric induced by J is of Morse-Smale type.

These conditions can be achieved by a generic perturbation of the almost complex strucrue. For (iii) this follows from the results of Section 2 and for (iv) from Theorem 8.1 in [21]. Theorem 7.3 ensures that this perturbation of J can be chosen without destroying conditions (i) and (ii).

Proposition 7.4. *Let (M,ω) be a compact symplectic manifold of dimension $2n$. Assume either that M is monotone or $c_1(\pi_2(M)) = 0$ or the minimal Chern number is $N \geq n$. Assume that the Morse function $H : M \to \mathbb{R}$ and the almost complex strucrue $J \in \mathcal{J}(M,\omega)$ satisfy the conditions (i) and (ii) in the statement of Theorem 7.3. Then there exists a number $m_0 = m_0(H) > 0$ such that every solution u of (10) and (7) with $\mu(u) \leq 1$ and $\tau^{-1} \in \{m_0, m_0+1, \ldots\}$ is independent of t.*

Proof: Choose $\tau_0 > 0$ as in Lemma 7.1 and let $u(s,t) = u(s,t+1)$ be a solution of (10) and (7) with $0 < \tau < \tau_0$, $\tau^{-1} \in \mathbb{Z}$, and

$$\mu(u) = \mathrm{ind}_H(x^+) - \mathrm{ind}_H(x^-) + 2\int u^*c_1 \leq 1.$$

Assume, by contradiction, that $u(s,t)$ is not independent of t. Then it follows from Lemma 7.1 that

$$\int u^*\omega > 0.$$

If u is simple then u must be independent of t since otherwise the functions $(s,t) \mapsto u(s_0+s, t_0+t)$ form a 2-dimensional family of simple solutions in contradiction with the dimension formula of statement (4) in the definition of $\mathcal{H}^{\mathrm{aut}}_{\mathrm{reg}}(J,\delta)$. If u is not simple then there exists an integer $m > 1$ such that

$$u(s, t + 1/m) \equiv u(s,t).$$

Let m be the largest such integer. (If there is no largest integer with this property then $u(s,t)$ is independent of t.) Then the function

$$v(s,t) = u(s/m, t/m) = v(s,t+1)$$

is a simple solution of (10) with $H\tau$ replaced by τ/m and index

$$\mu(v) = \mathrm{ind}_H(x^+) - \mathrm{ind}_H(x^-) + 2\int v^*c_1.$$

If

$$\int u^*c_1 \geq 0$$

then

$$\int v^*c_1 = \frac{1}{m}\int u^*c_1 \leq \int u^*c_1.$$

and hence $\mu(v) \leq 1$. If

$$\int u^*c_1 < 0.$$

then M is not monotone and hence must have minimal Chern number $N \geq n$ or $N = 0$. In the former case

$$\int v^*c_1 \leq -N \leq -n$$

and hence $\mu(v) \leq 0$. In the latter case $\mu(v) = \mu(u) \leq 1$. In all three cases v is a simple solution of (10) and (7) with $\tau = 1/m$ and $\mu(v) \leq 1$. But this contradicts the dimension formula of (ii) because the solution v belongs to a 2-dimensional family. This proves the proposition. □

Remark 7.5. The argument in the proof of Proposition 7.4 fails for $2n$-dimensional manifolds if the minimal Chern number is $n-1$. In this case there might be a sequence of connecting orbits u_ν of period $1/2$ with $\mathrm{ind}_H(x^+) - \mathrm{ind}_H(x^-) = 2n$ and $\int u_\nu{}^*c_1 = 2-2n$. Such connecting orbits would have index $\mu(u_\nu) = 4-2n \leq 0$. If the energy $E(u_\nu)$ converges to infinity then we cannot employ a compactness argument to arrive at a contradiction. Also considering the connecting orbits $v_\nu(s,t) = u_\nu(s/2, t/2)$ does not help since $\mu(v_\nu) = 2$ and therefore the dimension argument does not apply.

Proof of Theorem 6.1: In view of Theorem 5.2 the Floer cohomology groups are independent of the choice of J and H up to natural isomorphisms. Hence fix $J \in \mathcal{J}_{\mathrm{reg}}(M,\omega)$ and choose a time independent Hamiltonian $H \in \mathcal{H}^{\mathrm{aut}}_{\mathrm{reg}}(J,\delta)$ with $\delta > 0$ sufficiently small. Replace H by $H_0 = H/m$ with $m > 0$ sufficiently large. Then every solution u of (6) and (7) with $H = H_0$ and $\mu(u) \leq 1$ is independent of t. Hence the coboundary of the Floer chain complex agrees with the coboundary operator of the Morse complex of H.

Note that our time independent Hamiltonian H need not be regular in the sense of Section 3. However, by condition (iv) the gradient flow of H is of Morse-Smale type, and hence, by Corollary 4.3 and Theorem 7.3 in [21], the linearized operator D_u is onto for every solution u of (6) and (7) which is independent of t, and in particular for every solution with $\mu(u) \leq 1$. Using this one can prove directly that the continuation argument of Theorem 5.2 remains valid in the case where $H(s,t,x)$ is a homotopy from a time independent Hamiltonian function $H^\alpha = H_0$ as above to a regular Hamiltonian $H^\beta \in \mathcal{H}_{\mathrm{reg}}(J)$ in the sense of Section 3. It is here that the condition (iii) ie required to obtain compactness. The details are left to the reader. □

8 Examples

The product of n spheres

$$M = S^2 \times \cdots \times S^2$$

with the symplectic form $\omega = \omega_1 \times \cdots \times \omega_n$ is monotone if and only if all forms ω_j have the same volume. On the other hand the minimal Chern number of (M,ω) is 2 and hence the manifold is weakly monotone whenever $n \leq 4$. Our general theory only applies in the case $n = 1$ or $n = 2$. However, for arbitrary n, an almost complex structure of the form $J = J_1 \times \cdots \times J_n$ is generic in the sense of Section 2 and does not admit any J-holomorphic spheres of negative Chern number. Since there exists a Morse function whose critical points have only even indices our methods give the following refinement of the Lefschetz fixed point theorem.

Theorem 8.1. *Let (M,ω) be the n-fold product $S^2 \times \cdots \times S^2$ with any product symplectic structure. Then every 1-periodic Hamiltonian system on M with non-degenerate 1-periodic solutions has at least 2^{n-1} such solutions x with Conley-Zehnder index $\mu(x) = 0 (\bmod 4)$ and the same number with Conley-Zehnder index $\mu(x) = 2 (\bmod 4)$.*

Example 8.2. A 4-dimensional example of a Calabi-Yau manifold is the quartic surface

$$X = \left\{ [z_0 : \ldots : z_3] \in \mathbb{C}P^3 : \sum_{j=0}^{3} {z_j}^4 = 0 \right\}.$$

This is a compact, connected, simply connected 4-dimensional Kähler manifold with $c_1 = 0$. All 4-manifolds with these properties are diffeomorphic and they are called $K3$**-surfaces**. Their second Betti number is $b_2 = 22$. $K3$-surfaces have played an important role in 4 dimensional topology.

Example 8.3. A similar example in 3 complex dimensions is the hypersurface of degree d in $\mathbb{C}P^4$

$$Z_d = \left\{ [z_0 : \ldots : z_4] \in \mathbb{C}P^4 \mid \sum_{j=0}^{4} {z_j}^d = 0 \right\}.$$

This manifold is simply connected and has Betti numbers

$$b_2 = 1, \qquad b_3 = d^4 - 5d^3 + 10d^2 - 10d + 4.$$

In particular $\pi_2(Z_d) = \mathbb{Z}$ and the symplectic form ω does not vanish over $\pi_2(Z)$. Moreover the first Chern class of Z_d is given by

$$c_1 = (5-d){\iota_d}^* h$$

where $h \in H^2(\mathbb{C}P^4, \mathbb{Z})$ is the standard generator of the cohomology of $\mathbb{C}P^4$ and $\iota_d : Z_d \to \mathbb{C}P^4$ is the natural embedding of Z_d as a hypersurface in $\mathbb{C}P^4$. In particular the quintic hypersurface Z_5 is an example of a Calabi-Yau manifold.

Now let $A \in \pi_2(Z_d)$ be the generator of the homotopy group with $\omega(A) > 0$. An explicit representative of A is given by the holomorphic curve

$$[z_0 : z_1] \mapsto [z_0 : z_1 : iz_0 : iz_1 : 0]$$

when d is even and by $[z_0 : z_1] \mapsto [z_0 : z_1 : -z_0 : -z_1 : 0]$ when d is odd. Evaluating the first Chern class on this generator gives

$$\langle c_1(TZ_d), A \rangle = 5 - d.$$

So for $d \leq 4$ the manifold Z_d is monotone. For $d = 5$ and $d \geq 8$ the Arnold conjecture holds by Theorem 6.3 above. The manifold Z_d is always weakly monotone but our methods do not apply to the cases $d = 6$ and $d = 7$. In these cases, however, the Arnold conjecture follows from the recent work of Ono [18].

Example 8.4. An interesting example from the point of view of the Arnold conjecture is the 6-dimensional manifold

$$M = \mathbb{T}^2 \times X$$

This manifold has Euler characteristic 0, whereas the sum of the Betti numbers is 96. Thus the Lefschetz fixed point theorem does not give any periodic solution whereas our results show that a time dependent Hamiltonian flow on $\mathbb{T}^2 \times X$ with 1-periodic coefficients must have at least 96 contractible periodic solutions of period 1 provided that they are all nondegenerate.

Appendix A. Bubbling off analysis

Throughout let (M, ω) be a compact symplectic manifold with an ω-tame almost complex structure J. Consider the Riemann sphere $S^2 = \mathbb{C} \cup \{\infty\}$ covered by two charts with transition map $z \mapsto z^{-1}$. The group of biholomorphic maps of S^2 is $G = \mathrm{PSL}(2, \mathbb{C})$. It acts on S^2 by fractional linear transformations

$$\phi_A(z) = \frac{az+b}{cz+d}, \qquad A = \begin{pmatrix} a & b \\ c & d \end{pmatrix} \in \mathrm{SL}(2, \mathbb{C}).$$

We shall use the notation $\phi \in G$ for $\phi = \phi_A$ with $A \in \mathrm{SL}(2, \mathbb{C})$.

Theorem A.1. *For every sequence $u_\nu : S^2 \to M$ of J-holomorphic spheres representing a fixed homotopy class $A \in \pi_2(M)$ there exist a subsequence (still denoted by u_ν), sequences $\phi_\nu^j \in G$ for $j = 1, \dots, \ell$, and J-holomorphic spheres $v^1, \dots, v^\ell$ such that the following holds.*

(i) *The reparametrized curves $u_\nu \circ \phi_\nu^j$ converge to v^j with all derivatives uniformly on every compact subset of $S^2 \setminus C^j$ where $C^j \subset S^2$ is a finite set.*
(ii) *The connected sum $v^1 \# v^2 \# \cdots \# v^\ell$ represents the homotopy class A. In particular*

$$\sum_{j=1}^{\ell} E(v^j) = \omega(A), \qquad \sum_{j=1}^{\ell} c_1(v^j) = c_1(A).$$

(iii) *The set $\Sigma = \bigcup_{j=1}^{\ell} v^j(S^2)$ is connected.*
(iv) *For every neighbourhood U of Σ there exists a $\nu_0 > 0$ such that $u_\nu(S^2) \subset U$ for every $\nu \geq \nu_0$.*

The importance of J-holomorphic curves in symplectic geometry was discovered by Gromov in his seminal paper [11]. He also discussed in detail the phenomenon of bubbling and gave a geometric proof of Theorem A.1. A more analytical proof was recently given by Parker and Wolfson [19]. We sketch here the main ideas of the proof. As a first step we state an a-priori estimate for the derivatives of a nonconstant J-holomorphic curve with sufficiently small energy.

Lemma A.2. (A-priori estimate) *Assume that (M, ω) is a compact symplectic manifold and J is a smooth almost complex structure on M which is ω-tame. Then there exists a constant $\hbar > 0$ such that the following holds. If $r > 0$ and $v : B_r \to M$ is a J-holomorphic curve such that*

$$E(u) = \int_{B_r} |dv|^2 < \hbar$$

then

$$|dv(0)|^2 \leq \frac{8}{\pi r^2} \int_{B_r} |dv|^2.$$

The proof relies on a partial differential inequality of the form $\Delta e \geq -Ae^2$ for the *energy density* $e = |dv|^2$. The details are carried out in [20] for example.

Remark A.3.

(i) If $\hbar$ is as in Lemma A.2 then

$$E(v) = \int_{S^2} v^*\omega \geq \hbar$$

for every non-constant holomorphic sphere $v : S^2 \to M$. To see this identify $S^2 = \mathbb{C} \cup \{\infty\}$ and assume $E(v) < \hbar$. Since the mean value of $|dv|^2$ on the ball $B_r(z)$ converges to zero as $r \to \infty$ we have, by Lemma A.2, $dv(z) = 0$ for every $z \in \mathbb{C}$ and hence v is constant.

(ii) Given any number $c > 0$, there are only finitely many homotopy classes $A \in \pi_2(M)$ with $\omega(A) \leq c$ which can be represented by a holomorphic sphere.

The next lemma is a technical result about J-holomorphic annuli. It asserts that if the energy of the annulus is sufficiently small then it cannot be spread out uniformly but must be concentrated near the boundary circles. We denote $A(r, R) = B_R \setminus B_r \subset \mathbb{C}$ for $0 < r < R$.

Lemma A.4. *Let (M, ω) be a compact symplectic manifold and J be an ω-tame almost complex structure. Then there exist constants $c > 0, \hbar > 0$, and $T_0 > 0$ such that the following holds. If $u : A(r, R) \to M$ is a J-holomorphic curve such that $E(u) < \hbar$ then*

$$E_{A(e^T r, e^{-T} R)}(u) \leq \frac{c}{T} E_{A(r,R)}(u)$$

and

$$\int_0^{2\pi} d\left(u(re^{T+i\theta}), u(Re^{-T+i\theta})\right) \leq c\sqrt{\frac{E_{A(r,R)}(u)}{T}}$$

for $T \geq T_0$.

Proof: Choose the constant $\hbar > 0$ as in Lemma A.2 and consider the J-holomorphic curve $v(\tau + i\theta) = u(e^{\tau+i\theta})$ for $\log r < \tau < \log R$ and $\theta \in S^1 = \mathbb{R}/2\pi\mathbb{Z}$. Then for $\log r + T < \tau < \log R - T$ we have $E_{B_T(\tau+i\theta)}(v) \le TE(v) = TE(u)$ and hence, by Lemma A.2,

$$|\dot{\gamma}(\tau + i\theta)|^2 \le \frac{8E(u)}{\pi T}, \qquad \log r + T < \tau < \log R - T.$$

If T is sufficiently large then the loop $\gamma_\tau(\theta) = v(\tau + i\theta)$ is sufficiently short. Now for sufficiently short curves γ there is a well-defined symplectic action

$$a(\gamma) = -\int_\gamma \lambda$$

where λ is a 1-form on M such that $d\lambda = \omega$ on a geodesically convex neighbourhood of γ. This definition is independent of the choice of λ. Moreover, choosing local co-ordinates with $\gamma(0) = 0$ it is easy to see that

$$a(\gamma) \le c_0 \int_0^{2\pi} |\dot{\gamma}(\theta)|^2 \, d\theta.$$

Hence for the above loops $\gamma_\tau(\theta) = u(e^{\tau+i\theta})$ we obtain

$$|a(\gamma_\tau)| \le c_1 \, \ell(\gamma_\tau)^2 \le c_2 \frac{E(u)}{T}$$

for $\log r + T < \tau < \log R - T$. Since $\tau + i\theta \mapsto \gamma_\tau(\theta)$ is a J-holomorphic curve we have

$$\frac{d}{d}a(\gamma_\tau) = \int_0^{2\pi} |\dot{\gamma}_\tau(\theta)|^2 \, d\theta \ge \frac{\|\dot{\gamma}_\tau\|_{L^2}}{\sqrt{2\pi}} \ell(\gamma_\tau) \ge \frac{\|\partial_\tau \gamma_\tau\|_{L^2}}{\sqrt{2\pi c_1}} \sqrt{|a(\gamma_\tau)|}.$$

In particular, the function $\tau \mapsto a(\gamma_\tau)$ is strictly increasing. If $a(\gamma_\tau) > 0$ then

$$\frac{d}{d\tau}\sqrt{a(\gamma_\tau)} \ge c_3^{-1} \, \|\partial_\tau \gamma_\tau\|_{L^2}$$

and a similar inequality holds when $a(\gamma_\tau) < 0$. Integrating these from $\tau_0 = \log r + T$ to $\tau_1 = \log R - T$ (after splitting this interval into two according to the sign of $a(\gamma_\tau)$ if necessary) we obtain

$$\int_{\tau_0}^{\tau_1} \|\partial_\tau \gamma_\tau\|_{L^2} \, d\tau \le c_3 \left(\sqrt{|a(\gamma_{\tau_0})|} + \sqrt{|a(\gamma_{\tau_1})|} \right) \le c_4 \sqrt{\frac{E(u)}{T}}.$$

(A similar argument was used in [13].) Since $\|\partial_\tau \gamma_\tau\|_{L^2} \le c_5 \sqrt{E(u)/T}$ for $\tau_0 \le \tau \le \tau_1$ this implies

$$E_{A(e^T r, e^{-T} R)}(u) = \int_{\tau_0}^{\tau_1} \|\partial_\tau \gamma_\tau\|_{L^2}^2 \, d\tau \le c_6 \frac{E(u)}{T}.$$

Moreover,

$$\begin{aligned}\int_0^{2\pi} d(\gamma_{\tau_0}(\theta), \gamma_{\tau_1}(\theta))\, d\theta &\le \int_{\tau_0}^{\tau_1}\int_0^{2\pi} |\partial_\tau \gamma_\tau|\, d\theta d\tau \\ &\le \sqrt{2\pi}\int_{\tau_0}^{\tau_1} \|\partial_\tau\gamma_\tau\|_{L^2}\, d\tau \\ &\le c_7\sqrt{\frac{E(u)}{T}}\end{aligned}$$

and this proves the lemma. □

Let $u_\nu : S^2 = \mathbb{C}\cup\{\infty\} \to M$ be a sequence of J-holomorphic curves. A point $z \in \mathbb{C}$ is called **singular** for the sequence u_ν if there exists a sequence $z_\nu \to z$ such that $|du_\nu(z_\nu)|$ converges to ∞. The point $z = \infty$ is called singular if 0 is a singular point for the sequence $z \mapsto u_\nu(z^{-1})$. A singular point z for u_ν is called **tame** for u_ν if it is isolated and the limit

$$m_\varepsilon(z) = \lim_{\nu\to\infty}\int_{B_\varepsilon(z)} u_\nu^*\omega$$

exists for every sufficiently small $\varepsilon > 0$. In this case the **mass** of the singularity is defined to be the number

$$m(z) = \lim_{\varepsilon\to 0} m_\varepsilon(z).$$

This number exists because the function $\varepsilon \mapsto m_\varepsilon(z)$ is non-decreasing. It is always positive and in fact $m(z) \geq \hbar$.

The usual compactness argument shows that every sequence u_ν has a subsequence (still denoted by u_ν) with only finitely many singular points $z^1, \dots, z^\ell$ which are all tame. By definition the derivatives of u_ν are uniformly bounded in every compact subset in the complement of the singular set $C = \{z^1, \dots, z^\ell\}$. Hence it follows from elliptic bootstrapping that a further subsequence of u_ν converges with all derivatives uniformly on every compact subset of $S^2 \setminus S$ to a J-holomorphic sphere $u : S^2 \to M$. The energy of this limit satisfies the identity

$$E_{S^2\setminus B_\varepsilon(C)}(u) + \sum_{j=1}^{\ell} m_\varepsilon(z^j) = \omega(A)$$

for every sufficiently small $\varepsilon > 0$. Take the limit $\varepsilon \to 0$ to obtain

$$E(u) + \sum_{j=1}^{\ell} m(z^j) = \omega(A).$$

Soft rescaling

We examine the behaviour of the sequence u_ν near a singularity z in more detail. Composing u_ν with a suitable element of G (independent of ν) we may assume without loss of generality that $z = 0$ and denote $m = m(z)$. For every ν there exists a unique number $\delta_\nu > 0$ such that

$$\int_{B_{\delta_\nu}} u_\nu^*\omega = m - \frac{\hbar}{2}$$

where $B_\delta = B_\delta(0)$ denotes the ball of radius delta centered at 0. By definition of the mass m the sequence $\delta_\nu > 0$ converges to 0. Consider the sequence of J-holomorphic spheres $v_\nu : S^2 \to M$ defined by

$$v_\nu(z) = u_\nu(\delta_\nu z).$$

Lemma A.5. *There exists a subsequence (still denoted by v_ν) such that the following holds.*

(i) *The singular set $C' = \{w^1, \ldots, w^k\}$ of the subsequence v_ν is finite and tame and is contained in $\mathrm{cl}(B_1) \cup \{\infty\}$.*

(ii) *The subsequence v_ν converges with all derivatives uniformly on every compact subset of $S^2 \setminus C'$ to a non-constant J-holomorphic sphere $v : S^2 \to M$.*

(iii) *The energy of v and the masses of the singularities $w^1, \ldots, w^k$ satisfy*

$$\int_{S^2} v^*\omega + \sum_{j=1}^{k} m(w^j) = m.$$

(iv) $v(\infty) = u(0)$.

Proof: If follows immediately from the definitions that for every $R > 1$ and every $\varepsilon > 0$ there exists a $\nu_0 = \nu_0(R, \varepsilon) > 0$ such that

$$m - \frac{\hbar}{2} = E_{B_1}(v_\nu) \le E_{B_R}(v_\nu) = E_{B_{R\delta_\nu}}(u_\nu) \le m + \varepsilon$$

for $\nu \ge \nu_0$. Hence there is no bubbling outside the unit ball and this proves statement (i). We shall now prove that the limit curve $v : \mathbb{C} \to M$ satisfies

$$E_{\mathbb{C}\setminus B_R}(v) = \frac{\hbar}{2}.$$

We have already seen that $E_{\mathbb{C}\setminus B_R}(v) \le \hbar/2$. To prove the converse choose a sequence $\varepsilon_\nu > 0$ such that $E(u; B_{\varepsilon_\nu}) = m_0$. Then it follows again from the definition of m_0 that $\varepsilon_\nu \to 0$. Now consider the sequence $w_\nu(z) = u_\nu(\varepsilon_\nu z)$. It follows as above that $E(w_\nu; B_R - B_1)$ converges to zero for any $R > 1$. This implies that $E(w_\nu; B_1 - B_\delta)$ must also converge to zero for any $\delta > 0$ since otherwise a subsequence of w_ν would converge to a nonconstant J-holomorphic curve which is constant for $|z| \geq 1$ but such a curve does not exist. Since

$$E(w_\nu; B_1 - B_{\delta_\nu/\varepsilon_\nu}) = E(u_\nu; B_{\varepsilon_\nu} - B_{\delta_\nu}) = \frac{\hbar}{2}$$

it follows that $\delta_\nu/\varepsilon_\nu$ converges to 0. Now, by Lemma A.4, there exists a $T_0 > 0$ such that for $T > T_0$

$$E(u_\nu; B_{e^{-T}\varepsilon_\nu} - B_{e^T\delta_\nu}) \leq \frac{c}{T} E(u_\nu; B_{\varepsilon_\nu} - B_{\delta_\nu}) = \frac{c}{T}\frac{\hbar}{2}.$$

Pick any number $\alpha < 1$ and choose T so large that $1 - c/T > \alpha$. Then the energy of u_ν in the union of the annuli $A(\delta_\nu, e^T\delta_\nu)$ and $A(e^{-T}\varepsilon_\nu, \varepsilon_\nu)$ must be at least $\alpha\hbar/2$. But the energy of u_ν in $A(e^{-T}\varepsilon_\nu, \varepsilon_\nu)$ converges to 0 while the energy of u_ν in $A(\delta_\nu, e^T\delta_\nu)$ converges to $E(v; A(1, e^T))$. Hence $E(v; A(1, e^T)) \geq \alpha\hbar/2$. Since $\alpha < 1$ was chosen arbitrarily it follows that $E(v, \mathbb{C} - B_1) = \hbar/2$ as claimed. In particular this implies that v is nonconstant and thus we have proved statement (ii). Statement (iii) now follows from the usual bubbling argumant already used above.

To prove statement (iv) fix $\varepsilon > 0$ and $R > 0$ and define

$$E(\varepsilon, R) = \lim_{\nu\to\infty} \int_{B_\varepsilon \setminus B_{R\delta_\nu}} u_\nu^*\omega.$$

This limit exists for $\varepsilon > 0$ sufficiently small and $R > 0$ arbitrarily large. To see this consider the annuli $B_\varepsilon \setminus B_{\delta_\nu}$ and $B_{R\delta_\nu} \setminus B_{\delta_\nu}$. The limit on the former is $\hbar/2 + E_{B_\varepsilon}(u)$ while the limit on the latter is $E_{B_R\setminus B_1}(v) = \hbar/2 - E_{\mathbb{C}\setminus B_R}(v)$. Hence

$$E(\varepsilon, R) = E_{B_\varepsilon}(u) + E_{\mathbb{C}\setminus B_R}(v), \qquad \lim_{\substack{\varepsilon\to 0\\ R\to\infty}} E(\varepsilon, R) = 0.$$

Now $E(\varepsilon, R)$ converges to zero as $\varepsilon \to 0$ and $R \to \infty$. Hence it follows from Lemma A.4 that for $T > 0$ and $\nu > 0$ sufficiently large we have

$$\int_0^{2\pi} d(u_\nu(R\delta_\nu e^{T+i\theta}), u_\nu(\varepsilon e^{-T+i\theta}))\, d\theta \leq c\sqrt{E(\varepsilon, R)}.$$

Taking the limit $\nu \to \infty$ we obtain

$$\int_0^{2\pi} d(v(Re^{T+i\theta}), u(\varepsilon e^{-T+i\theta}))\, d\theta \leq c\sqrt{E(\varepsilon, R)}.$$

Here the constants T and c are independent of ε and R. Since $E(\varepsilon, R) \to 0$ for $\varepsilon \to 0$ and $R \to \infty$ we obtain $v(\infty) = u(0)$ as required. This completes the proof of Lemma A.5. □

Remark A.6. All the singularities of the sequence u_ν other than $z = 0$ are pushed to ∞ in the rescaled sequence v_ν. Lemma A.5 may be viewed as a *resolution* of the singularity $z = 0$.

Proof of Theorem A.1: Apply Lemma A.5 to each singular point for the sequence u_ν passing to a suitable subsequence. Now apply Lemma A.5 again to each singularity for each of the renormalized sequences v_ν obtained from the previous application of Lemma A.5 and proceed by induction. This process will terminate after finitely many steps since in each step the energy decreases by at least $\hbar$.

To prove convergence of u_ν to the collection of all bubbles take any sequence $z_\nu \in S^2$. If z_ν is uniformly bounded away from the singular set of u_ν then $u_\nu(z_\nu)$ accumulates on the limit curve of u_ν. Hence assume that z_ν converges to a singular point which we assume to be $z = 0$. There are two possibilities. If there exists a constant $\varepsilon > 0$ such that

$$\varepsilon \leq \frac{|z_\nu|}{\delta_\nu} \leq \frac{1}{\varepsilon}$$

where $\delta_\nu > 0$ is defined as in Lemma A.5 then $u_\nu(z_\nu)$ converges to the bubble point $u(0) = v(\infty)$. If

$$\lim_{\nu \to \infty} \frac{|z_\nu|}{\delta_\nu} = 0$$

then proceed by induction with u_ν replaced by the rescaled sequence $v_\nu(z) = u_\nu(\delta_\nu z)$.

A close examination of Lemma A.5 and the inductive procedure shows that the connected sum of all bubbles represents the original homotopy class A. This proves the theorem. □

References

[1] V.I. Arnold, Sur une propriété topologique des applications globalement canoniques et à mecanique classique, *C. R. Acad. Sci. Paris* **261** (1965), 3719–3722.

[2] V.I. Arnold, *Mathematical Methods in Classical Mechanics*, Springer-Verlag, 1978.

[3] P. Candelas and X.C. de la Ossa, Moduli space of Calabi-Yau manifolds, University of Texas Report UTTG-07, 1990

[4] P. Candelas, X.C. de la Ossa, P.S. Green, and L. Parkes, A pair of Calabi-Yau manifolds as an exactly soluble superconformal field theory, University of Texas Report UTTG-25, 1990

[5] C. Conley and E. Zehnder, The Birkhoff-Lewis fixed point theorem and a conjecture of V.I. Arnold, *Invent. Math.* **73** (1983), 33–49.

[6] A. Floer, Morse theory for Lagrangian intersections, *J. Diff. Geom.* **28** (1988), 513–547.

[7] A. Floer, Wittens complex and infinite dimensional Morse theory, *J. Diff. Geom.* **30** (1989), 207–221.

[8] A. Floer, Symplectic fixed points and holomorphic spheres, *Comm. Math. Phys.* **120** (1989), 575–611.

[9] A. Floer and H. Hofer, Coherent orientations for periodic orbit problems in symplectic geometry, *Math. Zeit.* **212** (1994), 13–38.

[10] A. Floer, H. Hofer, and D.A. Salamon, Transversality in elliptic Morse theory for the action functional, ETH Zürich preprint April 1994

[11] M. Gromov, Pseudo holomorphic curves in symplectic manifolds, *Invent. Math.* **82** (1985), 307–347.

[12] H. Hofer, Ljusternik-Schnirelman theory for Lagrangian intersections, *Ann. Henri Poincaré – analyse nonlinéaire* **5** (1988), 465–499.

[13] H. Hofer and C. Viterbo, The Weinstein conjecture in the presence of holomorphic spheres, *Comm. Pure Appl. Math.* **Vol. XLV** (1992), 583–622.

[14] D. McDuff, Examples of symplectic structures, *Invent. Math.* **89** (1987), 13–36.

[15] D. McDuff, Elliptic methods in symplectic geometry, *Bull. A.M.S.* **23** (1990), 311–358.

[16] D. McDuff, The local behaviour of holomorphic curves in almost complex 4-manifolds, *J. Diff. Geom.* **34** (1991), 143–164.

[17] S.P. Novikov, Multivalued functions and functionals — an analogue of the Morse theory, *Soviet Math. Dokl.* **24** (1981), 222–225.

[18] K. Ono, The Arnold conjecture for weakly monotone symplectic manifolds, to appear *Inv. Math.*

[19] T.H. Parker and J.G. Wolfson, A compactness theorem for Gromov's moduli space, Preprint, Michigan State University, East Lansing, 1991.

[20] D. Salamon, Morse theory, the Conley index and Floer homology, *Bull. L.M.S.* **22** (1990), 113–140.

[21] D. Salamon and E. Zehnder, Morse theory for periodic solutions of Hamiltonian systems and the Maslov index, *Comm. Pure Appl. Math.* **45** (1992), 1303–1360.

[22] J.-C. Sikorav, Homologie de Novikov associée à une classe de cohomologie réelle de degré un, thèse, Orsay, 1987.

[23] S. Smale, An infinite dimensional version of Sard's theorem, *Am. J. Math.* **87** (1973), 213–221.

[24] E. Witten, Supersymmetry and Morse theory, *J. Diff. Geom.* **17** (1982), 661–692.

Helmut Hofer
Departement Mathematik
ETH-Zentrum
CH-8092 Zürich

Dietmar A. Salamon
Mathematics Institute
University of Warwick
Coventry CV4 7AL
Great Britain

Symplectic invariants and Hamiltonian dynamics

H. Hofer and E. Zehnder

1 Symplectic capacities

There is a mysterious relation between rigidity phenomena of symplectic geometry and global periodic solutions of Hamiltonian dynamics. One of the links is provided by a special class of symplectic invariants discovered by I. Ekeland and H. Hofer in [2], [3] called symplectic capacities. We first recall this concept in a more general setting from [26] and consider the class of all symplectic manifolds (M, ω) possibly with boundary, but of fixed dimension $2n$. Here ω is a symplectic structure, i.e. a two-form on M which is closed and nondegenerate. A symplectic capacity is a map

$$(M, \omega) \mapsto c(M, \omega)$$

which associates with every symplectic manifold (M, ω) a nonnegative number or ∞, satisfying the following axioms (A1)–(A3):

(A1) Monotonicity: $c(M, \omega) \leq c(N, \tau)$,
if there is a symplectic embedding $\varphi : (M, \omega) \to (N, \tau)$;

(A2) Conformality: $c(M, \alpha\omega) = |\alpha| \; c(M, \omega)$
for all $\alpha \in \mathbb{R}$ with $\alpha \neq 0$;

(A3) Nontriviality: $c(B(1), \omega_0) = \pi = c(Z(1), \omega_0)$
for the open unit ball $B(1)$ and the open cylinder $Z(1)$ in the standard space $(\mathbb{R}^{2n}, \omega_0)$.

Here ω_0 denotes the standard symplectic structure in $\mathbb{R}^{2n} = \mathbb{R}^2 \oplus \ldots \oplus \mathbb{R}^2$ represented by the matrix J,

$$J = \begin{pmatrix} \begin{pmatrix} 0 & 1 \\ -1 & 0 \end{pmatrix} & & \\ & \ddots & \\ & & \begin{pmatrix} 0 & 1 \\ -1 & 0 \end{pmatrix} \end{pmatrix} . \tag{1}$$

Denoting by (x_j, y_j) the symplectic coordinates in $\mathbb{R}^2, 1 \leq j \leq n$, the ball $B(r)$ and the symplectic cylinder $Z(r)$ of radius $r > 0$ are defined by

$$B(r) := \left\{ \, |x|^2 + |y|^2 < r^2 \right\} \quad \text{and} \quad Z(r) := \left\{ x_1^2 + y_1^2 < r^2 \right\}.$$

In view of the monotonicity axiom, the invariants $c(M, \omega)$ represent, in particular, obstructions to symplectic embeddings. For example, an immediate consequence of the axioms is the celebrated squeezing theorem of M. Gromov [20]. It gave rise to the concept of capacity and asserts that there exists a symplectic embedding $\varphi : B(r) \to Z(R)$ if and only if $R \geq r$. Another consequence of the axioms is the stability of symplectic mappings under topological limits. For example, one readily deduces from the axioms that the group of symplectic diffeomorphisms of a compact symplectic manifold (M, ω) is C^0-closed in the group of all diffeomorphisms of M [4], [19], a statement which goes back to Ya. Eliashberg.

It is, of course, not surprising that in the special case $(\mathbb{R}^2, \omega_0)$ of dimension 2, the capacity of sets is related to the area. If, for example, $D \subset \mathbb{R}^2$ is a connected and bounded region with smooth boundary, then

$$c(D, \omega_0) \;=\; \text{area }(D)\,, \quad D \subset \mathbb{R}^2,$$

so that the capacity agrees with the Lebesgue measure of D. This can be deduced from the axioms, see [33] and [27]. In higher dimensions, the symplectic invariant c is quite different from the volume; for example in $\mathbb{R}^4 = \mathbb{R}^2 \oplus \mathbb{R}^2$,

$$c\Big(B^2(r) \times B^2(\frac{1}{r})\Big) = \pi r^2 \longrightarrow 0$$

$$\text{vol}\Big(B^2(r) \times B^2(\frac{1}{r})\Big) = \text{const}$$

as $r \to 0$. The capacity extends a familiar linear symplectic invariant defined for ellipsoids. Indeed representing an ellipsoid E as

$$E \;=\; E(r) \;=\; \Big\{(x,y) \in \mathbb{R}^{2n} \;\Big|\; \sum_{j=1}^{n} \frac{x_j^2 + y_j^2}{r_j^2} < 1\Big\}$$

in symplectic normal form, with $0 < r_1 \leq r_2 \leq \ldots \leq r_n$, then

$$c(E, \omega_0) \;=\; \pi r_1^2,$$

as one concludes immediately from the axioms. It should be said that the axioms do not determine the capacity function uniquely. An early example of a capacity deduced from Gromov's paper [18] is the so-called Gromov width $D(M, \omega)$. By Darboux's theorem there is always a symplectic embedding $(B(\varepsilon), \omega_0) \to (M, \omega)$ for an $\varepsilon > 0$. Trying to increase such a Darboux chart, one is lead to define $D(M, \omega)$ to be the supremum over all numbers such that there exists a symplectic embedding $\varphi : (B(r), \omega_0) \to (M, \omega)$.

Taking the existence of a capacity for granted, one verifies readily that $D(M, \omega)$ is a capacity, which is, moreover the smallest capacity,

$$D(M, \omega) \;\leq\; c(M, \omega),$$

for every c. There are, however, capacities which are very different from $D(M,\omega)$ and are dynamically defined by means of periodic orbits of Hamiltonian equations on M. An example is the distinguished capacity c_0 introduced in [26] which we recall next.

If (M,ω) is a symplectic manifold, one associates with a smooth function $H : M \to \mathbb{R}$ a unique vector field X_H on M by

$$\omega(X_H, \cdot) = -dH(\cdot).$$

A periodic solution $x(t)$ of the Hamiltonian equation $\dot{x} = X_H(x)$ on M is a solution satisfying the boundary condition $x(0) = x(T)$ for some $T > 0$. We single out a distinguished set $\mathcal{H}(M,\omega)$ of Hamiltonian functions H, satisfying the following properties (P1)–(P3):

(P1) There is a compact set $K \subset M$ (depending on H) satisfying $K \subset (M \backslash \partial M)$ and

$$H(M \backslash K) = m(H) \quad \text{i.e. is constant}\ .$$

(P2) There is an open set $U \subset M$ (depending on H) satisfying $U \subset (M \backslash \partial M)$ and

$$H(U) = 0.$$

(P3) $0 \le H(x) \le m(H)$ for all $x \in M$.

A function $H \in \mathcal{H}(M,\omega)$ is called admissible if it has the property that all T-periodic solutions of $\dot{x} = X_H(x)$ on M with periods $0 < T \le 1$ are constant solutions : $x(t) \equiv x(0)$ for all t. Abbreviating the set of admissible functions by $\mathcal{H}_a(M,\omega) \subset \mathcal{H}(M,\omega)$ we now define

$$c_0(M,\omega) = \sup \left\{ m(H) \;\middle|\; H \in \mathcal{H}_a(M,\omega) \right\}.$$

Therefore, if $c_0(M,\omega) < \infty$, then this number is characterized by the property that, for every H in $\mathcal{H}(M,\omega)$ whose oscillation satisfies $m(H) > c_0(M,\omega)$, the vector field, X_H possesses a nonconstant T-periodic solution for some $0 < T \le 1$ and $c_0(M,\omega)$ is the infinimum of the real numbers having this property.

It is proved in [26] that c_0 is a symplectic capacity. The existence proof is not intuitive and is based on the well known Hamiltonian variational principle of classical mechanics. It is interesting to know that in the special case of a convex and bounded domain $M \subset (\mathbb{R}^{2n}, \omega_0)$ with smooth boundary ∂M, the capacity c_0 is represented by a distinguished closed characteristic of ∂M,

$$c_0(M,\omega_0) = |A(\gamma^*)|,$$

where $|A(\gamma^*)| = \inf\{|A(\gamma)| : \gamma \subset \partial M \text{ is a closed characteristic }\}$. Here

$$A(\gamma) = \int_\gamma \lambda$$

is the action of the loop γ and λ is the Liouville form satisfying $d\lambda = \omega_0$. The proof can be found in [26].

It is difficult, in general, to compute the symplectic invariant c_0. Again, the 2-dimensional case is special; here one knows that

$$c_0(M,\omega) = \Big|\int_M \omega\Big|, \quad \dim M = 2,$$

provided M is connected, so that c_0 agrees with the total volume, which classifies 2-dimensional symplectically diffeomorphic manifolds. In addition, if $D \subset \mathbb{R}^2$ is open, bounded and path connected, one deduces from the regularity properties of c_0 that $c_0(D,\omega_0) = \text{area}(D)$. For proofs we refer to K. Siburg [33] and independently to Jiang [27]. In higher dimensions H. Hofer and C. Viterbo [24] computed the capacity c_0 for the projective space

$$c_0(\mathbb{C}P^n,\omega) = \pi,$$

where ω is the standard symplectic structure related to the Fubini-Study metric. The proof analyzes the structure of holomorphic spheres in $\mathbb{C}P^n$ and is technically based on a Fredholm-theory for first order elliptic systems. The same technique allows us to compute the capacity c_0 for other examples. Assume (M,ω) is a connected closed compact symplectic manifold and denote by $\pi_2(M)$ the second homotopy group of M. Then there is a map $\pi_2(M) \to \mathbb{R}$ defined by

$$u \mapsto \langle \omega, u\rangle = \int_{S^2} u^*\omega.$$

Define $\alpha^* = \alpha^*(M) \in \mathbb{R} \cup \{\infty\}$ by $\alpha^* = \inf\{\langle\omega,u\rangle | \langle\omega,u\rangle > 0 \text{ and } u \in \pi_2(M)\}$ and set $\alpha^* = \infty$ if $\pi_2(M) = 0$. Then the following result combines the computations of A. Floer, H. Hofer and C. Viterbo [10] and R. Ma [30]. If $(M \times \mathbb{R}^{2n}, \omega \oplus \omega_0)$, $n \geq 1$ satisfies $0 < \alpha^* \leq \infty$, then

$$c_0\Big(M \times B(r)\Big) = c_0\Big(M \times Z(r)\Big) = \pi r^2$$

for all r satisfying $0 < \pi r^2 < \alpha^*$. The cotangent bundle $T^*(T^n) \cong T^n \times \mathbb{R}^n$ with the standard symplectic structure $d\lambda$ plays a prominent role in classical mechanics. From the axioms one deduces readily that, for every capacity function c,

$$c\Big(T^n \times [-a,a]^n, d\lambda\Big) = c\Big(T^n \times [-a,a] \times \mathbb{R}^{n-1}, d\lambda\Big) = 4\pi a,$$

if $a > 0$. This result, due to Jiang [27] turns out to be very useful for the solution of the Weinstein conjecture for $T^*(T^n)$ as explained below. It should be pointed out that it is not known whether bounded sets, in general cotangent bundles have finite capacities c_0. Another intriguing case is the torus $T^{2n} = \mathbb{R}^{2n}/\mathbb{Z}^{2n}$. Here one finds symplectic structures ω^* by slightly modifying the standard structure ω_0 inherited from that of $\mathbb{R}^{2n}$, for which $c_0(T^{2n}, \omega^*) = +\infty$, [40]. This is in contrast to $D(T^{2n}, \omega^*) < \infty$. These examples are intimately related to M. Herman's celebrated counterexamples to the C^k-Closing Lemma for k large [39]. Amazingly, it is still not known whether $c_0(T^{2n}, \omega_0)$ is finite for the standard symplectic structure ω_0 if $n > 1$.

We finally would like to describe how the existence of the special capacity c_0 can be deduced from an interesting recent inequality for functions in $\mathbb{R}^{2n}$. To do so we first observe that it follows readily from the definition that c_0 satisfies the axioms (A1) and (A2). An example then shows that $\pi \leq c_0(B(1))$ and since, by (A1), $c_0(B(1)) \leq c_0(Z(1))$, it remains to verify the nontrivial estimate $c_0(Z(1)) \leq \pi$. Here we make use of the following inequality from [23]. If $U \subset (\mathbb{R}^{2n}, \omega_0)$ is an open set, then

$$\begin{aligned} &\sup \left\{ \|H\| \;\middle|\; H \in \mathcal{H}_a(U) \right\} \\ &\leq \inf \left\{ \|H\| \;\middle|\; H \in C_c^\infty([0,1] \times \mathbb{R}^{2n}) \text{ and } \psi_H(U) \cap U = \emptyset \right\}. \end{aligned} \tag{2}$$

The mapping ψ_H is the time-1 map of the flow of the Hamiltonian vector field X_H and

$$\|H\| := \int_0^1 \left(\sup_x H_t - \inf_x H_t \right) dt,$$

where $H_t(x) = H(t, x)$. The inequality (2) compares the special capacity $c_0(U, \omega_0)$ on the left hand side with the so-called displacement energy $e(U)$ on the right hand side. The latter is a symplectic invariant different from c_0. It is monotone: if $U \subset V$ then $e(U) \leq e(V)$. This follows from the definition. By an easy example one verifies that $e(Z(1)) \leq \pi$. Consequently, by (2), $c_0(Z(1)) \leq e(Z(1)) \leq \pi$ and hence $c_0(B(1)) = c_0(Z(1)) = e(B(1)) = e(Z(1)) = \pi$ and we see, in particular, that c_0 satisfies also the axiom (A3) as we wanted to show.

As an aside we would like to recall from [22] that the inequality (2) allows us to establish an intrinsic bi-invariant metric on the subgroup $\mathfrak{D} = \{\varphi \mid \varphi = \varphi_H$ and $H \in C_c^\infty([0,1] \times \mathbb{R}^{2n})\}$ of compactly supported symplectic mappings of $\mathbb{R}^{2n}$. Introducing the energy of $\varphi \in \mathfrak{D}$ by

$$E(\varphi) := \inf \left\{ \|H\| \;\middle|\; \varphi = \varphi_H \right\},$$

the bi-invariant metric $d : \mathfrak{D} \times \mathfrak{D} \to \mathbb{R}$ is defined by

$$d(\varphi, \psi) = E(\varphi^{-1}\psi).$$

It satisfies

(i) $d(\varphi,\psi)\geq 0$ and $d(\varphi,\psi)=0$ if and only if $\varphi=\psi$
(ii) $d(\varphi,\psi)\leq d(\varphi,\vartheta) + d(\vartheta,\psi)$
(iii) $d(\varphi\circ\vartheta,\psi\circ\vartheta) = d(\vartheta\circ\varphi,\vartheta\circ\psi) = d(\varphi,\psi)$.

All these properties of d are easily concluded from the definition for except the nontrivial statement that $d(\varphi,\psi)>0$ if $\varphi\neq\psi$, which, however, follows from (2). Indeed in order to prove that $E(\varphi)>0$ if $\varphi\neq id$, we pick a point x satisfying $\varphi(x)\neq x$ and hence find an open ball $U=B(x,\varepsilon)$ of radius $\varepsilon>0$ which is displaced from itself by φ such that $\varphi(U)\cap U=\emptyset$. Since $c_0(U)=\pi\varepsilon^2$, we conclude from the inequality (2) that $E(\varphi)\geq c_0(U)=\pi\varepsilon^2>0$ as claimed. We point out that the proof of (2) is again based on the Hamiltonian variational principle for periodic orbits.

There are other functions $(M,\omega)\mapsto c(M,\omega)$ which satisfy the axioms (A1), (A2) and, instead of (A3), they satisfy $c(B(1),\omega_0)>0$ and $c(Z(1),\omega_0)<\infty$, see I. Ekeland and H. Hofer [3]. Using the information given by all these invariants, one can show for example that if $\varphi: E(r)\to E(r')$ is a symplectic diffeomorphism between open ellipsoids as defined above, then there is a linear symplectic map $A\in Sp(n)$ such that $A(E(r))=E(r')$ and hence $r=r'$. Or, as in another application, one can prove that the polydisc $B^2(r)\times\ldots\times B^2(r)$ embeds symplectically into the open round ball $B^{2n}(R)$ if and only if $\sqrt{n}\,r\leq R$.

2 The special capacity c_0 and the existence of periodic orbits

We shall use now the dynamically defined symplectic invariant $c_0(M,\omega)$ in order to establish the existence of distinguished periodic orbits having prescribed energy.

Consider a compact and oriented hypersurface S contained in a symplectic manifold (M,ω) of dimension $2n$. Then the symplectic structure ω determines a canonical line bundle $\mathcal{L}_S\subset TS$ as follows. The restriction of ω to the odd dimensional tangent space $T_xS, x\in S$, is necessarily degenerate and has a kernel which is 1-dimensional. Hence we can define

$$\mathcal{L}_S = \left\{(x,\xi)\in T_xS\;\middle|\;\omega_x(\xi,\eta) = 0 \quad\text{for all}\;\; \eta\in T_xS\right\}.$$

This line bundle determines the direction of every Hamiltonian vector field X_H which has S as a regular energy surface. Indeed if $S=\{x\in M|H(x)=1\}$ and $dH(x)\neq 0$ for $x\in S$, then

$$X_H(x)\in\mathcal{L}_S(x)\quad\text{for all}\;\; x\in S.$$

The characteristics of $\mathcal{L}_S$ on S represent, therefore, the unparameterized solutions of every Hamiltonian vector field having S as a regular energy surface. The question arises, whether every hypersurface admits a closed characteristic, which then carries a periodic solution for every Hamiltonian vector field having S as a regular energy surface. It can be shown by means of the examples of a nonstandard torus (T^{2n}, ω^*) that not every hypersurface admits a closed characteristic [40].

Prompted by the seminal existence results due to P. Rabinowitz [29] and A. Weinstein [36] for closed characteristics on compact convex and starlike hypersurfaces S in $(\mathbb{R}^{2n}, \omega_0)$, Weinstein introduced in [37] the restricted class of hypersurfaces of contact type. A compact and orientable hypersurface $S \subset (M, \omega)$ is of contact type if there exists a vector field X on an open neighbourhood U of S satisfying

$$\begin{aligned} L_X\omega = \omega \qquad & \text{on } U \\ X(x) \notin T_xS \qquad & x \in S. \end{aligned}$$

The vector field X is, therefore, transversal to the hypersurface. Starlike hypersurfaces in $(\mathbb{R}^{2n}, \omega_0)$ are examples of this type.

A. Weinstein-Conjecture 1978: A hypersurface S of contact type and satisfying $H^1(S) = 0$ carries a closed characteristic.

The first landmark in the long history of this conjecture is the celebrated solution by C. Viterbo in 1987 [35], who showed that every compact hypersurfaces of contact type $S \in (\mathbb{R}^{2n}, \omega_0)$ carries a closed characteristic. Here the requirement $H^1(S) = 0$ is not necessary. Viterbo's proof prompted the existence proof of the capacity $c_0(M, \omega)$. This invariant allows us in turn to conclude that every compact hypersurface $S \subset (M, \omega)$ of contact type carries a periodic orbit provided only S has an open neighbourhood $U \subset M$ of finite capacity c_0:

$$c_0(U, \omega) < \infty .$$

We briefly sketch the argument from [26]. The trick is to thicken the given hypersurface S first and to find closed characteristics in a neighbourhood of S. If S is any compact and oriented hypersurface, there exists a parameterized family of hypersurface S_ε modelled on S, i.e. a diffeomorphism

$$\psi : S \times I \to U \quad \subset M,$$

where I is an open interval containing 0 and U an open bounded neighbourhood of S, such that

$$\psi(x, 0) = x \quad \text{for} \quad x \in S.$$

We shall abbreviate in the following the hypersurface $S_\varepsilon = \psi(S \times \{\varepsilon\})$ and denote by $\mathcal{P}(S_\varepsilon)$ the set of closed characteristics of S_ε.

If now $c_0(U, \omega) < \infty$, one concludes readily, using the definition of c_0 and the fact that we have the freedom to represent the hypersurface by convenient functions that the set of parameters

$$\sum = \left\{ \varepsilon \in I \;\middle|\; \mathcal{P}(S_\varepsilon) \neq \emptyset \right\}$$

is dense in I. This shows, in particular, that every compact hypersurface gives rise to an abundance of closed orbits. If now S is, in addition, of contact type, then the flow of the distinguished vector field X allows us to define a very distinguished parameterized family of hypersurfaces modelled on S. This family is defined by $\psi(x,t) = \varphi^t(x)$ for $x \in S$ and $|t| < \varepsilon$. It has the property that $T\varphi^t : \mathcal{L}_S \to \mathcal{L}_{S_t}$ is an isomorphism of vector bundles. Hence φ^t induces a one-to-one correspondence $\mathcal{P}(S) \longleftrightarrow \mathcal{P}(S_t)$. This finishes the sketch of the proof that the Weinstein conjecture holds true for all symplectic manifolds (M,ω) in which bounded open sets have finite capacity, such as for example $T^*(T^n)$ and $\mathbb{CP}^n$ and $(\mathbb{R}^{2n}, \omega_0)$.

We should point out that contact type hypersurfaces can topologically be very complicated. This has been illustrated by Ya. Eliashberg [5] and A. Weinstein [38] by using the technique of attaching symplectic handles. However, not every hypersurface is of contact type and we shall single out next a different class of hypersurfaces for which the Weinstein conjecture holds true.

Consider a compact oriented hypersurface $S \subset (M,\omega)$ which bounds a symplectic manifold (B,ω) so that $S = \partial B$. We then take a parameterized family (S_ε) modelled on S such that S_ε bounds the symplectic manifold (B_ε, ω), i.e. $S_\varepsilon = \partial B_\varepsilon$. The parameterization is chosen such that

$$B_\varepsilon \subset B_{\varepsilon'}, \quad \text{if } \varepsilon \leq \varepsilon'.$$

Then the function

$$C(\varepsilon) := c_0(B_\varepsilon, \omega)$$

is monotonically increasing.

Definition. The hypersurface S_{ε^*} is called of c_0-Lipschitz type if there are positive constants M and μ so that

$$C(\varepsilon) \leq C(\varepsilon^*) + M(\varepsilon - \varepsilon^*)$$

for every ε in $\varepsilon^* \leq \varepsilon \leq \varepsilon^* + \mu$.

Using the monotonicity property of c_0, one verifies readily that the definition does not depend on the choice of the parameterized family S_ε modelled on S_{ε^*}.

Theorem 1. *Assume $c_0(M,\omega) < \infty$. If the compact hypersurface $S \subset M$ bounds a compact symplectic manifold and is, in addition, of c_0-Lipschitz type then*

$$\mathcal{P}(S) \neq 0.$$

Proof a). By assumption we find a parameterized family S_ε with $S = S_0$ such that

$$C(\varepsilon) \leq C(0) + M\varepsilon \tag{3}$$

for $0 \leq \varepsilon \leq \mu$. If $0 < \tau < \mu$ we introduce the set $\mathcal{F}_\tau$ consisting of smooth functions $f : \mathbb{R} \to [C(0) - \tau M, \infty]$ satisfying

$$\begin{aligned} f(s) &= a && \text{if } s \leq 0 \\ f(s) &= b && \text{if } s \geq \frac{\tau}{2} \\ 0 < f'(s) &\leq c && \text{if } 0 < s < \frac{\tau}{2} \end{aligned}$$

where the constants a, b, c are restricted by the conditions

$$\begin{aligned} C(0) - M\tau \leq\ & a \leq C(0) \\ C(0) + 2M\tau \leq\ & b \leq C(0) + 3M\tau \\ c =\ & 10M. \end{aligned}$$

In view of the assumption (3) we have $\mathcal{F}_\tau \neq \emptyset$.
b) Fixing now τ and recalling the definition of the capacity $c_0(B_0, \omega)$, we find an admissible function $H \in \mathcal{H}_a(B_0, \omega)$ satisfying $C(0) - M\tau \leq m(H) < C(0)$. We can choose $f \in \mathcal{F}_\tau$ with $a = m(H)$ and define the Hamiltonian function F by

$$\begin{aligned} F(x) &= H(x) && \text{if } x \in B_0 \\ F(x) &= f(\varepsilon) && \text{if } x \in S_\varepsilon,\ 0 \leq \varepsilon \leq \tau \\ F(x) &= b && \text{if } x \notin \overline{B_\tau}. \end{aligned}$$

Clearly $F \in \mathcal{H}(B_\tau, \omega)$ and, in view of (3) $m(F) = b \geq C(0) + 2M\tau > C(0) + M\tau \geq C(\tau)$. Since $m(F) > C(\tau) = c_0(B_\tau, \omega)$ we deduce from the definition of the capacity c_0 a nonconstant periodic solution $x(t)$ having period $0 < T \leq 1$ of the Hamiltonian system

$$\dot{x} = X_F(x) \quad \text{on } B_\tau.$$

By construction, this solution cannot be contained B_0, since the restriction of F onto B_0 is an admissible function. Since B_0 is invariant under the flow of X_F, the solution must be contained in $B_\tau \backslash \overline{B_0}$. As the solution is not constant we conclude from the properties of f the existence of an ε in $0 < \varepsilon < \frac{\tau}{2}$ such that $x(t) \subset S_\varepsilon$.
c) This argument works for every $\tau > 0$. Hence we can choose a sequence $\tau_j \to 0$ and thus find sequences F_j and ε_j and a sequence $x_j(t)$ of periodic orbits of X_{F_j} satisfying

$$x_j(t) \subset S_{\varepsilon_j}, \quad \varepsilon_j \to 0$$

and having periods $0 < T_j \leq 1$. It follows from the construction that these solutions satisfy, in addition, some estimates independent of j. Define on U defined as the union of S_ε for $\varepsilon \in I$, a fixed Hamiltonian K having the hypersurfaces S_ε as energy surfaces by setting

$$K(x) = \varepsilon \quad \text{if } x \in S_\varepsilon .$$

If $x \in S_\varepsilon$ and $0 \le \varepsilon \le \mu$, then for every T_j, we have $F_j(x) = f_j(K(x))$ and consequently $X_F(x) = f'(K(x))X_K(x)$ where we did not indicate the dependence of the functions f and F on j. By construction,

$$\dot{x}_j(t) = f'(\varepsilon_j)\, X_K(x_j(t))$$
$$x_j(0) = x_j(T_j)$$

and $0 < T_j \le 1$. Reparameterizing, the functions

$$y_j(t) = x_j\Big(\frac{t}{f'(\varepsilon_j)}\Big)$$

solve the Hamiltonian equations $\dot{y}_j(t) = X_K(y_j)$ and $K(y_j(t)) = \varepsilon_j$. The crucial observation now is that the periods of y_j are given by $f'(\varepsilon_j)T_j$. Hence, in view of the construction $f'(\varepsilon_j) \le 10M$ so that the periods are bounded. We therefore find, by Arzelà-Ascoli's theorem a periodic solution $x(t)$ of X_K on the energy surface $K(x) = 0$, which is the hypersurface $S = S_0$. This periodic solution parameterizes the desired closed characteristic on S and the proof of Theorem 1 is finished. □

We should mention an immediate consequence of this existence result. Recall that the function $C(\varepsilon) = c_0(B_\varepsilon, \omega)$ is monotone; hence by a theorem due to Lebesgue, it is differentiable almost everywhere and we have proved

Theorem 2. *Assume the compact hypersurface $S \subset (M, \omega)$ bounds a compact symplectic manifold. If $S_\varepsilon, \varepsilon \in I$ is a parameterized family of hypersurfaces modelled on S, then (m denoting the Lebesgue-measure)*

$$m\Big\{\varepsilon \in I \;\Big|\; \mathcal{P}(S_\varepsilon) \neq \emptyset\Big\} = m(I),$$

provided $c_0(M, \omega) < \infty$.

A compact hypersurface $S \subset \mathbb{R}^{2n}$ separates the space into a bounded and an unbounded component and we therefore conclude for the standard manifold $(\mathbb{R}^{2n}, \omega_0)$:

Corollary: *For every compact hypersurface $S \subset (\mathbb{R}^{2n}, \omega_0)$ and every parameterized family $(S_\varepsilon), \varepsilon \in I$, of hypersurfaces modelled on S:*

$$m\Big\{\varepsilon \in I \;\Big|\; \mathcal{P}(S_\varepsilon) \neq \emptyset\Big\} = m(I).$$

This result was proved by M. Struwe in [34] who cleverly modified the proof of H. Hofer and E. Zehnder [25] of the weaker result that $P(S_\varepsilon) \neq \emptyset$ for all ε contained in a dense subset of I. The Corollary cannot be improved (at least in dimension $2n \ge 8$). Recent examples by V. Ginzburg and M. Herman show that there are compact energy surfaces in $\mathbb{R}^{2n}$ with $2n \ge 8$, which do not carry a periodic orbit, see [17], [21].

3 Floer homology and symplectic homology

The symplectic invariant $c_0(M, \omega)$ is intimately related to a distinguished critical point of a variational functional and represents a distinguished periodic orbit of a Hamiltonian system X_H on M. The question arises, whether the study of the structure of all critical points leads to more subtle symplectic invariants, which are not necessarily numbers anymore. In a joint effort with H. Hofer, A. Floer used his construction of the Floer homology on symplectic manifolds in order to establish a symplectic homology theory. Our next aim is to sketch this theory, which is, however, not yet in its final form.

We start by recalling very briefly Floer's analysis of the structure of periodic orbits of a time-dependent Hamiltonian vector field X_H on a compact symplectic manifold (M, ω). For simplicity we pose the topological restriction $\pi_2(M) = 0$. To a time-dependent Hamiltonian function $H : S^1 \times M \to \mathbb{R}$, there is an associated time-dependent vector field $X_H = X_t$ on M defined by

$$\omega\Big(X_t(x), \cdot\Big) = -dH(t, x). \tag{4}$$

We are interested in the structure of all periodic solutions of X_t, i.e.

$$\dot{x} = X_t(x)\,, \quad x(0) = x(1). \tag{5}$$

These are the Euler equations of the classical principle of least action. Denote by $\Omega(M)$ the set of parameterized contractible loops $x : S^1 \to M$, and define the functional $\varphi_H : \Omega \to \mathbb{R}$ by

$$\varphi_H(x) = -\int_D \bar{x}^* \omega + \int_0^1 H\Big(t, x(t)\Big)\, dt.$$

Here $\bar{x} : D \to M$ is the extension of x to the closed unit disc D, with $\bar{x}|_{\partial D} = x$. The critical points of φ_H are precisely the periodic solutions (5) we are looking for. We choose now an almost complex structure J on M compatible with ω, i.e. $J^2 = -\mathbb{1}$ and $\omega(\xi, J(x)\eta) = g_J(\xi, \eta)$ for $\xi, \eta \in T_xM$, with a Riemannian metric g_J. Then the L^2-gradient of φ_H with respect to the associated L^2-inner product for sections along loops $x \in \Omega$ is computed to be

$$\varphi_H'(x)(t) = J(x)\,\frac{d}{dt}\,x(t) + \nabla H\Big(t, x(t)\Big),$$

where ∇H is the g_J-gradient of H with respect to the second variable. In a series of papers [7], [9] Floer, inspired by C. Conley's homotopy index theory, studied the bounded solutions of the «flow equation» on Ω:

$$\frac{d}{ds}\,x = -\varphi_H'(x)\,, \quad x \in \Omega, \tag{6}$$

which, due to the gradient structure, are the connections of critical points of φ_H. However, due to a loss of one derivative, there is no Banach space set up in which φ_H' can be considered as a vector field on a Banach manifold; indeed the initial value problem on the loop space associated to (6) is not well posed. Floer's approach, therefore, is to study the «bounded solutions» directly by treating them as solutions of a partial differential equation and applying nonlinear Fredholm theory. Viewing x as a map $u : \mathbb{R} \times S^1 \to M$, the equation (6) is written

$$\frac{\partial u}{\partial s} + J(u)\frac{\partial u}{\partial t} + \nabla H(t,u) = 0, \tag{7}$$

which is a perturbed nonlinear Cauchy-Riemann equation. Floer imposes the asymptotic boundary conditions

$$u(s,\cdot) \to x^{\pm} \in \mathcal{P}_H \tag{8}$$

as $s \to \pm\infty$, where x^+ and x^- are 1-periodic solutions of the Hamiltonian system $\dot{x} = X_t(x)$. Assume now, that all the contractible periodic orbits of X_H are nondegenerate (1 is not a Floquet-multiplier) and call this set $\mathcal{P} = \mathcal{P}_H$. Denote by $C = C(M,H)$ the vector space over $\mathbb{Z}_2$ generated by the finitely many elements of $\mathcal{P}$. Then

$$C = \bigoplus_k C_k\,,\ C_k(M,H) = \operatorname{span}_{\mathbb{Z}_2}\{x \in \mathcal{P}_H \,|\, m(x) = k\}$$

where the grading is defined by the Maslov index $m(x) \in \mathbb{Z}$, which is associated intrinsically to the nondegenerate periodic orbit $x \in \mathcal{P}$, see [32]. Finding solutions of (7) and (8) turns out to be equivalent to finding zeroes of a Fredholm section of a Banach space bundle over a Banach manifold. It follows from Gromov's compactness result that the solution set, which we denote by $\mathcal{M}(x^-,x^+)$, consists of finitely many orbits (modulo time shift in s), whenever $m(x^-) - m(x^+) = 1$, provided the data are generic. This observation is used by Floer to construct a boundary operator $\partial_k = \partial_k(M,H,J) : C_k \to C_{k-1}$ via the formula

$$\partial y = \sum_{m(x)=k-1} \langle \partial y, x\rangle x$$

for $y \in \mathcal{P}$ having index $m(y) = k$. The matrix element $\langle \partial y, x\rangle$ is defined to be the number of components of $\mathcal{M}(y,x)$ modulo 2. Floer proves that $\partial \circ \partial = 0$ so that (C,∂) defines a chain complex. Its homology

$$F_*(M,H,J) := \frac{\ker \partial}{\operatorname{im} \partial}$$

is called the Floer homology of the generic pair (H,J). Note that this chain complex is determined by the one-dimensional components of the solution space $\mathcal{M}(x^-,x^+)$

of bounded solutions of (7) and (8). It is constructed in analogy with the Morse complex for Morse-Smale gradient flows on finite dimensional manifolds: the transversality condition corresponds to the property that for generic H, J, the linearized Fredholm operator along a solution u of (7) and (8) is onto while the Morse index of a critical point corresponds to the Maslov index of the periodic solutions $\mathcal{P}_H$.

The crucial observation by Floer is now, that changing the data $(H, J) \equiv (H_\alpha, J_\alpha)$ to another pair (H_β, J_β) of generic data, there exists a natural homotopy class of chain homotopic maps $\psi_{\beta\alpha}$:

$$[\psi_{\beta\alpha}] : (C_{H_\alpha, J_\alpha}, \partial_\alpha) \to (C_{H_\beta, J_\beta}, \partial_\beta).$$

Moreover his construction is functorial. In particular

$$\begin{aligned} [\psi_{\alpha\alpha}] &= id \\ [\psi_{\gamma\beta}] \circ [\psi_{\beta\alpha}] &= [\psi_{\gamma\alpha}]. \end{aligned}$$

Hence we obtain a connected simple system, which shows that the Floer homology $F_*(M, H, J)$ is up to natural isomorphisms independent of the date (H, J) and is therefore an invariant of the underlying manifold M. Choosing very special data (H^*, J^*), Floer established a natural isomorphism

$$\psi_\alpha : F_k(M, H_\alpha, J_\alpha) \to H_{k+n}(M, \mathbb{Z}_2)$$

between the Floer homology of the pair (H_α, J_α) and the singular homology of M. If, moreover, (H_β, J_β) is another pair, then $\psi_\alpha = \psi_\beta \circ \psi_{\beta\alpha}$. Hence, in the nondegenerate case, the set of contractible periodic solutions of a time-dependent Hamiltonian system X_H on M serves as a model of the singular homology of the symplectic manifold. One immediately deduces the Morse inequalities

$$\sum_{x \in \mathcal{P}_H} t^{m(x)} = P(t, M) t^{-n} + (1+t) Q(t),$$

where $P(t, M)$ is the Poincaré polynomial of the manifold M determined by the Betti numbers and $Q(t)$ has nonnegative integer coefficients. The left hand side is determined by all the contractible 1-periodic solutions $x \in \mathcal{P}_H$ of the Hamiltonian equation $\dot{x} = X_H(x)$, where $m(x) \in \mathbb{Z}$ denotes the Maslov index of the solution x. For the restricted class of symplectic manifolds under consideration, this answers a conjecture by V. Arnold about the number of fixed points of Hamiltonian mappings. The conjecture is prompted by old questions in celestial mechanics.

For the above variational functional, the Floer homology agrees with a familiar invariant of the underlying manifold M. Floer applied his construction, as is well known, also to the Chern-Simons functional on the space of $SU(2)$ connections on a homology-3-sphere Y. In that context the connecting orbits are interpreted as self-dual Yang-Mills connections of $Y \times \mathbb{R}$ having finite action. There, the associated Floer homology groups define new invariants of the underlying homology 3-sphere.

We turn now to the construction of a symplectic homology and start with the observation that, for a fixed generic pair (H, J), there exists an additional grading of the Floer homology defined by the values of the functional φ_H on Ω. We define, for $a \in \mathbb{R} \cup \{\infty\}$, the graded free groups

$$C^a(H,J) = \bigoplus_k C_k^a , \quad C_k^a(H) = \text{span}_{\mathbb{Z}_2} \{x \in \mathcal{P} \mid m(x) = k, \quad \varphi_H(x) < a\}. \tag{9}$$

Note that if u is a solution of (7) and (8) with asymptotic data $x^+, x^- \in \mathcal{P}$, then necessarily $\varphi_H(x^+) \leq \varphi_H(x^-)$ due to the variational structure of the problem. Consequently $\partial = \partial_{(H,J)}$ satisfies

$$\partial : C_k^a \to C_{k-1}^a.$$

Therefore, for $-\infty < a \leq b \leq \infty$ defining the quotient group

$$C^{[a,b)}(H,J) = \frac{C^b(H,J)}{C^a(H,J)}, \tag{10}$$

the boundary map induces a boundary operator $\partial_k : C_k^{[a,b)} \to C_{k-1}^{[a,b)}$, and we can define

$$F_k^{[a,b)}(H,J) = \frac{\ker(\partial_k)}{\text{im }(\partial_{k+1})}. \tag{11}$$

This Floer homology group depends, however, on the choice of a generic pair (H, J), and the aim is to construct a directed set of very special generic pairs, so that the limit of the corresponding groups defines a symplectic invariant of M.

We sketch this construction for compact symplectic manifolds (M, ω) possibly having a boundary ∂M. In the case that $\partial M \neq \emptyset$ we shall assume that ∂M is of contact type as defined above. Equivalently this requires that there exists a 1-form λ on ∂M satisfying $d\lambda = \omega$ on ∂M and $\lambda(x, \xi) \neq 0$ for all nonvanishing $\xi \in T_x(\partial M)$ for which $\omega(\xi, \eta) = 0$ for all $\eta \in T_x(\partial M)$. Moreover, we assume for simplicity, as above, that $\pi_2(M) = 0$. By $\mathcal{H}$ we denote the collection of all smooth and time-periodic Hamiltonian functions $H : S^1 \times M \to \mathbb{R}$ satisfying

$$\text{supp } (H) \text{ is compact in } S^1 \times (M \backslash \partial M) \text{ and } H(S^1 \times M) \subset (-\infty, 0]. \tag{12}$$

Clearly this condition implies that there are many periodic solutions x of X_H on M which are degenerate, and satisfy $\varphi_H(x) = 0$. In addition to (12) we, therefore, require that all contractible periodic solutions x of X_H satisfying $\varphi_H(x) < 0$ are nondegenerate. Consequently, if $-\infty \leq a \leq b \leq 0$, we can define as above the relative Floer groups $F^{[a,b)}(H, J)$. Here, however, not every J is admissible. The J

has to be chosen in such a way that every solution $u : \mathbb{R} \times S^1 \to M$ of (7) and (8) stays away from the boundary, i.e. $u(\mathbb{R} \times S^1) \subset M \backslash \partial M$, provided x^- and x^+ are contained in $M \backslash \partial M$. Such admissible J's do exist; the boundary ∂M is then called J-pseudoconvex in the sense of D. McDuff [28]. We shall call a pair (H, J) satisfying the above admissible conditions and introduce in the set of admissible pairs a partial ordering by defining (this is not a typing mistake)

$$(H,J) \leq (K,\tilde{J}) \Longleftrightarrow H(t,x) \geq K(t,x)$$

for all $(t,x) \in S^1 \times M$. For such pairs one can define a monotone homotopy. It consists of a pair $(L, \hat{J})$, where $L : \mathbb{R} \times S^1 \times M \to \mathbb{R}$ and where the almost complex structure $\hat{J}(s,t,x)$ is adapted to the symplectic form ω. Moreover $L = H, \hat{J} = J$ for $s \to -\infty$ while $L = K, \hat{J} = \tilde{J}$ for $s \to +\infty$, and $\frac{\partial L}{\partial s} \leq 0$. Given a monotone homotopy for $(H,J) \leq (K,\tilde{J})$, one considers the partial differential equations for $u : \mathbb{R} \times S^1 \to M$:

$$\begin{aligned} \frac{\partial u}{\partial s} + \hat{J}(s,t,u) \frac{\partial u}{\partial t} + (\nabla_{\hat{J}} L)(s,t,u) &= 0 \\ u(s,\cdot) \to x \in \mathcal{P}_H(s \to -\infty) & \\ u(s,\cdot) \to y \in \mathcal{P}_K(s \to +\infty). & \end{aligned}$$

Proceeding as in Floer's construction, the combinatorics of the set of solutions is used to construct a natural homomorphism

$$F^{[a,b)}(H,J) \to F^{[a,b)}(K,\tilde{J})$$

of $\mathbb{Z}_2$ vector spaces, which is independent of the chosen generic monotone homotopy. The partial ordering $\leq$ turns the set of admissible pairs (H,J) into a directed set, since given (H_1, J_1) and (H_2, J_2) there exists an admissible pair (H,J) satisfying

$$(H_j, J_j) \leq (H,J) \text{ for } j = 1,2.$$

Hence we may pass to the limit obtaining the symplectic homology group

$$S^M_{[a,b)} = \text{direct } \lim F^{[a,b)}(H,J).$$

In addition, if $U \subset M$ is an open subset, and if $\mathcal{H}_U$ consists of those admissible pairs (H,J) which are restricted by the requirement that supp $(H) \subset S^1 \times (U \backslash \partial M)$, we may take the directed limit over $\mathcal{H}_U$ and obtain a symplectic homology group of U,

$$S^M_{[a,b)}(U) := \text{direct } \lim_{\mathcal{H}_U} F^{[a,b)}(H,J).$$

We next list some of the properties about these groups known so far. There are natural homomorphisms of $\mathbb{Z}_2$ vector spaces

$$S^M_{[a,b)}(U) \to S^M_{[a,b)}.$$

Moreover, the short exact sequence $0 \to F^{[a,b)}(H,J) \to F^{[a,c)}(H,J) \to F^{[b,c)}(H,J) \to 0$, for $-\infty \leq a \leq b \leq c \leq 0$, gives rise to an exact triangle $\Delta_{a,b,c}(U)$ of symplectic homology groups:

$$\begin{array}{ccc} S^{[a,b)}(U) & \longrightarrow & S^{[a,c)}(U) \\ \partial \nwarrow & & \swarrow \\ & S^{[b,c)}(U). & \end{array}$$

Given a symplectic diffeomorphism $\psi : M \to M$ we can define for an admissible pair (H,J), where $H \in \mathcal{H}_U$, a new admissible pair (H_ψ, J_ψ) by setting

$$\begin{aligned} H_\psi(t,x) &= H\Big(t, \psi^{-1}(x)\Big) \\ J_\psi(x) &= T\psi\Big(\psi^{-1}(x)\Big) J\Big(\psi^{-1}(x)\Big) T\psi^{-1}(x). \end{aligned}$$

It can be shown that ψ induces a natural isomorphism

$$\psi_\# : S^M_{[a,b)}(U) \xrightarrow{\cong} S^M_{[a,b)}\Big(\psi(U)\Big).$$

If $U \subset V$ then the inclusion map induces a homomorphism

$$\sigma_{V,U} : S^M_{[a,b)}(U) \to S^M_{[a,b)}(V),$$

and if $\psi : M \to M$ is a symplectic diffeomorphism satisfying $\psi(U) \subset V$, we define the induced map ψ_* by

$$\psi_* = \sigma_{V,\psi(U)} \circ \psi_\# : S^M_{[a,b)}(U) \to S^M_{[a,b)}(V).$$

If then $\varphi : M \to M$ is another symplectic diffeomorphism satisfying $\varphi(V) \subset W$, one can prove

$$\begin{aligned} (\varphi \circ \psi)_* &= \sigma_{W,\varphi\circ\psi(U)} \circ (\varphi \circ \psi)_\# \\ &= \Big(\sigma_{W,\varphi(V)} \circ \varphi_\#\Big) \circ \Big(\sigma_{V,\psi(U)} \circ \psi_\#\Big) \\ &= \varphi_* \circ \psi_*. \end{aligned}$$

Moreover, the identity map of M induces the identity map *id*: $S^M_{[a,b)}(U) \to S^M_{[a,b)}(U)$. The most useful property is the isotopy invariance: if $\psi_s : M \to M$ for $0 \leq s \leq 1$ is a smooth family of symplectic diffeomorphisms satisfying $\psi_s(U) \subset V$ for all s, then $(\psi_s)_* : S^M_{[a,b)}(U) \to S^M_{[a,b)}(V)$ is independent of s.

It is rather difficult to compute the symplectic homology groups sketched above and we refer to [15], [16] for computations and applications. We would like to point out that the homology groups are again related to distinguished periodic orbits, which sit on or near the boundary of a domain. This is illustrated by the theorem below.

We consider a compact symplectic manifold (M,ω) with boundary $\partial M \neq \emptyset$ which is exact symplectic, i.e. $\omega = d\lambda$ for a 1-form λ. We assume the boundary to be of contact type with respect to λ. Then we can define the action spectrum $\sigma(\partial M) \subset \mathbb{R}$ as follows: $\sigma(\partial M) = \Big\{ \big| \int_x \lambda \big|$, where x is a periodic solution of the Reeb vector field on $\partial M \Big\}$.

Theorem 3. *Consider two exact symplectic and compact manifolds $(M, d\lambda)$ and $(N, d\tau)$ with boundaries ∂M and ∂N of contact type with respect to λ and τ. Assume there exists an exact symplectic diffeomorphism*

$$\varphi : \overset{\circ}{M} \to \overset{\circ}{N}$$

(i.e. $\varphi^*\tau - \lambda = dF$ for a smooth function $F : \overset{\circ}{M} \to \mathbb{R}$). If the periodic solutions on ∂M and ∂N and all their iterates are nondegenerate, then

$$\sigma(\partial M) = \sigma(\partial N).$$

Examples by Ya. Eliashberg and H. Hofer [6] demonstrate that under the assumptions of the theorem, the flows on ∂M and ∂N are not necessarily conjugated so that φ cannot be extended to a symplectic diffeomorphism $M \to N$.

The proof of Theorem 3 is based on the following nontrivial fact [16]. Let $\alpha < 0$ and define

$$d_M(\alpha) = \frac{1}{2} \lim_{\varepsilon \to 0} \dim S_M^{[\alpha-\varepsilon,\alpha+\varepsilon)}(\overset{\circ}{M}).$$

Then $d_M(\alpha) = 0$ if $-\alpha \notin \sigma(\partial M)$ and

$$d_M(\alpha) = \# \Big\{ \text{periodic solutions } x \text{ on } \partial M \text{ with action } \big| \int_x \lambda \big| = -\alpha \Big\}.$$

Since $S_M^{[a,b)}(\overset{\circ}{M}) = S_N^{[a,b)}(\overset{\circ}{N})$ for all $a < b$, we conclude $d_M(\alpha) = d_N(\alpha)$ so that the theorem follows. For the intricate details of the proof and for applications of this result, we refer to the forthcoming paper [16].

By computing the homology groups one also proves a conjecture by Gromov on the symplectic classification of open polydiscs. Define for $r = (r_1, r_2, \ldots, r_n)$, where $0 < r_1 \leq r_2 \leq \cdots \leq r_n$, the polydisc $D(r) = B^2(r_1) \times \cdots \times B^2(r_n)$.

Theorem 4. *$D(r)$ and $D(r')$ are symplectomorphic if and only if $r = r'$, assuming r and r' are ordered.*

For the proof we refer to [15].

References

[1] V.I. Arnold: Mathematical methods of classical mechanics, Appendix 9, Springer (1978).

[2] I. Ekeland and H. Hofer: Symplectic topology and Hamiltonian dynamics I. Math. Zeit. 200 (1990), 355–378.

[3] I. Ekeland and H. Hofer: Symplectic topology and Hamiltonian dynamics II. Math. Zeit. 203 (1990), 553–567.

[4] Ya. M. Eliashberg: A Theorem on the structure of wave fronts and its application in symplectic topology. Functional Analysis and its Applications, Vol. 21 (1987), 227–232.

[5] Ya. Eliashberg: Topological characterization of Stein manifolds of dimension > 2. Intern. J. Math. (1990), 29–46.

[6] Ya. Eliashberg, H. Hofer: Unseen Symplectic Boundaries, Preprint Stanford University 1993.

[7] A. Floer: Morse theory for Lagrangian intersections. J. Diff. Geom. 28 (1988), 513–547.

[8] A. Floer: Witten's complex and infinite dimensional Morse theory. J. Diff. Geom. 30 (1989), 207–221.

[9] A. Floer: Symplectic fixed points and holomorphic spheres. Comm. Math. Phys. 120 (1989), 575–611.

[10] A. Floer, H. Hofer and C. Viterbo: The Weinstein conjecture in $P \times \mathbb{C}^d$. Math. Zeitschrift 203 (1989), 355–378.

[11] A. Floer and H. Hofer: Coherent orientations for periodic orbit problems in symplectic geometry. Math. Zeit. 212 (1993), 13–38.

[12] A. Floer and H. Hofer: Symplectic homology I: Open sets in $\mathbb{C}^n$. Math. Zeit. 215 (1994), 37–88.

[13] A. Floer and H. Hofer: Symplectic homology II: General symplectic manifolds, to appear Math. Zeit.

[14] A. Floer, H. Hofer and D. Salamon: Transversality results in the elliptic Morse theory for the action functional, to appear Duke.

[15] A. Floer, H. Hofer and K. Wysocki: Applications of symplectic homology I, to appear Math. Zeit.

[16] A. Floer, H. Hofer and K. Wysocki: Applications of symplectic homology II, to appear Math. Zeit.

[17] V. Ginzburg: An embedding $S^{2n-1} \to \mathbb{R}^{2n}$, $2n-1 \geq 7$, whose Hamiltonian flow has no periodic trajectories, to appear Int. Math. Res. Notes.

[18] M. Gromov: Pseudo holomorphic curves in symplectic manifolds. Inv. Math. 82 (1985), 307–347.

[19] M. Gromov: Partial Differential Relations. Springer-Verlag (1986).

[20] M. Gromov: Soft and hard differential geometry. In: Proceedings of the ICM at Berkeley 1986, (1987), 81–89.

[21] M. Herman: Examples of compact hypersurfaces in $\mathbb{R}^{2p}$, $2p \geq 6$, with no periodic orbits, preprint.

[22] H. Hofer: On the topological properties of symplectic maps. Proceeding of the Royal Society of Edinburgh 115 A (1990), 25–38.

[23] H. Hofer: Estimates for the energy of a symplectic map. Commentarii Math. Helv. 68 (1993), 48–72.

[24] H. Hofer and C. Viterbo: The Weinstein Conjecture in the Presence of Holomorphic Spheres. Comm. Pure Appl. Math., Vol. XLV (1992), 583–622.

[25] H. Hofer and E. Zehnder: Periodic solutions on hypersurfaces and a result of C. Viterbo. Inv. Math. 90 (1987), 1–7.

[26] H. Hofer and E. Zehnder: A new capacity for symplectic manifolds. In: Analysis et cetera, Academic Press (1990), 405–428. Edited by P. Rabinowitz and E. Zehnder.

[27] Mei-Yue Jiang: Hofer-Zehnder capacity for 2-dimensional manifolds. Preprint, Peking University (1992)

[28] D. McDuff: Elliptic methods in symplectic geometry. Bull. A.M.S. 23 (1990), 311–358.

[29] P. Rabinowitz: Periodic solutions of Hamiltonian systems. Comm. Pure Appl. Math. 31 (1978), 157–184.

[30] Renyi Ma: A remark on the Hofer-Zehnder symplectic capacity in $M \times \mathbb{R}^{2n}$. Preprint (1991) Tsinghua University, Beijing.

[31] D. Salamon and E. Zehnder: In: Analysis et cetera, Floer homology, the Maslov index and periodic orbits of Hamiltonian equations. Academic Press (1990), 573–600. Edited by P. Rabinowitz and E. Zehnder

[32] D. Salamon and E. Zehnder: Morse Theory for Periodic Solutions of Hamiltonian Systems and the Maslov-Index. Comm. on Pure and Appl. Math., Vol. XLV (1992), 1303–1360.

[33] K.F. Siburg: Symplectic Capacities in Two Dimensions. Manuscripta math., 78 (1993), 149–164.

[34] M. Struwe: Existence of periodic solutions of Hamiltonian systems on almost every energy surface. Bol. Soc. Bras. Mat. 20 (1990), 49–58.

[35] C. Viterbo: A proof of the Weinstein conjecture in $\mathbb{R}^{2n}$. Ann. Inst. Henri Poincaré, Analyse nonlinéaire 4 (1987), 337–357.

[36] A. Weinstein: Periodic orbits for convex Hamiltonian systems. Ann. Math. 108 (1978), 507–518.

[37] A. Weinstein: On the hypothesis of Rabinowitz's periodic orbit theorems. J. Diff. Equations 33 (1979), 353–358.

[38] A. Weinstein: contact surgery and symplectic handlebodies. Preprint UC Berkeley (1990).

[39] J.C. Yoccoz: Travaux de Herman sur les tores invariants. Séminaire BOURBAKI, 44ème année, 1991–92, no. 754, février 1992.

[40] E. Zehnder: Remarks on Periodic Solutions on Hypersurfaces, in: NATO ASI, Series C Vol 209 (1987), 267–279.

H. Hofer
E. Zehnder
Dept. of Mathematics
ETH-Zentrum
CH-8092 Zürich
Switzerland

An irrational ruled symplectic 4-manifold

D. McDuff*

1 Introduction

Ruled surfaces form an important and fairly well understood class of symplectic 4-manifolds. They are orientable 2-sphere bundles V over a Riemann surface M which have a symplectic form ω that is non-degenerate on each fiber of the projection $\pi : V \to M$ and is compatible with the obvious orientations on the base and fiber. Such forms are said to be **π-compatible**. One of the main results of [3] is that every minimal symplectic 4-manifold that contains a symplectic 2-sphere with trivial normal bundle is π-compatible for a suitable fibration π. (A symplectic manifold is said to be minimal if it contains no symplectic 2-spheres with self-intersection -1.) We showed in [4] that when V is rational (that is, if the base M is a 2-sphere) or if V is the trivial bundle over T^2, then in each possible cohomology class the symplectic structure is unique up to isotopy. What happens in the general case is not yet clear, though one can show that uniqueness continues to hold in a certain range of cohomology classes.

In this paper we study one manifold in detail, namely the non-trivial bundle over the torus, and prove that in this case also uniqueness holds in all cohomology classes. Our arguments do not extend to base manifolds M of higher genus. It would be very interesting to know what happens in the general case: as yet, there are no known examples of bifurcating families of symplectic forms on a compact 4-manifold.

Although the subject of this paper is not directly related to any of Floer's work, the work presented here certainly owes a great deal to him, since it was Floer who taught me the analysis that is needed to put the theory of J-holomorphic curves on a sound footing.

2 Statement of results

Recall that for each base M, there are exactly two S^2-bundles up to fiber-wise diffeomorphism: the trivial bundle $\pi : M \times S^2 \to S^2$ and the non-trivial one $\pi : V_M \to M$. (If we think of V as the projectivization of a rank 2 complex vector bundle E over M, then these bundles may be distinguished by the Stiefel-Whitney

---

*) Partially supported by NSF grant DMS 9103033.

class $w_2(E)$.) Denote by M_+, resp M_-, sections of $\pi : V_M \to M$ with self-intersection numbers $+1$, resp. -1. The dual basis for $H^2(V_M; \mathbf{R})$ will be written b_+, b_-. Thus $b_+(M_+) = b_-(M_-) = 1$. Observe that because $[M_+].[M_-] = 0$ the following relations hold:

$$b_+ \cup b_+(V_M) = 1, \quad b_- \cup b_-(V_M) = -1, \quad b_+ \cup b_- = 0.$$

If $a = \mu_+ b_+ + \mu_- b_-$ is the class of a symplectic form we must have

$$0 < a^2(V_M) = (\mu_+)^2 - (\mu_-)^2,$$

so that $|\mu_+| > |\mu_-|$. Since we are also assuming that a is positive on the class $[F] = [M_+] - [M_-]$ of the fiber, we in fact need

$$\mu_+ > |\mu_-|.$$

However, note that μ_- need not be positive here. It turns out that when $M = S^2$, these classes a with $a^2 > 0$ do not all have compatible symplectic representatives, although they do for all other M.

Recall that two forms ω_0 and ω_1 are said to be **isotopic** if they may be joined by a family of cohomologous symplectic forms, or equivalently, if there is a family of diffeomorphisms g_t with $g_0 = id$ and such that $g_1^*(\omega_1) = \omega_0$. Note that even when ω_0 and ω_1 are π-compatible, one does not assume here that the isotopy g_t preserves the fibers of π, though we can assume that it preserves one fiber: see [3, Lemma 4.4]. Two forms are said to be **pseudo-isotopic** (or **deformation equivalent**) if they may be joined by a family of not necessarily cohomologous symplectic forms. Here is the main result of [4]. (There is an analogous result for the trivial bundle.)

Theorem 1

(i) *When $M = S^2$, the class $a = \mu_+ b_+ + \mu_- b_-$ may be represented by a π-compatible symplectic form on V_M only if $\mu_+ > \mu_- > 0$. Moreover, up to isotopy there is a unique such form in each class.*

(ii) *If M has genus $g > 0$, every class with $\mu_+ > |\mu_-| > 0$ has a π-compatible symplectic representative. There is a unique such form up to isotopy in each class such that $q\mu_- > (q-1)\mu_+$, where $q = [\frac{g+1}{2}]$.*

(iii) *All these forms are pseudo-isotopic.*

In this note, we prove:

Proposition 2 *Cohomologous π-compatible symplectic forms on V_T are isotopic.*

Before stating any other results, let me remind the reader that every symplectic ruled surface (V, ω) has a natural connection in which the horizontal spaces are just the symplectic orthogonals of the fibers. Thus, for each $x \in V$, the horizontal subspace is

$$\mathrm{Hor}_x = \{v \in T_x(V) : \omega(v, w) = 0, \text{ for all } w \text{ tangent to the fiber}\}.$$

(This observation applies much more widely: see [2].) Moreover, the holonomy of this connection consists of symplectomorphisms. Indeed, another way to describe this connection is to note that for every path α in the base, the inverse image $Q_\alpha = \pi^{-1}\alpha$ is a hypersurface in V. It is easy to check that the horizontal lifts of α are precisely the leaves of the characteristic foliation on Q_α. (Recall that these are given by the Hamiltonian flow of any function which has Q_α as a regular level set. They are transverse to the fibers because $\omega|_{\text{fiber}}$ is non-degenerate.)

Proposition 3 *Let ω be a π-compatible form on V_T in the class $a = \mu_+ b_+ + \mu_- b_-$. Then ω is isotopic to a π-compatible form which has trivial holonomy around some essential loop in T^2 iff $\mu_- > 0$.*

A similar argument shows:

Proposition 4 *Let M have genus 2, and suppose that ω_i, $i = 0, 1$, are π-compatible forms on V_M each of which has trivial holonomy around some essential loop. Then ω_0 and ω_1 are isotopic.*

The proof of Proposition 2 is much the same as the proof of uniqueness for the trivial bundle over T^2 except that the role of the section is played by a double cover of a section. The statements about holonomy follow from a new wrapping technique which trivialises holonomy around a loop. We will assume that the reader knows a little about the technique of J-holomorphic curves. Suitable references are [3], [4] and [5]. I wish to thank F. Lalonde for making some useful comments about wrapping.

3 Uniqueness for V_T

Here is a procedure (suggested to me by Hitchin) which constructs π-compatible symplectic forms on V_T in all classes a with $\mu_+ > |\mu_-|$. Identify T^2 with the quotient $\mathbf{R}^2/\mathbf{Z}^2$, and consider the action of $\mathbf{Z}^2$ on S^2 in which the two generators τ_1, τ_2 of $\mathbf{Z}^2$ act by half-turns h_1 and h_2 around two mutually orthogonal axes. It is easy to check that the quotient of $\mathbf{R}^2 \times S^2$ by the diagonal action is the non-trivial S^2-bundle over T^2. (This holds because the given representation of $\mathbf{Z}^2$ on $SO(3)$ does not lift to $SU(2)$.) Because $\mathbf{Z}^2$ acts by elements of order 2, the square

$$I_z = [0,2] \times [0,2] \times \{z\} \subset \mathbf{R}^2 \times S^2, \quad z \in S^2$$

descends to a closed submanifold X_z in V_T which lies in the class $4[M_+] - 2[F] = 2[M_+] + 2[M_-]$. Thus, if ρ_0 is an area form on S^2 with total area 1, the symplectic form $\lambda_1 ds \wedge dt \oplus \lambda_2 \rho_0$ on $\mathbf{R}^2 \times S^2$ descends to a form on V_T in the class $a = \mu_+ b_+ + \mu_- b_-$ where $\mu_+ + \mu_- = \lambda_1$ and $\mu_+ - \mu_- = \lambda_2$. By correctly choosing the parameters λ_1 and λ_2, one can obtain a form in any class a.

This construction may be adapted to the case when the base M has genus > 1. Here, think of M as $\mathcal{H}/\Gamma$, where $\mathcal{H}$ is the Siegel upper half plane with the area form induced by the Poincaré metric, and consider the action of Γ on S^2 in which a pair of conjugate generators act by the half-turns h_1 and h_2, and the others act trivially. (Two elements of a standard set of generating loops α_i, $1 \le i \le 2g$ for Γ are said to be **conjugate** if they intersect each other exactly once.) The bundle so obtained is the pull-back by a map of degree 1 of the above bundle on T^2, and all the above remarks apply.

Of course, one could use an arbitrary representation of $\pi_1(M)$ into $SO(3)$, but since the space of conjugacy classes of such representations is connected, all cohomologous symplectic forms obtained in this way are isotopic. Notice also that in all cases one can construct these standard forms on V_M from a form on $C \times S^2$ which is split near the boundary, by a gluing procedure which depends only on the genus of M. (Here C is a closed 2-cell. Observe also that a form on a product space is said to **split** if it is the sum of forms pulled back from the factors.) For example, when $M = T^2$, we may think of C as the square $[0,1] \times [0,1]$ and use the gluing maps:

$$(0,t,z) \mapsto (1,t,h_1(z)), \quad (s,0,z) \mapsto (s,1,h_2(z)).$$

We will call this gluing procedure the «standard procedure». A basic uniqueness theorem, essentially due to Gromov [1], but see also [3, Corollary 4.8], states that any form on $C \times S^2$ which splits near the boundary $\partial C \times S^2$ is isotopic to a split form. Hence,

Lemma 5 *Any symplectic form on V_T which has holonomy equal to h_1 and h_2 around a pair of generators for $\pi_1(T^2)$ is isotopic to a standard form.*

Let us look at this standard symplectic structure on V_T in more detail. It is a Kähler form with respect to the complex structure J induced by the product structure on $\mathbf{R}^2 \times S^2$. The manifolds X_z clearly form a holomorphic foliation of V_T. The generic leaf lies in class $4T_+ - 2F = 2T_+ + 2T_-$ and has trivial holomorphic normal bundle. Therefore, these leaves are not regular in the sense of Fredholm theory. In particular, a generic complex (or almost complex) structure close to J does not have such a holomorphic foliation. However, there are three special leaves corresponding to the three pairs of fixed points of the half-turns h_1, h_2 and $h_3 = h_1 h_2$. These leaves lie in the homology class $[T_+ + T_-]$ and are doubly covered by the nearby leaves. It follows that their normal bundles ν are not trivial as holomorphic bundles: instead they are elements of order two, that is, $\nu \otimes \nu$ is trivial. Thus $H^1(T^2, \nu) = 0$, which is the condition for regularity. Moreover, the moduli space of these curves has dimension 0, which means that they persist as J moves in the space $\mathcal{J}$ of almost complex structures which are tamed by some symplectic form which is pseudo-isotopic to ω.

Let α_i, $i = 1, 2, 3$, be three loops in the base T^2 corresponding to the translations $\tau_1, \tau_2, \tau_1\tau_2$, which intersect only at the point x. Then it is easy to check that, for each i, the holonomy of the fibration $V_T \to T^2$ over α_i is exactly h_i, and so it interchanges the two points in $X_j \cap F_x$ for $j \neq i$ and fixes the two points in $X_i \cap F_x$, where F_x is the fiber $\pi^{-1}x$. Thus, although the three manifolds X_i represent the same element, A say, in homology they are images of non-homotopic inclusions $\iota_i : T^2 \to V_T$. Let us denote by ξ_i, $i = 1, 2, 3$, these three components of the mapping space $\mathrm{Map}(T^2, V_T)$. We aim to show that as we move J in $\mathcal{J}$, there is at least one J-holomorphic torus $X_i(J)$ in each of these homotopy classes ξ_i. This is true for small deformations because our original tori are regular, but when we make a large deformation we have to ensure that the tori do not degenerate or disappear.

Lemma 6 *For «regular» J in $\mathcal{J}$ there are no J-holomorphic A-cusp-curves.*

Proof As shown in [3, Proposition 4.1], each $J \in \mathcal{J}$ gives rise to a fibration $\pi_J : V_T \to T^2$ whose fibers are J-holomorphic spheres. Moreover, just as in the case of spheres, there is a path-connected open subset $\mathcal{U}$ of $\mathcal{J}$ consisting of almost complex structures which do not admit any J-holomorphic tori C of negative self-intersection lying in a homology class such that $\omega(C) \leq \omega(X_i)$: see [3, (2.4)], or [5]. We will call J regular if it lies in $\mathcal{U}$. In particular, for these J there are no J-holomorphic tori in class T_-. Note that the standard integrable complex structure discussed above is in $\mathcal{U}$ by inspection: the only holomorphic tori in V_T are images of the flat curves $\mathbb{C} \times pt$ or, more generally, of graphs of suitable holomorphic maps $\mathbb{C} \to S^2$, and none of these have negative self-intersection number.

Suppose that $S = S_1 \cup \ldots \cup S_k$ is a J-holomorphic A-cusp-curve. The compactness theorem implies that a J-holomorphic A-cusp-curve S consists either of a torus plus some spherical bubbles or consists entirely of spheres. The only J-holomorphic spheres are the fibers of π_J. Thus, because π_J must map S onto the base T^2 by a map of degree 2, there must be a torus component, say S_1. Moreover, if S is a cusp-curve there must be some bubbles, and so $[S_1] = A - kF$ where $k > 0$. But then $S_1 \cdot S_1 = -2k < 0$ which contradicts regularity. □

Lemma 7 *Let ω be any π-compatible symplectic form on V_T. Then, for $1 \leq i \leq 3$, there is a symplectically embedded torus X_i in the class ξ_i which doubly covers the base.*

Proof Let $\omega_t, 0 \leq t \leq 1$ be a pseudo-isotopy from $\omega_0 = \omega$ to a standard form ω_1. Let J_0 be an ω_0-tame element of $\mathcal{J}$ such that the fibers of π are J_0-holomorphic. As in [3], Proposition 4.1, we may assume that $J_0 \in \mathcal{U}$. Further, let J_1 be the standard integrable complex structure on V_T, which, as we have already noted, lies in $\mathcal{U}$.

Then we may join J_0 to J_1 by a generic path J_t in $\mathcal{U}$. The set of tori in class ξ_i which are J_t-holomorphic for some $t \in [0, 1]$ forms a compact 1-dimensional manifold, which is non-empty and has exactly one endpoint over $t = 1$. Thus there must be at least one other endpoint X_i over $t = 0$.

Because $X_i \cdot F = 2$, X_i must intersect each fiber of π transversally in exactly two points. (To see this, note that the adjunction formula implies that X_i must be embedded. Thus, by positivity of intersections, the only other possibility is that the intersection of X_i with some fiber consists of a single point of tangency, but this cannot happen for topological reasons.) □

We will use these submanifolds X_i to modify the holonomy of ω. To do this, we must understand the relation between the group $\mathrm{Symp}(S^2, \mathrm{rel}\, p^{\pm})$ of symplectomorphisms of S^2 which are the identity near the two poles $p^{\pm}$ and its subgroup $\mathrm{Ham}(S^2, \mathrm{rel}\, p^{\pm})$ of Hamiltonian symplectomorphisms.

Lemma 8 *Let β be a geodesic arc joining p^+ to p^- and $\phi \in \mathrm{Symp}(S^2, \mathrm{rel}\ p^{\pm})$. Then ϕ is in the identity component of $\mathrm{Symp}(S^2, \mathrm{rel}\ p^{\pm})$ if and only if $\phi(\beta)$ is homotopic to β through embedded arcs which are fixed near their endpoints. Moreover $\phi \in \mathrm{Ham}(S^2, \mathrm{rel}\, p^{\pm})$ if and only if in addition there is zero area between β and $\phi(\beta)$.*

Proof Observe first that the inclusion of $\mathrm{Symp}(S^2, \mathrm{rel}\, p^{\pm})$ into the group G of all symplectomorphisms of S^2 with trivial 1-jet at the points $p^{\pm}$ is a homotopy equivalence. Since the corresponding group with trivial 1-jet at the single point p^+ is connected, the first statement follows by considering the long exact homotopy sequence of the fibration given by the evaluation map which takes $\phi \in G$ to its 1-jet at p^-. The second statement holds because the area between $\phi(\beta)$ and β measures the flux of ϕ. □

Proposition 9 *Any π-compatible symplectic form on V_T is isotopic to a standard form.*

Proof Our aim is to isotop ω to make its holonomy round the loops α_1 and α_2 equal to the half-turns h_1 and h_2. This proves the proposition by Lemma 5. We will assume that the cohomology class $\mu_+ b_+ + \mu_- b_-$ of ω is such that $\mu_- \leq 0$ since otherwise the result was proved in [4].

Let F be the fiber over the base point $x = \alpha_1 \cap \alpha_2$ and let $X = X_2$ be as constructed above. We will first put X into standard position relative to ω and π, so that the holonomy f_1 round the loop α_1 is equal to the half-turn h_1 near the points $p^{\pm} = F \cap X$. Thus f_1 interchanges the points p^- and p^+. It is essential to do this in such a way that that the element $f_1^{-1} h_1$ of $\mathrm{Symp}(F, \mathrm{rel}\, p^{\pm})$ is Hamiltonian. We then "inflate" V_T simultaneously along X and the fiber F, creating enough space in which to isotop f_1 to h_1. Finally, we adjust the holonomy round α_2 by a similar process, being careful not to mess up what we have already done.

Let us begin by looking at the problem of correcting the holonomy f_1 round α_1 near $p^{\pm}$. The first step is to isotop ω so it splits in neighbourhoods $D \times F$ of F and $X \times U'$ of X, where these product structures agree in the intersection $(D \times F) \cap (X \times U')$. We then isotop the fibration near X so that $f_1 = h_1$ near $p^{\pm}$. The final step is to arrange that $f_1^{-1}h_1$ is Hamiltonian. The first two steps may be accomplished by the methods developed in [4, §3]: see in particular Lemmas 3.9 and 3.11. For the final step we first arrange that $f_1^{-1}h_1$ is in the identity component of $\mathrm{Symp}(F, \mathrm{rel}\, p^{\pm})$ by isotoping ω in $D \times F$ so that the new holonomy is $f_1' = \psi \circ f_1$, for suitable ψ. The arguments below show that we may take ψ to be any Hamiltonian symplectomorphism of S^2 which is generated by a sufficiently small Hamiltonian H (the size is bounded by the area of D). But, using Lemma 8, one easily checks that ψ can be a twist about the point p^- generated by arbitrarily small H. However, there is no way to make $(f_1')^{-1}h_1$ Hamiltonian by changes with small H. Instead, we replace h_1 by a conjugate $h_1' = R_1^{-1}h_1R_1$, where R_1 is chosen to commute with h_2 and be such that, for some geodesic arc β joining p^- to p^+, the area between β and $f_1(\beta)$ is the same as that between $R_1(\beta)$ and $h_1R_1(\beta)$. It is easy to see that there is a suitable R_1 which is the identity near $p^{\pm}$ and which rotates the circles equidistant from these points by varying amounts. Then there is $U \subset U'$ so that $f_1 = h_1 = h_1'$ in $F \cap (X \times U)$.

We now need to isotop ω in $\pi^{-1}D = D \times F$ to make the holonomy equal to h_1' everywhere. As in [3, Lemma 4.14], this is possible if the disc $D \subset T^2$ is sufficiently large. The idea is the following. Identify D with a disc $D(R)$ of radius R in $\mathbf{R}^2$ so that $\alpha_1 \cap D$ is part of the arc $\{(s,0) : s \in \mathbf{R}\}$, with $x = (0,0)$, and $\alpha_2 \cap D \subset \{(0,t) : t \in \mathbf{R}\}$. Assume also that $\omega = ds \wedge dt + \rho_0$ in $D \times F \subset \mathbf{R}^2 \times F$, where ρ_0 is the standard form of area 1 on $F \equiv S^2$. Let ϕ_t be an isotopy of $\mathbf{R}^2 \times S^2$ supported in $D(R) \times S^2$ so that

$$\phi_1(s,0,y) = (s, H(s,y), y) \quad \text{for } (s,y) \in \mathbf{R} \times F.$$

We will suppose that $H(s,y) = 0$ for $s \leq 0$ and for $y \in F \cap X$, so that we may choose $\phi_t = \mathrm{id}$ on $\pi^{-1}(\alpha_2) \cup X$. The holonomy of $\phi_1^*(\omega)$ along $\alpha_1 \cap D$ is given by the characteristic flow of $\phi_1^*(\omega)$ on $\pi^{-1}(\alpha_1 \cap D)$, or equivalently, by the characteristic flow of ω on $\phi_1(\pi^{-1}(\alpha_1 \cap D))$. Thus, it is given by the flow of H. Moreover, we may choose $H(s,y)$ freely for $s > 0$ and $y \neq p^{\pm}$, adjusting it by a function of s alone in order to satisfy the boundary conditions. Thus, provided that we have enough space, we can arrange for the holonomy to be any Hamiltonian map of F which fixes the two points $p^{\pm}$. In particular, it can cancel out the holonomy of the rest of α_1, provided this is Hamiltonian.

In order to create the necessary space we follow the procedure of Lemma 4.15 in [3], isotoping ω so that it splits over a larger proportion of T^2. Consider the forms $\omega_t = \omega + 2t\pi^*(\sigma) + t\pi_X{}^*(\rho)$ where σ, ρ are non-negative 2-forms with supports in D and U, respectively, and total area 1, and where π_X is the projection $X \times U \to U$. Because ω splits with respect to the product structure $X \times U$, the forms ω_t are symplectic for all $t \geq 0$. We assume that ρ is invariant under h_1', so that the form τ_t induced by ω_t on F is invariant under h_1'. Clearly, we can choose σ and ρ so that $[\omega_t] = \kappa_t[\omega]$.

Since multiplication by a scalar does not affect holonomy, it suffices to show how to isotop ω_t for some large t so that its holonomy is h_1'. Because the product structure on $X \times U$ is compatible with π over the loop α_1, the holonomy of ω_t round α_1 does not change as t varies, even though the induced form τ_t on F does change. (Note that the set U on which this form changes is the set on which the holonomy is already correct.) Further, as t increases the "size" of D increases, i.e. $(D \times F, \omega_t)$ is symplectomorphic to some subset $D(R_t) \times F$ of $(\mathbf{R}^2 \times F, ds \wedge dt + \tau_t)$, where R_t becomes arbitrarily large. Thus, we can create as much space as we need by taking t large enough.

This process corrects the holonomy to $h_1' = R_1^{-1} h_1 R_1$ round α_1 by inflating along X_2, where R_1 commutes with h_2. We now correct the holonomy round α_2 by inflating along X_1. We show in the next paragraph that we may assume that X_1 intersects $\pi^{-1}\alpha_1$ along the pair of closed characteristics through the fixed points of h_1'. The reader may check that, given this, when we adjust ω near X_1 and then inflate along X_1 we will not alter the holonomy round α_1. Thus as before we may correct the holonomy round α_2 to $R_2^{-1} h_2 R_2$ where R_2 commutes with h_1'.

To construct X_1, take the double cover of V_T which is nontrivial over α_1. Cut it open along α_1 to get a symplectic form on $I \times S^1 \times S^2$ which splits near the boundary, and then identify each boundary component to single fibers F_0, F_1. This gives a symplectic form ω' on $S^2 \times S^2$ which is compatible with the projection π' onto the first factor and is invariant under the obvious extension ψ of the deck transformation of the 2-fold cover. Take a ψ-invariant ω'-tame almost complex structure J. As in Lemma 6, we may assume that the fibers of π' are J-holomorphic. Note also that by assumption on $[\omega]$, the integral of ω' on the fiber (the second factor) is at least as big as its integral over the base. By the results of [3], it follows that through each point of F_0 there is a unique J-holomorphic curve in the class $[S^2] \times$ pt. This is necessarily a section of π'. Consider the curves through the two fixed points of ψ on F_0. By uniqueness, these are both ψ-invariant, and hence they meet F_1 also in fixed points of ψ. Further, they descend to V_T to form a (possibly non-smooth) submanifold which double covers the base and intersects $\pi^{-1}(\alpha_1)$ along the closed characteristics through the fixed points of h_1'. For topological reasons it has to be connected. Therefore it is homotopic to X_1. We can smooth it out near its intersection with $\pi^{-1}(\alpha_1)$ using the techniques of [4, Lemma 3.13]. □

4 Wrapping

There is another procedure for modifying holonomy which, instead of using a submanifold such as X, uses one loop of trivial holonomy to create others.

Lemma 10 *Consider a symplectic S^2 bundle $\pi : V \to M$ with a compatible form ω. Let γ and δ be simple closed curves in M which intersect once transversally, and suppose that the holonomy around δ is trivial. Then there is an isotopy ϕ_t with support in a neighbourhood $N(\delta)$ of $\pi^{-1}(\delta)$ such that $\phi_1^*(\omega)$ has trivial holonomy around both γ and δ.*

Proof The symplectic neighbourhood theorem implies that, because the holonomy around δ is trivial, there is a neighbourhood $N(\delta)$ of $\pi^{-1}(\delta)$ that is symplectomorphic to the quotient

$$[-1,1] \times \mathbf{R} \times S^2/\sim, \quad (t,\lambda,y) \sim (t,\lambda+n,y),$$

where $\pi^{-1}(\delta) = \{t = 0\}$ and where the strip $[-1,1] \times \mathbf{R} \times S^2$ has a product symplectic form. By slightly perturbing π and γ, we may further suppose that π restricts to the obvious projection on $N(\delta)$ and that γ intersects $N(\delta)$ in the strip $\lambda = 0$.

As in Proposition 9, we choose $\phi_t : N(\delta) \to N(\delta)$ so that ϕ_1 lifts to a map of the form

$$(t,0,y) \mapsto (t, H(t,y), y)$$

on $\pi^{-1}(\gamma) \cap N(\delta)$, where

$$H = 0 \quad \text{for} \quad t \quad \text{near } [-1,0] \cup \{1\}.$$

Note that because the holonomy around δ is trivial, $H(t,y)$ may be as large as we want here. (If the holonomy were not trivial, the equivalence relation on $[-1,1] \times \mathbf{R} \times S^2$ would be more complicated and would impose restrictions on H.) Therefore, as before, we may arrange that the monodromy of $\phi_1(\gamma)$ in $N(\delta)$ cancel out the monodromy in the rest of γ without affecting the holonomy around δ. □

Proof of Proposition 3

The above lemma shows that if the form ω on V_T has trivial holonomy around one essential loop, say α_1, it is isotopic to a form which has trivial holonomy around a pair of loops α_1, α_2 which generate $\pi_1(T^2)$. Let $\Lambda = \alpha_1 \cup \alpha_2$. We may suppose that ω splits in some neighbourhood of Λ, and hence, if we identify Λ to a point we obtain a symplectic form on the bundle V_S, the non-trivial bundle over S^2 in a class $a = \mu_+ b_+ + \mu_- b_-$ with the same $\mu_\pm$ as for ω. We show in [3] that this bundle must have a section in class S_-, i.e. a section of self-intersection -1. (This holds because this section is an exceptional sphere.) Therefore, as stated in Theorem 1, we must have $\mu_- > 0$.

The converse statement follows from the uniqueness of forms on V_T and the fact that whenever $\mu_- > 0$ there is at least one form on V_T with trivial holonomy around some essential loop, namely one constructed as above from a bundle over S^2. □

Proof of Proposition 4

Let M be a Riemann surface of genus 2, and suppose that ω is a π-compatible form on V_M which has trivial holonomy around the essential loop α_1. It suffices to prove that ω is isotopic to one of the standard forms constructed in §3, since these are all isotopic. First suppose that α_1 separates M. Then, as in the previous proposition, we may identify $\pi^{-1}\alpha_1$ to a single fiber F_0, and hence construct from each half of $M - \alpha_1$ a form on a bundle over T^2. Note that one of these bundles will be trivial, and the other will be non-trivial. (Which is which depends on the characteristic foliation on $\pi^{-1}\alpha_1$ which determines the way in which it is identified with F_0.) But both are isotopic to a standard form. Moreover, inspection of the proofs shows that we may assume that these isotopies fix a neighbourhood of the fiber F_0. It is now not hard to check that they fit together to give an isotopy of ω to a standard form.

If α_1 does not separate M, then it is part of a standard set of generators, and so Lemma 5 implies that we may isotop ω so that it has trivial holonomy around a conjugate loop α_2 as well. Then, by identifying $\alpha_1 \cup \alpha_2$ to a single point, we may again reduce to a bundle over the torus, and the result follows much as before. □

Question 11 If ω is a π-compatible form on V_M where M has genus > 1, must ω have trivial holonomy around some essential loop? If so, the above reasoning would imply that all symplectic ruled surfaces are standard. But why should genus 2 be different in this respect from genus 1?

References

[1] M. Gromov: Pseudo-holomorphic curves in symplectic manifolds, Invent. Math. **82** (1985), 307–347.

[2] V. Guillemin, E. Lerman, and S. Sternberg: *Symplectic fibrations and multiplicity diagrams*, Camb. Univ. Press, to appear.

[3] D. McDuff: The Structure of Rational and Ruled Symplectic 4-manifolds, Journ. Amer. Math. Soc. 3 (1990), 679–712, Erratum 5 (1992), 987–988.

[4] D. McDuff: Notes on Ruled Symplectic 4-manifolds, Trans. Amer. Math. Soc. **345** (1994), 623–639.

[5] D. McDuff and D. Salamon: J-holomorphic curves and Quantum Cohomology, University Lecture Series, Amer. Math. Soc. (1994).

Dusa McDuff
Mathematics Dept.
SUNY at Stony Brook
Stony Brook NY 11794
USA

Floer cohomology of Lagrangian intersections and pseudo-holomorphic disks, III: Arnold-Givental Conjecture*

Y.-G. Oh

1 Introduction

It is an immediate corollary from the classical Morse inequality that on any compact manifold M, any Morse function, i.e., a nondegenerate smooth function has at least $\mathrm{SB}(M, \mathbb{Z})$ number of critical points where

$$\mathrm{SB}(M, \mathbb{Z}) := \dim H^*(M, \mathbb{Z}).$$

The critical points of f are the points q which satisfy the equation $df(q) = 0$, and therefore we have

$$\mathrm{Crit}(f) = M_0 \cap M_f$$

where

$$\begin{aligned} \mathrm{Crit}(f) &= \text{ set of critical points of } f \\ M_0 &= \text{ the zero section of } T^*M \\ M_f &= \text{ Graph } df = \{(q, df(q)) \in T^*M \mid q \in M\}. \end{aligned}$$

The cotangent bundle T^*M has the canonical symplectic structure $\omega_0 = -d\theta$ where the one-form θ is defined by

$$\theta(p)\xi = p(T\pi\xi)$$

for $\xi \in T_p(T^*M)$ where $\pi: T^*M \to M$ denotes the projection. The two submanifolds M_0 and M_f are *Lagrangian submanifolds*, i.e., n-dimensional submanifolds where the symplectic form ω_0 vanishes. Indeed, if we denote by $i_f: M_f \hookrightarrow T^*M$ the inclusion map, then we have

$$i_f^*\theta = df.$$

It can be shown that the graph M_f can be connected to the zero section M_0 via an *exact* (or *Hamiltonian*) isotopy on T^*M: A symplectic diffeomorphism ϕ on

*) Supported by NSF Grant DMS 9100383 and DMS 9012367.

(P, ω) is called *exact* if ϕ can be connected to the identity through an isotopy ϕ_t such that $\phi_0 = id, \phi_1 = \phi$ and there exists a smooth function $H : P \times \mathbb{R} \to \mathbb{R}$ such that ϕ_t is the flow of the equation

$$\dot{x} = X_{H_t}(x), \quad H_t(x) = H(x,t).$$

Here X_f denotes the Hamiltonian vector field associated to the smooth function f which is defined by the equation $X_f \rfloor \omega = df$. We call the isotopy ϕ_t of the above type an *exact* (or *Hamiltonian*) isotopy. We denote by $\mathfrak{D}_\omega$ the set of exact symplectic diffeomorphisms on (P, ω).

From the above observations on T^*M, Arnold [A] proposed a conjecture that if a Lagrangian submanifold $L \overset{i}{\hookrightarrow} T^*M$ is *Hamiltonian isotopic* to the zero section M_0, then

$$\#(L \cap M_0) \geq \mathrm{SB}\,(M, \mathbb{Z})$$

provided L meets M_0 transversely. This conjecture was solved by Hofer [Ho1] (see also [LS]) inspired by Conley and Zehnder's proof for the torus T^{2n} of another conjecture by Arnold which concerns the number of fixed points of exact symplectic diffeomorphisms on (P, ω). (See [F1] for a more detailed history and description of the conjecture.)

Since all of these notions like Lagrangian submanifolds and exact symplectic diffeomorphisms make sense on general symplectic manifolds (P, ω), it is natural to generalize the conjecture as follows: Given a compact Lagrangian submanifold $L \subset (P, \omega)$, is it true that

$$\#(L \cap \phi(L)) \geq \mathrm{SB}\,(L, \mathbb{Z}) \quad or \quad \mathrm{SB}\,(L, \mathbb{Z}_2) \tag{1.1}$$

provided that $\phi \in \mathfrak{D}_\omega$ and L intersects $\phi(L)$ transversely?

This, however, obviously fails in the present form as one can easily see that a «small» circle in a surface or more generally a «small» torus in general P can be separated from itself by $\phi \in \mathfrak{D}_\omega$. On the other hand, in the original conjecture on T^*M, M_0 is not an arbitrary Lagrangian submanifold but an *exact* Lagrangian submanifold in the sense that

$$\omega[w] = 0 \tag{1.2}$$

for all $w \in \pi_2(T^*M, M_0)$. In a seminal paper [Gr], Gromov introduced the important concept of *pseudo-holomorphic curves* on symplectic manifolds and using this, among other things, obtained some intersection theorem for the Lagrangian submanifolds which satisfy the condition (1.2).

The ideas of Conley-Zehnder and Gromov were culminated in the work by Floer [F1]. Under the hypothesis $\pi_2(P, L) = \{e\}$ which automatically implies (1.2), Floer introduced the so-called by now, *Floer cohomology* $I^*(L : P)$ of Lagrangian intersections of L on P, and proved that $I^*(L : P)$ is isomorphic to the singular cohomology $H^*(L, \mathbb{Z}_2)$. This, in particular, proves the estimate (1.1) in

$\mathbb{Z}_2$-coefficients. However, the condition $\pi_2(P, L) = \{e\}$ is such a strong restriction that it does not include the case of equators on S^2 for which the estimates (1.1) have been known to be true since Poincaré. Indeed, Givental (later also Chang-Jiang [CJ] by a different method) proved using the «non-linear Maslov index» [Gi] that (1.1) holds for the pair $(\mathbb{C}P^n, \mathbb{R}P^n)$. From this, Givental made the following conjecture (see [Gi]):

Conjecture (Givental). *Let (P, ω) be a symplectic manifold and let $\tau : P \to P$ be an anti-symplectic involution, i.e., $\tau^*\omega = -\omega$. Assume that the fixed point set $L = \operatorname{Fix} \tau$ is non-empty and compact. Then the estimate (1.1) holds for L.*

This conjecture includes the case of (T^*M, M_0), $(\mathbb{C}P^n, \mathbb{R}P^n)$ and the fixed point problem as special cases. The conjecture for the fixed point case was solved by Floer [F3] for the case of «monotone» symplectic manifolds.

The aim of the present paper is to resolve this conjecture for the real forms of compact Hermitian symmetric spaces under the assumption

$$\Sigma_L \geq 3$$

where Σ_L is the positive generator of the subgroup,

$$\{\mu(w) | w : (D^2, \partial D^2) \to (P, L)\} \subset \mathbb{Z}.$$

Here $\mu(w)$ denotes the Maslov index of w. (See [O2] for the definition).

Main Theorem. *Let (P, ω) be a compact symplectic manifold such that there exists an integrable almost complex structure J for which the triple (P, ω, J) becomes a compact Hermitian symmetric space. Let $L = \operatorname{Fix} \sigma$ where σ is an anti-holomorphic involutive isometry and assume that $\Sigma_L \geq 3$. Then the estimate (1.1) holds in $\mathbb{Z}_2$-coefficients, i.e.,*

$$\#(L \cap \phi(L)) \geq \operatorname{SB}(L, \mathbb{Z}_2)$$

provided $\phi \in \mathfrak{D}_\omega$ and L intersects $\phi(L)$ transversely.

Although the restriction $\Sigma_L \geq 3$ excludes the case $(\mathbb{C}P^1, \mathbb{R}P^1)$ (which was treated separately in [O3]), it is not a very strong restriction for the real forms of Hermitian symmetric spaces. Indeed if we denote

$$\Gamma_{c_1} = \text{ the positive generator of the subgroup } \{c_1|_{\pi_2(P)}\} \subset \mathbb{Z},$$

where c_1 is the first Chern class of (P, J), then it is proven in [O2] that

$$\Gamma_{c_1} | \Sigma_L | 2\Gamma_{c_1}$$

when L is simply connected, $\Sigma_L = 2\Gamma c_1$ for any $L = \operatorname{Fix} \tau, \tau$ an anti-symplectic involution (see Lemma 2.1 and Examples later in [O2]). Moreover, the computation of c_1 in [BH] for an irreducible compact Hermitian symmetric space shows that $\Gamma_{c_1} = 2$ for $\mathbb{C}P^1$ and $\Gamma_{c_1} \geq 3$ for all the other cases. Therefore the above theorem, together with Theorem 5 [O3], proves the estimate (1.1) for all irreducible compact Hermitian symmetric spaces:

Corollary. *Let (P,ω,J) be an irreducible compact Hermitian symmetric space and $L = \text{Fix}\,\sigma$, σ an anti-holomorphic involutive isometry. Assume that L is non-empty. Then*

$$\#(L \cap \phi(L)) \geq \text{SB}\,(L,\mathbb{Z}_2)$$

provided $\phi \in \mathfrak{D}_\omega$ and L meets $\phi(L)$ transversely.

In any case, the above theorem is much more general than the corollary. The method to prove the theorem is to use the author's refinement [O2] of Floer cohomology for *monotone Lagrangian submanifolds* introduced in [O2]: A Lagrangian submanifold $L \subset (P,\omega)$ is called *monotone* if

$$I_{\mu.L} = \lambda I_\omega \qquad \text{for some } \lambda \geq 0$$

where

$$I_{\mu,L} : \pi_2(P,L) \to \mathbb{Z} \text{ and } I_\omega : \pi_2(P,L) \to \mathbb{R}$$

are homomorphisms defined by

$$I_{\mu,L}(w) = \mu(w), \quad I_\omega(w) = \int_w \omega.$$

For monotone Lagrangian submanifolds, the following theorem is proved in [O2]:

Theorem A [O2]. *Let L be a compact monotone Lagrangian submanifold in (P,ω) and let $\phi = \{\phi_t\}_{0\leq t\leq 1}$ be an exact isotopy of P such that L meets $\phi(L)$ transversely. Suppose $\Sigma \geq 3$. Then there exists a homomorphism $\delta : \mathscr{C}^* \to \mathscr{C}^*$ with $\delta \circ \delta = 0$ such that the quotients*

$$I^*(L,\phi : P) = \text{Ker}\,\delta / \text{Im}\,\delta$$

are isomorphic as $\mathbb{Z}_2$-modules with $\mathbb{Z}/\Sigma$-grading for any exact isotopy ϕ provided L meets $\phi(L)$ transversely. We denote the common group by $I^*(L : P)$.

The construction of δ and so $I^*(L,\phi : P)$ involves almost complex structures J such that the bilinear form $g(\cdot,\cdot) = \omega(\cdot,J\cdot)$ becomes a Riemannian metric, but $I^*(L,\phi : P)$ does not depend on J, and so we omit the dependence of J in the above statement (see [O2] for more details). By the invariance of $I^*(L,\phi : P)$, we carry out the computation of $I^*(L : P)$ for an appropriate choice of (J,ϕ) which is regular in the sense described in [O2]. Here comes the special geometry of (P,ω) which is required in the main theorem. The same scheme of the proof was used by the author [O3] to compute $I^*(\mathbb{R}P^n : \mathbb{C}P^n)$ and to prove $I^*(\mathbb{R}P^n : \mathbb{C}P^n) \cong H^*(\mathbb{R}P^n,\mathbb{Z}_2) \cong \mathbb{Z}_2^{n+1}$ which in turn proves the estimate (1.1) for the case $(\mathbb{C}P^n,\mathbb{R}P^n)$.

The main theorem is an immediate corollary of the following stronger theorem.

Theorem B. *Let* (P,ω) *and* L *be as in the Main Theorem. Then we have the isomorphism*

$$I^*(L:P) \cong H^*(L,\mathbb{Z}_2)$$

as $\mathbb{Z}/\Sigma$*-graded* $\mathbb{Z}_2$*-vector spaces.*

It still remains to be seen whether the Floer cohomology $I^*(L:P)$ is isomorphic to $H^*(L,\mathbb{Z}_2)$ in the context of Theorem A. And there are degenerate versions of the estimate (1.1), which was proved for the case of $\pi_2(P,L)=\{e\}$ in [F4], [Ho2] and for the case $(\mathbb{C}P^n,\mathbb{R}P^n)$ in [CJ] and [Gi]. Our method in the present form does not give rise to any result on the degenerate case. We hope to investigate the degenerate case in the future.

Finally, as indicated in [O1], the estimate obtained in the Main Theorem should have some application to the Hamiltonian volume minimization problem as in the case of $(\mathbb{C}P^n,\mathbb{R}P^n)$, where the estimate (1.1) for $\mathbb{R}P^n$ in $\mathbb{C}P^n$ implies, via Crofton's formula, that $\mathbb{R}P^n$ is volume minimizing under its Hamiltonian deformations. We will investigate this application in the future.

We would like to thank C. LeBrun for bringing our attention to the curvature properties of compact Hermitian symmetric spaces. The present research is supported by NSF Grant DMS 9100383 and DMS 9012367. We would like to thank the Institute for Advanced Study for its nice research environment.

2 Geometry of compact Hermitian symmetric spaces

In this section we briefly review some curvature properties of compact Hermitian symmetric spaces and refer readers to [MZ] or references in it for more details.

Definition 2.1 A complex manifold $\overline{M}$ with a Hermitian structure is said to be a Hermitian symmetric space if each point $p \in \overline{M}$ is an isolated fixed point of an involutive holomorphic isometry s_p of $\overline{M}$.

We denote by $I(\overline{M})$ (respectively $A(\overline{M})$) the set of isometrics (respectively holomorphic isometries), and by $I^0(\overline{M})$ and $A^0(\overline{M})$ their identity components respectively. Then the group $G_u := A^0(\overline{M})$ acts transitively on $\overline{M}$ and if we denote by K_u the set of elements in G_u which fix a given point $0 \in \overline{M}$, then $\overline{M}$ is isometric to the quotient space G_u/K_u with the canonical G_u-invariant metric and the G_u-invariant complex structure J which reduce to the given ones on $(\overline{M}_0,\overline{g}_0,J_0)$. Here we use the notations G_u and K_u to match the notations in Section 3.

It turns out that the Hermitian structure of a Hermitian symmetric space is Kählerian, i.e., the J is parallel with respect to the Levi-Civita connection associated to the Riemannian metric $\overline{g}$. Therefore the bilinear form

$$\omega(\cdot,\cdot) = g(J\cdot,\cdot)$$

is a closed two form which in turn defines a symplectic structure. We denote the corresponding symplectic manifold by (P, ω) and the Kähler manifold by (P, ω, J).

Now, we recall the definition of the holomorphic bisectional curvature of a Hermitian manifold $(\overline{M}, \overline{g}, J)$.

Definition 2.2 Let R be the Riemann curvature tensor of $(\overline{M}, \overline{g})$ and extend it complex linearly to $T_{\mathbb{C}}\overline{M} = T\overline{M} \otimes \mathbb{C} = T^{(1,0)}\overline{M} \oplus T^{(1,0)}\overline{M}$. For each given $\xi, \eta \in T^{(1,0)}\overline{M}$, the holomorphic bisectional curvature in the complex plane spanned by ξ, η is defined by

$$\langle R(\xi, \bar{\xi})\bar{\eta}, \eta \rangle.$$

The following is a well-known fact (See [MZ]).

Lemma 2.3. *Let $(\overline{M}, \overline{g}, J)$ be a compact Hermitian symmetric space. Then it is an Einstein-Kähler manifold with positive Ricci curvature and with non-negative holomorphic bisectional curvature.*

Indeed, the above curvature properties characterize the compact Hermitian symmetric spaces (see [MZ]). By Kobayashi's theorem that a positive Einstein-Kähler manifold is simply connected, compact Hermitian symmetric spaces are also simply connected. From the above curvature property, we can prove the following proposition which will be essential for a later discussion. As usual, we will identify the tangent bundle $T\overline{M}$ with $T^{(1,0)}\overline{M}$ as a holomorphic vector bundle.

Proposition 2.4. *Let $(\overline{M}, \overline{g}, J)$ be a compact Hermitian symmetric space, and let $v : \mathbb{C}P^1 \to (\overline{M}, J)$ be a non-constant holomorphic map. Then the Grothendieck splitting*

$$E := v^* T\overline{M} = L_1 \oplus \cdots \oplus L_n$$

where L_k's are holomorphic line bundles, has the property

$$c_1(L_k) \geq 0$$

for all $k = 1, \dots, n$.

Proof By Lemma 2.3,

$$\langle R(\xi, \bar{\xi})\bar{\eta}, \eta \rangle \geq 0$$

for all $\xi, \eta \in T^{(1,0)}\overline{M}$. In particular, for any $\xi_k \in L_k$ and $s \in T\mathbb{C}P^1$, we have

$$\langle R(\xi_k, \bar{\xi}_k)\overline{Tv(s)}, Tv(s) \rangle \geq 0$$

since v is holomorphic. However, $\langle R(\xi_k, \bar{\xi}_k)\overline{Tv(s)}, Tv(s) \rangle$ for $\|\xi_k\| = 1$ is the induced curvature of the line bundle L_k since the second fundamental form of L_k in E vanishes due to the splitting. Therefore

$$c_1(L_k) = \frac{1}{2\pi} \int_{\mathbb{C}P^1} \langle R(\xi_k, \bar{\xi}_k)\overline{Tv(.)}, Tv(.) \rangle \geq 0$$

□

From this, we have the following immediate corollary.

Corollary 2.5. *Under the same hypothesis* $(\overline{M}, \overline{g}, J)$ *and* v *as in Proposition 2.4, there is no non-trivial global holomorphic sections of* $E^* \simeq \overline{E}$ *that vanishes at some point.*

Proof Since $c_1(L_k^*) = -c_1(L_k)$, we have

$$c_1(L_k^*) \leq 0$$

for all $k = 1, \ldots, n$. When $c_1(L_k^*) < 0$, there is no non-trivial global section whatsoever by the vanishing theorem. When $c_1(L_k^*) = 0$, the conclusion follows from the fact that $c_1(L_k^*)$ is the Euler characteristic of a non-trivial section and that the local degree of a sections of holomorphic line bundle over a surface is positive. □

3 Geometry of real forms

In this section, we will prove the following analogue to Proposition 2.1 [O3].

Proposition 3.1. *Let* (P, ω, J) *be a compact Hermitian symmetric space and let* σ *be an anti-holomorphic involutive isometry of* (P, ω, J). *Denote*

$$L = \operatorname{Fix} \sigma$$

and assume that L *is non-empty. Then there exists an exact isotopy* $\phi = \{\phi_t\}_{0 \leq t \leq 1}$ *such that*

$$\sigma \phi_t \sigma = \phi^{-1} \tag{3.1}$$

and

$$\#(L \cap \phi_t(L)) = \operatorname{SB}(L, \mathbb{Z}_2)$$

for all $0 < t \leq 1$. *Moreover,* ϕ_t *is the flow of a Killing vector field associated to an element in* $\mathfrak{g}_u$ *and it satisfies*

$$(\phi_1)^{2^N} = id \tag{3.2}$$

for a sufficiently large N.

To prove this, we first briefly summarize the geometry of the real forms of compact Hermitian symmetric spaces for the reader's convenience, following verbatim the description by Takeuchi [T]. Then in the end of this section, we will provide the proof of the above proposition.

Definition 3.2 Let $(\overline{M}, \overline{g}, J)$ be a Hermitian symmetric space and σ an anti-holomorphic involutive isometry. We call the fixed point set $M = \operatorname{Fix} \sigma$ a *real form* of $(\overline{M}, \overline{g}, J)$.

These real forms are completely classified by Takeuchi [T]. In [T] it is shown that the set of pairs $((\overline{M},\overline{g}),M)$ is in one-to-one correspondence with the set of pairs $(\mathfrak{g},\tau)$ where $(\mathfrak{g},\tau)$ is a *positive definite symmetric graded Lie algebra*, i.e.,

$$\mathfrak{g}=\mathfrak{g}_{-1}+\mathfrak{g}_0+\mathfrak{g}_1,\quad [\mathfrak{g}_p,\mathfrak{g}_q]\subset\mathfrak{g}_{p+q},$$

is a real semi-simple graded Lie algebra such that $\mathfrak{g}_{-1}\neq 0$ and $\mathfrak{g}_0$ acts effectively on $\mathfrak{g}_{-1}$, and τ a Cartan involution of $\mathfrak{g}$ with $\tau\mathfrak{g}_p=\mathfrak{g}_{-p}$. The correspondence between $(\mathfrak{g},\tau)$ and $((\overline{M},\overline{g}),M)$ goes as follows:

Let $\overline{G}=\mathrm{Aut}^0(\overline{M})$ which is a connected complex semi-simple Lie group with the trivial center, and let $G_u=I^0(\overline{M},\overline{g})$ which is a subgroup of $\overline{G}$ because $(\overline{M},\overline{g})$ is a compact Kähler manifold (cf. Kobayashi [Ko]). Let J denote the complex structure tensor of $\overline{M}$. We identify $\bar{\mathfrak{g}}=\mathrm{Lie}\,\overline{G}$ (resp. $\mathfrak{g}_u=\mathrm{Lie}\,G_u$) with the Lie algebra of all smooth vector fields X on $\overline{M}$ such that the Lie derivative of J with respect to X vanishes (resp. of all Killing vector fields on $(\overline{M},\overline{g})$) with Lie product $[X,Y]=YX-XY$. Then by Matsushima's theorem on compact Kähler-Einstein manifolds, we have

$$\bar{\mathfrak{g}}=\mathfrak{g}_u+J\mathfrak{g}_u,\quad \mathfrak{g}_u\cap J\mathfrak{g}_u=0. \tag{3.3}$$

Let $\mathfrak{g}(M)$ be a real subalgebra of $\bar{\mathfrak{g}}$ consisting of all $X\in\bar{\mathfrak{g}}$ such that the restriction $X|M$ is tangent to M, and $\mathfrak{k}(M)$ the Lie algebra of all Killing vector fields on M with respect to the Riemannian metric g induced from $\overline{g}$. We put

$$\mathfrak{k}=\mathfrak{g}(M)\cap\mathfrak{g}_u,\quad \mathfrak{p}=\mathfrak{g}(M)\cap J\mathfrak{g}_u,$$

and

$$\mathfrak{g}=\mathfrak{k}+\mathfrak{p}. \tag{3.4}$$

Then $[\mathfrak{k},\mathfrak{p}]\subset\mathfrak{p}$ and $[\mathfrak{p},\mathfrak{p}]\subset\mathfrak{k}$, and hence $\mathfrak{g}$ is a real subalgebra of $\bar{\mathfrak{g}}$. We need here the following:

Lemma 3.3.
(1) *The map* $\mathfrak{k}\to\mathfrak{k}(M)$ *defined by* $X\mapsto X|M(X\in\mathfrak{k})$ *is a Lie isomorphism.*
(2) *We have*

$$\mathfrak{g}_u=\mathfrak{k}+J\mathfrak{p},\quad \mathfrak{k}\cap J\mathfrak{p}=0. \tag{3.5}$$

Now, it follows from (3.3), (3.4) and (3.5) that $\mathfrak{g}$ is a real form of $\bar{\mathfrak{g}}$. Let σ and τ denote the complex conjugation of $\bar{\mathfrak{g}}$ with respect to $\mathfrak{g}$ and $\mathfrak{g}_u$, respectively. Then

$$\sigma JX=-J\sigma X\qquad \text{for } X\in\bar{\mathfrak{g}}, \tag{3.6}$$

$$\sigma\mathfrak{g}_u=\mathfrak{g}_u. \tag{3.7}$$

We fix a point $o\in M$ and put

$$K_u=\{a\in G_u;a(o)=o\},$$

which is known to be connected. (See Helgason [He] for fundamental results on symmetric spaces.) Then $\overline{M} = G_u/K_u$ as smooth manifold. Let $\mathfrak{k}_u = \mathrm{Lie}\, K_u$ and $\mathfrak{g}_u = \mathfrak{k}_u + \mathfrak{m}_u$ be the associated Cartan decomposition. Let H_0 be the unique element of the center of $\mathfrak{k}_u$ such that $J_o = \mathrm{ad}\,(H_o)|\mathfrak{m}_u$. Putting $Z = JH_o \in \bar{\mathfrak{g}}$, we define

$$\begin{aligned} \bar{\mathfrak{g}}_{\mathfrak{p}} &= \{X \in \bar{\mathfrak{g}}; [Z,X] = pX\} \quad (p = -1, 0, 1), \\ \bar{\mathfrak{u}} &= \bar{\mathfrak{g}}_0 + \bar{\mathfrak{g}}_1, \\ \overline{U} &= \{a \in \overline{G};\ \mathrm{Ad}\,(a)\bar{\mathfrak{u}} = \bar{\mathfrak{u}}\}. \end{aligned}$$

Then $\mathrm{Lie}\,\overline{U} = \bar{\mathfrak{u}}, \bar{\mathfrak{g}} = \bar{\mathfrak{g}}_{-1} + \bar{\mathfrak{g}}_0 + \bar{\mathfrak{g}}_1$ and $\overline{M} = \overline{G}/\overline{U}$ as a complex manifold. Note here that $\bar{\mathfrak{g}}_0$ acts on $\bar{\mathfrak{g}}_{-1}$ effectively. We define an involutive automorphism θ of $\overline{G}$ by

$$\theta(a) = \exp(\pi JZ)a(\exp(\pi JZ))^{-1} \quad \text{for} \quad a \in \overline{G}.$$

Then $\theta(G_u) = G_u$ and hence the differential of θ, denoted also by θ, satisfies $\theta\mathfrak{g}_u = \mathfrak{g}_u$. Moreover we have

$$\mathfrak{k}_u = \{X \in \mathfrak{g}_u; \theta X = X\}, \tag{3.8}$$

$$\mathfrak{m}_u = \{X \in \mathfrak{g}_u; \theta X = -X\}. \tag{3.9}$$

A diffeomorphism θ of $\overline{M} = G_u/K_u$ is defined by the correspondence $a \cdot o \mapsto \theta(a) \cdot o (a \in G_u)$ because K_u is connected. It is the symmetry of $(\overline{M}, \bar{g})$ at o. Since M is totally geodesic in $(\overline{M}, \bar{g})$ we have $\theta(M) = M$, and hence $\theta\mathfrak{g}(M) = \mathfrak{g}(M)$. Therefore we have $\theta\mathfrak{k} = \mathfrak{k}$ and $\theta\mathfrak{p} = \mathfrak{p}$, and hence $\theta\mathfrak{g} = \mathfrak{g}$. Thus (3.7), (3.8) and (3.9) imply

$$\sigma\mathfrak{k}_u = \mathfrak{k}_u, \tag{3.10}$$

$$\sigma\mathfrak{m}_u = \mathfrak{m}_u. \tag{3.11}$$

Now it follows from (3.6) and (3.11) that $\sigma J_o = -J_o\sigma$ on $\mathfrak{m}_u = T_o(\overline{M})$, and thus $[\sigma H_o, \sigma X] = J_o\sigma X$ for each $X \in \mathfrak{m}_u$,, where σH_0 is an element of the center of $\mathfrak{k}_u$ by (3.10). Therefore the uniqueness of H_0 implies that $\sigma H_0 = -H_0$, and so $\sigma Z = Z$, that is, $Z \in \mathfrak{g}$. Thus, putting $\mathfrak{g}_p = \bar{\mathfrak{g}}_p \cap \mathfrak{g}(p = -1, 0, 1)$ we get $\mathfrak{g} = \mathfrak{g}_{-1} + \mathfrak{g}_0 + \mathfrak{g}_1$. Moreover τ restricted to $\mathfrak{g}$ is a Cartan involution with $\tau Z = -Z$, and thus $\tau\mathfrak{g}_p = \mathfrak{g}_{-p}(p = -1, 0, 1)$. The effectiveness of $\mathfrak{g}_0$ on $\mathfrak{g}_{-1}$ follows from that of $\bar{\mathfrak{g}}_0$ on $\bar{\mathfrak{g}}_{-1}$. Therefore $(\mathfrak{g}, \tau)$ is a positive definite symmetric graded Lie algebra.

The group G_u acts on $\overline{M}$ as holomorphic isometries and in particular as symplectic diffeomorphisms with respect to the symlplectic structure $\omega(\cdot,\cdot) = g(J\cdot,\cdot)$. Due to the simple connectedness of $\overline{M}$ and the semi-simplicity of G_u, the action is a Hamiltonian action in the sense that there exists an equivariant moment map

$$\Phi : \overline{M} \to \mathfrak{g}_u^*.$$

Let $\mathfrak{h}^- \subset \mathfrak{p}$ be a maximal abelian subalgebra in $\mathfrak{p}$ with $Z \in \mathfrak{h}^-$ and take an abelian subalgebra $\mathfrak{h}^+$ of $\mathfrak{k}$ such that $\mathfrak{h} = \mathfrak{h}^+ + \mathfrak{h}^-$ is a Cartan subalgebra of

$\bar{\mathfrak{g}}$ and $\mathfrak{h}_R = \mathfrak{h}^+ + \sqrt{-1}\mathfrak{h}^-$ is a Cartan subalgebra of $\mathfrak{g}_u$. Since the vector field corresponding to an element $\sqrt{-1}\xi \in \sqrt{-1}\mathfrak{h}^-$ is given by $J\xi_{\overline{M}}$ where $\xi_{\overline{M}}$ is the vector field associated to ξ, $J\xi_{\overline{M}}$ is perpendicular to $M = \operatorname{Fix}\sigma$. By choosing an appropriate $\xi \in \mathfrak{h}^-$, we can make the flow ψ_t of $J\xi_{\overline{M}}$ is periodic with period 1. Now by the tightness of the moment map (see [D]), we have

$$M \cap \psi_t(M) = \operatorname{Crit}(f_{\sqrt{-1}\xi}) = \operatorname{SB}(M, \mathbb{Z}_2) \tag{3.12}$$

provided M meets $\psi_t(M)$ transversely.

Now we are ready to prove Proposition 3.1.

Proof of Proposition 3.1: We choose an element $\sqrt{-1}\xi \in \sqrt{-1}\mathfrak{h}_- \subset \mathfrak{g}_u$ and let ψ_t be the flow of $J\xi_{\overline{M}}$. The identity (3.1) follows from

$$f_{\sqrt{-1}\xi}(\sigma(x) = f_{\sqrt{-1}\xi}(x) \tag{3.13}$$

where

$$f_{\sqrt{-1}\xi}(x) := \langle \Phi(x), \sqrt{-1}\xi \rangle.$$

Indeed if we have proved (3.13), then we have

$$\sigma^*(J\xi_{\overline{M}}) = -J\xi_{\overline{M}} \tag{3.14}$$

since $\sigma^*\omega = -\omega$ and $J\xi_{\overline{M}}$ is the Hamiltonian vector field associated to $\sqrt{-1}\xi$. (3.1) is just the integrated version of (3.14). To prove (3.13), it is enough to show that

$$\sigma^* df_{\sqrt{-1}\xi} = df_{\sqrt{-1}\xi}$$

since $\operatorname{Fix}\sigma$ is non-empty. However,

$$\begin{aligned}
\sigma^* df_{\sqrt{-1}\xi}(x)(Y(x)) &= df_{\sqrt{-1}\xi}(\sigma(x)(T\sigma Y)) = (J\xi_{\overline{M}}(\sigma(x)) \rfloor \omega)(T\sigma Y)\\
&= \omega(J\xi_{\overline{M}}(\sigma(x)), T\sigma Y)\\
&= \omega(T\sigma T\sigma^{-1} J\xi_{\overline{M}}(\sigma(x)), T\sigma Y)\\
&= \omega(T\sigma(-J)T\sigma^{-1}\xi_{\overline{M}}(\sigma(x)), T\sigma Y)\\
&\qquad \text{since } \; T\sigma^{-1}J = -JT\sigma^{-1}\\
&= \omega(T\sigma(-J)\xi_{\overline{M}}(x), T\sigma Y) \;\; \text{since } \; \sigma^*\xi_{\overline{M}} = \xi_{\overline{M}}\\
&= -\omega(-J\xi_{\overline{M}}(x), Y(x)) \;\; \text{since } \; \sigma^*\omega = -\omega\\
&= \omega(J\xi_{\overline{M}}(x), Y(x)) = (J\xi_{\overline{M}}(x) \rfloor \omega)(Y(x))\\
&= df_{\sqrt{-1}\xi}(x)(Y(x))
\end{aligned}$$

which finishes the proof of (3.13). Now, (3.1) can be obtained by taking $\phi = \{\phi_t\}_{0 \le t \le 1}$

$$\phi_t = \psi_{t/2^N}$$

where ψ_t is the flow of $J\xi_{\overline{M}}$ and N is so large that (3.10) holds for all $0 < t \le 1$. □

4 Lagrangian intersection cohomology

In this section, we briefly summarize the construction of $I^*(L, \phi : P)$ in [O2] (see also [F1]). We follow the notations in [O2] below, fix the exponent $p = 2$ for the Sobolev exponetnt and take k so that $k > n/2$. We also denote $\Theta = \mathbb{R} \times I$, $I = [0, 1]$.

Definition 4.1 (1) $\Omega_\phi = \Omega(L, \phi) := \{z : I \to P | z(0) \in L, z(1) \in \phi_1(L)$ and the path $t \mapsto \phi_t^{-1} z(t)$ defines the zero element in $\pi_1(P, L)\}$.
(2) $\mathcal{P}_\phi = \mathcal{P}_k(L, \phi : P) := \{u \in L_k^2(\Theta, P) | u(\tau, 0) \subset L, u(\tau, 1) \subset \phi_1(L)$ and $u(\tau, \cdot) \in \Omega_\phi$ for all $\tau\}$.
(3) $\mathcal{M}_{J,\phi} = \mathcal{M}_J(L, \phi : P) := \{u \in \mathcal{P}_\phi \mid \overline{\partial}_J u := \frac{\partial u}{\partial \tau} + J\frac{\partial u}{\partial \tau} = 0, \int_{\mathbb{R}} \int_I |\frac{\partial u}{\partial \tau}|^2 dt d\tau < \infty\}$.
(4) $\mathcal{M}_{J,\phi}(x, y) := \{u \in \mathcal{M}_{J,\phi} | \lim_{\tau\to\infty} u = x, \lim_{\tau\to-\infty} u = y, x, y \in I(L, \phi)\}$.
(5) $\widehat{\mathcal{M}}_{J,\phi}(x, y) = \mathcal{M}_{J,\phi}(x, y)/\mathbb{R}$.

We also denote by

$$E_u = D\overline{\partial}_J(u) : T_u\mathcal{P}_\phi \to \mathcal{L}_\phi$$

the covariant linearization of $\overline{\partial}_J$ at u, where

$$\mathcal{L}_\phi := \{\xi \in L_{k-1}^2(\Theta, TP) | \xi(\theta) \in T_{u(\theta)}P\}.$$

Definition 4.3 We call $u \in \mathcal{M}_{J,\phi}(x, y)$ *regular* if $\mathrm{Coker} E_u = 0$ and we call (J, ϕ) *regular* if u is regular for all $u \in \mathcal{M}_{J,\phi}(x, y)$ and for all $x, y \in I(L, \phi)$.

It was shown (see Theorem 3 [F2].) that if x and y are transverse intersections of L and $\phi_1(L)$ and if u is regular, then $\mathcal{M}_{J,\phi}(x, y)$ is a smooth manifold near u, whose tangent space at u is isomorphic to $\mathrm{Ker} E_u$.

Definition 4.4 Assume that

(1) The pair (J, ϕ) is regular.
(2) The number of zero dimensional components of $\widehat{\mathcal{M}}_{J,\phi}(x, y)$, denoted by $\#(\widehat{\mathcal{M}}_{J,\phi}(x, y))$, is finite.
(3) The $\mathbb{Z}_2$-numbers defined by

$$\langle x, \delta y\rangle = \#(\widehat{\mathcal{M}}_{J,\phi}(x, y)) \bmod 2 \tag{4.1}$$

satisfy

$$\sum_{y \in I(L,\phi)} \langle x, \delta y\rangle \langle y, \delta z\rangle = 0$$

for any $x, z \in I(L, \phi)$. We then define the cohomology of (L, ϕ) as

$$I_J^*(L, \phi : P) = \mathrm{Ker}\, \delta / Im\, \delta$$

where $\delta : \mathcal{C}^* \to \mathcal{C}^*$, $\mathcal{C}^* =$ the free $\mathbb{Z}_2$-module over $I(L, \phi)$, is the operator defined by the matrix elements $\langle x, \delta y\rangle$ in (4.1).

One main theorem in [O2] can be stated as that for a monotone L with $\Sigma_L \geq 3$, the hypotheses (1), (2) and (3) in Definition 4.2 hold for generic J's in a certain sense provided L meets $\phi_1(L)$ transversely. Moreover, the resulting groups $I_J^*(L, \phi : P)$ are isomorphic as $\mathbb{Z}/\Sigma$-graded $\mathbb{Z}_2$-modules, as long as L meets $\phi_1(L)$ transversely and J satisfies (1), (2) and (3) at its end points, i.e., the initial and final points (see [O2] for details).

Now we specialize to the case of a real form L in a compact Hermitian symmetric space (P, ω, J). By Proposition 3.1, we have

$$\#\big(L \cap \phi_t(L)\big) = \#\,(\mathrm{Crit}\, f_\xi) = \mathrm{SB}\,(L, \mathbb{Z}_2)$$

for all $0 < t \leq 1$. Therefore to prove our main theorem, it is enough to prove the following:

Proposition 4.4 [Regularity]. *Let (P, ω, J) and L be as above and $\phi = \{\phi_t\}_{0 \leq t \leq 1}$ as in Proposition 3.1. Then the pair (J_0, ϕ) is regular i.e., the linearization*

$$E_u = D\bar{\partial}_J(u) : T_u \mathcal{P}_\phi \to \mathcal{L}_\phi$$

is surjective for all $u \in \mathcal{M}_{J,\phi}$.

Proposition 4.5 [Compactness]. *Under the same hypotheses as in Proposition 4.4, the zero dimensional component of $\widehat{\mathcal{M}}_{J,\phi}$ is compact and the one dimensional component $\widehat{\mathcal{M}}_{J,\phi}$ is compact up to the splitting of 2-trajectories.*

Proof (Assuming Proposition 4.4): This is an immediate consequence of the above regularity and Proposition 3.7 [O2]. Indeed, in the formula (3.3) in [O2]

$$I = \sum_{i=1}^{k} \mathrm{Index}(u_i) + 2\sum_j c_1(v_j) + \sum_l \mu(w_l),$$

each term in the right hand side is positive if the corresponding set is non-empty. The first term is positive due to the regularity proven above in Proposition 4.4. The second and the third terms are positive by the monotonicity of $(\mathbb{C}P^n, \mathbb{R}P^n)$. Therefore if we assume $I = 1$ or 2, then the sets $\underline{v} = \{v_j\}$ and $\underline{w} = \{w_l\}$ must be empty (Recall that we assume that $\Sigma_L \geq 3$.) and hence the proposition. Indeed if u_α be a sequence in $\mathcal{M}_{J,\phi}(x,y)$ with $I = 1$ or 2, then its weak limit in the sense of Proposition 3.7 [O2] is a genuine trajectory or a 2-trajectory respectively since the weak limit has the same index I and so bubbling cannot occur by the above observation. □

Propositon 4.6 [Vanishing]. *Under the same hypotheses as in Proposition 4.4, the number $\#\big(\widehat{\mathcal{M}}_{J,\phi}(x,y)\big)$ is even for all x, y and so the coboundary operator $\delta : \mathcal{C}^* \to \mathcal{C}^*$ vanishes.*

In the next section, we will prove Proposition 4.4 and 4.6 and so Theorem B in the introduction.

5 Proof of the main theorem

In this section, we prove Thorem B by proving Proposition 4.4 and 4.6.

Theorem 5.1. *Let* (P, ω) *as in Theorem B and* L *be a real form of the Hermitian symmetric space* (P, ω, J)*. Then*

$$I^*(L : P) \cong H^*(L, \mathbb{Z}_2) \cong \mathbb{Z}_2^{n+1}.$$

Corollary 5.2. *For any exact isotopy* $\phi = \{\phi_t\}_{0 \le t \le 1}$ *of* (P, ω) *such that* L *meets* $\phi_1(L)$ *transversally, we have*

$$\#\big(L \cap \phi_1(L)\big) \ge \mathrm{SB}\,(L, \mathbb{Z}_2).$$

The remaining section will be spent to prove Theorem 5.1.

We fix the pair (J_0, ϕ) as in Proposition 4.4. Note that

$$\phi_1(L) = \mathrm{Fix}\,(\phi_1 \sigma \phi_1^{-1}).$$

Due to the choice of ϕ, $\phi_1 \sigma \phi_1^{-1}$ is again an anti-homomorphic involutive isometry. We denote it by

$$\sigma_1 = \phi_1 \sigma \phi_1^{-1}.$$

For any given $u \in \mathcal{M}_{J_0, \phi}(x, y)$, we define

$$u_1(\tau, t) = \sigma_1 u(\tau, 1 - t).$$

This u_1 is J_0-holomorphic and

$$\begin{aligned} u_1(\tau, 0) &= \sigma_1 u(\tau, 1) = u(\tau, 1) \in \phi_1(L) \\ u_1(\tau, 1) &= \sigma_1 u(\tau, 0) = \phi_1 \sigma \phi_1^{-1}\big(u(\tau, 0)\big) \\ &= \phi_1 \sigma (\sigma \phi_1 \sigma)\big(u(\tau, 0)\big) \quad \text{from} \quad (3.1) \\ &= \phi_1^2\big(u(\tau, 0)\big) \end{aligned}$$

since $u(\tau, 0) \subset \mathrm{Fix}\,\sigma$. Now denote

$$\sigma_2 = \phi_1^2 \sigma (\phi_1^2)^{-1}.$$

We reflect u_1 along $\phi_1^2(L)$ using σ_2 and define

$$\begin{aligned} u_2(\tau, t) &:= \sigma_2 u_1(\tau, 1 - t) = \sigma_2 \sigma_1 u\big(\tau, 1 - (1 - t)\big) \\ &= \sigma_2 \sigma_1 u(\tau, t) = \phi_1^2 \sigma \phi_1^{-1} \sigma \phi_1^{-1}\big(u(\tau, t)\big) \\ &= \phi_1^2 \phi_1 \phi_1^{-1}\big(u(\tau, t)\big) = \phi_1^2 u(\tau, t). \end{aligned}$$

Here we used (3.1) for the 5-th identity. By repeating this procedure, we get

$$u_{2k} = \phi_1^{2k} u \quad \text{and} \quad u_{2k+1} = \phi_1^{2k} u_1$$

for $k = 0, 1, \ldots$. And we have from this expression.

$$u_{2k}(\tau, 1) = u_{2k+1}(\tau, 0).$$

Since both u_{2k} and u_{2k+1} are J_0-holomorphic, they can be smoothly glued together across $u_{2k}(\tau, 1) = u_{2k+1}(\tau, 0)$. By gluing $u, u_1, \ldots, u_{2^N} = u$ consecutively together, we have constructed a J_0-holomorphic map f from $\mathbb{R} \times [0, 2^N]$ to P such that

$$f(\tau, 0) = f(\tau, 2^N)$$

and

$$\int |\nabla f|^2 = 2^N \int |\nabla u|^2 < \infty. \tag{5.1}$$

We can consider f as a J_0-holomorphic map from the cylinder C_{2^N} to P where C_{2^N} has the complex structure induced from $\mathbb{R} \times i[0, 2^N] \subset \mathbb{C}$. Since C_{2^N} is conformally equivalent to $\mathbb{C}P^1 \backslash \{0, \infty\}$, (5.1) makes the J_0-holomorphic map $f \circ \chi : \mathbb{C}P^1 \backslash \{0, \infty\} \to P$ extended smoothly to a J_0-holomorphic map $v : \mathbb{C}P^1 \to P$ by the removal singularity theorem (see e.g., [O1]) since $\int |\nabla (f \circ \chi)|^2 = \int |\nabla f|^2 < \infty$. Here χ denotes a conformal map from $\mathbb{C}P^1 \backslash \{0, \infty\}$ to C_{2^N}.

Now, we are ready to prove Proposition 4.4. However, before proving it, we first prove Proposition 4.6 sssuming Proposition 4.4. For the simplicity of notation, we will omit the subscript 0 from J_0 from now on.

Proof of Proposition 4.6 (Assuming Proposition 4.4): By Proposition 5.1, the number $\langle x, \delta y \rangle = \#$ (isolated trajectories) in $\mathcal{M}_{J,\phi}(x, y)$ is finite. We shall show that any $u \in \widehat{\mathcal{M}}_{J,\phi}(x, y)$ comes in a pair $(u, \overline{u})$ where

$$\overline{u} = \phi^{2^l} u(z) \qquad 1 < l \leq N - 1,$$

and in particular $\langle x, \delta y \rangle \equiv 0 \mod 2$.

First, note that

$$\begin{aligned} (\phi_1^{2^{N-1}} u)(\tau, 0) &= \phi_1^{2^{N-1}} \big(u(\tau, 0)\big) \subset \phi_1^{2^{N-1}}(L) \\ (\phi_1^{2^{N-1}} u)(\tau, 1) &= \phi_1^{2^{N-1}} \big(u(\tau, 1)\big) \subset \phi_1^{2^{N-1}+1}(L) \end{aligned} \tag{5.2}$$

since $\phi^{2^N} = id$ (and so $\phi^{2^{N-1}+1}(L) = \phi_1(L)$). However, we have

$$\phi_1^{2^{N-1}}(L) = L.$$

Indeed, for any $p \in L$,

$$\sigma\phi_1^{2^{N-1}}(p) = \sigma\phi_1^{2^{N-1}}\sigma(p) = (\phi_1^{2^{N-1}})^{-1}(p) = \phi_1^{2^{N-1}}(p)$$

and so $\phi_1^{2^{N-1}}(p) \in \operatorname{Fix}\sigma = L$ for all $p \in \operatorname{Fix}\sigma = L$. Therefore we have

$$\phi_1^{2^{N-1}}u \in \mathcal{M}_{J,\phi}(x,y).$$

If $\phi_1^{2^{N-1}}u \not\equiv u$, then we choose $\phi_1^{2^{N-1}}u$ for $\overline{u}$ since $\phi_1^{2^{N-1}}u$ cannot be a translation of u (recall that the flow ψ_t is perpendicular to L and so $\psi_t(L)$ for any t). If $\phi_1^{2^{N-1}}u \equiv u$, then it can be easily shown similarly as before

$$\phi_1^{2^{N-2}}u \in \mathcal{M}_{J,\phi}(x,y).$$

If $\phi_1^{2^{N-2}}u \not\equiv u$, then we take $\phi_1^{2^{N-2}}u$ for $\overline{u}$. If $\phi_1^{2^{N-2}}u \equiv u$, then repeat the above procedure.

By choosing N sufficiently large in the beginning, we may assume that

$$\phi_1^2 u \neq u$$

and so the above procedure terminates at some $1 < l \leq N-1$ and hence we can take $\phi_1^{2^l}u$ for $\overline{u}$ for the l. □

Now we prove Proposition 4.4.

Proof of Proposition 4.4: Let us first explicitly write down the linearization E_u exploiting the fact that the almost complex structure J integrable. We denote by ∇ the Levi-Civita connection of the Fubini-Study metric $g(\cdot,\cdot) = \omega(\cdot, J\cdot)$. Then this connection is a torsion-free Hermitian connection with respect to the almost complex structure J. Let $u \in \mathcal{M}_{J,\phi}(x,y)$, i.e.,

$$\begin{cases} \dfrac{\partial u}{\partial \tau} + J(u)\dfrac{\partial u}{\partial t} = 0 \\ u(\tau,0) \subset L, u(\tau,1) \subset \phi_1(L) \quad \text{and} \quad \displaystyle\int |\nabla u|^2 < \infty. \end{cases}$$

We know by the elliptic regularity, u is C^∞. For each $\xi \in T_u\mathcal{P}$, we compute

$$E_u(\xi) = \nabla_s\big|_{s=0}\overline{\partial}_J(u_s)$$

where $\overline{\partial}_J = \frac{\partial}{\partial\tau} + J\frac{\partial}{\partial t}$, and $u_0 = u$ and $\frac{d}{ds}|_{s=0}\, u_s = \xi$.

$$\begin{aligned} \nabla_s\big|_{s=0}\overline{\partial}_J(u_s) &= \nabla_s\big|_{s=0}\left(\frac{\partial u_s}{\partial\tau} + J(u_s)\frac{\partial u_s}{\partial t}\right) \\ &= \nabla_s\frac{\partial u_s}{\partial\tau}\Big|_{s=0} + \nabla_s\left(J(u_s)\frac{\partial u_s}{\partial t}\right)\Big|_{s=0} \\ &= \nabla_s\frac{\partial u}{\partial\tau}\Big|_{s=0} + \nabla_s J(u_s)\big|_{s=0}\frac{\partial u}{\partial t} + J(u)\nabla_s\frac{\partial u}{\partial t}\Big|_{s=0} \\ &= \nabla_\tau\frac{\partial u}{\partial s}\Big|_{s=0} + J(u)\nabla_t\frac{\partial u}{\partial s}\Big|_{s=0} \text{ since } J \text{ is parallel} \\ &= \nabla_\tau\xi + J(u)\nabla_t\xi = (\nabla_\tau + J(u)\nabla_t)\xi. \end{aligned}$$

Hence, we have proved

$$E_u(\xi) = \big(\nabla_\tau + J(u)\nabla_t\big)\xi. \tag{5.3}$$

Recall that our eventual goal is to prove

$$\operatorname{Coker} E_u = \{0\} \quad \text{i.e.,} \quad \operatorname{Ker}\,(E_u)^* = \{0\}$$

where $(E_u)^*$ is the adjoint of E_u. Therefore, we want to characterize the elements in Ker $(E_u)^*$. The operator $(E_u)^*$ is characterized by the equation

$$\langle \xi, (E_u)^*\eta\rangle_2 = \langle E_u\xi, \eta\rangle_2$$

for all $\xi \in T_u\mathcal{P}_\phi$ and $\eta \in \mathcal{L}_\phi(u)$. For notational convenience, we denote $\Theta = \mathbb{R}\times I$ and then we have

$$\begin{aligned}
\langle E_u\xi, \eta\rangle_2 &= \int_\Theta \langle \nabla_\tau\xi + J(u)\nabla_t\xi, \eta\rangle \\
&= \int_\Theta \langle \nabla_\tau\xi, \eta\rangle + \int_\Theta \langle J(u)\nabla_t\xi, \eta\rangle \\
&= -\int_\Theta \langle \xi, \nabla_\tau\eta\rangle + \int_\Theta \omega(\nabla_t\xi, \eta)
\end{aligned}$$

by integration by parts and the identity $\omega(\cdot,\cdot) = \langle J\cdot,\cdot\rangle$.

The second term becomes

$$\begin{aligned}
\int_\Theta \omega(\nabla_t\xi, \eta) &= \int_{-\infty}^{\infty}\int_0^1 \frac{d}{dt}\omega(\xi,\eta) - \omega(\xi, \nabla_t\eta)dtd\tau \\
&= \int_{-\infty}^{\infty} \omega(\xi,\eta)|_0^1 d\tau - \int_\Theta \omega(\xi, \nabla_t\eta) \\
&= \int_{-\infty}^{\infty} \omega(\xi,\eta)|_{t=1} - \omega(\xi,\eta)|_{t=0} d\tau + \int_\Theta \langle \xi, J(u)\nabla_t\eta\rangle.
\end{aligned}$$

Therefore,

$$\langle E_u\xi, \eta\rangle_2 = \int_{-\infty}^{\infty} \omega(\xi,\eta)|_{t=1} - \omega(\xi,\eta)|_{t=0} d\tau + \int_\Theta \langle \xi, (-\nabla_\tau + J(u)\nabla_t)\eta\rangle dtd\tau.$$

If $\eta \in$ Coker $E_u =$ Ker $(E_u)^*$, then $\langle E_u\xi, \eta\rangle = 0$ for all $\xi \in T_u\mathcal{P}$ and hence we have

$$(-\nabla_\tau + J(u)\nabla_t)\eta = 0 \tag{5.4}$$

and

$$\begin{aligned}
&\eta(\tau, 1) \subset T\big(\phi_1(L)\big) = T\phi_1 \cdot T\big(\mathbb{R}P^n\big) \\
&\eta(\tau, 0) \subset TL.
\end{aligned} \tag{5.5}$$

Here for the boundary condition, we used the fact that L and $\phi_1(L)$ are Lagrangian and that

$$\xi(\tau,0) \subset TL, \xi(\tau,1) \subset T(\phi_1(L))$$

for $\xi \in T_u\mathcal{P}$ by Definition 4.1 (2) of $\mathcal{P}$. Hence we have

$$\begin{aligned} \text{Coker } E_u = \{\eta \in L^2_{k-1}(\Theta, u^*TP) | - \nabla_\tau\eta + J(u)\nabla_t\eta = 0 \\ \eta(\tau,1) \subset T\phi_1(TL) \\ \eta(\tau,0) \subset TL\}. \end{aligned}$$

Again by the elliptic regularity, any $\eta \in$ Coker E_u is smooth. Recall the construction of the J-holomorphic map $f : C_{2N} \to P$ in the proof of Proposition 4.4.

By the same reason as for the case of f, for each element $\eta \in$ Coker E_u, we can reflect and glue together to get a smooth vector field

$$V : C_{2N} \to TP$$

with $\pi \circ V(\theta) = f(\theta)$, and with $||V||_{1,2} < \infty$. It follows from the Cauchy-Riemann equation which V satisfies and from $||V||_{1,2} < \infty$ that

$$\lim_{\tau\to\pm\infty} |V(\tau)| = 0. \tag{5.6}$$

Now consider the complexified tangent bundle $T_\mathbb{C}P = TP \otimes \mathbb{C}$. Due to the complex structure on P, $T_\mathbb{C}P$ is decomposed into

$$T_\mathbb{C}P = E \oplus \overline{E}$$

where E is a holomorphic vector bundle which is isomorphic to TP, and $\overline{E}$ is the conjugate bundle to E i.e., the holomorphic vector bundle with opposite complex structure to E. Due to the Hermitian structure to TP, $\overline{E}$ is isomorphic to $E^* \simeq T^*P$ as a holomorphic vector bundle. Since f is J-holomorphic, the pullback of $T_\mathbb{C}P$ by f preserves the splitting i.e.,

$$f^*T_\mathbb{C}P = f^*E \oplus f^*\overline{E}.$$

Here f^*E is isomorphic to f^*TP and $f^*\overline{E}$ is isomorphic to f^*T^*P.

Under the identification of E and TP, the equation

$$-\nabla_\tau\eta + J(u)\nabla_t\eta = 0$$

becomes

$$0 = -\nabla_\tau\eta + i\nabla_t\eta = \nabla_{(-\frac{\partial}{\partial\tau}+i\frac{\partial}{\partial t})}\eta$$

and so η defines an anti-holomorphic section on f^*E i.e., a holomorphic section on $f^*\overline{E}$, and so on f^*T^*P. Using the conformal equivalence $\chi : \mathbb{C}^1\backslash\{0,\infty\} \to C_{2N}$, $V \circ \chi : \mathbb{C}P^1\backslash\{0,\infty\} \to TP$ defines an anti-holomorphic section on $(f \circ \chi)^*(TP)$ with $||V \circ \chi||_{1,2} < \infty$. Therefore, by the removal of isolated singularity, $V \circ \chi$ can be smoothly extended to an anti-holomorphic section $\Xi : \mathbb{C}P^1 \to v^*TP$ by defining

$$\Xi(0) = \Xi(\infty) = 0.$$

If η is non-trivial, then Ξ is non-trivial and so we have produced a non-trivial holomorphic section with some zeros on $v^*(T^*P)$ via the isomorphism between $v^*\overline{E}$ and $v^*(T^*P)$. Since v is non-trivial, this contradicts Lemma 2.4. Therefore there is no non-trivial element η in Coker E_u, which finishes our proof that Coker $E_u = \{0\}$. □

This finishes the proof of Theorem 5.1.

References

[A] Arnold, V.I., *Sur une propriété topologique des applications globallmant canoniques de la mécanique classique*, C. R. Acad. Sci. Paris **261** (1965), 3719–3722.

[BH] Borel, A., Hirzebruch, F., *Characteristic classes and homogeneous spaces I*, Amer. J. Math. **80** (1958), 458–538.

[CJ] Chang, K-C., Jiang, M.Y., *The Lagrange intersections for* ($\mathbb{C}P^n$, $\mathbb{R}P^n$), Manuscripta Math. **68** (1990), 89–100.

[D] Düistermaat, J., *Convexity and tightness for restrictions of hamiltonian functions to fixed points of an anti-symplectic involution*, Am. Math. Soc. Trans. **275** (1983), 417–429.

[F1] Floer, A., *Morse theory for Lagrangian intersections*, J. Diff. Geom. **28** (1988), 513–547.

[F2] Floer, A., *The unregularized gradient flow of the symplectic action*, Comm. Pure Appl. Math. **41** (1988), 775–813.

[F3] Floer, A., *Symplectic fixed points and holomorphic spheres*, Commun. Math. Phys. **118** (1988), 215–240.

[F4] Floer, A., *Cuplength estimates for Lagrangian intersections*, Comm. Pure Appl. Math. **42** (1989), 407–462.

[Gi] Givental, A., *Periodic maps in symplectic topology*, Funct. Anal. Appl. 23:4 (1989), 37–52.

[Gr] Gromov, M., *Pseudo-holomorphic curves in symplectic manifolds*, Invent. Math. **81** (1985), 307–347.

[GH] Griffiths, R., Harris, J., *Principles of Algebraic geometry*, Wiley, New York, 1978.

[He] Helgason, S., *Differential Geometry, Lie Groups and Symmetric Spaces*, Academic Press, New York, 1978.

[Ho1] Hofer, H., *Lagrangian embeddings and critical point theory*, Ann. Inst. H. Poincaré, Analyse Nonlinéaire **2** (1985), 407–462.

[Ho2] Hofer, H., *Lusternik-Schnirelman theory for Lagrangian intersections*, Ann. Inst. H. Poincaré, Analyse Nonlinéaire **51** (1988), 465–499.

[Ki] Kirwan, F., Cohomology of Quotients in Symplectic and Algebraic Geometry, Math. Notes 31, Princeton University Press, Princeton, 1984.

[Ko] Kobayashi, S., *Transformation Groups in Differential Geometry*, Academic Press, New York, 1962.

[MZ] Mok, N., Zhong, J.-G., *Curvature characterization of compact Hermitian symmetric spaces*, J. Diff. Geom. **23** (1986), 15–67.

[O1] Oh, Y.-G., *Removal of boundary singularities of pseudo-holomorphic curves with Lagrangian boundary conditions*, Comm. Pure Appl. Math. **45** (1992), 121–139.

[O2] Oh, Y.-G., *Floer cohomology of Lagrangian intersections and pseudo-holomorphic disks I*, Comm. Pure Appl. Math. **46** (1993), 949–994.

[O3] Oh, Y.-G., *Floer cohomology of Lagrangian intersections and pseudo-holomorphic disks II:* $(\mathbb{C}P^n, \mathbb{R}P^n)$, Comm. Pure Appl Math. **46** (1993), 995–1012.

[T] Takeuchi, M., *Stability of certain minimal submanifolds of compact Hermitian symmetric spaces*, Tohoku Math. J. **36** (1984), 293–314.

Yong-Geun Oh
Department of Mathematics
University of Wisconsin
Madison
Wisconsin 53706, USA

An obstacle to non-Lagrangian intersections*

L. Polterovich

1 Introduction and main results

In the sixties V.I. Arnold made a conjecture that (under certain assumptions) the number of intersection points of a Lagrangian submanifold with its image under an arbitrary Hamiltonian isotopy is not less than the sum of its Betti numbers. This conjecture became one of the main forces stimulating the development of Symplectic Topology. Though the complete solution is not yet found, the confirmative answer was obtained in various important cases. Omitting here the historical survey (see e.g. [McD2]) we mention only that one of the most remarkable contributions was made by Andreas Floer [F].

The aim of the present paper is to explain why the assertion of Arnold's conjecture does not in general remain true if, instead of Lagrangian submanifolds, *arbitrary middle-dimensional* ones are considered. More precisely, let L be a Lagrangian submanifold of a symplectic manifold M and let ξ be the Hamiltonian field generated by a smooth function $H : M \to \mathbb{R}$. The following local observation provides in fact the main motivation for the Lagrangian intersections conjecture:

There is a one-to-one correspondence between critical points of the restriction H on L and points where ξ is tangent to L.

In particular, if L is *closed* then *every Hamiltonian field on* M *is somewhere tangent to* L.

In the present paper we show that this property breaks for a wide class of middle-dimensional closed *non-Lagrangian* submanifolds of a symplectic manifold (see Theorem 1.2 below). In particular this class includes non-Lagrangian submanifolds which are sufficiently C^1-close to a Lagrangian one (see Corollary 1.6 below). As an immediate consequence we obtain that such a non-Lagrangian submanifold can be pushed away from itself by a certain arbitrary small Hamiltonian isotopy (see Theorem 1.12 below where this phenomenon is described in terms of Hofer's displacement energy). Therefore non-Lagrangian intersections already disappear on the infinitesimal level.

Our strategy is the following. We reduce the search of a transversal Hamiltonian vector field to the investigation of some partial differential relation (see §2) and prove the existence of a solution using certain h-principles (see §3).

---

*) A preliminary version of this paper appeared as a preprint (IHES/M/91/61, September 1991) entitled «Transversal Hamiltonian fields and differential relations».

Acknowledgements. I am deeply grateful to H. Hofer for many stimulating discussions during my stay at Ruhr-Universität, Bochum (this visit was partially supported by DSF-SFB 237). I appreciated the hospitality of Université Louis Pasteur (Strasbourg) and IHES (Bures-sur-Yvette) where the work on this paper was continued. I am thankful to V.I. Arnold, M. Audin, M. Gromov, D. McDuff, D. Salamon, J.-C. Sikorav and D. Spring for useful discussions.

2° Main theorem

We start with some basic notions of symplectic geometry (see [A], [A-G] for the detailed exposition). Symplectic structure on an even dimensional manifold M^{2n} is a closed differential 2-form, say ω, whose top power ω^n does not vanish at every point. A symplectic form defines an isomorphism between the tangent and cotangent bundles of the manifold as follows:

$$I : TM \to T^*M, \quad \xi \mapsto i_\xi \omega ,$$

where i denotes the interior product.

Therefore every smooth function $H : M \to \mathbb{R}$ generates a vector field

$$\xi = -(I^{-1})(dH)$$

which is called a *Hamiltonian field.*

A linear subspace $\ell \subset T_{(\cdot)}M^{2n}$ is called *Lagrangian* if $\dim \ell = n$ and the symplectic form ω vanishes on ℓ. A middle-dimensional submanifold $L \subset M$ is called *Lagrangian* if its tangent spaces are Lagrangian at every point. We shall say that a middle-dimensional submanifold is *non-Lagrangian* if its tangent space is not Lagrangian at least at one point.

In order to formulate our main result we need the following notion which was introduced by J.-C Sikorav [S1, Th. 2.4].

Definition 1.1. Let M be a symplectic manifold and let $L \subset M$ be a submanifold of middle dimension: $\dim L = \dim M/2$. The submanifold L is called *normal* if there exists a field of Lagrangian subspaces along L which is transversal to L.

Theorem 1.2. *Let M be a symplectic manifold and let L ⊂ M be a closed normal submanifold satisfying the following conditions:*

1) *L is non-Lagrangian;*
2) *Euler characteristic of L vanishes.*

Then there exists a Hamiltonian field on M which at every point is non-tangent to L.

A convention on terminology

We shall say that a vector field, say v, is *transversal* to a submanifold L if at every point it is non-tangent to L:

$$v(x) \notin T_x L \quad \text{for all} \quad x \in L .$$

Remark 1.3 Condition 2) of Theorem 1.2 cannot be removed since if L is normal and the Euler characteristic of L is not zero, then there is no vector field along L transversal to L.

Remark 1.4 We believe that the normality assumption can be removed in the following way.

Conjecture
Suppose that a closed middle-dimensional non-Lagrangian submanifold of a symplectic manifold admits a transversal vector field. Then it admits a transversal *Hamiltonian* vector field. (*Added on April, 1994* : This conjecture was recently proved in a paper F. Laudenbach, J.-C. Sikorav, Disjonction Hamiltonienne et limites de sous-variétés Lagrangiennes, Int. Math. Research Notices No. 4 (1994).)

Theorem 1.2 is proved in 4° §1.

3° **Examples of normal submanifolds**
Let M^{2n} be a symplectic manifold endowed with the symplectic structure ω. We say that an almost complex structure J on M is *compatible* with ω if the bilinear form

$$(\xi,\eta) \mapsto \omega(\xi, J\eta), \quad \xi,\eta \in T_{(\cdot)}M$$

is symmetric and positive definite at every point of M. It is well known that every symplectic manifold admits a compatible almost complex structure.

Proposition 1.5 *Every Lagrangian submanifold of M is normal.*

Proof For a Lagrangian submanifold $L \subset M$ the field $JT_{(\cdot)}L$ is a field of transversal Lagrangian subspaces (here J is a compatible almost complex structure on M). □

Corollary 1.6 *Let $L \subset M$ be a Lagrangian submanifold. Then every submanifold of M which is sufficiently C^1-close to L is normal.*

Another class of normal submanifolds can be constructed as follows. Let J be an almost complex structure on M which is compatible with the symplectic structure. Recall that a submanifold $L \subset M$ is called *totally real* if $\dim L = \dim M/2$ and subspaces $T_{(\cdot)}L$ and $JT_{(\cdot)}L$ are transversal at every point of L.

Proposition 1.7 *Let (M,ω) be a symplectic manifold. Let J be an almost complex structure on M compatible with ω. Suppose that $L \subset M$ is a totally real submanifold of M. Moreover assume that L is parallelizable. Then L is normal.*

Remark 1.8 It is not clear if the proposition remains true for non-parallelizable L.

Proof of Proposition 1.7 The almost complex structure J on M allows us to consider every tangent space $T_{(\cdot)}M$ as a complex one.

Define a Hermitian structure on M as follows:

$$\langle \xi, \eta \rangle = \omega(\xi, J\eta) + i\omega(\xi, \eta), \quad \xi, \eta \in T_{(\cdot)}M .$$

Since L is parallelizible one can choose a field of tangent frames $(e_1, \dots, e_n)$ on L (here $n = \dim L$). Since L is totally real then $(e_1, \dots, e_n)$ is a *complex* frame at every point of L. Let $(f_1, \dots, f_n)$ be a new complex frame which is obtained from $(e_1, \dots, e_n)$ by the standard orthogonalization procedure:

$$f_1 = e_1, \quad f_k = e_k - \sum_{j=1}^{k-1} \frac{\langle e_k, f_j \rangle}{\langle f_j, f_j \rangle} f_j \quad \text{for } k \geq 2 .$$

Obviously the subspace $\mathrm{Span}_{\mathbb{R}}(Jf_1, \dots, Jf_n)$ is Lagrangian. Moreover one can easily check that it is transversal to L at every point. The proof is complete. □

4° A normal form for a normal submanifold

Consider the cotangent bundle T^*L endowed with the standard symplectic structure Ω (see [A-G] for the definition). Denote by $\pi : T^*L \to L$ the natural projection. Let σ be a closed 2-form on L. Define a new symplectic form Ω_σ on T^*L as follows:

$$\Omega_\sigma = \Omega + \pi^*\sigma .$$

Proposition 1.9 (cf. [S1, Th. 2.4]). *Let L be a closed normal submanifold of a symplectic manifold (M, ω). Denote by σ the restriction $\omega|_L$. Then there exists a neighbourhood $\mathcal{U}$ of L in M and an embedding $\varphi : \mathcal{U} \to T^*L$ such that $\varphi(L)$ coincides with the zero section of T^*L and $\varphi^*\Omega_\sigma = \omega$.*

Proof We identify L with the zero section of T^*L. Since L is normal in M there exists a Lagrangian subbundle $\mathcal{L}$ of T_LM which is transversal to TL. Consider the following diagram:

$$\begin{array}{ccccccc}
T_LM & \xrightarrow{i_1} & \mathcal{L} \oplus TL & \xrightarrow{\Phi} & T^*L \oplus TL & \xrightarrow{i_2} & T_LT^*L \\
\downarrow & & \downarrow & & \downarrow & & \downarrow \\
L & \xrightarrow{1} & L & \xrightarrow{1} & L & \xrightarrow{1} & L
\end{array}$$

Here vertical arrows are natural projections, i_1 and i_2 are canonical identifications and an isomorphism Φ is defined as follows:

$$\Phi(\xi \oplus \eta) = i_\xi \omega \oplus \eta .$$

Here $i_{\cdot}\cdot$ denotes the interior product. Note that $\Phi^*\Omega_\sigma = \omega$. Indeed,

$$\begin{aligned}\Omega_\sigma\big(\Phi(\xi_1 \oplus \eta_1), \Phi(\xi_2 \oplus \eta_2)\big) &= \Omega_\sigma(i_{\xi_1}\omega \oplus \eta_1, i_{\xi_2}\omega \oplus \eta_2) = \\ &= \omega(\xi_1, \eta_2) - \omega(\xi_2, \eta_1) + \sigma(\eta_1, \eta_2) = \\ &= \omega(\xi_1 \oplus \eta_1, \xi_2 \oplus \eta_2)\ .\end{aligned}$$

Note that $\Phi|_{TL} = \mathbf{1}$. We conclude that there exists a diffeomorphism, say Ψ from a neighbourhood of L in M to T^*L such that $\Psi|_L = \mathbf{1}$ and $(\Psi^*\Omega_\sigma)\big|_{T_LM} = \omega\big|_{T_LM}$ (choose Ψ such that $\Psi_* = \Phi$ on T_LM). Thus the Darboux-Weinstein theorem (see [A-G]) implies that symplectic structures ω and Ω_σ are diffeomorphic in the neighbourhood of L by a diffeomorphism preserving L. This implies the proposition. □

Theorem 1.10 *Let L be a closed manifold whose Euler characteristic vanishes. Let σ be a closed 2-form on L which does not vanish identically. Then there exists a Hamiltonian vector field on (T^*L, Ω_σ) which is transversal to the zero section.*

Theorem 1.10 is proved in §3.

Proof of Theorem 1.2 Theorem follows immediately from Theorem 1.10 and Proposition 1.9. □

5° Symplectic displacement energy

Recently H. Hofer [H] introduced an interesting quantity, the so-called displacement energy, associated with every subset of a symplectic manifold. This quantity measures how large the variation of a compactly supported non-autonomous Hamiltonian function has to be in order to push this subset away from itself by a time-one map of the corresponding Hamiltonian flow. Let us give the precise definition. Consider a symplectic manifold (M, ω). Let $\mathcal{C}$ be the space of all compactly supported smooth functions on $M \times [0; 1]$. Assume that the segment $[0;1]$ is endowed with the time-coordinate t. For every function $H \in \mathcal{C}$ define the time-dependent Hamiltonian vector field ξ_H^t as follows:

$$\xi_H^t = -(I^{-1})(dH(\cdot, t))\ .$$

The flow g_H^t generated by the field ξ_H^t is called the *Hamiltonian flow*.

Definition 1.11 (H. Hofer). *Displacement energy* $e(A)$ of a subset $A \subset M$ is equal to

$$\inf\{\max_{M\times[0;1]} H - \min_{M\times[0;1]} H \,\big|\, \text{all } H \in \mathcal{C} \text{ such that } g_H^1 A \cap A = \emptyset\}\ .$$

Note that in general $e(A)$ belongs to $[0; +\infty]$.

Theorem 1.12 *Let M be a symplectic manifold and let $L \subset M$ be a closed normal non-Lagrangian submanifold whose Euler characteristic vanishes. Then displacement energy of L vanishes: $e(L) = 0$.*

Proof Let H be a function on M whose Hamiltonian field is transversal to L (see Theorem 1.2). One can choose such H compactly supported on M. Obviously for every positive sufficiently small $\delta > 0$ the following holds:

$$g^1_{\delta \cdot H} L \cap L = \emptyset .$$

Thus $e(L) = 0$. □

Remark 1.13 Given a closed Lagrangian submanifold $L \subset M$ with vanishing Euler characteristics, one can C^k-approximate it by normal non-Lagrangian submanifolds (here $k \geq 1$). Suppose that the displacement energy of L is positive. We conclude that L is a *point of discontinuity* of displacement energy which is considered as a functional $e : 2^M \to [0; +\infty]$. Note that there exist closed Lagrangian submanifolds with positive displacement energy (e.g. the standard Lagrangian torus in $\mathbb{C}^n$, see [S2]). We refer the reader to [P] for the detailed discussion.

6°

The rest of the paper is devoted to the proof of Theorem 1.10. In §2 we find a differential relation corresponding to the existence of a transversal Hamiltonian field. In §3 we prove the existence of a solution using certain h-principles.

2 Transversality relation

Let L be a closed manifold endowed with a closed differential 2-form σ. Consider the cotangent bundle T^*L with the symplectic structure Ω_σ (see 4° §1). We denote a point of T^*L by a pair (p, q) where $q \in L$ and $p \in T^*_q L$. Given a function $H : L \to \mathbb{R}$ and a vector field v on L, define function $H^v : T^*L \to \mathbb{R}$ as follows:

$$H^v(p, q) = H(q) + \langle p, v(q) \rangle ,$$

where $\langle . , . \rangle$ is the natural pairing.

The main result of this section is the following

Theorem 2.1 *The Hamiltonian field of H^v with respect to the symplectic structure Ω_σ is transversal to the zero section of the cotangent bundle if and only if a 1-form*

$$dH + i_v \sigma$$

on L is non-singular (i.e. at every point it does not vanish).

Here and below $i_{\cdot}\,\cdot$ denotes the interior product.

Proof Let (p, q) be canonical local coordinates on T^*L. In these coordinates the standard symplectic form Ω on T^*L can be written as $dp \wedge dq$. We shall identify vectors and covectors using the Euclidean scalar product. For a 2-form, say β, we denote by $\hat{\beta}$ the skew-symmetric matrix such that $\beta(\xi, \eta) = (\hat{\beta}\xi, \eta)$ for every vector ξ, η. Note that with this language $i_\xi \beta = \hat{\beta}\xi$.

Obviously the following relation holds:

$$\hat{\Omega}_\sigma = \begin{pmatrix} 0 & -\mathbf{1} \\ \mathbf{1} & \hat{\sigma} \end{pmatrix}$$

(we use here (p, q) coordinates).

Let $\eta = (\eta_p, \eta_q)$ be the Hamiltonian field of the function H^v with respect to the symplectic structure Ω_σ. By definition, $dH^v = -i_\eta \Omega_\sigma$. Our aim is to compute η on the zero section $\{p = 0\}$. Using the previous equation we obtain

$$\begin{pmatrix} v \\ \frac{\partial H}{\partial q} \end{pmatrix} = - \begin{pmatrix} 0 & -1 \\ \mathbf{1} & \hat{\sigma} \end{pmatrix} \begin{pmatrix} \eta_p \\ \eta_q \end{pmatrix} .$$

Thus $-\eta_p = \frac{\partial H}{\partial q} + \hat{\sigma} v$.

Note that the Hamiltonian field η is transversal to the zero section if and only if η_p at every point does not vanish. Thus we obtain (in local coordinates) that for every point q

$$\frac{\partial H}{\partial q}(q) + \hat{\sigma} v(q) \neq 0 .$$

Globally this means that the form $dH + i_v \sigma$ is non-singular. □

3 Solution of the transversality relation via h-principles

1°

The main result of this section is the following

Theorem 3.1 *Let L be a closed manifold whose Euler characteristic vanishes. Let σ be a closed 2-form on L which does not vanish identically. Then there exists a function $H \in C^\infty(L)$ and a smooth vector field v on L such that the 1-form*

$$dH + i_v \sigma$$

is non-singular on L.

Note that Theorem 1.10 follows immediately from Theorems 3.1 and 2.1. The rest of this section is devoted to the proof of Theorem 3.1.

2° Localisation of the problem

Step 1. Recall that σ is a closed 2-form on L. For a point $q \in L$ denote by E_q the kernel of σ:

$$E_q = \{\xi \in T_qL \mid i_\xi \sigma = 0\} .$$

The codimension of E_q is called the *rank* of σ at the point q. Note that the rank of σ is an *even* number since σ defines a symplectic form on T_qL/E_q. Since σ does not vanish identically on L one can choose a point, say $\mathbb{O} \in L$, such that rank of σ is constant and positive in some neighbourhood, say K of $\mathbb{O}$. We shall denote this rank by r. Note that $r \geq 2$ due to the previous discussion.

It is well known that distribution E_q is integrable in K provided the form σ is closed (see e.g. [A-M-R], p. 439). Denote by $\mathscr{E}$ the foliation defined by E_q, and set

$$E_q^{\perp} = \{\xi \in T_q^*L \mid \xi \text{ vanishes on } E_q\} .$$

The following (obvious) property of the foliation $\mathscr{E}$ is useful for our purposes.

Proposition 3.2 *The image of the linear map*

$$T_qL \to T_q^*L, \quad v \mapsto i_v\sigma$$

coincides with $E_q^{\perp}$. Restriction of this map to every r-dimensional subspace in T_qL transversal to E_q is an isomorphism.

Denote by n the dimension of L. Decreasing, if necessary, the neighbourhood K, one can introduce local coordinates $q_1 \dots q_n$ in the neighbourhood of $\mathbb{O}$ such that the following holds:

- The neighbourhood K is a cube

$$\{|q_j| < 1 \ , \ j = 1 \dots n\}$$

  with the center at $\mathbb{O}$.
- The foliation $\mathscr{E}$ is defined by the equations

$$q_1 = \text{const} , \dots q_r = \text{const} .$$

Step 2. We claim that there exists a function $h : L \backslash \mathbb{O} \to \mathbb{R}$ with the following properties.

- The 1-form dh does not vanish on $L \backslash \mathbb{O}$.
- The restriction of dh on the boundary ∂K can be smoothly extended to a non-singular (not closed !) 1-form on K.

In order to prove this recall that smooth functions with non-vanishing differential on an open manifold satisfy the h-principle (see [G], p. 45). In our situation this means the following. Given a non-singular 1-form, say α on $L \backslash \mathbb{O}$, one can deform it in the class of non-singular forms to the differential dh of some function $h : L\backslash\mathcal{O} \to \mathbb{R}$. Since the Euler characteristic of L vanishes, there exists a non-singular 1-form, say α' on L. Choose now the 1-form α as the restriction of α' on $L\backslash\mathcal{O}$ and deform it to dh according to the previous discussion. Note that the forms $dh\big|_{\partial K}$ and $\alpha'\big|_{\partial K}$ are homotopic in the class of non-singular 1-forms. Thus $dh\big|_{\partial K}$ admits a non-singular extension into K since α' does. Our claim is proved.

Step 3. Denote by $\mathcal{U}$ a neighbourhood of ∂K in Closure (K). The purpose of our localization program is to reduce Theorem 3.1 to the Proposition 3.3 below. We shall use notations of steps 1 and 2.

Proposition 3.3 *There exists a smooth function $\tilde{h} : K \to \mathbb{R}$ and a 1-form α on K such that the following holds:*

1) $\tilde{h}\big|_{\mathcal{U}} = h\big|_{\mathcal{U}}$;
2) $\alpha \in E_q^{\perp}$ *at every point* $q \in K$;
3) $\alpha\big|_{\mathcal{U}} \equiv 0$;
4) *the 1-form* $d\tilde{h} + \alpha$ *is non-singular on* K.

Proposition 3.3 is proved below in 3°.

Proof of Theorem 3.1 Let $\tilde{h}$ and α be as in Proposition 3.3. Proposition 3.2 implies that there exists a vector field v on K such that $\alpha = i_v\sigma$ and v identically vanishes in $\mathcal{U}$. Extend v outside K by the zero field. Let H be a smooth function on L such that

$$H = h \quad \text{outside} \quad K;$$
$$H = \tilde{h} \quad \text{in} \quad K.$$

We claim that $dH + i_v\sigma$ is a non-singular 1-form on L. Indeed,

$$dH + i_v\sigma = dh \quad \text{outside} \quad K;$$
$$dH + i_v\sigma = d\tilde{h} + \alpha \quad \text{in} \quad K.$$

The right-hand sides in both expressions are non-singular due to our construction. This completes the proof. □

3° Solution of the localized problem

Below we prove Proposition 3.3. We shall use notations of 2°. Note that condition 2) of Proposition 3.3 means that in local coordinates the 1-form α can be written as follows

$$\alpha = \sum_{i=1}^{r} \alpha_i(q)dq_i \ ,$$

where r is the rank of σ. Recall that $r \in [2; n]$. Without loss of generality we can assume that $r = 2$.

Let us emphasize now the following feature of our problem. The relation $d\tilde{h} + \alpha \neq 0$ gives us a restriction on the functions $\{\alpha_i\}$ themselves but not on their 1-jets. Therefore the techniques of [G] cannot be applied directly to the solution of this relation. It seems to be an interesting question how to formulate an h-principle which works in such a situation.

However we circumvent this difficulty by searching α_1 and α_2 in some special form, namely

$$\alpha_j = \sum_{i=1}^{n} \frac{\partial y_j}{\partial q_i} \quad \text{for} \quad j = 1, 2 \ ,$$

where y_1 and y_2 are smooth functions on K.

Proposition 3.4 *Let K be a cube $\{|q_j| < 1,\ j = 1, \dots, n\}$. Let $\mathcal{U}$, $\mathcal{U}'$ be neighbourhoods of ∂K in Closure (K) such that $\mathcal{U}' \supset$ Closure $(\mathcal{U})$. Suppose that h is a smooth function on $\mathcal{U}'$ with the following properties:*

- *the 1-form dh is non-singular on $\mathcal{U}'$;*
- *the 1-form dh admits a non-singular extension into K in the class of non-singular 1-forms.*

Then there exist smooth functions y_1, y_2, y_3, on K such that

$$y_3|_{\mathcal{U}} = h, \quad y_1|_{\mathcal{U}} = y_2|_{\mathcal{U}} \equiv 0$$

and the following relation holds:

$$\begin{pmatrix} \frac{\partial y_3}{\partial q_1} & + & \sum_{i=1}^{n} \frac{\partial y_1}{\partial q_i} \\ \frac{\partial y_3}{\partial q_2} & + & \sum_{i=1}^{n} \frac{\partial y_2}{\partial q_i} \\ & \frac{\partial y_3}{\partial q_3} & \\ & \vdots & \\ & \frac{\partial y_3}{\partial q_n} & \end{pmatrix} \neq 0$$

at every point $q \in K$.

Proof of Proposition 3.3 Let y_1, y_2, y_3 be as in Proposition 3.3. Set $\tilde{h} = y_3$ and $\alpha_j = \sum_{i=1}^{n} \frac{\partial y_j}{\partial q_i}$ for $j = 1, 2$. Set $\alpha = \alpha_1 dq_1 + \alpha_2 dq_2$. Obviously $\tilde{h}$ and α satisfy conditions 1)-4) of Proposition 3.3. This completes the proof. □

Proof of Proposition 3.4 Let $\theta : X \to K$ be a trivial fibration with the fiber $\mathbb{R}^3(y_1, y_2, y_3)$. Denote by X^1 the 1-jets bundle associated to θ. Every point of X^1 is represented by a triple (a, y, q) where a is a matrix (a_{ij}), $i = 1 \dots n$; $j = 1, 2, 3$. Each number a_{ij} corresponds to the partial derivative $\frac{\partial y_j}{\partial q_i}$.

Denote by $\widetilde{\Sigma}$ the subset of $\mathbb{R}^{3n}$ which is defined by the following equations:

$$\begin{cases} a_{13} + \sum_{i=1}^{n} a_{i1} = 0 \\ a_{23} + \sum_{i=1}^{n} a_{i2} = 0 \\ a_{33} = 0 \\ \dots \\ a_{n3} = 0 \end{cases}$$

where a_{ij}, $i = 1 \dots n$; $j = 1, 2, 3$ are coordinates in $\mathbb{R}^{3n}$. Define a subset $\Sigma \subset X^1$ as follows:

$$\Sigma = \{(a, y, q) \in X^1 \mid a \in \widetilde{\Sigma}\} .$$

For a section $y : K \to X$ denote by j^1_y its 1-jet extension. Note that j^1_y is a section $K \to X^1$. Let y^0 be a section $\mathcal{U}' \to X$ such that $y^0 = (0, 0, h)$. We can reformulate our problem as follows:

$$\left.\begin{array}{c} \text{To find a section } y : K \to X \text{ such that} \\ y = y^0 \text{ on } \mathcal{U} \text{ and } j^1_y(K) \subset X^1 \backslash \Sigma \end{array}\right\} \quad (\star)$$

In order to solve this problem we use the method of convex integration invented by M. Gromov (see [G]). One can find a brief and very clear exposition of this method in a paper by D. McDuff (see [McD1, §2]). We refer the reader to [McD1] for necessary notions and definitions. The method of convex integration prescribes the following procedure for establishing the existence of a solution for the problem $(\star)$.

Step 1. Existence of a formal solution. We have to check that there exists a map $\Psi : K \to X^1 \backslash \Sigma$ such that $\Psi|_{\mathcal{U}} = j^1_{y^0}$. Existence of such a map follows from assumptions on the function h.

Step 2. Ampleness of the relation $X^1 \backslash \Sigma$.

Denote by S_i, $i = 1, \dots, n$ the family of parallel 3-dimensional affine subspaces in $\mathbb{R}^{3n}$ which is determined by equations:

$$a_{jk} = \text{const} \quad \text{where} \quad j \neq i \ ; \ k = 1, 2, 3 \ .$$

In other words, coordinates a_{i1}, a_{i2}, a_{i3} are free, while all others are fixed. In order to check *ampleness* of the relation $X^1 \backslash \Sigma$ it is sufficient to prove that for every $i = 1, \dots, n$ and for every subspace $\ell \in S_i$ the intersection $\ell \cap \widetilde{\Sigma}$ has codimension more than 1 in ℓ:

$$\text{Codim}_\ell (\ell \cap \widetilde{\Sigma}) > 1$$

(see [McD1], p. 112]).

Consider two cases

1) $i = 1$ (the case $i = 2$ is analogous). Then either

$$\ell \cap \tilde{\Sigma} = \emptyset \text{ or } \ell \cap \tilde{\Sigma} = \left\{ (a_{11}, a_{12}, a_{13}) \;\middle|\; a_{13} + a_{11} = \text{const}, \; a_{12} = \text{const} \right\} .$$

Thus $\mathrm{Codim}_\ell(\ell \cap \tilde{\Sigma}) = 2$.

2) $i \geq 3$. Then either

$$\ell \cap \tilde{\Sigma} = \emptyset \text{ or } \ell \cap \tilde{\Sigma} = \left\{ (a_{i1}, a_{i2}, a_{i3} \;\middle|\; a_{i1} = \text{const}, \; a_{i2} = \text{const}, a_{i3} = 0 \right\} .$$

Thus $\mathrm{Codim}\,(\ell \cap \tilde{\Sigma}) = 3$.

We conclude that the relation $X^1 \backslash \Sigma$ is ample. Then Gromov's theorem (see [McD1, Th. 2.1]) implies solvability of the problem $(\star)$. This completes the proof. □

References

[A] V.I. Arnold, *Mathematical methods of classical mechanics*, Grundlehren 250, Springer, Berlin-Heidelberg-New-York, 1978.

[A-G] V.I. Arnold, A.B. Givental, «Symplectic geometry», in *Dynamical systems* **4**, *Encyclopedia of Math. Sciences*, Springer, 1990, 1–136.

[A-M-R] R. Abraham, J.E. Marsden, T. Ratiu, Manifolds, tensor analysis and applications, Springer, Berlin-Heidelberg-New-York, 1988.

[F] A. Floer, Morse theory for Lagrangian intersections, J. Diff. Geometry 28 (1988), 513–547.

[G] M. Gromov, *Partial differential relations*. Springer Ergebnisse 9, 1986.

[H] H. Hofer, On the topological properties of symplectic maps, Proceedings of the Royal Society of Edinburgh, 115A (1990), 25–38.

[McD1] D. McDuff, Application of convex integration to symplectic and contact geometry, Ann. Inst. Fourier 37 (1987), 107–133.

[McD2] D. McDuff, Elliptic methods in Symplectic Geometry, Bulletin A.M.S 23 (1990), 311–358.

[P] L. Polterovich, Symplectic displacement energy for Lagrangian submanifolds, Ergod. Th. & Dynam. Syst. 13 (1993), 357–367.

[S1] J.-C. Sikorav, Quelques propriétés des plongements Lagrangiennes, Preprint, Orsay (1990).

[S2] J.-C. Sikorav, Systèmes Hamiltoniens et topologie symplectique, ETS Editrice, Pisa, 1990.

Leonid Polterovich
Raymond and Beverly Sackler Faculty of Exact Sciences
School of Mathematical Sciences, Tel Aviv University
Ramat Aviv, 69978, Israel

A Mayer-Vietoris model for Donaldson-Floer theory

C.H. Taubes*)

Introduction

Andreas Floer [F] introduced a novel cohomology theory for a compact, oriented, 3-manifold with the homology groups of S^3. (A homology sphere.) Subsequently, Donaldson [D1] (see [A], [B], [DFK]) discovered that Floer cohomology has an intimate relationship to his gauge theoretic 4-manifold invariants [D2], [DK], [FM]. More precisely, Donaldson originally defined integer-valued invariants for compact, oriented 4-manifolds. He then found that his invariants naturally generalize to invariants of 4-manifolds with homology sphere boundary with one large difference: The generalized invariant takes values in the Floer cohomology of the boundary 3-manifold.

The purpose of this article is to present a reasonably simple model where the basic relationship between Floer cohomology and Donaldson invariants can be studied in abstraction. For example, the setting produces the famous $\mathbb{Z}/8$-grading and also the novel phenomena which arise when 3-manifolds with rational homology are considered.

The basic premise will be that Floer cohomology should be derived from considerations of the Donaldson invariant. Floer cohomology arises as the answer to a particular question. Meanwhile, this question is naturally posed by the use of a certain strategy to compute Donaldson invariants. All of this is explained below.

If the actual gauge theoretic constructions can be framed in terms of the abstract model in this article, then the definitions of Floer cohomology, the relative Donaldson invariants, and their relation to the Donaldson invariant for a boundary free 4-manifold, all will follow as essentially algebraic consequences. The question, of course, is whether such a framing is possible.

On one level, the answer to this question is no, mainly because the model below assumes that various spaces are compact, while the spaces which arise in gauge theory are emphatically not compact. This non-compactness stems from the famous «bubbling off» phenomenon that was first described by [U]. However, it is a fundamental fact (discovered by Donaldson and formalized in [FM]) that this sort of non-compactness is, in fact, irrelevant to many of the (essentially homological) considerations which appear.

The point is that the compactness assumptions below are, for the most part, only technical simplifications; the basic structures found in the model below persist in the gauge theoretic context.

---

*) Supported in part by the NSF under DMS-89-07995

In fact, on another level, the answer to the question above is yes. In a separate article [T1], the author describes Floer cohomology and its relation to Donaldson invariants for 3-manifolds which are topologically non-trivial circle bundles over Riemann surfaces. The description in [T1] succeeds in framing the discussion in what are essentially the terms below (modulo the aforementioned compactness issues). The analysis in [T1] is for non-trivial circle bundles only, but with just minor (and simplifying) notational changes, the presentation works for the homology sphere case as well. (The basic strategy in [T1] should work for the general 3-manifold, but only a careful attention to the details will tell for sure.)

The ensuing discussion presents a model of Floer cohomology which is different from the by now standard model of finite dimensional Morse theory. (See [W1], [C]. However, the approach parallels developments in [CJS].) There are also models of Floer cohomology and the Donaldson invariant which are based on topological field theory. (See [W2], [AJ], [BS].) The view below does not stem from field theory considerations either. (Although it is most probably true that the field theory language can be made to accommodate the constructions that follow.)

The model below also has some intellectual roots with ideas of Austin and Braam [AB] and Austin, Braam and Keck [ABK] for generalizing Floer cohomology to general 3-manifolds. While this paper was in preparation, the author received a preprint from Donaldson [D3] presenting ideas which are related to those reported here.

The description of the model starts in Section 2 below. Section 1 describes some of the basic phenomena which appear in Donaldson/Floer theory for non-trivial circle bundles over Riemann surfaces. (See [T1] for details). The Floer cohomology for these manifolds enjoys all of the properties of Floer cohomology for homology spheres plus some novel features which arise because they have rational first homology. It is expected that these novel features will occur generally for 3-manifolds with homology.

1 Nontrivial circle bundles

A non-trivial circle bundle, M, over a Riemann surface is determined up to oriented diffeomorphism by a non-zero integer k and a non-negative integer g. The latter is the genus of the surface M/S^1 and the former is the degree of M as a circle bundle over M/S^1. Change the orientation and k changes to $-k$ but g stays the same.

Unless $g = 0$, a circle bundle M will have non-trivial rational homology in dimensions 1 and 2 (the first and second Betti numbers are $2 \cdot g$), and the existence of this homology makes for a surprising generalization of the basic picture of Floer homology as a $\mathbb{Z}/8$-graded vector space. (Other authors have suggested generalizations of Floer's ideas to manifolds with non-trivial first homology, see [AB], [ABK], [Fuk] and [Fur]. Gauge theoretic aspects of non-trivial circle bundles are also considered in [T2])

Here is a taste of the most radical new phenomenon: Floer cohomology for a non-trivial circle bundle M is not a single vector space (as it is for a homology 3-sphere). Rather, there is a Floer cohomology for each point in a product of two Grassmannians. The first Grassmannian is the space of linear subspaces of $H_1(M;\mathbb{R})$, and the second Grassmannian is the space of linear subspaces of $H_2(M;\mathbb{R})$.

The Floer cohomology for the pair of subspaces $(0,0)$ most closely corresponds to the Floer cohomology for homology 3-spheres. This $(0,0)$ Floer cohomology is that of a square zero differential on a certain $\mathbb{Z}/8\mathbb{Z}$-graded $\mathbb{R}$-module, $\mathcal{C}$, of (Alexander) cocycles for the space of representation of $\pi_1(M)$ into $SU(2)$. (This $(0,0)$ Floer cohomology realizes, in a sense, some suggestions of Austin, Braam and Keck [ABK] for generalizing Floer cohomology to 3-manifolds with homology.)

In general, when V_1 is a linear subspace of $H_1(M;\mathbb{R})$ and V_2 is a linear subspace of $H_2(M;\mathbb{R})$, the Floer cohomolgy for (V_1,V_2) is computed from a differential on

$$\mathcal{C}\otimes \mathrm{Sym}^*(V_1)\otimes \Lambda^*(V_2). \tag{1}$$

(Here, $\mathrm{Sym}^*(V)\equiv \oplus_{p\geq 0}\mathrm{Sym}^p(V)$ and $\Lambda^*(V)\equiv \oplus_{p\geq 0}\Lambda^p(V)$ are the symmetric and antisymmetric powers of V.)

By the way, these Floer cohomologies do not appear to make a natural module for the group $Diff_+(M)$ of orientation preserving diffeomorphisms of M. Rather, there is a natural extension of $Diff_+(M)$ by the vector space

$$\mathrm{Hom}(H_1(M),H_2(M))\oplus \mathrm{Hom}(H_2(M),H_3(M)) \tag{2}$$

(real coefficients please) which naturally acts on these Floer cohomologies to cover the action of $Diff_+(M)$ on the product of the Grassmannians for $H_1(M)$ and $H_2(M)$.

This sort of generalization of Floer cohomology arises naturally if one takes as a postulate that the Floer cohomology should be the receptacle for Donaldson's invariant on a 4-manifold with boundary. (This is an inversion of the historical perspective.)

These generalized Floer cohomologies for a non-trivial circle bundle relates to Donaldson's invariants for a boundary-free manifold when the circle bundle (call it M) appears embedded in a 4-manifold W so as to split W in two as

$$W=W_-\cup W_+. \tag{3}$$

Here $W_\pm$ are 4-manifolds with M as boundary and which intersect as M. (Embeddings of non-trivial circle bundles appear ubiquitously in 4-manifolds as boundaries of neighborhoods of embedded surfaces with homologically non-trivial self-intersection.)

Donaldson's invariant for W can be computed from the Floer cohomology for a certain (V_1, V_2) in the product of Grassmannians. The subspaces (V_1, V_2) are

$$\begin{aligned} V_1 &\equiv \ker(H_1(M) \longrightarrow H_1(W_-) \oplus H_1(W_+)), \\ V_2 &\equiv \ker(H_2(M) \longrightarrow H_2(W_-) \oplus H_2(W_+)). \end{aligned} \tag{4}$$

The point is that there is a «relative» Donaldson invariant for W_- with values in the Floer cohomologies over

$$Gr(\ker(H_1(M) \to H_1(W_-)) \times Gr(\ker(H_2(M) \to H_2(W_-)). \tag{5}$$

(Here, $Gr(V)$ denotes the Grassmannian of all linear subspaces of V.) And, there is a «relative» Donaldson invariant for W_+ with values in the duals $(\mathrm{Hom}(\cdot; \mathbb{R}))$ to the Floer cohomologies over

$$Gr(\ker(H_1(M) \to H_1(W_+)) \times Gr(\ker(H_2(M) \to H_2(W_+)). \tag{6}$$

Then, the Donaldson invariant for W is obtained by pairing the cohomology with its dual over the pair (V_1, V_2) of (4). (This (V_1, V_2) sits in both (5) and (6).)

Note that there is a natural isomorphism between the (V_1, V_2) Floer cohomology for a non-trivial circle bundle with the dual to said Floer cohomology for M's orientation reversed twin.

The details for the preceding discussion will appear in [T1].

2 The Donaldson invariant

The Donaldson invariant arises in the following idealized context: An oriented manifold $\mathcal{M}$ is given with a canonical (up to homotopy) map, φ, to some fiducial space $\mathcal{B}$. This $\mathcal{B}$ has known cohomology. Using pullback by φ, the fundamental class of $\mathcal{M}$ pairs with $\mathcal{B}$'s cohomology. With this understood, the Donaldson invariant is the homomorphism

$$\mathcal{D} : H^*(\mathcal{B}) \longrightarrow \mathbb{R} \tag{7}$$

given by

$$\mathcal{D}(\omega) \equiv \langle \varphi^* \omega, \mathcal{M} \rangle \tag{8}$$

where $\langle \ , \ \rangle$ is the natural pairing between cohomology and homology.

Comment: In the context of a 4-manifold, W, the space $\mathcal{M}$ is defined as the space of equivalence classes of solutions to a certain differential equation on W. (This is the anti-self dual equation for a connection on a principle bundle over W.) The space $\mathcal{B}$ is the space of equivalence classes of all connections on bundles over W, and φ is the inclusion. The equivalence relationship is that induced by bundle equivalence.

As given, $\mathcal{M}$ has components $\{\mathcal{M}_m\}_{m\geq 0}$ of increasing dimensions. All but a few of the components are non-compact, so it is a subtle matter to prove that $H^*(\mathcal{B})$ pairs naturally with $\mathcal{M}$. The standard approach involves a natural compactification of $\mathcal{M}$ (as in [DK] or [FM]). Pretend here that each component of $\mathcal{M}$ is compact.

By the way, in the 4-manifold context, $H^*(\mathcal{B})$ is a free algebra which is isomorphic to $\mathrm{Sym}^*(H_{\mathrm{even}}(W)) \otimes \Lambda^*(H_{\mathrm{odd}}(W))$ with the generators graded as their Poincaré duals (see [DK].)

3 The fibered product

The baby model space for the discussion of the Donaldson invariant is the fibered product. Consider manifolds Q, P and Z with maps

$$f : Q \longrightarrow Z \quad \text{and} \quad g : P \longrightarrow Z. \tag{9}$$

The fibered product, $Q \times_Z P$ sits in $Q \times P$ as the inverse image by $(f \times g)$ of the diagonal in $Z \times Z$. That is,

$$Q \times_Z P \equiv \{(q,p) \in Q \times P : f(q) = g(p)\}. \tag{10}$$

Comment: The fibered product is relevant in the 4-manifold context when W is split as in (3). Here is why: Remember that each point in $\mathcal{M}$ is a solution to a differential equation on W. With respect to (3), any $A \in \mathcal{M}$ can be thought of as a pair (A_-, A_+), with A_- solving an equation (the anti-self dual equation) on W_- and with A_+ solving the equation on W_+. Of course, the solutions must agree in the middle on M. This view presents $\mathcal{M}$ as the fibered product of (10) where Q is the space of solutions on W_- and P is the space of solutions on W_+, and Z is the space of boundary values on M. The maps f and g are simply restriction to M, with M viewed, respectively, as a subset of W_- and W_+.

The fibered product of (10) will be a manifold when the maps f and g are transversal to each other, i.e., when the diagonal in $Z \times Z$ is a regular submanifold for the map $(f \times g)$ from $Q \times P$.

4 Push-forward and pull-back

Assume now that $\mathcal{M}$ in (8) is a fibered product as in (10) where f and g are transversal. Let ω be a closed form on $\mathcal{M}$ which can be written as

$$\omega = \omega_1 \wedge \omega_2 \tag{11}$$

with ω_1 a cocycle on Q and ω_2 a cocycle on P.

The calculation of (8), an integral over $\mathcal{M}$, can be simplified (in principle) when ω is given by (11). To simplify (8), integrate over the Q coordinates first, and then over P. The result is

$$\mathcal{D}(\omega) \equiv \langle g^* f_*(\omega_1) \wedge \omega_2, P \rangle, \tag{12}$$

where $f_*(\cdot)$ is push-forward of cohomology from Q and g^* is the usual pull-back of cohomology by g from P. (Thus, $f_*(\omega_1)$ is the result of integrating over Q first.)

Comment: When $f : Q \to Z$ is a fibration, cohomology push-forward, f_*, is simply integration over the fibers of f. Integration over the fibers also defines form push-forward for fibrations [BT]. Section 1 of [T1] is dedicated to the general subject of pushing forward forms by smooth maps.

Borrowing from [T1], remark that (12) makes sense when P is oriented and when f's differential, a section of $\mathrm{Hom}(TQ, f^*TZ)$, is everywhere Fredholm with oriented index bundle in the real K-theory of Q. This insures that the fiber of f over a regular value in Z is an oriented, finite dimensional manifold.

By the way, an orientation for said index bundle and an orientation for P produces an orientation for the fibered product in (10).

Assume for the sake of simplicity that P is compact and that f's fibers in Q are also compact. In the 4-manifold context, this assumption holds in spirit but not in body.

5 Floer cohomology at its simplest

Floer cohomology is the answer to the following question: What sort of beast is $f_*(\omega_1)$ when ω_1 is a cocyle on Q? Put differently, integration over the fibers of f takes a cocycle on Q and produces one on Z. In its simplest incarnation, Floer cohomology should be defined as the receptacle for the push-forward of closed cocycles.

Comment: In the 4-manifold context, the push-forward of closed cocycles will result in the $(0,0)$-Floer cohomology. The cohomologies for a pair (V_1, V_2) in the product of Grassmannians for $H_1(m)$ and $H_2(m)$ will be discussed shortly.

Now, it is a standard canard that push-forward commutes with the coboundary, d:

$$f_*(d\omega) = df_*(\omega). \tag{13}$$

But, this canard is true if f's fibers are compact manifolds. If not, then (13) cannot be guaranteed.

When (13) holds, the Floer cohomology of Z will be the ordinary cohomology of Z. (And, if Floer homology is defined as the cohomology's dual, then it is the usual homology of Z.)

6 A full grown model

In the 4-manifold context, the decomposition of $\mathcal{M}$ in (10) does not prove useful for a number of reasons. (Although, perhaps (10) can be exploited using [D4].)

A more realistic model for Floer cohomology starts by assuming that

$$\mathcal{M} \equiv \bigcup_{n\geq 0} \mathcal{M}_n. \tag{14}$$

Comment: In the 4-manifold context, a useful decomposition of $\mathcal{M}$ can be had by using an «energy» function on $\mathcal{M}$ that can be associated to an open subset of W. This function measures the amount of curvature (in an L^2 sense) on the subset. In certain contexts, this energy function takes discrete values; this occurs, for example, when the subset is the whole (compact) manifold. In this case, the energy comes in multiples of a fixed unit, a unit whose value is fixed by a normalization for $W = S^4$.

The energy «quantization» is modeled by (14). But agree now to change the dictionary between Donaldson/Floer theory and (14) as follows: Introduce a sequence (indexed by integer $n \geq 0$) of manifolds

$$W_n \equiv W_- \cup ([0,n] \times M) \cup W_+, \tag{15}$$

where the boundary of W_- is identified with $\{0\} \times M$ and the boundary of W_+ is identified with $\{n\} \times M$. (Note that all W_n are mutually diffeomorphic.) The dictionary between (14) and Donaldson/Floer theory should now make $\mathcal{M}_n$ in (14) correspond to the union of components of $\mathcal{M}$ on W_n with n units of energy. The use of (15) is a key part of the strategy in [T1].

The goal now is a useful decomposition of $\mathcal{M}_n$ in (14). In this regard, suppose that one is given spaces

$$Q \equiv \bigcup_{n\geq 0} Q_n, \quad P \equiv \bigcup_{n\geq 0} P_n, \quad X \equiv \bigcup_{n\geq 1} X_n. \tag{16}$$

Suppose that these spaces come with maps

$$f \equiv \{f_n : Q_n \to Z\},\ g \equiv \{g_n : P_n \to Z\} \quad \text{and a pair } r_\pm \equiv \{r_{\pm n} : X_n \to Z\}. \tag{17}$$

Make the following transversality assumptions: Each $\{f_n\}$ is transversal to each $\{g_m\}$ and to each $\{r_{-n}\}$, while each $\{r_{+n}\}$ is transversal to each $\{r_{-m}\}$ and to each $\{g_m\}$.

Now, each component of Q, P and X will be closed, and the interior of each will be a smooth manifold. In general, each component will be a generalized simplex (a manifold with corners, like a high dimensional cube or simplex), in the sense of having faces of various codimensions.

Start with X. The space X_1 should be a compact manifold without boundary. More generally, the codimension 1 face of X_n should be given by

$$\partial^1 X_n = \Big(\bigcup_{p+q=n} (X_p \times_Z X_q)\Big)/\sim . \tag{18}$$

Here, the fiber product uses r_+ on the left side (X_p) and r_- on the right side (X_q). The equivalence $\sim$ identifies $(X_p \times_Z X_q) \times_Z X_r \subset \partial^1 X_{p+q} \times_Z X_r$ with the analogous $X_p \times_Z (X_q \times_Z X_r) \subset X_p \times_Z \partial^1 X_{q+r}$.

Implicit in (18) is a compatibility of the maps $r_\pm$. That is, $r_{(p+q)+}$ should restrict to $X_p \times_Z X_q$ in (18) as the restriction of r_{q+} from the product $X_p \times X_q$. Likewise, $r_{(p+q)-}$ should restrict as the restriction or r_{-p} from $X_p \times X_q$.

With X understood, here is Q: The space Q_0 should be a compact manifold without boundary. For $n > 0$, the codimension 1 face of Q_n should be given by

$$\partial^1 Q_n = \Big(\bigcup_{p+q=n} (Q_p \times_Z X_q)\Big)/\sim . \tag{19}$$

Here, the fiber product uses f on the left side (Q_p) and r_- on the right side (X_q). The equivalence $\sim$ identifies $(Q_p \times_Z X_q) \times_Z X_r \subset \partial^1 Q_{p+q} \times_Z X_r$ with the analogous $Q_p \times_Z (X_q \times_Z X_r) \subset Q_p \times_Z \partial^1 X_{q+r}$.

Implicit in (19) is a compatibility of the maps f. That is, $f_{(p+q)}$ should restrict to $Q_p \times_Z X_q$ in (19) as the restriction of r_{q+} from the product $Q_p \times X_q$.

Consider, finally, P: The space P_0 should be a compact manifold without boundary. For $n > 0$, the codimension 1 face of P_n should be given by

$$\partial^1 P_n = \Big(\bigcup_{p+q=n} (X_p \times_Z P_q)\Big)/\sim . \tag{20}$$

Here, the fiber product uses r_+ on the left side (X_p) and g on the right side (P_q). The equivalence $\sim$ identifies $(X_p \times_Z X_q) \times_Z P_r \subset \partial^1 X_{p+q} \times_Z P_r$ with the analogous $X_p \times_Z (X_q \times_Z P_r) \subset X_p \times_Z \partial^1 P_{q+r}$.

Implicit in (20) is a compatibility of the maps g. That is, $g_{(p+q)}$ should restrict to $X_p \times_Z P_q$ in (18) as the restriction of r_{q-} from the product $X_p \times P_q$.

With Q, P and X now understood, here is how they enter the discussion: Suppose that $\mathcal{M}_n$ can be decomposed as

$$\mathcal{M}_n \equiv \Big(\bigcup_{q+p=n} (Q_q \times_Z P_p) \Big) / \sim, \tag{21}$$

where the fiber product uses f on the left side (Q_q) and g on the right side (P_p). The equivalence $\sim$ identifies $(Q_q \times_Z X_p) \times_Z P_r \subset \partial^1 Q_{q+p} \times_Z P_r$ with the analogous $Q_q \times_Z (X_p \times_Z P_r) \subset Q_q \times_Z \partial^1 P_{q+r}$.

It is a consequence of the equivalences in (18–21) that (21) is a manifold.

Comment: In the 4-manifold context, here is how to think of Q, P and X: The energy function on an arbitrary open subset of some W_n will not, in general, take on discrete values. However, the values will be approximately discrete for the open subset if its value is nearly zero on a neighborhood of said subset's boundary.

With this understood, Q_n should correspond to part of $\mathcal{M}$ for the space $W_- \cup ([0, n] \times M)$; the part with practically no energy near $\{n\} \times M$ but with approximately n units of energy total. Meanwhile, P_n corresponds to part of $\mathcal{M}$ for the space $([0, n] \times M) \cup W_+$, the part with practically no energy near $\{0\} \times M$ but with approximately n units of energy total.

As for X_n, it should correspond to a part of $\mathcal{M}$ for the space $[0, n] \times M$. A codimension 1 slice of that part with little energy near $\{0\} \times M$ and near $\{n\} \times M$, but with roughly n units of energy in total.

With these correspondences understood, the boundary behavior in (18–20) corresponds to the ways in which the energy can redistribute itself on $W_- \cup ([0, n] \times M)$ or $([0, n] \times M) \cup W_+$ as it nears a violation of the (implicit) constraint that only a little energy sit near $\{n\} \times M$ or $\{0\} \times M$, respectively. For instance, near the boundary of Q_n, one can have the n units of energy distributed with approximately p units on $W_- \cup ([0, p] \times M)$, the remaining (approximately) $(n - p)$ units situated on $[p - n, n] \times M$, and very little near $\{p\} \times M$. This distribution corresponds to the part $Q_p \times_Z X_{n-p}$.

For a second instance, near the boundary of X_n, one can have the n units of energy distributing as, roughly, p units on $[0, p] \times M$, the remainder on $[p - n, n] \times M$ with very little near $\{p\} \times M$. The interpretation for $\partial^1 P_n$ is analogous.

With the preceding understood, the decomposition of $\mathcal{M}_n$ in (21) describes the ways in which the n units on $W_- \cup ([0, n] \times M) \cup W_+$ can distribute itself. For example, $Q_p \times_Z P_{n-p}$ consists of that part of $\mathcal{M}_n$ where roughly p units of energy sit on $W_- \cup ([0, p] \times M)$ while roughly $(n - p)$ units sit on $([p, n] \times M) \cup W_+$ and almost no energy is near $\{p\} \times M$.

By the way, the existence of such a «simplicial» decomposition of $\mathcal{M}_n$ as in (21) is by no means obvious. Its existence when M is a non-trivial circle bundle is the main technical construction in [T1].

7 Some simplifying assumptions

There are some assumptions about Q, P and X which simplify the ensuing discussion. Some are violated in the actual 4-manifold situation, but the violations are not relevant to the basic picture. In particular, a lot of headaches disappear if one assumes that each P_n is compact, and that f_n and r_{+n} are proper.

Assume that each P_n is oriented and that the differentials of the maps f_n (as bundle maps from TQ_n to f_n^*TZ) have oriented index bundles in the real K-theory of Q_n. Also, assume that the index bundle of the differential of r_{n+} is oriented in the real K-theory of X_n. Of course, all of these orientations should be compatible with (18) and (19).

Here are some dimension assumptions: Assume that the numerical index of the differential of r_{n+} is a constant on X_n. Require that this index on X_1, say s, be positive. And, assume that on X_n, this index has value

$$\text{index}(dr_{n+}) \equiv n \cdot (s+1) - 1. \tag{22}$$

(This is compatible with (18).) Require as well that the numerical indices of the differentials of the f_n be constant on Q_n. Require that $\text{index}(df_n)$ increase with n. Require that the dimension of the components of P_n be constant and increase with n. These last assumptions imply that the dimension of $\mathcal{M}_n$ increase with n. Assume that the dimension the components of $\mathcal{M}_n$ is constant.

With these assumption understood, remark that

$$\begin{aligned} \dim(\mathcal{M}_{q+p}) &= \text{index}(df_q) + \dim(P_p), \\ \text{index}(df_p) &= p \cdot s + \text{index}(df_0), \\ \dim(P_p) &= p \cdot s + \dim(P_0). \end{aligned} \tag{23}$$

8 A realistic Floer cohomology

Certain cohomology classes on $\mathcal{M}_n$ in (21) will arise by pull-back from certain spaces. In particular, assume that Q maps to a space $\mathcal{B}_-$ and that P maps to a space $\mathcal{B}_+$ and require that these maps be compatible with (19) and (20). That is, write the map φ_- as $\{\varphi_{n-} : Q_n \to \mathcal{B}_-\}$ and require that φ_{n-} on $Q_p \times_Z X_{n-p}$ should be the restriction of φ_p from $Q_p \times X_{n-p}$. Make the analogous requirement for $\varphi_+ \equiv \{\varphi_{n+} : P_n \to \mathcal{B}_+\}$.

The preceding requirements insure that $\varphi_\pm$ restrict from $Q \times P$ to define maps (by the same name) from $\mathcal{M}$ to $\mathcal{B}_\pm$.

Comment: In the 4-manifold context, the maps $\varphi_\pm$ are defined by restricting to $W_\pm$.

In the 4-manifold context, one meets «Donaldson invariants» of the form

$$\langle \varphi_-^* \omega_- \wedge \varphi_+^* \omega_+, \mathcal{M}_n \rangle. \tag{24}$$

These can be evaluated with the help of (21) by integrating first over Q and then over P. The result is given for (21)

$$\sum_{q+p=n} \langle g_p^* f_{q*}(\mu_-) \wedge \mu_+, P_p \rangle, \tag{25}$$

where $\mu_- \equiv \varphi_-^* \omega_-$ and $\mu_+ \equiv \varphi_+^* \omega_+$.

Comment: Since P_p is not a manifold, one must interpret it suitably as a cycle. One must also be careful about defining the push-forward f_{q*} since the fibers of f_q in Q_q are not manifolds either. These issues are dealt with in Section 1 of [T1]. Suffice it to say that one should think first of fibrations and integrating differential forms over the fibers.

Comment: The fact that neither P_p nor f_q's fibers are closed cycles is very relevant to all that follows.

Floer cohomology arises as the answer to the question: On Z, what sort of cocycle is

$$f_*(\mu_-) \equiv \sum_{q \geq 0} f_{q*}(\mu_-)? \tag{26}$$

Likewise, Floer homology answers the question: What sort of cycle is the linear functional (on Z's cocycles) which sends ω to

$$\langle g^* \omega \wedge \mu_+, P \rangle \equiv \sum_{p \geq 0} \langle g_p^* \omega \wedge \mu_+, P_p \rangle? \tag{27}$$

(Note that both sums contain only finitely many non-zero terms because of the dimension restrictions from the preceding subsection.)

Fundamental to the answer for (26) is the fact that f_* *does not* commute with the exterior derivative, d. This is because the fibers of f are not closed cycles. However, suppose that μ is pulled back from $\mathcal{B}_-$ by φ_-, or, more generally:

Proposition. *Let μ be a cocycle on Q whose restriction to $Q_q \times_Z X_p \subset \partial^1 Q_{q+p}$ equals the restriction from $Q_q \times X_p$ of μ's pull back by projection to Q_q. For such μ, one has*

$$f_*(d\mu) = Df_*(\mu). \tag{28}$$

Here, D is an operator on a subcomplex, $C^*(Z)$, of Z's cocycles which sends ω to

$$D\omega \equiv d\omega - \sum_{p \geq 1} r_{+p*}(r_{-p}^* \omega). \tag{29}$$

Proof. The derivation of (29) is a version of Stokes' theorem.

Note that the sum in (29) has only finitely many non-zero terms because of the positivity of s in (22). In fact, the integer n term in (29) decreases the degree of a form by $n \cdot (s+1) - 1$. Thus, the integer n term in (29) acts as zero on forms of degree smaller than $n \cdot (s+1) - 1$.

The fact that the integer n term decreases degree by $n \cdot (s+1) - 1$ implies that D only preserves the grading in $\mathbb{Z}/(s+1) \cdot \mathbb{Z}$.

Now, D is an operator on Z's cocycles, and another version of Stokes' theorem is that

$$D^2 = 0. \tag{30}$$

Comment: Technically, one should be careful in one's interpretation of (30), for there are issues about D's range having to do with the definition of push-forward. See Section 2 of [T1] for the details. If r_+ is a fibration, no technicalities occur. In any event, the technicalities can all be absorbed by an appropriate definition of C*(Z) and do not affect the basic structure of the arguments below. For this reason, it is safe to ignore them in the subsequent discussion, and for ease of exposition, this will be done.

Equation (30) asserts that d-closed forms on Q (subject to compatibility on $\partial^1 Q$) push forward by f_* to D-closed forms on Z. And, d-exact forms on Q (again, subject to $\partial^1 Q$ compatibility) push forward to D-exact forms on Z.

The cohomology of this D can be rightly called Floer cohomology. With this understood, the answer to (26) is that the image of f_* is Floer cohomology.

By the way, since D only preserves the grading in $\mathbb{Z}/(s+1) \cdot \mathbb{Z}$, so D's cohomology only has a $\mathbb{Z}/(s+1) \cdot \mathbb{Z}$-grading.

Comment: In the 4-manifold context where M is a non-trivial circle bundle, the analog of D's cohomology is the $(0,0)$-Floer cohomology that was alluded to in the beginning of the introduction.

When M is a homology 3-sphere, the space Z is a finite set of points and so the first term on (29)'s left is missing. In either case, the appropriate value of s in (22) is 7, so that Floer theory is $\mathbb{Z}/8\mathbb{Z}$-graded.

Here is a final remark about this D: Note that the change in grading for the n'th term in (29) increases with n. This allows for the construction of a spectral sequence (of a filtered compex as in [BT]) which computes D's cohomology from d's (the E_2-term.) The r'th differential is obtained by truncating (29) at $p = r - 2$. (This is, more or less, the spectral sequence which appears in [ABK].)

9 Floer homology

If Floer cohomology is the answer to (26), then Floer homology is the answer to (27); that is, if one defines Floer homology as the Abelian group of linear maps from D's cohomology to $\mathbb{R}$. The basic point is that (27) obeys

$$\langle g^* D\omega \wedge \mu, P\rangle \equiv -\langle g^*(-1)^{\bullet}\omega \wedge d\mu, P\rangle \tag{31}$$

if μ is a form on P whose restriction to $X_q \times_Z P_p \subset \partial^1 P_{q+p}$ is the restriction from $X_q \times P_p$ of the pull-back of μ via projection to P_p.

Equation (31) is another version of Stokes' theorem.

10 Extending d-cohomology

Section 1 alluded to an extension of Floer cohomology as an assignment of vector spaces to points in a product of Grassmannians. This extended version of Floer cohomology has its origins already in the simple context of the fiber product in (10) where Floer cohomology is simply the cohomology of Z. The extended d-cohomology will be defined abstractly here; its relevance to (10) will be explained in the next section.

Let $H_1 \subset H^3(Z)$ and $H_2 \subset H^2(Z)$ be linear subspace.

Comment: More generally, one could replace H^3 with H^{odd} and H^2 with H^{even}, but $H^{3,2}$ are relevant in the 4-manifold context.

Choose a basis $\{z_{1\alpha}\}$ of closed forms which generate H_1, and likewise a basis $\{z_{2\alpha}\}$ of closed forms to generate H_2.

In terms of the basis $\{z_{1\alpha}\}$, identify $\text{Sym}^*(H_1)$ with the polynomials in the indeterminants $\{s_\alpha\}$. Use $\{\partial_\alpha\}$ to denote differentiation by s_α. Use $\eta_\alpha : \Lambda^p(H_2) \to \Lambda^{p-1}(H_2)$ to denote interior multiplication along the direction defined by $z_{2\alpha}$. Let $\Theta : \Lambda^*(H_2) \to \Lambda^*(H_2)$ denote act as $(-1)^p$ on Λ^p. Let $C^*(Z)$ denote the forms on Z. Let $\{z_{1\alpha}\}$ and $\{z_{2\alpha}\}$ act on $C^*(Z)$ by wedge product. A differential on $C^*(Z) \otimes \text{Sym}^*(H_1) \otimes \Lambda^*(H_2)$ is given by

$$\underline{d} \equiv \Theta \cdot (d - z_{1\alpha} \cdot \partial_\alpha) - z_{2\alpha} \cdot \eta_\alpha. \tag{32}$$

An easy calculation sees $\underline{d}^2 = 0$.

Note that $\underline{d}$ preserves

$$C^*(Z) \otimes \text{Sym}^*(V_1) \otimes \Lambda^*(V_2) \tag{33}$$

whenever $V_{1,2} \subset H_{1,2}$ are linear subspaces. The cohomology of $\underline{d}$ on (33) will be denoted by $H^*_{\underline{d}}(V_1, V_2)$. It is the simplest analog of the (V_1, V_2) cohomology to which the introduction alluded.

Now, $\underline{d}$ depends on the choice of closed forms $(\{z_{1\alpha}\}, \{z_{2\alpha}\})$. A different choice produces a $\underline{d}'$ which differs from $\underline{d}$ by

$$\underline{d}' = U^{-1} \cdot d \cdot U \tag{34}$$

where U is an invertible operator on (33). Thus, the cohomologies of $\underline{d}$ and of $\underline{d}'$ are isomorphic. However, the isomorphism is not generally canonical, for U (on $H^*_{\underline{d}}(V_1, V_2)$ is unique up to an action (as a vector space) of

$$\text{Hom}(H_1; H^2(Z)) \oplus \text{Hom}(H_2; H^1(Z)). \tag{35}$$

Let $Gr(H_1) \times Gr(H_2)$ denote the product of the Grassmannians of linear subspaces of H_1 and H_2, respectively. The association to (V_1, V_2) in this product of $H^*_{\underline{d}}(V_1, V_2)$ defines an assignment,

$$H^*_{\underline{d}} \to Gr(H_1) \times Gr(H_2) \tag{36}$$

of a vector space to every point in $Gr(H_1) \times Gr(H_2)$. This assignment is analogous to the assignment of Floer cohomologies. The vector space over the point $(0,0)$ is the d-cohomology of Z. The group in (35) acts naturally as a group of automorphisms of this $H^*_{\underline{d}}$.

By the way, the fact that (35) acts as a group of automorphisms of $H^*_{\underline{d}}$ is related to the observation that the $(H_1 \times H_2)$-preserving subgroup of Diff(Z) does not act naturally on $H^*_{\underline{d}}$. (The group acts naturally on $Gr(H_1) \times Gr(H_2)$). Rather, an extension of this group by (35) acts.

11 The geometry of the extended cohomology

The cohomology of $\underline{d}$ on (33) arises in the following context: Return to the simple fiber product $\mathcal{M} = Q \times_Z P$ of (10). To begin the exposition, consider a class $\underline{z} \in H_1$ with the property that $f^*\underline{z} = 0 = g^*\underline{z}$. Represent $\underline{z}$ by a closed form, z, on Z and there are 2-forms, α on Q and β on P such that

$$f^*z = d\alpha \qquad \text{and} \qquad g^*z = d\beta. \tag{37}$$

Equation (37) implies that

$$\tau = \alpha - \beta \tag{38}$$

restricts to $\mathcal{M}$ as a *closed* 2-form. (Remember: $f = g$ on $\mathcal{M}$.)

With the preceding understood, consider how one might calculate the «Donaldson invariant» given by

$$\langle(\alpha - \beta)^n \wedge \omega_1 \wedge \omega_2, \mathcal{M}\rangle \tag{39}$$

by exploiting the decomposition of (10). Here, ω_1 is a closed form on Q and ω_2 is a closed form on P.

Clearly, one should split the $(\alpha - \beta)^n$ in (39) into a part from Q and a part from P; and here is how: Introduce an indeterminant s and let ∂ denote differentiation by s. Then, write (39) as

$$\langle \exp(-\beta \cdot \partial)(\alpha + s)^n \wedge \omega_1 \wedge \omega_2, \mathcal{M} \rangle|_{s=0}. \tag{40}$$

(This is Taylor's theorem.) Finally, evaluate (40) by first integrating over Q and then P to give

$$\langle \exp(-\beta \cdot \partial) \cdot g^* f_*((\alpha + s)^n \wedge \omega_1) \wedge \omega_2, P \rangle|_{s=0}. \tag{41}$$

Here, the result of Q integration is

$$f_*((\alpha + s)^n \wedge \omega_1), \tag{42}$$

an element in $C^*(Z) \otimes \mathbb{R}[s]$.

The cocycle in (42) is not closed unless $z \equiv 0$ since α is not a closed cocyle on Q. However, (42) is annihilated by

$$d - z \cdot \partial. \tag{43}$$

Furthermore, change α to $\alpha + d\nu$, and change ω_1 to $\omega_1 + d\nu'$ and (42) changes by $(d - z \cdot \partial)$·(something). Thus, (42) defines an element in $H^*_{\underline{d}}(V_1, V_2)$ for any V_1 which contains z and for any V_2.

Meanwhile, a linear functional on $C^*(Z) \otimes \mathbb{R}[s]$ is defined by sending μ to

$$\langle \exp(-\beta \cdot \partial) \cdot g^* \mu \wedge \omega_2, P \rangle|_{s=0} \tag{44}$$

A version of Stokes' theorem will prove that a changing β to $\beta + d\nu$ and changing ω_2 to $\omega_2 + d\nu'$ will have no effect on (44) when $\underline{d}\mu = 0$. And, if μ is changed to $\mu + (d - z \cdot \partial)$·(anything), then (44) will not change.

In summary, (44) defines an element of $\mathrm{Hom}(H^*_{\underline{d}}(\mathbb{R} \cdot \underline{z}, 0); \mathbb{R})$. Furthermore, (42) defines an element in $H^*_{\underline{d}}(\mathbb{R} \cdot \underline{z}, 0)$ and the value of (39) is equal to the value of the former on the latter.

12 Relative Donaldson invariants in $\underline{d}$-(co)homology

The push-forward in (42) is a model for a «relative Donaldson invariant» with values in Floer cohomology; while the linear functional in (44) is a model for a relative Donaldson invariant with values in Floer homology. These relative invariants have an intricate algebraic structure and the reader is referred to [T1]'s sequel. (To frame the issue, note that the restriction in (8) to the symmetric algebra generated by $H^2(\mathcal{B})$ makes $\mathcal{D}$ an element in Hom(Sym$^*(H^2(\mathcal{B}); \mathbb{R})$. With this understood, the question is one of interpreting the relative Donaldson invariants in a similar vein, as Hom of something to something else.)

Here is a rough summary: Push-forward from Q (as in (42)) produces elements in $H^*_{\underline{d}}|_{S_1}$ for $S_1 \equiv Gr(\ker(f^*|_{H_1})) \times Gr(\ker(f^*|_{H_2}))$. Pull-back to P and then integration (as in (44)) produces elements in $\mathrm{Hom}(H^*_{\underline{d}}|_{S_2}; \mathbb{R})$ for $S_2 \equiv Gr(\ker(g^*|_{H_1})) \times Gr(\ker(g^*|_{H_2})$. Let $V_{1,2}$ denote the intersections of the kernels of f^* and g^* on H_1 and H_2, respectively. Then, the pairing between $H^*_{\underline{d}}|_{(V_1,V_2)}$ and its dual gives the value of «Donaldson invariants» of the form

$$\langle (\bigwedge_i (\alpha_i - \beta_i) \wedge \omega_1 \wedge \omega_2 \rangle \tag{45}$$

for ω_1 a closed form on Q, ω_2 a closed form on P, and (α_i, β_i) as in (37) but with $z \equiv z_i$ in V_1 or V_2.

13 The extended D-cohomology

The cohomology of D in (29) has an extension which is completely analogous to the extension of d to $\underline{d}$ that was just described. The extension will be described in the following circumstances: Suppose that $\mathcal{B}_0$ is a smooth manifold and that

$$\psi : Z \to \mathcal{B}_0 \tag{46}$$

is a given map.

The space X maps to $\mathcal{B}_0$ by $\psi or_\pm$. Assume that these two maps are homotopic as maps from X to $\mathcal{B}_0$. One must assume that the homotopy, say h, is natural with respect to the face structure on X. That is, write h as a set of maps $\{h_n : [0, n] \times X_n \to \mathcal{B}_0\}$. Then h_{q+p} on $X_q \times_Z X_p \subset \partial^1 X_{q+p}$ must be the concatenation of h_q and h_p. (So, $h_{q+p}(t, x, y) = h_q(t, x)$ if $0 \le t \le q$ and otherwise, it equals $h(t - q, y)$.)

Let $H_1 \subset H^3(\mathcal{B}_0)$ and $H_2 \subset H^2(\mathcal{B}_0)$ be given subspaces. The extended operator $\underline{D}$ is defined on

$$C^*(Z) \otimes \mathrm{Sym}^*(H_1) \otimes \Lambda^*(H_2). \tag{47}$$

Here is how to define $\underline{D}$: Let $\{z_{1\alpha}\}$ be a basis of closed forms for H_1 and let $\{z_{2\alpha}\}$ be a like basis for H_2. If z is one of $(\{z_{1\alpha}\}\{z_{2\alpha}\})$, write

$$s(z) \equiv \pi_*(h^* z), \tag{48}$$

where $\pi \equiv \{\pi_n : [0, n] \times X_n \to X_n\}$ is the projection. Note that $s(z)$ is a form on X of degree one less than z's. Also

$$ds(z) = (\psi \circ r_+)^* z - (\psi \circ r_-)^* z. \tag{49}$$

Furthermore, this s is consistent with (18) in the sense that s on X_{p+q} (call it s_{p+q}) restricts to $X_p \times_Z X_q \subset \partial^1 X_{p+q}$ as the restriction of $s_p + s_q$ from the product $X_p \times X_q$.

Introduce on $C^*(X) \otimes \mathrm{Sym}^*(H_1) \otimes \Lambda^*(H_2)$ the operator

$$\underline{s} \equiv \Sigma_\alpha(s(z_{1\alpha}) \cdot \partial_\alpha + s(z_{2\alpha}) \cdot \Theta \cdot \eta_\alpha). \tag{50}$$

With $\underline{s}$ understood, define $\underline{D}$ on (47) by

$$\underline{D} = \Theta \cdot (d - \Sigma_\alpha z_{1\alpha} \cdot \partial_\alpha) - \Sigma_\alpha z_{2\alpha} \cdot \eta_\alpha - \Theta \cdot r_{+*}(\exp(\underline{s}) \cdot r_{-*}((\cdot))). \tag{51}$$

A version of Stokes' theorem shows that $\underline{D}^2 = 0$. This $\underline{D}$ preserves

$$C^*(Z) \otimes \mathrm{Sym}^*(V_1) \otimes \Lambda^*(V_2) \tag{52}$$

whenever $(V_1, V_2) \subset Gr(H_1) \times Gr(H_2)$. The assignment to such (V_1, V_2) of the $\underline{D}$-cohomology on (52) defines an assignment,

$$H^*_{\underline{D}} \to Gr(H_1) \times Gr(H_2), \tag{53}$$

of vector spaces to points in $Gr(H_1) \times Gr(H_2)$ which is a realistic analog of the Floer cohomologies. (Note that the vector space of (53) over $(0,0)$ is the cohomology of (29).)

As with $H_{\underline{d}}$, the assignment $H^*_{\underline{D}}$ depends on the choice of representative forms $\{z_{1\alpha}, z_{2\alpha}\}$. Change the forms and the new assignment is naturally isomorphic to the old modulo the action of the vector space in (35) as a group of automorphisms.

14 Donaldson invariants and extended Floer cohomology

The extended Floer cohomology appears in the context of $\mathcal{M} = \{\mathcal{M}_n\}$ of (21). This cohomology makes its appearence in the following circumstances: Let Q, P and X be as described subsequent to (16). Let $\underline{z}$ be a class in H_1 or H_2 with the property that $f^*\psi^*\underline{z}$ and $g^*\psi^*\underline{z}$ both vanish in the cohomology of Q and P, respectively. If z is a form which represents $\underline{z}$, then

$$f^*\psi^* z = d\alpha \qquad \text{and} \qquad g^*\psi^* z = d\beta \tag{54}$$

with α a form on Q and β one on P. Now, suppose that α and β can always be chosen consistently with (19) and (20) in the following sense: On Q_{q+p}, the form α (call it α_{q+p}) must restrict to $Q_q \times_Z X_p$ as the restriction of $\alpha_q + s(z)_p$ from $Q_q \times X_p$. Likewise, on P_{q+p}, the form β (call it β_{p+q}) must restrict to $X_q \times_Z P_p$ as the restriction of $s(z)_q + \beta_p$.

Given the preceding proviso, observe that the form $\alpha - \beta$ on the product $Q \times P$ restricts to $\mathcal{M}$ in (21) as a closed form. With this understood, one can imagine being asked to compute a «Donaldson invariant» of the form

$$\langle(\alpha - \beta)^n \wedge \varphi_-^* \omega_- \wedge \varphi_+^* \omega_+, \mathcal{M}\rangle \tag{55}$$

with $\varphi_\pm^* \omega_\pm$ as in (24).

The strategy for (55) is analogous to that for (39). That is, introduce an indeterminant, s. Let ∂ denote differentiation in the s direction and write (55) as

$$\langle \exp(-\beta \cdot \partial) \cdot g^* f_*((\alpha + s)^n \wedge \varphi_-^* \omega_-) \wedge \varphi_+^* \omega_+, P\rangle|_{s=0}, \tag{56}$$

where f_* is defined as in (26) and P is given as in (16). (Note that (56) is a sum of infinitely many terms with only finitely many non-zero.)

Of course, (56) is a calculation of (55) which does the Q integration first and the P integration second. The result of doing the Q integration gives

$$f_*((\alpha + s)^n \wedge \varphi_-^* \omega_-), \tag{57}$$

which is in (52) for any $V_{1,2} \subset \ker(f^* \psi^*|_{H_{1,2}})$ which contains $\underline{z}$. (Technically, $V_{1,2}$ must lie in that subset of the kernel which satisfies the Q-proviso after (54). This may be all of the kernel.)

The extended Floer cohomology of $\underline{D}$ arises as the answer to the question: What kind of beast is (57)? For, (57) is $\underline{D}$-closed, and any change of ω_- and α by d(something) changes (57) by the image of $\underline{D}$. Here, the change in α must obey the proviso after (54). (The proof of this last statement is, as usual, a version of Stokes' theorem.)

Meanwhile, consider the linear functional on (52) which sends μ to

$$\mu \to \langle \exp(-\beta \cdot \partial) \cdot g^* \mu \wedge \varphi_+^* \omega_+, P\rangle|_{s=0}. \tag{58}$$

If $\underline{z}$ is in H_1, set $V_1 = \operatorname{Span}(\underline{z})$ and $V_2 = 0$, otherwise, set $V_1 = 0$ and $V_2 = \operatorname{Span}(\underline{z})$. Then, Stokes' theorem shows that (58) descends to a linear functional on $H^*_{\underline{D}}(V_1, V_2)$.

By construction, the linear functional in (58) and the $\underline{D}$-closed cocycle in (57) pair to give (55).

15 Relative Donaldson invariants

As with the $\underline{d}$-cohomology, there are relative Donaldson invariants for the $\underline{D}$-cohomology and for the $\underline{D}$-homology, $\operatorname{Hom}(H^*_{\underline{D}}; \mathbb{R})$. Their pairing computes

$$\langle(\bigwedge_i (\alpha_i - \beta_i)) \wedge \varphi_-^* \omega_- \wedge \varphi_+^* \omega_+, \mathcal{M}\rangle \tag{59}$$

when (α_i, β_i) are now a set as in (54) and the subsequent discussion. Their formal algebraic properties are described in [T1]. (See Section 12 above, but make the obvious notational changes: Replace $\underline{d}$ by $\underline{D}$ and (45) by (59)).

16 Group actions

There is an important sense in which (21) is unrealistic in the context of the 4-manifold invariants. This is because group actions have been neglected. In the 4-manifold context, a connected, compact Lie group, G acts on $\mathcal{M}$. When the G action on $\mathcal{M}$ is free, and ω is a closed cocycle on $\mathcal{M}/G$, then

$$\langle \omega, \mathcal{M}/G \rangle \tag{60}$$

is a better model than (8) for a Donaldson invariant. The question arises as to the method for calculating (60) when $\mathcal{M}$ is given by (21).

The group G will be assumed to act on Q, P, and X in (16) and also on Z, with all maps, $f, g, r_\pm$, being equivariant. When G acts freely on Z, then the complication is moot, for one can replace every space by its G-quotient and repeat verbatim the preceeding discussion.

However, in the 4-manifold context, the group G does not act freely on Z and so complications will arise. These complications can be dealt with, at least in the following context: Assume that the G action is free on X, P and on $\mathcal{M}$. The case where the action is free on X, Q and $\mathcal{M}$ and other scenarios will not be considered here.

Comment: In the 4-manifold context considered in [T1], the group G is $SO(3)$. In [T1], the action is only assumed free on the analogs of $\mathcal{M}_{n>0}$, on X and on $P_{n>0}$. Even so, the discussion follows the strategy which is outlined below.

By the way, these conditions for the free action of G in [T1] are guaranteed by imposing restrictions on the intersection pairing for $H_2(W_\pm)$. Said restrictions are either $b_2^+(W_+) > 1$ and $b_2^+(W_-) \geq 0$, or $b_2^+(W_+) = 1$ and $b_2^+(W_-) \geq 1$. (Thus, the discussion in [T1] handles the case where W_- is a tubular neighborhood of an embedded surface with negative homological self-intersection, and also the case where W_+ is a tubular neighborhood of an embedded surface with positive homological self-intersection.) If $b_2^+(W_+) = 0$ and $b_2^+(W) > 1$, then $G = SO(3)$ will not act freely on finitely many of the P_n.

Let EG be a separable, Hilbert manifold model for the universal G bundle. Its quotient is BG. When Y is a space on which G acts, write Y^G for $Y \times_G EG$. This is a fiber bundle over BG with fiber Y. If G acts freely on Y, then Y^G is also a fiber bundle over Y/G with fiber EG. In the latter case, sections exist, and any two sections are homotopic. This is because EG is contractible. By the way, a section of Y^G over Y/G is the same thing as a G-equivariant map from Y to EG. The two notions will be used interchangebly.

In the context of (24), suppose that the group G acts on $\mathcal{B}_\pm$ and that the maps $\varphi_\pm$ are G-equivariant. Then these maps descend as maps from Q^G and P^G, respectively, to $\mathcal{B}_\pm^G$ (without change of notation). Then, they give maps $\varphi_\pm : \mathcal{M}^G \to \mathcal{B}_\pm^G$ and a typical Donaldson invariant is

$$\langle \underline{\sigma}^*(\varphi_-^* \omega_- \wedge \varphi_+^* \omega_+), \mathcal{M}/G \rangle, \tag{61}$$

where $\underline{\sigma} : \mathcal{M}/G \to \mathcal{M}^G$ is any section. There is an analogous generalization of (59) which will be left unstated. The question is: How can one use (21) to compute (61) (or its (59) analog)?

Here is how: To begin, note that the construction of Y^G is natural with respect to fiber products in the sense that the fibered product of Q^G and P^G over Z^G is the same as

$$(Q \times_Z P)^G. \tag{62}$$

Under the assumption that G acts freely on P, one can choose a section, $\sigma : P/G \to P^G$. Write $\sigma \equiv \{\sigma_n\}$, with σ_n defined on P_n. Demand that σ_{q+p} restrict to $(X_q \times_Z P_p)/G \subset \partial^1 P_{q+p}/G$ in the following way: As a G-equivariant map from $X_q \times_Z P_p$ to EG, it should be the restriction from the product $X_q \times P_p$ of the map which is given by projection onto P_p followed by σ_p.

Note now that this σ defines a section, $\underline{\sigma}$, of $\mathcal{M}^G$ over $\mathcal{M}/G$: As a G-equivariant map from $\mathcal{M}_n$ in (21) to EG, it is given simply by restriction from $Q \times P$ of the map which is projection to P followed by σ.

With the preceeding understood, remark that (61) is identical to

$$\langle \sigma^* g^* f_*(\varphi_-^* \omega_-) \wedge \varphi_+^* \omega_+, P/G \rangle, \tag{63}$$

where f_* is push-forward by f of cocycles from Q^G to Z^G, and where g^* is the pull-back of cocycles from Z^G to cocycles on P^G. This last formula suggests a generalization for D's cohomology: Simply replace D by the operator on $C^*(Z^G)$ which is given by (29) save that $r_\pm$ should be interpreted as maps to Z^G from X^G.

Make this change. Then, Stokes' theorem shows that, $f_*(\varphi_-^* \omega_-)$ defines a D-cohomology class, and $\mu \to \langle \sigma^* g^* \mu \wedge \varphi_+^* \omega_+, P/G \rangle$ defines a D-homology class. Furthermore, the value of (61) is computed from the natural pairing between D-homology and D-cohomology.

Comment: If push-forward is defined for a G-equivariant map from a space Y to Z, then it will be defined using the induced map from Y^G to Z^G. For example, the latter has Fredholm differential if the former does. Also, the index bundle of the latter is oriented if the former is. See [T1].

There is an analogous G-equivariant generalization of $\underline{D}$ of (51). See Section 6 of [T1].

Under the assumption that G acts freely on Q but not on P, there is an alternate strategy which can be pursued. See [T1].

References

[A] M.F. Atiyah, «New invariants in three and four dimensions», Proc. Symps. Pure Mathematics (The Mathematical Heritage of Hermann Weyl) **48** (1985) 285–299.

[AJ] M.F. Atiyah and L. Jeffries, «Topological lagrangians and cohomology», J. Geom. Phys. 7 (1990) 120.

[AB] D.M. Austin and P.J. Braam, «Morse-Bott theory and equivariant Floer Cohomology for rational homology spheres», this volume.

[ABK] D.M. Austin, P.J. Braam and Keck, «Equivariant Floer theory for 3-manifolds», preprint.

[BS] L. Balieu and I.M. Singer, «Topological Yang-Mills symmetry», Nucl. Phys. Proc. suppl. **5b** (1988) 12.

[BT] R. Bott and L. Tu, Differential forms in algebraic topology, Springer Verlag, New York, 1982.

[B] P.J. Braam, «Floer homology groups for homology 3 spheres», Adv. in Math. 88 (1991) 131–144.

[C] R.L. Cohen, «Topics in Morse theory», Lecture Notes, Stanford University (1991).

[CJS] R.L. Cohen, J.D.S. Jones and G.B. Segal, «Morse theory and classifying spaces», this volume.

[D1] S.K. Donaldson, private communication.

[D2] S.K. Donaldson, «Polynomial invariants for smooth 4-manifolds», Topology **29** (1990) 257–315.

[D3] S.K. Donaldson, «Gluing techniques in the cohomology of moduli spaces», preprint.

[D4] S.K. Donaldson, «Boundary value problems for Yang Mills fields», preprint.

[DFK] S.K. Donaldson, M. Furuta and D. Kotschick, Floer homology groups in Yang-Mills theory», to appear.

[DK] S.K. Donaldson and P.B. Kronheimer, The geometry of 4-manifolds, Oxford University Press 1990

[F] A. Floer, «An instanton invariant for three-manifolds», Commun. Math. Phys. **118** (1987) 215–240.

[FM] R. Friedman and J. Morgan, Smooth 4-manifolds and complex surfaces, preprint.

[Fuk] K. Fukaya, «Instanton homology for oriented 3-manifolds», preprint.

[Fur] M. Furuta, «An analogue of Floer homology for lens spaces», preprint.

[T1] C.H. Taubes, «Donaldson-Floer theory for non-trivial circle bundles», in preparation.

[T2] C.H. Taubes, L2 moduli spaces on 4-manifolds with cylindrical ends, International Press, Hong Kong, 1993.

[U] K.K. Uhlenbeck, «Connections with L2 bounds on curvature» Commun. Math. Phys. **83** (1982) 31–42.

[W1] E. Witten, «Supersymmetry and Morse theory», J. Diff. Geom. **17** (1982) 661–692.

[W2] E. Witten, «Topological Quantum Field Theory», Commun. Math. Phys. **117** (1988) 353–386.

Clifford Henry Taubes
Department of Mathematics
Harvard University
Cambridge, MA 02138

The cup-product on the Thom-Smale-Witten complex, and Floer cohomology

C. Viterbo

Introduction

Let M be a finite dimensional smooth manifold, f a C^2 Morse function on M satisfying the Palais-Smale condition

(PS) Any sequence $(x_n)_{n\geq 1}$ such that $df(x_n) \to 0$ and $f(x_n)$ is bounded has a converging subsequence.

In the sequel we shall consider a «nice» pseudogradient vector field, X, that is

(i) $df(x) \cdot X(n) \geq \varepsilon |df(x)|^2$

(ii) X is a Morse-Smale vector field, that is, the stable manifold of a critical point is transverse to the unstable manifold of another critical point.

The (PS) assumption guarantees that the flow of X is well defined.

Given two critical points x and y with $f(x) < f(y)$, we denote by $\mathcal{P}(x,y)$ the set of maps $\gamma : \mathbf{R} \to M$ such that $\dot\gamma(s) = X(\gamma(s))$ and

$$\lim_{s\to+\infty} \gamma(s) = x \qquad \lim_{s\to+\infty} \gamma(s) = y .$$

As we have an obvious $\mathbf{R}$-action on $\mathcal{P}(x,y)$ defined by $(t*\gamma)(s) = \gamma(s+t)$, it is natural to define

$$\widehat{\mathcal{P}}(x,y) = \mathcal{P}(x,y)/\mathbf{R} .$$

Finally we denote by $\mathcal{M}(x,y)$ the image of $\mathcal{P}(x,y)$ by the evaluation map $\gamma \mapsto \gamma(0)$. In other words $\mathcal{M}(x,y)$ is the set of points on some curve of $\mathcal{P}(x,y)$.

Cauchy's uniqueness theorem tells us that the evaluation map is a diffeomorphism from $\mathcal{P}(x,y)$ onto $\mathcal{M}(x,y)$.

Let us denote by $i(x)$ the Morse index of the critical point x. We then have by our transversality assumptions

$$\dim \mathcal{P}(x,y) = i(y) - i(x)$$
$$\dim \widehat{\mathcal{P}}(x,y) = i(y) - i(x) - 1 .$$

Now for any $a < b$ regular values for f, we may construct the Thom-Smale-Witten complex by defining $W^k(f;a,b)$ to be the free R-module with one generator for each critical point of f with index k and critical value in $[a,b]$

$$\delta : W^k(f;a,b) \to W^{k+1}(f;a,b)$$
$$\delta(x) = \Sigma n(x,y) \cdot y ,$$

where $n(x,y)$ is the number of points in $\widehat{\mathcal{P}}(x,y)$ counted with the proper sign (i.e. as an intersection point of $W^u(x)$ and $W^s(y)$). As we are not particularly interested about this aspect, we shall just say that at each critical point, we may orient arbitrarily $W^u(x)$. This gives an orientation of $W^s(x)$ if we require that $W^s(x) \cap W^u(x) = (+1) \cdot x$.

It has been proved by several authors that the cohomology of (W^k, δ^k) is isomorphic to $H^*(M^b, M^q)$. (cf. [T], [M], [Fl 4], etc...)

The above isomorphism is an isomorphism of R-modules, but there is more structure in the $H^*(M)$ module in particular through the cup-product operation

$$H^*(M) \otimes H^*(M^b, M^a) \to H^\alpha(M^b, M^a)$$
$$\omega \otimes \alpha \to \omega \cup \alpha \,.$$

We now would like to read this structure on the complex $W^k(f; a, b)$ directly.

In Section 1, we define an $H^*(M)$-module structure on the cohomology of $W^k(f; a, b)$. In Section 3, we shall show that our argument extends to the more general framework of Floer cohomologies (in the symplectic or instanton case) with minor modifications, to yield an $H^*(\Lambda M)$ module structure (in the symplectic case) or an

$$H^*(BG)/\Big(\bigoplus_{k\geq 8} H^k(BG)\Big)$$

in the instanton case (BG is the classifying space of the gauge group).

In Section 2 we prove, returning to the case of a finite dimensional manifold, that the module structure, defined in Section 1, coincides with the cup-product. We then use this result to prove that the $H^*(\Lambda M)$, (M is the free loop space of M) structure defined on the symplectic Floer cohomology, in the case where this Floer cohomology may be identified with $H^*(M)$, reduces to the composition of cup-product and $i^* : H^*(\Lambda M) \to H^*(M)$ ($i : M \to \Lambda M$) is the inclusion of constants).

The existence of the product is sketched in [Fl 3], where the associativity result of Lemma 4 is conjectured. If M has holomorphic spheres, things are slightly more complicated (see Section 3). In particular we compute our product for complex projective spaces.

Let us point out that according to ideas of Cohen, Jones and Segal ([CJS]), one may define a kind of «stable Floer-Witten homotopy type», associated to the set of all bounded trajectories of X, such that its holomogy is just $H^*(M^b, M^a)$ (or the Floer cohomology).

However, on stable objects, there is no cup-product, thus there seems to be no way to reduce our cup-product to the above object, if we require a construction relying only on the spaces of bounded trajectories (this is the only thing we may hope to extend in the infinite dimensional case).

The author would like to thank Prof. J-M. Bismut, R. Cohen, F. Laudenbach and G. Segal, for useful discussions.

1 An $H^*(M)$-module structure on the cohomology of (W, δ)

From now on, we assume that $R = \mathbb{R}$. Let $(W(f;a,b), \delta)$ be the Thom-Smale-Witten chain complex defined in the introduction, taking real coefficients, and let $[\omega] \in H^*(M)$ be a de Rham cohomology class.

Theorem 1. *Let ω be a closed form on M, and let $\pi(\omega)$ be the map*

$$\pi(\omega) : W^k(f;a,b) \to W^{k+d}(f;a,b)$$
$$x \to \sum_y \left(\int_{\mathcal{M}(x,y)} \omega \right) \cdot y .$$

Then $\pi(\omega)$ commutes with δ, inducing a map in cohomology

$$P(\omega) : HW^k(f;a,b) \to HW^{k+d}(f;a,b) .$$

This map itself only depends on the cohomology class of ω in $H^*(M)$, moreover we have

$$P(\omega)P(\omega') = P(\omega \cup \omega') \quad (associativity) .$$

As a result P defines an $H^*(M)$ module structure on $HW^*(f;a,b)$.

Remark. In the above theorem, it is understood that $P(\omega)$ does not depend on the choice of the pseudo gradient vector fiels X.

The proof of the above theorem rests on several lemmas. We start by describing the closure of $\mathcal{M}(x,z)$ (or $\mathcal{P}(x,z)$).

Lemma 1. (cf. [Fl 1], [Fl 3]). *The closure of $\mathcal{M}(x,z)$ may be described as*

$$\overline{\mathcal{M}}(x,z) = \bigcup \mathcal{M}(x,y_1) \cup \quad \cup \mathcal{M}(y_q,z)$$

the union being over all sequences $y_1, \dots, y_q$ of critical points such that $\mathcal{M}(x,y_1)$, $\mathcal{M}(y_1,y_2)$, ..., $\mathcal{M}(y_q,z)$ are **all non empty**.

Moreover for any such sequence $(y_1, \dots, y_q)$, there is a map

$$G : \widehat{\mathcal{P}}(x,y_1) \times \dots \times \widehat{\mathcal{P}}(y_q,z) \times \Delta^{q+1} \to \overline{\mathcal{M}}(x,z)$$

where

$$\Delta^{q+1} = \{(\lambda_0, \dots, \lambda_q) \in [-\infty, +\infty]^{q+1} \mid \forall j, 1 + \lambda_j \le \lambda_{j+1}\}.$$

(1) *The image of G is a neighbourhood of $\mathcal{M}(x,y_1) \cup \quad \cup \mathcal{M}(y_q,z)$ in $\overline{\mathcal{M}}(x,z)$.*

(2) *The restriction of G to $\widehat{\mathcal{P}}(x_1,y_1) \times \quad \times \widehat{\mathcal{P}}(y_q,z) \times \mathring{\Delta}^{q+1}$ is a diffeomorphism onto its image.*

(3) $G(a_0, \dots, a_q, -\infty, \dots, -\infty, \mu_j, \dots, \mu_{j+p}, +\infty, \dots, +\infty) = G(a_j, \dots, a_{j+p}, \mu_j, \dots, \mu_{j+p})$.

Proof: We shall only sketch the construction of G (the first part of the theorem is standard). An essentially equivalent construction may be found in [Fl 3] or [C-J-S]. It is based on the *gluing* of flow trajectoires of X that one may find in [Fl 3].

Finally, we point out that we only need the lemma for $q = 1$.

We find a section of the natural projection $\mathcal{M}(a,b) \to \widehat{\mathcal{P}}(a,b)$ to identify $\widehat{\mathcal{P}}(a,b)$ to a submanifold of $\mathcal{M}(a,b)$. Using a tubular neighbourhood of $\mathcal{M}(a,b)$, we may construct a submanifold $\Sigma(a,b)$ such that

(i) $\Sigma(a,b) \cap \mathcal{M}(a,b) = \widehat{\mathcal{P}}(a,b)$

(ii) $T_u\Sigma(a,b) \cap T_u\mathcal{M}(a,b) = T_u\widehat{\mathcal{P}}(a,b)$

(iii) $T_u\Sigma(a,b) + T_u\mathcal{M}(a,b) = T_uM$
(One usually says that $\Sigma(a,b)$ and $\mathcal{M}(a,b)$ have «clean intersections» along $\widehat{\mathcal{P}}(a,b)$).

(iv) The projection $x : \Sigma(a,b) \to \widehat{\mathcal{P}}(a,b)$ makes $\Sigma(a,b)$ into a disk bundle.

Now let γ be a trajectory in $\mathcal{P}(a,b)$ near $\mathcal{M}(x,y_1) \cup \quad \cup \mathcal{M}(y_q,z)$. Then $\gamma(\mathbf{R})$ intersects all the $\Sigma(y_j,y_{j+1})$ (we set $y_0 = x$, $y_{q+1} = z$) at times λ_j $(j = 0,\dots,q)$, and $\pi(\gamma(\lambda_j)) = a_j \in \widehat{\mathcal{P}}(y_j,y_{j+1})$.

Conversely, given a sequence (a_j,λ_j) with $\lambda_{j+1} > 1+\lambda_j$, there is a unique γ in $\mathcal{P}(a,b)$ such that $\gamma(\lambda_j) \in \Sigma(y_j,y_{j+1})$ and $\pi(\gamma(\lambda_j)) = a_j$. This follows from the transversality of stable and unstable manifolds, and the implicit function theorem.

So our map G is defined on

$$\widehat{\mathcal{P}}(x,y_1) \times \quad \times \widehat{\mathcal{P}}(y_q,z) \times \overset{\circ}{\Delta}{}^{q+1} .$$

and is a diffeomorphism onto its image.

We now extend G to the values of λ_j in $\{\pm\infty\}$.

If $\lambda_0,\dots,\lambda_j$ go to $-\infty$, then for any $t \in \mathbf{R}$, $\gamma(t)$ goes to a point in $\overline{\mathcal{M}(y_j,z)}$. Indeed, if γ is a trajectory, and $\gamma(\lambda_{j-1}) \in \Sigma(y_{j-1},y_j)$, $\gamma(\lambda_j) \in \Sigma(y_j,y_{j+1})$, then $\frac{1}{|\lambda_{j+1}-\lambda_j|}$ measures how close γ passes y_j (because going near y_j slows down the trajectory), so that as $\lambda_j \to -\infty$ while $\lambda_{j+1} > -C$, γ converges to a trajectory starting from y_j.

The same argument shows that if $\lambda_{k+1},\dots,\lambda_q$ go to $+\infty$, $\gamma(0)$ converges to a point in $\overline{\mathcal{M}(x,y_k)}$.

Moreover as $\lambda_0,\dots,\lambda_j$ got to $-\infty$, $\lambda_{k+1}, \dots, \lambda_q$ to $+\infty$, we have that the definition of γ, only depending on $(a_{j+1}, \dots, a_k, \lambda_{j+1}, \dots, \lambda_k)$, will eventually be given by $G((a_{j+1}, \dots, a_k, \lambda_{j+1}, \dots, \lambda_k)$. This proves (3) and concludes the proof.

Lemma 2. *Let x,z with $i(x) - i(z) = k + 1$. Define*

$$\partial\mathcal{M}(x,z) = \sum_{i(y)=i(x)+1} n(x,y)\mathcal{M}(y,z) + \sum_{i(y)=i(z)-1} n(y,z)\mathcal{M}(x,y) .$$

Then we have for any k-form φ

$$\int_{\partial\mathcal{M}(x,z)} \varphi = \int_{\mathcal{M}(x,z)} \varphi .$$

Remarks. This is our version of Stoke's formula. Hence the «sum» in the definition of $\partial\mathcal{M}(x,z)$ should be understood as a summation of integration currents.

Proof: (see also [L], Prop. 2). As in the proof of Stoke's theorem, we may, using a partition of unity, restrict our attention to a neighbouhrood of $\mathcal{M}(x,y_1) \cup \quad \cup$ $\mathcal{M}(y_q,z)$.

Since G^{-1} is a diffeomorphism of such a neighbourhood into $\widehat{\mathcal{P}}(x,y_1)\times\ldots\times \widehat{\mathcal{P}}(y_q,z)\times \overset{\circ}{\Delta}_{q+1}$, we may apply the change of variable formula (to carry back our integration on the latter set) the Stoke's formula for manifolds with corners (see the Appendix for details). We see that $G^*\varphi$ must be integrated only on

$$\widehat{\mathcal{P}}(x,y_1)\times\ldots\times\widehat{\mathcal{P}}(y_q,z)\times\{-\infty\}\times \overset{\circ}{\Delta}_q \bigcup \widehat{\mathcal{P}}(x,y_1)\times\ldots\times\widehat{\mathcal{P}}(y_q,z)\times \overset{\circ}{\Delta}_q \times\{+\infty\},$$

and since $G(a_0,a_1,\ldots,a_q,-\infty,\lambda_1,\ldots,\lambda_q)$ does not depend on a_0, integration is really only $\widehat{\mathcal{P}}(y_1,y_2)\times\ldots\times\widehat{\mathcal{P}}(y_q,z)\times \overset{\circ}{\Delta}_q \cup \widehat{\mathcal{P}}(x,y_1)\times\ldots\times\widehat{\mathcal{P}}(y_{q-1},y_q)\times \overset{\circ}{\Delta}_q$. Moreover this set has dimension $> k$ unless $i(x)-i(y_1)=1$ (or $i(y_q)-i(z)=1$).

Thus the only integrals of φ that appear, are these on $\mathcal{M}(y,z)$ (resp. $\mathcal{M}(x,y)$) with $i(x)-i(y)=1$ (resp. $i(y)-i(z)=1$) and the integral appears once for each element in $\mathcal{P}(x,y)$ counted with the proper sign.

This concludes the proof.

Consider now the map $\pi(\omega)$. It is a linear map, and depends linearly on ω.

Lemma 3.

$$\pi(d\omega) = \delta\pi(\omega) \pm \pi(\omega)\delta$$

where δ *is the coboundary on* $W^*(f;a,b)$.

Proof:

$$\begin{aligned}
\pi(d\omega)x &= \sum_t \left(\int_{\mathcal{M}(x,t)} d\omega\right) t \\
&= \sum_t \left(\int_{\partial\mathcal{M}(x,t)} \omega\right) t \\
&= \sum_t \left(\sum_{i(y)=i(x)+1} n(x,y) \int_{\mathcal{M}(y,t)} \omega + \sum_{i(z)=i(t)-1} n(z,t) \int_{\mathcal{M}(x,z)} \omega\right) t \\
&= \sum_{i(y)=i(x)+1} n(x,y) \sum_t \left(\int_{\mathcal{M}(y,t)} \omega\right) t + \sum_{i(z)=i(t)-1} n(z,t) \left(\int_{\mathcal{M}(x,z)} \omega\right) t \\
&= \big(\delta\pi(\omega) + \pi(\omega)\delta\big)x
\end{aligned}$$

Remark. This is the analogue of the formula $d(\varphi\wedge\alpha) = d\varphi\wedge\alpha+(-1)^{\deg(\varphi)}\varphi\wedge d\alpha$ for $\varphi \in \Omega(M)$, $\alpha \in \Omega(M^b, M^a)$.

Corollary. *If ω is closed, $\pi(\omega)$ induces a map $P(\omega)$,*

$$HW^k(f;a,b) \to HW^{k+d}(f;a,b)$$

which depends only on the cohomology class of ω.

Proof: If $d\omega = 0$, $\delta\pi(\omega) = -\pi(\omega)\delta$, hence $(-1)^{\deg} \cdot \pi(\omega)$ is a chain map, hence define a map in cohomology. If $\omega = d\varphi$, $\pi(d\varphi) = \delta\pi(\varphi) + \pi(\varphi)\delta$ hence $\pi(\delta\varphi)$ sends cycles to boundaries: it induces the zero map in cohomology. We now prove

Lemma 4. *Let ω_1, ω_2 be closed forms. Then we have $P(\omega_1 \wedge \omega_2) = P(\omega_1) \cdot P(\omega_2)$.*

Remark. It is understood that$P(\omega_1), P(\omega_2)$ and $P(\omega_1 \wedge \omega_2)$ only depend on the cohomology classes of ω_1 and ω_2.

Proof: We have

$$P(\omega_1 \wedge \omega_2) \cdot x = \sum_z \left(\int_{\mathcal{M}(x,z)} \omega_1 \wedge \omega_2 \right) z$$

$$P(\omega_1) \cdot P(\omega_2) \cdot x = P(\omega_1) \cdot \sum_y \left(\int_{\mathcal{M}(x,y)} \omega_1 \right) y$$

$$= \sum_z \left(\sum_y \left(\int_{\mathcal{M}(x,y)} \omega_1 \int_{\mathcal{M}(y,z)} \omega_2 \right) \right) z .$$

It will then be sufficient to prove that adding a suitable exact form to ω_1 and ω_2, we have

$$(*) \qquad \sum_y \int_{\mathcal{M}(x,y)} \omega_1 \int_{\mathcal{M}(y,z)} \omega_2 = \int_{\mathcal{M}(x,z)} \omega_1 \wedge \omega_2 .$$

Let $k_1 = \deg \omega_1$, $k_2 = \deg \omega_2$; then if $\dim \mathcal{M}(x,y) = i(x) - i(y) < k_1$, we may chane ω_1 by an exact form, so that it vanishes in a neighbourhood of $\mathcal{M}(x,y)$. The same holds for ω_2 near $\mathcal{M}(y,z)$ whenever $i(y) - i(z) < k_2$. Thus we only have on the left hand side of (*) terms with $i(y) = i(x) + k_1 = i(z) - k_2$.

In other terms, if we also use the $\mathbb{R}$ action (i.e. the flow) to enlarge the region where ω_1 and ω_2 vanish, we may assume that ω_1 vanishes on any (given) compact subset of $\mathcal{M}(x,z)$ and $\mathcal{M}(y,z)$, and ω_2 vanishes on any compact subset of $\mathcal{M}(x,z)$ and $\mathcal{M}(x,y)$.

Now, we see that ω_1 vanishes outside a neighbourhood of the $\overline{\mathcal{M}(x,y)}$, and ω_2 outside a neighbourhood of the $\overline{\mathcal{M}(y,z)}$. Thus we may assume in proving $(*)$

that ω_1 and ω_2 vanish outside neighbourhood of $\overline{\mathcal{M}(x,y)}$ and $\overline{\mathcal{M}(y,z)}$, i.e. prove $(*)$ for forms supported near $\overline{\mathcal{M}(x,y)} \cup \overline{\mathcal{M}(y,z)}$.

Now we claim that this neighbourhood is the image $C\widehat{\mathcal{P}}(x,y) \times C\widehat{\mathcal{P}}(y,z)$, where CA is the cone over A. Here, we identify $C\widehat{\mathcal{P}}(x,y)$ with $\widehat{\mathcal{P}}(x,y) \times [-\infty, +\infty[/\widehat{\mathcal{P}}(x,y) \times \{+\infty\}$and $C\widehat{\mathcal{P}}(y,z)$ with $\widehat{\mathcal{P}}(y,z) \times]-\infty, +\infty]/\widehat{\mathcal{P}}(y,z) \times \{+\infty\}$.

According to Lemma 1, a neighbourhood of $\mathcal{M}(x,y) \cup \mathcal{M}(y,z)$ is described as the image of $\widehat{\mathcal{P}}(y,y) \times \widehat{\mathcal{P}}(y,z) \times \overset{\circ}{\Delta}_2$

by the glueing map. We then associate to $(a_1, t_1, a_2, t_2) \in C\widehat{\mathcal{P}}(x,y) \times C\widehat{\mathcal{P}}(y,z)$ the image by the glueing map of $(a_1, a_2, t_1, t_1 + t_2)$, and $(a_1, -\infty, a_2, +\infty)$ goes to $\{y\}$.

We now pull back ω_1 and ω_2 on $C\mathcal{P}(x,y) \times C\widehat{\mathcal{P}}(y,z)$, and we get two forms φ_1, φ_2. From the fact that ω_1 vanishes away from $\mathcal{M}(x,y)$, we get that in fact

$$\omega_1 \in H^*(C\widehat{\mathcal{P}}(x,y) \times C\widehat{\mathcal{P}}(y,z), D_1)$$
$$\omega_2 \in H^*(C\widehat{\mathcal{P}}(x,y) \times C\widehat{\mathcal{P}}(y,z), D_2)\,.$$

where $D_i = \{(a_1, t_1, a_2, t_2) \mid t_i \geq C\}$ In other words, if $CA = A \times [0,1]/A \times \{0\}$, A is identified with $A \times \{1\}$, we have, given $A = \widehat{\mathcal{P}}(x,y)$, $B = \widehat{\mathcal{P}}(y,z)$,

$$\varphi_1 \in H^*(CA \times CB, A \times CB)$$
$$\varphi_2 \in H^*(CA \times CB, CA \times B)$$
$$\varphi_1 \wedge \varphi_2 \in H^*(CA \times CB, A \times CB \cup CA \times B)\,.$$

The fact that

$$\int_{CA \times CB} \varphi_1 \wedge \varphi_2 = \int_{CA} \varphi_1 \int_{CB} \varphi_2$$

is given by the following:

Sublemma. *Let A, B be manifolds without boundary, of dimension a, b. Let*

$$\int_{CA} : H_c^{a+1}(CA, A) \to \mathbf{R}$$

be the isomorphism given by integration of forms (note that $H^i(CA,A) \simeq H^{i-1}(A)$ for $i \neq 0$). Then for $\alpha \in H_c^{a+1}(CA \times CB, A \times CB)$, $\beta \in H_c^{b+1}(CA \times CB, CA \times B)$, we have

$$\int_{CA \times CB} \alpha \wedge \beta = \int_{CA \times \{b_0\}} \alpha \int_{\{a_0\} \times CB}\,.$$

Proof: Note that $\alpha \wedge \beta \in H^*(CA \times CB, A \times CB \cup CA \times B) \simeq H^*(CA, A) \otimes H^*(CB, B)$ (Künneth isomorphism). Now, because $H_c^*(CB) = 0$, α is always cohomologous to $\overline{\alpha} \otimes 1$, β to $1 \otimes \overline{\beta}$, with $\overline{\alpha} \in H_c^*(CA, A)$, $\overline{\beta} \in H_c^*(CB, B)$.

Then

$$\int_{CA\times\{b_0\}} \alpha = \int_{CA} \overline{\alpha} \qquad \int_{\{a_0\}\times CB} \beta = \int_{CB} \overline{\beta}$$

and $\alpha \wedge \beta$ is the image by Künneth of $\overline{\alpha} \otimes \overline{\beta}$, hence

$$\int_{CA\times CB} \alpha \wedge \beta = \int_{CA\times CB} \overline{\alpha} \otimes \overline{\beta} = \int_{CA} \overline{\alpha} \int_{CB} \overline{\beta} = \int_{CA\times\{b_0\}} \alpha \int_{\{a_0\}\times CB} .$$

This concludes our proof.

We still want to prove that

$$\int_{C\widehat{\mathcal{P}}(x,y)} \varphi_1 = \int_{\mathcal{M}(x,y)} \omega_1 ,$$

but $C\widehat{\mathcal{P}}(x,y) \times \{(a_1, +\infty)\}$ goes to $\mathcal{M}(x,y)$; thus this follows from the change of variable formula.

We still want to prove that $P(\omega)$ does not depend on the choice of (X, f). In other words, let (X_t, f_t) be a family of data, such that $M_t^b = M_0^b$, $M_t^a = M_0^a$.

Then there is a chain map $K : W^*(f_0, X_0; a, b) \to W^*(f_1, X_1; a, b)$ inducing as an isomorphism

$$\overline{K} : HW^*(f_0, X_0; a, b) \to HW^*(f_1, X_1; a, b) .$$

Proposition.

$$\overline{K}P(\omega) = P(\omega)\overline{K} .$$

Proof: Assume

$$\begin{aligned} F_t(x) &= f_0(x) + \frac{\varepsilon}{2}t^2 \qquad 0 \le t \le \delta \quad \varepsilon > 0 \\ &= f_1(x) - \frac{\varepsilon}{2}t^2 \quad 1 - \delta \le t \le 1 \end{aligned}$$

and set $F(t, x) = f_t(w)$.

Then $F : M \times [0, 1]$ has the following pseudogradient vector field

$$\begin{aligned} X(t, X) &= X_0(x) + \frac{\varepsilon}{2}t\frac{\partial}{\partial t} \qquad 0 \le t \le \delta \\ &= X_t(x) + \frac{\partial}{\partial t}f_t(x) \qquad \delta \le t \le 1 - \delta \\ &= X_1(x) - \frac{\varepsilon}{2}t\frac{\partial}{\partial t} \quad 1 - \delta \le t \le 1 \end{aligned}$$

Thus the flow preserves $M \times \{0\}$ and $M \times \{1\}$. Moreover if x is a critical point of f_0 (resp. f_1) with index i, $(x,0)$ (resp. $(x,1)$) is a critical point of F with index i (resp. $i+1$).

For x a critical point of f_0, we set

$$K(x) = \sum_y n(x,y) \cdot y$$

where y is a critical point of f_1, and $n(x,y)$ is the number of trajectories of X from $(x,0)$ to $(y,1)$.

Thus K defines a map $W^*(f_0, X_0, a, b) \mapsto W^*(f_1, X_1, a, b)$.

Note that if we consider $W^*(f_0, X_0, a, b)$ and $W^{*-1}(f_1, X_0, a, b)$ as subcomplexes of $W^*(F, X, a, b)$, the boundary map Δ on this complex may be written as a matrix

$$\Delta = \begin{pmatrix} \delta_0 & 0 & \vdots & 0 \\ K & \delta_1 & \vdots & ? \\ \dots & \dots & \vdots & \dots \\ ? & 0 & \vdots & ? \end{pmatrix}$$

where the first and second rows (resp. column) correspond to $W^*(f_0, X_0, a, b)$ and $W^*(f_1, X_1, a, b)$. Since $\Delta^2 = 0$, we get $\delta_0 K + K\delta_1 = 0$ and K is (up to sign) the required chain map.

Now let $\widetilde{\omega}$ denote the obvious extension of ω to $M \times [0,1]$. Then, since we proved that $\Delta \circ P(\widetilde{\omega}) = P(\widetilde{\omega}) \circ \Delta$, this implies $K \circ P(\widetilde{\omega}) = P(\widetilde{\omega}) \circ K$; but $P(\widetilde{\omega})$ corresponds to $P(\omega)$ on $W^*(f_0, X_0, a, b)$ and $W^*(f_1, X_1, a, b)$.

This proves our proposition.

Remark. It would be interesting to understand what are the other cohomology operations that we may define using only trajectory spaces. In particular, in view of our paper [V 2], one would expect to be able to define Massey products $< \omega, x, \omega' >$, but we were unable to prove this, as the definition of Massey products makes crucial use of the existence of a product, defined on cochains, which is already associative on the cochain level.

Remark. It seems that all this may also be done with any coefficients ring. For this one would have to triangulate the spaces $\mathcal{M}(x,y)$, and define a «fundamental class», (that should be the sum of all oriented simplexes in $\mathcal{M}(x,y)$ so that for any cocycle c in M, we may define $< c, \mathcal{M}(x,y) >$, and the equivalent of Lemma 2 is satisfied.

2 Identification with the «usual» cup-product structure

We consider for example the case where f_0 has a nondegenerate critical manifold N, assumed to be compact (if M is compact we could have $M = N$). Assume that all critical points in $f^{-1}([a,b])$ are on N.

Proposition 2.1. *With the above assumptions, let $j : N \to M$ be the inclusion, i the index of N. Then $H^*(M^b, M^a) \simeq H^{*-i}(N)$, and under this identification $P(\omega)\alpha = j^*(\omega) \cdot \alpha$. In particular, when M is compact, $a = -\infty, b = +\infty$, $P(\omega)$ induces the usual cup-product on M.*

Proof: We start with the case $M = N$. Even in this case we have to explain the meaning of our proposition, since the identification $HW^*(M) \to H^*(M)$ has not been explicitly made.

In this case, if we moreover assume f to be an ordered function, that is, all critical values are distinct, and for x, y two critical points of f, $f(x) < f(y)$ implies $i(x) \leq i(y)$.

Let $(a_i)_{i=0,\dots,n}$ be real numbers separating the critical values, so that $f^{-1}([a_i, a_{i+1}])$ contains all critical points of index i.

Since

$$\begin{aligned} H^j(M^{a_{i+1}}, M^{a_i}) &= 0 && j \neq i \\ &= R^{m_i} && j = i \text{ if } f \text{ has } m_i \\ & && \text{critical points with index } i \end{aligned}$$

we see that the map

$$d_i : H^i(M^{a_i}, M^{a_{i-1}}) \to H^{i+1}(M^{a_{i+1}}, M^{a_i})$$

may be identified with the map

$$\delta^i : W^i(f, X) \to W^{i+1}(f, X)$$

as follows: each class in $H^i(M^{a_{i+1}}, M^{a_i})$ may be identified to a sum of duals of $W^s(x)$ for x critical point of index i, that is, to x we associate $\alpha \mapsto \int_{W^s(x)} \alpha$. In other words, we get a map

$$H^i(M^{a_{i+1}}, M^{a_i}) \to W^i(f, X) \qquad \omega \mapsto \sum_x \left(\int_{W^s(x)} \omega \right) x .$$

Let us check that this identification is compatible with the differentials d_i and δ_i.

Now we know from [M], that d_i corresponds to counting the number of points of $W^s(x) \cap W^u(y) \cap M^{a_i}$. This is exactly how δ_i is computed.

Now let $\omega \in H^d(M)$, we may assume that ω vanishes on M^{a_d} since $H^d(M^{a_d}) = 0$. Then $P(\omega) \cdot 1 = \sum_x \left(\int_{W^s(x)} \omega \right) x = \mu(\omega)$ where μ is the map $H^*(M) \to HW^*(f, X)$ defined above. This proves our claim for $N = M$.

The general case follows from this, since N has a tubular neighbourhood ν, which may be decomposed as $\nu = \nu^+ \oplus \nu^-$, positive and negative eigenspaces for $d^2 f$.

Let g be a function on N, and set

$$f_\varepsilon(x,y) = f(y) + \varepsilon(|y|^2)g(x)$$

(here $y \in \nu_x$ and $f(y) \simeq |y_+|^2 - |y_-|^2$). Then

$$df_\varepsilon(x,y) = \left(df(y) + \varepsilon'(|y|^2)\cdot yg(x), \varepsilon(|y|^2)dg\right)$$

provided $\left|\varepsilon'(|y|^2)\right| < 1$ and $|g(x)| < 1$

$$df_\varepsilon(x,y) = 0 \Leftrightarrow y = 0dg(x) = 0$$

and we may choose $X_\varepsilon(x,y) = (0, \nabla g(x))$ on $\{y = 0\}$.

Thus the trajectories of X_ε are in one-to-one correspondence with those of g. Choosing g to be well ordered, we are reduced to the case treated.

3 Generalizations to Floer cohomology

Here we shall only deal with the case of symplectic Floer homology, even though there is no obstruction to applying the same methods for the instanton case.

Let (P, ω) be a symplectic manifold, H a C^2 function on $P \times S^1$. We choose an almost complex structure on P compatible with ω, that is,

$$\omega(J\xi, \eta) = g(\xi, \eta),$$

where g is a Riemannian metric. Note that for fixed ω, the set of almost complex structures compatible with ω make a contractible space.

Let ΛP be the free loop space of P, i.e. the set of maps $S^1 \to P$. Floer associates to the action the functional on ΛP that is defined as

$$A_H(x) = \int_{D^2} u^*\omega - \int_{S^1} H(\theta, x)d\theta$$

where $u : D^2 \to P$ satisfies $uQ|\partial D = x$. Note that

(i) A_H is only defined on the connected component $\Lambda_0 P$ of the set of constant maps (otherwise u does not exist)

(ii) u is not unique. In fact A_H is multivalued (i.e. defined on some covering space of $\Lambda_0 P$) unless $[\omega]\pi_2(P) = 0$.

We first deal with the classical case $[\omega]\pi_2(P) = 0$. Then, the L^2 gradient flow of A_H, studied by Floer, is given by the equation

$$\frac{\partial u}{\partial t} = -J\frac{\partial u}{\partial \theta} - \nabla H(\theta, u)$$

that is,

$$\bar{\partial} u = -\nabla H(\theta, u)\,. \tag{$*$}$$

Moreover, the critical points of A_H are the 1-periodic solutions of

$$\text{(H)} \qquad \dot{x} = J\nabla H(\theta, x) = X_H(\theta, x)$$

so that the connecting trajectories are solutions of

$$\text{(E)} \qquad \begin{cases} \bar{\partial} u = -\nabla H(n) \\ \lim_{t\to+\infty} u(t,\theta) = y(\theta) \qquad \lim_{t\to-\infty} u(t,\theta) = x(\theta)\,. \end{cases}$$

The set $\mathcal{P}(x, y)$ of solution for the above equation then behaves exactly as in the finite dimensional case. In particular, we have both the compactness modulo break up, and the glueing of solutions ([Fl 3], p. 584, prop. 1b), as well as a homeomorphism $\mathcal{P}(x, y) \to \mathcal{M}(x, y)$ obtained from the evaluation map.

The only difference comes from the Morse index. Despite the fact that the Morse index is alway infinite, we may define a finite number, still called an index, that measures the size of the negative eigenspace of $D^2 A_H$. For the definition we refer to [V 1], [Fl], [S-Z]. It is now dependent on v, in fact if $u, v : D^2 \to P$ are such that $v = w$ on ∂D^2, $i(v) - i(w) = 2c_1(P)(v - w)$ where $(v - w) : S^2 \to P$ is the map obtained by glueing w and v along ∂D^2, $c_1(P)$ is the first Chern class of (P, J) (cf. [V]).

Let Γ be the subgroup of $\mathbf{Z}$ generated by the values of $2c_1(P)$ over all spheres. Then we may define $i(x)$ in $\mathbf{Z}/\Gamma$ for x a solution of (H).

We first assume $\Gamma = \{0\}$. Then one has a well defined index. The Floer complex is then constructed like the Thom-Smale-Witten complex. We denote its cohomology by $HF^*(P, H; a, b)$. Then we get a module structure of $HF^*(P, H; a, b)$ over the sign $H^*(\Lambda P)$.

Note that for P compact, $a = -\infty$, $b = +\infty$, HF^* does not depend on H. If we take H to be independent of θ and C^2 small, the solutions of (E) will be $u(\theta, t) = u(t)$ with $\dot{u}(t) = -\nabla H(u)$, that is u is a path of constant loops given by the flow trajectories of $-\nabla H$ on P.

Thus $HF^*(P, H) \simeq H^*(P)$, and according to proposition 1 of section 2, the $H^*(\Lambda P)$ structure is given by $\omega \cdot \alpha = c^*(\omega) \cdot \alpha$ where $c : P \to \Lambda P$ is the inclusion of constant loops.

Now let us deal with the case $[\omega]\pi_2(P) \neq 0$ and/or $c_1(P)\pi_2(P) \neq 0$.

Let $\widetilde{\Lambda} P$ be the covering of $\Lambda_0 P$ defined as

$$\widetilde{\Lambda} P = \left\{ v : D^2 \to P \right\} / \sim$$

where

$$v \sim w \Leftrightarrow v = w \text{ on } \partial D^2$$

and

$$\langle \omega, v - \omega \rangle = \langle c_1(P), v - \omega \rangle = 0\,.$$

Then, A_H is well defined on $\widetilde{\Lambda}P$, and each critical point has a well defined index.

Now let u be a solution of (E); it will then be a trajectory from (x, v) to (y, w) where w is homotopic to $v\#u$, that is, we glue v and u along ∂D^2 to get a map from the disk obtained by glueing D^2 and $S^1 \times \mathbf{R}$ (along $\partial D^2 \simeq S^1 \times \{-\infty\}$).

Thus, to each solution of (E) corresponds a set of trajectories, each of them starting from (x, v) with different v. However, if we require $A_H(x, v) \in [a, b]$, $i(x, v) \in [d_1, d_2]$ then we get only finitely many v's.

Now, due to the possible interference of holomorphic spheres in P, the space of trajectories $\mathcal{P}((x, v); (y, w))$ is subject to a more complicated failure of compactness than described by Lemma 1. Indeed, as it is now classical, a holomorphic sphere may «bubble off». However, if we denote by α the sphere that bubbles off from a trajectory with $i(y, w) - i(x, v) = d$, we must have $1 + 2\langle c_1(P), \alpha\rangle \leq d$, since what is left after bubbling off is still a trajectory $\tilde{u}$ from $(x, \tilde{v})$ to $(y, \tilde{w})$ with $\tilde{w} - \tilde{v} = \tilde{u}$ and $u = \tilde{u}\#\alpha$; thus

$$\begin{aligned} 2\langle c_1(P), \alpha\rangle = 2\langle c_1(P), u - \tilde{u}\rangle &= (i(y, w) - i(x, v)) - i(y, \tilde{w}) - i(x, \tilde{v}) \\ &= d - \dim \mathcal{P}\,(x, \tilde{v}); (y, \tilde{w}))\ . \end{aligned}$$

Since $\dim \mathcal{P}((x, \tilde{v}), (y, \tilde{w}))$ must be greater than one, we get the desired inequality.

Note: Our argument is not complete for two reasons. First together with bubbling off, we may have break up of the trajectory as in Lemma 1. However, this is not a problem, since the two phenomena may be treated separately.

Second we have to worry about orientations. This is more tricky than in the finite dimensional case, and we refer to [Fl-H] for this.

As a result, and as remarked by Floer, if $c_1(P) > 0$ on holomorphic spheres (this is in particular true for P monotone, i.e. $c_1(P) = \lambda\omega, \lambda > 0$) we have that bubbling off does not occur on one dimensional trajectories. Thus the δ operator is well defined and $\delta \circ \delta = 0$ (this is based on the compactness of trajectory spaces of dimension 1, but $1 + 2c_1(P)$ is at least 3 on a holomorphic sphere).

Note that we may also define the maps $P(\sigma)$ for $\sigma \in H^k(\tilde{\Lambda}P)$ with $k < 2\inf_\alpha \langle c_1(P), \alpha\rangle = d(P)$ (we need an extra dimension to apply Lemma 3, and show that $P(\sigma)$ only depends on the cohomology class of σ thus $k < 1 + 2\inf_\alpha \langle c_1(P), \alpha\rangle$ would not suffice).

Moreover, we have $P(\sigma_1, \sigma_2) = P(\sigma_1) \cdot P(\sigma_2)$ provided $\deg(\sigma_1) + \deg(\sigma_2) < d(P)$.

Remark. It is interesting to note that if $\deg(\sigma_1)$ and $\deg(\sigma_2)$ are $< d(P)$, but $\deg(\sigma_1) + \deg(\sigma_2) > d(P)$, $P(\sigma_1) \cdot P(\sigma_2)$ is well defined. However we cannot prove that it depends only on $\sigma_1 \cdot \sigma_2$. We cannot even prove that $P(\sigma_1)P(\sigma_2) =$

$\pm P(\sigma_2)P(\sigma_1)$. Thus if $R_* = H^*(M)$ is defined as a ring though generators and relations, we get that $R_*^{(d(P))}$ operates on HF^*, where $R_*^{(k)}$ is defined as the ring obtained from R_* by forgetting all generators and relations of degree $\geq k$.

Theorem. *Let $d(P) = 2\inf\{\langle c_1(P), \alpha\rangle \mid \alpha$ is a holomorphic sphere $\}$ and R be the ring $H^*(\tilde{\Lambda}P)$. Then provided $d(P) > 0$, we may define $HF^*(\tilde{\Lambda}P, H; a, b)$. It has an $R^{(d(P))}$ module structure.*

Remark. If we want everything to be independent of the almost complex structure J, we essentially need to replace $d(P)$ by

$$\tilde{d}(P) = 2\inf\{\langle c_1(P), \alpha\rangle \mid \alpha \in \pi_2(P), \langle \omega, \alpha\rangle > 0\} .$$

However, there may be a big difference between d and $\tilde{d}$ ($\tilde{d}$ may even vanish); thus it may sometimes be convenient to work with an almost given complex structure, and try to estimate d for this structure

Remark. Instanton Floer cohomology. In [Fl 5], Floer defined the instanton cohomology $IH^*(M)$ for a three dimensional homology sphere M. This is obtained by constructing the «TSW complex associated to the Chern-Simons functional»

$$CS(A) = \int_M \left(\frac{1}{2}A \wedge dA + \frac{1}{3}A \wedge A \wedge A\right)$$

where $A \in \mathcal{A}(M)$ the space of $SU(2)$ connections for the trivial bundle over M (note that every $SU(2)$ bundle over a three dimensional manifold is trivial).

The space of maps from M to $SU(2)$ is a group g, called the gauge group, acting on $\mathcal{A}(M)$ though

$$A \mapsto gAg^{-1} + dg \cdot g^{-1} .$$

Then $CS(g * A) = CS(A) + 2\pi \deg(g)$, ($g : M \to SU(2)$ is a map between three dimensional manifolds, : $\deg(g)$ is its degree). Thus CS is not well defined on $\mathcal{A}(M)/G$, but on its universal cover $\mathcal{A}(M)/\mathcal{G}_0$ with $\mathcal{G}_0 = \{g \in \mathcal{G} \mid \deg(g) = 0\}$. At any rate the gradient flow of A is well defined, by

$$\frac{d}{dt}A(t) + *F_{A(t)} = 0 \text{ on } M \times \mathbf{R}$$

where $*$ is the Hodge duality from two forms to one forms, and F_A is the curvature of $A(t)$.

Since this is a Fredholm equation, the set of trajectories displays the usual compactness, except for «bubbling off» of a «point instanton». But this can only happen on a family $\mathcal{M}(A, B)$ of dimension at least 8.

Now for any two flat connections on $M : A, B, \mathcal{M}(A, B)$ has connected components of different dimensions, but they are all equal modulo 8 (in other words $i(A)$ is only defined in $\mathbf{Z}/8\mathbf{Z}$.

Thus there is no problem in defining $IH^*(M)$ since this only involves 1 and 2 dimensional trajectory spaces, as well as $P(\omega)$ for $\omega \in H^k(\mathcal{A}/\mathcal{G}_0) = H^k(\mathcal{B}\mathcal{G}_0)$ for $k < 8$ (since $\mathcal{A}(M)$ is contractible $\mathcal{A}/\mathcal{G}_0$ has the homotopy type of the classifying space $\mathcal{B}\mathcal{G}_0$). Here we totally ignored all difficulties in the construction of $IH^*(M)$, due in particular to the existence of reducible connections, and non-transversality of $\mathcal{M}(A, B)$.

However, it seems to us that there is no obstruction to defining an $R_*^{(8)}$ module structure on $IH^*(M)$, where

$$R_* = H^*(\mathcal{B}\mathcal{G}_0)\,.$$

Appendix.
About the use of Stokes' formula

Lemma *Let X be a compact manifold with corners, and $G : X \to M$ a map such that*

(1) *G is Lipschitz;*

(2) *G is C^2 on $\overset{\circ}{X}$, and $G_{|\overset{\circ}{X}}$ is a diffeomorphism on its image;*

(3) *G is C^2 on the regular part of $X - \overset{\circ}{X} = \partial X$ and $G_{|\partial X}$ is a diffeomorphism on its image.*

Then we have, denoting $G(X) = W$, $G(\partial X) = \partial W$

$$\int_W d\alpha = \int_{\partial W} \alpha$$

for any form α.

Proof: Let G_n be a sequence of C^2 maps such that $G_n \to G$ in the Lipschitz topology; in other terms

$$|[G_n(x) - G(x)] - [G_n(y) - G(y)]| < \varepsilon_n \, |y - x|\ .$$

The sequence G_n may be obtained by convolution. Then, for any form φ on M, and Z a submanifold of X,

$$\int_Z G_n^* \varphi \to \int_Z G^* \varphi\,.$$

Indeed, φ is a sum of forms of the type $a(x) dx^1 \wedge \quad \wedge dx^p$; then

$$\int G_n^* \varphi = \int a(G_n(x)) dG_n^1 \wedge \quad \wedge dG_n^p\,.$$

Now $|a(G_n(x))dG_n^1 \wedge \quad \wedge dG_n^p| \leq M(K+\varepsilon)^p$ where $M = \sum |a|$, K is the Lipschitz constant for G. Since on Z, $G_n^*\varphi \to G^*\varphi$ pointwise, we have by Lebesgue's dominated convergence that $\int_Z G_n^*\varphi$ goes to $\int_Z G^*\varphi$.

The result is now clear, since

$$\begin{aligned} \int_W d\alpha = \int_X G^*(d\alpha) &= \int_X d(G^*\alpha) \quad \text{using (2)} \\ &= \lim \int_X d(G_n^*\alpha) \\ &= \lim_n \int_{\partial X} G_n^*\alpha \qquad \text{Stokes} \\ &= \int_{\partial X} G^*\alpha = \int_{\partial W} \alpha \qquad \text{by (3)} \end{aligned}$$

References

[C-J-S] Cohen, R., Jones, J., Segal, G., Personnal communication, see also this proceeding (Floer Memorial volume).

[Fl 1] Floer, A., The unregularized Gradient Flow of the Symplectic Action; Comm. Pure Appl. Math., **41**, (1988), pp. 775–813.

[Fl 2] Floer, A., Morse Theory for Lagrangian Intersections; Journal of Differential Geometry, **28**, (1988), pp. 513–547.

[Fl 3] Floer, A., Symplectic fixed points and Holomorphic spheres; Commun. in Math. Phys., **120**, (1988), pp. 575–611.

[Fl 4] Floer, A., Witten's complex and infinite-dimensional Morse theory; J. of Differential Geom. **30**, (1989), pp. 207–221.

[Fl 5] Floer, A., An instanton invariant for three manifolds; Communication in Math. Phys., **118**, (1988), pp. 215–240.

[Fl 6] Floer, A., Cuplength estimates on Lagrangian intersections; Comm. Pure Appl. Math., **42**, (1989), pp. 335–357.

[Fl-H] Floer, A., Hofer, H., Coherent orientation for periodic orbits in symplectic geometry; Math. Zeitschrift, 1993, pp. 13–38.

[H-S] Hofer, H., Salamon, D.A., Floer Homology and Novikov rings, in this volume.

[L] Laudenbach, F., On the Thom-Smale complex, Appendix to «An extension of a theorem by Cheeger and Müller» by J.M. Bismut; Astérisque, Soc. Math. France, (1992).

[M] Milnor. «Lectures on the h-cobordism theorem», Notes by L. Siebenmann and J. Sondow; Princeton University Press, 1965.

[S-U] Sacks, J. Uhlenbeck, K., The existence of minimal immersions of 2-spheres; Ann. of Math, **113** , (1981), pp. 1–24.

[S-Z] Salamon, D. and Zehnder, E. Floer Homology, Maslov index and periodic orbits of Hamiltonian equations; Analysis et cetera, Papers in the honor of J. Mosers 60th birthday, Academic Press 1987, pp. 573–600.

[Th] Thom, R., Sur une partition en cellules associée à une fonction sur une variété. Comptes Rendus Acad. Sci. Paris Série I, (1949), pp. 973–975 .

[V 1] Viterbo, C., Intersections de sous-variétés lagrangiennes, fonctionelles d'action et indice des systèmes hamiltoniens; Bull. Soc. Math. France, **115**, (1987), pp. 361–390.

[V 2] Viterbo, C., Some remarks on Massey products and Lusternik-Shnirelman category; in preparation.

[Wi] Witten, E., Supersymmetry and Morse theory; J. Diff. Geometry, **17**, (1982), pp. 661–692.

U.R.A. 1169
Département de mathématique
Bâtiment 425 Université de Paris-Sud
F-91405 Orsay Cedex
France

DMI
Ecole Normale Supérieure
45 rue d'Ulm
F-75230 Paris Cedex 05
France

Added in proof: We refer to the paper by Braam and Donaldson in this volume for more information on the Instanton Floer homology applications of the cup product.

The symplectic structure on moduli space

A. Weinstein[1])

1 Introduction

The «moduli space» referred to in the title of this paper is the space $\mathcal{M}$ of gauge equivalence classes of flat connections on a principal G bundle over a compact Riemann surface Σ. Atiyah and Bott [1] constructed a symplectic structure $\omega_{\mathcal{M}}$ on $\mathcal{M}$ by symplectic reduction from the infinite dimensional symplectic manifold of *all* connections. Since $\mathcal{M}$ is a finite-dimensional object, it seems desirable to have a finite-dimensional construction of its symplectic form. In fact, $\mathcal{M}$ can also be realized as the representation space $\mathrm{Hom}(\pi, G)/G$, where π is the fundamental group of Σ and G acts by conjugation. Using the resulting identification of the tangent spaces of $\mathcal{M}$ with cohomology spaces of π with suitable coefficients, Goldman [9] gave a direct construction of $\omega_{\mathcal{M}}$ as a nondegenerate 2-form, but he was unable to prove that this form is closed without recourse to the infinite-dimensional picture. This gap in Goldman's approach was filled recently by Karshon [10], who showed that $d\omega_{\mathcal{M}} = 0$ by methods of group cohomology, without using the space of connections.[2])

The purpose of this paper is to place Karshon's computations in an appropriate general setting, namely the double complex of forms on cartesian powers of G [5] [14], or more precisely the corresponding complex of equivariant forms [2] [3] with respect to the action of G by conjugation.

Andreas Floer's refinement of the Casson invariant for homology 3-spheres was constructed in terms of the gradient «flow» of the Chern-Simons invariant on a space of connections. Floer himself remarked that there should be an alternative construction of his invariant which, like Casson's own, takes place on the finite-dimensional space $\mathcal{M}$.[3]) Such a construction would use the symplectic structure of $\mathcal{M}$ in an essential way. We hope that the new description of the symplectic structure here may be of some assistance in the realization of Floer's hope.

I would like to thank S. Bates, R. Bott, A. Givental, Y. Karshon, and J. Weitsman for their valuable help in the preparation of this paper.

2 The de Rham-bar bicomplex

The standard «bar» complex for computing the cohomology of a group G [6] involves functions on cartesian powers of the group. When G is a Lie group, the

---

1) Research partially supported by NSF Grant DMS-90-01089.

2) L. Guillou and J. Huebschmann also inform me that they have found finite-dimensional proofs that the form is closed.

3) Some progress in this direction is reported in [7].

functions can be replaced by differential forms, and one obtains a double complex $C^{**}(G)$, with $C^{p,q}(G)$ defined to be the space $\Omega^q(G^p)$ of q forms on G^p. The second coboundary operator is just the exterior differential d, while the first, δ, is defined by the usual formula from group cohomology, with some reinterpretation of the notation:

$$\delta\alpha(a_0,\dots,a_p) = \alpha(a_1,\dots,a_p) + \sum_{i=1}^{p}(-1)^i\alpha(a_0,\dots,a_{i-1}a_i,\dots,a_p)$$
$$+(-1)^{p+1}\alpha(a_0,\dots,a_{p-1}).$$

Here, when ϕ is a differential form on Y and $y = f(x)$ is the notation «in variables» for a mapping $f : X \to Y$, the expression $\phi(f(x))$ is interpreted to mean the pullback of ϕ to X by f, just as in the case of functions.

The importance of the de Rham-bar bicomplex revolves around the fact that the cohomology of the corresponding total complex (with differential $(-1)^p d + \delta$) is naturally isomorphic to the real cohomology $H^*(BG)$ of the universal classifying space of G. Furthermore, there is a natural mapping Φ from the ring of invariant polynomials on the Lie algebra $\mathfrak{g}$ to the total cocycles in $C^{**}(G)$ (see [14]).

Application of the de Rham-bar bicomplex $C^{**}(G)$ to the moduli space comes about through the following construction. Let π be a discrete group. We consider the space $\mathrm{Hom}(\pi, G)$ as a singular «manifold» as is usual in this subject [9]. The evaluation mappings $E : \pi^p \times \mathrm{Hom}(\pi, G) \to G^p$ induce pullbacks $E^* : C^{p,q}(G) \to \Omega^q(\pi^p \times \mathrm{Hom}(\pi, G))$. Since π is discrete, the last space can be identified with $C^p(\pi) \otimes \Omega^q(\mathrm{Hom}(\pi, G))$, where $C^p(\pi)$ is the usual space of (real-valued) p-cochains on π, i.e. the functions on π^p.

Our key observation is that E^* is a mapping of bicomplexes, where $C^*(\pi) \otimes \Omega^*(\mathrm{Hom}(\pi, G))$ is considered as the tensor product of the bar complex of π and the de Rham complex of $\mathrm{Hom}(\pi, G)$. Composing E^* with Φ, we get a map from the invariant polynomials on $\mathfrak{g}$ to the tensor product of the cocycles in $C^*(\pi)$ and the closed forms on $\mathrm{Hom}(\pi, G)$. Applying this map to the quadratic polynomial on $\mathfrak{g}$ given by an invariant symmetric bilinear form, and pairing the result with the «fundamental cycle» for π, when π is the fundamental group of a Riemann surface, we obtain a closed 2-form $\tilde{\omega}$ on $\mathrm{Hom}(\pi, G)$.

It turns out that $\tilde{\omega}$ is just the pullback of the symplectic structure on moduli space. It is clear from our discussion so far that $\tilde{\omega}$ is invariant under the adjoint action of G, but it is not so clear that $\tilde{\omega}$ is horizontal. Horizontality is, however, analogous to closedness where the operator d is replaced by the operators of interior product with the infinitesimal generators of the adjoint action. The sum of these two operators is essentially the equivariant differential [2] [3], so it will be natural to extend all of our bicomplexes and complexes to their equivariant versions.

3 Maurer-Cartan calculus

Given a Lie algebra $\mathfrak{g}$ and a manifold M, the $\mathfrak{g}$-valued differential forms on M form a super-Lie algebra; i.e. the algebraic identities defining a Lie algebra are satisfied by $\mathfrak{g}$-valued forms with the appropriate inclusion of signs [11]. In particular, $[\eta, \eta]$ is not necessarily zero when η is a $\mathfrak{g}$-valued 1-form, but the super Jacobi identity implies that $[\eta, [\eta, \eta]] = 0$. For further properties of vector-valued differential forms, we refer to [4].

On any Lie group G, we denote by ω the $\mathfrak{g}$-valued left invariant 1-form which maps each tangent vector to the left invariant vector field having that value. The corresponding right invariant form will be denoted by $\overline{\omega}$. They are related by the equation

$$\overline{\omega} = \phi\omega,$$

where ϕ is the adjoint representation, considered as a 0-form on G with values in $\mathrm{Hom}(\mathfrak{g}, \mathfrak{g})$.

The forms ω and $\overline{\omega}$ satisfy the structure equations

$$d\omega = -\frac{1}{2}[\omega, \omega]$$

and

$$d\overline{\omega} = \frac{1}{2}[\overline{\omega}, \overline{\omega}].$$

Suppose now that $\mathfrak{g}$ has an Ad-invariant inner product (not necessarily positive definite), which we denote simply by $\cdot$. Ad-invariance means that $[x, y] \cdot z = -y \cdot [x, z]$ for any three elements of $\mathfrak{g}$, so that the triple product

$$[x, y] \cdot z \tag{1}$$

is completely antisymmetric. For $\mathfrak{g}$-valued 1-forms, it is completely symmetric.

We consider the 3-form

$$\lambda = \frac{1}{6}\omega \cdot [\omega, \omega].$$

Lemma 3.1 $d\lambda = 0$.

Proof. $[\omega, \omega]$ is closed by the structure equation, so

$$d\lambda = -\frac{1}{12}[\omega, \omega] \cdot [\omega, \omega].$$

By Ad-invariance, the last expression equals $-\frac{1}{12}\omega \cdot [\omega, [\omega, \omega]]$, which vanishes by the super Jacobi identity. □

We will use the following notation for forms and vector fields on $G \times G$. If α is any differential form on G, we denote by α_i the pullback of α to $G \times G$ by the projection p_i onto the i'th component. If X is a vector field on G, we denote by X_i the vector field on $G \times G$ which projects to X under p_i and to zero under the other projection. $m : G \times G \to G$ is group multiplication.

In terms of this notation, we have the following lemma, whose proof is left to the reader:

Lemma 3.2 $m^*\omega = \phi_2^{-1}\omega_1 + \omega_2$.

The object which will eventually give us the symplectic structure on moduli space is the 2-form

$$\Omega = \omega_1 \cdot \overline{\omega}_2$$

on $G \times G$. Using the structure equations and the relation between ω and $\overline{\omega}$, we compute the exterior derivative

$$d\Omega = -\frac{1}{2}([\omega_1, \omega_1] \cdot \phi_2\omega_2 + \omega_1 \cdot [\phi_2\omega_2, \phi_2\omega_2]).$$

The form $d\Omega$ is evidently nonzero (unless G is abelian). However, we have:

Lemma 3.3 $d\Omega = \delta\lambda$.

Proof.

$$\begin{aligned} -\delta\lambda &= m^*\lambda - \lambda_1 - \lambda_2 \\ &= \frac{1}{6}(\phi_2^{-1}\omega_1 + \omega_2) \cdot [\phi_2^{-1}\omega_1 + \omega_2, \phi_2^{-1}\omega_1 + \omega_2] \\ &\quad - \frac{1}{6}\omega_1 \cdot [\omega_1, \omega_1] - \frac{1}{6}\omega_2 \cdot [\omega_2, \omega_2]. \end{aligned}$$

Using the fact that the values of ϕ are automorphisms of both the bracket and inner product structures on $\mathfrak{g}$, we may simplify the expression above to:

$$\frac{1}{2}([\omega_1, \omega_1] \cdot \phi_2\omega_2 + \omega_1 \cdot [\phi_2\omega_2, \phi_2\omega_2]).$$ □

Another simple computation, which we leave to the reader, shows:

Lemma 3.4 $\delta\Omega = 0$.

Combining Lemmas 3.1, 3.3, and 3.4, we conclude:

Corollary 3.5 *$\Omega - \lambda$ is a cocycle of degree 4 in the total complex of the double complex $C^{**}(G)$.*

Remark The forms Ω and λ were originally found by analysis of objects in [10], but in fact $\Omega - \lambda$ is precisely the result of applying the Bott-Shulman map Φ to the invariant polynomial $X \cdot X$ on $\mathfrak{g}$.

4 The equivariant theory

The equivariant cohomology of a G-manifold M can be computed using the complex of *equivariant differential forms* on M [2] [3]. By definition, these are polynomial maps from $\mathfrak{g}$ to $\Omega^*(M)$ which are equivariant with respect to the adjoint representation. The equivariant differential d_G is defined by $(d_G\alpha)(X) = (d - i_{X_M})(\alpha(X))$. Here, X_M denotes the vector field given by the Lie algebra element X acting on M.

In the setting of the previous section, if a group H acts on G by automorphisms, it is natural to extend the double complex $C^{**}(G)$ to its equivariant version $C_H^{**}(G)$, where $C_H^{p,q}(G)$ consists now of the H-equivariant q-forms on G^p, and the second differential is replaced by d_H.[4]) In what follows, we will restrict our attention to the special case where $H = G$, acting by inner automorphisms.

The form $\Omega - \lambda$, which is closed for the operator $\delta + (-1)^p d$, is *not* equivariantly closed. The extra part of the equivariant differential is calculated in the next two lemmas, in which the X on the right hand side is to be interpreted as a constant $\mathfrak{g}$-valued 0-form on G.

Lemma 4.1 $i_{X_G}\lambda = -d(X \cdot (\omega + \overline{\omega}))$.

Proof. We begin by noting that X_G for the adjoint representation is just $X - \overline{X}$, where $\overline{X}$ is the right-invariant vector field which agrees at the identity with the left invariant vector field X. Then we have

$$6 i_{X_G}\lambda = i_{X-\overline{X}}(\omega \cdot [\omega, \omega]).$$

Since λ is also equal to $\overline{\omega} \cdot [\overline{\omega}, \overline{\omega}]$, the last expression can be rewritten as

$$i_X(\omega \cdot [\omega, \omega]) - i_{\overline{X}}(\overline{\omega} \cdot [\overline{\omega}, \overline{\omega}]).$$

Using the equations $i_X\omega = X$ and $i_{\overline{X}}\overline{\omega} = X$ and the symmetry properties of the triple product ((1)), the last expression simplifies to

$$3X \cdot ([\omega, \omega] - [\overline{\omega}, \overline{\omega}]),$$

which, by the structure equations, is just

$$-d(6X \cdot (\omega + \overline{\omega})). \qquad \square$$

Lemma 4.2 $i_{X_{G\times G}}\Omega = X \cdot (\omega_1 - \phi_2^{-1}\omega_1 + \overline{\omega}_2 - \phi_1\overline{\omega}_2)$.

4) We expect that the cohomology of the corresponding total complex should be isomorphic to the H-equivariant cohomology of BG, which is in turn isomorphic to the cohomology of BK, where K is the semidirect product of H with G.

Proof. Since $X_{G\times G} = X_1 - \overline{X}_1 + X_2 - \overline{X}_2$, we have

$$i_{X_{G\times G}}\Omega = i_{X_1-\overline{X}_1+X_2-\overline{X}_2}(\omega_1 \cdot \overline{\omega}_2)$$

Using the fact that $\overline{X} = \phi^{-1}X$, one can easily transform the last expression into the right hand side of the statement of the lemma. □

Lemmas 4.1 and 4.2 suggest that a piece of the puzzle is missing, namely the equivariant form

$$\theta(X) = X \cdot (\omega + \overline{\omega}).$$

The calculations above show that

$$(\delta + (-1)^p d_G)(\Omega - \lambda + \theta)(X) = -i_{X_{G\times G}}\Omega + \delta\theta(X) + i_{X_G}\theta(X).$$

The next two lemmas now complete the picture.

Lemma 4.3 $i_{X_G}\theta(X) = 0.$

Proof.

$$i_{X_G}\theta(X) = i_{X-\overline{X}}(X \cdot (\omega + \overline{\omega})) = X \cdot (i_X\omega - i_{\overline{X}}\overline{\omega}) = 0.$$ □

Lemma 4.4 $\delta\theta(X) = i_{X_{G\times G}}\Omega$

Proof.

$$\begin{aligned}\delta\theta(X) &= \theta(X)_1 + \theta(X)_2 - m^*\theta(X)\\ &= X \cdot (\omega_1 + \overline{\omega}_1 + \omega_2 + \overline{\omega}_2 - (\phi_2^{-1}\omega_1 + \omega_2 + \overline{\omega}_1 + \phi_1\overline{\omega}_2))\\ &= X \cdot (\omega_1 + \overline{\omega}_2 - (\phi_2^{-1}\omega_1 + \phi_1\overline{\omega}_2)).\end{aligned}$$

Now compare with Lemma 4.2. □

As a consequence of the preceding calculations, we have

$$(\delta + (-1)^p d_G)(\Omega - \lambda + \theta) = 0.$$

We state our main results below.

Theorem 4.5 *$-\lambda + \theta$ is an equivariantly closed form of degree 3 on G, and $Q_4 = \Omega - \lambda + \theta$ is an equivariantly closed element of degree 4 in the total complex of the bicomplex $C_G^{**}(G)$.*

5 Forms on representation spaces

To complete our work, we have little to do but to pull back the 4-form Q_4 of Theorem 4.5 by the evaluation maps $E : \pi^p \times \mathrm{Hom}(\pi, G) \to G^p$. Since each E is obviously equivariant, E^*Q_4 is a cocycle in the total complex of the bicomplex $C^*(\pi) \otimes C^*_G(\mathrm{Hom}(\pi, G))$. By pairing with cycles in $C_*(\pi)$, we obtain equivariant forms on $\mathrm{Hom}(\pi, G)$.

If $c \in C_2(\pi)$ is a 2-cycle, then

$$\begin{aligned} d_G\langle E^*\Omega, c\rangle &= d_G\langle E^*Q_4, c\rangle = \langle E^*d_GQ_4, c\rangle \\ &= E^*\langle -\delta Q_4, c\rangle = -E^*\langle Q_4, \partial c\rangle = 0, \end{aligned}$$

so $\langle E^*\Omega, c\rangle$ is an equivariantly closed 2-form on $\mathrm{Hom}(\pi, G)$. A similar computation using $\Omega - \lambda$ and the ordinary differential shows that $\langle E^*\Omega, c\rangle$ is closed. It follows that $\langle E^*\Omega, c\rangle$ is a G-invariant closed form on $\mathrm{Hom}(\pi, G)$ which is annihilated by $i_{X_{\mathrm{Hom}(\pi,G)}}$ for every $X \in \mathfrak{g}$, so that it is the pullback of a closed 2-form $\omega_{\mathcal{M}}$ on (at least the smooth part of) the moduli space $\mathcal{M}$.

If we pair with a 2-boundary ∂b, then we have

$$\langle E^*\Omega, \partial b\rangle = \langle \delta E^*\Omega, b\rangle = \langle E^*\delta\Omega, b\rangle,$$

which is zero by Lemma3.4. Thus we have proven:

Theorem 5.1 *Pairing $E^*\Omega$ with 2-cycles defines a natural homomorphism from $H_2(\pi)$ to the space of closed 2-forms on $\mathcal{M}$.*

Following [10], one may show easily that if π is the fundamental group of a closed Riemann surface and c is the fundamental cycle of that surface, then $\omega_{\mathcal{M}}$ is the usual symplectic structure on moduli space.

It is also possible to pair $E^*(\lambda)$ with a class of degree 1 in $H_*(\pi)$ (i.e. a conjugacy class in π). The resulting object is a closed, invariant, 3-form on $\mathrm{Hom}(\pi, G)$. This form does not, however, push down to $\mathcal{M}$. In fact, if we subtract $E^*\theta$ from $E^*\lambda$, the result is an equivariantly closed form on $\mathrm{Hom}(\pi, G)$, but it can be pushed down as a class to $\mathcal{M}$ only where the G action is free, and as a form only with the aid of a connection for that action.

A reader familiar with other work on the cohomology of moduli spaces may recognize that we have found, in a new guise, the generating classes of Newstead [12]. (Actually, these classes apply directly only to the «twisted» moduli spaces; see [1].) One of Newstead's generators is, however, missing, namely the class of degree 4 obtained by pairing with the fundamental 0-cycle of π. In fact, we can obtain this class as well by considering the element of $C^{0,4}_G(G)$ given simply by the invariant quadratic polynomial $\Phi(X) = X \cdot X$. This pulls back to $\Omega^*_G(\mathrm{Hom}(\pi, G))$ as the same polynomial, which is an equivariant form of degree 4. The limitations on pushing down to $\mathcal{M}$ are the same as those just discussed for the classes of degree 3.

It is perhaps useful to describe the symplectic form on $\mathcal{M}$, as we have constructed it, in more concrete terms. A 2-cycle for π is just a (finite) formal linear combination of elements of $\pi \times \pi$. Pairing $E^*(\Omega)$ with such a cycle gives a linear combination of pullbacks of Ω from G^2 to $\mathrm{Hom}(\pi, G)$ by evaluation on pairs of elements of π. Now choose a finite subset γ of π which contains a generating set and is large enough so that our cycle is supported in $\gamma \times \gamma$. We consider the singular space $\mathrm{Hom}(\pi, G)$ as embedded into the smooth manifold G^γ of maps from γ to G. The evaluation map E extends in an obvious way to a map from $\pi^2 \times G^\gamma$ to G^2, so when we pair with the cycle we get a smooth 2-form on G^γ which extends our form on $\mathrm{Hom}(\pi, G)$. It is important to note, though, that the extended evaluation map is no longer compatible with the two bicomplex structures, so *the extended 2-form is not necessarily closed.* We do not know whether a smooth closed extension can be found; nevertheless, it is reassuring to see the form on the singular space $\mathrm{Hom}(\pi, G)$ extended to a smooth form on a smooth manifold.

6 Discussion

Our construction of the symplectic structures on moduli spaces raises several questions.

First of all, it would be nice to prove with our formalism that the closed 2-form on moduli space is nondegenerate. One way to do so would be to construct the corresponding Poisson structure by an analogous procedure. This approach leads to the idea of applying our method to the moduli spaces of flat connections on manifolds with boundary, which are Poisson manifolds in which the symplectic leaves are given by specifying the conjugacy class of the holonomy on each boundary circle.

Secondly, we have dealt with only the characteristic class of degree 4 for the group G. It turns out that this leads to several equivariant cohomology classes on $\mathrm{Hom}(\pi, G)$, only one of which passes to a canonical form on $\mathcal{M}$. It would be interesting to carry out a similar analysis for the entire characteristic ring of G.

Although the forms Ω and λ were seen to arise directly by the Bott-Shulman construction, the equivariant form $\theta(X)$ was put in «by hand». It should be possible to find a «grand unified theory» encompassing all three of these forms, as well as the degree 4 class mentioned near the end of Section 5, by using a suitable de Rham model for $H^*_G(BG)$.

The geometric quantization of $\mathcal{M}$, which is so important for topological quantum field theory, begins with the construction of a line bundle whose curvature is $\omega_{\mathcal{M}}$. The construction of this line bundle in [13] relies on the Chern-Simons functional. Is it possible to construct the line bundle from an object analogous to the de Rham-bar complex?

We hope to return to these points in a sequel to this paper.

Note added in proof (February 28, 1994): significant developments along the lines initiated in this paper have been made (independently) by J. Huebschmann and L. Jeffrey, in work soon to be published.

References

[1] Atiyah, M.F. and Bott, R., The Yang-Mills equations over Riemann surfaces, *Phil. Trans. R. Soc. Lond.* **A 308** (1982), 523–615.

[2] Atiyah, M.F. and Bott, R., The moment map and equivariant cohomology, *Topology* **23** (1984), 1–28.

[3] Berline, N., Getzler, E., and Vergne, M., *Heat Kernels and Dirac Operators*, Berlin/New York, Springer-Verlag, 1992.

[4] Bishop, R.L., and Crittenden, R.J., *Geometry of Manifolds*, Academic Press, New York, 1964.

[5] Bott, R., On the Chern-Weil homomorphism and the continuous cohomology of Lie groups, *Advances in Math.* **11** (1973), 289–303.

[6] Brown, K.S., *Cohomology of groups*, Springer-Verlag, New York, 1982.

[7] Cappell, S.E., Lee, R., and Miller, E.Y., A symplectic geometry approach to generalized Casson invariants of 3-manifolds, *Bull. Amer. Math. Soc.* **22** (1990), 269–275.

[8] Floer, A., An instanton invariant for three manifolds, *Comm. Math. Phys.* **118** (1988), 215–240.

[9] Goldman, W., The symplectic nature of fundamental groups of surfaces, *Adv. in Math.* **54** (1984), 200–225.

[10] Karshon, Y., An algebraic proof for the symplectic structure of moduli space, *Proc. Amer. Math. Soc.* **116** (1992), 591–605.

[11] Manin, Y.I., Gauge field theory and complex geometry, Berlin/New York, Springer-Verlag, 1988.

[12] Newstead, P.E., Characteristic classes of stable bundles over an algebraic curve, *Trans. Amer. Math. Soc.* **169** (1972), 337–345.

[13] Ramadas, T.R., Singer, I.M., and Weitsman, J., Some coments on Chern-Simons gauge theory, *Commun. Math. Phys.* **126** (1989), 409–420.

[14] Shulman, H.B., Characteristic classes and foliations, Ph.D. Thesis, University of California, Berkeley (1972).

Alan Weinstein
Department of Mathematics, University of California
Berkeley, CA 94720 USA
(alanw@math.berkeley.edu)

Chern-Simons gauge theory as a string theory

E. Witten*

Abstract. Certain two dimensional topological field theories can be interpreted as string theory backgrounds in which the usual decoupling of ghosts and matter does not hold. Like ordinary string models, these can sometimes be given space-time interpretations. For instance, three-dimensional Chern-Simons gauge theory can arise as a string theory. The world-sheet model in this case involves a limiting case of Floer/Gromov theory of symplectic manifolds. The instantons usually considered in Floer theory give rise to Wilson line insertions in the space-time Chern-Simons theory. A certain holomorphic analog of Chern-Simons theory can also arise as a string theory.

1 Introduction

In this paper, I will describe how Chern-Simons gauge theory in three dimensions can be viewed as a string theory. The string theory in question will be constructed using a topological sigma model [1] associated with a degenerate case of Floer/Gromov theory of symplectic manifolds [2], [3], in which the target space is T^*M, M being a three-manifold. The perturbation theory of this string theory will coincide with Chern-Simons perturbation theory, in the form that this has been presented by Axelrod and Singer [4] and further studied by Kontsevich [5]. Mathematically, the idea is roughly that there are no instantons with target space T^*M and boundary values in M, so in an appropriate topological field theory, the usual counting of instantons is vacuous. However, there are virtual instantons at infinity; their proper counting leads to Chern-Simons perturbation theory. Chern-Simons theory enters in this particular string theory in much the same way that ordinary space-time physics (with general relativity as the long wavelength limit) arises in conventional string theory.

Physically, one might take the following as the starting point. String theorists usually construct two dimensional field theories describing particular classical solutions of string theory by constructing a «matter» system, of total central charge 26, and coupling it to the ghosts, of central charge -26. The vanishing total central charge ensures the existence of a BRST operator Q, obeying $Q^2 = 0$, and playing a crucial role in world-sheet and space-time gauge invariance.

One knows, however [7], that Q can be interpreted as a generator of linearized gauge transformations, mixing ghosts and matter, so the assumption that the matter and ghosts are decoupled cannot be valid as a fundamental principle; it is merely a (partial) gauge condition and very likely cannot be imposed at all in some situations, perhaps time dependent ones.

---

*) Research supported in part by NSF Grant PHY91-06210.

So in §2, we will ask what is left of the standard structure if one does not assume decoupling of matter from ghosts. From the discussion, it will become obvious that exotic realizations of the same basic structure can be constructed using topological sigma models; we therefore do this in §3. The resulting models can be considered for either open or closed strings, but the open string case is in some ways easier to understand. In §4, we use open-string field theory as a short-cut to determining the space-time interpretation of the open-string version of our models, with the result — alluded to above — that one type of model is equivalent in perturbation theory to three dimensional Chern-Simons gauge theory. This is possibly the first time that the background independent, gauge invariant space-time interpretation of a string theory has been completely determined. (But see [8] for a previous investigation of the space-time interpretation of some topological string theories of a rather different flavor.) As one application, our result explains certain observations by Kontsevich about Chern-Simons gauge theory, and we will make a small digression on that account. Another version of the theory has for its space-time interpretation a sort of holomorphic version of Chern-Simons theory. In §5, we attempt to discuss the closed string sector (which among other things should be more closely related to usual manifestations of mirror symmetry). This concluding section is brief since I do not understand it. (However, see [32], which was written subsequently to the present work and shed considerable light on the closed string sector.)

To keep this paper within reasonable length, it has not been possible to give a self-contained explanation of all of the relevant background. The relevant material on topological sigma models can be found in [6]; for their coupling to topological gravity see [1]. The relation of Riemann surfaces to gauge theories will be briefly recalled presently. Apart from this, it is helpful to have some familiarity with the construction of string models, the relation between world-sheet and space-time physics, string field theory, and Chern-Simons perturbation theory. The main point of the paper is the unexpected relation of those latter topics to each other and to Floer theory.

Riemann surfaces and gauge theories

Riemann surfaces enter our story in two ways. On the one hand, when we study two dimensional quantum field theory, Riemann surfaces are present at the beginning. Then we wish to show how gauge theory in the target space emerges.

On the other hand, oriented two dimensional manifolds arise in gauge theories with $U(N)$ gauge group in the following fashion, due originally to 't Hooft [9]; see also [10], §4 for a recent explanation. ($SO(N)$ and $Sp(N)$ can be considered similarly and lead to not necessarily orientable Riemann surfaces.) Though the following general comments also go through for ordinary Yang-Mills theory in any dimension, let us to be definite consider the Chern-Simons action for a $U(N)$-valued connection A on a three manifold M:

$$I = \frac{k}{4\pi} \int_M \mathrm{Tr} \left(A \wedge dA + \frac{2}{3} A \wedge A \wedge A \right). \tag{1.1}$$

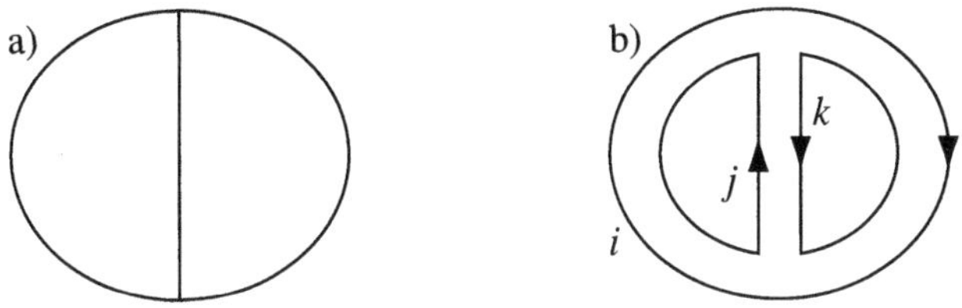

Fig. 1 A two loop Feynman diagram (a) and its thickening (b), in which the boundary components are labeled by gauge indices.

Consider expanding this theory in perturbation theory in $1/k$, say around the trivial connection. The successive terms are all topological invariants. The coefficient of $1/k^r$ comes from Feynman diagrams with r loops, and is a function of N. For given r, by considering the dependence on N, how many invariants can be extracted? This question can be conveniently answered as follows. The gauge field A is essentially a one-form valued in hermitian $N \times N$ matrices. Write A as $A^i{}_j$, $i, j = 1, \ldots, N$, making the matrix indices explicit while leaving implicit the fact that A is a one-form. The interaction

$$\mathrm{Tr}\, A \wedge A \wedge A = A^i{}_j \wedge A^j{}_k \wedge A^k{}_i \tag{1.2}$$

involves sewing the «lower» index of one A field to the «upper» index of the next. To exhibit this in drawing Feynman diagrams, represent the A propagator not as a line but as a slightly thickened band in which one edge represents the «upper» index and the other line represents the «lower» index. The index flow at the cubic vertex (1.2) is then neatly incorporated by smoothly joining the bands, as in Figure 1(b). In the process, the Feynman diagram of Figure 1(a) has been replaced by the Riemann surface Σ of Figure 1(b). Each boundary component is labeled by a gauge index i running from 1 to N. If there are h boundary components, the sum over gauge indices gives a factor of N^h. The coefficients of N^h/k^r are the three-manifold invariants that can be extracted from Chern-Simons theory with gauge group $U(N)$.

The genus of Σ is $g = (r - h + 1)/2$. As r and h vary, g and h vary independently, and Σ varies over all topological types of oriented two dimensional surface with boundary. Thus, $U(N)$ gauge theory gives us a three manifold invariant $\Gamma_{g,h}(M)$ for every topological type of oriented two dimensional surface with boundary. On the other hand, in Sections 3–4 we will consider a two dimensional topological field theory (closely related to the counting of almost holomorphic curves) in which one can conveniently take the target space to be T^*M, the cotangent bundle of a three-manifold M. The partition function $Z_{g,h}(M)$ of this theory formulated on a Riemann surface of genus g with h holes is, again, a three-manifold invariant depending on g and h. Our main conclusion is that $\Gamma_{g,h}(M) = Z_{g,h}(M)$. In reaching this conclusion, the link between the Riemann surfaces of the two dimensional field theory and those of $U(N)$ gauge theory is provided by the fatgraph [11]–[13] or open string field theory [14]–[15] description of the moduli space of Riemann surfaces with boundary; in this description, complex Riemann surfaces are built up from pictures similar to that of Figure 1(b), as we will recall in more detail in §4.

2 Axioms

In this section, we will discuss what remains of the usual structures in string theory if one does not assume that ghosts and matter are decoupled. Σ will be a Riemann surface of genus g with local coordinates x^α, $\alpha = 1,2$. The symbols J and h will be used to denote a complex structure and a metric on Σ; of course a metric determines a complex structure. The space of all complex structures will be called $\mathcal{J}$; the group of diffeomorphisms of Σ will be called $\mathcal{G}$. The quotient $\mathcal{J}/\mathcal{G}$ is the moduli space of complex structures on Σ. The space of metrics on Σ will be called $\mathcal{H}$.

Usually one considers conformally invariant world-sheeet theories consisting of «matter», of central charge $c = 26$, and «ghosts», of $c = -26$. As explained in the introduction, we want to drop the assumption of this decoupling; so if conformal invariance is retained, we preserve the fact that the total central charge is $c = 0$, but no longer build this from a cancellation between different contributions. We will, however, not necessarily retain the assumption of conformal invariance.

One part of the structure that must be kept for all that follows is the existence of a BRST operator Q with $Q^2 = 0$, and with the further property that the stress-tensor $T_{\alpha\beta}$ can be written as

$$T_{\alpha\beta} = \{Q, b_{\alpha\beta}\}, \tag{2.1}$$

for some field $b_{\alpha\beta}$ — which in the usual case is called the antighost field. This ensures general covariance — or topological invariance — of the two dimensional theory. In the conformally invariant case, b and T are traceless. The stress tensor is conserved, $D_\alpha T^\alpha{}_\beta = 0$, and so we will assume that the b field obeys

$$D_\alpha b^\alpha{}_\beta = 0 \tag{2.2}$$

which when b is traceless is its usual equation of motion. In addition, in the usual constructions, there are no short-distance singularities in operator products of b fields, and there are only the standard short-distance singularities in products of b and T. Concerning the generalization, topological sigma models are an example showing that these assumptions are too strong. In those models, there is a delta function contact term in the $b \cdot T$ product. (This results in the $\rho\rho\psi\psi$ term in the coupling to topological gravity, eqn. (4.18) of [1].) I do not know what sort of singularities should be allowed in $b \cdot b$ and $b \cdot T$ operator products, in general, except to say that they must be such that a coupling of the model to topological gravity must be possible.

Another feature of the usual case that we wish to preserve is that there is a «ghost number» operator G (mathematically it would usually be called the dimension), with Q and b having $G = 1$ and $G = -1$, respectively. Moreover, we wish to preserve the usual fact that the «ghost number of the vacuum» is $-3\chi(\Sigma)$ where $\chi(\Sigma)$ is the Euler characteristic of Σ; this means that only an operator product of total ghost number $3\chi(\Sigma)$ can have a non-zero vacuum expectation value. For instance, for an orientable Riemann surface of genus g, the case that we will consider in this section for definiteness, $3\chi(\Sigma) = -(6g-6)$.

A particularly important case (in genus ≥ 2 for simplicity, to avoid ghost zero modes) is the expectation value of a product of $6g-6$ b's. Let $\delta^{(k)}h$, $k = 1, \ldots, 6g-6$ be $6g-6$ variations of h (that is, $6g-6$ tangent vectors to the space $\mathcal{H}$ of metrics). Let

$$b^{(k)} = \int_{\Sigma} d^2x \sqrt{h} h^{\alpha\alpha'} h^{\beta\beta'} \delta^{(k)} h_{\alpha\beta} \cdot b_{\alpha'\beta'} \tag{2.3}$$

be the corresponding modes of b, and let

$$\Theta(\delta^{(1)}h, \ldots, \delta^{(6g-6)}h) = \langle b^{(1)} \ldots b^{(6g-6)} \rangle, \tag{2.4}$$

with $\langle\ \ \rangle$ the expectation value in the «measure» determined by the genus g world-sheet path integral. By fermi statistics, Θ is skew-symmetric in the $\delta^{(k)}h$'s and so can be interpreted as a $6g-6$ form on $\mathcal{H}$. In the usual case, one proves that Θ is closed starting with

$$0 = \langle \{Q, b^{(1)} \ldots b^{(6g-5)}\} \rangle = \sum_{j=1}^{6g-5} (-1)^{j-1} \langle b^{(1)} \ldots \{Q, b^{(j)}\} \ldots b^{(6g-5)} \rangle. \tag{2.5}$$

Then using (2.1) to write $\{Q, b^{(j)}\}$ as a moment of T, and interpreting the insertion of T as a derivative on $\mathcal{H}$, one can interpret (2.5) as a first order differential equation for Θ which amounts to

$$d\Theta = 0. \tag{2.6}$$

This proof goes through in general if the $b \cdot b$ and $b \cdot T$ operator products are standard; however, topological sigma models are an example in which the definition of Θ must be modified by addition of contact terms to ensure $d\Theta = 0$. A successful coupling of the model to topological gravity will always ensure the existence of a suitable modification (by addition of contact terms) of the definition of Θ.

Θ is obviously diffeomorphism invariant. Moreover, in the usual case it is «basic»; that is, it vanishes if any $\delta^{(k)}h$ is of the form $\delta^{(k)}h_{\alpha\beta} = D_\alpha v_\beta + D_\beta v_\alpha$ induced by an infinitesimal diffeomorphism $x^\alpha \to x^\alpha + \epsilon v^\alpha$. This is proved by integrating by parts, using (2.2) and the absence of short distance singularities of the b's. In general (because of the relation of the topological gravity multiplet to the equivariant cohomology of the diffeomorphism group), a successful coupling to topological gravity will always ensure that Θ is basic.

The fact that Θ is basic and diffeomorphism invariant means that it can be interpreted as the pullback of a closed $6g-6$ form ν on the space $\mathcal{H}/\mathcal{G}$ of metrics modulo diffeomorphisms. In the conformally invariant case, Θ is actually a pullback from a form μ on the moduli space $\mathcal{M}$ of complex structures; as $\mathcal{M}$ is $6g-6$ dimensional, μ is a top form or measure on $\mathcal{M}$. If conformal invariance does not hold, there is no natural way to construct from μ a measure on $\mathcal{M}$, but the fact that ν is closed, $\mathrm{d}\nu = 0$, is almost as good. It means that if s is a section of the bundle $\mathcal{H}/\mathcal{G} \to \mathcal{M}$, and $\mu = s^*(\nu)$, then μ is independent of the choice of s, modulo an exact form. μ is the desired $6g-6$-form on $\mathcal{M}$.

Vertex operators and ghosts

The most traditional string theory vertex operators can be written $W = c\bar{c}V$ where c and $\bar{c}$ are the ghosts and V is a spin $(1,1)$ conformal field constructed from matter fields only. It is well known in the operator formalism [16] that these conditions can be relaxed. The only really essential properties of W in conventional string theory are that it have ghost number 2, spin zero, and be annihilated by Q, and that the Fock space state corresponding to W should be annihilated by an operator usually written as ${b_0}^- = b_0 - \bar{b}_0$ (here b_0 and $\bar{b}_0$ are the zero modes of left- and right-moving ghosts). The requirement that W have spin zero and be annihilated by ${b_0}^-$ ensures that the analogs of Θ defined with an insertion of W have the essential properties that entered above (diffeomorphism invariant and basic). The ghost number two condition shifts the ghost number of the vacuum, leading in a natural way to measures on the moduli space of Riemann surfaces with marked points.

These conditions make sense in our abstract setting, and can be borrowed bodily, except that if we do not assume conformal invariance, the definition of ${b_0}^-$ must be written ${b_0}^- = \oint \mathrm{d}x^\alpha b_{\alpha 0}$. (The integration is over a parametrized circle and «0» is the normal direction to the circle.) Note, therefore, that in including vertex operators, we need make no explicit mention of the ghosts; only the product $W = c\bar{c}V$, and not the separate factors in that product, needs to be generalized. Where, therefore, do the ghosts appear in the discussion? Of course, the antighosts entered in the definition of the measure for $g \geq 2$; but what about the ghosts?

We perhaps could «find» the ghosts by trying to generalize the usual definition of the genus one measure — as c and $\bar{c}$ each has a zero mode on a surface of genus one. To get a different perspective, I will instead discuss the issue from the standpoint of closed string field theory.

In closed string field theory as we know it today (see [17] for a review), the string field Ψ is a vector in a string Hilbert space obtained by quantization on a circle S. Picking a parametrization of the circle, let ${L_0}^- = L_0 - \bar{L}_0$ be the generator of a rotation of the circle, and let ${c_0}^- = (c_0 - \bar{c}_0)/2$. Ψ is required to obey

$$L_0{}^-\Psi = b_0{}^-\Psi = 0. \tag{2.7}$$

The quadratic part of the Lagrangian is

$$\left(\Psi, c_0{}^- Q\Psi\right), \tag{2.8}$$

and the linearized gauge invariance is

$$\delta\Psi = Q\epsilon, \tag{2.9}$$

where ϵ has ghost number 1 and

$$b_0{}^-\epsilon = 0. \tag{2.10}$$

This is proved to be a gauge invariance as follows. In the usual theory, (2.10) implies that

$$\epsilon = b_0{}^-\alpha \tag{2.11}$$

for some α. Inserting this in (2.8), and using

$$\{b_0{}^-, c_0{}^-\} = 1, \tag{2.12}$$

together with

$$\{Q, b_0{}^-\} = L_0{}^- \tag{2.13}$$

(which is a consequence of (2.1)) along with (2.7), one verifies the invariance of (2.8).

What are the essential points that should be retained if the ghosts and matter fields are not decoupled? In conventional string theory, one has

$$\left(b_0{}^-\right)^2 = 0, \tag{2.14}$$

a consequence of the absence of singularities in the $b \cdot b$ operator products. This property also holds in topological sigma models, and it seems reasonable to insist on it in general even if one permits some kind of $b \cdot b$ singularities. The ability to write ϵ as in (2.11) is then the assertion that the cohomology of $b_0{}^-$ vanishes in ghost number 1. If the cohomology of $b_0{}^-$ is altogether zero, then there exists an operator $c_0{}^-$ obeying (2.12). Moreover, $c_0{}^-$ is uniquely determined modulo $c_0{}^- \to c_0{}^- + \{b_0{}^-, f\}$ for some f; using (2.7) and (2.13), one sees that the Lagrangian (2.8) is invariant under such a shift of $c_0{}^-$. Thus, (2.14) and the trivial cohomology of $b_0{}^-$ are sufficient requirements for free closed string field theory (and I believe also for the interacting case).

What if the cohomology of $b_0{}^-$ is not zero? Then using (2.13) and the fact that the cohomology can be represented by states invariant under the compact group generated by $L_0{}^-$ (and in any case since the physical field Ψ is required to have this invariance), we see that Q generates a linear transformation of the cohomology of b_0^-. If this linear transformation is zero, there will be an operator c_0^- obeying not (2.12) but

$$\{b_0^-, c_0^-\}Q = Q\{b_0^-, c_0^-\} = Q. \tag{2.15}$$

This actually is enough to ensure the gauge invariance of the free Lagrangian (2.8). (If the cohomology of b_0^- is nonzero in ghost number one, one cannot use (2.11); but the fact that Q annihilates the cohomology of b_0^- means that for ϵ annihilated by b_0^- and L_0^-, one can write $Q\epsilon = b_0^-\beta$ for some β; this is good enough.) Topological sigma models give an example in which the cohomology of b_0^- is non-zero, but annihilated by Q. We therefore can write down closed string field theory in this case. (Understanding it involves puzzles mentioned in §5 and partly clarified in [32].)

A conundrum

The conundrum that I want to state is obvious. The properties of conventional string backgrounds that I have cited are the only ones I know of that make sense in the general case in which ghosts and matter are not decoupled. Yet they are so general as to permit bizarre realizations, like the topological sigma models that we will consider presently. Are they adequate and if not how should they be supplemented?

The following example will perhaps serve to sharpen the puzzle. In Type II superstring theory, it is usually supposed that a classical solution is described by a superconformal world-sheet theory. One therefore might expect and hope to supplement the above discussion with general properties of superconformal symmetry. Yet it has been pointed out [18] that Ramond-Ramond vertex operators do not commute with the left- or right-moving superconformal currents, and consequently that when such vertex operators are added to the world-sheet Lagrangian (as one would expect in a generic time-dependent situation), the world-sheet theory is not superconformal. Therefore, the structure of Type II backgrounds, at the general level of our above discussion, looks hard to distinguish from that of bosonic string theory. If we do not wish to claim that bosonic strings and Type II superstrings are the same theory (and I would be skeptical of that interpretation), we apparently must find general properties of the world-sheet theory that go well beyond the ones I have cited.

3 Realizations via topological field theories

We will now use topological sigma models to make some realizations of the structure just explained. There are two classes of such models [6], [19], [20], which I will call the **A** and **B** models following some of the papers just cited.

These models govern maps from a Riemann surface Σ to a target space X, which must be presented with an almost complex structure, in the case of the **A** model, or an actual complex structure, for the **B** model. The **A** model is concerned with almost holomorphic maps from Σ to X, while the **B** model is related to periods of differential forms on X. The **B** model is only well-defined for X obeying the Calabi-Yau condition $c_1(X) = 0$. The **A** model is defined without that condition, but obeys the axioms of §2 (definite ghost number of the vacuum) only if $c_1(X) = 0$. Therefore, we will limit ourselves to Calabi-Yau target spaces. This is, of course, also the case in which the sigma model is conformally invariant.

For either the **A** or **B** models, the ghost number of the vacuum is $-d \cdot \chi(\Sigma)$, with $d = \dim_{\mathbb{C}}(X)$. This coincides with the desired valued $-3\chi(\Sigma)$ that played such an important role in §2 precisely if $d = 3$, and therefore we will limit ourselves to this case. It is curious to note that this is the same value of d that arises in the usual «physical» applications of Calabi-Yau manifolds in superstring compactification.

General tangent space indices to X will be written as I, J, K, while indices of type $(1,0)$ or $(0,1)$ will be written as i, j, k or $\bar{i}, \bar{j}, \bar{k}$, respectively. We consider X as a Kahler manifold endowed with a Ricci-flat Kahler metric $g_{i\bar{j}}$. The bosonic fields of the **A** or **B** models are simply a map $\Phi : \Sigma \to X$; if we pick local coordinates ϕ^I on X, this map can be described by giving functions $\phi^I(x^\alpha)$. It is also convenient to pick a local complex coordinate z on Σ.

The A model

The detailed construction of the **A** and **B** models is explained in [6]. In the **A** model, the fermi fields are a section χ^I of $\Phi^*(TX)$, and a one-form ψ with values in $\Phi^*(TX)$. ψ obeys a self-duality condition which says that its $(1,0)$ part $\psi_z^{\bar{i}}$ has values in $\Phi^*(T^{0,1}X)$, and its $(0,1)$ form $\psi_{\bar{z}}^{i}$ has values in $\Phi^*(T^{1,0}X)$. The BRST transformation laws are

$$\begin{aligned} \delta\phi^I &= i\alpha\chi^I \\ \delta\chi^I &= 0 \\ \delta\psi_z^{\bar{i}} &= -\alpha\partial_z\phi^{\bar{i}} - i\alpha\chi^{\bar{j}}\Gamma^{\bar{i}}_{\bar{j}\bar{m}}\psi_z^{\bar{m}} \\ \delta\psi_{\bar{z}}^{i} &= -\alpha\partial_{\bar{z}}\phi^{i} - i\alpha\chi^{j}\Gamma^{i}_{jm}\psi_{\bar{z}}^{m} \end{aligned} \tag{3.1}$$

with α an anticommuting parameter and Γ the affine connnection of X. The BRST operator Q is defined by writing $\delta\Lambda = -i\alpha\{Q, \Lambda\}$ for any field Λ. The Lagrangian can be written in the form $L = i\{Q, V\}$ with any V such that L is nondegenerate. A suitable choice of V is

$$V = t\int_\Sigma \mathrm{d}^2z \quad g_{\bar{i}j}\left(\psi_z^{\bar{i}}\partial_{\bar{z}}\phi^{j} + \partial_z\phi^{\bar{i}}\psi_{\bar{z}}^{j}\right) \tag{3.2}$$

(with $\mathrm{d}^2z = |\mathrm{d}z \wedge \mathrm{d}\bar{z}|$) and gives

$$L = 2t\int_\Sigma \mathrm{d}^2z\left(\frac{1}{2}g_{IJ}\partial_z\phi^I\partial_{\bar{z}}\phi^J + i\psi_z^{\bar{i}}D_{\bar{z}}\chi^i g_{\bar{i}i} + i\psi_{\bar{z}}^{i}D_z\chi^{\bar{i}} g_{\bar{i}i} - R_{i\bar{i}j\bar{j}}\psi_{\bar{z}}^{i}\psi_z^{\bar{i}}\chi^j\chi^{\bar{j}}\right) \tag{3.3}$$

with t a constant.

The fact that $L = i\{Q, V\}$ means that the t dependence (and dependence on the target space metric) of (3.3) is of the form $\{Q, \ldots\}$ and so does not affect the BRST invariant physics. As weak coupling means large t, the fact that there is no t dependence means that weak coupling is exact.

The dependence on the metric of Σ is likewise of the form $\{Q, \ldots\}$, so we can write $T_{\alpha\beta} = \{Q, b_{\alpha\beta}\}$ where at the classical level

$$b_{\alpha\beta} = itg_{IJ}\left(\psi^I_\alpha\partial_\beta\phi^J + \psi^I_\beta\partial_\alpha\phi^J - h_{\alpha\beta}h^{\sigma\tau}\psi^I_\sigma\partial_\tau\phi^J\right). \tag{3.4}$$

Quantum mechanically, the formula for b requires modification except in the Calabi-Yau case (T and b will no longer be traceless), but an appropriate b still exists, as the theory admits a higher derivative regularization preserving the fermionic symmetry.

The B model

In the **B** model, the fermi fields include first of all sections $\eta^{\bar{i}}$, $\theta^{\bar{i}}$ of $\Phi^*(T^{0,1}X)$; actually, it is convenient to write the formulas in terms of not $\theta^{\bar{i}}$ but $\theta_j = g_{j\bar{i}}\theta^{\bar{i}}$. The other fermi fields are a one-form ρ^i with values in $\Phi^*(T^{1,0}X)$. The transformation laws are

$$\begin{aligned} \delta\phi^i &= 0 \\ \delta\phi^{\bar{i}} &= i\alpha\eta^{\bar{i}} \\ \delta\eta^{\bar{i}} &= \delta\theta_i = 0 \\ \delta\rho^i &= -\alpha\, \mathrm{d}\phi^i \end{aligned} \tag{3.5}$$

and the Lagrangian is again of the form $L = i\{Q, V\}$; for a suitable V, one gets

$$\begin{aligned} L =& t\int_\Sigma d^2z \left(g_{IJ}\partial_z\phi^I\partial_{\bar{z}}\phi^J + i\eta^{\bar{i}}(D_z\rho^i_{\bar{z}} + D_{\bar{z}}\rho^i_z)g_{i\bar{i}}\right. \\ &\left. + i\theta_i(D_{\bar{z}}\rho_z{}^i - D_z\rho_{\bar{z}}{}^i) + R_{i\bar{i}j\bar{j}}\rho^i_z\rho^j_{\bar{z}}\eta^{\bar{i}}\theta_k g^{k\bar{j}}\right). \end{aligned} \tag{3.6}$$

The physics is independent of the coupling parameter t for the same reason as in the **A** model.

In this case, we can write $T_{\alpha\beta} = \{Q, b_{\alpha\beta}\}$ with

$$b_{\alpha\beta} = itg_{IJ}\left(\rho^I_\alpha\partial_\beta\phi^J + \rho^I_\beta\partial_\alpha\phi^J - h_{\alpha\beta}h^{\sigma\tau}\rho^I_\sigma\partial_\tau\phi^J\right). \tag{3.7}$$

3.1 Boundary conditions

In our applications, we will want to formulate these theories on Riemann surfaces Σ that may have non-empty boundary. In doing so, we require a boundary condition.*) We want local, elliptic boundary conditions that preserve the fermionic symmetry. I will explain natural boundary conditions for both the **A** and **B** model, but there may very well be other natural boundary conditions — perhaps required for mirror symmetry — that should be considered.

The **A** model is closely related to Floer theory, and the boundary condition that I want to consider for the **A** model is the one that was studied by Floer.†) For each component C_i of $\partial\Sigma$, we pick a Lagrangian submanifold M_i of the Kahler manifold X. For instance, in the nicest case, there may be a real involution τ_i of X, and M_i may be a component of the fixed point set of τ_i. Let TM_i and NM_i be the tangent and normal bundles to M_i in X; we regard them as the real and imaginary subbundles of $TX|_{M_i}$. Then regarding C_i as a real locus in the complex Riemann surface Σ, we require that the boundary values should be real in the following sense: $\Phi|_{C_i}$ is real, that is it maps C_i to M_i; the normal derivative to Φ at C_i is imaginary (it takes values in $\Phi^*(NM_i)$); χ and the pullback of ψ to M_i are real (they take values in $\Phi^*(TM_i)$). These boundary conditions make sense even if M_i does not arise as the fixed point set of a real involution.

For the **B** model, we pick instead «free» boundary conditions that do not require anything analogous to the choice of the M_i. We require that the normal derivative of Φ vanishes on $\partial\Sigma$, that θ vanishes on $\partial\Sigma$, and that the pullback to $\partial\Sigma$ of $\star\rho$ vanishes ($\star$ being here the Hodge star operator).

3.2 Boundary terms

Now, we further wish to couple this system to target space gauge fields in the sense of string theory, that is, to gauge fields on X. This will be done by introducing Chan-Paton factors, that is, by coupling the gauge fields to «charges» that propagate on $\partial\Sigma$. These charges are string theory analogs of the labels on the boundaries in Figure 1(b), and will reduce to the latter when we make contact with target space Chern-Simons theory. Just as in the construction leading to Figure 1(b), we must take the gauge group to be $U(N)$ if we want to consider oriented Riemann surfaces only; $SO(N)$ and $Sp(N)$ are possible if one wishes to permit unoriented surfaces. The restriction on the gauge group is not needed in the preliminary discussion of the present section but is essential in §4.

The quantum version of either of the models we have introduced is described by a Feynman path integral

$$\int D\Psi_i \ \ \exp(-L(\Psi_i)), \tag{3.8}$$

where Ψ_i are the various fields and L is the Lagrangian. Let now $A = A_I d\phi^I$ be a connection with structure group $U(N)$ on a rank N complex vector bundle

---

*) For some generalities about topological field theories on Riemann surfaces with boundary, and some specifics about the **A** model, see [21].

†) Readers to whom these boundary conditions may seem strange at first sight are invited to try to find other local, elliptic, Q-invariant boundary conditions for the **A** model.

E over X. Let Σ be an oriented Riemann surface whose boundary is a disjoint union of circles C_i. The orientation of Σ induces an orientation of each C_i. Given $\Phi : \Sigma \to X$, for each C_i we can take the trace of the holonomy of $\Phi^*(A)$ around C_i; we write this as

$$\mathrm{Tr}\, P \exp \oint_{C_i} \Phi^*(A). \tag{3.9}$$

(The trace is taken in the defining N dimensional representation of $U(N)$.) Then, tentatively, the coupling to gauge fields in the target space is accomplished by replacing (3.8) by

$$\int D\Psi_i \quad \exp(-L(\Psi_i)) \cdot \prod_i \mathrm{Tr}\, P \exp \oint_{C_i} \Phi^*(A). \tag{3.10}$$

(For the **B** model, this will require some modification.) If A is trivial, this is simply a factor of N for each boundary component, just as in the evaluation of the Feynman amplitude of Figure 1(b).

We must determine whether this modification of the path integral preserves the fermionic symmetry. In general the variation of the trace of the holonomy about a circle C is

$$\delta\, \mathrm{Tr}\, P \exp \oint_C \Phi^*(A) = \mathrm{Tr} \oint_C \delta\phi^I \frac{\mathrm{d}\phi^J}{\mathrm{d}\tau} F_{IJ}(\tau)\mathrm{d}\tau \cdot P \exp \oint_{C;\tau} \Phi^*(A). \tag{3.11}$$

Here τ is a coordinate on C, and $\exp \oint_{C;\tau} \Phi^*(A)$ is the holonomy of $\Phi^*(A)$ around C, starting and ending at τ; and F_{IJ} (or more fastidiously $\Phi^*(F_{IJ})$) are the components of the pullback by Φ of the space-time curvature $F = \mathrm{d}A + A \wedge A$.

The A model

In the case of the **A** model, since $\Phi^*(C_i) \subset M_i$, the bundle E and connection A actually need only be defined on the union of the M_i, not on all of X.

Inserting the **A** model transformation law $\delta\phi^I = i\alpha\chi^I$ in (3.11), we see that the holonomy factors in (3.10) are invariant under the fermionic symmetry if and only if the space-time curvature F vanishes. Thus, in the **A** model, it is possible to couple only to *flat* connections on the M_i. We will give this a more intuitive explanation in §4, where we will see that the target space physics of the **A** model is Chern-Simons gauge theory on the M_i. The classical solutions of Chern-Simons gauge theory are precisely the flat connections. Thus, the requirement of the **A** model that the target space connection must be flat is a special case of the fact that in string theory, the background fields that can be incorporated in the world-sheet theory are always classical solutions of the space-time theory.

Since A is flat, its role is from some points of view almost trivial. The holonomy factors in (3.10) depend only on the topology of Σ and the choice of a homotopy class of maps $\Phi : \Sigma \to X$, in which one is evaluating the functional integral. But deeper aspects of the theory involve summing over the topology of Σ and the homotopy type of Φ, and then the factors coming from holonomies of a flat connection on X fit together coherently (to give, as we will see, Chern-Simons theory expanded around the given flat connection A).

The B model

The situation in the **B** model is similar but more subtle. Plugging the transformation laws $\delta\phi^i = 0$, $\delta\phi^{\bar{i}} = i\alpha\eta^{\bar{i}}$ into (3.11), we find that the holonomy is invariant under the fermionic symmetry if and only if the $(1,1)$ and $(0,2)$ parts of the curvature vanish. The vanishing $(0,2)$ curvature asserts that the $(0,1)$ part of A defines a holomorphic structure on the bundle E; the additional vanishing of the $(1,1)$ curvature says that A is a holomorphic connection (locally it can be represented by a holomorphic one-form of type $(1,0)$). The latter condition is rather restrictive and we wish to eliminate it. This can be done as follows. Replace $\Phi^*(A)$ by the following «improved» connection on the bundle $\Phi^*(E)$ over Σ:

$$\widetilde{A} = \Phi^*(A) - i\eta^{\bar{i}} F_{\bar{i}j}\rho^j. \tag{3.12}$$

Then using the transformation laws of the **B** model, one readily sees that for any circle $C \subset \Sigma$, the trace of the holonomy

$$\operatorname{Tr} P \exp \oint_C \widetilde{A} \tag{3.13}$$

of $\widetilde{A}$ is invariant under the fermionic symmetry, provided that the $(0,2)$ part of the curvature of A vanishes, that is provided A determines a holomorphic structure on E. Thus, the formula (3.10) for coupling external gauge fields to the **A** model should be modified, in the case of the **B** model, to

$$\int D\Psi_i \quad \exp(-L(\Psi_i)) \cdot \prod_i \operatorname{Tr} P \exp \oint_{C_i} \widetilde{A}. \tag{3.14}$$

I will leave it to the interested reader to verify that up to terms of the form $\{Q, \dots\}$, the coupling to gauge fields in (3.14) depends only on the holomorphic structure of E (i.e., the $(0,1)$ part of A), not on the $(1,0)$ part of A. This is in accord with the fact that, as we will see in §4, the **B** model has for its classical solutions in space-time the holomorphic vector bundles. As always, the classical solutions are the objects to which the world-sheet theories can be coupled.

3.3 Large t limit

To set up the Hamiltonian version of these theories, for open strings, we take Σ to be an infinite strip $0 \le \sigma \le \pi$, $-\infty \le \tau \le \infty$ with metric $\mathrm{d}s^2 = \mathrm{d}\sigma^2 + \mathrm{d}\tau^2$. We consider the **A** and **B** models with the boundary conditions just described.

In the case of the **A** model, we use the same Lagrangian submanifold M at the two ends of the strip. A quantum Hilbert space $\mathcal{H}$ is introduced in the usual way by quantizing on the initial value surface $\tau = 0$. We want to compute the cohomology of Q — the analogs of the Floer groups — and certain aspects of the large t behavior that will be essential in the next section.

The A model

If we write $\psi = \psi_\sigma \mathrm{d}\sigma + \psi_\tau \mathrm{d}\tau$, then, as the self-duality condition determines ψ_σ in terms of ψ_τ, we can regard χ and ψ_τ as the independent fermi variables. The canonical commutation relations are

$$\begin{aligned}\left[\frac{\mathrm{d}\phi^I}{\mathrm{d}\tau}(\sigma), \phi^J(\sigma')\right] &= -\frac{i}{t} g^{IJ} \delta(\sigma - \sigma') \\ \{\psi_\tau(\sigma), \chi(\sigma')\} &= \frac{1}{t}\delta(\sigma - \sigma').\end{aligned} \tag{3.15}$$

The Hilbert space $\mathcal{H}$ consists of functionals $\mathcal{A}(\Phi, \ldots)$ where now Φ is a map of the interval $I = [0, \pi]$ to X mapping ∂I to M, and «...» is a subset of half the fermi variables, depending on a choice of representation of the canonical anticommutators.

The Hamiltonian is

$$L_0 = \int_0^\pi \mathrm{d}\sigma \; T_{00}. \tag{3.16}$$

Using the canonical commutation relations to write $\mathrm{d}\phi/\mathrm{d}\tau$ in terms of $\delta/\delta\phi$, this can be written

$$L_0 = \frac{1}{2}\int_0^\pi \left(-\frac{1}{t} g^{IJ} \frac{\delta^2}{\delta\phi^I \delta\phi^J(\sigma)} + t g_{IJ} \frac{\mathrm{d}\phi^I}{\mathrm{d}\sigma}\frac{\mathrm{d}\phi^J}{\mathrm{d}\sigma}\right) + \text{terms with fermions.} \tag{3.17}$$

The fundamental relation $T_{\alpha\beta} = \{Q, b_{\alpha\beta}\}$ implies that if we introduce the zero mode of the b field,

$$b_0 = \int_0^\pi \mathrm{d}\sigma \; b_{00} \tag{3.18}$$

then

$$L_0 = \{Q, b_0\}. \tag{3.19}$$

(3.19) implies that the Q cohomology can be computed in the subspace of $\mathcal{H}$ annihilated by L_0. Since the Q cohomology (and everything else of essential interest) is independent of t, it is enough to study the kernel of L_0 for large t. Actually, with other applications in mind, we want to understand not just the kernel of L_0 but all eigenvalues that are of order $1/t$ for large t.

Looking at (3.17), we see that such eigenfunctions must be localized near the region with $\mathrm{d}\phi^I/\mathrm{d}\sigma = 0$, that is, the space of constant maps $\Phi : I \to X$. Because of the boundary condition that $\partial\Sigma$ is mapped to M, the constant map in question must in fact map I to M. The non-zero modes of the fermions make contributions of order 1 to L_0. Combining these observations, the low-lying eigenfunctions of L_0 can be described by a functional $\mathscr{A}$ of the bose and fermi zero modes, with other modes in their Fock vacuum. (One must check that the energy of this Fock vacuum is zero, but this follows from Q-invariance or better from the supersymmetry of the untwisted theory.) The bose and fermi zero modes are all tangent to M; denote them as $q^a, \chi^a, \psi_\tau{}^a$, $a = 1 \dots 3$, with q^a being coordinates on M. (We reserve the letter ϕ for coordinates on X.) The canonical commutation relations (3.15) show that $\psi_\tau{}^a$ can be represented as $\partial/\partial\chi^a$; then $\mathscr{A}$ reduces to a function $\mathscr{A}(q^a, \chi^a)$. This has an expansion in powers of χ^a

$$\mathscr{A} = c(q) + \chi^a A_a(q) + \chi^a \chi^b B_{ab}(q) + \dots \tag{3.20}$$

in which the successive terms can be interpreted as p-forms on M of degrees $0 \leq p \leq 3$. If Chan-Paton factors are included in the discussion, these differential forms take values in $\mathrm{End}(E)$, the endomorphisms of some flat vector bundle E over M.

The commutation relations $[Q, \phi^I] = -\chi^I$, $\{Q, \chi^J\} = 0$ show that, if we interpret χ^I as $-\mathrm{d}\phi^I$, we can identify Q with the exterior derivative d on M (with values in $\mathrm{End}(E)$). The Q cohomology is thus $H^*(M, \mathrm{End}(E))$.

Moreover, (3.17) shows that acting on differential forms on M, L_0 reduces to

$$L_0 = \frac{\pi}{2t}\Delta, \tag{3.21}$$

with $\Delta = \mathrm{dd}^* + \mathrm{d}^*\mathrm{d}$ the usual Laplacian on forms. The underlying relation $L_0 = \{Q, b_0\}$ leads one to guess that in the same approximation

$$b_0 = \frac{\pi}{2t}\mathrm{d}^*, \tag{3.22}$$

and this can be verified using the commutation relations.

The B model

For the **B** model we can be brief, as the arguments are so similar. In the large t limit, the wave-functional $\mathscr{A}$ reduces to a function of the zero modes of ϕ^I and $\eta^{\bar{i}}$ (the zero mode of ρ being represented as $\partial/\partial\eta$). Expanding

$$\mathscr{A}(\phi^I, \eta^{\bar{i}}) = c(\phi^I) + \eta^{\bar{i}} A_{\bar{i}}(\phi^I) + \eta^{\bar{i}}\eta^{\bar{j}} B_{\bar{i}\bar{j}}(\phi^I) + \dots, \tag{3.23}$$

we see that for the low-lying states, $\mathscr{A}$ reduces to a sum of $(0, q)$ forms on X (valued in $\mathrm{End}(E)$, E being a holomorphic vector bundle on X), for $0 \leq q \leq 3$. The commutation relations $[Q, \phi^i] = 0$, $[Q, \phi^{\bar{i}}] = -\eta^{\bar{i}}$, $\{Q, \eta^{\bar{i}}\} = 0$ show that, if

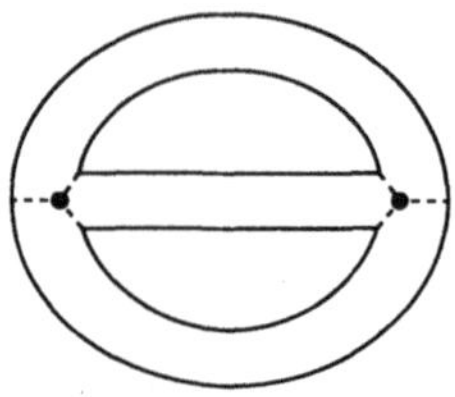

Fig. 2 The cell decomposition of open string moduli space depends on building Riemann surfaces by gluing together flat strips S_i of width π and variable length T_i, $0 \leq T_i \leq \infty$. The strips are glued together in groups of three along the dotted lines; their midpoints meet in conical singularities of deficit angle $-\pi$ that are marked as solid dots. The resemblance of this figure to Figure (1) is central in this paper.

we interpret $\eta^{\bar{i}}$ as $-\mathrm{d}\phi^{\bar{i}}$, we can identify Q with the $\overline{\partial}$ operator (with values in $\mathrm{End}(E)$). So the cohomology of Q is $H^{0,*}(X, \mathrm{End}(E))$. Likewise, by arguments similar to those for the **A** model, L_0 and b_0 reduce to

$$\begin{aligned} L_0 &= \frac{\pi}{2t}\left(\overline{\partial}\,\overline{\partial}^* + \overline{\partial}^*\overline{\partial}\right) \\ b_0 &= \frac{\pi}{2t}\overline{\partial}^*. \end{aligned} \tag{3.24}$$

4 Space-time interpretation via string field theory

In this section we will determine the space-time interpretation of the models just constructed in the case of open strings, that is in the case of orientable world-sheets with boundary. The reason for focussing on this case is that a simple and beautiful answer will arise, as we will see. The more perplexing closed string case (Riemann surfaces without boundary) is the subject of §5.

Usually, closed strings are inescapable even when one tries to do open string physics, for elementary topological reasons. The reader may therefore be surprised at the lengthy discussion of open strings in this section ignoring closed strings. In fact, as we discuss in §4.2, our main points can be formulated at the classical level, where closed strings cannot appear; also, we argue in §5 that in these topological theories, the open and closed strings are decoupled.

The simplicity of the open string topological theories is closely related to the comparative simplicity of open string field theory [14], which in turn is intimately tied up [15] with the existence of a simple cell decomposition of the moduli space of Riemann surfaces with boundary [11]–[13]. This cell decomposition involves the construction of Riemann surfaces by gluing together flat strips S_i of width π and length T_i; they are glued, as sketched in Figure 2, at vertices that generically (unless some $T_i = 0$) are trivalent. The resulting pictures have been called «fat-graphs» by Penner. In the limit when the T_i are all large, or equivalently when the widths of the strips can be neglected, the fatgraphs of the open-string field theory

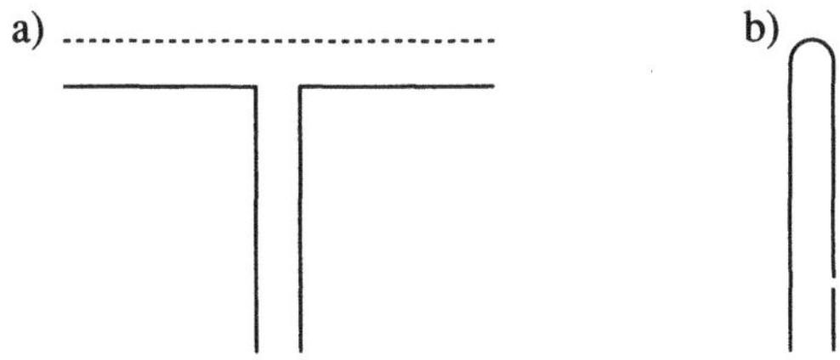

Fig. 3 The multiplication law (a) and integration law (b) of open string field theory are defined by pictures involving gluing of strings.

reduce to ordinary graphs corresponding to Feynman diagrams of ordinary field theory.

Construction of open string field theory

Now we recall the highlights of the construction of open string field theory. A key part of the structure is a $\mathbb{Z}$-graded associative algebra $\mathcal{B}$, with a multiplication law that we denote as $\star$, and a derivation Q of degree 1 with $Q^2 = 0$. The degree is usually called a ghost number.*) There is a linear functional $\int : \mathcal{B} \to \mathbb{C}$, of degree -3 (that is, if b is an element of $\mathcal{B}$ of definite ghost number, then $\int b$ vanishes unless b has ghost number three), obeying $\int a \star b = (-1)^{\deg a \deg b} \int b \star a$, $\int Qb = 0$, for all $a, b \in \mathcal{B}$. The string field is a ghost number 1 element $\mathcal{A} \in \mathcal{B}$. The Lagrangian is

$$L = \frac{1}{2} \int \left(\mathcal{A} \star Q\mathcal{A} + \frac{2}{3} \mathcal{A} \star \mathcal{A} \star \mathcal{A} \right). \tag{4.1}$$

This is invariant under gauge transformations generated by

$$\delta\mathcal{A} = Q\epsilon - \epsilon \star \mathcal{A} + \mathcal{A} \star \epsilon. \tag{4.2}$$

Chan-Paton factors are introduced as follows (in the case of a trivial rank N bundle in space-time with flat connection; the generalization is discussed presently). Let $M_N(\mathbb{C})$ be the associative algebra of $N \times N$ complex matrices. One simply replaces $\mathcal{B}$ with $\mathcal{B} \otimes M_N(\mathbb{C})$ (and $\int$ by $\int \otimes \mathrm{Tr}$, Tr being the usual trace on $M_N(\mathbb{C})$). This preserves the basic structures. In this process, $\mathcal{A}$ acquires matrix indices. If a suitable reality condition is imposed, $\mathcal{A}$ takes values in $N \times N$ hermitian matrices — the Lie algebra of $U(N)$.

Conventionally, $\mathcal{B}$ is taken to be the space of open string states in some critical string theory; the multiplication and integration operations $\star$ and $\int$ are defined by operations of gluing strings that we recall in Figure 3. Q is the BRST operator of the critical string theory. The integration law is of ghost number -3

*) The ghost number as I will count it here is the ghost number as defined in [14] plus 3/2; thus, the $SL(2, \mathbb{R})$ invariant vacuum has ghost number zero.

since the Euler characteristic of a disc is -1, and in critical string theory the ghost number of the vacuum is $-3\chi(\Sigma)$.

Now we want to consider open string field theory with the conventional string models replaced by the more exotic ones discussed in Sections 2–3. In particular, for the world-sheet theory we take a topological sigma model with a Calabi-Yau target space X of complex dimension three; this ensures the correct ghost number of the vacuum — or of the integration law. Multiplication and integration are defined by the standard gluing operations; Q will now be the BRST operator of whatever topological field theory we consider. This framework enables us to construct an open-string field theory from any world-sheet theory (obeying the usual axioms). For instance, instead of tensoring with $M_N(\mathbb{C})$ to introduce trivial Chan-Paton factors, we can use the world-sheet theories with boundary interactions constructed in §3.2 to couple to a flat bundle on M in the case of the **A** model, or a holomorphic bundle on X in the case of the **B** model.

4.1 The A model

We want to understand the physical content of these models. First we consider the **A** model, which is closely related to Floer theory. It is necessary to pick boundary conditions, and as in §3, we follow Floer and pick boundary conditions associated with Lagrangian submanifolds of X. In §3, in studying a particular surface Σ, we introduced a separate Lagrangian submanifold M_i for each boundary component C_i of Σ. In string field theory, one generates all possible Σ's via a Feynman diagram expansion, and the M_i's must be built in universally at the outset. We will do this in the simplest way by picking a single M once and for all. Thus the free boundaries of all our strings and surfaces will be mapped to $M \subset X$. (Generalizations exist; by correlating the choice of M with the Chan-Paton factors one could make a gauge invariant string field theory with more than one M.)

Since X is of complex dimension three, M is a three-manifold. A neighborhood of M in X is equivalent topologically and even symplectically to a neighborhood of M in its cotangent bundle T^*M. The topological string theory with target space X involves, roughly, two ingredients. One is the instantons with target X (and boundary values in M); these are the usual subjects of study in Floer theory. The other side of the story, as we will see, involves Chern-Simons theory with target space M. I want first to isolate this «new» ingredient — new in the sense that it is not usually coupled with Floer theory. The instantons can be suppressed and the new ingredient isolated by replacing X by T^*M, since a simple vanishing theorem (discussed presently) shows that there are no non-constant instantons mapping Σ to T^*M. Later we will generalize to arbitrary X and determine the instanton corrections to the space-time Chern-Simons theory.

So until further notice, in discussing the **A** model, our target space will be the cotangent bundle T^*M of an oriented three-manifold M. *) Like any symplectic

*) The orientation is needed to consistently define the sign of the fermion determinant in the world-sheet theory of §3.

manifold, T^*M can be given an almost complex structure such that the symplectic structure is positive and of type $(1,1)$; indeed, this is essential in Floer/Gromov theory of symplectic manifolds. This is good enough for formulating the **A** model with target space T^*M; though the transformation laws and Lagrangian of the **A** model were written in §3 in a way that assumed the integrability of the almost complex structure; this assumption can be relaxed, as explained in detail in [1].

The vanishing theorem

Now let us briefly explain the vanishing theorem, which asserts that instantons mapping Σ to T^*M and mapping $\partial\Sigma$ to M are necessarily constant.

Consider temporarily a general symplectic manifold X with symplectic form ω and an almost complex structure J such that ω is of type $(1,1)$ and positive. Positivity means that the metric $g_{IK} = J^S{}_I\omega_{SK}$ is positive definite; the $(1,1)$ condition means that if i,j,k and $\bar{i},\bar{j},\bar{k}$ are indices of types $(1,0)$ and $(0,1)$, respectively, then $g_{ij} = g_{\bar{i}\bar{j}} = 0$, and $g_{i\bar{j}} = g_{\bar{j}i} = -i\omega_{i\bar{j}}$. If Σ is a Riemann surface, an instanton or almost holomorphic map is a map $\Phi : \Sigma \to X$ with $\bar{\partial}\phi^i = 0$. Consider the bosonic sigma model action

$$I = i\int_\Sigma \mathrm{d}z \wedge \mathrm{d}\bar{z} \quad g_{IJ}\partial_z\phi^I\partial_{\bar{z}}\phi^J. \tag{4.3}$$

Instantons minimize this action for a given homotopy class since

$$\begin{aligned} I =& 2i\int_\Sigma \mathrm{d}z\wedge\mathrm{d}\bar{z} \quad g_{\bar{i}j}\partial_z\phi^{\bar{i}}\partial_{\bar{z}}\phi^j - i\int_\Sigma \mathrm{d}z\wedge\mathrm{d}\bar{z} \quad g_{\bar{i}j}\left(\partial_z\phi^{\bar{i}}\partial_{\bar{z}}\phi^j - \partial_z\phi^j\partial_{\bar{z}}\phi^{\bar{i}}\right) \\ =& 2i\int_\Sigma \mathrm{d}z\wedge\mathrm{d}\bar{z} \quad g_{\bar{i}j}\partial_z\phi^{\bar{i}}\partial_{\bar{z}}\phi^j + \int_\Sigma \Phi^*(\omega). \end{aligned} \tag{4.4}$$

The first term on the right hand side of (4.4) is positive semi-definite and vanishes precisely for instantons; so for instantons, the action reduces to $I = \int_\Sigma \Phi^*(\omega)$. The vanishing theorem comes by showing that if $X = T^*M$, and $\Phi(\partial\Sigma) \subset M$, then

$$\int_\Sigma \Phi^*(\omega) = 0. \tag{4.5}$$

If this is known, then I vanishes for instantons that map $\partial\Sigma$ to M; but from the definition (4.3) it is clear that I vanishes only for constant maps.

To justify (4.4), pick on M local coordinates q^a, $a = 1\ldots3$. The symplectic structure of T^*M can be written as $\omega = \sum_{a=1}^3 \mathrm{d}p_a \wedge \mathrm{d}q^a$, with p_a linear coordinates in the fibers that vanish on M. This is $\omega = \mathrm{d}\rho$, where $\rho = \sum p_a \mathrm{d}q^a$ vanishes on M. So $\int_\Sigma \Phi^*(\omega) = \int_{\partial\Sigma}\Phi^*(\rho) = 0$, if $\Phi(\partial\Sigma) \subset M$.

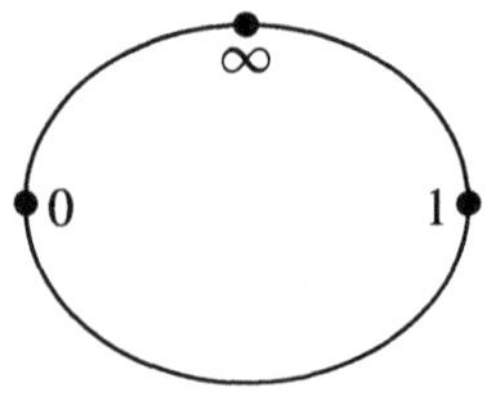

Fig. 4 The coupling of three gauge fields on a disc.

Low energy expansion

As explained following equation (3.3), the key simplification of the **A** model is that the essential physics is independent of the coupling parameter t. As t and the target space metric g_{IJ} appear only in the combination tg_{IJ}, large t is simply the limit in which the target space metric is scaled up; it is the limit of large distances or long wavelengths. This is the limit in which ordinary string theory reduces approximately to field theory. Since ordinary string theory is t-dependent, the large t behavior is only an approximation. The topological string theories that we are studying are t-independent, so we can hope for an exact description by looking at the large t behavior.

We have analyzed the large t behavior of the string states in §3.3. In particular, as we saw there, the low-lying modes can be described as functions $\mathcal{A}(q^a, \chi^b)$ of the center of mass variables. In §3.3, we computed the cohomology for all ghost numbers, but in the string field theory context, we want to insist that $\mathcal{A}$ have ghost number 1. This means that it must be linear in χ, so we write

$$\mathcal{A} = \chi^a A_a(q), \tag{4.6}$$

with $A_a dq^a$ a one-form on M, valued possibly in $\mathrm{End}(E)$, the endomorphisms of a flat vector bundle E.

In §3.3, we identified Q, in the same large t limit, with the exterior derivative d. Then the quadratic part of the string field Lagrangian $\frac{1}{2}\int \mathcal{A} \star Q\mathcal{A}$ reduces (for the low-lying modes in the large t limit) to

$$\frac{1}{2}\int_M \mathrm{Tr}\, A \wedge dA, \tag{4.7}$$

with Tr a trace on the Chan-Paton factors. This is the quadratic part of the ordinary Chern-Simons action for a gauge field A.

What about the cubic part of the string field action? If $A^{(j)}$ is a mode of the gauge field A, the corresponding vertex operator is $V^{(j)} = \chi^a A_a^{(j)}(q)$. Let us evaluate by world-sheet path integrals the coupling of three such modes on the disc. This is done by inserting three vertex operators $V^{(i)}$, $i = 1 \ldots 3$ at boundary points of the disc, as in Figure 4. Because of $SL(2,\mathbb{R})$ symmetry, all that matters in the choice of the three points is the cyclic order. So we can take

$\langle V^{(1)}(0)V^{(2)}(1)V^{(3)}(\infty)\rangle$. In the large t limit, the path integral reduces to an integral over the constant zero modes; this integral gives

$$\int \mathrm{d}q^1 \ldots \mathrm{d}q^3 \mathrm{d}\chi^1 \ldots \mathrm{d}\chi^3 \mathrm{Tr}\chi^a A_a^{(1)}(q)\chi^b A_b^{(2)}(q)\chi^c A_c^{(3)}(q) = \int_M \mathrm{Tr} A^{(1)} \wedge A^{(2)} \wedge A^{(3)}. \tag{4.8}$$

This is simply a matrix element of the cubic part of the usual Chern-Simons action. So putting the pieces together, the string field action (4.1) reduces to the ordinary Chern-Simons Lagrangian

$$L_T = \frac{1}{2}\int_M \mathrm{Tr}\left(A \wedge \mathrm{d}A + \frac{2}{3}A \wedge A \wedge A\right). \tag{4.9}$$

The string field gauge invariances can similarly be seen in the same approximation to reduce to conventional gauge invariance. So, under the conditions that we have formulated, the abstract Chern-Simons Lagrangian of string field theory reduces to the ordinary Chern-Simons Lagrangian.

Now string theories usually reduce to field theories at low energy only approximately. For instance, ordinary string theories reduce approximately at low energies to general relativity plus massless matter fields, but there are $O(\alpha')$ corrections.

In the case of topological string theories (such as the **A** model considered here) the situation is quite different. The analog of the usual $O(\alpha')$ corrections would be perturbative corrections in $1/t$. These are absent, because the model is t-independent. Instanton corrections are also absent, because for target space T^*M, there are no non-constant instantons, as we have seen. So the reduction of the **A** model to (4.9) is exact. If we replace T^*M by some Calabi-Yau threefold X (having M as a Lagrangian subspace), there are instanton corrections to (4.9) that we will discuss later.

4.2 Reconsideration in the light of Floer theory

Now I want to reconcile what we have said above with previous discussions of the **A** model such as [6] (or [1], §4.1 where the coupling to topological gravity is considered).

Formally, the **A** model is a quantum field theory that counts holomorphic curves obeying various conditions. (This is familiar in applications of the **A** model to mirror symmetry, for instance.) However, in the vanishing theorem that we have discussed above, we have seen that, for target space T^*M, there are no non-constant instantons. How therefore can the model possibly be non-trivial, and in particular equivalent as we have claimed to perturbative Chern-Simons gauge theory?

To answer this question, it is necessary to recall some background. Formally, in topological quantum field theories of this general nature («cohomological field

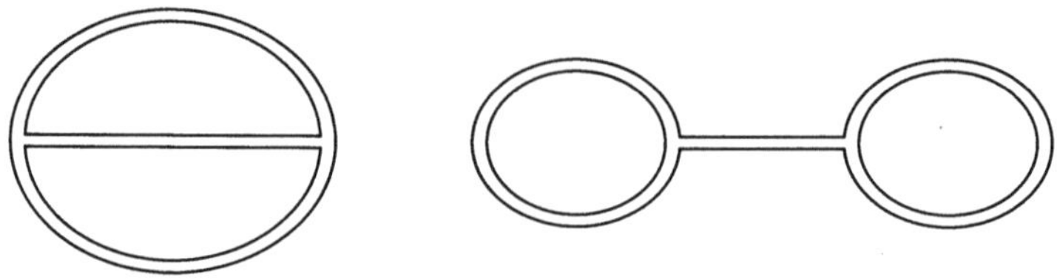

Fig. 5 Degeneration of a three-holed sphere to a trivalent graph can occur in two ways.

theories»), one has an infinite dimensional function space $\mathcal{S}$, and over it an infinite dimensional vector bundle $\mathcal{W}$; one wishes to make sense of the Euler class of this bundle. To tame the situation, one has a section w of $\mathcal{W}$. In finite dimensions, if $\mathcal{S}$ is compact and without boundary, the Euler class of $\mathcal{W}$ would be Poincaré dual to a homology class $[w]$ supported on the zeros of w. The Euler class of $\mathcal{W}$, integrated over $\mathcal{S}$, can be represented by various integral formulas; one can construct an integral formula which is localized near the zeros of w [22]. The Feynman path integral of a cohomological field theory is a function space version of such a formula. This point of view is developed (in an analogous four dimensional gauge theory) in [23].

To identify $\mathcal{S}$, $\mathcal{W}$, and w in the case at hand, let Σ be a compact oriented two dimensional surface, perhaps with boundary, not endowed with any *a priori* complex structure. Let X be a symplectic manifold, with an almost complex structure obeying the usual conditions, and with a Lagrangian submanifold M. Fix a homotopy class $\mathcal{H}$ of maps of $\Phi : \Sigma \to X$ with $\Phi(\partial\Sigma) \subset M$. Such a map can be described locally by complex functions ϕ^i on Σ (corresponding to local complex coordinates on X). Then we set $\mathcal{S}$ to be the space of pairs (J, Φ), where J is a complex structure on Σ, and Φ is a map $\Sigma \to X$ in the class $\mathcal{H}$, with two such pairs considered equivalent if they differ by a diffeomorphism of Σ. For $\mathcal{W}$ we take the vector bundle over $\mathcal{S}$ whose fiber is the space of sections of $\Omega^{0,1}(\Sigma) \otimes \Phi^*(T^{1,0}X)$, where $\Omega^{0,1}(\Sigma)$ is the space of $(0,1)$-forms on Σ in the complex structure J, and $T^{1,0}X$ is the holomorphic tangent bundle of X. For w, we pick a natural section of $\mathcal{W}$ given by the equation for an almost holomorphic curve; it can be written $w = \overline{\partial}_J \phi^i$, with $\overline{\partial}_J$ the $\overline{\partial}$ operator on Σ determined by J.

Now, let us focus on the case $X = T^*M$, endowed with an almost complex structure of the usual type and so in particular with a metric. The vanishing theorem discussed earlier implies that for H a non-trivial homotopy class, there are no instantons; w has no zeros. If $\mathcal{S}$ were compact, we would conclude that the Euler class of $\mathcal{W}$ is zero.

However, $\mathcal{S}$ is non-compact, and its non-compactness is essential. In fact, in a certain sense, there are instantons or zeros of w at infinity. These arise in a limit in which the Riemann surface degenerates to a graph Γ. This possibility is sketched in Figure 5; from the point of view of open-string field theory, the degeneration arises in the limit in which all T_i go to infinity in Figure 2.

In a given homotopy class of maps of the graph Γ to X, there will be a map Φ for which $\Phi(\Gamma)$ is a «geodesic graph» of least total length. Φ can be regarded as a limit of maps from almost degenerate Riemann surfaces (obtained by slight thickening of Γ) that almost obey the instanton equation. It is an «instanton at infinity». The sigma model action (4.3) (which should vanish for instantons obeying our boundary conditions, as we saw in the proof of the vanishing theorem) vanishes in the limit of such a geodesic graph.

The existence of zeros at infinity means that there is no straightforward evaluation of the world-sheet path integral by counting zeros of w. There are no real zeros, but there are virtual zeros at infinity. While the contributions of real zeros are integers that can be counted topologically, there is no apparent reason for this to be true for the virtual zeros at infinity. In any event, the contributions of the virtual zeros, summed over all homotopy classes, can be evaluated by computing the measure on the moduli space (of complex structures on Σ) that comes from the world-sheet path integral. *)

The essential simplification of the **A** model always arises because, for large t, this measure is concentrated near zeros of w. In the case that the only zeros are virtual zeros at infinity, the measure is concentrated for large t on the region near infinity; but this is the limit in which string theory reduces to field theory, and world-sheet path integrals reduce to Feynman graphs. So the non-vanishing contributions to the world-sheet path integral come entirely from the regions in the moduli space (of complex structures on Σ) in which the string theory reduces to the evaluation of Feynman graphs. This is the equivalence of the string theory to a field theory that was claimed earlier. Of course, one must remember that (as in the figure, where there are two possible degenerations) one Riemann surface will in general be capable of degenerating to several possible Feynman graphs.

Perturbation theory

To make this a little more explicit, we will now discuss how to extract the field theoretic Feynman rules from the string theory.

We want to describe the measure on the moduli space of Riemann surfaces that follows from the description of moduli space via fatgraphs. This can be worked out systematically by quantizing open string field theory. The necessary techniques were given by Thorn and Boccicchio [25], [26] and involve an elegant application of the Batalin-Vilkovisky approach to quantization. Instead of following that road,

---

*) When $\pi_1(M)$ has infinitely many conjugacy classes, so that the number of homotopy classes is infinite, it is scarcely plausible that the full answer can be written as a sum over homotopy classes with each class contributing an integer. For convergence, the contributions would have to be almost all zero. The possible noncommutativity of the sum over homotopy classes and the integration over moduli space may also mean that the final answer (which should be obtained by summing over homotopy classes first) cannot be written as a sum of contributions of homotopy classes, though the measure on moduli space can be so written. In this case, the question of whether the individual contributions are integers does not even make sense. This seems to be the case for the one loop contributions, as one can see by comparing to D. Fried's formulas expressing the logarithm of analytic torsion as a sum over geodesics [24].

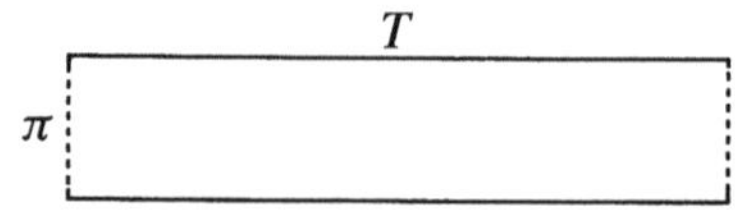

Fig. 6 A strip S of width π with one real modulus T.

I will merely use the fatgraph description of moduli space and the definition of the measure in equation (2.4).

The basic fact that we will use is therefore the fact that every Riemann surface Σ with boundary has a canonical flat metric (with some conical singularities) built as in Figure 2. This metric is obtained by gluing together some standard strips. The strips are flat rectangles of width π in the $\sigma - \tau$ plane, say $0 \leq \sigma \leq \pi$, $0 \leq \tau \leq T$, with metric $\mathrm{d}s^2 = \mathrm{d}\sigma^2 + \mathrm{d}\tau^2$. Such a strip, say S (Figure 6), has one real modulus, namely T. A deformation of this modulus from T to $T + \delta T$ can be made by changing the metric to

$$\mathrm{d}s^2 = \mathrm{d}\sigma^2 + \mathrm{d}\tau^2 \left(1 + 2f(\tau)\right), \tag{4.10}$$

with $f(\tau)$ any function with $\int_0^T \mathrm{d}\tau f(\tau) = \delta T$. It is convenient to take $f(\tau) = \delta T \cdot \delta(\tau - \tau_0)$, for some τ_0. If one of the $\delta^{(k)} h$ in equation (2.4) is the deformation just described of the length of the strip S, then the corresponding mode $b^{(k)}$ of equation (2.3) is simply

$$b^{(k)} = \int_0^\pi \mathrm{d}\sigma \; b_{00}(\sigma, \tau_0). \tag{4.11}$$

This is in other words an insertion of the zero mode b_0 at $\tau = \tau_0$.

We also have to worry about the propagation of the string through the proper time T; this gives a factor of $\exp(-TL_0)$, with L_0 the string Hamiltonian discussed in §3.3. Allowing for the integral over T and the insertion of b_0, we get for each strip a factor

$$\int_0^\infty \mathrm{d}T \; b_0 \exp(-TL_0). \tag{4.12}$$

Let $\mathcal{V}$ be the kernel of L_0, and let $1/L_0$ be the inverse of L_0 in the subspace orthogonal to $\mathcal{V}$. Since b_0 annihilates the kernel of L_0,*) the integral in (4.12) can

*) The topological sigma model in question was constructed by twisting an ordinary supersymmetric model (as explained in detail in [6]) by a power of the canonical bundle. On the flat strip under discussion here, the canonical bundle is naturally trivialized, and the twisting does nothing. In the underlying supersymmetric model, b_0 is the adjoint of Q, and as in Hodge theory, the relation $L_0 = \{Q, b_0\}$ implies that Q and b_0 annihilate the kernel of L_0.

be evaluated to give

$$\int_0^\infty \mathrm{d}T\; b_0 \exp(-TL_0) = b_0 \frac{1}{L_0}. \tag{4.13}$$

In equations (3.21) and (3.22), we determined the large t limits of L_0 and b_0 to be $\pi\Delta/2t$ and $\pi \mathrm{d}^*/2t$. (Thus, b_0 vanishes for large t, but b_0/L_0 has a non-vanishing limit because of the small eigenvalues of L_0.) So the large t limit of (4.13) is

$$\mathrm{d}^* \frac{1}{\Delta}. \tag{4.14}$$

Now, (4.13) is the open string propagator, while (4.14) is the propagator of Chern-Simons field theory.

In the fatgraph approach to construction of moduli space, the strips we have just analyzed are glued together at cubic vertices to make Riemann surfaces. The structure of the interaction vertices is the same as in the open string multiplication law, (Figure 3(a)). We have already seen in connection with Figure 4 how these vertices reduce at large t to wedge products of differential forms, as in Chern-Simons field theory.

While the underlying string field $\mathscr{A}$ had ghost number 1, and the underlying Chern-Simons gauge field A is a one-form, in (4.13) and (4.14) states of all ghost number — or differential forms of all degrees — are propagating. In either case, the vertices of the gauge fixed theory have in a natural sense the same structure (gluing of strings or wedge products) as that in the classical theory, though the constraint on the ghost number or degree is absent. In the case of string theory, this structure was explained by Thorn and Boccicchio [25], [26] by solving the master equation and gauge fixing of string field theory. In Chern-Simons theory, the analogous formulation of the gauge fixed theory, combining the fields of different ghost number, is due to Axelrod and Singer [4].

The tree level S-matrix

I would now like to make this more explicit in a particular example. I have picked the example to anticipate the following question that may perplex some readers. We have extracted Chern-Simons theory from the open string degenerations of moduli space, but are there additional closed string contributions? It will be clear in §5 that the role of the closed strings is not fully understood, though I will argue there that the closed and open strings are decoupled. I therefore want to demonstrate that the statement «the topological string theory of the **A** model is equivalent to Chern-Simons field theory» can be tested in a situation in which closed strings are definitely not relevant. This is the case for scattering amplitudes (or the analog of what in ordinary string theory would be scattering amplitudes) for the case that Σ is a disc.

As we know from §3.3, the «physical states» of the open string field theory are elements of $H^1(\Sigma, \mathrm{End}(E))$, E being a flat vector bundle that we will keep

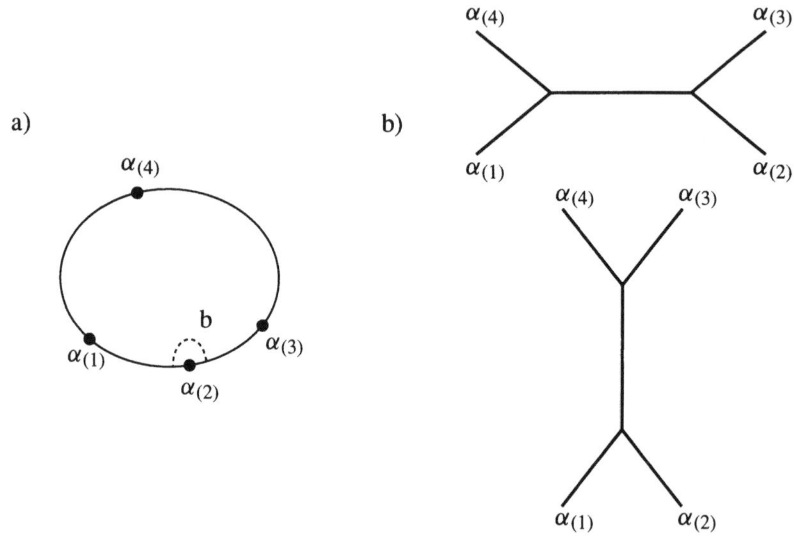

Fig. 7 Coupling of four states $\alpha_{(1)}, \ldots, \alpha_{(4)}$ on the boundary of a disc (a) with an antighost insertion around $\alpha_{(2)}$; the Feynman diagrams corresponding to the two possible «channels» (b).

fixed in the discussion. The exterior derivative with values in that bundle will be denoted as d. Let $\alpha_{(1)}, \ldots, \alpha_{(4)}$ be four such states. We consider their coupling on the disc in that cyclic order, as in Figure 7. To calculate the four point function in field theory, all that we really need to know is that there are two possible Feynman diagrams, indicated in Figure 7, and that the propagator is d^*/Δ. The amplitude is therefore

$$I(\alpha_{(1)},\ldots,\alpha_{(4)})=\int_M \mathrm{Tr}\left(\alpha_{(1)}\wedge\alpha_{(2)}\frac{\mathrm{d}^*}{\Delta}\alpha_{(3)}\wedge\alpha_{(4)}+\alpha_{(2)}\wedge\alpha_{(3)}\frac{\mathrm{d}^*}{\Delta}\alpha_{(4)}\wedge\alpha_{(1)}\right). \tag{4.15}$$

To do the same calculation in string theory, we introduce vertex operators $V_{(i)} = \alpha_{(i)\,a}X^a$ for the external states, and one antighost insertion corresponding to the one real modulus of the disc. We can map $V_{(1)}$, $V_{(3)}$, and $V_{(4)}$ to $0, 1$, and ∞, and take the modulus to be the position of $V_{(2)}$. In that case, the antighost insertion can be taken to be a small contour integral around the position of $V_{(2)}$, as in the figure. The contour integral can be evaluated by thinking of $V_{(2)}$ as an operator inserted on the end of an open string in a Hamiltonian formulation; the antighost contour computes the commutator of this operator with the antighost zero mode b_0, and so replaces the zero form operator $V_{(2)}$ by its one-form descendant

$$W_{(2)} = [b_0, V_{(2)}] = A_a(q)\frac{\mathrm{d}q^a}{\mathrm{d}\tau}. \tag{4.16}$$

The string theory amplitude is therefore

$$\int_0^1 \mathrm{d}\sigma \langle V_{(1)}(0) \cdot A_a \frac{\mathrm{d}q^a}{\mathrm{d}\tau}(\sigma) \cdot V_{(3)}(1) V_{(4)}(\infty) \rangle . \tag{4.17}$$

Now we wish to evaluate this, as usual, in the limit of $t \to \infty$. If the integral were limited to a range of σ bounded away from 0 and 1, it would vanish as $1/t$ for $t \to \infty$. This is so because the ϕ and $\psi - \chi$ propagators are both proportional to $1/t$. A non-zero result arises only because if the result is blindly expanded in powers of $1/t$, the endpoint contributions are infinite. This is because the term of lowest order in $1/t$ in the operator product $V_{(1)}(0) \cdot A_a \frac{\mathrm{d}q^a}{\mathrm{d}\tau}(\sigma)$ is proportional to $1/\sigma$, coming from $\phi - \partial\phi$ contractions. A simple-minded $1/t$ expansion of (4.17), evaluating those contractions (and discarding a term proportional to $\{Q, \ldots\}$), leads to a divergent contribution near $\sigma = 0$ that looks like

$$\frac{\pi}{2t} \int_0^\epsilon \mathrm{d}\sigma \frac{1}{\sigma} \langle \chi^a \left(\mathrm{d}^* (\alpha_{(1)} \wedge \alpha_{(2)}) \right)_a (0) \;\; V_{(3)}(1) \;\; V_{(4)}(\infty) \rangle , \tag{4.18}$$

with ϵ an arbitrary fixed positive number.

To cure the infinity, we must remember the anomalous dimensions of the various operators, which are given by the eigenvalues of L_0. To order $1/t$, we had $L_0 = \pi\Delta/2t$ in §3.3. The physical states have harmonic representatives, and if we pick such representatives (discarding terms of the form $\{Q, \ldots\}$), the $V_{(i)}$ have $L_0 = 0$ in order $1/t$. However, the vertex operator $\chi^a \left(d^* (\alpha_{(1)} \wedge \alpha_{(2)}) \right)_a$ in (4.18) corresponds to a state not annihilated by L_0. Including powers of σ coming from the anomalous dimensions, and replacing the expectation values $\langle \;\; \rangle$ by zero mode integrals, (4.18) is replaced by

$$\frac{\pi}{2t} \int_0^\epsilon \mathrm{d}\sigma \frac{1}{\sigma} \int_M \mathrm{Tr} \left(\mathrm{d}^* (\alpha_{(1)} \wedge \alpha_{(2)}) \sigma^{\pi\Delta/2t} \wedge \alpha_{(3)} \wedge \alpha_{(4)}(\infty) \right) \tag{4.19}$$

The endpoint contribution of the σ integral is now finite as $t \to \infty$, and reproduces one of the two Feynman diagram contributions in equation (4.15). The other arises from the endpoint at $\sigma = 1$.

Now let us discuss briefly the physical significance of these couplings. Given a flat bundle E corresponding to a solution $A_{(0)}$ about which one wishes to expand, the space of linearized deformations is $H^1(M, \mathrm{End}(E))$. Writing $A = A_{(0)} + B$, impose the gauge condition $\mathrm{d}^* B = 0$ (with d^* being defined with respect to $A_{(0)}$). In turn, write $B = \sum_i u_i \alpha_{(i)} + \sum_\sigma v_\sigma \beta_{(\sigma)}$, where $\alpha_{(i)}$ are a basis of zero modes of the Laplacian Δ and $\beta_{(\sigma)}$ are the non-zero modes; and u_i, v_σ are real coefficients. The underlying Lagrangian

$$L = \int_M \mathrm{Tr} \left(A \wedge \mathrm{d}A + \frac{2}{3} A \wedge A \wedge A \right) \tag{4.20}$$

can be regarded as a function $L(u_i, v_\sigma)$. The tree diagrams that we have discussed above are a calculational technique for «integrating out» the v_σ to get an «effective potential» $V(u_i)$. It can be written in terms of the n-point functions on the disc discussed above for $n = 4$ as

$$V(u_i) = \sum_{i,j,k} \frac{u_i u_j u_k}{3!} \langle \alpha_{(i)} \alpha_{(j)} \alpha_{(k)} \rangle + \sum_{i,j,k,l} \frac{u_i u_j u_k u_l}{4!} \langle \alpha_{(i)} \alpha_{(j)} \alpha_{(k)} \alpha_{(l)} \rangle + \dots. \tag{4.21}$$

Since V is obtained from the underlying Lagrangian L by «integrating out the massive modes», its stationary points, i.e., solutions of $\partial V / \partial u_i = 0$, are in one-to-one correspondence with the stationary points of the underlying Lagrangian.

The process of «integrating out the massive modes» can be described as follows. Let L be a function with a critical point at the origin, and let n be the dimension of the kernel of the Hessian or matrix of second derivatives of L. Then by a generalization of the Morse lemma, it is possible to pick coordinates $u_i, i = 1 \dots n$ and v_σ in a neighborhood of the origin so that

$$L(u_i, v_\sigma) = V(u_i) + \sum_\sigma a_\sigma {v_\sigma}^2, \tag{4.22}$$

with non-zero constants a_σ; $V(u_i)$ is then the «effective potential with massive modes integrated out». The u_i are of course only uniquely determined up to diffeomorphism. Describing the possible canonical forms of $V(u_i)$ up to a reparametrization of the variables is more or less the problem of singularity theory.

Since the u_i can be described in first order as coordinates near the origin on the naturally defined vector space $H^1(M, \mathrm{End}(E))$, it is natural to think of the effective potential V as a function defined on that space, and as such it is uniquely defined up to a coordinate transformation

$$u_i \to \widetilde{u}_i = u_i + c_{ijk} u_j u_k + \dots \tag{4.23}$$

that is trivial in first order. The quartic coupling $\langle \alpha_{(i)} \alpha_{(j)} \alpha_{(k)} \alpha_{(l)} \rangle$, for particular values of i, j, k, and l, is invariant under such transformations if and only if the cubic coupling $\langle \alpha_{(m)} \alpha_{(n)} \alpha_{(p)} \rangle$ vanishes for m, n equal to two consecutive values of i, j, k, l;*) by Poincaré duality, this is so precisely if $\alpha_{(m)} \wedge \alpha_{(n)}$ vanishes in $H^2(M, \mathrm{End}(E))$ whenever m and n equal two consecutive values of i, j, k, and l.

Under such conditions, the four point coupling can be given the following interpretation. Since $\alpha_{(i)} \wedge \alpha_{(j)}$ and $\alpha_{(j)} \wedge \alpha_{(k)}$ vanish in cohomology, there are $\mathrm{End}(E)$-valued one-forms y, z with

$$\mathrm{d}y = \alpha_{(i)} \wedge \alpha_{(j)}, \quad \mathrm{d}z = \alpha_{(j)} \wedge \alpha_{(k)}. \tag{4.24}$$

*) One considers i, j, k, l arranged on a circle in that order, so l and i are consecutive.

In fact, we can set

$$y = \mathrm{d}^* \frac{1}{\Delta} \left(\alpha_{(i)} \wedge \alpha_{(j)}\right)$$
$$z = \mathrm{d}^* \frac{1}{\Delta} \left(\alpha_{(j)} \wedge \alpha_{(k)}\right) \tag{4.25}$$

(Because $1/\Delta$ is the inverse of Δ only in the orthocomplement of the kernel of Δ, in proving that (4.25) obeys (4.24), one needs the fact that $\alpha_{(i)} \wedge \alpha_{(j)}$ and $\alpha_{(j)} \wedge \alpha_{(k)}$ vanish in cohomology and so are orthogonal to the kernel.) Let

$$w(\alpha_{(i)}, \alpha_{(j)}, \alpha_{(k)}) = y \wedge \alpha_{(k)} + \alpha_{(i)} \wedge z. \tag{4.26}$$

Then $\mathrm{d}w = 0$, so w determines an element of $H^2(M, \mathrm{End}(E))$. Considering how w transforms under a shift in y and z, one sees that it is not well-defined as an element of $H^2(M, \mathrm{End}(E))$ but only as an element of the quotient group

$$H^2(M, \mathrm{End}(E))/(H^1(M, \mathrm{End}(E)) \wedge \alpha_{(l)} \oplus \alpha_{(i)} \wedge H^1(M, \mathrm{End}(E))). \tag{4.27}$$

The element w of that quotient space is known as the Massey triple product of $\alpha_{(i)}, \alpha_{(j)}$, and $\alpha_{(k)}$. Under the further hypothesis that $\alpha_{(k)} \wedge \alpha_{(l)}$ and $\alpha_{(l)} \wedge \alpha_{(i)}$ vanish in cohomology,

$$\int_M \mathrm{Tr}\, w(\alpha_{(i)}, \alpha_{(j)}, \alpha_{(k)}) \wedge \alpha_{(l)} \tag{4.28}$$

is well-defined and in fact equals coefficient of the quartic term in the effective potential as computed above. The contribution of the quartic coupling to $\partial V/\partial u_l$ is therefore

$$\frac{1}{3!} \sum_{i,j,k} u_i u_j u_k \int_M \mathrm{Tr}\, w(\alpha_{(i)}, \alpha_{(j)}, \alpha_{(k)}) \cdot \alpha_{(l)}. \tag{4.29}$$

As a flat connection is a critical point of the Chern-Simons action, this is in keeping with the observation [27] that the Massey triple product and its higher order analogs are the obstructions to deformation of a flat connection.

4.3 Comparison to remarks of Kontsevich

What do we learn by interpreting Chern-Simons gauge theory as a string theory? Here I will only point out one obvious consequence. Other applications may involve, for instance, a deeper relation of Chern-Simons theory to mirror manifolds.

The fact that Chern-Simons theory can be interpreted as a topological sigma model coupled to topological gravity means that one can introduce the usual observables of topological gravity — the stable cohomology classes on the moduli space of Riemann surfaces introduced by Mumford, Morita, and Miller. In the present context, this would amount to the following. As was sketched in §2, the topological sigma model determines a differential form Θ on moduli space; for open strings, the relevant moduli space is the moduli space $\mathcal{M}_{g,s,h}$ of Riemann surfaces of genus g with s punctures (where vertex operators are inserted) and h boundary components. Instead of integrating the closed form Θ over $\mathcal{M}_{g,s,h}$, we can integrate it over a homology cycle C in $\mathcal{M}_{g,s,h}$ of codimension, say, n.

The trouble is that we have fixed X to have complex dimension three precisely to ensure that Θ is a top form. It therefore cannot be integrated over a cycle of positive codimension unless the definition is modified somehow. The following way of doing this will enable us to make contact with observations by Kontsevich.

Instead of considering a single target space X with a fixed metric, we will consider a family of X's with a family of metrics. The partition function of the theory will be not a number but a closed differential form on the space of metrics on X. To implement this idea, introduce along with the metric $g_{IJ}(\phi)$ of X a fermionic variable $\zeta_{IJ}(\phi)$ (of ghost number one) which one can think of as (up to a constant) the exterior derivative of $g_{IJ}(\phi)$. Take the new transformation laws

$$\delta g_{IJ} = i\alpha\zeta_{IJ}, \quad \delta\zeta_{IJ} = 0. \tag{4.30}$$

Recalculating the Lagrangian starting from (3.2), we find that (3.3) is replaced by $\widetilde{L} = L + L'$, with

$$L' = -t\int_\Sigma d^2z\zeta_{\bar{i}j}\left(\psi_z^{\bar{i}}\partial_{\bar{z}}\phi^j + \partial_z\phi^{\bar{i}}\psi_{\bar{z}}^j\right). \tag{4.31}$$

Since L' is of ghost number -1 in the «matter fields» ϕ, χ, ψ, every insertion of L' shifts the ghost number by one unit. If we want to integrate not over moduli space but over a cycle C of codimension n, the non-vanishing contributions will be precisely n^{th} order in L'. As L' is linear in ζ_{IJ}, these contributions will be n^{th} order in ζ_{IJ} and so will define an n-form Ω on the space $\mathcal{R}$ of metrics on X. Q-invariance means that Ω is closed. Let $\mathcal{F}$ be the group of diffeomorphisms of X. Ω is invariant under the natural action of $\mathcal{F}$ on $\mathcal{R}$ and moreover is basic (to show the later one notes that if $\zeta_{IJ} = D_I v_J + D_J v_I$ for some v^I, then L' is of the form $\{Q,\ldots\}$ up to terms that vanish by the equations of motion). So if one has a fiber bundle

$$\begin{array}{ccc} X & \to & Y \\ & & \downarrow \\ & & B, \end{array} \tag{4.32}$$

with an arbitrary base B, then, by picking a metric on the total space Y, one gets a family of metrics on X, parametrized by B, and Ω determines an n-dimensional cohomology class of B.

We have found, therefore, a map from the codimension n homology of $\mathcal{M}_{g,s,h}$ to the n-dimensional cohomology of B. This map was described by Kontsevich [5] by examining Chern-Simons perturbation theory. The considerations just explained give a more conceptual explanation for its existence.

A digression

Let us now make a small digression to examine some related observations by Kontsevich in the light of standard quantum field theory ideas. We want to consider the standard Fadde'ev-Popov-BRST quantization of three dimensional Chern-Simons gauge theory, with gauge group G, on a three-manifold M. In doing so, in addition to the gauge field A_a, one introduces a ghost field c (anticommuting, of ghost number one, transforming in the adjoint representation). The usual BRST transformation laws are

$$\begin{aligned} \delta A_a &= -D_a c \\ \delta c &= \frac{1}{2}[c, c]. \end{aligned} \tag{4.33}$$

We temporarily postpone introducing the antighosts and auxiliary fields that enter the gauge fixing.

Just from the $A - c$ system, it is possible to construct new observables (discussed, for instance, in [28], where more information can be found). Let T be an invariant, antisymmetric polynomial on the Lie algebra $\mathcal{G}$ of G. If $P \in M$ is any point, let

$$\mathbb{O}_T^{(0)}(P) = T(c(P)). \tag{4.34}$$

It is evident that $\mathbb{O}^{(0)}$ is BRST invariant and cannot be written in the form $\delta(\ldots)$. By solving the «descent equations»

$$\begin{aligned} d\mathbb{O}^{(0)} &= \delta\mathbb{O}^{(1)} \\ d\mathbb{O}_T^{(1)} &= \delta\mathbb{O}^{(2)} \\ d\mathbb{O}_T^{(2)} &= \delta\mathbb{O}^{(3)} \\ d\mathbb{O}_T^{(3)} &= 0, \end{aligned} \tag{4.35}$$

one finds for each i, $0 \leq i \leq 3$, an operator-valued i-form $\mathbb{O}_T^{(i)}$ that is BRST invariant up to $\delta(\ldots)$. Hence, if $Y \subset M$ is a i-dimensional cycle, then

$$\int_Y \mathbb{O}_T^{(i)} \tag{4.36}$$

is BRST invariant (and is easily seen, by virture of (4.35), to depend only on the homology class of Y). In particular, setting $i = 3$ and $Y = M$, we get new terms that can be added to the Chern-Simons Lagrangian. To be precise, if L is the usual Chern-Simons Lagrangian and T_α are the antisymmetric invariants on the Lie algebra, we can take

$$L \to L + \sum_\alpha t_\alpha \int_M d^3x \ \mathbb{O}_{T_\alpha}^{(3)}. \tag{4.37}$$

This modification of the standard Chern-Simons theory is implicit in the work of Kontsevich, who describes the situation in terms of a certain class of homotopy Lie algebras.

One could also, as in Donaldson theory, pick closed submanifolds M_α of M, of dimension d_α, and generalize (4.37) to

$$L \to L + \sum_\alpha t_\alpha \int_{M_\alpha} d^3x \ \mathcal{O}_{T_\alpha}^{(d_\alpha)}. \tag{4.38}$$

As it stands, (4.37) is not very useful. Since the usual Chern-Simons theory conserves ghost number, and the $\mathcal{O}^{(3)}$'s all have ghost number $+3$, the results will be independent of the t_α unless we also introduce some interaction of negative ghost number. To get something interesting, we will now modify the usual gauge fixing.

Gauge fixing requires the introduction of antighost and auxiliary fields. The standard procedure is to introduce the antighost $\bar{c}$ (anticommuting, of ghost number -1, in the adjoint representation) and the auxiliary field w (commuting, of ghost number 0, in the adjoint representation), with

$$\delta\bar{c} = iw, \quad \delta w = 0. \tag{4.39}$$

Gauge fixing is then carried out by

$$L \to L + \delta\Gamma, \tag{4.40}$$

with any convenient Γ. A standard choice involves picking a metric g_{ab} on M. Writing also $A = A_{(0)} + B$, where $A_{(0)}$ is a solution of the classical equations about which one wishes to expand, and denoting the covariant derivative with respect to $A_{(0)}$ as $D^{(0)}$, we take

$$\Gamma = -\int_M d^3x \ \sqrt{g}g^{ab} \operatorname{Tr} \bar{c} D_a^{(0)} B_b. \tag{4.41}$$

Computing $\delta\Gamma$, one gets the usual gauge fixing Lagrangian

$$L_{GF} = \int_M d^3x \ \sqrt{g} \operatorname{Tr} \left(iwD_b^{(0)} B^b - \bar{c} D_b^{(0)} D^b c \right). \tag{4.42}$$

Just as in our previous discussion, to modify this, we will consider not a fixed target M but a family of M's with a variable metric. Instead of considering the metric g_{ab} of M to be «inert» under the BRST transformations, we introduce corresponding fermi variables ζ_{ab} (of ghost number 1), with

$$\delta g_{ab} = \zeta_{ab}, \qquad \delta\zeta_{ab} = 0. \tag{4.43}$$

Then (4.42) is replaced by $\widetilde{L}_{GF} = L_{GF} + \Delta L_{GF}$ with

$$\Delta L_{GF} = -\int_M d^3x \sqrt{g}\left(\zeta^{ab} - \frac{1}{2}g^{ab}\zeta^c{}_c\right) \operatorname{Tr}\bar{c}D_a^{(0)}B_b. \qquad (4.44)$$

Since ΔL_{GF} is of ghost number -1 in the matter fields, insertions of $\mathcal{O}^{(3)}$'s can be balanced by insertions of ΔL_{GF}. As ΔL_{GF} is linear in ζ, the resulting amplitudes, just as in our string theoretic discussion, will be naturally not numbers but differential forms on the base space of a fibration.

4.4 General target spaces

So far we have only considered the **A** model with target space $X = T^*M$. Now we want to generalize the discussion to consider an arbitrary symplectic target manifold X (of $c_1 = 0$), with M as a Lagrangian submanifold.

The first consequence of replacing T^*M with a more general X is that there may be nonconstant instantons. The same argument that we used in proving the vanishing theorem for T^*M shows that a nonconstant instanton would necessarily have a positive value of the instanton number

$$q = \int_\Sigma \Phi^*(\omega). \qquad (4.45)$$

To improve the convergence of our formulas, we pick a positive number θ, and weight instantons of instanton number q with a factor of $\exp(-\theta q)$. This can be naturally built into the formulas by adding to the Lagrangian a suitable multiple of q:

$$L \to L + \theta \int_\Sigma \Phi^*(\omega). \qquad (4.46)$$

In the absence of non-constant instantons, the space-time theory of the **A** model is ordinary Chern-Simons theory. We want to determine the corrections to this coming from the nonconstant instantons. In doing so, our goal is to find the classical Lagrangian underlying the space-time physics of the **A** model. To this end, we concentrate on the case that the world-sheet is a disc Σ. (However, corrections due to higher topologies can be described similarly.) The target space Lagrangian L_T is equal to a world-sheet path integral on the disc:

$$L_T = \int DX \ldots D\psi \ \exp(-L) \qquad (4.47)$$

We already know that the contribution to (4.47) of instanton number $q = 0$ is the ordinary Chern-Simons action. We want to determine the contribution for some non-zero value of q.

Consider the moduli space of holomorphic maps $\Phi : \Sigma \to X$ with $\Phi(\partial\Sigma) \subset M$ and with two such maps identified if they differ by an $SL(2, \mathbb{R})$ transformation. The fact that $c_1(X) = 0$ and $\dim_{\mathbb{C}}(X) = 3$ means that in the moduli problem, the dimensions of the appropriate H^0 and H^1 are zero. «Generically» this means that there are only finitely many such instantons for each value of q. For simplicity, we will consider only this case. If Φ is such an instanton, let $C = \Phi(\partial\Sigma)$. Generically, C is a knot in the three-manifold M. Let us work out the contribution of Φ to the path integral (4.47). If we are expanding around a background connection A in M, the contribution of the Chan-Paton factors is

$$\operatorname{Tr} P \exp \int_C A. \tag{4.48}$$

Note that though C bounds a disc in X, it may not do so in M, so (4.48) can be non-trivial even if A is flat. We also get a factor of $\exp(-\theta q)$ from the instanton-counting term in (4.46). The remaining contributions are nearly trivial since (i) they are independent of t; (ii) they reduce in the large t limit to a ratio of determinants; (iii) except for a possible sign, the boson and fermion determinants cancel because of the Q symmetry. The contribution of an instanton is hence

$$\eta \exp(-\theta q) \operatorname{Tr} P \exp \int_C A \tag{4.49}$$

where $\eta = \pm 1$ is the ratio of determinants.

The total action is therefore easy to evaluate. If Φ_i, $i = 1, 2, 3, \ldots$ are the instantons of non-zero instanton number, with instanton numbers q_i, boundaries C_i, and determinant factors η_i, then the action is

$$L_T = \frac{1}{2} \int_M \operatorname{Tr}\left(A \wedge dA + \frac{2}{3} A \wedge A \wedge A\right) + \sum_{i=1}^{\infty} \eta_i \exp(-\theta q_i) \operatorname{Tr} P \exp \int_{C_i} A. \tag{4.50}$$

For instance, for $\theta >> 0$, the factors $\exp(-\theta q_i)$ are small, and the instanton corrections to amplitudes can be evaluated perturbatively. Their evaluation would involve calculating expectation values of products of Wilson lines on the three-manifold M.

4.5 The B model

Now, we would like, in a similar spirit, to identify the space-time field theory that is equivalent to the **B** model, with the «free» boundary conditions of §3.1. To be more precise, we consider the open string sector of the **B** model, and we use the same general framework of open string field theory as in §4.1. We can be brief, as the arguments are so similar.

As in the case of the **A** model, the main simplification comes from the invariance under rescaling the metric of the target space X by an arbitrary factor t.

In §3.3, we saw that the low-lying modes of the string are functions $\mathcal{A}(\phi^I, \eta^{\bar{i}})$ of the zero modes. As we now wish $\mathcal{A}$ to have ghost number 1, we take it to be linear in η, so in fact

$$\mathcal{A} = \eta^{\bar{i}} A_{\bar{i}}(\phi^I). \tag{4.51}$$

So the physical field A is a one-form of type $(0, 1)$. If gauge fields are included via Chan-Paton factors, then A takes values in $N \times N$ matrices or more generally in the endomorphisms of some holomorphic vector bundle E.

The linearized gauge transformation law $\delta\mathcal{A} = Q\epsilon$ reduces for large t to $\delta A = \overline{\partial}\epsilon$, so A must be interpreted as the $(0, 1)$ part of a connection on E. What should be the field equation for A? This can be anticipated from the discussion at the end of §3.2, where we showed that the background connections to which the **B** model can be coupled are precisely those for which the $(0, 2)$ part of the curvature vanishes, in other words those that define holomorphic structures on E. To write a Lagrangian from which this equation can be derived, let λ be an everywhere non-zero holomorphic three-form on X. Then up to an undetermined constant, the Lagrangian whose solutions are connections of vanishing $(0, 2)$ curvature is

$$L_T = \frac{1}{2} \int_X \lambda \wedge \mathrm{Tr}\left(A \wedge \overline{\partial} A + \frac{2}{3} A \wedge A \wedge A \right). \tag{4.52}$$

Arguments similar to those that we have given for the **A** model show that in the large t limit, the open string field theory of the **B** model reduces to (4.52).

The quantum field theory with Lagrangian (4.52) is unrenormalizable by power counting. However, it has the following all-but-unique property: there are no possible counter-terms that respect the classical symmetries. The symmetries of (4.52) include *complex* gauge transformations, $\overline{\partial}_A \to g \overline{\partial}_A g^{-1}$, with g an arbitrary gauge transformation of E not respecting any reality or unitarity condition; and local complex changes of coordinates that preserve λ. There is no local density constructed from A that is invariant under these symmetries. (Even L_T itself, though possessing these invariances at least if one considers only gauge transformations that are connected to the identity, is not the integral of an invariant local density.) Infinities in quantum field theory are ordinarily integrals of invariant local densities, so if (4.52) could be quantized preserving the symmetries, one would expect this theory to be *finite*, though superficially unrenormalizable.

Relying only on usual field theory arguments, it is not at all clear that (4.52) can be quantized preserving its symmetries. However, the equivalence of (4.52) to a string theory strongly suggests that it in fact is finite. One might worry about whether closed string poles can ruin the finiteness; at the end of §5 we will argue that this does not occur. One might also wonder whether the finite theory given by the string theory is really (4.52) or some more elaborate theory with (4.52) coupled to closed strings. In §5 we argue that the closed strings are decoupled.

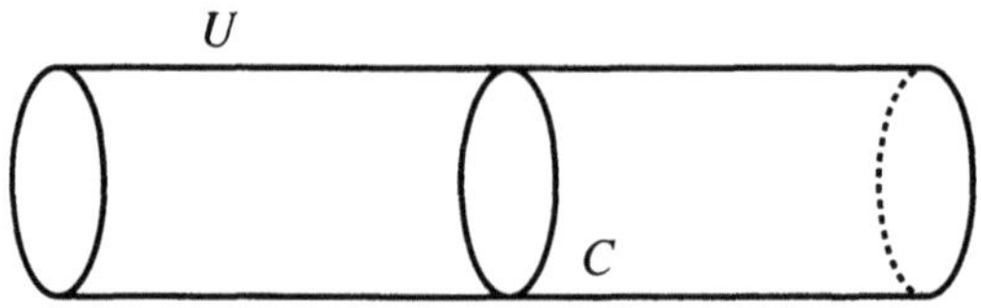

Fig. 8 A cylinder U and an embedded circle C. Any diagram containing this embedded cylinder has two moduli naturally associated with the cylinder; one is the length of the cylinder and the second comes from the possibility of «cutting» the cylinder on C, and twisting about C before regluing.

5 The closed string sector

We have discovered that, with some reasonable boundary conditions, the open string sector of the topological **A** and **B** string theories has an elegant interpretation in terms of a space-time field theory. The extension to closed strings does not work so nicely; the first point of this section is to sketch what the problem is. After doing this, I will conclude by trying to show that open and closed strings are decoupled in these models; this is intended as partial justification for studying the open strings separately in §4.

For open strings, the propagator is

$$\frac{b_0}{L_0}, \tag{5.1}$$

with L_0 the Hamiltonian of the string and b_0 the antighost zero mode. For closed strings, one has separate zero modes b_0 and $\overline{b}_0$ for right- and left-moving antighosts. It is convenient to set $b_0^{\pm} = b_0 \pm \overline{b}_0$. The formula analogous to (5.1) is that the closed string propagator is

$$\frac{b_0^- \overline{b}_0{}^+}{2L_0^+}\Pi = \frac{b_0\overline{b}_0}{L_0^+}\Pi. \tag{5.2}$$

Here $L_0^+ = L_0 - \overline{L}_0$ is the total string Hamiltonian, and Π is the projection operator on states invariant under rotation of the circle. This formula is fairly well known, and in any case can be derived similarly to (5.1), replacing the strip in Figure 6 with the cylinder of Figure 8; the extra ghost field in the numerator and the projection on rotation-invariant states come from the twist symmetry of the cylinder, indicated in the figure.

For the **A** model, for instance, the long wavelength limits of b_0 and $\overline{b}_0$ are the ∂^* and $\overline{\partial}^*$ operators of the target space X. With L_0 reducing at long wavelengths to the Laplacian Δ, the string propagator looks like

$$\frac{\partial^*\overline{\partial}^*}{\Delta}. \tag{5.3}$$

This propagator, however, does not seem to arise by gauge fixing of any local Lagrangian. It is a pseudodifferential operator of degree zero, so for it to arise as the inverse of a differential operator, that operator would have to be of degree zero, that is a constant.

In fact, by repeating for the closed string the analysis of §3.3, one finds that the low energy modes of the closed string sector of the **A** model are naturally represented by a two-form h in space-time. *) (For large t, winding sectors of the closed string cannot have small eigenvalues of L_0.) The free Lagrangian for h that one might guess by analogy with our open string results would be

$$L = \int_X h \wedge \partial\overline{\partial} h. \tag{5.4}$$

This has

$$\frac{\partial^* \overline{\partial}^*}{\Delta^2} \tag{5.5}$$

for a gauge fixed propagator. The extra factor of Δ in the denominator, which is in sharp variance with (5.3), of course, makes (5.5) a pseudodifferential operator of degree -2, in keeping with the fact that the kinetic operator in (5.4) is second order.

What sort of Lagrangian do we get by taking the low energy limit of closed string field theory? As we discussed in §2 in a related context, the free part of the closed string Lagrangian is

$$(\Psi, c_0^- Q\Psi), \tag{5.6}$$

where ideally

$$\{b_0{}^-, c_0{}^-\} = 1. \tag{5.7}$$

Such a $c_0{}^-$ does not exist, since $b_0{}^-$, whose field theory limit is $\partial^* - \overline{\partial}^*$, has a non-trivial cohomology. However, one can pick a $c_0{}^-$ such that $\{b_0{}^-, c_0{}^-\} = 1 - T$, where T is the projection operator onto a subspace (say the kernel of L_0) annihilated by Q; this is good enough to ensure gauge invariance of (5.7). Such a $c_0{}^-$ is, in the field theory limit,

$$\frac{\partial - \overline{\partial}}{2\Delta}. \tag{5.8}$$

With this choice of $c_0{}^-$, and recalling that the field theory limit of Q is $\partial + \overline{\partial}$, the field theory limit of the Lagrangian is not (5.4) but

$$\int_X h \wedge \frac{1}{\Delta} \partial\overline{\partial} h. \tag{5.9}$$

When this is gauge-fixed and inverted to get a propagator, the factor of Δ^{-1} migrates to the numerator, canceling a factor of Δ^{-1} in (5.5) and reproducing the field theory limit of the closed string propagator (5.3).

---

*) For the **B** model, one gets instead a sum of $(0, i)$ forms with values in $\wedge^j T^{(1,0)}$, with $i + j = 2$.

A general puzzle

The conclusion seems to be that closed string field theory of the **A** model, or similarly of the **B** model, would be non-local in space-time. Sometimes such apparent non-localities can be eliminated by introducing additional fields (such that the apparent non-locality arises in integrating them out). I have no evidence that that can be done here. In any event, certain puzzling arguments seem to show that the closed string **A** and **B** theories do not behave as one would expect of space-time field theories.

In either the **A** or the **B** model, there are non-trivial cubic and higher order couplings of the physical modes. In studying the open string sector of the **A** model, there was at the classical level (the world-sheet being a disc) a cubic coupling of three physical fields given by the classical formula $\int_M \mathrm{Tr}\, A \wedge A \wedge A$, and various higher interactions involving Massey products, as we saw in §4.2. As always in field theory, these couplings are Taylor series coefficients of a natural potential $V(t_i)$ for sources t_i coupling to the physical modes. The classical solutions of the space-time theory — or equivalently the possible world-sheet theories — are in one-to-one correspondence with the critical points of V.

What about the closed string sectors? There are analogous cubic and higher couplings of physical modes. For instance, for the closed string sector of the **A** model, the cubic coupling in the large volume limit is $\int_X h \wedge h \wedge h$. Experience with both field theory and with the open string sectors of the **A** and **B** models lead us to form a generating function $V(t_i)$ from these couplings, with the expectation that the allowed world-sheet theories will correspond to the critical points of V. This latter expectation proves to be false. For the **A** model, for instance, the part of the h field that is a $(1,1)$ form in space-time represents a displacement in the Kahler class of the metric of X. The **A** model makes sense for any choice of this Kahler class, so in contrast to what one would have anticipated, having $V'(t_i) \neq 0$ is not an obstruction to being able to define the world-sheet theory. Likewise, for the **B** model, the low energy modes include a displacement in the complex structure of X. The **B** model makes sense for any complex structure on X, even though there are non-zero cubic and higher order couplings for the fields representing a displacement in the complex structure.

5.1 Closed string contributions to amplitudes

Leaving this puzzle as food for thought, I will conclude by making some simple comments about the closed string contributions to open string amplitudes of the **A** and **B** models.

Let us look briefly at the couplings of open and closed strings in these models. To do so we will use the fatgraph or open string field theory description of external open and closed strings. An external open string (a marked point on the boundary of a Riemann surface) should have one real modulus, while an external closed string (a marked point in the interior) should have two. In the fatgraph description, as one might expect, external open strings are represented by open string propagators going off to infinity, as for O_1 and O_2 in Figure 9(b). The one

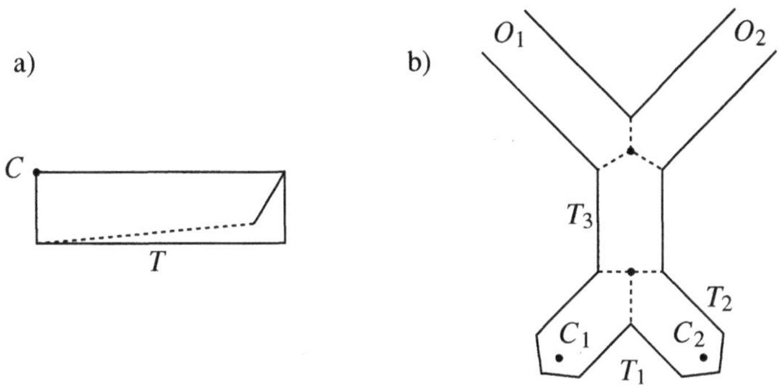

Fig. 9 In the fatgraph description of a Riemann surface coupling open and closed strings, the closed strings are incorporated as in (a). A standard open string strip of width π and length T, open at one end (where it attaches to the rest of the diagram) is closed at the other end by folding it over on itself; the closed string is attached at the resulting conical singularity, as shown. A typical fatgraph coupling open and closed strings is shown in (b). It consists of five flat strips glued on the dotted lines, with conical singularities shown as solid dots. The open strings O_1 and O_2 are attached at the ends of infinite flat strips, while the closed strings appear as conical singularities on strips of finite lengths T_1 and T_2. This particular world-sheet has three real moduli: T_1, T_2, and the length T_3 of the one internal propagator.

real modulus of the open string is, roughly, the location at which its propagator is attached to the rest of the figure. The proper coupling of external closed strings (which can be deduced by seeing how closed string poles arise in open string diagrams, as in [29]) is as follows. One attaches to the open string diagram an external open string propagator of *finite* length T, closes it up by folding together its free end, and inserts the closed string at the resulting conical singularity. The two real moduli of the closed string are, roughly, T and the position at which the propagator is attached to the rest of the diagram. This is sketched in Figures 9(a,b).

To compute the classical couplings of n open strings and m closed strings, we consider the moduli space $\mathcal{D}_{n,m}$ of a disc with n marked points on the boundary and m in the interior; its real dimension is $n+m-3$. (The analysis goes through the same way in higher genus.) Comparing the dimension of $\mathcal{D}_{n,m}$ to that of $\mathcal{D}_{n+m,0}$, there is one extra modulus for each marked point in the interior. As explained in the last paragraph, the fatgraph description of $\mathcal{D}_{n,m}$ (Figure 9(b)) is similar to the fatgraph description of $\mathcal{D}_{n+m,0}$ except that, while the external open strings are attached to outgoing propagators of infinite length, the external closed strings are attached to propagators of variable length; the one extra real modulus for each external closed string is precisely the length of the propagator to which it is attached.

In the **A** model, for instance, an external open string is represented by an (End(E)-valued) one-form α, while an external closed string is represented by a two-form h. Integration over the extra modulus of the propagator by which the closed string is attached to the rest of the diagram multiplies h by the *open string* propagator b_0/L_0. In the large t limit, this simply turns h into a one-form

$$\frac{\mathrm{d}^*}{L_0} \cdot h. \tag{5.10}$$

This one-form then couples as just one more external open string state (which happens to be valued in the center of $\mathrm{End}(E)$). But the one-form in (5.10) is exact (since, for instance, the open string propagator annihilates harmonic forms), so the corresponding open string state decouples. Consequently the on-shell couplings of open and closed strings are all zero — in either the **A** or **B** model.

If the closed strings are, then, decoupled from the open strings, what is their role? The following conjecture seems natural to me. The field theories we have extracted from the **A** and **B** models all have c-number anomalies, analogous to the central charge in two-dimensional conformal field theory. For instance, the anomalies of Chern-Simons theory are connected with framings of three-manifolds and of knots [30]. One-loop anomalies of the field theory related to the **B** model were calculated long ago by Ray and Singer [31]. These anomalies are possible only because of the ultraviolet difficulties of quantum field theory, which of course are greatly ameliorated in string theory. It therefore seems reasonable to suspect that the closed string contributions in the **A** and **B** theories cancel the anomalies of these theories, without, in view of the decoupling argued above, having much effect on the open string «physics».

In any event, whatever the closed string contributions, they are finite. In usual string theory, possible infinities come from physical closed string poles, but in these topological models, there are no such poles since the closed string propagator annihilates the physical states or harmonic forms; this is because (in contrast to conventional string backgrounds) the b_0 operator annihilates the kernel of L_0. Thus, for instance, in the case of the **B** model, whose finiteness perhaps comes as a surprise (since the field theory related to the low energy limit of the open strings is superficially unrenormalizable), the closed strings will not ruin this finiteness.

For further information on the closed string version of these theories, see [32].

Acknowledgements: The discussion of closed string field theory in §2 benefitted considerably from discussions with B. Zwiebach. Several relevant references were pointed out by L. Jeffrey.

References

[1] E. Witten. «Topological Sigma Models», Comm. Math. Phys. **118** (1988), 411.

[2] M. Gromov, «Pseudo-Holomorphic Curves In Symplectic Manifolds», Invent. Math. **82** (1985), 307.

[3] A. Floer, «Symplectic Fixed Points And Holomorphic Spheres», Comm. Math. Phys. **120** (1989), 575.

[4] S. Axelrod and I.M. Singer, «Chern-Simons Perturbation Theory», MIT preprint (1991).

[5] M. Kontsevich, lecture at Institute for Advanced Study (January, 1992).

[6] E. Witten, «Mirror Manifolds and Topological Field Theory», in *Essays On Mirror Manifolds*, ed. S.-T. Yau (International Press, 1992).

[7] W. Siegel, «Covariantly Second-Quantized Strings, II,III», Phys. Lett. **151B** (1985) 391, 396.

[8] S. Elitzur, A. Forge, and E. Rabinovici, «On Effective Theories Of Topological Strings», preprint CERN-TH.6326 (1991).

[9] G. 't Hooft, Nucl. Phys. **B72** (1974) 461.

[10] E. Witten, «Two Dimensional Gravity And Intersection Theory On Moduli Space», Surv. Diff. Geom. **1** (1991) 243.

[11] R. Penner, «The Teichmuller Space Of A Punctured Surface», Comm. Math. Phys. (1987), «Perturbative Series And The Moduli Space Of Riemann Surfaces», J. Diff. Geom. **27** (1988) 35.

[12] J. Harer, «The Virtual Cohomological Dimension Of The Mapping Class Group Of Orientable Surfaces», Inv. Math. **84** (1986) 157.

[13] B.H. Bowditch and D.B.A. Epstein, «Natural Triangulations Associated To A Surface», Topology **27** (1988) 91.

[14] E. Witten, «Non-Commutative Geometry And String Field Theory», Nucl. Phys. **B**268 (1986) 253.

[15] S. Giddings, E. Martinec, and E. Witten, «Modular Invariance In String Field Theory», Phys. Lett. **176B** (f1986) 362.

[16] J. Distler and P. Nelson, «Topological Couplings And Contact Terms In 2d Field Theory», Comm. Math. Phys. **138** (1991) 273.

[17] B. Zwiebach, «Closed String Field Theory: Quantum Action And The BV Master Equation», preprint IASSNS-HEP-92/41 (June, 1992).

[18] L. Dixon, lecture at Princeton University, ca. 1987.

[19] C. Vafa, «Topological Mirrors And Quantum Rings», in *Essays on Mirror Manifolds*, ed S.-T. Yau (International Press, 1992).

[20] J.M.F. Labastida and P.M. Llatas, «Topological Matter in Two Dimensions» (preprint, 1991).

[21] P. Horava, «Equivariant Topological Sigma Models», preprint (1991).

[22] V. Mathai and D. Quillen, «Superconnections, Thom Classes, and Equivariant Differential Forms», Topology **25** (1986) 85.

[23] M.F. Atiyah and L. Jeffrey, «Topological Lagrangians And Cohomology», J. Geom. Phys. **7** (1990) 119.

[24] D. Fried, Inv. Math. **84** (1986) 523.

[25] C.B. Thorn, «Perturbation Theory For Quantized String Fields», Nucl. Phys. **B287** (1987) 61.

[26] M. Bochicchio, «Gauge Fixing For The Field Theory Of The Bosonic String», Phys. Lett. **193B** (1987) 31.

[27] D. Johnson and J. Millson, «Deformation Spaces Associated To Compact Hyperbolic Manifolds», in *Discrete Groups In Geometry And Analysis: Papers In Honor Of G.D. Mostow*, ed R. Howe (Birkhäuser, Boston, 1987).

[28] O. Piguet and S.P. Sorella, «On The Finiteness Of BRS Modulo-d Cocycles», Univ. of Geneva preprint UGVA-DPT 1992/3-759.

[29] D.Z. Freedman, S.B. Giddings, J.A. Shapiro, and C.B. Thorn, Nucl. Phys. **B298** (1988) 253.

[30] E. Witten, «Quantum Field Theory And The Jones Polynomial», Comm. Math. Phys. **121** (1989) 351.

[31] D. Ray and I.M. Singer, «Analytic Torsion And The Laplacian On Complex Manifolds» Ann. of Math. **98** (1973) 154.

[32] M. Bershadsky, S. Cecotti, H. Ooguri and C. Vafa, «Holomorphic Anomalies in Topological Field Theories», Nucl. Phys. **B405** (1993) 279, «Kodaira-Spencer Theory of Gravity and Exact Results for Quantum String Amplitudes», Nucl. Phys. B405 (1993) 279.

Edward Witten
School of Natural Sciences
Institute for Advanced Study
Olden Lane
Princeton, NJ 08540

Index

Progress in Mathematics

Progress in Mathematics is a series of books intended for professional mathematicians and scientists, encompassing all areas of pure mathematics. This distinguished series, which began in 1979, includes authored monographs, and edited collections of papers on important research developments as well as expositions of particular subject areas.

We encourage preparation of manuscripts in such form of TeX for delivery in camera-ready copy which leads to rapid publication, or in electronic form for interfacing with laser printers or typesetters.

Proposals should be sent directly to the editors or to: Birkhäuser Boston, 675 Massachusetts Avenue, Cambridge, MA 02139, U.S.A.

1 GROSS. Quadratic Forms in Infinite-Dimensional Vector Spaces
2 PHAM. Singularités des Systèmes Differentiels de Gauss-Manin
4 AUPETIT. Complex Approximation
5 HELGASON. The Radon Transform
6 LION/VERGNE. The Weil representation Maslov index and Theta series
7 HIRSCHOWITZ. Vector bundles and differential equations
10 KATOK. Ergodic Theory and Dynamical Systems I
11 BALSLEY. 18th Scandinavian Congress of Mathematicians
12 BERTIN. Séminaire de Théorie de Nombres, Paris 79-80
13 HELGASON. Topics in Harmonic Analysis on Homogeneous Spaces
14 HANO. Manifolds and Lie Groups
15 VOGAN JR. Representations of Real Reductive Lie Groups
16 GRIFFITHS/MORGAN. Rational Homotopy Theory and Differential Forms
17 VOVSI. Triangular Products of Group Representations and Their Applications
18 FRESNEL/VAN DER PUT. Géometrie Analytique Rigide et Applications
19 ODA. Periods of Hilbert Modular Surfaces
20 STEVENS.Arithmetic on Modular Curves
21 KATOK. Ergodic Theory and Dynamical Systems II
22 BERTIN. Séminaire de Théorie de Nombres, Paris 80-81
23 WEIL. Adeles and Algebraic Groups
24 LE BARZ. Enumerative Geometry and Classical Algebraic Geometry
25 GRIFFITHS. Exterior Differential Systems and the Calculus of Variations
27 BROCKETT. Differential Geometric Control Theory
28 MUMFORD. Tata Lectures on Theta I
29 FRIEDMANN. The Birational Geometry of Degenerations
30 YANO/KON. Submanifolds of Kaehlerian and Sasakian Manifolds
31 BERTRAND. Approximations Diophantiennes et Nombres Transcendant
32 BROOKS. Differential Geometry
33 ZUILY. Uniqueness and Non-Uniqueness in the Cauchy Problem
34 KASHIWARA. Systems of Microdifferential Equations
35/36 ARTIN/TATE. Vol. 1 Arithmetic. Vol. 2 Geometry
37 BOUTET. Mathématique et Physique
38 BERTIN. Séminaire de Théorie de Nombres, Paris 81-82
39 UENO. Classification of Algebraic and Analytic Manifolds

40 TROMBI. Representation Theory of Reductive Groups
41 STANLEY. Combinatorics and Commutative Algebra
42 JOUANOLOU. Théorèmes de Bertini et Applications
43 MUMFORD. Tata Lectures on Theta II
45 BISMUT. Large Deviations and the Malliavin Calculus
47 TATE. Les Conjectures de Stark sur les Fonctions L d'Artin en s=0
48 FRÖHLICH. Classgroups and Hermitian Modules
49 SCHLICHTKRULL. Hyperfunctions and Harmonic Analysis on Symetric Spaces
50 BOREL ET AL. Intersection Cohomology
51 Séminaire de Théorie de Nombres, Paris 82-83
52 GASQUI/GOLDSCHMIDT. Déformations Infinitésimales desStructures Conformes Plates
53 LAURENT. Théorie de la 2ième Micro-localisation dans le Domaine Complexe
54 VERDIER. Module des Fibres Stables sur les Courbes Algébriques
55 EICHLER/ZAGIER. The Theory of Jacobi Forms
56 SHIFFMAN/SOMMESE. Vanishing Theorems on Complex Manifolds
57 RIESEL. Prime Numbers and Computer Methods for Factorization
58 HELFFER/NOURRIGAT. Hypoellipticité Maximale pour des Opera-teurs Polynomes de Champs de Vecteurs
59 GOLDSTEIN. Séminaire de Théorie de Nombres, Paris 83-84
60 ARBARELLO. Geometry Today
62 GUILLOU. A la Recherche de la Topologie Perdue
63 GOLDSTEIN. Séminaire de Théorie des Nombres, Paris 84-85
64 MYUNG. Malcev-Admissible Algebras
65 GRUBB. Functional Calculus of Pseudo-Differential Boundary Problems
66 CASSOU-NOGUES/TAYLOR. Elliptic Functions and Rings of Integers
67 HOWE. Discrete Groups in Geometry and Analysis
68 ROBERT. Autour de l'Approximation Semi-Classique
69 FARAUT/HARZALLAH. Analyse Harmonique: Fonctions Speciales et Distributions Invariantes
70 YAGER. Analytic Number Theory and Diophantine Problems
71 GOLDSTEIN. Séminaire de Théorie de Nombres, Paris 85-86
72 VAISMAN. Symplectic Geometry and Secondary Characteristic Classes
73 MOLINO. Riemannian Foliations
74 HENKIN/LEITERER. Andreotti-Grauert Theory by Integral Formulas
75 GOLDSTEIN. Séminaire de Théorie de Nombres, Paris 86-87
76 COSSEC/DOLGACHEV. Enriques Surfaces I
77 REYSSAT. Quelques Aspects des Surfaces de Riemann
78 BORHO/BRYLINSKI/MCPHERSON. Nilpotent Orbits, Primitive Ideals, and Characteristic Classes
79 MCKENZIE/VALERIOTE. The Structure of Decidable Locally Finite Varieties
80 KRAFT/ SCHWARZ/PETRIE (eds.) Topological Methods in Algebraic Transformation Groups
81 GOLDSTEIN. Séminaire de Théorie des Nombres, Paris 87–88
82 DUFLO/PEDERSEN/VERGNE (eds.) The Orbit Method in Representation Theory
83 GHYS/DE LA HARPE (eds.) Sur les Groupes Hyperboliques d'après M. Gromov
84 ARAKI/KADISON (eds.) Mappings of Operator Algebras
85 BERNDT/DIAMOND/HALBERSTAM/ HILDEBRAND (eds.) Analytic Number Theory
89 VAN DER GEER/OORT/STEENBRINK (eds.) Arithmetic Algebraic Geometry
90 SRINIVAS. Algebraic K-Theory
91 GOLDSTEIN. Séminaire de Théorie des Nombres, Paris 1988-89

92 CONNES/DUFLO/JOSEPH/RENTSCHLER. Operator Algebras, Unitary Representations, Enveloping Algebras, and Invariant Theory. A Collection of Articles in Honor of the 65th Birthday of Jacques Dixmier
93 AUDIN. The Topology of Torus Actions on Symplectic Manifolds
94 MORA/TRAVERSO (eds.) Effective Methods in Algebraic Geometry
95 MICHLER/RINGEL (eds.) Representation Theory of Finite Groups and Finite–Dimensional Algebras
96 MALGRANGE. Equations Différentielles à Coefficients Polynomiaux
97 MUMFORD/NORMAN/NORI. Tata Lectures on Theta III
98 GODBILLON. Feuilletages, Etudes géométriques
99 DONATO/DUVAL/ELHADAD/ TUYNMAN. Symplectic Geometry and Mathematical Physics. A Collection of Articles in Honor of J.-M. Souriau
100 TAYLOR. Pseudodifferential Operators and Nonlinear PDE
101 BARKER/SALLY. Harmonic Analysis on Reductive Groups
102 DAVID. Séminaire de Théorie des Nombres, Paris 1989-90
103 ANGER/PORTENIER. Radon Integrals
104 ADAMS/BARBASCH/VOGAN. The Langlands Classification and Irreducible Characters for Real Reductive Groups
105 TIRAO/WALLACH. New Developments in Lie Theory and Their Applications
106 BUSER. Geometry and Spectra of Compact Riemann Surfaces
107 BRYLINSKI. Loop Spaces,Characteristic Classes and Geometric Quantization
108 DAVID. Séminaire de Théorie des Nombres, Paris 1990-91
109 EYSSETTE/GALLIGO. Computational Algebraic Geometry
110 LUSZTIG. Introduction to Quantum Groups
111 SCHWARZ. Morse Homology
112 DONG/LEPOWSKY. Generalized Vertex-Algebras and Relative Vertex Operators
113 MOEGLIN/WALDSPURGER. Décomposition Spectrale et Series d'Eisenstein
114 BERENSTEIN/GAY/VIDRAS/YGER. Residue Currents and Bezout Identities
115 BABELON/CARTIER/KOSMANN-SCHWARZBACH (eds.) Integrable Systems. The Verdier Memorial Conference
116 DAVID (ed.) Séminaire de Théorie des Nombres, Paris, 1991-1992
117 AUDIN/LAFONTAINE (eds.) Holomorphic Curves in Symplectic Geometry
118 VAISMAN. Lectures on the Geometry of Poisson Manifolds
119 JOSEPH/MIGNOT/MURAT/PRUM/ RENTSCHLER (eds.) First European Congress of Mathematics (Paris, July 6-10, 1992). Volume I: Invited Lectures (Part 1)
120 JOSEPH/MIGNOT/MURAT/PRUM/ RENTSCHLER (eds.) First European Congress of Mathematics (Paris, July 6-10, 1992). Volume II: Invited Lectures (Part 2)
121 JOSEPH/MIGNOT/MURAT/PRUM/ RENTSCHLER (eds.) First European Congress of Mathematics (Paris, July 6-10, 1992). Volume III: Round Tables
122 GUILLEMIN. Moment Maps and Combinatorial Invariants of Hamiltonian T^n-spaces
123 BRYLINSKI/BRYLINSKI/GUILLEMIN/KAC (eds.) Lie Theory and Geometry: In Honor of Bertram Kostant
124 AEBISCHER/BORER/KÄLIN/ LEUENBERGER/REIMANN. Symplectic Geometry. An Introduction based on the Seminar in Bern, 1992
125 LUBOTZKY. Discrete Groups, Expanding Graphs and Invariant Measures
126 RIESEL. Prime Numbers and Computer Methods for Factorization
127 HÖRMANDER. Notions of Convexity
128 SCHMIDT. Dynamical Systems of Algebraic Origin
129 DIJKGRAAF/FABER/VAN DER GEER. The Moduli Space of Curves
130 DUISTERMAAT, Fourier Integral Operators

MIX
Papier aus verantwortungsvollen Quellen
Paper from responsible sources
FSC® C105338
FSC
www.fsc.org

If you have any concerns about our products,
you can contact us on
ProductSafety@springernature.com

In case Publisher is established outside the EU,
the EU authorized representative is:
Springer Nature Customer Service Center GmbH
Europaplatz 3, 69115 Heidelberg, Germany

Printed by Libri Plureos GmbH
in Hamburg, Germany